COMPUTATIONAL MATERIALS

COMPUTATIONAL MATERIALS

WILHELM U. OSTER
EDITOR

Nova Science Publishers, Inc.
New York

For permission to use material from this book please contact us:
Telephone 631-231-7269; Fax 631-231-8175
Web Site: http://www.novapublishers.com

LIBRARY OF CONGRESS CATALOGING-IN-PUBLICATION DATA

Computational materials / Wilhelm U. Oster [editor].
p. cm.
Includes bibliographical references and index.
ISBN 978-1-60456-896-7 (hardcover)
1. Materials science--Mathematical models. 2. Materials science--Computer simulation. I. Oster, Wilhelm U.
TA403.6.C6375 2008
620.1'1015118--dc22

2008026219

Published by Nova Science Publishers, Inc. ✢ New York

CONTENTS

Preface		**vii**
Chapter 1	Quantitative Description of the Morphology Evolution and Crystallization Kinetics in Thin Films *Vladimir I. Trofimov*	**1**
Chapter 2	Overview of the Effects of Shot Peening on Plastic Strain, Work Hardening and Residual Stresses *Baskaran Bhuvaraghan, M.S.Sivakumar and Om Prakash*	**49**
Chapter 3	Vibrational Dynamics of Bulk Metallic Glasses Studied by Pseudopotential Theory *Aditya M. Vora*	**119**
Chapter 4	A Computational Study of the Phonon Dynamics of Some Complex Oxides *Prafulla K. Jha and Mina Talati*	**177**
Chapter 5	Diffusion Mechanisms Near Tilt Grain Boundaries in Ni, Cu, Al and Ni_3Al *G.M. Poletaev, M.D. Starostenkov and S.V. Dmitriev*	**265**
Chapter 6	Computational Materials: III-V Semiconductor Clusters *Hamidreza Simchi*	**289**
Chapter 7	Computational Prognoses of Damage Growth and Failure in Some Steel Elements of a Liquefied Natural Gas Terminal Covering the Temperature Range of the Ductile-to-Brittle Transition Region *J. Jackiewicz*	**319**
Chapter 8	Porosity and Mechanical Properties of Cement Mortars Based on Microstructural Investigation *Ali Ugur Ozturk and Bulent Baradan*	**351**
Chapter 9	Cleavage Fracture Crystallography *J. Flaquer and A. Martín-Meizoso*	**369**

Chapter 10 Theoretical Simulation on Molecular Electronic Materials and Molecular Devices **385**
Jianwei Zhao, Yanwei Li and Hongmei Liu

Chapter 11 Numerical Simulation of the Curing Process of Rubber Articles **445**
Mir Hamid Reza Ghoreishy

Chapter 12 Relaxation Element Method in Mechanics of Deformed Solid **479**
Ye.Ye. Deryugin, G. Lasko and S. Schmauder

Index **547**

PREFACE

This book provides up-to-date research on field computational materials. This field of study consists of the construction of mathematical models and numerical solution techniques with the use of computers to analyze and solve scientific, social scientific and engineering problems.

Chapter 1 - A quantitative description of the morphology evolution in thin film growing via 3D island mechanism and crystallization kinetics of amorphous film is presented. The main concepts underlying a treatment are the survival probability for any point on (above) a substrate during the growth process, a 'feeding zone' and a 'dangerous zone' around a growing island or crystallite in amorphous matrix. The growth law for two island forms (hemisphere and paraboloid) in different condensation regimes is derived in self-consistent manner at all successive deposition stages, which shows growth acceleration at late stages due to island collisions. The kinetics of a lateral growth front perimeter allowing experimental determination of the growth law is deduced. At paraboloid growth, a steady (in time) and random (in space) surface relief is eventually formed. For this relief, composed of growth hillocks, the height-height autocorrelation function (ACF) is derived for various hillock shape and space distribution. The latter influences on both the ACF form and the roughness and the hillock shape strongly affects on the surface roughness at their random space distribution. For hemisphere growth, analytical expressions are derived for a variety of the average surface roughness parameters (rms roughness, the roughness coefficient, and the package density factor) and the surface height distribution providing a rather complete quantitative description of the evolving surface morphology. The surface height distribution is a non-Gaussian and the average surface roughness parameters kinetics are universal functions of coverage or film thickness with a maximum just prior the completed layer formation. A model for crystallization kinetics of amorphous film is developed by extension familiar Kolmogorov-Johnson-Mehl-Avrami (KJMA) model to take into account a finite film thickness. Two model versions: volume induced crystallization (VIC) and surface induced crystallization (SIC) are explored. Finite film thickness effects lead to important consequences in the VIC: the crystallization profile reaches maximum in film middle, the crystallites population is always higher than in bulk material, the thinner the film the slower it crystallizes and a spatially inhomogeneous crystalline structure with a fine-grained subsurface layer is formed. A VIC – kinetics follows a generalized KJMA equation with parameters depending on a film thickness and a SIC-kinetics obeys 2D KJMA equation in thin enough film and is almost linear in a thick film. The model is extended for the case of non-constant growth rate and crystallization of subsurface film layer, which is shown to be consistent with experimental data.

Chapter 2 - Residual compressive surface stresses are found to enhance the fatigue life of components, as fatigue cracks originate mostly from surfaces. Mechanical surface enhancement processes develop such residual compressive stresses. This review paper focuses on shot peening method and attempts to understand the underlying mechanism. This will help to quantify and optimize the material response due to shot peening computationally and help the designer to evolve better design. Starting with an introductory overview of shot peening, this paper describes the experimental and theoretical studies performed in the residual stress development to mitigate various damaging mechanisms that include SCC, corrosion, fretting etc. The paper also covers the development of cold work due to peening and related modifications in the surface microstructure of different materials. The material response remains complex due to the stochastic nature of several variables. Theoretical studies involving Finite Element method, and/or Discrete Element method are currently employed to understand the physics of shot peening. Shot peening produces different amounts of near surface plastic deformation for the same level of residual stresses and vice versa for different combinations of input parameters. The paper covers the efforts that overcome these challenges in optimizing the peening parameters. It also provides a glimpse of the research in the area of crack growth reduction due to shot peening. The residual stress that is developed relaxes due to service environment. In all the above phenomenon, dislocations play a vital role and therefore a multi-scale approach involving dislocations can provide a common platform in explaining them. In addition, this paper points to some future directions into which the research that can possibly explore to quantify the material response more accurately.

Chapter 3 - In this chapter, the author discusses a vibrational dynamics of some bulk metallic glasses (BMG) in terms of the phonon eigen frequencies of the localized collective excitations using model potential formalism at room temperature for the first time. The theoretical effective atom model (EAM) with Wills-Harrison (WH) form are used to compute the interatomic pair potential and pair correlation function (PCF) for the glassy systems. The phonon dispersion curves are computed from the three approaches proposed by Hubbard-Beeby (HB), Takeno-Goda (TG) and Bhatia-Singh (BS). Various elastic and thermodynamic properties have been studied from the elastic limit of the dispersion relations. Different types of the local field correction functions are used for the first time in the present investigation to study the screening influence on the aforesaid properties. The pseudo-alloy-atom (PAA) model is applied for the first time instead of Vegard's Law.

Chapter 4 - Understanding the mechanism of strong correlation effects of spin, charge and lattice responsible for the fascinating properties such as large Resistivity changes, huge volume changes, high T_c superconductivity, strong thermoelectric response, gigantic non linear optical effect and finally colossal magnetoresistance (CMR) in the transition metal oxides is an important challenge. One of the central issues of concern in many manganites is to understand the occurrence of colossal magnetoresistance around the insulator (I) to metal (M) transition along with the paramagnetic to ferromagnetic transition at same temperature. Double exchange interaction and electron-phonon interaction triggered by Jahn-Teller (JT) distortions of the MnO_6 octahedral have been proposed. In this situation, correlated information on the phonon properties is essential for a detailed understanding of the lattice distortion in these technological and fundamentally important compounds. The present chapter focuses on recent advances in understanding the phonon properties at ambient condition and far from the ambient to correlate the JT distortions and magnetoresistance in manganites. LaMnO3, which undergoes phase transition at high temperature, is considered in its cubic and rhombohedral phase. The chapter also considers the study of the effect of Sr-doping at La-site. The difference in structural symmetry of cubic and rhombohedral manganites is manifested in their phonon spectra and Phonon density of states. In addition, to understand the effect of pressure and temperature, 30% Sr-doped $LaMnO_3$ *i.e.* $La_{0.7}Sr_{0.3}MnO_3$ (LSMO) is considered at different applied pressure and temperature. The phonon properties are also reported for the $NaCoO_2$ compound in its two different geometry positions.

Chapter 5 - With the use of the molecular dynamics technique the diffusion mechanisms along tilt grain boundaries <111> and <100> are investigated in pure metals Ni, Cu, Al and in Ni3Al intermetallide. The following three basic mechanisms of grain boundary diffusion were revealed: migration of atoms along the cores of grain boundary dislocations, cyclic mechanism near the core, and the formation of the chains of atoms displaced from the core of one dislocation to the core of the other one. The density of steps at grain boundary dislocations strongly affects the probability of the realization of all three mechanisms. Temperature and misorientation angle also were found to be important factors. Main peculiarity of grain boundary diffusion in Ni3Al intermetallide is related to the fact that it occurs mainly by the displacements of Ni atoms over their sublattices in L12 superstructure. As a result, the short-range order is almost preserved. At high temperature Al atoms start to participate in the migration process. Although their displacements are small in comparison to the displacements of Ni atoms, they cause an essential reduction of a superstructural order in the alloy.

Chapter 6 - Semiconductor materials have been playing an important role in daily life. Among them, the narrow gap III-V semiconductor materials InAs, GaSb, InSb and their alloys are particularly interesting and useful materials since they offer the promise of being able to access the 2-10 microns wavelength region and should provide the next generation of LEDs lasers and photo detectors for applications such as sensors, molecular spectroscopy and thermal imaging. On the other hand the discovery of fullerenes and nanotubes of carbon have led to a wide spread interest to understand their properties, applications and development of unconventional forms (i.e Clusters) of materials. Therefore it is expected, that, the clusters of III-V semiconductors have been widely studied. One of the main interests in semiconductor clusters is the variation of band gap with size which affects the photoluminescence properties. Here the author reviews the progress in understanding the structure and electronic spectra of InAs, GaAs, InSb and CdSe clusters. First principles approaches based on Hartree-Fock and Density Functional Theory have become central to such studies and a brief overview of them is given.

Chapter 7 - The paper discusses a method for numerical simulation of the transition from ductile to brittle cleavage behavior in some steel elements of a liquefied natural gas terminal. To determine an onset of material failure, a coupled mechanistic model is developed for the effect of preceding ductile damage on cleavage fracture. The aim of the research is to improve our understanding of transferable and applicable, formulated fracture criterion when the cleavage crack initiation is preceded by ductile crack growth giving a large scatter to values of fracture toughness. For grains and interfaces, results of micromechanical simulations refer to macroscopic predictions, which can be obtained by means of a version of the Gurson type model incorporating effects of a microvoid aspect ratio.

An additional motivation of the presented work is further development of combined method of contour elements because the standard finite element method has some limitations in several applications. In the paper implemented, new numerical formulations, that are built-up on analytical integration methods, are briefly described and implemented.

The validity of the developed fracture model is checked by comparisons of results of standardized experimental tests of fracture mechanics with results of micromechanical simulations obtained by means of the combined method of contour elements.

Index Terms—Contour subdomain method, Ductile to brittle transition, Mesomechanics behavior of the polycrystalline microstructure, Weibull statistics, Methods of elastic-plastic analysis, Liquefied natural gas, Ferritic and duplex steels.

Chapter 8 - Microstructure and mechanical behaviour of cement based mortars has become a more important issue due to the recent developments in microstructural investigations by using computational analysis techniques. Microstructural investigations have a considerable capacity to define the inner structural formation of cementitious materials. A relationship between pore structures of cement mortars and their mechanical properties can be determined by studies based on image processing and analysis.

In the scope of this investigation, the effects of retarders have been investigated by implementation of these methods as a case study. Standard cement mortars were prepared by incorporation of retarders with various ratios in order to obtain different pore formations. Microstructures of different cement mortars were investigated by using optical microscope. Micrographs of polished sections of cement mortar samples were taken to determine the area ratio values of pore formations. The pore area ratio values represent total pore area amount in a polished section. The development of pore area ratio values for 1, 2, 7 and 28-day old cement mortars have been determined. The test results indicate that mechanical properties of cement mortars increase as the pore area ratio values decrease. The relationship between pore structures of cement mortars and their mechanical properties has been established by various mathematical models. The chapter shows that image analysis techniques have a remarkable potential to define the relationship between pore area ratios obtained by microstructural investigations and mechanical properties.

Chapter 9 - Cleavage fracture of cubic materials takes place usually along {001} planes. In a cubic crystal there are three sets of <001> orientations available for cleavage. The crystal will cleave along the best oriented among the three. The distribution of the best oriented cleavage plane is computed by simulation for a random polycrystalline material. It is proved that cleavage fracture is intrinsically unstable. Beside a simple and effective procedure is described for computing the distribution of observed tilt (dihedral) angles in 2D sections and/or projections distributed randomly in 3D.

Chapter 10 - Over the past decade, there has been remarkable progress in the studies of molecular electronics. In this chapter, the authors will survey the recent theoretical research in this field. In principle, two theoretical strategies, namely static and dynamic approaches are employed in the theoretical modeling. For static approach, an *in-situ* static theoretical calculation by considering the influence of electric field is introduced. The *in-situ* simulation results displayed that both the geometric and electronic structures of the conjugated molecular materials are sensitive to the electric field. On the other hand, the electronic transportation through molecular wires has been studied intensively in these years by using the non-equilibrium Green's functional formulism combined with density functional theory (NEGF-DFT), which may directly give the current-voltage character to compare with the experimental measurement. A series of typical molecular wires have been studied to compare with the experimental transportation behavior. Based on the chain-length dependence of conductance for each series of molecular wires, the attenuation factor, β, has been obtained and compared with experimental data where applicable. The β value has also been quantitatively correlated to the molecular HOMO-LUMO gap. Following the study of molecular wires, molecular rectification has become the focus in the field of molecular devices. A series of asymmetrically substituted conducting molecular wires have been studied with the same method. The results demonstrated that the fully-conjugated molecular wire with asymmetric substitution has minor rectification. Therefore, other rectification mechanisms are essentially required.

Chapter 11 - This work is devoted to the numerical modeling of the curing (vulcanization) process of rubber articles. The main aim is to give an overall prospectus of the research works carried out in this field as well as practical approaches to solve an industrially scaled problem. The chapter begins with an introduction in which the aim and scope of the work is briefly described. A comprehensive review of the related published works can be found in the next section. It tried to cover all research papers from the beginning of the research in this area up to recent publications. The governing equations of the heat transfer and kinetics of the rubber curing reaction as well as finite element method used for the solution of these equations are then given and discussed in detail. The main focus is on the derivation of the working equations in conjunction with boundary conditions encountered in a typical rubber curing process. The solution algorithm and the commercial and in house developed computer codes are drawn in the next part. In order to give a better understanding of topics presented in this chapter, two practical examples of the finite element simulation for the curing process of rubber goods including a metal reinforced rubber slab and a truck tire in the mold are presented and finally the conclusion as well references are given.

Chapter 12 - In this chapter an original method of calculation and modelling of plastic strain localization in a loaded solid — the relaxation element method (REM) — is represented. The fundamental property of solids: "plastic deformation is accompanied by stress relaxation in local volumes" is the basis of the method. The theoretical foundation of the method is derived from the basic equations of elasticity and continuum theory of defects. For the plane-stress state, the technique of the construction of local sites of plastic deformation at the mesoscopic scale level is introduced. Examples and results of modelling the process of plastic strain localization, accompanied by the effect of Lüders band propagation and Portevin Le Chatelier effect are presented. The difficulty of the traditional description of strain localization phenomena lies in the fact that it is not possible to formulate a universal physical law of the connection between plastic deformation and stresses in the solid due to the relaxation nature of plastic deformation. Application of the REM allows to overcome this difficulty. By methods of mechanics of deformed solid it was shown that stress relaxation inside the structural element on a definite value is unambiguously connected with the change of its external shape, which according to the physical sense is not-elastic, but plastic. As a result, the stress field changes accordingly outside the considered structural element and is defined unambiguously as well. So, the structural element, having undergone plastic deformation becomes a relaxation element with its own fields of internal stresses. This stress field can be connected with the specific value of plastic deformation, ensuring corresponding changes of the external shape of the structural element. Applications of REs as mesoscopic defects, defining the relation of plastic deformation inside the RE with the stresses outside this element allow to simulate the processes of strain localization and to obtain the dependencies of the flow stress from the consequent involvement of separate structural elements of the solid into plastic deformation. The developed REM model for the plastic deformation localization operates on the principles of cellular automata. The calculation area is divided into a number of cells, playing the role of structural element (for example, grains in polycrystals). In the model each element of the simulated medium possesses the ability to switch its state by discrete steps of plastic deformation. Thus, the elements of the medium are able to increase the degree of plastic deformation one after the other and as a stress concentrator to effect the change of the stress field in the whole volume of the solid. The involvement of a structural elements into plastic deformation occurs when under the influence of external applied stress the shear stress at the center of this element achieves a critical value (e.g., according to the Mises-Tresca criterion). The interaction of the fields of internal stresses from different relaxation elements, which have undergone plastic deformation takes place automatically.

In: Computational Materials
Editor: Wilhelm U. Oster
ISBN: 978-1-60456-896-7

Chapter 1

QUANTITATIVE DESCRIPTION OF THE MORPHOLOGY EVOLUTION AND CRYSTALLIZATION KINETICS IN THIN FILMS

Vladimir I. Trofimov*
Institute of Radioengineering and Electronics of the Russian Academy of Sciences,
11/7 Mokhovaya Street, 125009 Moscow, Russia

ABSTRACT

A quantitative description of the morphology evolution in thin film growing via 3D island mechanism and crystallization kinetics of amorphous film is presented. The main concepts underlying a treatment are the survival probability for any point on (above) a substrate during the growth process, a 'feeding zone' and a 'dangerous zone' around a growing island or crystallite in amorphous matrix. The growth law for two island forms (hemisphere and paraboloid) in different condensation regimes is derived in self-consistent manner at all successive deposition stages, which shows growth acceleration at late stages due to island collisions. The kinetics of a lateral growth front perimeter allowing experimental determination of the growth law is deduced. At paraboloid growth, a steady (in time) and random (in space) surface relief is eventually formed. For this relief, composed of growth hillocks, the height-height autocorrelation function (ACF) is derived for various hillock shape and space distribution. The latter influences on both the ACF form and the roughness and the hillock shape strongly affects on the surface roughness at their random space distribution. For hemisphere growth, analytical expressions are derived for a variety of the average surface roughness parameters (rms roughness, the roughness coefficient, and the package density factor) and the surface height distribution providing a rather complete quantitative description of the evolving surface morphology. The surface height distribution is a non-Gaussian and the average surface roughness parameters kinetics are universal functions of coverage or film thickness with a maximum just prior the completed layer formation. A model for crystallization kinetics of amorphous film is developed by extension familiar Kolmogorov- Johnson-Mehl-Avrami (KJMA) model to take into account a finite film thickness. Two model versions: volume induced crystallization (VIC) and surface

* Tel.: +7 495 629 3403, fax: +7 495 629 3678, E-mail: trof@mail.cplire.ru

induced crystallization (SIC) are explored. Finite film thickness effects lead to important consequences in the VIC: the crystallization profile reaches maximum in film middle, the crystallites population is always higher than in bulk material, the thinner the film the slower it crystallizes and a spatially inhomogeneous crystalline structure with a fine-grained subsurface layer is formed. A VIC-kinetics follows a generalized KJMA equation with parameters depending on a film thickness and a SIC-kinetics obeys 2D KJMA equation in thin enough film and is almost linear in a thick film. The model is extended to the case of non-constant growth rate and crystallization of subsurface film layer, which is shown to be consistent with experimental data.

1. Introduction

Modeling thin film growth and morphology evolution at successive deposition stages important from the viewpoint of both science and engineering presents significant challenge to theoretical physicists and materials scientists. The growth of a thin film in molecular-beam epitaxy via Volmer-Weber 3D island mechanism involves three main kinetic stages: 1) very short nucleation period, 2) a longer growth stage, where deposited atoms are mostly captured by growing islands, and 3) the longest coalescence stage, where the growing islands collide with each other and eventually form a continuous film. The growth of individual islands at second stage is rather well described by familiar 'diffusion model' based on the solution of a diffusion equation for adatom density in the island surroundings at appropriate boundary conditions with following calculation of an atomic current arriving to an island (e.g., [1, 2]. However, as islands grow they inevitably collide with each other with formation of island clusters, so that this approach becomes progressively wrong with time and after percolation threshold, marking the formation of semicontinuous film, fails at all. Some aspects of the growth and surface morphology evolution at these late stages have been addressed [3-10], but no general unifying methodology has been emerged. In this article we present a systematic quantitative analysis these issues basing on the statistic growth models approach [2] developed by the extension of the familiar Kolmogorov [11], Johnson and Mehl [12], and Avrami [13] (KJMA) theory of crystallization. The central concept of this approach is a survival probability for any randomly chosen point on (above) a substrate during film growth process. In section 2 we present and further develop a simple growth model [2] that takes into account island collisions and thus allows finding out *in a self-consistent manner* the growth law at all successive deposition stages, from the nucleation commence up to the formation of completed layer. In section 3 a variety of the average surface roughness parameters and height-height autocorrelation function are derived for different nucleation mode and island form, and thus the surface morphology evolution during film growth is analyzed. In section 4, using the same approach, a modified KJMA model of the crystallization kinetics in thin amorphous films and surface layers that takes into account a finite film thickness is presented.

2. GROWTH

2.1. Self-Consistent Growth Model

The model is based on a concept of a *feeding zone* (FZ) that is introduced as a join of substrate regions covered by islands and an annular-like strip of some width λ (feeding strip) adjoining to the islands perimeter (figure 1). It is supposed that all the atoms deposited onto a FZ remain there and are incorporated into a growing film, whereas those landing outside a FZ reevaporate from a substrate.

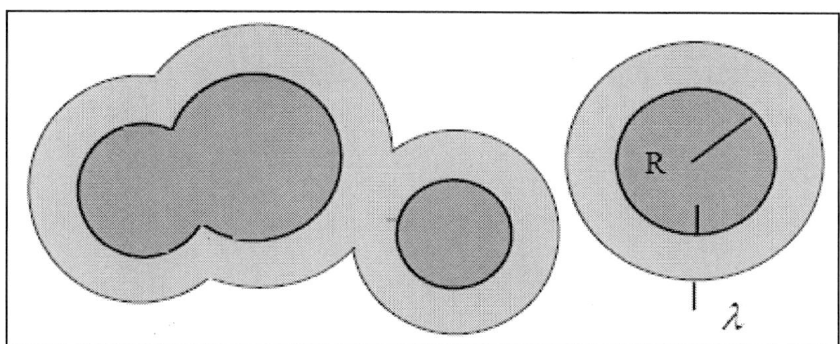

Figure 1. Schematic of a feeding zone.

To verify the applicability of the model let us compare it to the diffusion model. In both models an atomic flux onto islands from a vapor phase is the same and it is sufficient therefore to compare only diffusion currents. In the diffusion model the average current of adatoms into the island of radius R per unit time is [1, 2]

$$F_d^{(d)} = \pi X_a^{\,2} \cdot J \cdot f(\rho), \tag{1}$$

where $f(\rho) \equiv 2\rho K_1(\rho)/K_0(\rho)$; $\rho = R/X_a$, $X_a = (D\tau_a)^{1/2}$ is the mean diffusion length of adatom before desorption, $D\,(\mathrm{cm^2 s^{-1}})$ the surface diffusion coefficient of adatom, τ_a a mean lifetime of adatom on a substrate before desorption, $J\,(\mathrm{cm^{-2} s^{-1}})$ the deposition flux and $K_0(x)$ and $K_1(x)$ are modified Bessel functions.

In our model this current is proportional to a feeding strip area

$$F_d^{(fz)} = \pi X_a^{\,2} \cdot J \cdot \varphi(\rho, u), \tag{2}$$

where $\varphi(\rho,u) = 2u\rho + u^2$ with $u = \lambda / X_a$. Calculation results (figure 2) show that at $u = 1$ $\varphi(\rho)$ is very close to $f(\rho)$, except for very small ρ, and with taking into account asymptotic behaviour of Bessel functions ($K_1(x)/K_0(x) \xrightarrow[x\to\infty]{} 1 + 1/(2x) + \ldots$) they eventually coincide, and thus for individual islands a FZ description coincides with the diffusion one. We will set therefore $\lambda = X_a$.

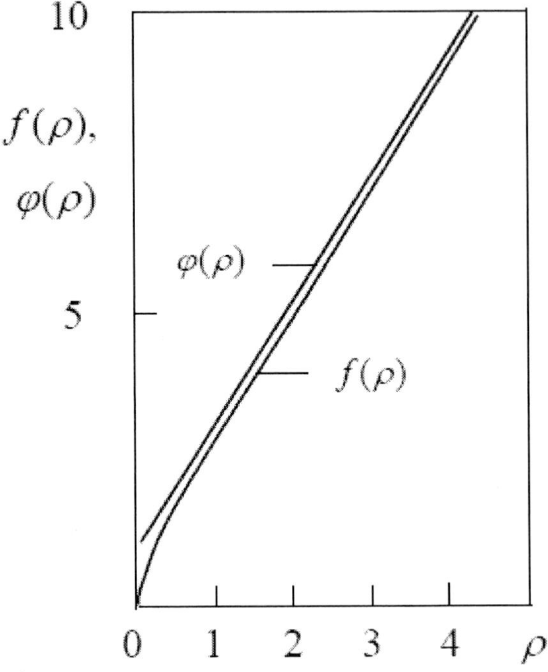

Figure 2. Functions $f(\rho)$ and $\varphi(\rho)$.

Let us designate a specific (per unit area) FZ area at some time moment t as $\xi_f(t)$, and a specific film volume as $V(t)$, which coincides with an average film thickness $\bar{h}$. Then, a simple kinetic equation may be written down

$$dV(t)/dt = d\bar{h}/dt = J\Omega \cdot \xi_f(t), \qquad (3)$$

where Ω is the atomic volume in the crystal. A specific FZ area obviously coincides with substrate coverage ξ by an imaginary film resulted from a real film by an instantaneous radius increment of all the islands by $\Delta R = X_a$, i.e., $\xi_f(t) = \xi(R + X_a)$. Now we need to specify a growth process. We consider two growth modes: a hemispherical growth, when the free islands grow in a hemisphere form and a 'paraboloid growth' when the lateral growth of an island occurs in a circle form via capture of migrating adatoms and its vertical growth proceeds thanks to direct impingement of deposited atoms with a constant velocity $v_n = J\Omega$, so that a freely growing island acquires a paraboloid form.

2.1.1. Hemispherical Growth

In this model the free islands grow in a hemisphere form with a growth law

$$R(t',t) = \int_{t'}^{t} v(\tau)d\tau\,, \tag{4}$$

where $R(t',t)$ is an island radius at the time t if it nucleated at some time $t' \le t$ and $v(\tau)$ is the growth rate. After collisions islands cease growth at their contacts, but continue to grow in all other available directions. Consider a film at some time moment t (figure 3). Let $\xi(z,t)$ be a specific coverage of a film section at some height z from a substrate; it's clear that $\xi(z,t)$ is an ever decreasing function of z: $\xi(0,t) = \xi(t)$ - substrate coverage and $\xi(z,t) = 0$ at $z \ge R_m$, where R_m is the radius of the largest island. Then, a specific film volume is written down as

$$V(t) = \int_{0}^{\infty} \xi(z,t)dz\,. \tag{5}$$

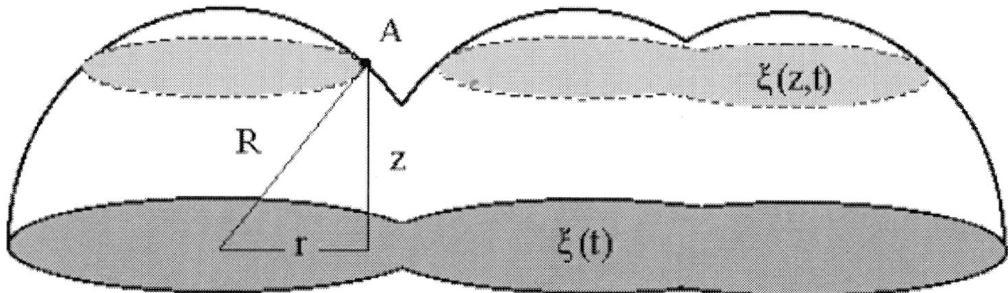

Figure 3. Schematic of a film at some time moment t.

To find a function $\xi(z,t)$ one needs to specify a nucleation mode. We consider two cases: continuous nucleation (CN) of islands during film growth with some intensity $I(t)$ ($cm^{-2}s^{-1}$) at random points of uncovered substrate and instantaneous nucleation (IN) of all the islands at the time $t = 0$ at random points of a substrate with density N (cm^{-2}); note that the IN-model is a special case of the CN-model at $I(t) = N(t)\cdot\delta(t)$ with $\delta(t)$ being the delta-function. Let A is some point at a height z from a substrate; the probability $q(z,t)$ that this point will survive by the time t, i.e., will not be captured by a growing film, may be written down as [2, 14]

$$q(z,t) = \exp\left(-\int_{0}^{t} \mu_A(\tau)d\tau\right),$$

where $\mu_A(\tau)d\tau$ is the probability of the formation of an 'aggressor'- a nucleus at the time $\tau, \tau + d\tau$ in a circle with a radius $r = \left[R^2(\tau,t) - z^2\right]^{1/2}$ centered under point A provided that before τ it was free of nuclei – that can capture point A by the time t. In the CN-model, $\mu_A(\tau)d\tau = \pi I\left[R^2(\tau,t) - z^2\right]\cdot d\tau$, so that taking into account that $q(z,t)$ coincides with the uncovered fraction at a level z at the time t, i.e., $q(z,t) = 1 - \xi(z,t)$, one gets

$$\xi(z,t) = 1 - \exp\left\{-\pi \int_0^{t^*} I(\tau)\left[R^2(\tau,t) - z^2\right]\cdot d\tau\right\}, \tag{6}$$

where t^* is defined by the relation $R(t^*,t) = z$.

In the IN-model $q(z,t)$ coincides simply with a probability of the nuclei absence in a circle of a radius $r = \left[R^2(0,t) - z^2\right]^{1/2}$, so that due to the Poisson distribution of nuclei

$$\xi(z,t) = 1 - \exp\left[-\pi N(R^2(t) - z^2)\right]. \tag{7}$$

Note that the latter formula is obtained also from (6) by setting $I(\tau) = N \cdot \delta(\tau)$. Insertion (6) and (7) into (5) yields

$$V(t) = \int_0^{R_m} \left\{1 - \exp\left[-\pi I \int_0^{t^*} \left(R^2(\tau,t) - z^2\right) d\tau\right]\right\} dz \quad \text{CN-model} \tag{8}$$

$$V(t) = \int_0^{R_m} \left\{1 - \exp\left[-\pi N\left(R^2(t) - z^2\right)\right]\right\} dz \quad \text{IN-model} \tag{9}$$

Now, noticing that the coverage $\xi(t)$ and FZ area $\xi_f(t)$ are given by

$$\xi(t) = 1 - \exp\left(-\pi I \int_0^t R^2(t,t')dt'\right) \tag{10}$$

$$\xi_f(t) = 1 - \exp\left(-\pi I \int_0^t (R(t,t') + X_a)^2\, dt'\right) \tag{11}$$

in the CN-model and

$$\xi(t) = 1 - \exp(-\pi NR^2) \tag{12}$$

$$\xi_f(t) = 1 - \exp(-\pi N(R + X_a)^2) \tag{13}$$

in the IN-model, for the second required equation from (3) one obtains

$$\frac{dV}{dt} = J\Omega\left\{1 - \exp\left[-\pi I\int_0^t \left(R(\tau,t) + X_a\right)^2 d\tau\right]\right\} \text{ CN-model} \tag{14}$$

$$\frac{dV}{dt} = J\Omega\left\{1 - \exp\left[-\pi N\left(R(t) + X_a\right)^2\right]\right\} \text{ IN-model} \tag{15}$$

Coupled equations (8) and (14), and (9) and (15) are desired ones that determine self-consistently growth law $R(t)$ in the CN- and IN-model, respectively.

Further on, we will concentrate on the IN-model as a simpler one and that serves as a good approximation of a nucleation process in thin films [1, 2]. In this model, $R_m(t) = R(0,t) = R(t)$ and introducing dimensionless variables $\rho = R/a$ and $\tau = t/t_0$ with natural scales of the model $a = (\pi N)^{-1/2}$, $t_0 = a/(J\Omega)$ from (9) and (15) after differentiation and processing one finally gets

$$\int_0^\rho \frac{2\rho' \int_0^{\rho'} \exp\left[-(\rho'^2 - y^2)\right]\cdot dy}{1 - \exp\left(-(\rho' + \sqrt{z})^2\right)} \cdot d\rho' = \tau \,. \tag{16}$$

Here, $z \equiv (X_a/a)^2 = \pi N X_a{}^2$ is the numerical parameter, characterizing a condensation regime: at $z << 1$ a so called extremely incomplete condensation (high substrate temperatures) takes place; with increasing z a crossover to incomplete $(z < 1)$ and in the limit $z \to \infty$ - to complete condensation occurs. As follows from (13), the parameter z is connected with the FZ area by a simple relationship $\xi_f(0) = 1 - e^{-z}$. Numerical computations with Eq. (16) at different z are shown in figure 4. As can be seen, there exist two limiting curves at $z = \infty$ and $z = 0$ corresponding to complete and extremely incomplete condensation regime, respectively. All curves tend with time toward the limiting curve for $z = \infty$, since with increasing substrate coverage any condensation regime changes toward a complete condensation. It's worthwhile to note that the curve with $z = 1$ nearly coincides with a limiting curve with $z = \infty$. It means that a complete condensation regime is practically realized at $z \geq 1$, what has a simple explanation: at $z \geq 1$ even the initial FZ area $\xi_f(0) \geq 0.63$, i.e., more than 63 % of the substrate surface are already covered by the capture zones of existing nuclei at the very commence of deposition. It is also seen that the growth accelerates at late stages because of the reduction of the growth front perimeter length at island collisions. To detect the effect of collisions in explicit form let us consider the growth of a free island. The latter occurs owing to the capture of atoms deposited into its feeding zone, a circle of a radius $R + X_a$ so that taking into account an overlapping of these circles, we can write down $2\pi R^2 dR/dt = J\Omega\left\{1 - \exp\left[-\pi N(R + X_a)^2\right]\right\}$, whence one finally gets

$$\int_0^{\rho} \frac{2\rho'^2}{1 - \exp\left(-(\rho' + \sqrt{z})^2\right)} \cdot d\rho' = \tau \qquad (17)$$

Calculation results by this formula, depicted in figure 4 by broken curves, confirm that the growth

acceleration at late stages are indeed caused by the island collisions and this effect is naturally stronger manifested in the complete condensation regime with its initial growth law $\rho \propto \tau^{1/3}$, since in this regime an average atomic flux into each island remains constant during all the growth process.

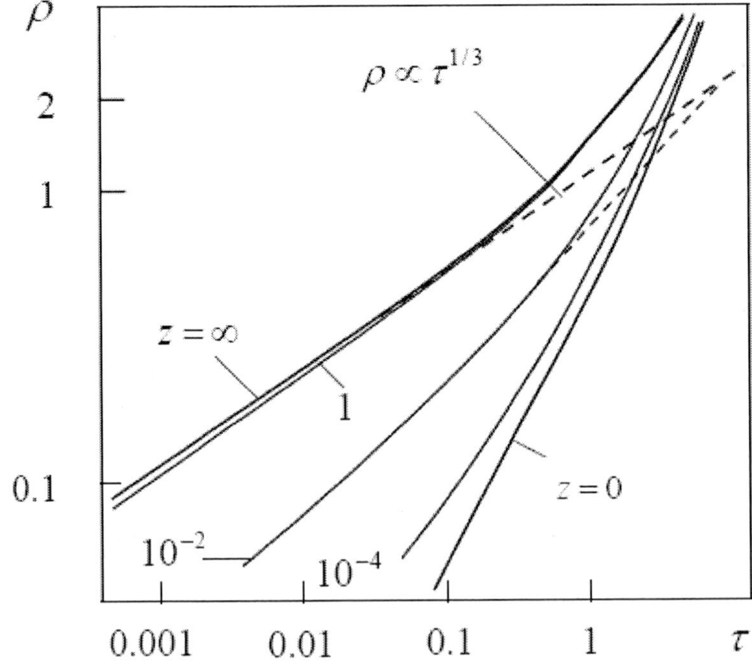

Figure 4. Hemispherical growth law in various regimes of condensation.

2.1.2. Paraboloid Growth

In this model the lateral growth of a free island occurs in a circle form due to capture of migrating adatoms from a ring $R, R+X_a$ and after collisions islands cease the growth at their contacts but continue to grow in all the other available directions via capture of adatoms from a feeding strip. Therefore, the substrate coverage kinetics follows a simple equation $d\xi / dt = J\Omega^{2/3}\left(\xi_f(t) - \xi(t)\right)$, whence noticing that in the considered case of the instantaneous nucleation of all the islands at $t = 0$ in random substrate points with density N, the substrate coverage $\xi(t)$ and the specific FZ area $\xi_f(t)$ are given by Eqns. (12) and (13), we obtain

$$\int_0^R \frac{2\pi N r dr}{1-\exp\left(-\pi N(X_a^{\,2} + 2rX_a)\right)} = J\Omega^{2/3} t, \qquad (18)$$

which, after introducing dimensionless size $\rho = R/a$ with the same $a = (\pi N)^{-1/2}$ and dimensionless time $\tau = J\Omega^{2/3} t$ having a simple meaning of a number deposited monolayers, becomes

$$\int_0^\rho \frac{2\rho' d\rho'}{1-\exp\left(-(z+2\sqrt{z}\rho')\right)} = \tau \qquad (19)$$

Calculation results with (19) depicted in figure 5 by solid curves show again that with increasing z the

$\rho(\tau)$ curves tend toward a limiting curve $\rho = \tau^{1/2}$ for the complete condensation regime and

practically coincide with the latter at $z = 1$. In this model, the second limiting curve for $z = 0$

coincides with an abscissa axis, since a feeding strip vanishes at $z = 0$. To detect in explicit form the

effect of collisions consider again the growth of a free island. Now, the latter occurs due to capture of adatoms from a feeding strip, an annular $R,\ R+X_a$, so that taking into account their overlapping, we can write down $2\pi N R dR / dt = J\Omega^{2/3}\left[e^{-\pi N R^2} - e^{-\pi N (R+X_a)^2}\right]$, hence the growth law is given by

$$\int_o^\rho \frac{2x dx}{\left[e^{-x^2} - e^{-(x+\sqrt{z})^2}\right]} = \tau . \qquad (20)$$

Calculation results with (20) demonstrate again (broken curves in figure.5) that island collisions lead to the growth acceleration at late stages and that this effect naturally weakens with decreasing z. A simple comparison with figure 4 shows also that at paraboloid growth this effect is much more pronounced than at hemispherical growth. This can easily be understood. Really, at paraboloid growth all atoms deposited onto a feeding strip are 'consumed' in the lateral growth process, whereas at hemispherical growth these are used for both lateral and vertical growth.

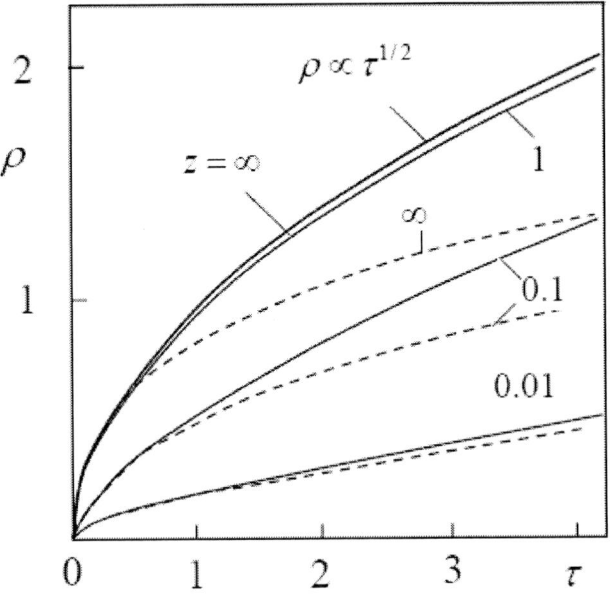

Figure 5. Paraboloid growth law in various regimes of condensation.

2.2 Growth Front Perimeter

For calculation of the lateral growth front perimeter (GFP) length, i.e., length of free island boundaries, consider a film in the substrate plane at some time moment t, figure 6a, where GFP is shown by bold circles. Let all the islands received instantaneous increment of their size, $\Delta R = r$ resulting in the increase of coverage, so that it becomes $\xi(r,t) = \xi(R+r)$ (grey in figure 6a).

Then, a specific (per unit area) GFP length (L) may be written down as [2, 14]

$$L = \partial\xi(r,t)/\partial r\big|_{r=0} \tag{21}$$

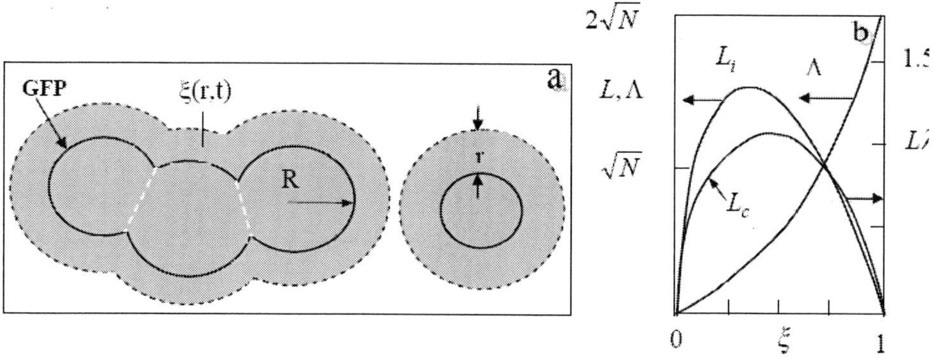

Figure 6. Schematic of GFP (a) and kinetics of GFP (at continuous (L_c) and instant (L_i) nucleation), and Λ (b).

Now we consider the same two nucleation modes, instantaneous nucleation (IN) and continuous nucleation (CN). In the CN-model according to (7), $\xi(r,t) = 1 - \exp\left[-\pi \int_0^t I(t')(R(t') + r)^2 dt'\right]$, so that at $I(t) = const. \equiv I$ and $v(t) = const. \equiv v$, introducing the dimensionless time $\tau = (Iv^2t^3)^{1/3}$ from (21) one gets the GFP kinetics

$$L(\tau) = \frac{\pi}{\lambda}\tau^2 \exp(-\frac{\pi}{3}\tau^3), \tag{22}$$

with a single parameter $\lambda = (v/I)^{1/3}$, which is depicted by the curve with a maximum $L_m = 1.03/\lambda$ at $\tau = 0.86$. Noticing that in the CN-model the coverage kinetics is given by $\xi(t) = 1 - \exp(-\pi I v^2 t^3 /3)$, Eq. (22) may then be rewritten in the form of universal function of coverage

$$L(\xi) = \frac{\pi}{\lambda}(1-\xi)\left[\frac{3}{\pi}\ln\frac{1}{1-\xi}\right]^{2/3}, \tag{23}$$

that is depicted by the curve with a maximum at coverage $\xi_m = 0.487$ (figure 6b). In the IN-model pursuant to (12), $\xi(r,t) = 1 - \exp\left[-\pi N(R(t) + r)^2\right]$, so in the dimensionless time $\tau = (\pi N)^{1/2} R(t)$ Eq. (21) yields the GFP kinetics

$$L(\tau) = 2(\pi N)^{1/2} \tau \exp(-\tau^2) \qquad (24)$$

with a single parameter, the island density, that is also represented by the curve with a maximum $L_m = 1.52N^{1/2}$ at $\tau = 0.707$. Taking into account that in this case $\xi(t)$ is given by (12), Eq. (24) can also be rewritten as universal function of coverage

$$L(\xi) = 2(\pi N)^{1/2}(1-\xi)[-\ln(1-\xi)]^{1/2} \qquad (25)$$

represented by the unimodal curve (figure 6b) with earlier ($\xi_m = 0.393$) maximum. A nonmonotonic GFP kinetics at any nucleation mode has a simple explanation: initially GFP naturally grows along with growing islands, but afterwards due to islands impingement and ceasing the lateral growth at interisland boundaries GFP reduces. In the CN-model, a ceaseless nucleation of new islands partially compensate an impingements induced GFP decrease - that is why in this case a maximum is achieved later ($\xi_m = 0.487$) than in the IN case ($\xi_m = 0.393$).

For more complete description of the film morphology a total interisland boundaries length (IBL), shown in figure 6a by white broken lines, should be included. For the IN-model a specific IBL, can be obtained following [14, 15] in explicit form and presented also as universal coverage function

$$\Lambda(\xi) = 2\sqrt{N}\left(erf(\sqrt{\frac{1}{1-\xi}}) - \frac{2(1-\xi)}{\sqrt{\pi}}\sqrt{\ln\frac{1}{1-\xi}} \right), \qquad (26)$$

where $erf(z) \equiv \frac{2}{\sqrt{\pi}} \int_0^z e^{-y^2} dy$. Figure 6b shows that IBL is naturally an ever-increasing function of coverage. It is noteworthy to notice that the summarized all boundaries length ($L_i + \Lambda$) kinetics is represented (not shown) by a curve with a later ($\xi_m = 0.747$) maximum.

An analysis of the GFP kinetics, which can be measured on the electron micrographs of the film, allows finding the growth law at all successive deposition stages. To that end rewrite (21a) as

$$L(t) = (1-\xi)L_0(t), \; L_0 = 2\pi NR(t), \qquad (27)$$

whence it is seen that a quantity L_0, having a meaning of the extended GFP and easily calculated from measured L and ξ, is proportional to R. As an example, figure 7 shows the GFP kinetics during growth of Au and $CrSi_2$ films on (001) KCl substrates at conditions close to complete condensation. It is seen that experimentally observed GFP kinetics are described by curves with a maximum near $\xi \cong 0.4$ similarly to the theoretical curve L_i in figure 6b. A quantitative analysis confirms that the experimental GFP kinetics curves are well described by (25) and the island number density determined from here agree well with that counted up directly on the micrographs. Calculated from here with using (27) growth law for Au and $CrSi_2$ films is presented in figure 8. As can be seen, the experimental growth law is described by a power law $R \propto t^p$ with an exponent varied with time. At early stages, $p = 1/3$, confirming the model predictions (figure 4) that the initial growth of hemispherical islands in the complete condensation regime obeys the simple law $R \propto t^{1/3}$. At further stages, the exponent p ceaselessly increases, also confirming the model predictions of the growth acceleration at later stages due to the islands impingement effect. Remarkably that the observed growth acceleration begins at coverage $\xi \approx 0.15$. Indeed, albeit at random islands space distribution their collisions occur from the very onset of the nucleation, they become significant, when the average islands size ($2\overline{R}$) becomes comparable to the average interisland distance $a = (2\sqrt{N})^{-1}$, i.e., at coverage $\xi \cong \pi\overline{R}^2 N \cong \pi(\frac{1}{4\sqrt{N}})^2 N = 0.196$.

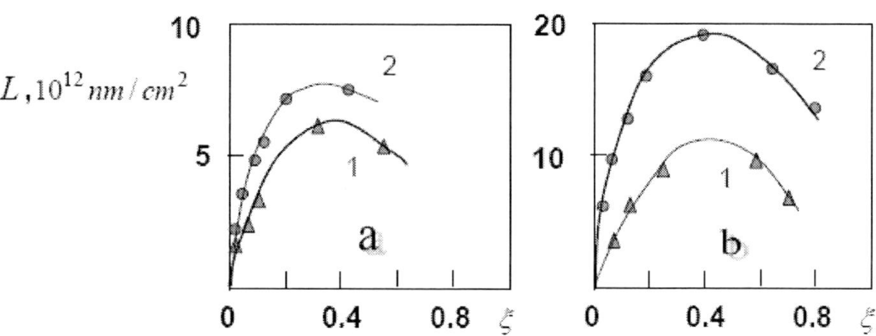

Figure 7. The GFP kinetics in Au (a) and $CrSi_2$ (b) films at different temperatures: a) 1 – 150^0C, 2 – 100^0C, b) 1 – 400^0C, 2 – 300^0C.

3. Surface Morphology

The most important quantitative characteristic of a rough surface is its height-height autocorrelation function (ACF) defined as (figure 9)

$$G(u) = \langle (h_A - \langle h \rangle)(h_B - \langle h \rangle) \rangle, \quad (28)$$

where h_A and h_B denote the surface height (or depth) at two randomly chosen points A and B separated by a distance u, $\langle h \rangle$ is the mean surface height (depth) and angle brackets denote the spatial averaging over a planar reference plane. However, a surface profile of a growing film is not *a priori* known and given only in probability terms, so it needs to rewrite (28) in the form

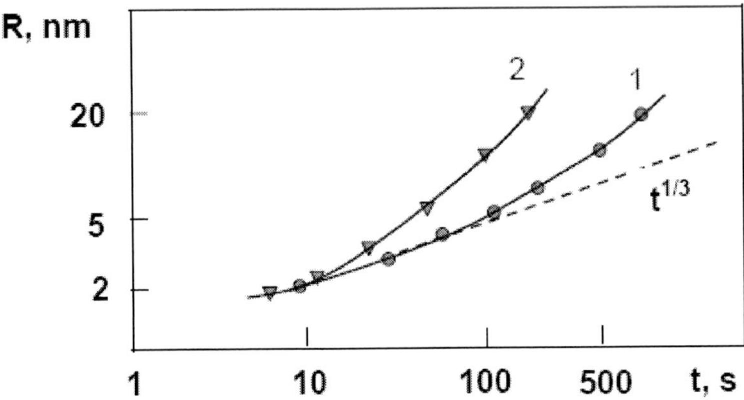

Figure 8. Growth law for Au (1) and $CrSi_2$ (2) films, curve 1 corresponds to curve 2 in figure 7a and curve 2 – to curve 2 in figure 7b.

$$G(u) = \iint_{\{h_1,h_2 \geq 0\}} h_1 h_2 f(h_1,h_2;u)dh_1 dh_2 - <h>^2 \tag{29}$$

and the problem is to find the conditional probability density

$$f(h_1,h_2;u) = P\{h_A = h_1, h_B = h_2; |AB| = u\}. \tag{30}$$

In subsection 3.1 we show that for the surface relief forming at paraboloid growth this function can be found out and hence, ACF can be calculated [16]. Then, in subsection 3.2 we analyze the surface morphology for hemispherical growth model in terms of the average surface roughness parameters.

3.1. Paraboloid Growth

In this model, islands nucleate at random points of a 2D flat substrate (model dimension d=2+1) and during the early stages of deposition they grow in a paraboloid form (figure 9). At later stages due to collisions and sticking of neighboring growing islands continuous film is eventually formed with a steady (in time) and random (in space) surface relief that afterwards simply lifts up with a constant velocity v_n. For this relief, consisting of the randomly distributed growth hillocks, generated by growing islands, the desired function (30) can be found as follows. First, it is more convenient to work on a profile depth (rather than a profile height) and place its reference line at a level of the highest hillock apexes (figure 9). In this model at each substrate point vertical growth with a constant velocity v_n starts immediately after its capture by a growing island. Therefore, the event: the profile depth at a randomly chosen point A at some time moment t equals to $h_A = h_1$ means that this point has been captured by a growing island just at the time moment $t_A = h_1 / v_n$ (figure 9), hence the nearest to the point A nucleus is located at a distance $r_A = R(t_A)$, where $R(t)$ is the lateral growth law.

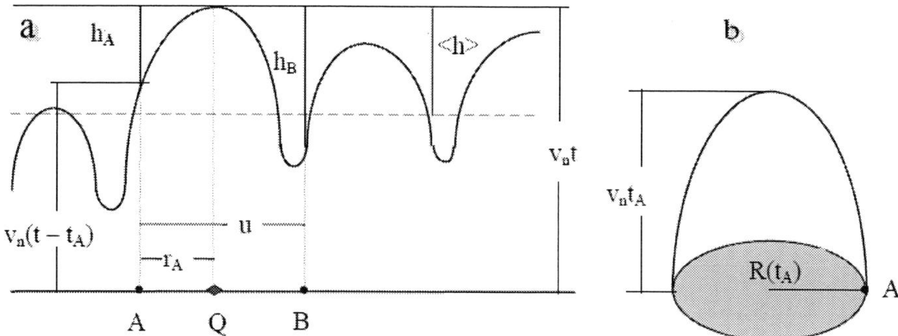

Figure 9. Schematic of a film surface profile (a) and an isolated island at some time moment t_A (b).

In other words, for any substrate point A at the time t there exists some 'dangerous zone' (DZ) around it such that the presence of at least one nucleus within this DZ leads inevitably to its capture by the time t. Due to isotropic lateral growth a DZ for a point A is a circle C_A with a radius $R(t)$ centered at A. The same is valid with respect to a point B, i.e., its DZ is a similar circle C_B centered at a point B. It is more convenient therefore to go over from a profile depth h_A (h_B) to a random distance r_A (r_B) from a point A (B) to its nearest nucleus and respectively to go over from the probability density $f(h_1, h_2; u)$ to the cumulative probability $F_H(h_1, h_2; u) = P\{h_A > h_1, h_B > h_2; |AB| = u\}$, and further to the cumulative probability $F_R(x, y; u) = P\{r_A > x, r_B > y; |AB| = u\}$. Then, the main formula (29) for ACF may be rewritten as

$$G(u) = -\iint_{\{x,y \geq 0\}} h(x)h(y)dF_R(x, y; u) - \langle h \rangle^2 \tag{31}$$

This formula allows calculating ACF if a depth profile $h(r)$, i.e., a hillock shape and hillock spatial distribution (HSD) are known. First we explore the ACF behaviour depending on a hillock shape at a fixed random HSD, and then we study the effect of the HSD at a fixed hillock shape.

Effect of a hillock shape. Let all the islands nucleated with some density N at random substrate points at the initial time t = 0 (IN- model). Then, due to the Poisson distribution of the island centers over a substrate, a probability function $F_R(x, y; u)$ is written down in a simple form

$$F_R(x, y; u) = \exp[-NS(x, y; u)], \tag{32}$$

where S is the area of the joined dangerous zone, $C = C_A \cup C_B$. Depending on the magnitudes of x, y and u there exist four types of the joined DZ shown in figure 10: 1) $x + y < u$, C_A and C_B don't intersect and so $S = \pi (x^2+y^2)$, 2) $x > y + u$, C_A envelops C_B and so $S = \pi x^2$, 3) $y > x + u$, C_B envelops C_A, hence $S = \pi y^2$ and 4) $x + y > u$, $x - y < u$, C_A and C_B are partially overlapped and therefore

$$S = \pi(x^2 + y^2) - \left(x^2 arcosine(\frac{u^2+\rho^2}{2ux}) + y^2 arcosine(\frac{u^2-\rho^2}{2u\rho}) \sqrt{u^2x^2 - (u^2+\rho^2)^2/4} \right),$$

$\rho^2 = x^2 - y^2$, and accordingly the integration field in (31) is divided into four domains (figure 10).

It can easily be shown that for the probability function (32) the average surface profile depth

$$< h >= v_n \int_0^t \exp\left(-\pi R^2(t) N\right) \cdot dt \tag{33}$$

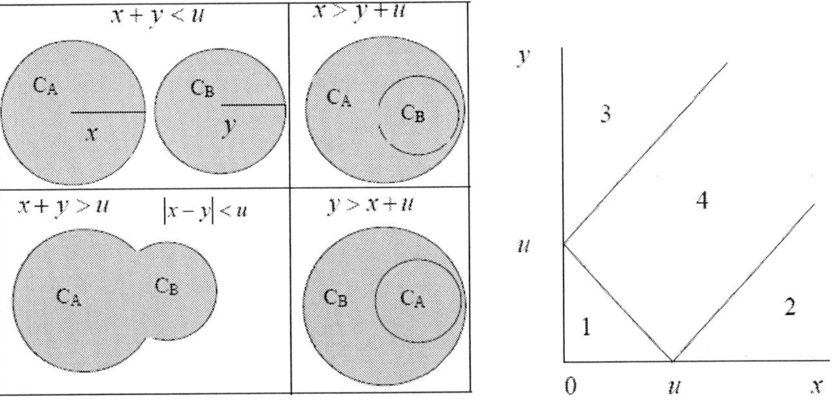

Figure 10. Four types of the joined DZ, $C = C_A \cup C_B$ (left) and corresponding subdivision of the integration field in (31) into four domains (right).

As we saw above, the lateral island growth obeys a simple power law $R(t) = At^p$ that generates a power hillock (depth) profile $h(r) = v_n (r / A)^{1/p}$, where the exponent p depends on the growth regime (substrate temperature) and varies during the growth process between $1/2$ and 1. In this case the average profile depth is calculated exactly

$$< h >= v_n (a / A)^{1/p} \Gamma(1 + 1/(2p)), \tag{34}$$

where $a = (\pi N)^{-1/2}$ and $\Gamma(...)$ the gamma-function. The second moment $<h^2>$, hence the rms roughness $\sigma = \sqrt{G(0)}$, are also obtained in exact analytical form

$$\sigma = v_n (a/A)^{1/p} \left[\Gamma(1+1/p) - \Gamma^2(1+1/(2p))\right]^{1/2} \tag{35}$$

The normalized surface roughness

$$\hat{\sigma} \equiv \frac{\sigma}{<h>} = \left[\frac{\Gamma(1+1/p)}{\Gamma^2(1+1/(2p))} - 1\right]^{1/2} \tag{36}$$

as a function of p is shown in figure 11 and numerically computed ACFs for two characteristic values of an exponent p are shown in figure 12, where $s = u/a$.

It is clearly seen that the normalized surface roughness increases with decreasing p, i.e., with changing a hillock shape from the conic shape $(p = 1)$ to the paraboloid one $(p < 1)$. This can easily be understood as follows. With decreasing p the hillocks become progressively steeper since their profile $h(r) \propto r^{1/p}$ and due to their random spatial distribution there inevitably exist hillocks with large spacing between them, where the deep canyons arise that contribute greatly into a second moment, hence to the roughness. The real roughness increases even stronger because the average profile depth $<h>$ also grows with reducing p.

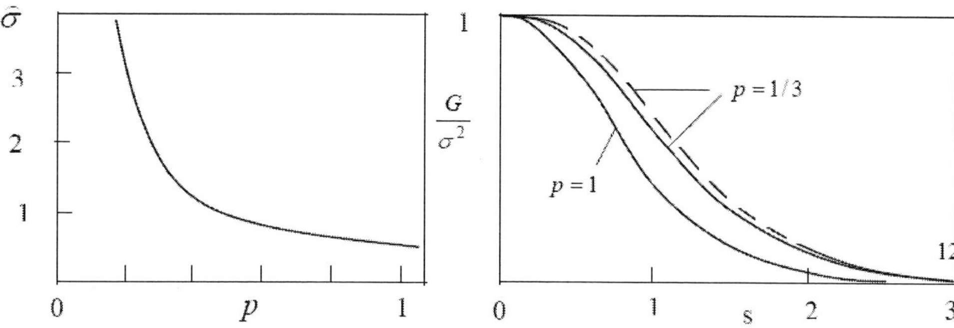

Figure 11. Normalized surface roughness

Figure 12. ACF for two values of p (solid curves) and as a function of p . its Gauss approximation for $p = 1/3$ (broken curve).

As figure 12 demonstrates, computed ACF may be rather well approximated by a Gauss function

$$G(u) = \sigma^2 \exp\left(-u^2 / \Lambda^2\right) \tag{37}$$

with the autocorrelation length

$$\Lambda = \left(\alpha\pi N\right)^{-1/2} \quad \alpha \approx 0.6 - 0.8, \tag{38}$$

hence the roughness spectrum (Fourier transform of ACF)

$$g(k) = \pi\sigma^2\Lambda^2 \exp\left(-k^2\Lambda^2 / 4\right), \tag{39}$$

where $k = 2\pi / L$ is the roughness vector modulus, or spatial frequency and L the spatial period.

These results may directly be compared with experimental data of Ref. [17], where the ACF of the growing surface of thin metallic (Mg, Cu, Ag and Au) films grown on a super smooth quartz substrate has been measured from electron micrographs. These films exhibit just a hillock-like surface morphology assumed in our model, and the main, initial part of the experimental ACF is well approximated by a Gaussian function (37), and the measured quantities σ and Λ are rather well described by a dependence $\sigma(\Lambda)$

$$\Lambda = \frac{A\sigma^p}{\alpha^{1/2} {v_n}^p [\Gamma(1+1/p) - \Gamma^2(1+1/2p)]^{p/2}} \tag{40}$$

resulting from (35) and (38), thus confirming that the lateral correlation length has a simple physical meaning of a characteristic inter-hillock distance.

Effect of the HSD. Due to the extreme difficulty of studying spatially inhomogeneous models in 2+1 dimensions, we turn to the case of 1+1 dimensions, where a DZ becomes a linear segment what significantly simplifies an analysis [18]. For a fixed triangle hillock shape (owing to the constant vertical and lateral growth velocities v_n and v_τ, respectively) we consider three different nucleation mode giving rise three different HSD: 1) All the nuclei instantly formed at random points with linear density N at the time $t = 0$: ideal random HSD (IR-HSD); 2) The nuclei are forming continuously with some constant rate I at random points of uncovered substrate; in this case of finite-size restricted random HSD (FSRR-HSD) the final island (hillock) density $N_\infty = \int_0^\infty I \exp\left(-Iv_\tau t^2\right) dt = \frac{1}{2}\sqrt{\frac{\pi I}{v_\tau}}$, and 3) All the nuclei instantaneously formed at the time $t = 0$ at 1D lattice sites with density N : regular or lattice HSD (L-HSD). With the aim of comparison these versions we suppose $N = N_\infty$, what allows to introduce a unique length scale $a = 1/N$, hence a unique dimensionless distance $s = u/a$. For all these versions ACFs are obtained from (31) in exact analytical form.

For IR-HSD

$$G(s) = \sigma^2 (1+2s)\exp(-2s)\,,\ \sigma = a\eta/2\ \ \eta \equiv v_n / v_\tau\,. \tag{41}$$

For FSRR-HSD

$$G(s) = \sigma^2 \exp(-4s^2/\pi) + \Delta G\,,\ \sigma = \pi\eta a\sqrt{(4-\pi)/\pi}\,/4\,, \tag{42}$$

where a Gauss term (specially picked out) is a main one and

$$\Delta G \equiv \left(\frac{\pi}{4}\eta a\right)^2 \left(\Phi^2\!\left(\frac{2s}{\sqrt{\pi}}\right) - [1-\exp(-\frac{4s^2}{\pi})] + 2\sqrt{2}[\exp(-\frac{2s^2}{\pi})\Phi(\frac{2s}{\sqrt{\pi}}) - \frac{2s}{\sqrt{\pi}}\sqrt{\frac{2}{\pi}}][1-\Phi(2s\sqrt{\frac{2}{\pi}})] \right)$$

($\Phi(z)$ is the error integral) is a small correction: $\Delta G(0) = 0$ and $(\Delta G / G)_{\max} \le 20\%$.

And for L-HSD one naturally gets a periodic function

$$G(s) = \sigma^2 \left[(4s-1)^3 - 3(4s-1)\right]/2\,,\ |s| \le 1/2\,,\ \sigma = \eta a/(4\sqrt{3}) \tag{43}$$

ACFs calculated by these formulas are depicted in figure 13. As expected, the highest roughness is achieved at an ideal random distribution of nuclei because in this case between far distant neighbor nuclei deep canyons arise that increase the roughness, and the smallest roughness is achieved at a regular space distribution of nuclei. In the case of finite-size restricted random nucleation an intermediate situation occurs. The ACF' behaviour with growth parameters is expressed by relations

$$\partial G/\partial v_n > 0\,,\ \partial G/\partial v_\tau < 0 \text{ and } \partial G/\partial N < 0\,,$$

which mean that the roughness augments with increasing the vertical growth rate and decreases with increasing the lateral growth velocity and nuclei density.

In 1+1 dimensions at IR-HSD a simpler form of the probability function (32) enables to obtain additional analytical results.

1) Besides a limiting ($t \to \infty$) ACF's form (41) it is possible to derive its exact time evolution:

$$G(s,\theta)/\sigma^2 = \left[(1+2s)e^{-2s} - e^{-4\theta}\right] + 4e^{-(s+\theta)}\left[e^{-\theta} - |s-\theta|e^{-|s-\theta|}\right],\ 0 \le s \le 2\theta$$

$$G(s,\theta) = 0\,,\ \ s \ge 2\theta \tag{44}$$

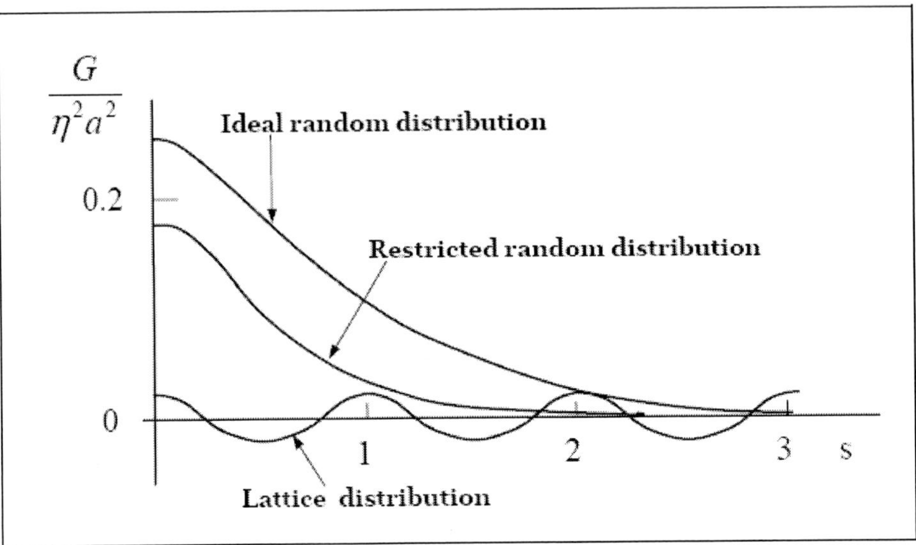

Figure 13. ACF for various hillock space distributions.

where $\theta = v_\tau t / a = v_\tau tN$ is the dimensionless radius or the dimensionless time. It is clear that the function $G(s,\theta)$ tends to its limiting form (41) as $\theta \to \infty$ and figure 14 demonstrates a fast convergence $G(s,\theta) \to G(s)$: they practically coincide at $\theta = 5$, when the normalized mean film thickness $\bar{h} / a = \eta(\theta - \pi / 4) = 4.2\eta$.

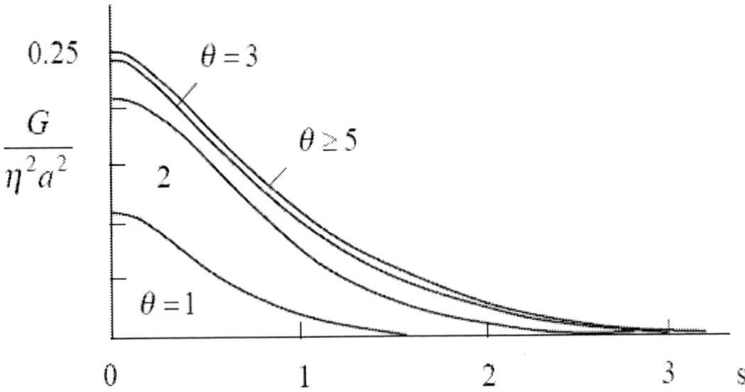

Figure 14. The time evolution of ACF.

2) The consideration of a lateral growth with a power law $R(t) = At^p$ (as in above case of 2+1 dimensions) lets to derive the limiting ACF at different values of an exponent p in exact form:

$$\frac{G(s)}{\sigma^2} = e^{-2s}\left[1 + 2s + \frac{2}{19}(13s^2 + \frac{14}{3}s^3 + 2s^4 + \frac{2}{15}s^5)\right] \quad , \quad \sigma = \frac{3 \cdot 19^{1/2} v_n}{4A^3N^3} \text{ at } p = 1/3 \tag{45}$$

$$\frac{G(s)}{\sigma^2} = e^{-2s}\left[1 + 2s + \frac{2}{5}(3s^2 + \frac{2}{3}s^3)\right], \ \sigma = \frac{5^{1/2} v_n}{2A^2N^2} \text{ at } p = 1/2 \tag{46}$$

and at $p = 1$ one gets naturally (41) with $A = v_\tau$. For the rms roughness an expression at any p can be obtained

$$\frac{\sigma}{<h>} = \left[\frac{\Gamma(1 + 2/p)}{\Gamma^2(1 + 1/p)} - 1\right]^{1/2} \tag{47}$$

with

$$<h> = v_n (a/(2A))^{1/p} \Gamma(1 + 1/p) \tag{48}$$

It is remarkable that introducing a combination $\mu = d'p$ with d' being the substrate dimension allows to get a general formula for σ

$$\frac{\sigma}{<h>} = \left[\frac{\Gamma(1 + 2/\mu)}{\Gamma^2(1 + 1/\mu)} - 1\right]^{1/2} \tag{49}$$

that reduces to (47) and (36) at $d' = 1$ and $d' = 2$, respectively.

3) As we saw above (Eq. (36) and figure 11), in the 2+1 dimension case of IR-HSD the rms roughness increases with decreasing p, i.e., with changing a hillock shape from the conic shape $(p = 1)$ to the paraboloid one $(p < 1)$. In order to gain additional evidence that such an effect of a hillock shape is just obliged to their random spatial distribution, turn back to just considered 1+1 dimension case of a lattice hillock space distribution. It can easily be shown that in this case σ is given by

$$\sigma = \frac{v_n}{p + 1} \cdot \left(\frac{1}{2AN}\right)^{1/p} \cdot \left(\frac{1}{1 + 2/p}\right)^{1/2}, \tag{50}$$

which at p=1 reduces to (43). Calculation results with Eqns. (50) and (47) are presented in figure 15, where for purposes of comparison the roughness in both cases has been normalized to the same quantity $B = v_n(a/(2A))^{1/p}$. These plots brightly demonstrate that the hillock shape greatly affects on the roughness only at a random hillock space distribution, especially at most relevant for growing films $p < 1$ (at $p = 1/3$ $\hat{\sigma} = 26$), whereas at the lattice hillock space distribution the roughness is very small and almost independent of the hillock shape.

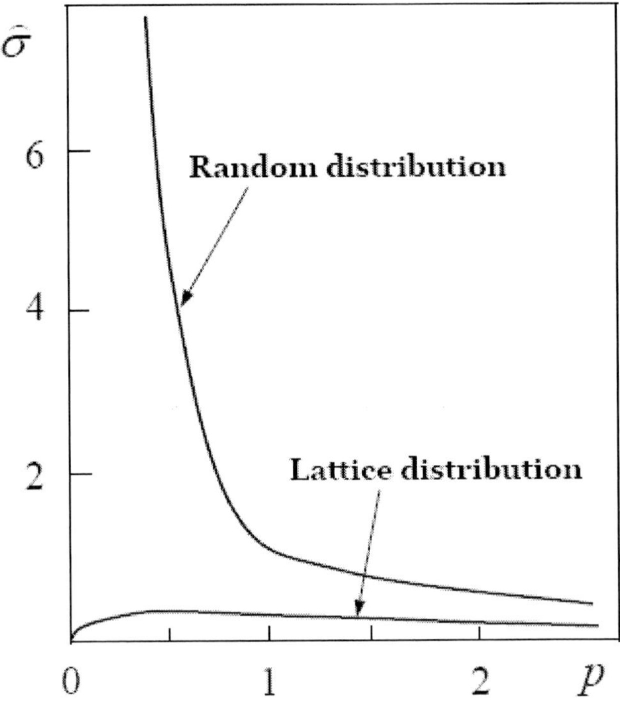

Figure 15. The rms roughness as a function of p at various hillock space distributions.

It's worth while to note that in [19] the model described has been extended to both the one-dimensional elliptical case and to the pyramidal two-dimensional one and the kinetics of σ and ACF have been obtained, and it has been shown that properly rescaled σ is universal function of coverage that obey a power law with the exponent depending upon island shape.

Effect of deposition conditions on the surface roughness of a growing film. Let us designate the coverage of a film section at some height h from a substrate as $\xi(h,t)$, which is obviously ever-decreasing function of h: $\xi(0,t) \equiv \xi(t)$ - substrate coverage and $\xi(h,t) \equiv 0$ at $h \geq h_{\max}$, $h_{\max} = b + v_n \cdot t$ - the maximum surface height. The quantity $\xi(h,t)$ coincides obviously with the probability that a surface height above randomly chosen substrate point at the time t is not less than h, so the probability distribution density of the surface height may be written down as

$$-\partial\xi(h,t)/\partial h \quad 0 \leq h \leq h_{\max}$$

$$f_t(h) = (1-\xi(t))\delta(h) + \{ \qquad (51)$$

$0,\ h_{max} \le h$

where the first term with the delta-function accounts for the contribution of uncovered parts of a substrate whose fraction is $(1-\xi(t))$. In this model at every substrate point vertical growth with a constant velocity $v_n = J\Omega$ starts just after its capture by a growing island, so the event: some point A at height h is captured by the time moment t is equivalent to capture of its projection on a substrate by the earlier time moment $t' = t - (h-b)/v_n$ and therefore

$\xi(t),\ h \le b$

$$\xi(h,t) = \{\ \ \xi(t - \frac{h-b}{v_n}),\ b \le h \le b + v_n t \qquad (52)$$

$0,\ h \ge b + v_n t$

Let all the islands nucleated instantly at the time $t = 0$ at random points with areal density N. Then, due to the Poisson distribution of island centers on a substrate, the coverage is given by (12) or

$$\xi(\tau) = 1 - \exp\left(-\rho^2(\tau)\right) \qquad (53)$$

in dimensionless variables $\rho = R/a$, $a = (\pi N)^{-1/2}$, $\tau = Jb^2 t$, $b = \Omega^{1/3}$, so for the rms roughness

$$\sigma^2 = \int_0^\infty (h-\overline{h})^2 f_t(h)dh \qquad (54)$$

calculations with (51)-(53) yield

$$\tilde{\sigma}^2(\tau) = \xi(\tau) - \overline{h'}^2 + 2\int_0^\tau \overline{h'}(\tau')d\tau',\ \tilde{\sigma} = \sigma/b. \qquad (55)$$

The mean dimensionless film thickness $\overline{h'} \equiv \overline{h}/b = \gamma_t \cdot \tau$, where the total condensation coefficient

$\gamma_t = \frac{1}{\tau}\int_0^{\tau} \gamma_i(\tau')d\tau'$ with the instantaneous condensation coefficient γ_i equal obviously to a feeding zone area, $\gamma_i = 1 - \exp\left[-(\rho + \sqrt{z})^2\right]$, i.e.,

$$\overline{h'}(\tau) = \tau - \int_0^{\tau} \exp\left[-(\rho(\tau') + \sqrt{z})^2\right] \cdot d\tau', \tag{56}$$

whence for the case of incomplete condensation (where according to (19), growth law has a form $\rho \cong \sqrt{z} \cdot \tau$) one finally gets

$$\overline{h'}(\tau) = \tau - \sqrt{\pi / 4z} \cdot \left[\Phi(\sqrt{2z} \cdot (\tau + 1)) - \Phi(\sqrt{2z})\right], \tag{57}$$

where $\Phi(y)$ is the error integral..

Differentiating (55) yields $2\tilde{\sigma} \cdot d\tilde{\sigma} / d\tau = d\xi / d\tau + 2\overline{h'} \cdot (1 - \gamma_i) \geq d\xi / d\tau \geq 0$, since $\gamma_i \leq 1$, hence $\tilde{\sigma}$ is a non-decreasing function of time.

In the complete condensation regime, $\gamma_i \equiv 1$, hence $\overline{h'} = \tau$, whence $\tilde{\sigma}^2(\tau) = \xi(\tau)$, i.e., $\tilde{\sigma} \xrightarrow[\tau \to \infty]{} 1$.

In the incomplete condensation regime, the roughness kinetics can be computed by (55) in association with (53) and (57). Results of such computations at several values of z are shown in figure 16, where one can see that the rms roughness at first increases with time and then saturates to some limiting value, because in the paraboloid growth model a steady in time relief is eventually formed.

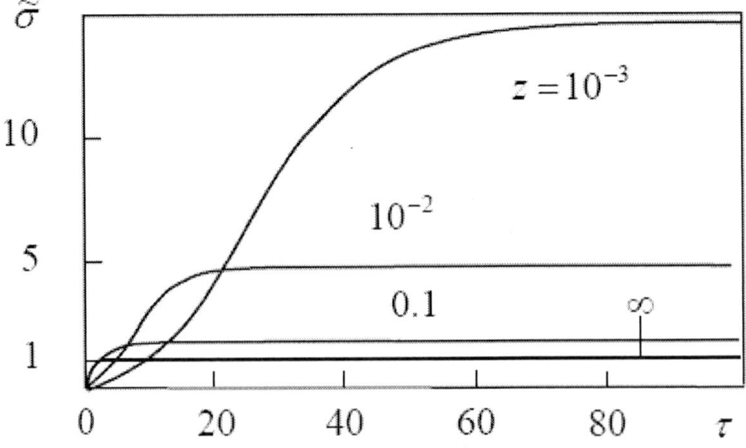

Figure 16. Roughness kinetics in various regimes of condensation.

The saturation rms roughness naturally increases with decreasing z, i.e., increasing substrate temperature. For description this steady relief that is formed at both complete and incomplete condensation it is convenient to go over to the deviation of the surface height from its average, $\widetilde{h} = h' - \overline{h'}$, then a film thickness has no importance and may be as much as desired. The distribution of $\widetilde{h}$ can be obtained from (52) by a passage to the limit $t \to \infty$. In limiting cases of extreme incomplete and complete condensation this distribution and thus the saturation rms roughness are derived exactly.

In the extreme incomplete condensation ($z << 1$)

$$f(\widetilde{h}) = \begin{cases} 2z(h_z^* - \widetilde{h}) \cdot \exp[-z(h_z^* - \widetilde{h})^2], \ \widetilde{h} < h_z^* \\ 0, \ \widetilde{h} > h_z^*, \end{cases} \tag{58}$$

where $h_z^* = 1 + \sqrt{\pi / 4z} \cdot [1 - \Phi(\sqrt{2z})]$ is a maximum value of $\widetilde{h}$. From here it follows

$$\widetilde{\sigma} \equiv \sqrt{< \widetilde{h}^2 >} = \left[(h_z^*)^2 + \frac{1}{z} - h_z^* \cdot \sqrt{\frac{\pi}{z}} \right]^{1/2} \tag{59}$$

Notice that at $z << 1$, $h_z^* \approx \sqrt{\pi / 4z}$, therefore (59) simplifies

$$\widetilde{\sigma} \approx \left[(1 - \frac{\pi}{4}) \cdot \frac{1}{z} \right]^{1/2} \tag{60}$$

Calculation results by (59) presented in figure 17 show on how the saturation rms roughness drops with increasing z.

In the complete condensation regime $\rho = \tau^{1/2}$ and $\overline{h'} = \tau$, so

$$f(\widetilde{h}) = \begin{cases} \exp\left[-(1 - \widetilde{h})\right], \ \widetilde{h} \le 1 \\ 0, \ \widetilde{h} \ge 1 \end{cases} \tag{61}$$

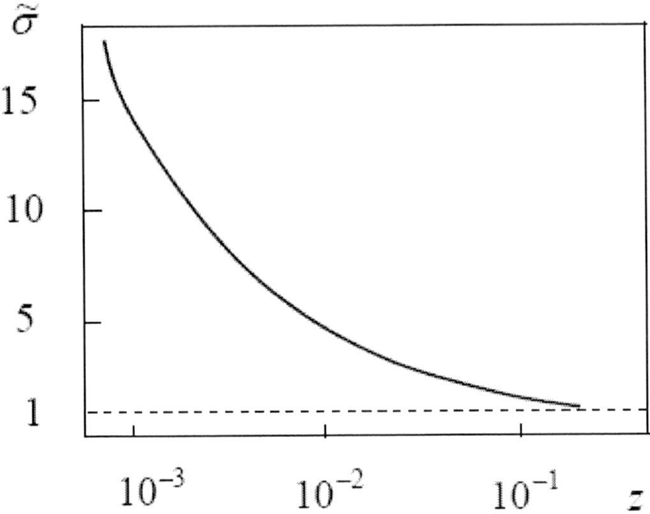

Figure 17. The saturation rms roughness as a function of the parameter z .

whence $\tilde{\sigma} = 1$, i.e., in this regime a film with very smooth surface ($\sigma = b \cong 0.3nm$) is formed. Turning back to figure 17 we see that (59) deduced under condition $z << 1$ predicts correct roughness values up to $z \approx 0.1$. Figure 18 visually demonstrates how the surface width narrows down with increasing z. This plot shows also that the height distribution is asymmetric, non-Gaussian with a negative skewness what is more consistent with available experimental data than a widely spread hypothesis of a Gaussian height distribution of the random rough surface[1] which in addition is often combined with a conjecture of a Gaussian correlation function. Generally speaking, the height distribution and correlation function may be independent and above results tell us that the random rough surface with undoubtedly non-Gaussian height distribution is characterized by close-to- Gaussian height-height correlation function.

In optical thin film applications, for characterizing a film microstructure is used the so-called packing density factor (PDF) defined as a ratio of a volume of the solid part of the film to the total volume occupied by a film [22]. In our model, this quantity is written down as

$$P = \overline{h'} / h_{\max}, \; h'_{\max} = \tau + 1 \qquad (62)$$

where $\overline{h'}$ is given by (56). Calculation results with (62) for several values of z are shown in figure 19.

As can be seen the higher the parameter z (i.e. the lower substrate temperature) the more compact film is formed. These plots may be used for estimation of a contribution of the surface roughness associated film porosity to an overall film porosity.

---

[1] Thus, until recent time all the diffraction theories from random rough surface had been based on the assumption of a Gaussian surface height distribution (e.g., [20]), and only recently a diffraction theory for non-Gaussian rough surfaces has become to developed [21].

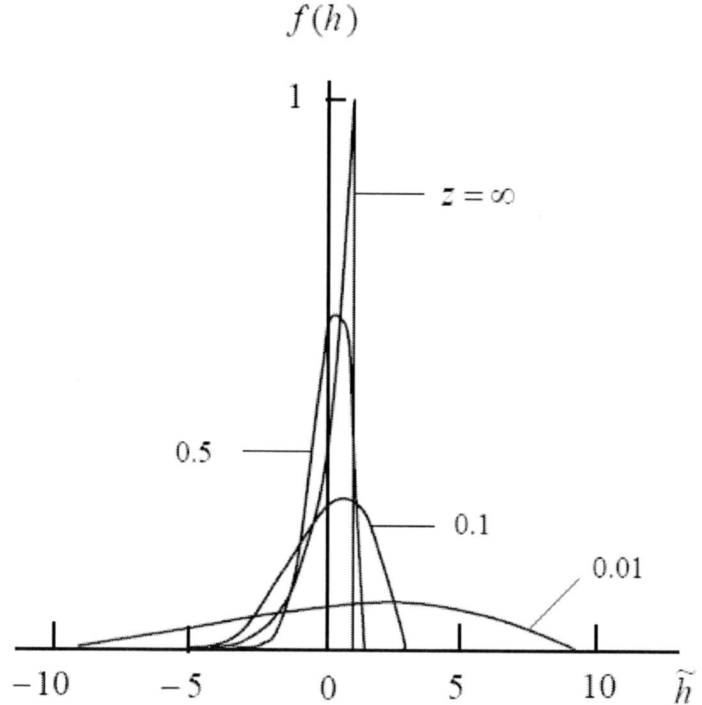

Figure 18. The surface height distribution in various regimes of condensation.

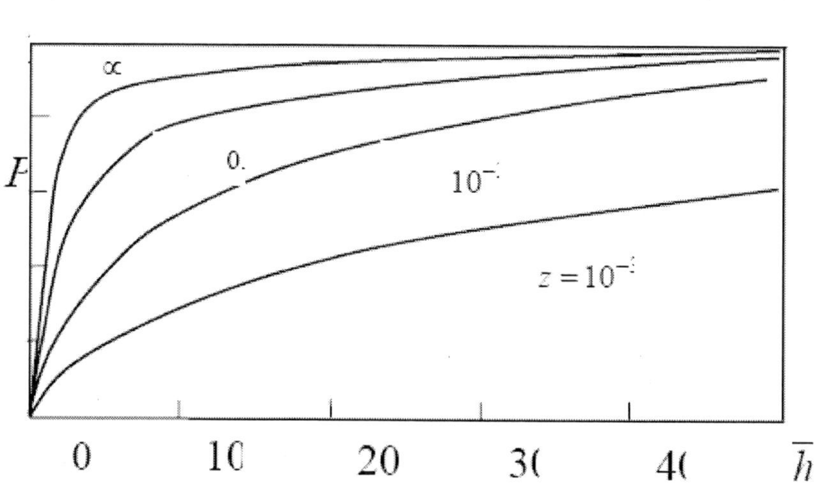

Figure 19. Packing density factor kinetics in various regimes of condensation.

3.2. Hemispherical Growth

In this model a freely growing island has a hemispherical shape; after collisions islands cease their growth where they meet, but continue to grow in all other available directions with a law $R(t)$ that has been found in section 1. Consider a film profile at some time moment t (figure 3) and designate as $\xi(h,t)$ the coverage of a film section at some height h from a substrate. Since $\xi(h,t)$ coincides with a probability that a profile height above a randomly chosen substrate point at time t is not less than h, the probability distribution density of a profile height is written down as

$$f_t(h) = (1-\xi(t))\delta(h) + \begin{cases} -\partial\xi(h,t)/\partial h & 0 \le h \le R \\ 0, & R \le h, \end{cases} \tag{63}$$

where the first term with the delta-function accounts for the contribution of the uncovered parts of a substrate, whose fraction is $(1-\xi(t))$. In the case of the instantaneous nucleation of all the islands at the time $t = 0$ at random points with surface density N, $\xi(h,t) = 1 - \exp[-\pi N(R^2 - h^2)]$ and $\xi(t)$ is given by (12), hence for a mean film thickness $\overline{h} = \int_0^\infty h f_t(h) dh$ one gets

$$\overline{h'} = \tau - \int_0^\tau \exp[-(\rho(\tau') + z^{1/2})^2] d\tau' \quad \overline{h'} \equiv \overline{h}/a \tag{64}$$

in dimensionless variables. Numerical computations with (64) show (figure 20) that similar to the growth law (figure 4), with increasing z kinetic curves $\overline{h'}(\tau)$ tend asymptotically towards a limiting law $\overline{h'} = \tau$ for a complete condensation and at $z = 1$ almost merge it, and any kinetic curve $\overline{h'}(\tau)$ tends with time to a linear law $\overline{h'} \propto \tau$, because, as already mentioned, any condensation regime with increasing substrate coverage changes towards the complete condensation.

Now we can calculate the variety of the surface morphology characteristics. Among them one of the most important is the rms roughness $\sigma \equiv \left[\int_0^\infty (h-\overline{h})^2 f_t(h) dh \right]^{1/2}$ for which using (63) one obtains

$$\tilde{\sigma}^2(\tau) = \rho^2(\tau) - \overline{h'}^2(\tau) - \xi(\tau), \; \tilde{\sigma}(\tau) \equiv \sigma / a \tag{65}$$

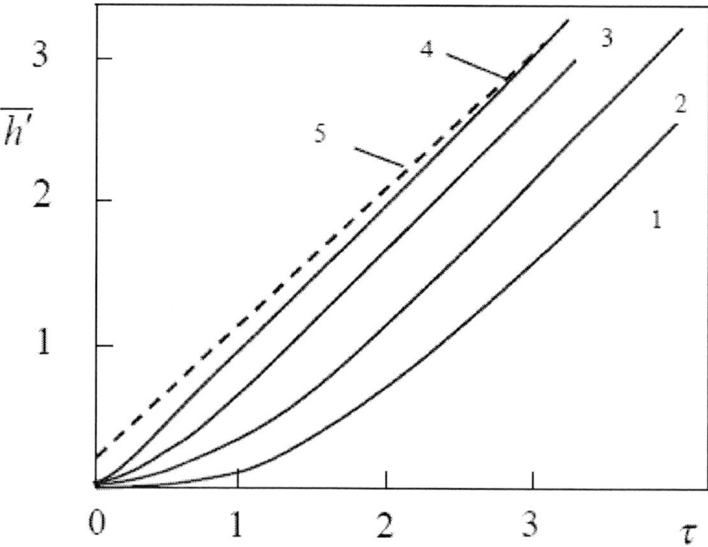

Figure 20. The time dependence of the mean film thickness at different values of the parameter z : 10^{-4}(1), 10^{-2}(2), 10^{-1}(3), 1(4) and ∞ (5).

Equation (65) combined with Eqs. (12), (16) and (64) yields the $\tilde{\sigma}(\tau)$ curve with a maximum (figure 21) what has a simple explanation: initially the nucleation and growth of islands increase the roughness but at later stages due to islands impingement and filling of interisland voids the roughness decreases with time. With decreasing z (i.e. increasing substrate temperature) the curve $\tilde{\sigma}(\tau)$ moves from one limiting curve ($z = \infty$) corresponding to complete condensation regime to another one ($z = 0$) corresponding to extremely incomplete condensation regime, because of the growth rate decrease caused by desorption of adatoms. While the maximum position (τ_m) varies with the regime of condensation: τ_m increases with a decrease of the parameter $z = \pi N X_a^{\ 2}$, characterizing the condensation regime, the maximum height σ_m doesn't depend on the growth regime, $\sigma_m \cong 0.457$ and to all τ_m values the same coverage value $\xi_m \cong 0.856$ corresponds. Indeed, in this model both quantities ρ and $\overline{h'}$ are universal functions of coverage: $\rho(\xi)$ is given by (12), and $\overline{h'}$ given by (64) may also be rewritten as a function of coverage Therefore $\tilde{\sigma}$ given by (65) may also be represented as a universal function of either coverage or film thickness. The first of them, $\tilde{\sigma}(\xi)$ is depicted by a curve (figure 22a) with a maximum $\tilde{\sigma}_m \cong 0.457$ just prior completed film formation, at $\xi_m \cong 0.856$. The second one, $\tilde{\sigma}(\overline{h'})$ that may be more useful in applications, especially when one deals with a continuous film, is shown in figure 22b, where as an example the absolute values of σ and $\overline{h}$ at typical for metallic films on insulator substrates interisland distance $a = 25nm$ are indicated.

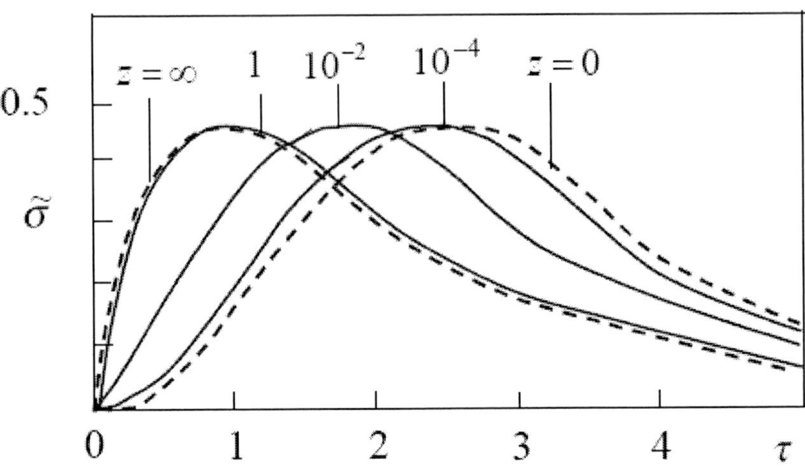

Figure 21. Surface roughness kinetics in different condensation regimes.

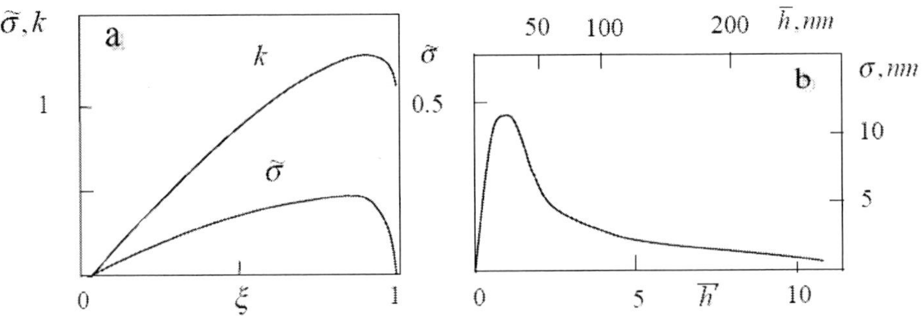

Figure 22. The rms roughness and the roughness coefficient as a function of coverage (a) and the rms roughness as a function of the mean film thickness (b).

The nonmonotonic σ behavior during film growth suggests that its physical properties sensible to the surface roughness should also display a similar behavior with changing film thickness. As an example it is reasonable to consider some of the so called surface enhanced phenomena, e.g., the surface enhanced infrared spectra, induced by the great enhancement of the incident electromagnetic field around the small metal islands through the plasmon excitation, resulting in the enhanced absorption spectra of adsorbates. Namely, in Ref. [23] the infrared absorption spectra of azobenzene monolayers adsorbed on silver island films of various thickness (10, 25 and 45 nm) on CaF_2 substrate were studied and it was revealed that the infrared absorption is greatly enhanced due to the presence of the silver island film, and the enhancement effect reaches its maximum just on the 25-nm thick silver film, which according to the atomic-force microscope observations has the 'roughest' surface.

At $\xi \cong 1$ (64) simplifies: $\overline{h'} \cong \rho - 1/2\rho$, i.e., $\rho \cong \overline{h'} + 1/2\overline{h'}$, so from (65) it follows that in the descending branch of the $\tilde{\sigma}(\overline{h'})$ curve in figure 22b, i.e., in a continuous film ($\overline{h'} > 3$) the rms roughness decreases with film thickness in a simple law

$$\tilde{\sigma} \approx 1/2\overline{h}', \tag{67}$$

i.e., rather slowly so that in a film of finite thickness a finite surface relief is formed, which depends on a scale $a = (\pi N)^{-1/2}$, i.e., on the island density. For example, at typical for metallic films on insulator substrates, $a = 25nm$ in $60-nm$ film $\sigma \cong 3nm$ in consistence with experimentally measured roughness (e.g., [17]). In real variables (67) becomes

$$\sigma = \frac{a^2}{2\overline{h}} \tag{68}$$

or, noticing that the average interisland distance a is equal to the average island size $2R$ that is proportional to the average film thickness $\overline{h}$,

$$\sigma \propto \overline{R} \tag{69}$$

Eqns. (68), (69) suggest that reducing the island size by increasing the nucleation density will result in smoother film surface. This was confirmed experimentally in [24], where a two-step growth process separating the nucleation and growth phases has been proposed and implemented by deposition either at a lower substrate temperature or higher deposition rate during the initial nucleation stage, and then reversing one or the other at the growth stage. Using this two-step deposition process the lithitium niobate films with smooth surface ($\sigma < 2nm$), excellent crystalline quality and optical losses below 1.8 dB/cm at a wavelength 632.8 nm were grown on the sapphire substrate.

Another important surface relief characteristic is the mean surface depth $\overline{h_d'} = \rho - \overline{h'}$. As can be seen in figure 23a, its kinetics behaviour depending on the condensation regime is similar to that of the surface roughness (figure 21). This quantity is also a universal function of either coverage or film thickness. The first of them, $\overline{h_d'}(\xi)$ is also depicted with the unimodal curve with an earlier ($\xi_m \cong 0.6$) maximum $\overline{h_{d_m}'} \cong 0.54$ (figure 23b).

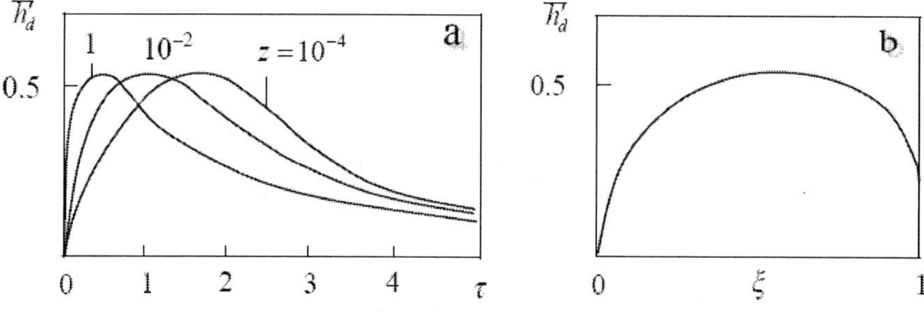

Figure 23. The average surface depth as a function of time in various regimes of condensation (a) and as a function of coverage (b).

The next significant surface morphology characteristic (e. g., for adsorption film properties) is a roughness coefficient k defined as a specific actual film surface. It coincides obviously with a specific growth front surface, i.e., equals to a derivative of a specific film volume (or average film thickness) with respect to a radius: $k = dV / dR = d\overline{h} / dR = d\overline{h'} / d\rho$, whence

$$k = 2\rho(\rho - \overline{h'}) . \qquad (70)$$

This quantity is also a universal function of ξ or $\overline{h'}$ and the former is depicted by a curve (figure 22a) similar to that of $\widetilde{\sigma}(\xi)$ with a slightly later ($\xi_m = 0.895$) maximum $k_m = 1.285$.

In continuous film, throwing off the first term in (62) for the probability distribution density of a deviation of a surface height from its average, $\widetilde{h} = h' - \overline{h'}$ one gets

$$f(\widetilde{h}) = \begin{cases} 2(\widetilde{h} + \overline{h'}) \cdot \exp\left[-\left(1 + \dfrac{1}{4\overline{h'}^2} - \widetilde{h}^2 - 2\widetilde{h}\overline{h'} \right) \right], & \widetilde{h} \le \dfrac{1}{2\overline{h'}} \\ 0, & \widetilde{h} > \dfrac{1}{2\overline{h'}} . \end{cases} \qquad (71)$$

Figure 24 shows that the surface width narrows down with film thickness and that the surface height distribution is asymmetric, non-Gaussian, with a negative skeweness, likewize the above considered case of the paraboloid growth (figure 18).

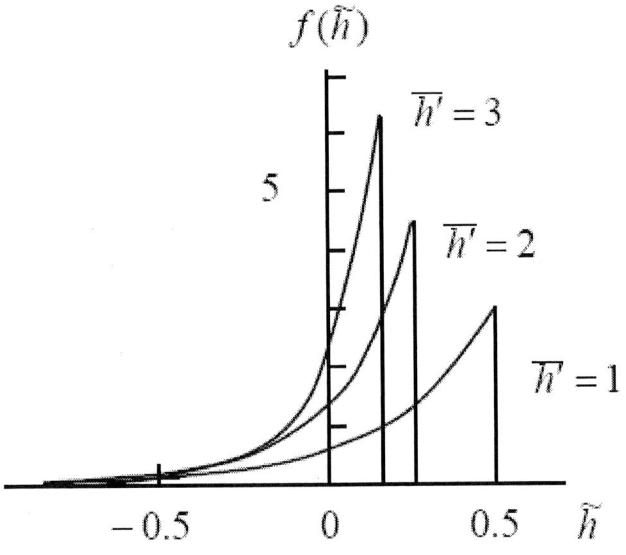

Figure 24. The time evolution of the surface height distribution.

Finally, the packing density factor (PDF), in this model is defined as

$$P = \bar{h} / h_g = \bar{h}' / h_g', \tag{72}$$

where the geometric film thickness h_g coincides with a maximum island radius $h_g = R_m(t)$. Calculation results with (72) show on (figure 25) how the PDF tends to unity with increasing film thickness because of the surface roughness decrease. This plot may be used for estimation of a contribution of surface roughness associated porosity to an overall film porosity.

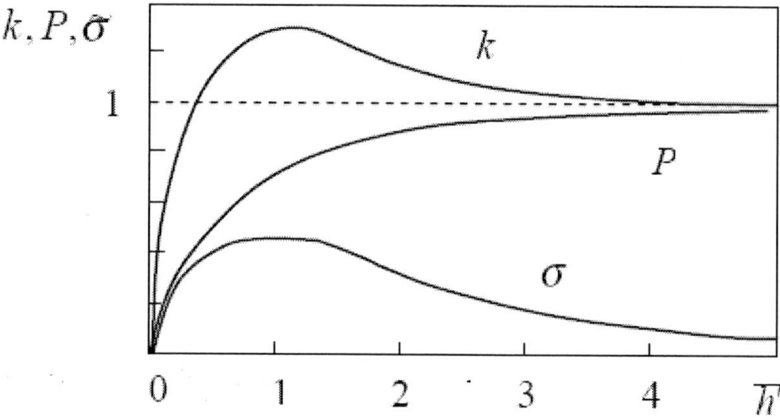

Figure 25. The roughness coefficient, packing density factor and rms roughness as a function of the mean film thickness.

4. Crystallization Kinetics

In analyses of experimental data on phase transformation kinetics in thin films, e.g., crystallization of amorphous films under annealing or aging, or high-energy irradiation until recent time the Kolmogorov – Johnson – Mehl - Avrami (KJMA) statistical model of crystallization [11-13] has been widely using [25-38], despite one of its main premises is that of an infinite medium. In this section, a modified KJMA model of crystallization kinetics in thin amorphous films and surface layers that takes into account a finite film thickness is presented.

4.1 Model

We study two variants of the crystallization process in a film: common volume induced crystallization (VIC) when the crystallites randomly nucleate over a whole film volume and surface induced crystallization (SIC) when they nucleate on a film/substrate interface (figure 26).

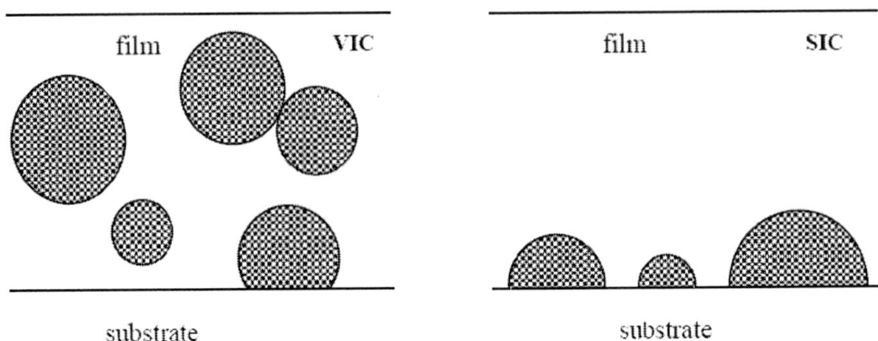

Figure 26. Schematic of a volume induced crystallization and a surface induced crystallization.

And two nucleation mode are considered: continuous nucleation during all the process with some intensity $I(t)$ at random points of untransformed volume (CN) and instantaneous random nucleation (IN) of all the crystallites at the onset of crystallization ($t = 0$) with density N; note that the IN-model is a special case of the CN-model at $I(t) = N(t) \cdot \delta(t)$ with $\delta(t)$ being the delta-function. The crystallites grow isotropically with a linear rate $v(t)$, so that by time t the radius of a crystallite nucleated at some time moment $t' \leq t$ becomes

$$R(t',t) = \int_{t'}^{t} v(\tau)d\tau \,. \tag{73}$$

After impingements crystallites cease growth at their contact boundaries but continue to grow in all other available directions.

Let A is some point at height z from a substrate (figure 27); the probability $q(z,t)$ that this point will survive by time t, i.e. will not be captured by a crystalline phase, may be written down as [39]

$$q(z,t) = \exp\left(-\int_{0}^{t} \mu_A(\tau)d\tau \right), \tag{74}$$

where $\mu_A(\tau)d\tau$ is the probability of formation of an "aggressor"- a nucleus in a "dangerous zone" (DZ) at time $\tau, \tau + d\tau$ – that can capture point A by time t. For a VIC-model a DZ is a sphere of radius $R = R(\tau,t)$ centered at point A with a volume $V(z,R)$ and for a SIC-model a DZ is an intersection of this sphere and a substrate with an area $S(z,R)$.

Thus $\mu_A(\tau)d\tau = I(\tau)V(z,R)d\tau$ for a CN-VIC, hence

$$q(z,t) = \exp\left(-\frac{I}{v} \int_0^{R_0} V(z,R)dR \right) \text{ CN-VIC} \tag{75}$$

where $R_0 = R(0,t) = vt$ and

$$q(z,t) = \exp\left(- NV(z,R_0)\right) \text{ IN-VIC} \tag{76}$$

In the SIC-model, in Eqns. (75) and (76) the quantity $V(z,R)$ should be merely substituted with $S(z,R)$ and the "volume" parameters I $(cm^{-3}s^{-1})$ and $N\,(cm^{-3})$ – with corresponding 'areal' ones $\widetilde{I}$ $(cm^{-2}s^{-1})$ and $\widetilde{N}\,(cm^{-2})$.

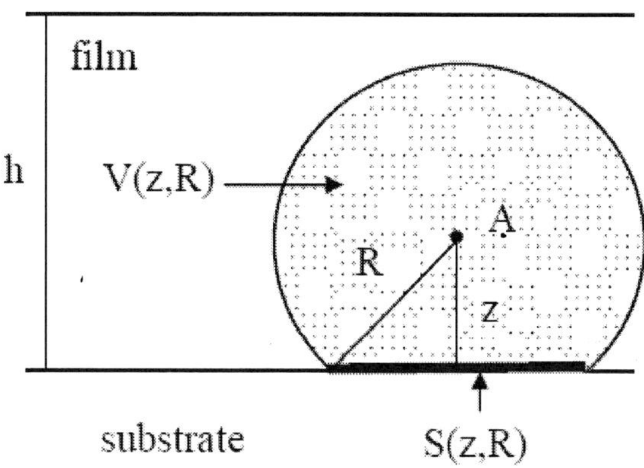

Figure 27. Schematic of a dangerous zone.

4.2. Crystallization Profile

In the VIC-model, the crystallization process is obviously symmetric relative to a central plane $z = h/2$, it is sufficient therefore to consider only $z \subset [0, h/2]$. A simple geometrical arguments (figure 27) show that a volume of a dangerous zone

$$V(z,R) = \frac{\pi}{3} \begin{cases} 4R^3, & R \le z \\ 2R^3 + 3R^2 z - z^3, & z \le R \le h-z \\ 3R^2 h - z^3 - (h-z)^3 & h-z \le R \end{cases} \tag{77}$$

Likewise, in the SIC-case a dangerous zone area

$$S(z,R)=\begin{cases}0, R\le z\\ \pi(R^2-z^2), z\le R\end{cases} \tag{78}$$

Hereafter we will suppose $I(\tau)=cons\tan t\equiv I$, $v(\tau)=cons\tan t\equiv v$ and go over to dimensionless variables: $\tau=t/t_0$ and $\tilde{z}=z/a$ with natural length and time scales: $a=(\pi I/3v)^{-1/4}$ for CN-VIC, $a=(4\pi N/3)^{-1/3}$ for IN-VIC, $a=(\pi\tilde{I}/3v)^{-1/3}$ for CN-SIC, $a=(\pi\tilde{N})^{-1/2}$ for IN-SIC, and $t_0=a/v$ for all versions. Then, inserting (77) and (78) into (75) and (76), respectively yields

$$q(\tilde{z},\tau)=\exp\begin{cases}-\left[\tau^4\right], \tau\le\tilde{z}\\ -\left[\tilde{z}^4+\frac{1}{2}(\tau-\tilde{z})(\tau+\tilde{z})^3\right], \tilde{z}\le\tau\le\lambda-\tilde{z} \\ -\left[\tau^3\lambda-\tau\left[\tilde{z}^3+(\lambda-\tilde{z})^3\right]+\frac{1}{2}\left[\tilde{z}^4+(\lambda-\tilde{z})^4\right]\right], \lambda-\tilde{z}\le\tau\end{cases} \quad \text{CN-VIC} \tag{79}$$

$$q(\tilde{z},\tau)=\exp\begin{cases}-\left[\tau^3\right], \tau\le\tilde{z}\\ -\left[\frac{1}{2}\tau^3+\frac{3}{4}\tau^2\tilde{z}-\frac{1}{4}\tilde{z}^3\right], \tilde{z}\le\tau\le\lambda-\tilde{z} \\ -\left[\frac{3}{4}\tau^2\lambda-\frac{1}{4}\tilde{z}^3-\frac{1}{4}(\lambda-\tilde{z})^3\right], \lambda-\tilde{z}\le\tau\end{cases} \quad \text{IN-VIC} \tag{80}$$

$$q(\tilde{z},\tau)=\begin{cases}1, \tau\le\tilde{z}\\ \exp\left[-(\tau-\tilde{z})^2(\tau+2\tilde{z})\right], \tilde{z}\le\tau\end{cases} \quad \text{CN-SIC} \tag{81}$$

$$q(\tilde{z},\tau)=\begin{cases}1, \tau\le\tilde{z}\\ \exp\left[-(\tau^2-\tilde{z}^2\right], \tilde{z}\le\tau\end{cases} \quad \text{IN-SIC} \tag{82}$$

Now the model contains a single parameter, dimensionless film thickness $\lambda=h/a$. Noticing that $q(\tilde{z},\tau)$ coincides simply with the parent phase volume fraction on a height $\tilde{z}$, hence Eqns. (79) - (82) allow calculating the time evolution of the crystalline volume fraction profile $\xi(\tilde{z},\tau)=1-q(\tilde{z},\tau)$.

Examples of such calculations are presented in figure 28. It is clearly seen a distinctly different profile distribution across a film in VIC- and SIC-models. In a VIC-model, the crystalline phase volume fraction reaches a maximum in a film middle, because inhere crystallization can proceed thanks to crystallites growing from all directions, whereas near a film/substrate interface or external film surface it can occur only due to the crystallites growing from within a film volume. In a SIC-model, the crystalline phase volume fraction naturally falls inwards a film because nuclei germinate only on a film/substrate interface.

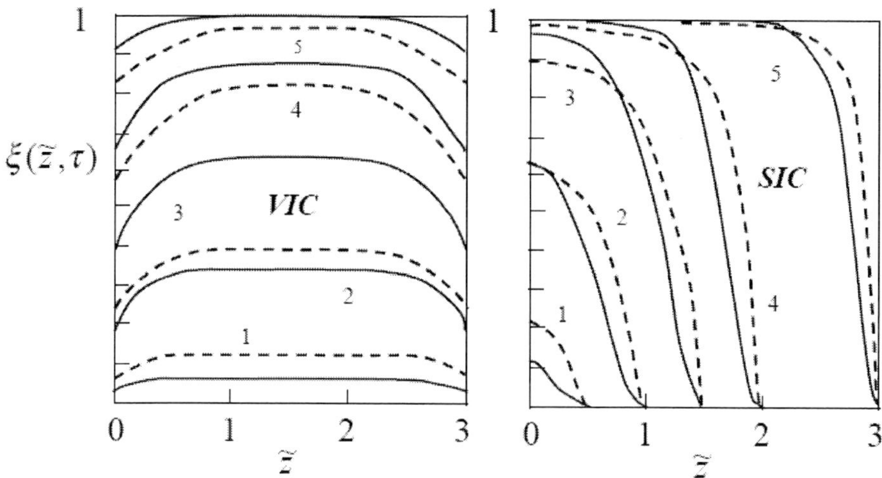

Figure 28. The temporal evolution of the crystallized volume fraction profile across a film of a thickness λ=3 for VIC and SIC at CN (solid curves) and IN (broken curves); at VIC, τ=: 0.5(1), 0.8(2), 1(3), 1.2(4) and 1.5(5), at SIC, τ=: 0.5(1), 1(2), 1.5(3), 2(4) and 3(5).

In the CN-VIC the nucleation frequency is proportional to the uncrystallized volume fraction at a given time moment, so a local crystallites population density at a distance $\tilde{z}(\leq \lambda/2)$ at the time τ is

$$N(\tilde{z},\tau) = I\int_0^\tau q(\tilde{z},\tau')d\tau' \tag{83}$$

And because the crystalline phase distribution profile $\xi(\tilde{z},\tau)$ achieves a maximum in a film center,

the crystallites concentration has a minimum in a film center and increases towards its borders. This spatially nonuniform microstructure remains in completely crystallized film $(\tau \to \infty)$ as is demonstrated in figure 29, where is represented the final crystallites density profile $N(\tilde{z},\infty) \equiv N(\tilde{z})$:

$$\frac{N(\tilde{z})}{N_\infty} = \frac{1}{\Gamma(5/4)} \cdot \int_0^\infty q(\tilde{z},\tau)d\tau \tag{84}$$

$N_\infty = \Gamma(5/4)\cdot\left(3I^3/\pi v^3\right)^{1/4}$ is the crystallites density in bulk material, $\Gamma(x)$ - the gamma-function.

Figure 29 shows also that the crystallites population in a film is always larger than in bulk material.

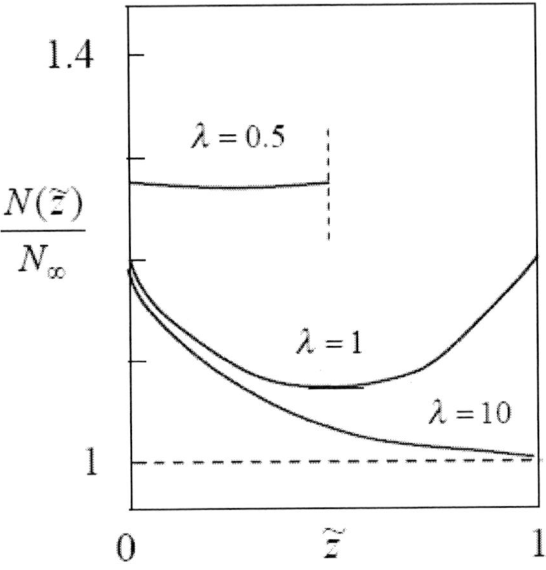

Figure 29. Final crystallites density profile in films of different thickness.

The average overall crystallites concentration in fully crystallized film

$$\overline{N} = \frac{2}{\lambda} \cdot \int_0^{\lambda/2} N(\tilde{z}) d\tilde{z} \tag{85}$$

drops with film thickness (figure 30) towards its bulk value N_∞ because a relative role of a film surfaces effect weakens with increasing its thickness, at small λ - in a simple power law

$$\frac{\overline{N}}{N_\infty} = \frac{\Gamma(4/3)}{\Gamma(5/4)} \cdot \lambda^{-\frac{1}{3}} \tag{86}$$

4.3. Crystallization Kinetics

The above-presented results allow calculating the crystallization kinetics

$$\xi(\tau) = \frac{1}{\Lambda} \int_0^{\Lambda} \xi(\tilde{z}, \tau) d\tilde{z} \ , \tag{87}$$

where $\Lambda = \lambda/2$ for VIC and $\Lambda = \lambda$ for SIC.

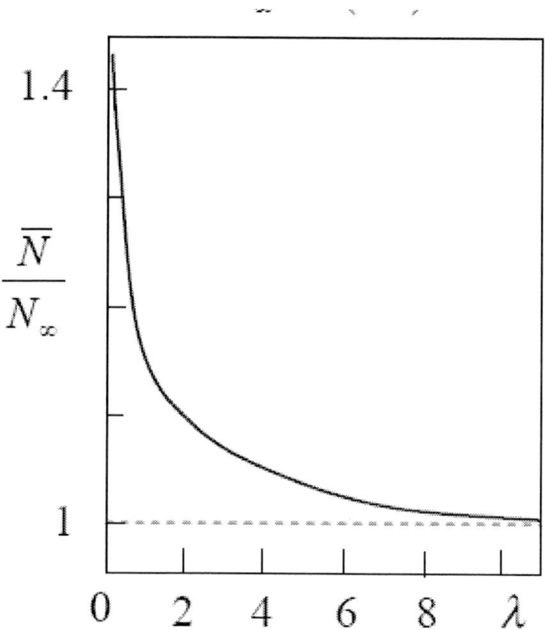

Figure 30. Final crystallite density as a function of film thickness.

VIC- kinetics. Examples of calculated VIC-kinetics are presented in figure 31. The key observation to make is that at any time the crystalline phase fraction in a thicker film is greater, i.e. the thicker film the faster it crystallizes, what for the first glance contradicts a common sense. But a careful consideration confirms that. Indeed, in an infinite medium or thick enough film any point can crystallize thanks to the crystallites growing from all directions, but in thin film only central points can crystallize in the same way, whereas those in the vicinity of the film surfaces can crystallize only owing to the crystallites growing from within film volume. It is evidenced quantitatively also. Namely, differentiating (87) with respect to λ yields

$\frac{\partial \xi(\tau)}{\partial \lambda} = \frac{2}{\lambda} \cdot \int_0^{\lambda/2} \left\{ \left[\xi(\lambda/2,\tau) - \xi(\tilde{z},\tau) \right] \cdot \frac{1}{\lambda} + \frac{\partial \xi(\tilde{z},\tau)}{\partial \lambda} \right\} d\tilde{z}$ and noticing that the term in square brackets is positive, because $\xi(\tilde{z},\tau) < \xi(\lambda/2,\tau)$ at $0 \le \tilde{z} \le \lambda/2$) and $\partial \xi(\tilde{z},\tau)/\partial \lambda \ge 0$, because $\xi(\tilde{z},\tau)$ increases with λ, hence $\partial \xi(\tau)/\partial \lambda \ge 0$, i.e. $\xi(\tau)$ is increasing function of λ. In other words, thin film crystallizes slower than a thick film (or an infinite medium) due to a slower crystallization of its subsurface layers. On the other side, it means that the untransformed (amorphous) phase volume fraction in the subsurface layers of a thin film, hence a nucleation rate is higher (see (83)) resulting in the higher crystallites density (figure 30). Thus a slower crystallization in the subsurface layers of a thin film leads to its overall slower crystallization, the overall higher crystallites density and the formation of a spatially nonuniform microstructure with fine-grained subsurface layers in the crystallized film. It should be noted also that in the CN-VIC the crystallization is faster than in the IN-VIC apparently due to the ceaseless appearance of new crystallites during the CN-VIC process.

The next important feature of plots in figure 31 is that with increasing film thickness the kinetic $\xi(\tau)$ curves move to the left, towards the limiting curve (broken one) for an infinite medium, described by the classical KJMA equation

$$\xi(\tau) = 1 - \exp(-\tau^n) \tag{88}$$

with an exponent $n = 4$ for VIC and $n = 3$ for SIC, and at $\lambda \approx 5$ merge it.

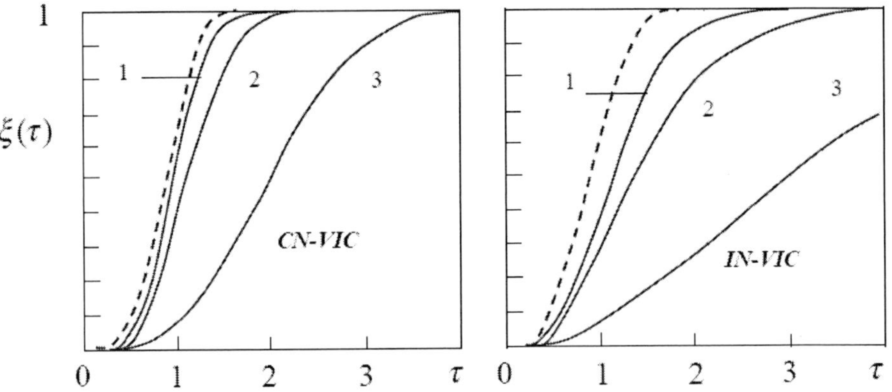

Figure 31. Crystallization kinetics at CN-VIC and IN-VIC for various film thickness λ: 1(1), 0.5(2) and 0.1(3); broken curves - λ = ∞ (infinite medium).

In another limiting case of extremely thin film a VIC may be represented as a 2D KJMA-process with intensity $\widetilde{I} = Ih$ at CN and density $\widetilde{N} = Nh$ at IN, then using Eqns. (79) and (80) we obtain

$$\xi_\lambda(\tau) = 1 - \exp\left(-K_\lambda \cdot \tau^n\right) \tag{89}$$

where $K_\lambda = \lambda$, $n = 3$ for CN-VIC and $K_\lambda = 3\lambda/4$, $n = 2$ for IN-VIC. The existence of analytic asymptotes (88) and (89) suggests that the crystallization kinetics for a film of arbitrary thickness may be approximated by a generalized KJMA equation

$$\xi_\lambda(\tau) = 1 - \exp\left(-K_\lambda \cdot \tau^{n(\lambda)}\right) \tag{90}$$

with parameters depending on a film thickness. This is confirmed by plots in figure 32, clearly demonstrating that the CN-VIC kinetic curves are well straightened in the appropriate to (90) coordinates $\ln[-\ln(1-\xi)]$ against $\ln\tau$, and table 1, indicating that parameters, founded from here, in limiting cases of extremely thin and thick film converge to their KJMA values.

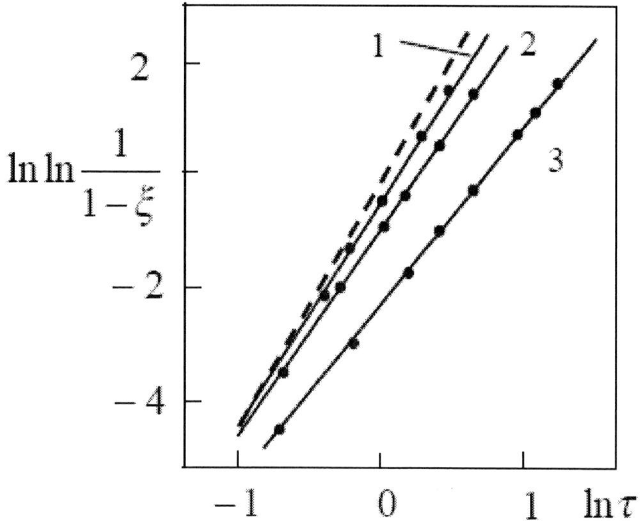

Figure 32. CN-VIC kinetic curves from figure 31 in the coordinates $\ln[-\ln(1-\xi)]$ vs $\ln\tau$.

Table 1. Parameter values in equation (90)

Parameter CN-VIC IN-VIC
λ 0.1 0.5 1.5 4.0 5 ∞ 0.15 0.3 1.2 4.3 8 ∞
$n(\lambda)$ 3 3.4 3.5 3.8 3.9 4 2 2.1 2.7 2.8 2.9 3
K_λ 0.1 0.4 0.8 0.9 0.95 1 0.12 0.25 0.6 0.8 0.9 1

The exponent n is often written down as [33, 34, 40] $n = a + Dp$, where a characterizes the nucleation process and varies from 1 (nucleation with constant rate) to zero (instantaneous nucleation), D is the process dimensionality and p is an exponent in the growth law $R \propto t^p$. Above results indicate that in thin film the crystallization process is characterized by a non-integer exponent n (2 - 3 at IN and 3 - 4 at CN) what may be treated as a fractional D between 2 and 3.

SIC - kinetics. In this case the crystalline nuclei form only on a film/substrate interface (or external film surface), so the thicker film the slower it crystallizes and kinetic curves move therefore to the right as film thickness increases (figure 33). In sufficiently thin film (at $\lambda \le 0.2$), when crystallites size is comparable with a film thickness a situation is equivalent to the plane, 2D KJMA-process, so crystallization kinetics $\xi(\tau)$ obeys Eq. (89) with $n = 3$ for CN-SIC and $n = 2$ for IN-SIC and $K_\lambda = 1$ for both variants. With increasing film thickness a situation is changed, because at any time moment τ crystallization front penetrates into a film only to a depth $\tilde{z} = \tau$ and ahead of it $\xi(\tilde{z}, \tau) = 0$. Setting $\xi(\tilde{z}, \tau) = 1$ in the crystallization rear we find $\xi(\tau) \cong \tau / \lambda$. That is why in thick enough film the crystallization kinetics is linear up to $\tau \approx \lambda$ with a slope just equal to $1/\lambda$ (curve 4 in figure 33).

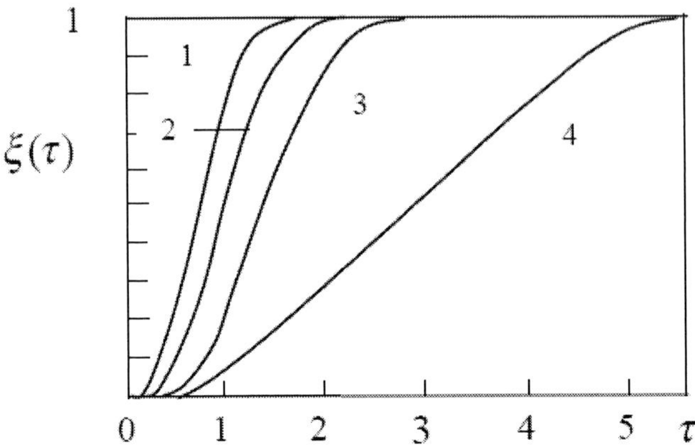

Figure 33. Crystallization kinetics for CN-SIC at different film thickness λ : 0.1 (1), 1 (2), 2 (3), 5 (4).

Crystallization kinetics with a time-dependent growth rate. So far, we restricted our analysis to a constant growth rate corresponding to the case of isothermal crystallization. Now we extend model to the case of a time-dependent growth rate what takes place at nonisothermal crystallization. As it is commonly accepted the growth rate exponentially depends on a temperature $v = v_0 \cdot \exp(-E/kT)$, where E is the activation energy, k the Boltzmann's constant, so under film heating at constant rate ω from T_0 to $T = T_0 + \omega \cdot t$, a crystallite grows up to a size

$$R = \int_0^t v(T(t'))dt' = \frac{v_0}{\omega}\int_{T_0}^{T} \exp\left(-\frac{E}{kT'}\right)dT' \tag{91}$$

At substitution $y = E/kT$ the integral in (91) is transformed into the Doyle p-function [31]

$p(y) = \int_y^{\infty} e^{-y} y^{-2} dy$, which at $y > 15$ is approximated by a simple relation $\lg p(y) = -2.315 - 0.4567y$, whence $R = C(E/\omega) \cdot \exp(-\varepsilon/T)$, $\varepsilon = 1.052 \cdot E/k$. This implies that at substitution $\tau = A \cdot e^{-\varepsilon/T}$ with $A = CE/a\omega$ Eq.(90) is valid at nonisothermal crystallization as well, i.e., its kinetics $\xi(\tau)$ is given by

$$\ln[-\ln(1-\xi)] = \ln(KA^n) - n\varepsilon/T \; . \tag{92}$$

4.4. Comparison with Experiment

4.4.1. Nonisothermal Crystallization

For model verification, experiments on annealing of amorphous $CrSi_2$ and resistive alloy RA-3710 (37 at. %Cr + 10 % Ni +53 %Si) films deposited onto (001) KCl and polished silica-glass substrates with *in situ* resistance monitoring by a four-probe method were carried out [41]. The typical resistivity variation curves during $CrSi_2$ film heating in a vacuum are shown in figure 34. Electron diffraction patterns and transmission electron microscopy (TEM) showed (figure 35) that as deposited (at 200^0C) films were amorphous, whereas after annealing they became crystalline and thus observed resistivity variations may be attributed to the amorphous-to-crystalline transition.

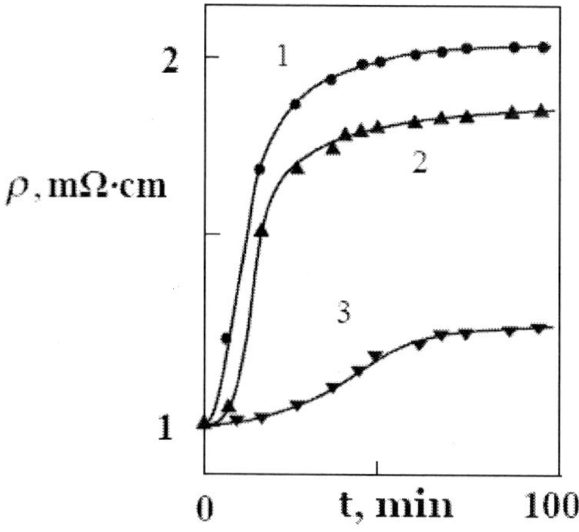

Figure 34. Resistivity variation curves of amorphous $CrSi_2$ films during heating at a rate $\omega, K/\min$: 4.4 (1), 2.6 (2) and 1.4 (3).

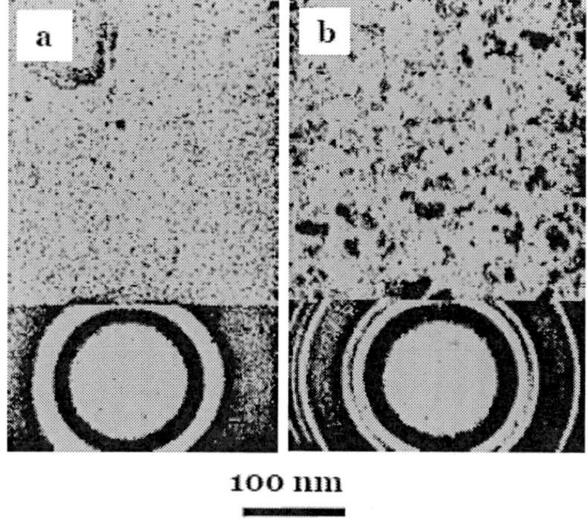

Figure 35. TEM images and corresponding electron diffraction patterns of the $CrSi_2$ film as deposited (a) and after annealing (b).

Using the relationship [26]

$$\rho/\rho_a = 4\cdot\left\{(3\xi-1)/\gamma+(2-3\xi)+\{(3\xi-1)/\gamma+(2-3\xi)^2+8/\gamma\}^{1/2}\right\}^{-1},$$
$$\gamma \equiv \rho_c/\rho_a \quad (93)$$

connecting the resistivity ρ of the mixture of amorphous and crystalline phases with volume fraction $(1-\xi)$ and ξ, and resistivity ρ_a and ρ_c, respectively, the resistivity variation curves $\rho(t)$ were converted to the crystallization kinetics curves $\xi(\tau)$, which pursuant to Eq. (92) would yield straight lines in a plot $\ln[-\ln(1-\xi)]$ against $1/T$ with a slope $n_\lambda\cdot\varepsilon$ and intercept KA^n. This is confirmed by plots in figure 36 showing that the crystallization kinetics of resistive films is depicted by two linear segments. In studied films of thickness $h = 25-50nm$ the characteristic intercrystallite distance, as measured on electron micrographs, $a = 25-100nm$, i.e. $\lambda = h/a \approx 0.5-1$, hence (according to the Table 1) $n(\lambda) = 2.5$. Thus, from the product $n_\lambda\cdot\varepsilon$ found from the slope of these lines one finally gets a reasonable estimate of the activation energy $E = 1.1eV$ and $E = 0.24eV$ at first and second stage, respectively for RA-3710, and $E = 0.55eV$ and $E = 0.15eV$ for $CrSi_2$.

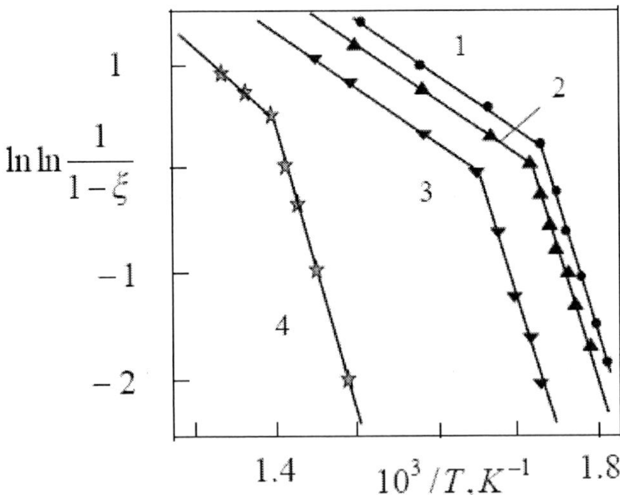

Figure 36. Crystallization kinetics of amorphous $CrSi_2$ (1- 3) and RA-3710 (4) films in coordinates $\ln[-\ln(1-\xi)]$ vs $1/T$.

4.4.2. Isothermal Crystallization

In Ref. [25] using the X-ray diffraction the quantitative data on the isothermal (at 400^0C) crystallization kinetics of 150-nm Ge film had been obtained and shown that it obeys Eq. (88) with an exponent $n = 4$. This means that the film has crystallized via CN-VIC mechanism and the condition of a 'thick' film $\lambda = h/a \geq 5$ allows evaluating the crystallites size $a \leq 30nm$. The isothermal (at 422^0C and 455^0C) crystallization kinetics of 1-μm thick amorphous Ge film obtained from electro conductivity measurements has been fitted to the KJMA Eq. (88) with an exponent $n = 1.2$ and $n = 1.5$, respectively [26]. According to TEM observations the large ($> 1\ \mu m$) spherical crystallites grow up from the preexisting nuclei and so these results may be explained by the IN-VIC model ($\lambda < 1$) with a growth exponent $p = 0.6 - 0.75$. The crystallization kinetics of amorphous Si_3N_4 and $Si_xN_yO_z$ films at 1100^0C obeys Eq. (88) with an exponent $n = 3$ [27]. The model predicts this magnitude for both CN-VIC in sufficiently thin film ($\lambda \leq 0.1$) and IN-VIC in sufficiently thick film ($\lambda \geq 5$). In the CN-VIC the kinetic coefficient $K = 0.1(\pi I v^3/3) = 0.1(v/a)^3$ and in the IN-VIC $K = 4\pi N v^3/3 = (v/a)^3$. From these K values and known film thickness we get estimates: $a = 1\mu m$, $v = 0.28nm/s$, $I = 2.8 \cdot 10^8 cm^{-3}s^{-1}$ at CN-VIC and $a = 20nm$, $v = 2.6 \cdot 10^{-3} nm/s$, $N = 3 \cdot 10^{16} cm^{-3}$ at IN-VIC for Si_3N_4 and $a = 1\mu m$, $v = 0.15nm/s$, $I = 1.5 \cdot 10^8 cm^{-3}s^{-1}$ at CN-VIC and $a = 20nm$, $v = 1 \cdot 10^{-3} nm/s$, $N = 3 \cdot 10^{16} cm^{-3}$ at IN-VIC for $Si_xN_yO_z$. The absence of additional structural data doesn't allow discriminating between these two mechanisms, but anyway it is evident that a faster crystallization of Si_3N_4 film is obliged to a quicker crystallites growth. The isothermal crystallization kinetics of 500 - nm thick amorphous $Se_{70}Te_{30}$ film has been fitted to the Eq. (88) what produced non-integer exponent $n = 2.2$, whereas for the bulk $Se_{70}Te_{30}$ an integer exponent $n = 3$ was obtained [38] . This implies that $Se_{70}Te_{30}$ film has crystallized via IN-VIC mechanism.

4.4.3. Subsurface Layer Crystallization.

In Ref. [30] the isothermal (at 540^0C) crystallization kinetics of amorphous Si films on Si(100) and Si(111) surfaces in UHV has been studied using LEED. In this case the crystalline structure of the topmost surface layer is registered and therefore analysis these data needs a knowledge crystallization kinetics of the subsurface layer. Designate this subsurface layer thickness as Δh and following [30] consider the SIC process, which begins on the crystalline Si/amorphous Si film interface. Up to the time moment $\tau^* = k\lambda$, when the crystallization front reaches a lower boundary of this layer at height $h^* = h - \Delta h \equiv kh$ the crystalline phase is absent in this layer and therefore its crystallization kinetics $\xi_s(\tau)$ should be written down as

$$\xi_s(\tau) = \frac{1}{\lambda(1-k)} \int_{k\lambda}^{\lambda} \xi(\tilde{z}, \tau) d\tilde{z}\ ,\ \tau > \tau^* = k\lambda \tag{94}$$

where the integrand is given by Eq. (81) or (82). Figure 37 shows results of the numerical integrating in (94) for films of various thickness with the coefficient k varied so that to keep $\Delta h = const.$: for films of thickness λ_1 and λ_2 its values are related by $k_2 = 1 - \lambda_1(1 - k_1)/\lambda_2$. The steeper IN – curves are more appropriate to experimental kinetic curves in [30]. In the IN-version Eq. (94) allows yielding an approximate analytical formula

$$\xi_s(\tau) = \frac{\tau - k\lambda}{\lambda(1-k)} - \frac{e^{-2\tau(\tau-\lambda)} - e^{-2\tau(\tau-k\lambda)}}{2\tau(\lambda-k)}, \ \tau < \lambda$$

$$\xi_s(\tau) = 1 - \frac{e^{-2\tau(\tau-\lambda)} - e^{-2\tau(\tau-k\lambda)}}{2\lambda\tau(1-k)}, \ \tau > \lambda \qquad (95)$$

that for $\lambda > 4$ provides an accuracy of 1%. In LEED studies a structure of a surface layer is actually registered, so setting $\Delta h \to 0$ from Eq. (95) one obtains in real variables

$$\xi_s(t) = 1 - \exp[-\pi N(v^2 t^2 - h^2)] \qquad (96)$$

Indeed, experimental kinetic curves from Ref. [30] are well straightened in respective coordinates $\ln[1/(1-\xi)] - t^2$ what allows determining the rate constant $K = \pi N v^2$ whence, using known growth rate $v = 0.2 nm/s$, one gets the estimates of the surface density of crystallization nuclei $N = 2.1 \cdot 10^{10} cm^{-2}$ and $N = 3.5 \cdot 10^{9} cm^{-2}$ in Si films of thickness 120 and 150 nm, respectively.

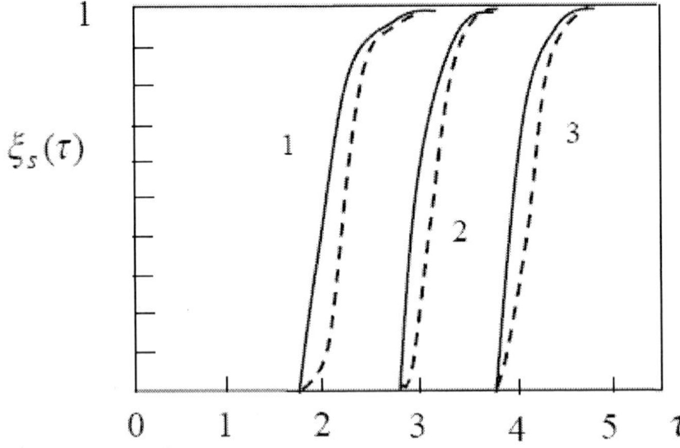

Figure 37. Crystallization kinetics of the subsurface layer at IN-SIC (solid curves) and CN-SIC (broken curves) in films of different thickness: 1 - $\lambda = 1$, $k = 0.9$, 2 - $\lambda = 3$, $k = 0.93$, 3 - $\lambda = 4$, $k = 0.95$.

In summary, a general methodology based on the extension and development of the KJMA statistical model of crystallization is presented, which provides in a unifying manner a quantitative description of the growth and surface morphology evolution in thin films growing via Volmer-Weber 3D island mechanism as well as the crystallization kinetics of amorphous film.

REFERENCES

[1] J. A. Venables, G. D.T. Spiller, and Hanbuken, *Rept. Progr. Phys*. 47 (1984) 399.

[2] V. I. Trofimov, V .A. Osadchenko, Growth and Morphology of Thin Films, Energoatompubl., Moscow, 1993, pp.272, in Russian.

[3] D. Kaschchiev, *J. Cryst. Growth* 40 (1977) 29.

[4] T.J.Newman and A.Volmer, *J.Phys*. A29 (1996) 2285.

[5] J.D.Aze and Y.Yamada, *Phys. Rev*. B34 (1986)1599.

[6] E.Ben-Naim, A.R.Bishop, I.Daruka and P.L.Krapivsky, *J. Phys. A:Math.Gen*. 31 (1998) 5001.

[7] D.W.Clinton, Van Siclen, *Phys. Rev*. B54 (1996)11845.

[8] Ge Yu, S.T.Lee, J.K.L.Lai and L.Ngai, *J. Appl. Phys*. 81 (1997) 89.

[9] M.Fanfoni, M.Tomellini, *Eur. Phys. J*. B34 (2003)331.

[10] M. F. Gyure, C. Ratsch, B. Merriman, R. E. Caflisch, S. Osher, J. J. Zinck, D. D. Vvedensky, *Phys. Rev*. E 58 (1998) R6927.

[11] A. N. Kolmogorov, Izv. Akad. Nauk SSSR, Ser. Mathem. No. 1 (1937) 355.

[12] W. A. Jonson and R. F. Mehl, Trans. *Am. Inst. Min. Metall. Eng*. 135 (1939) 416.

[13] M. Avrami, *J. Chem. Phys*. 7 (1939) 1103; 8 (1940) 212; 9 (1941) 177.

[14] V.Z.Belen'kii, *Geometric-probabilistic models of crystallization: phenomenological approach*, Nauka, Moscow, 1980, in Russian.

[15] J.L.Meijering, Philips Res.Rep. 8 (1953) 270.

[16] V.I. Trofimov, *Thin Solid Films* 428 (2003) 56.

[17] G. Rasigni, F. Varnier, M. Rasigni et al., Surface Sci. 162 (1985) 985, *J. Vac. Sci. Technol.* A10 (1992) 2869.

[18] V.I. Trofimov, H.S. Park, Appl. Surface Sci. 219 (2003) 93.

[19] B. Pacchiarotti, M. Fanfoni, M. Tomellini, Physica A 358 (2005) 379.

[20] H.-N. Yang, G.-C. Wang, and T.-M. Lu, Diffraction from Rough Surfaces and Dynamic Growth Fronts, *World Scientific*, Singapore, 1993.

[21] Y.-P. Zhao, G.-C. Wang, and T.-M. Lu, *Phys. Rev*. B55 (1997)13938.

[22] H. Angus Macleod, *J. Vac. Sci. Technol*. A4 (1986) 418.

[23] J. Zhang, J. Zhao, H. X. He, H. L. Li, Z. F. Liu, *Thin Solid Films* 327-329 (1998) 287.

[24] S.Y. Lee, R.S. Feigelson, *J. Cryst. Growth* 186 (1998) 594.

[25] N.N.Sirota, W.A.Denis, L.S.Unyarka, *Kristall und Technik* 11 (1976) 629.

[26] P.Germain, S.Squelard, J.Bourgoin, A.Gheorgiu, *J. Appl. Phys*. 48 (1977) 1909.

[27] L.N. Alexandrov, F.L. Edelman, Proc. 7th Vacuum Congr. , Vienna, vol. 2, 1977, p.1619.

[28] N.M.Mayer, H.Hoffman, A.Schafer, *Thin Solid Films* 88 (1982) 225.

[29] L.Shikmanten, M.Talianker, M.P.Dariev, *J. Phys. Chem. Solids* 44 (1983) 745.

[30] V.V.Korobtsov, V.G.Zavodinski, A.V.Zotov, *Surface Science* 130(1983) L325.

[31] K.Matusita, T.Komatsu, R.Yokota, *J. Mater. Sci.* 19 (1984) 291.

[32] M.C.Morilla, C.N.Afonso, A.K.Petford-Long, R.C.Doole, *Thin Solid Films* 275(1996)78

[33] D.Dimitrov, M.A.Ollacarizqueta, C.N.Afonso, N.Starbov, Thin Solid Films 280(1996)278

[34] N.Ohshima, *J. Appl. Optics*. 79 (1996) 8357.

[35] J. Pelleg, *Thin Solid Films* 325(1998) 60.

[36] J.Olivares, A.Rodriguez, J.Sangrader, T.Rodriguez, C.Ballesteros, A.Kling, *Thin Solid Films* 337(1999) 51.

[37] M.-Y. Hua, R.-Y. Tsai, *Thin Solid Films* 388 (2001) 165.

[38] A.H.Moharram, *Thin Solid Films* 392 (2001) 34.

[39] V.I.Trofimov, I.V.Trofimov, Jongil Kim, *Nuclear Instrum. and Methods in Phys. Res.* B 216 (2004) 334.

[40] V.R.V.Ramanan, G.E.Fish, *J. Appl. Phys*. 53(1982) 2273.

[41] V.I.Trofimov, I.V.Trofimov, Jongil Kim, *Thin Solid Films* 495 (2006) 398.

In: Computational Materials
Editor: Wilhelm U. Oster

ISBN: 978-1-60456-896-7

Chapter 2

OVERVIEW OF THE EFFECTS OF SHOT PEENING ON PLASTIC STRAIN, WORK HARDENING AND RESIDUAL STRESSES

Baskaran Bhuvaraghan[1], M.S.Sivakumar[2] and Om Prakash[1]
[1]GE Aviation;
[2]Professor, IIT Madras, Chennai, India

ABSTRACT

Residual compressive surface stresses are found to enhance the fatigue life of components, as fatigue cracks originate mostly from surfaces. Mechanical surface enhancement processes develop such residual compressive stresses. This review paper focuses on shot peening method and attempts to understand the underlying mechanism. This will help to quantify and optimize the material response due to shot peening computationally and help the designer to evolve better design. Starting with an introductory overview of shot peening, this paper describes the experimental and theoretical studies performed in the residual stress development to mitigate various damaging mechanisms that include SCC, corrosion, fretting etc. The paper also covers the development of cold work due to peening and related modifications in the surface microstructure of different materials. The material response remains complex due to the stochastic nature of several variables. Theoretical studies involving Finite Element method, and/or Discrete Element method are currently employed to understand the physics of shot peening. Shot peening produces different amounts of near surface plastic deformation for the same level of residual stresses and vice versa for different combinations of input parameters. The paper covers the efforts that overcome these challenges in optimizing the peening parameters. It also provides a glimpse of the research in the area of crack growth reduction due to shot peening. The residual stress that is developed relaxes due to service environment. In all the above phenomenon, dislocations play a vital role and therefore a multi-scale approach involving dislocations can provide a common platform in explaining them. In addition, this paper points to some future directions into which the research that can possibly explore to quantify the material response more accurately.

NOMENCLATURE

A	Area of the shot spread
a	Total area of indentations at time interval t
C[t]	Coverage during time t
D	Diameter of the shot
E	Young's Modulus
$E^{p}_{xx}(z)$	Plastic strain in x direction along depth 'z'
FWHM	Full Width at Half Maximum[used in X-Ray Diffraction]
H	Height[intensity]
$\overset{o}{m}$	Mass flow rate of shots
n	Hardening exponent
p	Cumulative Plastic strain
$P_{\max}$	Maximum Hydrostatic Pressure
R_1, R_2	Radius of spheres 1 and 2
R_t	Surface roughness
r	Indentation radius
$\bar{r}$	Shot radius=D/2
S	Total area
t	Time during which indentations are created
V	Velocity of the shot
σ_a	Applied stress amplitude
σ_e	Equivalent stress
σ_H	Hydrostatic Stress
σ_m [0]	Mean stress prior to fatigue
σ_m [N]	Mean Stress after N cycles including residual stress
σ_R [t], σ_{R0}	Residual Stress at time 't' and time=0
σ_y, p_y	Yield strength
ξ_a	$\sqrt{J_{2a}}$, the square root of amplitude from max and min deviatoric stresses
υ	Poisson ratio
ρ	Density of the shot
τ_a	Shear stress amplitude

1. Introduction

Metal fatigue is a major failure mode of many aircraft engine components due to the application of both mechanical and thermal cyclic loads[1]. Fatigue cracks originate mostly from surfaces for the following reasons:

- stresses are high at the surface and have a negative gradient in engine component features such as bores
- surfaces are subjected to machining and handling defects
- fatigue resistance of inner material is nearly one-third higher than that of the surface [2]

There are many ways to improve the fatigue strength. Generally, the methods to enhance the fatigue strength include use of expensive advanced materials having higher fatigue strength or making the parts larger enough. These options will make them costlier and heavier. Besides, surface machining such as polishing to reduce surface roughness and case hardening techniques to strengthen the surface are employed[figure 1]. These processes are more economical than using expensive materials or processes[3]. However, polishing has limited potential to improve fatigue strength. The case hardening techniques include mechanical, thermal and chemical treatments. The mechanical treatments mainly consist of Shot Peening[SP], Low Pressure Burnishing[LPB] and Hole Expansion[HE] while Laser Shock Peening[LSP] falls under thermal treatment. A historical overview of the development of the processes can be found in [4]. The above processes induce Residual Compressive Stress[RCS] and shift the crack initiation location to subsurface region [5] to mitigate the crack initiation due to fatigue loading.

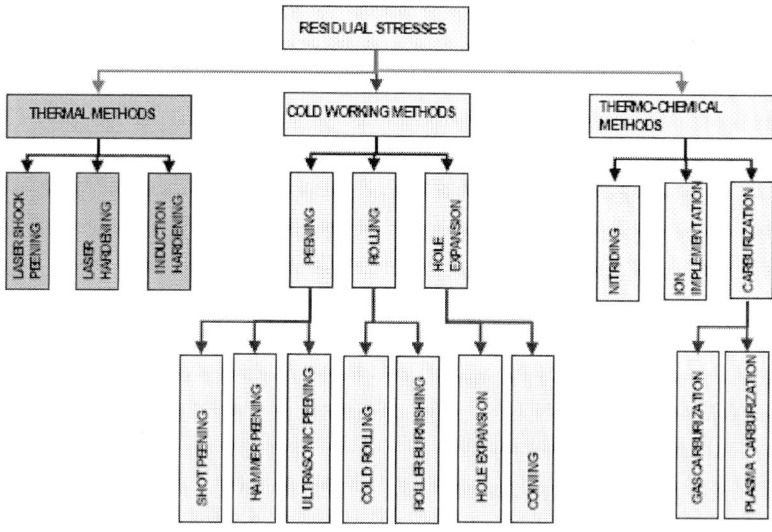

Figure 1. Methods of residuals generation.

Apart from service loads that cause fatigue failures, detrimental residual tensile stresses are induced on the metal surfaces due to different machining operations that affect the fatigue response, dimensional stability, further machining and assembly[6]. The RCS developed through Mechanical Surface Enhancement[MSE] processes reduces the adverse effects of such tensile stresses. It also helps to increase low crack-growth rate region and decrease the wear rate[7]. Bozdana has discussed the application of such MSE processes and the related forming processes in aircraft and engine manufacturing where expensive metallic alloys are used[8].

The designer wants to include the RCS effects in a reliable way during the design to avoid increased cost and size. The fatigue strength and the predicted crack growth life significantly increase if RCS effect is considered thus eliminating the conservatism in the design[9]. Each of these processes modifies the microstructure, cold work and the roughness differently[10]. The induced RCS relaxes due to mechanical and thermal loads to different levels due to such treatments. The relaxation can occur due to static load or cyclic loads. The RCS is modified even due to crack propagation. Also, over-exposure to surface enhancement processes such as SP can result in debit in fatigue strength. The above facts necessitate better understanding of the MSE processes to fully exploit their economic potential.

Among the four processes, SP is widely used and highly economical. Deriving its name from shot blasting[11], shot peening is a controlled cold working process involving multiple and progressively repeated impacts. Many authors have covered the history, evolution and details of SP process[12-16]. More information on different shot types and mechanisms of shot delivery can be found in [17]. Three important specifications that cover the peening aspects are SAE J442/443[18,19], AMS2432[20] and MIL-S-13165C[21]. Tosha has done a statistical analysis of the research work till 1999[22] and he has concluded that the enhancing fatigue strength is accomplished by mechanism of residual stresses.

Significant amount of research has been done in understanding the process of shot peening. Much of the work is experimental focusing on specific applications and materials. Theoretical studies that have been done use mostly Finite Element Method[FEM] involving unit cell approach. Recently some researchers have explored Discrete Element Method[DEM] coupled with FEM as an alternative method to increase the number of shots in the simulation to make it more realistic. But, the relationship between the RCS and fatigue performance is complex to model due to various factors such as stress distribution and its gradient, multi-axial loading, relaxation due to service loads and material structure. Therefore shot peening is still used as an additional safety measure[23] in the industry than being used directly in the design. In order to help the designer use RCS in the design, it needs to be quantified in a clear manner.

The main purpose of the paper is to make a systematic study of the current research to understand the progress as well as the issues in the shot peening process simulation. It focuses on current level of research on SP and its mechanics through the available literature. Figure 2 summarizes the scope of the study which aims to understand the different phenomena such as RCS development[with changes in cold work, surface properties], fracture, relaxation etc that are related to SP. Firstly the techniques and numerical tools that are used in the SP need to be known. Section 2 provides an overview of the experimental techniques used in the residual stress measurement and types of numerical methods that can be used in any process simulations. This is followed by section 3 which gives an overview of SP method and the related systems. The success of any process implementation depends on proper selection of measurement parameters. Section 4 covers the different aspects measuring shot peening parameters.

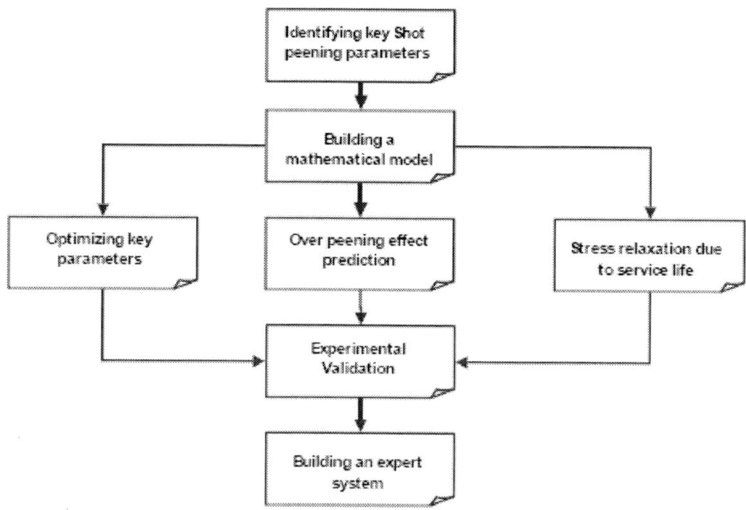

Figure 2. Mathematical modeling of shot peening process.

The generation of residual stresses and cold work in different materials such as steels, aluminum, titanium, superalloys etc. due to different SP parameters are reviewed in section 5. Subsequent to the understanding of mechanics of RCS development, the designer would like to optimize the peening parameters so that the RCS is maximized with fatigue crack growth and relaxation due to service loads are minimum. The studies related to optimization, crack growth and stress relaxation due to SP are briefly covered in sections 6 to 8 respectively. Many researchers have conducted both experimental and theoretical studies, but each of the sections from 5 to 8 has been classified under two different categories depending on the key conclusions. There has been also some research where the material behavior under SP, LSP or LPB is compared. Section 9 discusses SP process and the other MSE methods. Dislocations play vital role in plastic deformation of metals and hence in the residual stress development and all subsequent phenomena. Section 10 covers how dislocations can be used in understanding the SP process. The last section outlines the possible areas of future work.

2. Methods for Studying Process Simulations

Both experimental and theoretical studies are necessary to understand any process simulation. Raabe [24] in his book brings out many factors that are to be considered in a process simulation:

- is it to be solved through analytically or experimentally?
- is it to be done using empirical or phenomenological models?
- what are the physical scales in space and time?

- what simulation method has to be used for the scales used?
- is it necessary to include microstructure?
- is the phenomena deterministic or stochastic?
- should an atomistic or continuum model be used?
- are different spatial and/or time scales to be combined?
- what are the independent and dependent variables?
- do simulation codes exist?
- what data are available from testing?

2.1. Experimental Methods

Several measurement techniques exist to measure the residual stresses and it becomes important to measure non-destructively[25]. Withers [26] discusses these methods to measure the three levels[Type I, II and III] of residual stresses. The following three methods are used extensively:

- X-ray Diffraction (for type I and II residual stresses)[27-31]
- Neutron Diffraction [32]
- Hole Drilling method [33]

2.2. Theoretical Methods

SP is a process involving metal plasticity, contact mechanics, impact dynamics and fracture mechanics with probabilistic nature[figure 3]. Most of the researchers use continuum mechanics based approach for predicting the residual stresses and plastic strains. The constitutive equations may be obtained either from tests or from multiscale modeling. Several numerical methods have evolved[34] over the years to find use in various disciplines of engineering:

- finite element method[h version, p version, mixed formulation etc]
- finite difference method
- spectral method
- wavelet technique
- mesh-free method
- discrete element method
- finite-discrete element method
- boundary element method
- finite volume method
- arbitrary Lagrangian-Eulerian method
- multigrid method

Amongst them Finite Element Method[FEM] has been widely used in SP simulation. The method is explained in many text books[35-37]. FEM can accommodate plasticity, contact, impact and fracture effects well, but it mainly operates on continuum. To enhance the functionality of FEM, new methods have evolved into generalized finite element methods[38-40] that use non-polynomial shape functions. Examples are finite element partition of unity[41], cloud based hp finite element method[42] and spectral finite element method[43] etc. Each of these methods is used for special applications such as fracture mechanics[44], wave propagation[45] etc. As mentioned before, contact mechanics and metal plasticity play a key role in generating cold work and residual compressive stress in all the MSE processes. The computational aspects of contact mechanics are given in[46-48]. The different plasticity computational models can be seen in [49]. The SP process requires study of wave propagation as due to impacts[47] as both elastic and plastic stress waves are generated. Figure 4 shows different alternatives that need to be considered in SP simulation.

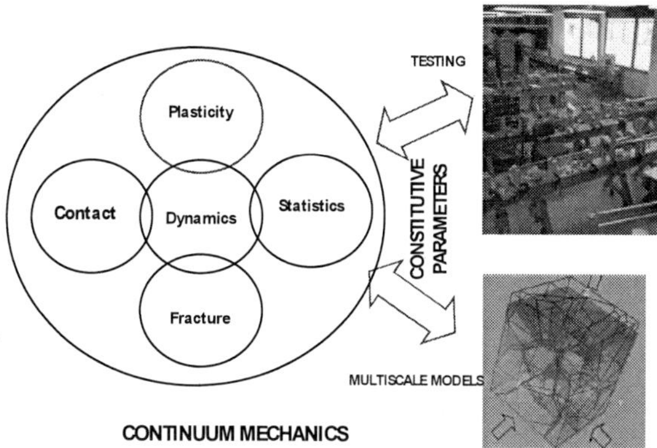

Figure 3. Multi-disciplinary aspects in shot peening.

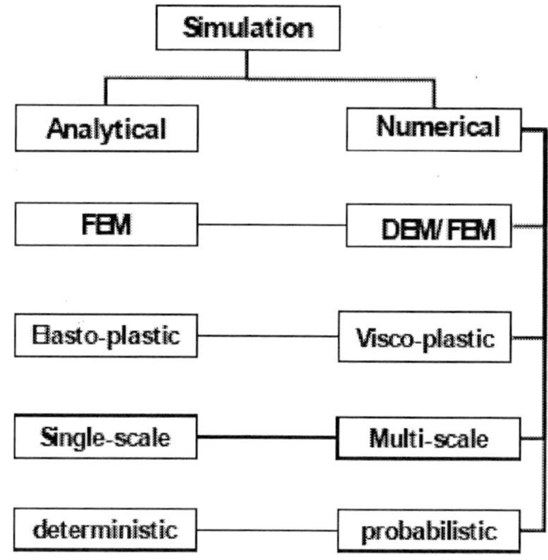

Figure 4. Simulation Strategies for Shot Peening Simulation.

2.2.1. Multiscale Methods

The multiscale methods range from atomistic to single crystals to poly-crystals to micromechanical to macro models[figure 5]. An overview of multiscale methods can be found in[50-57]. Different techniques exist at different spatial levels that also have a bearing on the temporal scales[figure 6]. In the area of plasticity, there have been multiscale simulations[58-62]. They couple atomistic level[63-65] to dislocation level [66] to crystal and poly-crystal levels[67-72].

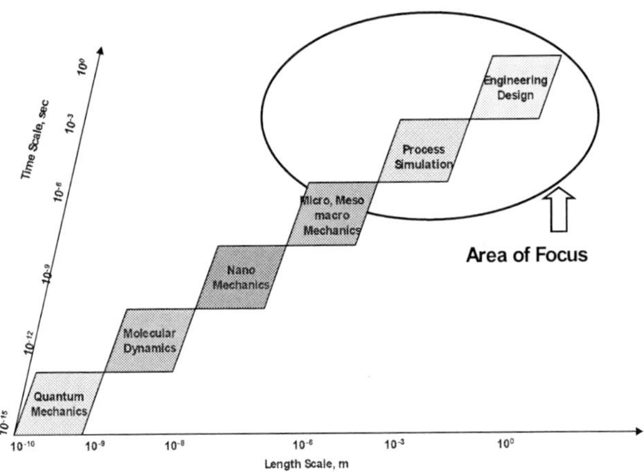

Figure 5. Focus Area for Multiscale Modeling.

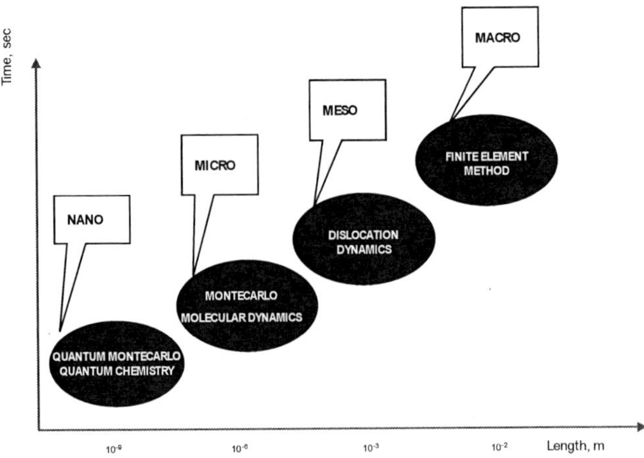

Figure 6. Tools for Multiscale Modeling.

The SP process involves processes whose spatial and temporal scales are many orders more than the area and time duration at which the stress wave is applied. Besides, the nano-crystalline structure, which is 10 times stronger and harder [73] develops due to peening[74,75]. It can be concluded that in future it will be necessary to use of multi-scale methods for shot peening simulations. Such simulations span from meso-scale to macro-scale involving dislocations.

3. Shot Peening Systems and Applications

The impacts result in plastic deformation of the surface layer causing them to elongate during the impact duration. Once the shot leaves the work piece, the surface material tries to become normal. However, the adjoining layer of material does not allow to contract thus causing RCS. It has been observed that the near-yield stress compressive stresses are balanced by sub-surface tensile stresses, resulting in a self-equilibrated material state.

3.1. Media

The shot is the most important factor in SP process. The shot must be at least as hard as the peened material and not leave much material residue[76]. It is made in different sizes and from different materials such as cast steel, carbon steel, ceramics, glass etc. Four types of shots made of wrought carbon steel, cast steel, ceramic and glass are compared in terms of size, shape, hardness, density and durability[77]. Shots are generally spherical in shape, but cut wire shots are also widely used due to their lower costs. These shots are found to change their shape from cylinder to oblong and finally to spherical shape during use[78]. They produce consistent depth and with improved conditioning better fatigue lives than cast steel shots[79]. Typically stainless steel shots are used for Titanium alloys to have better surface finish than glass beads that break during use[80]. In a study with steel shots, it has been found that effect of ageing through hardness reduction and cracks does not change the fatigue strength of the material[81]. Surface contamination on non-ferrous and stainless steel materials are observed when using steel shots. Zirconium shots produce better surface quality than glass beads[82] for the same Almen intensity[see Sec. 4.1] thus possibly reducing the fatigue strength of aluminium alloys. New type of glass beads are developed with alumina-silica material that have better ageing properties on aluminium and titanium alloys than conventional glass beads[83]. Ceramic shots produce good surface finish due to less breakage and intermediate intensity between steel and glass shots [84]. The variation of size and shape also important in the cold work and subsequent RCS development. Better control of the above parameters will result in less RCS variation. The shape and size of the shots can be monitored through image analysis before and after usage[85]. Conditioning of shots has reduced the variation in hardness and grain sizes in shots even below the specification limits, but it has less significant effect in the RCS[86].

It is very important to determine the shot velocity both theoretically and experimentally. Iida has calculated shot velocities from equations of motion in terms of air velocity and the velocity loss due to drag[87]. He has also calculated the velocity for centrifugal type delivery systems. A new method involving three disks[88] and another method with two disks are developed for measuring shot velocities[89]. Within the range of parameters measured, shot velocity is linear with nozzle pressure. Meguid and Duxbury [90] have explained the use of photo-transistors and strain gauges to measure the shot velocity and intensity. Kopp et al use photo-electric barriers [91] while Barker et al. have devised a non-laser based sensor [92].

3.2. Delivery Systems

Shots derive kinetic energy from pressurized air and are delivered by nozzles. Of late, centrifugal wheels have found use in aircraft industry due to better energy efficiency, velocity, coverage and flow rate over air nozzle systems[93]. Suction systems also are used for glass beads [94]. Different types of separation systems such as sieves, multi-step air separation systems are used to maintain the shot quality[95].

3.3. Other Peening Methods

Other peening methods include media such as water[96-100]. oil [101], ultrasound[102-105]. flap peening [106], cavitation jet and even dry ice [107]. The mechanism of developing RCS is the same as SP, but the difference is the type of media employed. Ultrasound peening produces less surface roughness as the shot velocity is lower than in conventional peening[108]. SP generates noise and dust and vacuum is used to enable peening to reduce dust and noise pollution effects[109]. Modifications to SP include Wet Shot peening[WSP] where glass beads are sent along with high pressure water jet[110]. Wet peening reduces dust and it is easy to separate undersize shots and impurities[111]. Needle peening with cylindrical head produces less surface roughness and is likely to be a candidate for commercial use[112]. But, it produces less RCS than conventional SP[113]. Micro-shot peening helps to improve the surface finish after micro-milling [114].

3.4 Applications

It is known that shot peening helps to improve the fatigue strength through high level of work hardening and residual compressive stress[115]. It protects against fretting, SCC [116], corrosion fatigue, galling[117], hydrogen embrittlement in high strength steels[118] and cavitation in ship propeller blades[119]. The improvement against different damaging mechanisms occurs primarily due to RCS. SP also helps in electroplating and testing the quality of electroplated parts.

The change in surface roughness due to SP is also helpful. For example, the surface texture modification due to SP with increased roughness is used to help lubrication. The SP produces less coefficient of drag due to small vertex flow in each concave dent similar to golf balls[120]. It also increases the thermodynamic efficiency of pipes[121] and boiling heat transfer due to increased surface roughness[122]. SP is also used to provide lining of soft metals with hard metals[123].

Metal forming using peening methods have been increasingly used [124] in forming sheet metals such as aircraft skins[125] replacing expensive methods. The survival probability has increased after SP is applied when compared to unpeened aircraft aluminium parts[126]. Peen forming reduces production steps and provides high flexibility in form and thickness[127]. SP is effective for thin sections and it causes double curvature first and then single curvature[128]. SP also is applied to modify part dimensions. More controlled forming with balls with specified energy and locations can replace the conventional peen forming[129]. Pre-stressing and straining have been applied to increase the effectiveness of SP process. The pre-stressing can be due to stretching or bending[130] and the resultant elastic stresses increases the depth and the average values of RCS[131]. This process eliminates tensile stresses being developed due to conventional bending operation. Deeper curvatures can be obtained using strain peening[132].

Variations to peen forming processes are applied for specific purposes. Kondo et al. analyze the bulging and sinking mechanisms occurring during peen forming[133] by means of needle indentation. Ball-drop plate bending is effected using bigger shots with less velocity[134]. Recently, multi-axis CNC machines are used in controlled shot peening process with many complex aerospace applications[135,136]. A special method called Flexible Peen Forming[FPF] is used to manufacture large parts through forming between a hard spherical punch and hard specially mounted balls[137]. Impact Metal Forming is a new process that uses guided flying punch on exact position[138]. Peening with 1.5 to 4mm diameter bearing balls and 4 to 9 m/s velocities with compressed air is also attempted[139] to produce optimum intensity in different metals. Such bearing balls also find application in peening large components such as steam turbine blades[140] and aircraft components[141].

4. Measuring Shot Peening

SP induces work hardening and RCS on the target material. The following parameters affect the work hardening and the compressive stresses[142]:

- shot: size, velocity, shape, material, hardness, angle of impact
- part: material and hardness
- process: exposure time, coverage

Simpson and Garibay classify the process variables under four categories, namely, workpiece related variables, workload related variables, energy related variables and saturation and process procedural variables in a comprehensive manner[143]. Intensity and coverage are the main shot peening measures. The effect of shot peening can be seen in [144] on different treatments and failure mechanisms.

4.1. Intensity

In the industry, the curvature of the Almen strip is used as a measure of the intensity of peening. It remains as the simplest, most flexible but least expensive mechanism to repeat the peening process after standardization[145]. The material and dimensions of Almen strips are standardized by American Military Standards. The details of calibration of Almen gages are given by Champaigne [146]. He has described the R&R study involving Almen gages[147] due to equipment and operator variations. The Almen strips produce good correlation with steel parts within the hardness range of the strips. Efforts are made to relate different parameters, such as shot stream energy with Almen strip etc[148]. Holes can be peened using deflector peening [149] and their peening intensity is measured through special calibration. Though widely used, the Almen system does not provide a very accurate prediction of the strain hardening due to various peening factors. The key issues are[150,151]:

- the same intensity can be produced with different combinations of input parameters such as shot sizes and velocities.
- for parts with different materials and/or hardness, the fatigue lives measured do not match even when the Almen intensities are same.
- inability to peen small features and components using Almen intensities.
- the inherent variation in the strips and gauges[152-154]
- the operator bias and integrity that increase the process variation during manual peening [155]

Therefore for production parts, limited testing with narrow intensity band may result in insufficient data[156].

4.2. Coverage

Coverage is measured as the ratio of the peened surface over the total area. To measure the peened area, the projected area of the dents is considered. When the area is completely peened it is called 100% coverage. Suppose the duration of peening is 2 times the time taken for 100% peening, then the coverage is called 200%. The coverage of peening process can be evaluated through Avrami equation [157]:

$$S = 100(1-\exp(\frac{-3r^2 \overset{o}{m} t}{4A\bar{r}^3 \rho}))$$

4.3. Other Measures

Due to the limitations of Almen strip, new measures have been evolved. Roth et al have invented alternate methods to measure the SP process through stress and depth parameters instead of using the conventional Almen strip height[158]. Some researchers have attempted to use the same components such as springs that are peened to measure the intensity [159]. Kirk has proposed the use of circular disks in place of Almen strips and LVDTs[160,161]. or strain gauges[162] to monitor the intensity interactively. More automated control has been used that results in savings[163] and better reliability[164-167]. As certain combination of parameters make the tracer removal correlation with coverage difficult, using microscopes with UV or white light may be considered [168].

5. Studies on Residual Stress and Cold Work

Generally, the fatigue life improvement depends on peening intensity and spread. The RCS is assumed to affect the mean stress due to the applied loads. However, the following parameters also need to be considered if one needs to accurately predict the fatigue strength improvement:

- residual stress gradient
- load stress gradient
- plastic deformation due to first cyclic loading and subsequent cyclic loadings
- subsequent hardness and RCS changes

Designers use Goodman's diagram to calculate the allowable alternating stress for a given mean stress. Compared to Goodman's equation used to represent such stress field, the biaxial stress state on the surface is better represented by Dang-Van criteria where hydrostatic pressure vs. shear stress amplitude is computed to evaluate fatigue limit[169]:

$\tau_a + \alpha P_{\max} = \beta$ where α and β are constants

SP is applied on many materials that include carbon steels, high alloy steels, nickel and cobalt alloys [170]. In most metals the RCS pattern is the same and the maximum tensile stresses are one-third of surface compressive stresses [171]. SP shifts the fatigue crack zone to the subsurface area where the fatigue strength is higher [172].

The peening parameters change the following material parameters[142]:

Metallurgical : structure, hardness

Mechanical : residual stresses, residual stress gradient, depth of plastic deformation

Geometrical : roughness

5.1. Experimenatl Studies

SP is investigated experimentally for decades by many researchers. The fundamental mechanisms of SP are reviewed by Lieurade and Bignonnet[173]. The fatigue performance not only depends on the RCS at the surface, but also on the gradient inside the material. The stresses developed due to the cold work as a result of shot impact and the Hertzian contact pressure get superimposed to provide a cumulative effect.

Kobayashi et al [174] have brought out the differences between static compressive test and dynamic tests with single and multiple impacts. The indentation in static test is quite different from dynamic test probably due to high strain rate effects. Also, the stress is tensile in single-ball dynamic test at the center of indentation compared to the compressive stress found in static test. As more impacts were performed around the first, due to superposition, the residual stress at the center has turned out to be compressive. Kirk discusses about four different influencing areas from experiments: shot properties, peening intensity, surface coverage and surface properties[175].

5.1.1. Shot Parameters

Shot size, velocity, material strength and hardness are the key parameters that affect the SP process. The effect of single-ball impact made of steel on the hardness and dent dimensions of different materials such as aluminum, brass and steel are studied experimentally[176]. It is obvious that the hardness change in materials bears direct relation with shot velocity. The RCS magnitude is found to be dependent on the shot hardness[78]. Though the shot is assumed to be rigid in simulations, the yield strength of the shot affects the dent size and the corresponding RCS field of the solid[177]. Saturation occurs quicker with smaller shots though of smaller dimensions. When smaller shots are used the maximum RCS occurs closer to the surface. As the shot size is increased, the maximum stress occurs farther away from the surface and the total affected depth also increases[178]. Maximum value of RCS depends on the peened material state while shot size and velocity influence the RCS layer thickness[179]. The plasticized depth increases with the hardness and diameter of the shot[180]. The effect of shot velocity on Almen intensity and depth of RCS are measured [181]. The hardness and dent depth increase with shot velocity. Also, higher velocity increases intensity levels. The shot with oblique velocity creates less plastic zone than the one with normal velocity. The depth of RCS field is more for 90 deg impact than say 45 deg impact and it helps to reduce the Fatigue Crack Growth[FCG][182]. Besides, the micro and macro surface irregularities are deformed plastically[180]. SP produces more RCS than grit blasting [183]. SP with steel grit produces higher RCS and lower specimen deflection than SP with spherical shot[184]. Even in grit blasting, the tangential component of the velocity reduced the RCS but increased the surface roughness[185]. Even the plasticized layer thickness is higher while the maximum hardness is lower[186].

5.1.2. Intensity

Almen intensity is measured using the permanent offset occurring due to its plastic deformation. The intensity determines the depth of RCS below the peened surface. Very low intensity does not increase the fatigue resistance as much as a higher intensity[187]. But increasing the intensity beyond certain value can be detrimental. Very high intensity is likely to have higher tensile stresses in the interior and the compressive stresses also will be smaller but spread deeper than in the case of low intensity peening. To increase the surface RCS value, secondary or even multiple peening can be effected[188].

5.1.3. Surface Coverage

It is always assumed that 100% coverage is essential for uniform development of RCS. In many cases, the coverage is more than 100%. Sometimes, 100% coverage is not possible while peening hard materials and complex features. The RCS value increases linearly with coverage until 80% then decreases[189]. In fact, the dent size is smaller than the residual stress field in single-shot impact, which implies that 100% coverage may not be required [190]. In a study to understand the incomplete coverage using punches, Meguid and Klair[191] have found that the residual stress remains compressive even when distance between punches is four times the diameter.

5.1.4. Surface Properties

MSE methods, SP in particular, alter the microstructure at the surface in terms of hardness, grain size and even phase-transformation due to deformation. The surface microstructural changes are covered in [192]. The recrystallization to a fine structure at the surface helps fatigue properties while the coarser grain at the bulk region improves creep performance.

Detailed experimental studies on different metals and alloys show increased fatigue strength of surfaces with good fatigue strength further than the surfaces with poor fatigue strength caused due to machining etc[193]. Wang et al [194] have conducted series of experiments on different materials and empirically determined the relationships between material properties and peening parameters through regression.

Material Hardness: The Hertzian effects dominate at higher hardness of material while plastic stretching effects influence at lower material hardness[192]. The fatigue strength improvement for high strength materials primarily depends on the residual stresses, but for low strength materials, the fatigue strength depends on the work hardening also[195]. When the material hardness is high, less energy is absorbed to deform the surface layer plastically, but more energy is used to deform deeper layers due to Hertzian loads[196]. The depth of the metal plasticized decreases as the material hardness increases. The fatigue strength improvements are lower for low strength materials than for high strength ones[197]. SP has improved fatigue strength by reducing notch factors in low hardness materials and by RCS in high hardness materials [198] Overall, different structural changes occur based on difference in the original structure due to SP and hence drawing conclusions based on increase in hardness is insufficient[199]. Failures can also occur in the zone below the hardened layer due to cyclic stresses or high mean stress[200].

The cold work topographies of typical materials are given in [201]. SP results in work-hardening, work-nonhardening or work-softening on the target material[202]. Work hardening occurs in the low-hardness target materials due to reduction in domain size, while work-softening happens in high hardness materials with no domain size change[203]. For example, work-softening can be seen on rolled sheet with pre-strain. But even work-softened materials exhibit fatigue strength increase due to SP[204,205]. This is because the RCS does not change, though there is recovery of crystal structure[206]. SP also produces cyclic plastic strains in the surface layer and the cyclic hardening increases the dislocation density[207]. The higher the cyclic yield strength and lower the cyclic strain hardening exponent, better the fatigue strength of materials[208]. The surface residual stresses are more when hard shots are used, but the depth of RCS is relatively small[209]. In gravity aided shot peening experiments with steel shots, SP increases the hardness such that the local yield strength is increased much above bulk yield strength[210].

Surface Roughness: The surface roughness factor R_a increases with shot velocity. It is also affected by the hardness of the material peened and the shot diameter. Loersch and Neal [211] have concluded that higher shot diameter causes better surface finish. High intensity peening followed by low intensity peening[HLSP] and mechanical polishing produces same residual stresses as HSP but low fatigue strength due to the surface roughness effect [212]. For identical peening conditions, roughness is more in 42CrMo4 due to lower yield strength than 54SiCr6. In a study with glass beads, it is found that the nozzle angle between 45 to 75 produces better surface finish[213]. The roughness of the surfaces can be measured by sharp surface sections. A new method where 10 sample points will be used to measure the surface roughness will eliminate the inadequacy of conventional roughness[214].

Studies on Steel: Extensive experimental research has been done on various steels. SP has improved the fatigue life due to surface cold-working than unpeened or bulk-formed steel specimen[215]. Iida has also mentioned that SP has improved the fatigue strength of carbon steels without work-hardening[216]. When glass beads are used to peen machined medium carbon steels, the surface roughness is reduced and the tensile stresses produced by machining are turned into compressive stresses[217]. In the same way, when SP is done as a pretreatment, it reduces the tensile stresses developed due to chromium plating [218]. SCC occurs due to the presence of tensile stresses, corrosive environment, temperature and material susceptibility. The shot peened EN8 plain carbon steel exhibited good resistance to SCC and peening coverage, intensity and time determine the improvement in SCC strength[219]. Contact failure is due to pitting, shallow spalling and deep spalling. The 40Cr steel shows higher contact fatigue strength after a compound heat treatment consisting of ion-nitriding and induction hardening followed by SP[220]. The 20CrMnTi steel, after heat-treatment and SP, shows increased hardness due to higher dislocation density[221]. The RCS and the hardened layer prevent crack formation in the surface, but in the sub-surface. Pre-annealing temperature decreases the hardness of the material. In a study on carbon steel, pre-annealing temperature increases hardness layer depth and surface roughness due to SP[222]. Transgranular slip decoherence and fine striations cause low crack propagation in hardened smooth specimens of 40Cr steel[223]. The cyclic stress-strain curves are smaller for 42CrMo4 steel in push-pull and bending fatigue tests resulting in increased fatigue life due to the different nature of stress distributions[224]. Specimens with different thicknesses have different fatigue strength depending on the Almen intensities[225]. At high stresses structural steels have even shown negative improvement to corrosion fatigue in both air and water to SP[226]. But after preceding treatments of toughening and induction hardening, SP is found to increase the fatigue strength[227]. In Ck45 steel subjected to bending, the notched fatigue strength improvement is notably higher than the improvement in unnotched specimens after SP[228]. The same effect is obtained even after high temperature fatigue testing[229]. Both austenitic and martenistic steels show improvement in SCC fatigue lives due to SP [230]. Pakrasi has reported decrease in retained austenite in carbonitrided 16MnCr5[231]. Changed microstructure like increased martensite in carburized gears and compressive stress field changed from tensile field due to carburizing[232] prevent initiation and propagation of cracks. Matsumoto et al have derived regressed the relationship between peening parameters such as shot velocity, intensity, residual stress etc[233]. For example, the intensity is defined as:

$$H \infty \log(CD^3V^2)$$

Wu et al breaks up the flow stress into five components including petch stress, macro and micro stresses, secondary particle stresses and conclude that the primitive dislocation distribution has to be controlled to avoid static stress relaxation to facilitate the rearrangement of dislocations[234,235]. The static and fatigue bending strengths of carburized steels increase after shot peening and polishing to remove internal oxidation and reduce surface roughness[236]. The combined effect of carburizing and SP depends on hardness, RCS and austenite[237]. Carburized 17NiCrMo4-6 steel used in gears has shown to have better fatigue strength without affecting fatigue contact strength[238]. Non-metallic inclusions such as Oxygen reduce the fatigue strength in carburized steels[239]. However, carburizing after peening has increased carbon retention, but has reduced the contact strength[240]. Fatigue strength development in bainitic-austenitic nodular cast iron is caused by work-hardening, martensite formation and RCS development[241]. Pitting formation due to inter-granular oxidation layer is suppressed in carburized steels[242]. After carburizing, hardening and tempering, application of SP and polymer coating have enhanced the fatigue strength of carbon steels proving that SP works well on hard surfaces[243]. Fatigue strengths of soft-nitrided components and austempered ductile cast irons increase with SP[244,245]. The intergranular corrosion of ferritic stainless steels is also reduced due to SP [246].

Several literature point towards the residual compressive stress as the primary reason for fatigue life improvement even in corrosive conditions. For example, Risch brings out the effect of SP on SCC and corrosion fatigue on austenitic steels[247]. The 12% Cr steel shows improvement against SCC due to the plasticized layer[248]. The intergranular corrosion resistance increases, but pitting resistance is unaffected[249,250]. The austenitic steel that is cold worked to represent neutron irradiation also derives fatigue strength and SCC benefits from SP[251]. SP not only retards but also prevents SCC as long as the compressive stress is not altered[252]. The 304 steel, used in nuclear reactors, with similar level of cold working also shows SCC strength improvement when shot peened with water jet[253]. The austenite in 304 steel transforms to martensite, which is magnetic[254]. The SP increases RCS in the martensite while the water peening reduces it compared to austenite in 316 steel[255]. Small features such as fillets, holes are peened locally either with masks or without masks[256]. Masking produces surface tensile stresses in the masked zone[257]. In a contact fatigue study, the hardness variation along depth has caused decrease in fatigue life after peening[258]. As the depth of decarburized layer increases, shot-peening effect has less effect on fatigue life[259]. Unlike other materials, work softening results due to shot peening in austenitic steels[260] resulting in insignificant fatigue strength change[261]. With two modifications of SP, viz., stress double shot peening and stress reflection double shot peening, residual stresses have been increased significantly in QT and IH steels[262,263] Forgings also show fatigue life improvement after airless blast cleaning[264]. Peening with super-hard fine particles improve the fatigue property of maraging steel[265]. Cemented carbide shots have produced larger RCS and depth thus enabling the cold forged dies to have better fatigue strengths[266]. Steels with and without surface structure anomalies have shown same level of fatigue strengths after SP[267]. Zinc coated steel has shown fatigue strength improvement after SP depending the load type and sequence of application[268]. In large specimens, SP has less pronounced effect due to the possibility of large inclusions[269]. Stainless steel welded plates show enhancement in fatigue strength after SP with combined bending and torsion loading[270]. The friction coefficient is found to decrease after peening [271]. Experiments with AISI 4340 steel indicate that even 20% coverage is adequate[272]. Peening before coating helps to prevent crack propagation from the coating into the base metal [273]. AISI 4340 with higher hardness experiences less RCS than the one with lower hardness for the same intensity [274].

Studies on Aluminium: Investigations have been done on 7075 alloys for T6[275]and T7351 conditions[276]. For example, in an experiment on 7075 alloy, it is found that the RCS and microstructure change help the fatigue strength while the roughness is unfavorable[277]. In another experiment, 7075-T6 alloy has not produced any significant fatigue strength increase due to peening with oil though higher RCS is expected due to hydrodynamic pressure and better surface finish by lubrication. The RCS does not seem to affect the crack growth rate of T7351 alloy. When peening is performed on partially fatigue-damaged 7075 alloy notched specimens, it shows improvement in fatigue strength than even peened specimens before fatigue loading though the RCS is not built to the original level[278]. Repeening fatigue damaged specimens produce a mixed set of results. Compressive preloading has insignificant effect on the RCS distribution while tensile prestress has a positive effect than peening without any prestress[279]. Jaensson and Magnusson also conclude that superposition of damage is applicable in 7050-T73651 specimens. Even in 7075-T3511 alloy, the fatigue strength improvement due to SP in notches is marginally more for specimens with higher SCF than for specimens with low SCF[280]. Similarly, welded 7075-T6 alloys show good improvement against SCC due to SP[281]. Under corrosive environment, SP produces better fatigue strength than unpeened condition for the 7075-T6 alloy[282]. Experiments on 6065-T651 alloy show that the RCS is related to air pressure through power law and to the exposure time through exponential law[283]. The secondary peening with glass beads is found to improve both surface finish and fatigue life[284]. When nozzle angles are shallower, even secondary peening produces higher surface roughness[285]. Another study with 2024 and 7075 alloys[286] concludes that the metallurgical and roughness changes and RCS improve the fatigue response. Apart from RCS and surface hardening, the closure and modification of pores cause improvement in fatigue strength of die-cast aluminium alloy[287]. SP treatment with aluminium shots produce higher RCS though the Almen intensities are lower compared to harder shots made of steel or ceramics probably due to a different mechanism[288]. Fretting caused by relative movement of surfaces under loading consists of corrosion and fatigue[289]. SP improves fretting strength due to more RCS and surface roughness and less tangential force [290]. The relationships for the damage due to pitting and hydrogen embrittlement are provided by de los Rios[291]. The alloys 7010 and 8090 have been tested for fatigue and fretting strength improvements after SP [292] and it is found that the 8090 alloy has less improvement in LCF and fretting due to high deformation of the surface. The desensitization of welds by TIG followed by SP has produced fatigue strength of welds[293] sometimes even almost equal to the base metal[294]. The delamination of layers due to exfoliation corrosion in aluminium rolled plates is suppressed by stretching and peening[295]. The Holdgate model seems to predict better than Avrami model for 2024 and 7150 aluminium alloys [296]. For single source of peening, it is given as:

$$C(t+\delta t)=1-[1-C(t)]\left[1-\frac{a}{S}\right]$$

Studies on Titanium: Different Titanium alloys show different levels of fatigue life enhancements[297]. The fatigue strength improvement also is different for notched and unnotched specimens[298]. Hanyuda et al conclude that the fatigue strength improvement depends mainly on the magnitude of surface RCS and not on the tensile stresses developed inside the material in their study on Ti-3Al-2V alloy[80]. But RCS magnitude first increases to a maximum value and then decreases with Almen intensity. With increasing coverage, the RCS rises and the maximum value of RCS shifts inside the material of Ti-6AL-4V plates[299]. The RCS even extends to the locations not covered by peening. Wagner and Luetjering[300] have reported that dislocation density, residual stress and plasticity depth increased while micro-hardness decreased during peening of Ti-6Al-4V. Double shot peening first with cast iron shot followed by glass shot improves both fatigue and surface qualities

[301]. The internal tensile residual tensile stresses can also cause crack nucleation after SP [302] thus providing a lower fatigue strength in vacuum fatigue test for Ti-6Al-4V alloy[303]. The fatigue strength of TIMETAL 1100 depends on the tensile stresses, the micro-structure and its sensitivity to the environment[304]. Titanium alloys with creep-resistant coarse grain structure are peened and annealed at high temperatures to form fatigue-resistant fine grain structure at the surface [305].

Studies on Superalloys: Superalloys and powder metals also show improvement in fatgue strength due to SP. The fine recrystallization of the peened layer in superalloys improves the fatigue strength even at high temperatures[306]. Inco-718 alloy shows improvement in fatigue strength possibly due to the elimination of small machining defects as well[307]. The Rene-95 powder superalloy undergoes reduction in micorposorsities due to SP and the fatigue strength improves due to RCS, smaller grain-size resulting in increased density [308]. Even in the case of ceramics, SP is able to induce RCS to a magnitude of 1Gpa[309, 310] apart from fracture toughness improvement [311]. In Ti-6AL-4V powder compacts, the blended element material has not shown any improvement for commonly employed peening intensities. In the prealloyed material even higher intensities have resulted in lower fatigue strength after recrystallization annealing[312]. The Al-SiC metal matrix composites also show improvement due to SP. The average RCS can be measured by incremental hole drilling method while the RCS in fibre/matrix can be measured by X-Ray diffraction[313]. The SiC particles and whiskers are used in 2024 Aluminium matrix[314]. The SiCp shows improvement due to SP while SiCw does not. SP creates surface roughness, RCS and cold-work[315]. The main difference from unreinforced material is the possibility of reinforcements fracture except for the possibility of reinforcements fracture.

Studies on Magnesium Alloys: SP is used to enhance corrosion strength of magnesium alloys. Ebihara et al have compared the acoustic, eddy-current, optical and X-Ray measurements of three magnesium alloys[316]. Zhang and Lindermann[317] have studied the high cycle fatigue behavior of AZ80 Magnesium alloy with shot peening. Increasing beyond certain peening intensity reduced the fatigue strength possibly due to increased crack size in the surface layer. This could be due to the hexagonal structure with limited deformability by slip[318]. Wang et al have calculated the material responses such as RCS, depth and hardness from yield and ultimate strengths and peening intensity[194]. In a SCC study of Al-Zn-Mg alloy, different types of peening processes are compared with vibratory and thermal stress relieving processes. The RCS developed due to various peening processes is lost due to subsequent welding[319].

Studies on Components: Peening also helps small components such as compression coil springs[320] with crack growth slowing down in the RCS field[321]. In components such as coil springs, the RCS distribution is not different than in plates[322]. Barrel springs used in automotives benefit more by coarser shot peening[323]. In leaf springs, the RCS distribution can be approximated to sine wave approximation[324]. Peening with shots with higher hardness has produced better fatigue strength in gears[325]. Similarly in welds, SP can increase the weld strength higher[326] even close to that of plain plate[327]. When peening is done on damaged parts due to fatigue loading, the fatigue strength can be improved by the rejuvenating factor which is the ratio of fatigue cycles of damaged specimens after peening to those of undamaged and unpeened specimens [328]. The corroded spring steel also has shown improvement after peening [329].

Thermal Effects: The heating due to peening occurs due to abrasion, which is due to difference in hardness of shot and material. The temperature rise increases with increase in the hardness and decrease of shot size[330]. The heat generated during SP may reduce the residual stresses[331]. Herzog et al have conducted a set of experiments exploring the influence of different peening parameters on steel and aluminium[332]. They have concluded that the RCS depends on the energy loss of the shots to create plasticity [neglecting thermal losses] as well as the Hertzian effect. SP at higher temperatures produces higher RCS[333]. Peening at high temperatures cause more RCS and stability than conventional peening[334]. Maximov uses Gouy-Stodola's theorem to minimize entropy generation[335]. However, heat generated is generally ignored in simulations as the energy loss is very small.

5.2. Theoretical Studies

The theoretical studies on shot peening have been analytical as well as numerical. The analytical solution proposed by Hertz is the expression for the maximum elastic compressive stress for two spheres in contact[336]:

$$\sigma_c = -0.62\sqrt[3]{\left(\left[\frac{P}{\Delta^2}\right]\left[\frac{1}{R_1}+\frac{1}{R_2}\right]^2\right)} \text{ where } \Delta = \frac{1}{E_1}(1-\upsilon_1)^2 + \frac{1}{E_2}(1-\upsilon_2)^2 \text{ and } P \text{ is the contact force}$$

Using the Hertzian contact and elastic-plastic theories, simplified formulae have been developed to predict the maximum value and the depth of the RCS field by Ogawa and Asano[337]. In his paper, Al-Obaid [338] has attempted to provide theoretical expressions for single-shot impact by assuming the peening to be quasi-static and compared with experimental results. The magnitude of the RCS depends on factors such as material, hardness, size and velocity of the shot [339]. Shot velocity is an important factor in the peening process. Shen and Atluri have evolved an analytical solution to calculate RCS field that considers shot velocity[340]. DeLitizia has come up with a cubical expression of depth for the residual stresses[341]. Al-Hassani has provided an overview of various aspects of the mechanics through analytical solutions[342,343]. He has calculated a damage number represented by $\rho V^2/\bar{p}$. The plasticized depth is calculated as a function of peening parameters which includes strain hardening as

$$\frac{h_p}{R} = 2.57\left[\frac{4+n}{6+2^{n/2}}\right]\left(\frac{\rho V_2}{\bar{p}_y}\right)^{1/(4+n)}$$

Analytically the evaluation of RCS is developed by Watanabe et al as a superposition of stresses. The nominal fatigue strength is evaluated based on the analytical expressions for the residual stresses and internal fatigue strength[344,345]. The tensile residual stress introduced by peening is given by

$$\sigma_t = \frac{(z-z_0)^{1.35}}{a(z-z_0)^2+b}$$ where a and b are constants based on RCS field.

The RCS which is elastic is calculated as the sum of stresses due to forces and temperatures [346]. An equivalent temperature load that produces the strain pattern as that of multiple shots has been used alternately[347] to predict SPF. Iida has calculated dent dimensions as a function of peening parameters[348]:

Dent diameter, $d = K_d DV^{1/2}$ and depth $h = K_h DV$ where K_d and K_h material constants.

An empirical model for weld strength is provided by Chang and Lawrence [349]. For machine elements a design procedure involving the surface characteristics have been evolved by IPM, Poland[350]. Collision energy is used as a measure to optimize the peening in coil springs[351]. When the stress range is below the endurance limit, the mean stress can be added to stresses due to shot peening to calculate the fatigue strength[352]. Multi-axial loading and surface roughness are used through Crossland model $\xi_a + \alpha P_{\max} \leq \beta(\alpha, \beta = cons\tan ts)$ and correction factor respectively to evaluate the fatigue strength taking into peening and machining stresses into account[353]. Guechichi and Castex conclude that the cold work has more beneficial effect than RCS due to SP[354].

In this work by Levers et al, pre-stress effects have been created using temperature loads to avoid the complexity of modeling of multiple impacts. This has been extended by Gardiner to simulate peen forming [355]. Planes parallel to the surface have been applied with eigen strains to impart the pre-stress conditions [356] This method has found match the cold work calculated through analytical methods [357].

FE Models: Meguid et al [358, 359] have simulated single-shot peening effect through an elastic-plastic three-dimensional impact analysis for different shot velocities, sizes and hardening characteristics. The results reveal that the depth of the compressed layer and the surface and sub-surface residual stresses are influenced by shot velocity, shot shape and to a lesser extent by the strain-hardening rate of the target[360]. Using ADINA, two-dimensional axis-symmetric analysis has been performed with dynamic effects by Schiffner et al[361]. The results from analyses for different shot velocities on two materials have compared well with theoretical results reported by Al-Hassani. Stress stabilization after a small number of impacts at the same location has been reported based on FE analysis[362]. Curtis et al [363] have determined the boundary conditions relating the roughness and RCS. Baragetti has used axi-symmetric and 3D analyses to determine the stress field[364], while a 3D elastic-plastic analysis with strain-hardening is considered by Kyriacou[365]. Single-shot simulation indicates that the dent form is influenced by the yield strengths of shot and material and the hardening effects[366]. Hassani has used ABAQUS to calculate stresses with strain-rate non-linear work hardening effects to compare with his theoretical results[367]. Guagliano et al have combined FEM and a set of non-dimensional parameters to relate the peening parameters and the stresses[368]. As Almen intensity continues to be the measure of peening process, Guagliano has come up with a process of relating it to FE results[369]. The multi-shot impingement produces more RCS than the single-shot impingement, while the plastic strain remains the same [370].

Webster [371] has indicated that apart from temperature effect, high cyclic plastic strains can also wipe out the fatigue strength improvement in the part due to peening. Han et al. [372], have calculated the residual stress distribution due to a simulated multi-impact and applied the results on a test specimen so that stresses due to real-life loads can be superimposed. Meo and Vignjevic[373] have analyzed the residual stress development in welded structures due to shot peening using transient method to account for elastic and plastic waves, inertia and strain rate effects. Deslaef et al[374]. Have measured the coverage rate and shot speed by FEM. It is shown that the stress filed created by impacts of first set of shots being is made inhomogeneous by subsequent impacts[375]. When multi-shot impacts occur, residual stresses prevent further plastic flow and after a few cycles the entire deformation will be elastic, thus causing shakedown[376].

Using LS/DYNA Meguid et al [377] have analyzed the effect of peening on AISI 4340 steel and have concluded that strain-rate is a key parameter that affects plastic strain and residual stresses along with shot velocity and hardness. The residual stresses depend on the coverage and it is uniform when subjected to multiple impacts. The friction effect on both plastic strain and compressive stresses is very minimal. Majzoobi et al [378] have simulated multiple shot impacts using LS/DYNA for different velocities. They have found good correlation with tests conducted by Torres and Voodwald. Beyond certain velocity, the maximum RCS value decreases[379]. Using ABAQUS explicit code, an elastic-plastic analysis has been performed to study the peening parameters for steel and aluminium [380]. Double-sided peening has been simulated with two-step process of explicit and implicit analyses to correlate with experiments[381]. Based on plasticity, coverage is estimated and using an explicit-implicit analysis is done to simulate the impact of nearly 1000 balls[382]. Peen forming has been simulated using FEM with temperature loads[383]. A 3D analysis is performed for simulating multiple shots using ABAQUS-explicit tool that uses infinite boundaries used to quieten the stress ransgranula[384]. A typical process map involving FEM is shown in figures 7 and 8.

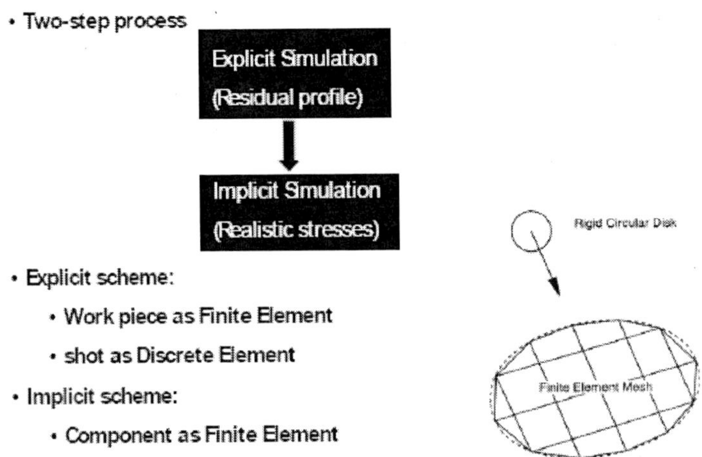

Figure 7. Two step process for FEM simulation.

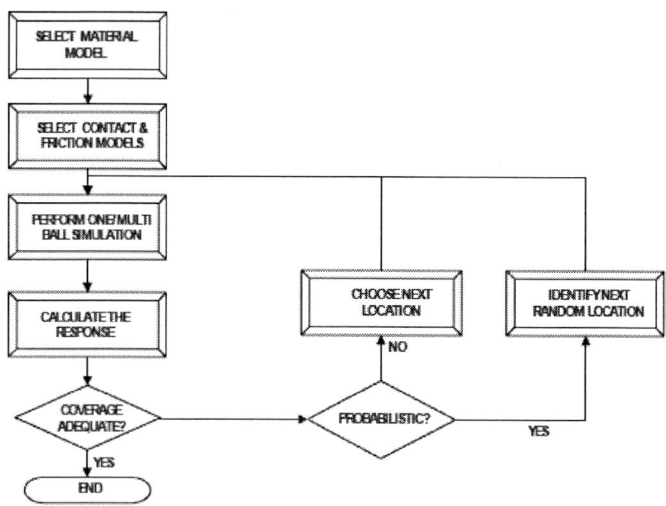

Figure 8. Process map for FEM analysis.

DEM-FEM Models: Han and Peric[385] has performed a two-dimensional analysis treating the sphere as a rigid circle. They have used different contact laws and included damping. In the DE/FE analysis the surface stresses have been evaluated as tensile in nature, necessitating three-dimensional analysis for more accurate stress prediction. The work has been extended with three-dimensional laws by the same authors[386]. They have also simulated multi-impact and found that the single-impact results have been significantly different from multi-impact results. The above studies using discrete element methods may preclude the effects of shots made from different materials and hardness. Figure 9 shows the flow-chart depicting FEM and DEM.

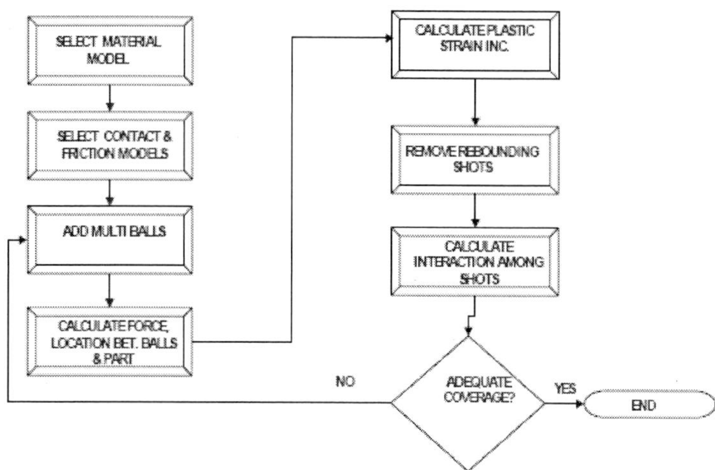

Figure 9. Process map for FEM/DEM coupled analysis.

Multi-Scale Models: Peric and Han have supported strongly the case for multiscale modeling due to the very small shot size when compared to the size of the component. The material microstructure at the surface is modified by SP [387] to nano-level[figure 7] which plays important role in the determination of the RCS distribution and fatigue strength[388]. For example, Lu and Lu report the formation of Nano-crystals at the surface of the shot-peened component by Surface Mechanical Attrition Treatment[SMAT] with multi-directional loading [389] due to dislocation formation, movement and annihilation. The nano-crystals formed at the surface are found to improve the fatigue strength [390]. Thus Roland mentions that nano-crystals improve the yield stress and fatigue strength in 316L stainless steel [391]. The nano-structure is found to enhance the fatigue characteristics of coil springs [392]. Xinling et al [393] has developed a method of using dislocation dynamics to evaluate plastic strains due to shot peening. A new process, in fact, is evolved called Surface Nano-crystallization and Hardening[SNH] method that uses higher kinetic energy generates manifolds of plastic strain and RCS than SP thus creating nano-crystals[394]. The formation of nano-layer depends on the shot size which determines the contact duration and strain-rate[395]. The nano-layer has very high hardness and thermal stability than the underlying hardened layer.

Constitutive Models: In continuum mechanics, the type of material models determine the material response. Elastic-plastic material model with strain rates, damping and deformable shot are considered to predict the stress field due to SP [396]. Fathallah has used Guechichi-Khabou model which does not include the tangential stretching but only Hertzian effect[397]. The constitutive relation between plastic strain and residual stress, say, in X direction is:

$$E_{xx}^{p}(z) = -\frac{1}{E}(R_{xx}(z) - \upsilon R_{yy}(z)) \text{ where } R_{xx}, R_{yy} \text{ are residual stress tensors.}$$

He has considered the effects of friction, inclined impact and hardness ratio [398]. Frija et al. have used a combined damage model of Chabache and Lemaitre for Waspalloy [399].

$$D \cong \frac{D_c}{\varepsilon_R - \varepsilon_D}\left(p\left[\frac{2}{3}(1+\upsilon)+3(1-2\upsilon)(\frac{\sigma_H}{\sigma_{eq}})^2\right] - \varepsilon_D \right) \text{ where } D_c,\ \varepsilon_R \text{ and } \varepsilon_D \text{ are constants.}$$

Lillamnad et al. also use Chabache model for engine disk components [400]. Al-Hassani has proposed that shakedown, reverse yielding, Bauschinger effects and strain-rate play a role in the accurate prediction of RCS [401]. Slim et al. use the elastic-plastic method proposed by Zarka and Inglebert to calculate RCS [402]. Another study using Zarka model finds that the RCS distribution over depth has matched well between theory and experiments[403]. The effect of temperature rise due to SP is coupled with mechanical effects through a thermo-elastic plastic model [404].

Stochastic Models: Some of the experiments indicate that the scatter in fatigue lives have not increased after SP thus making it a useful process in the industry [405]. Depending only on standards based on Almen intensity measurement can result in big variation [406]. The intensity measurement using Almen strips and coverage measurement are statistically inadequate[407]. On the contrary, moving towards SPC and computer control of key process parameters, the use Almen strip can be mostly avoided [408]. The SP is made statistically capable by relating the process variables statistically to the component strength. Ideally, all process variables should be treated as statistical with proper tolerances that can be controlled to make the SP statistically capable[409]. Equivalent diameter is calculated to represent the variation in shot size [410] in calculating the same coverage ratio. Even the measurement parameters are stochastic in nature. For example, the coverage is assumed to follow normal distribution below the nozzle [411]. By tightly controlling the tolerances of these variables, a statistically minimum intensity called true capability can be assured. Statistically the impacts follow random distribution and coverage is calculated accordingly for multiple impacts[412-414]. Some of the variables are controllable and a robust design method is proposed using them[415]. Using Monte-Carlo method, the surface topography is simulated using random simulations and found to match well with experiments[416]. The forming shape is predicted through regression of peening variables in ball shot forming with minimal statistical error[417]. Ford Corp. has proposed a statistical process control to define and monitor critical peening parameters than relying on Almen intensity [407]. GE has come up with Full Assurance Shot Peening covering different aspects that controls the robustness of quality [418]. The impact locations are specified by random numbers to specify the coverage [419].

6. Optimization

SP generates RCS along with cold work and surface roughness. In an optimum condition, the surface roughness should be as low as possible with the magnitude and depth of RCS should be as high as possible[420]. It is also desired by the designer to have less cold work to reduce the levels of relaxation. The factors to be considered for peening are given by Kiefer[421]. For example, the size of the shot is determined by the smallest radius of the part. Yu-Kui et al [422] have studied the effect of peening on 40Cr steel for different peening parameters. They have concluded that the maximum subsurface compressive stress for a given material is always the same while surface RCS depends on the peening parameters and the material properties. The optimum results depend on the ratio of compressive stress layer to the tensile stress layer. As the number of peening variables is large, a large data base is required to predict optimum process levels to use in the design of components[143]. The area under the RCS distribution curve between the surface and maximum values can be used as a measure of fatigue property[423]. Simpson and Probst [424] have concluded that the optimum intensity range is determined by the peening induced surface damage rather than by RCS.

6.1 Experimental Studies

Optimization of peening process depends on

- choosing the right set of parameters
- setting optimum values for them and
- controlling them effectively

Optimum values of peening intensity depend on the microstructure of the materials. The austenitic steels reach an optimum in fatigue strength beyond which they lose the improvement due to formation of deformation related martensite formation[425]. Even the SCC strength deteriorates due to such metallurgical change[426]. However, in 18Mn4Cr steel optimum peening parameters are experimentally determined to avoid martensite formation[427]. SP does not improve the strength against localized corrosion such as pitting[428]. In another optimization study, the life improvement in 2024-T351 alloy in LCF is due to slow crack growth and in HCF due to prolonged crack arrest[429]. Prevey and Cammett have concluded that 100% coverage is not required for an optimum fatigue performance for the typical load spectrum considered[430]. The process parameters can also be controlled through computer to improve the reliability of the peening[431-433] which reduces the variation of intensity and surface finish. Experiments conducted on peen formed components indicate reduction in curvature due to relaxation of RCS to a level induced by plastic deformation[434]. Through on-line control the peen forming by continuously measuring the shape and modifying the process parameters [435].

DoE studies are generally employed for optimizing responses of any process[figure 9]. Five responses have been measured with respect to six control parameters to optimize through a DoE study[436]. Figure 10 shows the methodology of Design of Experiments[DoE] in a schematic way. The shot hardness and exposure time are found to have influence over the RCS in a DoE study while stand-off distance and shot diameter have less influence[437]. A new carbon-steel alloy has been formulated with high Mo and Low Si that will enhance the RCS by its ability for higher peening intensity[438]. Tufft has conducted a DoE study of peening parameters and concluded that velocity is the key parameter influencing the intensity[439] and it influences the slip band development and transient temperature rise[440]. Similar DoE study has been conducted on SAE8620 to determine the optimum peening parameters[441]. Fatigue models based on the DoE study has been proposed to calculate fatigue life and crack growth [442]. A lower bound estimate on fatigue life has been found using the slip depth to find the threshold velocity for Rene88[443].

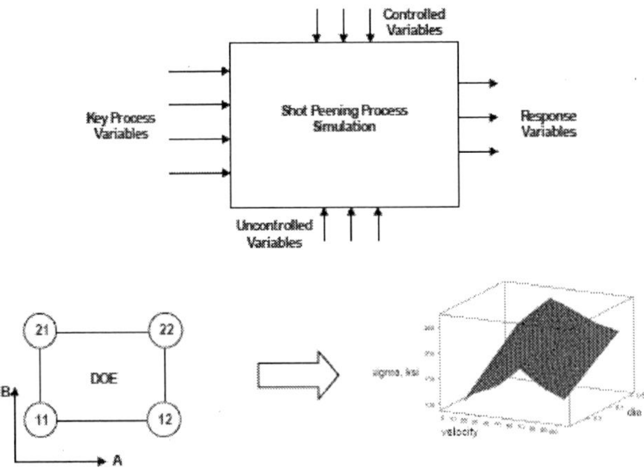

Figure 10. Optimization through DoE method.

Over-peening: Over-peening at highly localized areas can cause erosion of metals[444]. More cold work arising out of such over-peening can cause inversion of stress thus reducing the compressive stress [445]. But in harder materials, surface roughness due to over-peening is less pronounced[256]. When very high coverage, to 600%, is done through SP in carburized CrMo gear steels, white layer of adiabatic shear bands are formed in the subsurface layer which can cause tooth-chipping [446].

Stress and Strain Peening: Application of stress or strain increases the RCS field depth. Pre-stressing improves the fatigue life of components though its mechanism is not clearly known. In pre-stressed peening, stresses are applied in the lateral direction before the part is peened and then the applied load is removed. Among different types loads, the maximum benefit is derived for tensile load and for bigger specimens[447]. The elastic pre-stress applied increases the maximum values of RCS and its location than in free-state peened zones. The total RCS depth does not change in pre-stressed part or free-state peening[448]. It is also reported that the magnitude of RCS and its layer thickness increase with stress-peening[449]. Leaf springs are generally bent to have a preset of twice the yield strain and then strain-peened on the tensile side[450]. However, torsional prestraining is found to be detrimental [451]. Also, when peening follows plasticization/pre-setting, the RCS reaches the maximum upto yield strength[452]. The value of RCS increases with material hardness.

6.2. Theoretical Studies

Peenstress is a software developed by Metal Improvement Company that helps to choose the correct process variables and it predicts the velocity and RCS[453,454] for the given shot, material and intensity. The software has a library of materials and geometries to choose from. A two-dimensional FEM based parametric study has been done and the shot with hardness higher than the target provided better RCS than the softer shot[455]. Baragetti et al [456,457] have performed numerical simulations using Design of Experiments[DoE] and have identified a non-dimensional parameter which takes the ratio of RCS magnitude to yield point to express the shot peening conditions that can be applied to different materials with different treatments. Taguchi technique is used to optimize the shot peening parameters[458]. The change in improvement in alternating bending strength is expressed as a superposition of changes in roughness depth, RCS and hardening [459]:

$$\Delta\sigma_{w,b} = \alpha_1 \Delta R_t + \alpha_2 \Delta FWHM + \alpha_3 \Delta\sigma_R \text{ where } \alpha_1, \alpha_2, \alpha_3 \text{ are constants}$$

FEM in conjunction with statistical analysis of peening parameters is used to optimize the forming of sheet metals [460].

7. Crack Growth Studies

Figure 11 shows the crack propagation with and without the presence of RCS. As the RCS field is modified, the crack front shape is modified. The compressive stresses in the surface layer retard the crack propagation than in the bulk material.

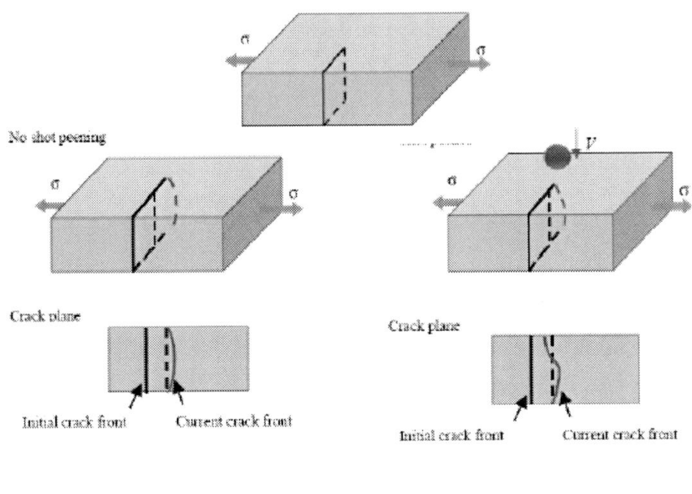

Figure 11. Crack propagation with and without Residual Stress.

7.1. Experimental Studies

SP induces RCS, cold work and surface roughness and each response parameter is critical in fracture mechanics studies. Surface roughness accelerates the crack initiation while cold work retards it. But, crack propagation is accelerated by cold work but is retarded by RCS[461]. Fine crystalline structure at the surface helps in controlling crack initiation while the coarser bulk material retards crack growth[192]. To have low Fatigue Crack Growth[FCG], the surface RCS should be as high as possible[462]. This occurs as RCS reduces the tensile stresses at the crack tip[463]. In high carbon steels, generally the crack grows from the SP dent which acts as a stress raiser[464]. In carburized notched steels, the fracture is a mixed mode of granular and transgranular one, whereas after SP it has only transgranular features[465]. The RCS and the dislocation density created by SP provide favorable conditions against the crack growth while the surface roughness provides a negative effect in Ti-6Al-4V alloy[466]. In 7075-T351[467] alloy, SP reduced crack growth at low R ratios and the SCF due to SP is insignificant[equal to 1.14]. The RCS and surface roughness play a role in improving fretting fatigue with RCS having more influence[468]. During fretting, the stress relaxation increases with increasing number of cycles, thus pushing the crack initiation zone towards the contact surface. The surface cracks have not grown as long as the superimposed[RCS and the applied] stress intensity is below the threshold value [469]. Waspalloy is tested on four-point bending machine to determine the crack initiation, stage 1 and 2 growth for finding the peening effect to provide healing effect after fatigue damage has occurred [470]. Crack initiation life has been predicted using equivalent strain energy density with Neuber's rule and Morrow's equation [471]. Jaensson has indicated that the crack initiation mechanisms, such as shear mode or tensile mode, are dependent on the preload type[472].

Among the three types of failures viz., deep spalling, shallow spalling from surface and shallow spalling from sub-surface, the shallow spalling is mainly active which consumes longest duration[221]. At temperatures of 250 and 450 deg C, Ti-6Al-4V shows lower FCG due to dynamic strain aging process which leads to local increase of yield strength[473]. The crack nucleation is primarily due to surface roughness and the propagation is controlled favorably by RCS and unfavorably by high dislocation density[474]. The crack nucleation points are pushed to the sub-surface due to SP in the coil springs [475]. In steel and magnesium alloy, the crack growth rate is influenced by RCS than by the hardened layer [476].

7.2. Theoretical Studies

Figure 12 gives the techniques employed in fracture mechanics analysis. A theoretical FE analysis calculating the stress intensity factor[SIF] has concluded that SP induced RCS has locally retarded the crack growth, but has not affected in an overall sense[477].

LEFM (Linear Elastic Fracture Mechanics)

- CTOD(Crack tip opening Displacements) method
- Virtual Crack Extension Method
- Strain Energy Release rate (G)
- J-Integral method
- Modified Crack Closure Integral Method

Figure 12. Fracture mechanics methods.

Analytically, it is shown that the crack growth is delayed in RCS environment by using fracture mechanics techniques along with S-N curves[478]. A local concept which predicts the crack initiation location makes use of the local stresses in case-hardened notched specimens[479]. The SCF due to SP has been analyzed by simplified models[one-pit, seven-pit and loop-pit] and the Goodman diagram is accordingly modified[480]. FE models with residual stresses, contact between crack faces and applied load have been used to predict SIF[481]. The conditions for crack arrest are presented in the form of Fatigue damage Map[482]. Romero et al proposes a suitable adaptation of Navarro-Rios model using microstructural fracture mechanics that takes microstructure, initial flaw size, residual stresses, surface hardness, work-hardened layer into account[483]. The surface roughness due to SP is analyzed as a local increase in the far-field stress [484].

8. Stress Relaxation

The MSE processes develop different levels of cold work on the surface. Therefore, when mechanical loads[static and cyclic] or thermal loads are applied, the RCS relaxes to different levels. An excellent review on the relaxation of residual stresses is given by McClung[485]. The mechanisms involved in relaxation due to static[or first cycle], fatigue, thermal loads and fracture are quite different. The author also points out that the RCS does not relax to zero due to these loadings. McClung also indicates that probabilistic aspects of residual stresses need to be considered due to its inherent variation. Schulze also mentions that the relaxation depends on the surface cold work, stress amplitude and the cycles[486]. The temperature, uni-axial or cyclic mechanical loading make both macro and micro residual stresses to relax[331]. The macrostresses relax less compared to microstresses that affects the crack behavior[487]. The mechanisms leading to relaxation are described by Schulze[488]. Niku-Lari also points out that due to cyclic loads and temperature, the residual stresses are reduced. The stress relaxation due to temperature is depicted in figure 13.

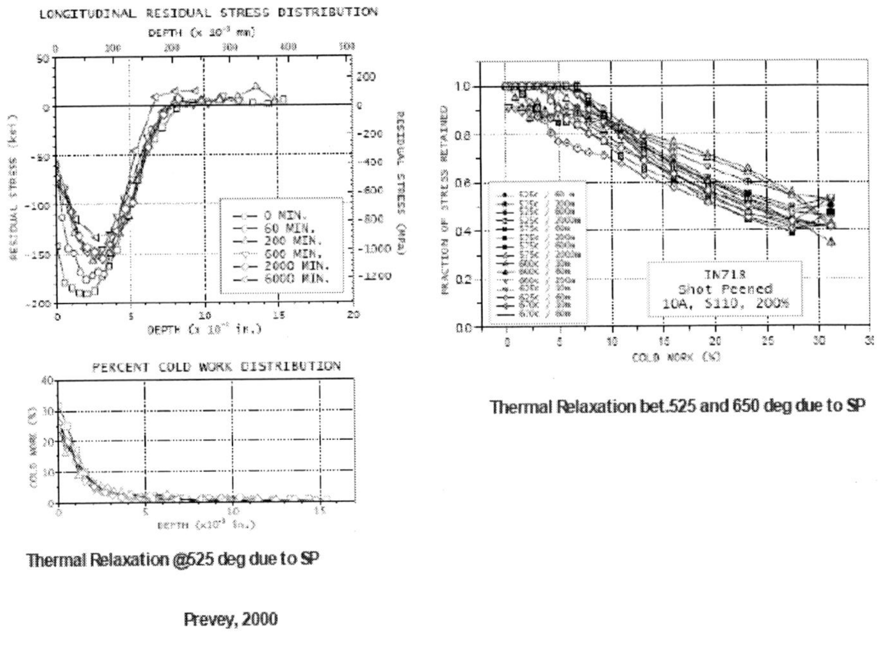

Figure 13. Thermal relaxation due to temperature (Prevey, 2000).

At high temperatures, the inhomogeneous micro residual stresses developed during MSE relax more than macro residual stresses[489]. The material behavior at high temperatures are strongly material dependent. For example, magnesium alloys such as AZ31 show poor fatigue properties due to annealing and recrystallization while AISI 304 steel shows stable microstructure except for small reduction of dislocation density. Higher the amount of dislocation density higher is the stress relaxation[461]. In other words the rate of relaxation is a function of cold work[490]. SP does not affect the Bauschinger effect in the static loading, but affects in cyclic loading [491]. The relaxation is more for residual stresses in the direction where the load is applied [492] and for stresses that are shallowly distributed [493].

8.1. Experimental Studies

Capello et al. report that stress relaxation starts from the very first cycle of mechanical loading [494]. The elastic shakedown occurs due to plastic deformation with macroscopic yielding[495] followed by relaxation with micro-plasticity[496]. The relaxation causes the residual stresses to diminish, the crack tip stresses increase steadily causing stress intensity factors to increase[497]. Notched and unnotched specimens relax to the same level. The relaxation occurs due to the super-position of RCS with applied compressive stress[205]. Applied tensile stresses do not cause any relaxation [498] and also there is no relaxation if the applied stresses are below fatigue limit [499].

Studies on Steels: Carbon steel loses its residual stresses when it is subjected to fatigue testing at high temperatures[500]. The high dislocation density in the hardened surface creates higher relaxation than the bulk material in plain Carbon steel[501]. The type of applied stresses plays a role in stress-relaxation. In 17Cr7Ni austenitic stainless steel, bending produced more relaxation in terms of maximum RCS as well as depth than uni-axial loading[502]. Kirk has indicated that higher alternating stresses cause increased relaxation of residual stresses in steels[113]. Experiments on AISI 4340 also point out that the relaxation depends on the applied stress and number of load cycles[503]. In AISI H11 steel used in tool making, thermal fatigue results in the relaxation of RCS in addition to work hardening[504]. Though the fatigue life improves more for low strength materials due to RCS, they have lower resistance against stress relaxation due to cyclic loads with mean stress moving to zero [505]. Stress peening produces higher compressive stresses while warm peening improves the stability of residual stresses[506]. Warm peening is found to retain the residual stresses[507] better than conventional peening due to dislocation stability[508]. Iida et al conclude that RCS with shallow distribution relaxes more than that of deeper distribution [509]. In 42CrMo4 steel, Avrami approach seems to fit the thermal relaxation data well [510]. The relaxation is linear with log[N] where N is the load cycles when $N < Ni$, the cycles for crack initiation[511]. When $N > Ni$, the relaxation is more pronounced. In 42 CrMo4 steel, the relaxation is different in tensile and compressive loading direction due to Bauschinger effect[512]. Stress relieving after shot peening has not affected the fatigue strength of martenistic steels indicating that the strength improvement is due to texture induced by rotation of surface crystals[513]. Tensile stresses decrease the surface hardness of the peened surface while compressive stresses increase the hardness.

Studies on Aluminium: The stress relaxation rate depends on surface yield strength, surface residual stress and stress amplitude[514]. Using acoustic emission studies on two aluminium alloys, the stages of residual stress creation and relaxation are explained [515]. They are

a. cyclic cold working with RCS evolution
b. plasticization of surface and stabilization of stresses
c. appearance of micro-cracks without any change in stresses and d]coalescence of cracks with stress relaxation.

The 7050-T451 alloy shows relaxation due to thermal exposure during adhesive bonding and subsequent spectrum loading[516].

Studies on Titanium: In cold worked Ti-6Al-4V alloy used in aircraft engines, fatigue loading with higher amplitudes cause more relaxation making it more susceptible for impact failure[517]. Lee et al [518] have established that stress relaxation occurs in the order of 20-50% due to repeated cyclic loading and high temperatures. The effect of high temperature compounds the stress relaxation effect occurring due to mechanical loads. The fatigue strength drops below the unpeened value at elevated temperatures because the surface roughness remains, but the residual stress relaxation occurs[519]. For Timetal also Avrami equation explains the relaxation effect[520].

Studies on Superalloys: Experimentally, Ofsthun[521] has proved that the typical aircraft engine materials[Nickel alloy 718 and Titanium 6Al 4V alloy] show no drop of fatigue strength when they are subjected to high temperatures when they are not peened. However, when they are shot peened and then exposed to high temperatures of the order of 200-300 deg C, the fatigue strengths significantly drop. Highly cold worked IN718 specimens caused by more coverage relax more at high temperatures than less cold worked specimens caused by less coverage [522]. In a typical turbine part, the creep phenomenon is even present that can nullify the peening effect. Buchanan et al [523] have determined through experiments the level of relaxation of both residual stresses and cold work for IN100 super alloy. They have concluded that the benefit of residual stresses can still be considered in the design despite the relaxation effects. Masmoudi and Castex [524] have used X-ray diffraction for analyzing the surface stresses of IN100 alloy and the incremental hole drilling method for the stress in depth. Khadhraoui et al.[525] have observed that a higher temperatures produce greater relaxation due to rapid annihilation of the unstable crystalline defects in highly deformed materials. Cao et al [526] also mention that pure thermal loads cause annihilation and reorganization of the crystalline defects induced by shot peening, whereas the mechanical relaxation is linked to cyclic plasticity of materials. In a study involving UDIMET 720Li alloy, the relaxation due to mechanical loads is anisotropic with relaxation higher along loading axis[527,528]. Belassel et al have studied the LCF effects on discs made of Nickel Base alloys and concluded that the RCS follows a linear trend based on the LCF cycles[529]. The surface roughness effect limits the effect of peening as the RCS field reaches saturation[530]. Introducing tensile plastic strain including accidental dents can cause reduction of RCS, sometimes even introduction of tensile stresses [531].

8.2. Theoretical Studies

An empirical Zener-Wert-Avrami equation is used to represent stress relaxation given by $\sigma^{RS}(T,t)/\sigma_0^{RS} = e^{(-At)^m}$ where m and A are constants. A depends on material defined by

$A = Ce^{(-Q/kT)}$ where C is a constant.

The relationship for stress relaxation due to cyclic loads is given by the following Morrow's equation[173]:

$$\frac{\sigma_m(N)}{\sigma_m(o)} = \frac{o_{ys} - \sigma_a}{\sigma_m(o)} - \left(\frac{\sigma_a}{o_{ys}}\right)^b$$

and due to temperature[T] and time[t] by the relationship:

$\sigma_R(t) = \sigma_{RO} - \frac{RT}{\beta}\log(\frac{t}{t_0}+1)$ where β and t_0 are material dependent constants

Using ABAQUS code, Dattoma et al. have simulated stress relaxation with welding as a pre-stress condition [532]. Analytical models using strip method [533] with non-linear kinematic hardening law show good correlation of relaxation rate with experimental results. In order to simulate the relaxation due to quasi-static loading at high temperatures, FE model with different material properties at different layers are used [493]. Including the cyclic properties of the surface layer after peening helps in better prediction of the relaxation[534]. Meguid et al. have developed a unit cell model involving Cowper and Symonds strain-rate material model. Johnson-Cook model is used to evaluate stresses at different temperatures. Multiple layers of impacts are employed to calculate the shot peening and subsequent relaxation effects[535]. Different hardening models[isotropic, kinematic and chaboche] have been tried. Thus, appropriate material models that can capture both development and relaxation of RCS are to be used to cover the reduction of yield strength due to temperature along with relaxation due to loads.

9. Comparison of SP with other MSE Methods

The key parameters of different Mechanical Surface Enhancement[MSE] Processes are compared in table.1. As can be seen, the number of influencing parameters are high in the case of SP than the other processes. The various phenomenon that need to be considered are listed in table.2. Again it can be seen that more phenomenon influence SP when compared with other processes. For example, statistical aspects play important role in SP simulation. The magnitudes of various mechanical responses are given in table.3. figure 14 shows the surface roughness attributes of SP, LSP and LPB. Figure 15 gives the schematic variation of residual stresses for the MSE processes.

9.1. SP and LSP

LSP is able to provide better surface finish along with more and deeper RCS than SP [536] in sheet metals. Also, LSP has resulted in better fatigue life than SP in 7075-T351 Friction Stir Welded specimens [537] with low fatigue crack growth marked by smaller striations due to deeper RCS[538]. LSP and a SP-LSP combination in 2024-T351 alloy provide better fatigue strengths in comparison with just SP alone as SP results in high loss of ductility [539]. Precipitates are formed in 2024-T3 alloy and in fact the nucleation and growth of such precipitates can be enhanced or suppressed by the magnitude of laser scan velocities. Formation of precipitates is observed after shot peening[540-542] irrespective of the magnitude of laser scan velocities applied before. SP also has been used to alleviate the tensile stresses created by the laser melting [543].

Table 1. Key Parameters in MSE Processes

Sl. No	Shot Peening	Laser Shock Peening	Roller Burnishing	Hole Expansion
1	Shot veleocity	Laser Power density	Ball Diameter	Tool Material
2	Shot Diameter	Repetitions	Ball Material	Hole Material
3	Shot Material	Target material	Force	Force
4	Target Material		Feed rate	
5	Angle of Impact		Speed of Ball	
6	Number of shots		Target Material	
7	Coverage		Lubrication	
8	Intensity			

Table 2. Modeling Phenomenon

Modeling Phenomenon	Shot Peening	Laser Shock Peening	Roller Burnishing	Hole Expansion
Plasticity	High	High	High	High
Contact mechanics	High	Low	High	Medium
Friction effects	High	Low	High	Low
Strain rate effects	High	High	Low	Low
Statistical effects	High	Medium	Low	Low
Thermal Effects	Medium	High	Medium	Medium/Low
Strainhardening/softening	High	High	High	High
Bauschinger Effect	High	High	High	High

Table 3. Response Parameters in MSE processes

	Shot Peening	Laser Shock Peening	Low Plasticity Burnishing	Hole Expansion
Surface RCS	Low	Medium	High	Medium
Max. RCS	Low	Medium	High	Medium
Depth of Max. RCS	Low	Medium	High	Medium
Cold Work	High	Medium	Low	Medium
Surface Roughness	High	Medium	Low	Low

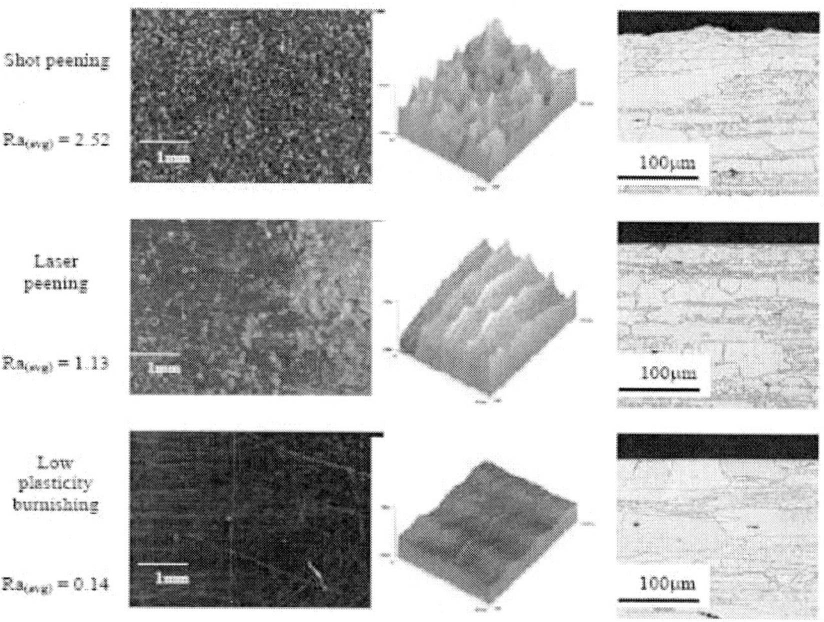

Figure 14. Surface roughness comparison among SP, LSP and LPB (Guimmara).

LSP has produced the same level of RCS of intense peening and even more increased intensity of plastic strain compared to SP in Titanium alloy [544]. In mild steel, the LSP has caused 80% increase in hardness due to dislocation density increase, but the RCS level is more in SP [545]. This can be due to the choice peening parameters. Rankin et al. found that the SP and LSP have same RCS till 0.1 mm depth, but the depth of RCS is more in LSP[546,547]. Thus LSP is likely to replace other conventional processes [548].

Comparison of different MSE methods is complex as the variation of input parameters can result in wide variation of material response spectrum. LSP is investigated for different laser intensities on two aluminium cast alloys to determine the effects of optimization, repetition, juxtaposing etc. and it is found that LSP with certain input parameter combination has even produced less maximum RCS than SP, but the depth is more[549]. In AISI 316L steel, LSP has created less RCS and cold work due to non-formation of martensite compared to SP [550].

9.2. SP and LPB

Deep rolling exhibited better fatigue strengths at all test temperatures than SP[489]. In a study involving magnesium alloy AZ80, Zhang et al also conclude that RB is found to be better than SP in fatigue strength at optimum conditions[551,552]. Both SP and LPB increase the fatigue strength of Titanium alloys[553], but sub-surface cracks emanated due to the presence of sub-surface tensile stresses. Lower deformation rate brings benefit in LPB in titanium alloy while polishing has shown improvement after SP[554]. When LPB is applied on peened aluminium and copper specimens, the surface finish as well as the hardness, fatigue strength and corrosion resistance are found to improve [555]. In Udimet 720 LI superalloy, the deep rolling improves the fatigue life and RCS more than SP[556]. In spring steel, deep rolling has produced higher and deeper RCS than stress peening[557]. Stress rolling of spring steel shows higher RCS values than stress peening[558]. The increased depth of compressive layer in LPB resists stress relaxation due to thermal loads, found in gas turbines, better than shot peening [559] with less cold work. Thus from many literature we can conclude that LPB is better than SP. However, sometimes SP parameters can be optimized such that it produces high RCS and low roughness so that even LPB can be replaced with SP [560]. The required magnitude and distribution of RCS for a targeted HCF life is estimated for LPB process with Haigh diagram using FEM, LEFM and X-Ray diffraction [561]. Critical distance models and fracture mechanics have been used to study both SP and LPB[562]. In a study involving different aged AA6610 Al alloy, the relaxation of RCS due to deep rolling is attributed to elastic strain converting to micro-plastic strains [563].

9.3. Other Methods

The variation of residual stresses for different MSE methods is shown in figure 15. It can be observed that RCS at the surface due to LPB is same as LSP. But LPB develops higher and deeper RCS than LSP below the surface. SP develops the smallest magnitudes of RCS at and below the surface. Amongst SP, LSP and LPB, the SP is the most economical and fastest while LSP is the opposite [490]. LPB produces same magnitude of RCS as LSP but is deeper than LSP. SP produces less deep RCS and cold work but the amount of cold work is higher compared to LPB and LSP. Altenberger [461] compares MSEs on microstructure, surface roughness, RCS, magnitude and thickness of cold work and concludes that strain-rate and glide behavior of dislocations determine the microstructure. The MSE processes develop high dislocation densities near the surfaces that result in plastic strains[564]. In AISI 304 steel, Nikitin and Altenberger have found that the LPB produces nano-crystalline structure with high dislocation density and martensite formation, but no such observations is found for LSP[565].

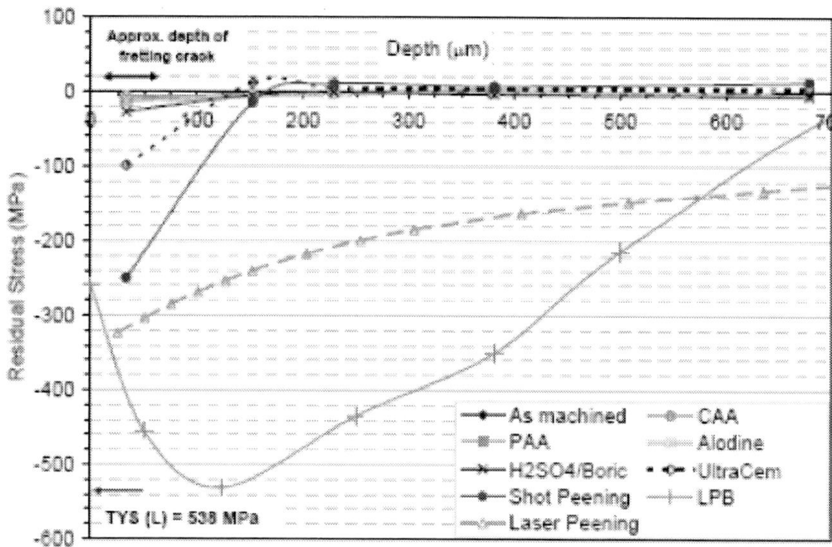

Figure 15. RCS comparison among SP, LSP and LPB.

10. Dislocation Dynamics

Dislocations cause the slip which causes for the plastic deformation[566]. Typically metals show very low dislocation density after annealing where as SP increases it by many folds. SP is a process that involves work hardening with high strain rate. In SP, as we know, the kinetic energy of the shots are used to create plastic deformation. The energy that needs to be applied to overcome the barriers during the slip depends on the temperature and strain rate. Work-hardening occurs when dislocations find increased glide resistance as they move, interact and change their density and distribution. Typically metals are poly-crystals. In poly-crystals, the yielding starts first with few grains and progresses to others. The grain boundaries act as barriers for dislocation motions. The piled-up dislocations at the boundaries of plastically deformed grains make dislocations in the adjoining grains to operate. Ductile fracture involves significant yielding before failure while brittle fracture has no or very little yielding. The type of fracture is again decided by the amount of dislocation generation and the way of propagation. Hence, the authors also feel that better understanding the dislocation dynamics with respect to specific materials through simulations can help linking the different phenomena such as cold work, RCS and fracture that occur during SP process. It also needs to be pointed out full-fledged simulation of dislocation dynamics demand huge computing resources.

11. Conclusions

When we look at the experimental and theoretical studies, we can conclude that current level of research has focused on one or a few of the phenomena such as RCS development, cold work, optimization, fracture or relaxation. Each study is applied only on specific materials depending on the industrial application and damaging mechanism such as fatigue or SCC. Therefore, drawing general conclusions based on such focused studies becomes difficult.

As SP involves millions of impacts, theoretical simulation on large real-life part with complex features such as holes, fillets has not been attempted till now. Rather, unit cell approach is followed with suitable simplifications in terms of boundary conditions. Some research has been done by using suitable material models to simulate the development of residual stresses and stress-relaxation. The material models need to be robust enough to cover different phenomena discussed in the paper. For example, FE model with unit cell approach can be modified with appropriate constitutive models capable of handling strain-rate dependent plasticity is required that can be used in optimization with respect to RCS, cold work and surface roughness. Further, the FE model can be augmented by numerical fracture mechanics and relaxation models to have a unified approach. In addition, the computational aspects of SP can be expanded to cover the following areas for enhanced understanding:

- Similar to mentioned in [567], creation of a knowledge base using FEM will be the first step in this multi-disciplinary work to predict the RCS and cold work due to SP. Information based solution approach can be used to analyze the complex shot peening process [568].
- Currently the shots are assumed to be spherical in the analysis. Due to wear, they lose the shape. DEM [569] can be easily used handle irregular shapes to calculate the contact forces.
- More shots can be simulated that considers the shot-shot interactions as more efficient algorithms are available [570].
- Parallelizing the DEM/FEM simulation also will help to handle complex systems [571].
- Adaptive meshing or even generalized finite element methods, mesh free methods can be employed to simulate the surface modification due to SP which causes high surface roughness and cold work.
- Sampling such as Latin Hypercube method and optimization based on genetic algorithm can be used to predict and optimize non-linear response with many variables[572] than Taguchi methods[figure 16]
- Multi-scale simulations that consider dislocation interactions at grain-level will provide more insight in the development of residual stresses and related cold work as well as relaxation[figure 17].

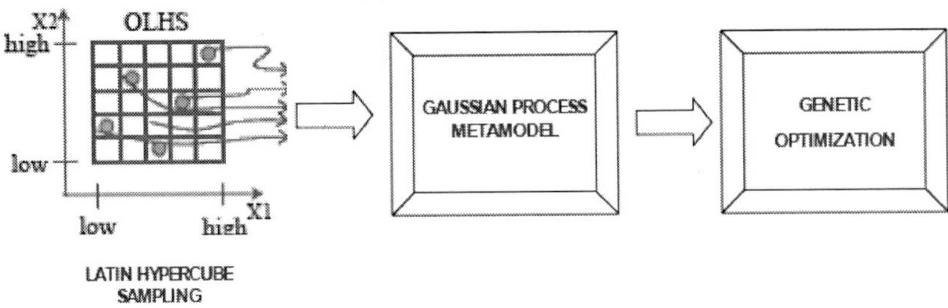

Figure 16. Optimization through Latin Hypercube method.

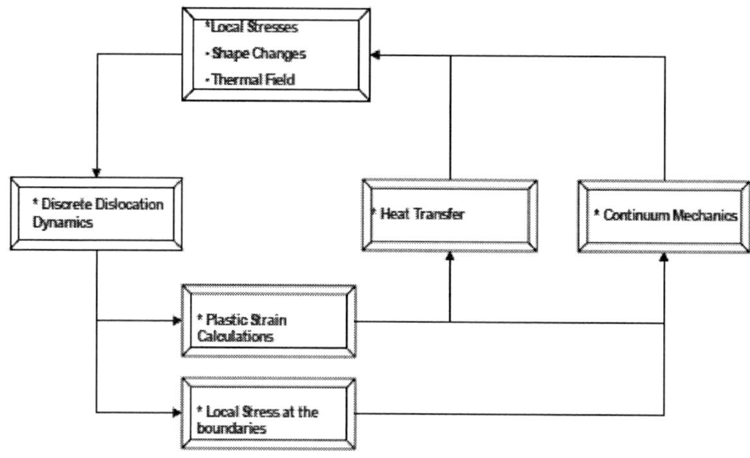

Figure 17. Multiscale modeling with dislocation dynamics.

REFERENCES

[1] König G. Life Enhancement of Aero Engine Components by Shot Peening: Opportunities and Risks. In: *ICSP-8*. Garmisch-Partenkirchen, Germany: 2002. p. 13-22.

[2] Peige S, Wang R, Jilai R, Li X, Mei Y. Qualitative Analysis about Effect of Shot Peening on Fatigue Limit of 300M Steel under the Rotating Bending Condition. In: *ICSP-6*. San Francisco, USA: 1996. p. 290-295.

[3] Burrell NK. Controlled Shot Peening to Increase the Fatigue Properties of Crankshafts. In: *ICSP-2*. Chicago: 1984. p. 361-364.

[4] Kloos KH, Macherauch E. Development of Mechanical Surface Strengthening Processes from the beginning until Today. In: *ICSP-3*. Garmisch-Partenkirchen, Germany : 1987. p. 3-27.

[5] Gabb TP, Telesman J, Kantzos P, Preve´y PS. Surface Enhancement of Metallic Materials. *Advanced Materials and Processes*. 2002; 160(1):69-72.

[6] Korsunsky AM, Regino GM, Latham D, Liu J, Nowell D, Walsh M. Eigenstrain Analysis of Synchrotron X-ray Diffraction Measurement of Residual Strains in Machined Nickel Alloy Plates. *Journal of Strain Analysis for Engineering Design*. 2006; 41(5):381-395.

[7] Jiawen H. On the Relation Between Delamination Wear and Microcrack Propagation. In: *ICSP-2*. Chicago: 1984. p. 266-268.

[8] Bozdana AT. On the Mechanical Surface Enhancement Techniques in Aerospace Industry - a Review of Technology. *Aircraft Engineering and Aerospace Technology*. 2005; 77(4):279-292.

[9] John R, Larsen, J, Buchanan D, Ashbough N. Incorporating Residual Stresses in Life Prediction of Turbine Engine Components. In: *Proceedings of the 8th International Fatigue Congress*. Stockholm: 2002. p. 1063-1070.

[10] Flavenot J. Residual Stresses Due to Mechanical Surface Treatment. *Revue de Metallurgie*. Cahiers D'Informations Techniques. 1997; 94(2):188-197.
[11] Cary P. History of shot peening. In: *ICSP-1*. Paris: 1981. p. 23-28.
[12] Champaigne J. History of Shot Peening Specifications. In: *ICSP-9*. Paris, France: 2005.
[13] Tufft M. *Development of a Fracture Mechanics/Threshold Behavior Model to assess the Effects of Competing mechanisms induced by Shot Peening on the Cyclic Life ofa Nickel Based Superalloy, Rene-88DT*. 1997.
[14] Kirk D. Shot peening. *Aircraft Engineering and Aerospace Technology*. 1999; 71(4):349-361.
[15] Leghorn G. The Story of Shot Peening. *A.S.N.E.Journal*. 1957.
[16] Burakowski, T., Nakonieczny, A. General Aspects of Shot Peening, Criteria of Parameters Selection. In: *ICSP-1*. Paris: 1981. p. 139-146.
[17] Plaster, H.J. Technical Aspects of Shot Peening Machinery and Media. In: *ICSP-1*. Paris: 1981. p. 83-94.
[18] *SAE J443*. Test Strip, Holder And Gage For Shot Peening. 1961.
[19] *SAE J442*. Procedures for Using Standard Shot Peening Test Strip. 1961.
[20] *AMS 2432*. Peening Media General Requirements. 2006.
[21] *MIL-S-13165*. Military Specification Peening of Metals. 1966.
[22] Tosha K. Papers on Shot Peening Published in the World for the Last Thirteen Years. In: *ICSP-7*. Warsaw, Poland: 1999. p. 5-10.
[23] Diepart CP, Burrell NK. Improved Fatigue Performance of Gears through Controlled Shot Peening. In: *ICSP-3*. Garmisch-Partenkirchen, Germany : 1987. p. 169-177.
[24] Raabe D. *Computational Materials Science: The Simulation of Materials, Microstructures and Properties*. Wiley-VCH Verlag Gmbh; 1998.
[25] Matzkanin GA, Yolken HT. A Review of Techniques for Nondestructively Characterizing Residual Stress in Metals. *American Society of Mechanical Engineers, Pressure Vessels and Piping Division* (Publication) PVP. 2000; 4291-6.
[26] Withers PJ, Bhadeshia HKDH. Residual Stress Part 1 - Measurement Techniques. *Materials Science and Technology*. 2001; 17(4):355-365.
[27] Martinez SA, Sathish S, Blodgett MP, Shepard MJ. Residual Stress Distribution on Surface-treated Ti-6al-4v by X-ray Diffraction. *Experimental Mechanics*. 2003; 43(2):141-147.
[28] P.S. Prevey. *X-Ray Diffraction Characterization of Residual Stresses produced by Shot Peening*. France: Lambda Research Inc; 1990.
[29] *P.S. Prevey*. X-ray Diffraction Residual Stress Techniques. Lambda Research Inc;
[30] Cheong WCD, Zhuang WZ, Zhang LC. Study of Residual Stress Relaxation Using X-ray Diffraction. *Key Engineering Materials*. 2004; 274-276(I):871-876.
[31] Noyan IC, Cohen J. *Residual Stress Measurement by Diffraction and Interpretation*. Springer-Verlog New York Inc.; 1987.
[32] Webster G, Wempory R. Non-Destructive Measurement of Residual Stress by Neutron Diffraction. *Journal of Materials Processing Technology*. 2001; 117395-399.
[33] Lord J. Hole Drilling Techniques[Internet]. 2000 Sep 12; Available from: *www.npl.co.uk/materials/residualstress/pdf_files/hole_drilling.pdf*
[34] Stein E, Borst R, Hughes T. *Encyclopedia of Computational Mechanics*. John Wiley and Sons Inc; 2004.

[35] Zienkiewicz, O.C., Taylor, R.L., Zhu, J.Z. *The Finite Element Method* (Vol.1-3). Butterworth-Heinemann; 2005.

[36] Bathe K. *Finite Element Procedures*. Printice-Hall; 1995.

[37] Cook R, Malkus D, Plesha M, Witt R. *Concepts and Application of Finite Element Analysis*. John Wiley and Sons Inc; 2002.

[38] Strouboulis, T. , Copps, K. , Babuška, I. Computational Mechanics Advances. the Generalized Finite Element Method. *Computer Methods in Applied Mechanics and Engineering*. 2001; 190(32-33):4081-4193.

[39] Belytschko, T., Krongauz, Y., Organ, D., Fleming, M., Krysl, P. Meshless Methods: An Overview and Recent Developments. *Computer Methods in Applied Mechanics and Engineering*. 1996; 139(1-4):3-47.

[40] Shaofan Li, Wing Kam Liu. Meshfree and Particle Methods and Their Applications. *Applied Mechanics Review*. 2002; 55(1):1-34.

[41] Melenk, J.M., Babuška, I. The Partition of Unity Finite Element Method: Basic Theory and Applications. *Computer Methods in Applied Mechanics and Engineering*. 1996; 139(1-4):289-314.

[42] Duarte, C.A., Oden, J.T. An h-p Adaptive Method Using Clouds. *Computer Methods in Applied Mechanics and Engineering*. 1996; 139(1-4):237-262.

[43] Bazant, Z.P., Wang, T. Spectral Finite Element Analysis of Random Shrinkage in Concrete. *Journal of Structural Engineering*. 1984; 110(9):2196-2211.

[44] Moës, N., Dolbow, J., Belytschko, T. A finite element method for crack growth without remeshing. *International Journal for Numerical Methods in Engineering*. 1999; 46(1):131-150.

[45] Kudela P, Krawczuk M, Ostachowicz W. Wave Propagation Modelling in 1D Structures Using Spectral Finite Elements. *Journal of Sound and Vibration*. 2007; 300(1-2):88-100.

[46] Wriggers P. *Computational Contact Mechanics*. John Wiley and Sons Inc; 2002.

[47] Laursen T. *Computational Contact and Impact Mechanics*. Springer-Verlog berlin Heidelberg; 2003.

[48] Jaeger J. *New Solutions in Contact Mechanics*. WIT Press; 2005.

[49] Dunne F, Petrinic N. *Introduction to Computational Plasticity*. Oxford University Press; 2006.

[50] Chong K, Sung L, Ho D, et al. Solid Micromechanics: Research and Challenges. In: *Proceedings of the 13th Micromechanics Europe (MME2002) Workshop*. Romania: 2002. p. 347-358.

[51] Lu G, Kaxiras E. *An Overview of Multiscale Simulations of Materials.*

[52] Gates T, Odegard G, Frankland S, Clancy T. Computational Materials: Multi-scale Modeling and Simulation of Nanostructured Materials. *Composites Science and Technology*. 2005; 652416-2434.

[53] Noor AK. *Perspectives on Multiscale Modeling, Simulation and Visualization*. NASA; 2002.

[54] Weinan E, Engquist B. Multiscale modeling and Computation. *Notices of AMS*. 2003; 50(9):1062-1070.

[55] Ghoniem N, Busso E, Kioussis N, Huang H. Multiscale Modelling of Nanomechanics and Micromechanics: An Overview. *Philosophical Magazine.* 2003; 83(31-34):3475–3528.

[56] Nakano A, et al. Scalable Atomistic Simulation Algorithms for Materials Research. Denver: 2001.

[57] Liu W, Karpov E, Zhang S, Park H. An Introduction to Computational Nanomechanics and Materials. *Computer Methods in Applied Mechanics and Engineering.* 2004; 1931529-1578.

[58] Mecking H. Syntheses: Modelling of Polycrystal Plasticity. *Modelling Simul. Mater. Sci. Eng.* 1999; 7669-673.

[59] Zbiba HM, Rubia T. A Multiscale Model of Plasticity. *International Journal of Plasticity.* 2002; 181133-1163.

[60] Kadowaki H, Liu W. Bridging Multi-scale Method for Localization Problems. *Computer Methods in Applied Mechanics and Engineering.* 2004; 1933267-3302.

[61] Needleman A. Computational Mechanics at the Mesoscale. *Acta Materialia.* 2000; 48105-124.

[62] Cuitino A, et al. A Multiscale Approach for Modeling Crystalline Solids. *Journal of Computer Aided Materials Design.* to appear;

[63] Shilkrot L, Miller R, Curtin W. Multiscale Plasticity Modeling: Coupled Atomistics and Discrete Dislocation Mechanics. *Journal of the Mecahanics and Physics of Solids.* 2004; 52755-787.

[64] Saiki I, Kanke S, Nakajima A, Terada K. Study on Multiscale Modeling of Plasticity by Atomic Model. *Structural Engineering/Earthquake Engineering JCSE.* 2005; 22(2):209-217.

[65] Ghoneim N, Cho K. The Emerging Role of Multiscale Modeling in Nano and Micro Mechanics of Materials. *Computer Modeling in Engineering and Sciences.* 2002; 3(2):147-173.

[66] Hartley C. Multi-scale Modeling of Dislocation Processes. *Materials Science and Engineering.* 2001; A319–321133-138.

[67] Y.Y.Li, J.Z.Qui. The Multi-scale Computational Method for the Mechanics Parameters of the Materials with Random Distribution of Multi-scale Grains. *Composites Science and Technology.* 2005; 651447-1458.

[68] Campbell G, Foiles S, Huang H, Hughes D, King W, Lassila D, et al. Multi-scale Modeling of Polycrystal Plasticity: A Workshop Report. *Materials Science and Engineering.* 1998; A2511-22.

[69] Clayton J, McDowell D. Homogenized Finite Elastoplasticity and Damage: Theory and Computations. *Mechanics of Materials.* 2004; 36799–824.

[70] Markovic D, Niekamp R, Ibrahimbegovic A, Matthies H, Taylor R. Multiscale Modeling of Heterogeneous Structures with Inelastic Constitutive Behaviour. *Engineering Computations.* 2005; 22(5):664-683.

[71] Owen D, Feng Y, Cottrel M, Yu J. Numerical Modelling of Continuum to Discontinuum/Discrete Transition Phenomena. In: *Multiscale Computational Mechanics for Material and Structures.* Cachan: 2002.

[72] Lebensohn R. Modelling the Role of Local Correlations in Polycrystal Plasticity Using Viscoplastic Self-consistent Schemes. *Modelling Simul. Mater. Sci. Eng.* 1999; 7739-746.

[73] Koch CC, Youssef KM, Scattergood RO, Murty KL. Breakthroughs in Optimization of Mechanical Properties of Nanostructured Metals and Alloys. *Advanced Engineering Materials*. 2005; 7(9):787-794.
[74] Umemoto M, Todaka Y, Tsuchiya K. Nanograined Steels Produced by Shot Peening and Drilling. In: *Proceeding of the 25th Riso International Symposium*. Denmark: 2004.
[75] Todaka Y, Umemoto M, Tsuchiya K. Comparison of Nanocrystalline Surface Layer in Steels Formed by Air Blast and Ultrasonic Shot Peening. *Materials Transactions*. 2004; 45(2):376-379.
[76] Gillespie RD. Shot Peening Media - Its Effect on Process Consistency and Resultant Improvement in Fatigue Characteristics. In: *ICSP-5*. Oxford, UK: 1993. p. 55-64.
[77] Gillespie RD, Gloerfeld H. An Investigation of the Durability and Breakdown Characteristics of Shot Peening Media. In: *ICSP-4*. Tokyo, Japan: 1990. p. 27-36.
[78] Flavenot J, Lu J. Influence of Ageing of a Cut Steel Wire Shot during Shot Peening. In: *ICSP-4*. Tokyo, Japan: 1990. p. 83-94.
[79] Edwards, PE. A Comparison of the Effect of Peening Using Cut Wire and Cast Steel Shot on Fatigue Performance of 7010-T7651 Aluminium Alloy Plate. In: *ICSP-6*. San Francisco, USA: 1996. p. 154-165.
[80] Hanyuda T, Nakamura M, Endo T, Shimizu H. Effect of Shot Peening on Fatigue Strength of Titanium Alloy. In: *ICSP-5*. Oxford, UK: 1993. p. 116-125.
[81] Bertoli A, Flavenot J. Influence of the Ageing of Cast Steel Shot on the Fatigue Strength of Shot-peened TiAl6V and 40 NSCD 7 Grade Steel. In: *ICSP-5*. Oxford, UK: 1993. p. 166-172.
[82] Sharma MC. Fatigue and Heat Transfer Behaviour of Shot Peened Aluminium Alloy. In: *ICSP-6*. San Francisco, USA: 1996.
[83] Arker P, Lu J, Flavenot J. Effect of Glass Beads Nature, Manufacturing and Ageing on the Behaviour of Shot Peened Aeronautical Materials. In: *ICSP-4*. Tokyo, Japan: 1990. p. 209-218.
[84] Barzoukas H, Jauffret J. Peening with Ceramic Shot. In: *ICSP-4*. Tokyo, Japan: 1990. p. 47-56.
[85] Gillespie B. Image Analysis of Shot Peening Media. In: *ICSP-6*. San Francisco, USA: 1996. p. 213-222.
[86] Marconi P. Correlation between Mechanical and Geometrical Characteristics of Shot and Residual Stress Induced by Shot Peening. In: *ICSP-8*. Garmisch-Partenkirchen, Germany: 2002. p. 167-172.
[87] Iida K. The Analyses of the Shot Velocity thrown from the Nozzle and the Bladed Wheel. In: *ICSP-5*. Oxford, UK: 1993. p. 40-45.
[88] Andziak J. A New Method of Measurement of the Velocity of Solid Particles and Their Mass for Air Blasting. In: *ICSP-7*. Warsaw, Poland: 1999. p. 308-312.
[89] Clausen R, Stangenberg J. Device for the Determination of Impact Velocities in Shot Peening. In: *ICSP-8*. Garmisch-Partenkirchen, Germany: 2002. p. 89-95.
[90] Meguid, SA, Duxbury, JK A Practical Approach To Forming And Strengthening of Metallic Components Using Impact Treatment. In: *ICSP-1*. Paris: 1981. p. 217-228.
[91] Linnemann W, Wustefeld F, Kopp R, Kittel S. Shot Velocity Measurement. In: *ICSP-6*. San Francisco, USA: 1996. p. 118-129.
[92] Barker B, Young K, Pouliot L. Particle Velocity Sensor for Improving Shot Peening Process Control. In: *ICSP-9*. Paris, France: 2005.

[93] Baughman, DL The Evolution of Centrifugal Wheel Shot Peening in Aerospace Industry and Recent Applications. In: *ICSP-1*. Paris: 1981. p. 101-108.
[94] Goldman D, Sutker B, Schmidt R, Suttmeier P. Comparative Performance of Glass Breads in Suction and Direct Pressure Shot peening Applications. In: *ICSP-4*. Tokyo, Japan: 1990. p. 9-18.
[95] Kreikenbaum H, Fischer G. About the Influence of Different Separation Systems on Shot Peening Treatment. In: *ICSP-3*. Garmisch-Partenkirchen, Germany : 1987. p. 55-62.
[96] Hashish M, Chillman A, Ramulu M. *Waterjet Peening at 600mpa: A First Investigation*. ASME Publications. 2005; 261 FED45-52.
[97] Mochizuki M, Enomoto K, Sakata S, Kurosawa K, Saito H, Tsujimura H, et al. A Study on Residual Stress Improvement by Water Jet Peening. In: *ICSP-5*. Oxford, UK: 1993. p. 246-255.
[98] Ishiguro T, Shimazaki T, Terayama K, Uchida S, Uno A. Effects of Stress Shot Peening with Back Water Jet Pressure on Surface Residual Stress Distribution of Carburized Steels. Tetsu-To-Hagane/*Journal of the Iron and Steel Institute of Japan*. 1996; 82(4):303-308.
[99] Kunaporn S, Ramulu M, Jenkins MG, Hashish M. Residual Stress Induced by Waterjet Peening: A Finite Element Analysis. *Journal of Pressure Vessel Technology*. 2004; 126333-340.
[100] Salko D. Peening by Water. In: *ICSP-2*. Chicago: 1984. p. 37-39.
[101] Grinspan AS, Gnanamoorthy R. Effect of Oil Jet Peening Duration on Surface Modification and Fatigue Behavior of Medium Carbon Steel, AISI 1040. *Materials Science and Engineering A*. 2007; 456(1-2):210-217.
[102] Duchazeuabeneix J. The Interest to Use Tungsten Carbide Balls with Stressonic Ultrasonic Shot-peening Process: Application to Aero Engines Parts. In: *ICSP-9*. Paris, France: 2005.
[103] Duchazeaubeneix JM. Stressonic Shot Peening (Ultrasonic Process). In: *ICSP-7*. Warsaw, Poland: 1999. p. 444-452.
[104] Watanabe Y, Hattori K, M.Handa, N.Yoshida , Taki T, Duchazeaubeneix J. Effect of Ultra Shot Peening on Contact Surface Strength. In: *ICSP-9*. Paris, France: 2005.
[105] Micoulaut M, Retraint D, Viot P, François M. Heterogeneous Ultrasonic Shot Peening: Experiment and Simulation. In: *ICSP-9*. Paris, France: 2005.
[106] Bailey PG. Manual Peening with the Rotary Flap Process. In: *ICSP-7*. Warsaw, Poland: 1999. p. 405-414.
[107] Uhlmann E, Krieg M. Shot Peening With Dry Ice. In: *ICSP-9*. Paris, France: 2005.
[108] Watanabe Y, Hattori K, Handa M. Effect of Ultrasonic Shot Peening on Fatigue Strength of High Strength Steel. In: *ICSP-8*. Garmisch-Partenkirchen, Germany: 2002. p. 305-310.
[109] Pieper G, Ruhland S. Vaccum-suction Peening:a Novel Method for Emission-free Shot Peening. In: *ICSP-8*. Garmisch-Partenkirchen, Germany: 2002. p. 73-77.
[110] Takemoto M, Shinohara T, Shirai M. Prevention of Stress Corrosion Cracking of Weldment by Wet Shot Peening. In: *ICSP-2*. Chicago: 1984. p. 39-42.
[111] Bosshard RG, Hornauer K. Controlled Shot Peening, Complete Plants and Examples of Application. In: *ICSP-3*. Garmisch-Partenkirchen, Germany : 1987. p. 63-70.

[112] Kirk D. The Development of Residual Stresses and Curvature in Mild Steel by Needle Peening. In: *ICSP-6*. San Francisco, USA: 1996. p. 312-319.

[113] Kirk D, Birch DG. Residual Stresses Induced by Peening Austenitic Ductile Cast -Iron. In: *ICSP-7*. Warsaw, Poland: 1999. p. 23-32.

[114] Horsch C, Schulze V, Löhe D. Deburring and Surface Conditioning of Micro Milled Structures by Abrasive and Non-Abrasive Micro Peening. In: *ICSP-9*. Paris, France: 2005.

[115] Niku-Lari A. *Overview on the Shot Peening Process*. 1987. p. 155-170.

[116] Feld P, Gillespie R. Controlled Shot Peening Reduces or Eliminates Stress Corrosion Cracking. In: *ICSP-2*. Chicago: 1984. p. 121-125.

[117] Nachman G. Modern Shot Peening Technology. In: *ICSP-3*. Garmisch-Partenkirchen, Germany : 1987. p. 37-47.

[118] Watanabe Y, Hasegawa N. Effect of Shot peening on Delayed Fracture of High Strength Steel. In: *ICSP-5*. Oxford, UK: 1993. p. 219-225.

[119] Salko D. The Protection of the Propeller Blades Surfaces by the Shot Peening. In: *ICSP-6*. San Francisco, USA: 1996. p. 56-60.

[120] Iida K, Miyazaki K. Peening Effect on Flow Resistance of Air. In: *ICSP-4*. Tokyo, Japan: 1990. p. 73-82.

[121] Salko D. Effect of Shot Peening on the Thermodynamic Properties of Piping. In: *ICSP-5*. Oxford, UK: 1993. p. 319-332.

[122] Sharma M C, Deepak. Shot Peened Surfaces and Boiling Heat Transfer In: *ICSP-5*. Oxford, UK: 1993.

[123] Harada Y, Mori K, Maki S. Lining of Metal Surface with Hard-Metal Foil using Shot Peening. In: *ICSP-8*. Garmisch-Partenkirchen, Germany: 2002. p. 200-207.

[124] Burmeister TJ. Peen Forming of Airfoil Shapes for Aircraft Wing Sections. In: *ICSP-2*. Chicago: 1984. p. 1-5.

[125] Baughman DL. An Overview of Peen Forming Technology. In: *ICSP-2*. Chicago: 1984. p. 28-32.

[126] Meyer R, Reccius H, Schulein R. Shot Peen-Forming of NC-Machined Parts with Integrated Stringers Using Large Balls. In: *ICSP-3*. Garmisch-Partenkirchen, Germany : 1987. p. 327-334.

[127] Wang JY, Kohler W. Development of the Peen Forming Process for Spherical Shaped Components. In: *ICSP-4*. Tokyo, Japan: 1990. p. 585-594.

[128] Tatton RJD. Shot Peen Forming -An Economical Solution. In: *ICSP-3*. Garmisch-Partenkirchen, Germany : 1987. p. 309-318.

[129] Clausen R, Bruder M. Peen Forming by Pointed Ball Shooting (Ball Shot Forming). In: *ICSP-3*. Garmisch-Partenkirchen, Germany : 1987. p. 319-326.

[130] Kopp R, Ball H. Recent Developments in Shot Peen Forming. In: *ICSP-3*. Garmisch-Partenkirchen, Germany : 1987. p. 297-308.

[131] Li K. Using Stress Peen Forming Process for Integrally Stiffened Wing Panels. In: *ICSP-1*. Paris: 1981. p. 555-564.

[132] O'Hara P. Peen-Forming – A Developing Technique. In: *ICSP-8*. Garmisch-Partenkirchen, Germany: 2002. p. 217-226.

[133] Kondo K, Tsuzuki S, Kato A. Investigations on Peen Forming. In: *ICSP-1*. Paris: 1981. p. 565-572.

[134] Ghosh S, Johnson W, Mamalis A. Ball-drop plate bending: An experimental study of some of its process Variables. In: *ICSP-1*. Paris: 1981. p. 573-582.
[135] Wustefeld F, Linnemann W, Kittel S. Towards Peen Forming Process Automation. In: *ICSP-8*. Garmisch-Partenkirchen, Germany: 2002. p. 44-52.
[136] Friese A, Lohmar J, Wustefeld F. Current Applications of Advanced Peen Forming Applications. In: *ICSP-8*. Garmisch-Partenkirchen, Germany: 2002. p. 54-61.
[137] Reccius H, Endemann D, Hornauer K. Sheet Forming by Producing the Coverage with a Simultaneous Working System of Balls. In: *ICSP-5*. Oxford, UK: 1993. p. 199-212.
[138] Reccius H. Impact Metal Forming. In: *ICSP-8*. Garmisch-Partenkirchen, Germany: 2002. p. 234-239.
[139] Lunarski J, Zielecki W. Possibilities of Improving Fatigue Properties of Machine Elements by Pneumatic Shot Peening. In: *ICSP-4*. Tokyo, Japan: 1990. p. 263-272.
[140] Korzyhski M. Ball Peening Machines and Some Aspects of Ball Peening of Steam Turbine Blades. In: *ICSP-4*. Tokyo, Japan: 1990. p. 19-26.
[141] Korzynski M. Some Properties of the Overheated W73-1 Titanium Alloy after Glass Bead Peening or Bearing Ball Peening. In: *ICSP-6*. San Francisco, USA: 1996. p. 282-289.
[142] Niku-Lari A. Shot-peening. In: *ICSP-1*. Paris: 1981. p. 1-21.
[143] Simpson RS, Garibay RP. Important Considerations for utilizing Shot Peening as a Design Tool. In: *ICSP-3*. Garmisch-Partenkirchen, Germany : 1987. p. 71-78.
[144] Champaigne J. Shot Peening Overview. Metal Improvement Co; 2001.
[145] Nachman G. Shot Peening - Past, Present and Future. In: *ICSP-7*. Warsaw, Poland: 1999. p. 1-4.
[146] Champaigne J. Almen Gage Calibration. In: *ICSP-8*. Garmisch-Partenkirchen, Germany: 2002. p. 108-113.
[147] Champaigne J. Almen Gauge Accuracy and Repeatability. In: *ICSP-5*. Oxford, UK: 1993. p. 16-27.
[148] Neelakantan M, Green R, Foo S, John R. Measure and Control of Impact Finishing Processes. In: *ICSP-1*. Paris: 1981. p. 147-158.
[149] Happ M. Shot Peening of Bolt Holes in Aircraft Engine Hardware. In: *ICSP-2*. Chicago: 1984. p. 43-49.
[150] Fuchs HO. Defects and Virtues of the Almen Intensity Scale. In: *ICSP-2*. Chicago: 1984. p. 74-78.
[151] Bosshard R. The Unsatisfactory Situation in Residual Stress Evaluation. In: *ICSP-8*. Garmisch-Partenkirchen, Germany: 2002. p. 65-72.
[152] Simpson RS, Clark O, Chiasson GL. The Effect of Inherent Tolerances in the Almen Test Strip on Shot Peening Process Reliability. In: *ICSP-3*. Garmisch-Partenkirchen, Germany : 1987. p. 101-107.
[153] Kirk D, Hollyoak R. Factors Affecting Almen Strip Curvature Readings. In: *ICSP-7*. Warsaw, Poland: 1999. p. 291-300.
[154] Bailey P, Champaigne J. Factors That Influence Almen Strip Arc Height. In: *ICSP-9*. Paris, France: 2005. p. 392-399.
[155] Clarke D, Birley S. The Control of Manual Shot Peening. In: *ICSP-1*. Paris: 1981. p. 167-174.
[156] Simpson R,. McCagg D. Effects of Peening on Fatigue Life of Wrought 5454 Aluminum Automotive Road Wheels. In: *ICSP-2*. Chicago: 1984. p. 58-62.

[157] Kirk, D., Abyaneh, M. Y.. Theoretical Basis of Shot Peening Coverage Control. In: *ICSP-5*. Oxford, UK: 1993. p. 175-183.

[158] Roth M, Bernasconi J, Strassler S. Residual Stress Distribution in Shot Peened Plates. In: *ICSP-2*. Chicago: 1984. p. 293-296.

[159] Sandeson A, Slingsby R. Improved Method of Shot Peening Control Examines Variables in Production Process. In: *ICSP-1*. Paris: 1981. p. 159-166.

[160] Sharma MC, H.Deo, Modi S, Bosshard R. Virtuesand Limitations of Almen Round. In: *ICSP-8*. Garmisch-Partenkirchen, Germany: 2002. p. 83-88.

[161] Kirk D. Interactive Shot Peening Control. In: *ICSP-5*. Oxford, UK: 1993. p. 10-15.

[162] Kirk D. Developments in Interactive Control of Shot Peening Intensity. In: *ICSP-6*. San Francisco, USA: 1996. p. 95-106.

[163] Gillespie R, Hasty Jr. W. Automated Shot Peen Quality Control through the Use of Parametric Sensors. In: *ICSP-3*. Garmisch-Partenkirchen, Germany : 1987. p. 93-99.

[164] Aoki H, Takeda H. Recent Developments in Shot Peening Machine. In: *ICSP-4*. Tokyo, Japan: 1990. p. 105-112.

[165] Bosshard RG, Fritchman TT, Badertscher A. Monitoring of Shotpeening Processes by Hybrid Recorders. In: *ICSP-4*. Tokyo, Japan: 1990. p. 1-8.

[166] Wüstefeld F, Linnemann W, Kittel S, Friese A. On-Line Process Control for Shot Peening Applications. In: *ICSP-9*. Paris, France: 2005.

[167] Haubold T, Hennig W, Wüstefeld F, Kittel S, Friese A. Implementing On-Line Process Control for Shot Peening. In: *ICSP*-9. Paris, France: 2005.

[168] Bailey P. Effective Use of Flourescent Tracers for Peening Coverage. In: *ICSP-8*. Garmisch-Partenkirchen, Germany: 2002. p. 96-100.

[169] Hu NS, Wang X, Zhang D, Xu K. Effect of Biaxial Stress by Shot Peening on Fatigue Limit. In: *ICSP-4*. Tokyo, Japan: 1990. p. 427-434.

[170] Nagel G. Application of Shot Peening and Blasting Methods in the Aircraft Industry. In: *ICSP-3*. Garmisch-Partenkirchen, Germany : 1987. p. 111-115.

[171] Ezeilo AN, Webster G, Webster P, Webster P. Comparison of Shot Peening Residual Stress Distributions in a Selection of Materials. In: *ICSP-5*. Oxford, UK: 1993. p. 274-281.

[172] Gao Y-, Yao M, Shao P-, Zhao Y-. Another Mechanism for Fatigue Strength Improvement of Metallic Parts by Shot Peening. *Journal of Materials Engineering and Performance*. 2003; 12(5):507-511.

[173] Lieurade HP, Bignonnet A. Fundamental Aspects of the Effect of Shot Peening on the Fatigue Strength of Metallic Parts and Structures. In: *ICSP-3*. Garmisch-Partenkirchen, Germany : 1987. p. 343-360.

[174] Kobayashi M, Matsui T, Murakami Y. Mechanism of Creation of Compressive Residual Stress by Shot Peening. International Journal of Fatigue. 1998; 20(5):351-357.

[175] Kirk D. Evolution of Shot Peening Experimental Techniques. In: *ICSP-9*. Paris, France: 2005.

[176] Chen L, Wang Q, Wang Y, Guo E. Measurement and Research of the Shot Peening Velocity and the Surface Hardness. In: *ICSP-4*. Tokyo, Japan: 1990. p. 127-132.

[177] Rouhaud E, Deslaef D. I*nfluence of Shots' Material on Shot Peening, a Finite Element Model*. 2002. p. 153-158.

[178] Robertson G. The Effects of Shot Size on the Residual Stresses resulting from Shot Peening. *The Shot Peener*. 1987; 11(3):46-48.

[179] Zinn W, Scholtes B. Influence of Shot Velocity and Shot Size on Almen Intensity and Residual Stress Depth Distributons. In: *ICSP-9*. Paris, France: 2005.
[180] Neema, ML., Pandey, PC. A Study of Controlled Peening Parameters on the Properties of Machined Surfaces. In: *ICSP-1*. Paris: 1981. p. 201-208.
[181] Zinn W, Schulz J, Kopp R, Scholtes B. The Influence of the Velocity of a Peening Medium on the Almen Intensities and Residual Stress States of Shot Peened Specimens. In: *ICSP-8*. Garmisch-Partenkirchen, Germany: 2002. p. 161-166.
[182] Ebenau A, Vohringer O, Macherauch E. Influence of the Shot Peening Angle on the Condition of Near Surface Layers in Materials. In: *ICSP-3*. Garmisch-Partenkirchen, Germany : 1987. p. 253-260.
[183] Tosha K, Lu J. Effect of Shot Peening and Grit Blasting on Surface Integrity: Influence on Residual Stresses. *Transactions of Nonferrous Metals Society of China* (English Edition). 2006; 16(SUPPL. 2):230-235.
[184] Kitamura T, Terasaki T, Akiyama T. Effect of Grain Shapes on Residual Stress Distribution Due to Shot Peening. *Zairyo/Journal of the Society of Materials Science*, Japan. 2000; 49(6):651-654.
[185] Tosha K, Iida K. Stock Removal and Surface Residual Stress of Grit Blasted Titanium. In: *ICSP-4*. Tokyo, Japan: 1990. p. 435-444.
[186] Tosha K, Lu J, Guelorget B, Nagashima E. Shot Peening and Grit Blasting. In: *ICSP-9*. Paris, France: 2005.
[187] Fuchs HO. Optimum Peening Intensities. In: *ICSP-3*. Garmisch-Partenkirchen, Germany : 1987. p. 639-646.
[188] Woelfel M. Maximization of Flexural Fatigue Strength of Heat-treated Steel Specimens Using Multiple Peening Techniques. In: *ICSP-3*. Garmisch-Partenkirchen, Germany : 1987.
[189] Kirk D, Jayaraman N. Relationship Between Coverage and Surface Residual Stress. In: *ICSP-9*. Paris, France: 2005.
[190] Ludian T, Wagner L. Coverage Effects in Shot Peening of Al 2024-T4. In: *ICSP-9*. Paris, France: 2005.
[191] Meguid S, Klair MS. Finite Element Studies into Incomplete Coverage in Shot Peening. In: *ICSP-2*. Chicago: 1984. p. 306-315.
[192] Schulze V. Characteristics of Surface Layers Produced by Shot Peening. In: *ICSP-8*. Garmisch-Partenkirchen, Germany: 2002. p. 145-160.
[193] Koster W, Gatto L, Cammett JT. Influence of Shot Peening on Surface Integrity of Some Machined Aerospace Materials. In: *ICSP-1*. 1981.
[194] Wang S, Li Y, Yao M, Wang R. Compressive Residual Stress Introduced by Shot Peening. *Journal of Materials Processing Technology*. 1998; 73(1-3):64-73.
[195] Lang K, Schulze V, Vöhringer O. Shot Peening and Fatigue Strength of Steels. In: *ICSP-8*. Garmisch-Partenkirchen, Germany: 2002. p. 281-294.
[196] Wohlfahrt H. The Influence of Peening Conditions on the Resulting Distribution of Residual Stress. In: *ICSP-2*. Chicago: 1984. p. 316-331.
[197] Fuchs, H.O. The Strength of Shot Peened Parts, Design Calculations and Specifications. In: *ICSP-1*. Paris: 1981.
[198] Lunarski J, Zielecki W. Fatigue Strength of Welded Joints after Pneumatic Shot Peening. In: *ICSP-4*. Tokyo, Japan: 1990. p. 263-272.

[199] Hu, N., Zhou, H. The Effect of Surface Plastic Deformation on Fatigue Strength and its Mechanism. In: *ICSP-1*. Paris: 1981. p. 351-358.

[200] Koibuchi K, Hayama T, Kawai S. Residual Stress and Fatigue Strength of Surface Hardened Components. In: *ICSP-1*. Paris: 1981. p. 413-422.

[201] Kohler W, Hornauer K. Selected Examples on the Topography of Shot Peened Metal Surfaces. In: *ICSP-3*. Garmisch-Partenkirchen, Germany : 1987. p. 269-277.

[202] Iida K, Tosha K. Fatigue Strength of Work Softening Layer Produced by Shot Peening. In: *ICSP-3*. Garmisch-Partenkirchen, Germany : 1987. p. 611-618.

[203] Burgahn F, Vohringer O, Macherauch E. Microstructural Investigations of the Shot Peened Steel 42 CrMo 4 in Different Heat Treatment conditions by the Aid of a X-Ray Profile Analysis. In: *ICSP-4*. Tokyo, Japan: 1990. p. 199-208.

[204] Iida K, Tosha K. Fatigue Strength and Residual Stress Distribution of the Work-Softened Steel by Shot Peening. In: *ICSP-4*. Tokyo, Japan: 1990. p. 233-242.

[205] Schulze V, Macherauch E, Vohringer O, Lang K. Relaxation of Shot Peening Induced Residual Stresses in Quenched and Tempered Steel AISI 4140 Due to Uniaxial Cyclic Deformation. In: *ICSP-6*. San Francisco, USA: 1996. p. 403-412.

[206] Iida K, Tosha K. Work-softening Induced by Shot Peening for Austenitic Stainless Steel by Shot Peening. In: *ICSP-5*. Oxford, UK: 1993. p. 311-318.

[207] Wang, R., Tan, Y., Li, X., Yan, M. Investigation on the Microstructure in Shot Peening Surface Straining Layer of Martensite. In: *ICSP-1*. Paris: 1981. p. 185-192.

[208] Qiu Q, Wang R. Influence of the Change of Microstructure and Residual Stress Field in the surface Shot-Peening Straining Layer on Fatigue Behavior. In: *ICSP-3*. Garmisch-Partenkirchen, Germany : 1987. p. 231-238.

[209] Sollich A, Wohlfahrt H. Optimization of the Fatigue Strength of Heat Treated Steels as a Consequence of an Optimum State of the Surface and Subsurface Layers after Shot Peening. In: *ICSP-6*. San Francisco, USA: 1996. p. 251-262.

[210] Nobre JP, Kornmeier M, Dias AM, Scholtes B. Comparative Analysis of Shot-peening Residual Stresses Using Hole-drilling and X-ray Diffraction Methods. *Materials Science Forum*. 2000; 347138-143.

[211] Joersch, J. F., Neal, J. W. A Simultaneous Method of Imparting compressive stresses on various substrates while maintaining surface integrity. In: *ICSP-1*. Paris: 1981. p. 649-662.

[212] Wierschowski D, Ostertag A, Wagner L. Fatigue Performance of Mechanically Surface Treated Steels 42CrMo4 and 54SiCr6: Shot Peening Vs Roller Burnishing. In: *ICSP-8*. Garmisch-Partenkirchen, Germany: 2002. p. 468-473.

[213] Ferrari V, Woelfel M. The Relationship Between Peening Intensity and Metal Erosion Rate with Respect to Nozzle Angle. In: *ICSP-2*. Chicago: 1984. p. 269-270.

[214] Clausen R, Stangenberg J. Roughness of Shot-Peened Surfaces - Definition and Measurment. In: *ICSP-7*. Warsaw, Poland: 1999. p. 69-77.

[215] Hornbogen E, Thumann M, Verpoort C. Influence of Shot Peening on the Fatigue Behavior High Cycle Fatigue Properties of a Precipitation Hardenable Austenitic Steel. In: *ICSP-1*. 1981. p. 381-387.

[216] Iida K, Hirose Y. The Residual Stress Distribution in Shot Peened Carburized Steel under Fatigue. In: *ICSP-7*. Warsaw, Poland: 1999. p. 96-101.

[217] Iida K, Ito J. Peening Effect on Machining of Steel. In: *ICSP-4*. Tokyo, Japan: 1990. p. 347-352.

[218] Lu J, Reby J, Vasseurr G, Flavenot JF, Sutte B. Modification of the Residual Stresses and Microstructures in Hard Chromium Plating by Shot Peening. In: *ICSP-4*. Tokyo, Japan: 1990. p. 179-190.

[219] Kirk D, Render PE. Effects of Peening on Stress Corrosion Cracking in Carbon Steel. In: *ICSP-7*. Warsaw, Poland: 1999. p. 167-176.

[220] Jingpu Z. The Effect of Shot Peening on Contact Fatigue Behavior of 40Cr Steel after Compouind Heat Treatment. In: *ICSP-2*. Chicago: 1984. p. 215-224.

[221] Zhu J, Guo P. The Research to the Contact Fatigue Behavior Influenced by the Residual Stresses on the Surface of the Shot-peened Specimens. In: *ICSP-3*. Garmisch-Partenkirchen, Germany : 1987. p. 525-532.

[222] Tosha K, Lu J, Retraint D, Guelorget B, Iida K. Influence of Pre-Annealing on Surface and Surface Layer Characteristics Produced by Shot Peening. In: *ICSP-8*. Garmisch-Partenkirchen, Germany: 2002. p. 173-180.

[223] He Jia-wen, Hu Zhi-zhong , Zhang Ding-quan. Influence of Residual Stress on Fatigue Limint of Various Carbon Steels. In: *ICSP-2*. Chicago: 1984.

[224] Ebenau A, Eifler D, Vohringer O, Macherauch E. Influence of Shot Peening on the Cyclic Deformation Behaviour of the Steel 42 CrMo 4 in a Normalized State. In: *ICSP-4*. Tokyo, Japan: 1990. p. 337-346.

[225] Woelfel MM. Experimental Investigation of Fatigue Strength as Related to Peening Intensity and Specimen Thickness. In: *ICSP-2*. Chicago: 1984. p. 365-368.

[226] Kirk D, Jarrett M. Influence of Shot-Peening on Fatigue and Corrosion-Fatigue Behavior of Structural Steel. In: *ICSP-2*. Chicago: 1984. p. 133-142.

[227] Burakowski T, Nakonieczny A. Influence of Shot Peening on the Fatigue Strength of Various Heat Treated Steels. In: *ICSP-2*. Chicago: 1984. p. 334-339.

[228] Hoffmann JE, Scholtes B, Vohringer O, Macherauch E. Influence of Shot Peening on the Bending Fatigue Behaviour of Notched Specimens of Ck 45. In: *ICSP-3*. Garmisch-Partenkirchen, Germany : 1987. p. 239-246.

[229] Chen Y, Wang R. Investigation on the Effect of Shot Peening on the Elevated Temperature Fatigue Behavior of Superalloy. In: *ICSP-3*. Garmisch-Partenkirchen, Germany : 1987. p. 655-658.

[230] Wang R, Li X, Yin Y-. The Effect of ShotPeening on Stress Corrosion Cracking Behavior of Stainless Steel. In: *ICSP-2*. Chicago: 1984. p. 143-151.

[231] Pakrasi, S., Betzold, J. Effect of Shot Peening on Properties of Carbonitrided Case with Retained Austenite. In: *ICSP-1*. Paris: 1981. p. 193-200.

[232] Hashimoto M, Shiratori M, Nagashima S. The Effects of Shot-Peening on Residual Stresses and Fatigue Strength of Carburized Gear Steels. In: *ICSP-4*. Tokyo, Japan: 1990. p. 495-504.

[233] Matsumoto T, Suzuki Y, Okada Y. The Relation between Shot Peening Conditions and Fatigue Strength of Carburized Gears. In: *ICSP-4*. Tokyo, Japan: 1990. p. 455-464.

[234] Wu D, Cai R, Teng F, Li J. The Microstructure and the Fatigue Properties of Shot-Peened Carburized Gear Steel 20CrMnTi. In: *ICSP-4*. Tokyo, Japan: 1990. p. 527-534.

[235] Nakonieczny, A, Szyrle W. Residual Stresses, Microstructure and Fatigue Behavior of Carburized Layers Before and After Shot Peening. In: *ICSP-6*. San Francisco, USA: 1996. p. 263-269.

[236] Hisamatsu S, Kanazawa T, Toyoda T. Influence of the Surface Properties on the Bending Strength of Shot Peened Carburized Steel. In: *ICSP-4*. Tokyo, Japan: 1990. p. 477-484.

[237] Peyrac C, Flavenot J, Convert F. Combining Case Hardening and Shot Peening for Gear Steels; Influence on Residual Stress. In: *ICSP-7*. Warsaw, Poland: 1999. p. 362-370.

[238] Peyrac C, Flavenot J. Optimisation of Carburising and Shot Peening, In Order To Improve Both Bending and Contact Fatigue Behaviour For Gear Application. In: *ICSP-9*. Paris, France: 2005.

[239] Toyoda T, Kanazawa T, Hisamatsu S, Ishiguro T. Influence of Inclusions on the Fatigue Strength of Shot Peened Carburized Steel. In: *ICSP-5*. Oxford, UK: 1993. p. 333-340.

[240] Nakonieczny A, Zolciak T, Monka G. Effects of Shot Peening on Process of Carburization and Selected Strength Properties of Steel 18HGT. In: *ICSP-7*. Warsaw, Poland: 1999. p. 135-144.

[241] Ebenau A, Lohe D, Vohringer O, Macherauch E. Influence of Shot Peening on the Microstructure and the Bending Fatigue Strength of Bainitic-Austenitic Nodular Cast Iron. In: *ICSP-4*. Tokyo, Japan: 1990. p. 389-398.

[242] Kobayashi M, Hasegawa K. Effect of Shot Peening on the Pitting Fatigue Strength of Carburized Gears. In: *ICSP-4*. Tokyo, Japan: 1990. p. 465-476.

[243] Sharma MC, Rao RNVDM. Effect of Shot Peening and Polymer Coating on Corrosion Fatigue Behaviour of Carbon Steels. In: *ICSP-4*. Tokyo, Japan: 1990. p. 551-560.

[244] Ohsawa M, Yonemura T. Improvement of Hardened Surface by Shot Peening. In: *ICSP-4*. Tokyo, Japan: 1990. p. 147-158.

[245] Satoru E, Muneto H, Ichiro A, Yasuo T. Effects of Shot Peening and Heat Treatment on Endurance Limits of Austempered Ductile Cast Iron Gears. In: *ICSP-6*. San Francisco, USA: 1996. p. 14-23.

[246] Lamaire E, Le Guernic, Y, Roch E, Scott C. Shot Peening to Prevent Intergranular Corrosion of Ferritic Stainless Steel. In: *ICSP-6*. San Francisco, USA: 1996. p. 61-72.

[247] Risch K. Shot Peening to Prevent Stress Corrosion Cracking and Corrosion Fatigue Failures of Austenitic Stainless Steels. In: *ICSP-3*. Garmisch-Partenkirchen, Germany : 1987. p. 141-148.

[248] Prabhugaunkar GV, Rawat MS, Prasad CR. Role of Shot Peening on Life Extension of 12% Cr Turbine Blading Martensitic Steel Subjected to SCC and Corrosion Fatigue. In: *ICSP-7*. Warsaw, Poland: 1999. p. 177-183.

[249] Iida K, Ohshima M. The Influence on Corrosion with Residual Stress and Transformation Induced in Shot - Peened Austenitic Stainless Steel. In: *ICSP-7*. Warsaw, Poland: 1999. p. 192-198.

[250] Rawat MS, Jayaraman TR, Prasad CRK, Kalpana Y. Influence of Shot Peening on Corrosion Properties of Steam Turbine Blading Steel. In: *ICSP-7*. Warsaw, Poland: 1999. p. 184-191.

[251] Obatu M, Sudo A. Effect of Shot Peening on Residual Stress and Stress Corrosion Cracking for Cold Worked Austenitic Stainless Steel ?? In: ICSP-5. Oxford, UK: 1993.

[252] Gillespie B. Controlled Shot Peening Can Help Prevent Stress Corrosion Cracking. In: *ICSP-3*. Garmisch-Partenkirchen, Germany : 1987. p. 149-156.

[253] Obata M, Sudo A, Matsumoto J. The Effect of Shot Peening on Residual Stress and Stress Corrosion Cracking for Austenitic Stainless Steel. In: *ICSP-6*. San Francisco, USA: 1996. p. 24-33.

[254] Kirk D, Payne R. Transformations Induced in Austenitic Stainless Steels by Shot Peening. In: *ICSP-7*. Warsaw, Poland: 1999. p. 15-22.

[255] Krull P, Nitschke-Pagel T, Wohlfahrt H. Stability of Residual Stresses in Shot Peened and High Pressure Water Peened Stainless Stees at Elevated Temperatures. In: *ICSP-7*. Warsaw, Poland: 1999. p. extra.

[256] Sharma MC. Effect of Work-piece Hardness on Peening Intensity Under Local Peening. In: *ICSP-3*. Garmisch-Partenkirchen, Germany : 1987. p. 205-212.

[257] Kirk D. Residual Stresses and Retained Austenite in Shot Peened Steels. In: *ICSP-1*. Paris: 1981. p. 271-278.

[258] Cheng Jia-xiang, Ao Bing-qiu. The Effect of Shot Peening on Contact Fatigue Life of Carburized Steel. In: *ICSP-1*. Paris, France: 1981. p. 333-339.

[259] S.Sato, K.Inoue, A.Ohno. The Effect of Shot Peening to Decarburized Spring Steel Plate. In: *ICSP-1*. Paris, France: 1981. p. 303-312.

[260] Iida K, Tosha K. Work-softening and Strain-induced Transformation Produced by Shot Peening and Grit Blasting for Austenitic Stainless Steel. In: *ICSP-6*. San Francisco, USA: 1996. p. 296-301.

[261] Muller M, Verpoort C, Gessinger G. Influence of Shot Peening on Fatigue and Corrosion Fatigue Behavior of an Austenitic-ferritic Stainlessn Steel. In: *ICSP*-1. Paris: 1981. p. 479-484.

[262] Ishigami H, Matsui K, Jin Y, Ando K. Study on Stress, Reflection and Double Shot Peening to Increase Compressive Residual Stress. *Fatigue and Fracture of Engineering Materials and Structures*. 2000; 23(11):959-963.

[263] Ando K, Matsui K, Ishigami H. Study on Double Shot Peening and Fatigue Limit of Gear. *Key Engineering Materials*. 2000; 183II/.

[264] Kloos KH, Kaiser B, Oppermann T. Improvement to the Fatigue Strength of Drop-forged Specimens by Airless Blast Cleaning. In: *ICSP-3*. Garmisch-Partenkirchen, Germany : 1987. p. 533-540.

[265] Kawagoishi N, Nagano T, Moriyama M, Kondo E. Improvement of Fatigue Strength of Maraging Steel by Shot Peening. *Key Engineering Materials*. 2007; 348-349537-540.

[266] Ito T, Mizuno T, Takahashi T, Kurosaki J. Peening Characteristics of Cemented Carbide Peening Shotand Life Improvement of Cold Forging Die. In: *ICSP-7*. Warsaw, Poland: 1999. p. 371-377.

[267] Kikuchi M, Ueda H, Hanai K, Naito T. The Improvement of Fatigue Durability of Carburized Steels with Surface Structure Anomalies by Shot Peening. In: *ICSP-2*. Chicago: 1984. p. 208-214.

[268] Burakowski T, Nakonieczny A, Babul T. The Effect of Shot Peening on Fatigue Strength of Zinc Coated Steels. In: *ICSP-3*. Garmisch-Partenkirchen, Germany : 1987. p. 603-610.

[269] Hisamatsu S, Kanazawa T. Size Effect on Fatigue Strength of Shot Peened Carburized Steel. In: *ICSP-3*. Garmisch-Partenkirchen, Germany : 1987. p. 517-524.

[270] Nadkarni VS, Sharma MC. Behaviour of Shot Peened Electron Beam Welded Stainless Steel Plate Samples under Combined Reverse Torsional and Bending Cyclic Stress. In: *ICSP-7*. Warsaw, Poland: 1999. p. 161-166.

[271] Adamovic D, Babic M, Jeremic B. Shot Peening Influence on Tribological Characteristics of surfaces. In: *ICSP-7*. Warsaw, Poland: 1999. p. 350-358.

[272] Prevey P, Cammett J. The Effect of Shot Peening Coverage on Residual Stress,Cold Work and Fatigue in a Ni-Cr-Mo Low Alloy Steel. In: *ICSP-8*. Garmisch-Partenkirchen, Germany: 2002. p. 295-304.

[273] Torres MAS, Baptista C, Nascimento M, Voorwald HJC. Analysis of the Influence of Shot Peening on the Fatigue Life of a Hard Chromium Electroplated AISI 4340 Steel Underplate. In: *ICSP-9*. Paris, France: 2005.

[274] Torres MAS, Nascimento N, Voorwald HJC. *Consideration of Shot Peening Treatment Applied to a High Strength Aeronautical Steel with Different Hardnesses*. Garmisch-Partenkirchen, Germany: 2002. p. 37-43.

[275] Was G, Pelloux R, Frabolot M. Effect of Shot Peening Methods on the Fatigue Behavior of Alloy 7075-t6. In: *ICSP-1*. 1981. p. 445-452.

[276] Honda T, Ramulu M, Kobayshi AS. Shot Peening and Fatigue Crack Growth in 7075-T7351 *Aluminum. Key Engineering Material*s. 2005; 297-300 172-77.

[277] Wang R, Li X, Wu H. Relationship between Shot Peening Surface Strain Layer and the Fatigue Strength of High-Strength Aluminium Alloy. In: *ICSP-3*. Garmisch-Partenkirchen, Germany : 1987. p. 417-422.

[278] Kyriacou S, Al-Khaja J. Investigation of Shot Peening and Re peening Effects on Partially Fatigued Notched Components. In: *ICSP-6*. San Francisco, USA: 1996. p. 233-242.

[279] Jaensson B, Magnusson L. An Investigation into the Influence on the Fatigue Strength of Aluminium Alloy Parts of Load Spectrum Base Level and Residual Stresses Induced shot Peening or Straightening. In: *ICSP*-3. Garmisch-Partenkirchen, Germany : 1987. p. 423-430.

[280] Akber J, Kyriacou S, El-Zafrany A. Effect of Shot Peening on the Fatigue Life of Axially Loaded Notched Components. In: *ICSP-5*. Oxford, UK: 1993. p. 349-358.

[281] Rahmatalla H, Al-Rimawi M, Al-Hadid T. Effect of Shot Peening on Stress Corrosion Cracking (SCC) 7075-T6Aluminum Alloy. In: *ICSP-6*. San Francisco, USA: 1996. p. 184-191.

[282] Cheong, Nam J, Lee J, Kim T, T. Taki. Effects of Shot Peening on the Corrosion Fatigue Life of Al 7075-T6. In: *ICSP-9*. Paris, France: 2005.

[283] Xiao Z, Fok WC, Lwin TD. Parametric Study of Residual Stress Due to Shot Peening. *Journal of Materials Processing Technology*. 1993; 39(3-4):469-483.

[284] Snowman, A. , Schmidt, R.G. The Improvement of Fatigue and Surface Characteristics of Alloy 7075-T6 by Secondary Peening with Glass Beads. In: *ICSP-1*. 1981. p. 313-322.

[285] Balcar, G.P., Maltby, F.C. Basic Curves of Surface Finish After Glass Bead Peening. In: *ICSP-1*. Paris: 1981. p. 295-302.

[286] Bathias C, Bonnafe JP. Shot Peening and Its Beneficial Effects. *New Metall Mater and New Fabr Processes*. 1988; 12. 0-12. 28.

[287] Kurth B, Bomas H , Mayr P. The Effect of Shot Peening on Fatigue Properties of a Die-cast Aluminium Alloy. In: *ICSP-5*. Oxford, UK: 1993. p. 154-165.

[288] Koehler W. Influence of Shot Peening with Different Peening Materials on the Stress Corrosion and Corrosion Fatigue Behavior of a welded AlZnMg-Alloy. In: *ICSP-2*. Chicago: 1984.

[289] Guernic Y. Shot Peening Retards "Fretting". In: *ICSP-4*. Tokyo, Japan: 1990. p. 281-296.
[290] Watanabe Y, Hasegawa N, Endo H, Marui E, Aoki Y. An Effect of Peening on Fretting Fatigue. In: *ICSP-7*. Warsaw, Poland: 1999. p. 127-134.
[291] De Los Rios ER. Mechanisms and Modelling of Cracking under Corrosion and Fretting Fatigue Conditions. In: *ICSP-8*. Garmisch-Partenkirchen, Germany: 2002. p. 243-254.
[292] Fair GH, Noble B, Waterhouse RB. The Effect of Shot-Peening on the Fatigue and Fretting Fatigue Behaviour of 8090 and 7010 Aluminium Alloys. In: *ICSP-3*. Garmisch-Partenkirchen, Germany : 1987. p. 431-438.
[293] Hamidi T, Castex L, Fathallah R, Sidhom H, Barrallier L. Behavior of Shot Peened Welded Steel Under Cyclic Loading. In: *ICSP-6*. San Francisco, USA: 1996. p. 130-141.
[294] Wohlfahrt H, Nitschke-Pagel T, Kopp R. Optimization of the Fatigue Behaviour of Welded Joints by means of Shot Peening -A Comparison of Results on Steel and Aluminium joints. In: *ICSP-6*. San Francisco, USA: 1996. p. 243-250.
[295] Wubbenhorst W. Exfoilation Corrosion Located by Search Peening. In: *ICSP-7*. Warsaw, Poland: 1999. p. 209-215.
[296] Karuppanan S, Romero J, De Los Rios ER, Rodopoulos C. A Theoretical and Experimental Investigation into the Development of Coverage in Shot Peening. In: *ICSP-8*. Garmisch-Partenkirchen, Germany: 2002. p. 101-107.
[297] Oberparleiter W, Schuetz W. *Shot Peening- A Choice to Improve Fatigue Life of Titanium Components*. 1984. p. 1803-1810.
[298] Schutz W. Improvement of Fatigue Strength of Titanium components under Variable Amplitude Loading. In: *ICSP-2*. Chicago: 1984. p. 166-174.
[299] Kussmaul K, Guth W, Meyer R. On the Effects of Shot Peening on the Fatigue Strength and the Residual Stresses of Parts Made of TiAl6V4. In: *ICSP-3*. Garmisch-Partenkirchen, Germany : 1987. p. 335-340.
[300] Wagner, L., Lutjering, G. Influence of Shot Peening on the Fatigue Behavior of Titanium Alloys. In: *ICSP-1*. Paris: 1981. p. 453-460.
[301] Gao Y-. Influence of Shot Peening on Tension-tension Fatigue Property of Two High Strength Ti Alloys. *Surface Engineering*. 2006; 22(4):299-303.
[302] Wagner L, Lutjering G. Influence of the Shot Peening Parameters on the Surface Layer Properties and the Fatigue Life of Ti-6Al-4V. In: *ICSP-2*. Chicago: 1984. p. 194-200.
[303] Wagner L, Lutjering G. Influence of Shot Peening Treatment on the Fatigue Limit of Ti-6Al-4V. In: *ICSP-2*. Chicago: 1984. p. 194-200.
[304] Dorr T, Wagner L. Influence of Stress Gradient on Fatigue Behavior of Shot Peened Timetal 1100. In: *ICSP-6*. San Francisco, USA: 1996. p. 223-232.
[305] Gray H, Wagner L, Lutjering G. Effect of Modified Surface Layer Microstructures through Shot Peening and Subsequent Heat Treatment on the Elevated Temperature Fatigue Behaviour of Ti Alloys. In: *ICSP-3*. Garmisch-Partenkirchen, Germany : 1987. p. 447-458.
[306] Wang, R., Zhang, D., Song, Y Yin. Shot Peening of Superalloys and Its Fatigue Properties at Elevated Temperature. In: *ICSP-1*. Paris: 1981. p. 395-404.

[307] Guedou J, Evry F. Effect of Shot Peening Conditions on Fatigue Strength of Two Alloys for Turboengines Disks. In: *ICSP-3*. Garmisch-Partenkirchen, Germany : 1987. p. 459-466.

[308] Ru J, Wang R, Li X. Investigation on Shot Peening Strengthening of Rene'95 Powder Alloy. In: *ICSP-6*. San Francisco, USA: 1996. p. 338-347.

[309] Pfeiffer W, Frey T. *Strengthening of Ceramics by Shot Peening*. 2002. p. 101-108.

[310] Frey T, Pfeiffer W. Shot Peening of Ceramics: Damage or Benefit? In: *ICSP-8*. Garmisch-Partenkirchen, Germany: 2002. p. 185-190.

[311] Pfeiffer W, Frey T. Advances in Shot Peening of Silicon Nitride Ceramics. In: ICSP-9. Paris, France: 2005.

[312] Cammett J, Eylon D, Cooke C, Froes F. Effects of Shot Peening on Fatigue of Ti-6Al-4V Powder Compacts. In: *ICSP-2*. Chicago: 1984. p. 181-193.

[313] Lu J, Miege B, Namiki K. Study on the Effect of Shot Peening on the Residual Stress in Al-SiC Composite. In: *ICSP-4*. Tokyo, Japan: 1990. p. 169-178.

[314] Tohriyama S, Kumano M, Hisamatsu S. Influence of Peening on Fatigue Life of SiC Reinforced Aluminum. In: *ICSP-4*. Tokyo, Japan: 1990. p. 307-316.

[315] Omam Nango C, Lu J, Flavenot J, Lieurade H, Bathias C. Study of Residual Stresses Produced in Metal Matrix Composite AI-SiCw by Shot Peening. In: *ICSP*-5. Oxford, UK: 1993. p. 290-299.

[316] Ebihara W, Fiore N, Angelini M. Plastically deformed depth in shot peened magnesium alloys. In: *ICSP-1*. Paris: 1981. p. 209-216.

[317] Zhang P, J. L. Influence of Shot Peening on High Cycle Fatigue Properties of the High-strength Wrought Magnesium Alloy. *Scripta Materialia*. 2005; 52485-490.

[318] Dorr T, Hilpert M, Beckmerhagen P, Kiefer A, Wagner L. Influence of Shot Peening on Fatigue Performance of High-Strength Aluminium and Magnesium Alloys. In: *ICSP-7*. Warsaw, Poland: 1999. p. 153-160.

[319] Birley SS. Shot Peening As a Stress Corrosion Preventive in Al-Zn-Mg Welded Joints. In: *ICSP-1*. Paris: 1981. p. 529-540.

[320] Baao W, Zhenhuan G, Zhunling W, Renzhi W. Effects of Shot-Peening on Fatigue of Compressive Coil Springs. In: *ICSP-2*. Chicago: 1984. p. 115-121.

[321] Kaiser B. Effects of Shot Peening on the Corrosion Fatigue Properties of Spring Steels. In: *ICSP-3*. Garmisch-Partenkirchen, Germany : 1987. p. 667-674.

[322] Hirose, S. Effect of Shot Peening on the Fatigue Strength of Small Spring Wire and Coil Springs. In: *ICSP-1*. Paris: 1981. p. 51-60.

[323] Hutter W, Kloos KH. Influence of Peening Parameters, Coating Temperatures and Operating Loads on Residual Stresses of Shot-Peened Barrel Springs. In: *ICSP-3*. Garmisch-Partenkirchen, Germany : 1987. p. 509-516.

[324] Kurihara Y, Shimoseki M, Sato K. Relationship between Change in Leaf Camber and Residual Stresses Produced by Shot Peening. In: *ICSP-4*. Tokyo, Japan: 1990. p. 353-362.

[325] Wohlfahrt H, Hirsch T, Macherauch E. Fatigue Strength of Case Hardened and Shot Peened Gears. In: *ICSP-3*. Garmisch-Partenkirchen, Germany : 1987. p. 547-560.

[326] Banas G, Lawrence Jr. FV. Improvement of Ni 18 (250) Maraging Steel Weldment Fatigue Strength Through Shot Peening. In: *ICSP-3*. Garmisch-Partenkirchen, Germany : 1987. p. 675-680.

[327] Chang, S., Lawrence Jr., F.V. Predicting the Effect of Shot Peening on Weld Fatigue Life. In: *ICSP-1*. Paris: 1981. p. 461-478.

[328] Oshida Y, Daly J. Effects of Shot Peening on Fatigue-Damaged High Strength Aluminum Alloy -Applicability of Rejuvenation on Fatigue-Damaged Components. In: *ICSP-4*. Tokyo, Japan: 1990. p. 317-326.

[329] Nakonieczny A, Monka G. Effect of shot peening on the Fatigue Strength of Spring Steel after exposure to Corrosion. In: *ICSP-9*. Paris, France: 2005.

[330] Hasegawa N, Hattori K, Watanabe Y, Fukuyama K, Fuwa T. The Temperature Change on Surface of Workpiece During Peening. In: *ICSP-7*. Warsaw, Poland: 1999. p. 54-59.

[331] Vohringer O. Changes in the State of the Material by Shot Peening. In: *ICSP*-3. Garmisch-Partenkirchen, Germany : 1987. p. 185-204.

[332] Herzog R, Wohlfahrt H, Scholtes B, Zinn W. The Significance of Almen Intensity for the Generation of Shot Peening Residual Stresses. In: *ICSP-6*. San Francisco, USA: 1996. p. 270-281.

[333] Schilling-Praetzel M, Hegemann F, Gomez P, Gottstein G. Influence of Temperature of Shot Peening on Fatigue Life. In: *ICSP-5*. Oxford, UK: 1993. p. 225-236.

[334] Wick A, Schulze V, Vohringer O. Shot Peening and Stress Peening of AISI 4140 at Increased Temperatures. In: *ICSP-7*. Warsaw, Poland: 1999. p. 110-116.

[335] Maximov JT. Thermodynamic Optimization of Mechanical Systems with Dissipative Processes. *International Journal of Mechanical Sciences*. 2006; 48(1):62-74.

[336] Timoshenko SP, Goodier JN. *Theory of Elasticity*. McGraw-Hill; 1970.

[337] Ogawa K, Asano T. Theoretical Prediction of Residual Stress Induced by Shot Peening and Experimental Verification for Carburized Steel. *Zairyo/Journal of the Society of Materials Science, Japan*. 1999; 48(12):1360-1366.

[338] AI-Obaid Y. Shot Peening Mechanics: Experimental and Theoretical Analysis. *Mechanics of Materials*. 1995; 19251-260.

[339] Levers A, Prior A. Finite Element Analysis of Shot Peening. J*ournal of Materials Processing Technology*. 1998; 80-81304-308.

[340] Shen S, Atluri SN. An Analytical Model for Shot-peening Induced Residual Stresses. Computers, *Materials and Continua*. 2006; 4(2):75-85.

[341] DeLitizia AT. Influence of Shot Peening on Residual Stresses in Spring Steel Plate. In: *ICSP-2*. Chicago: 1984. p. 237-240.

[342] Al-Hassani STS. Mechanical Aspects of Residual Stress Development in Shot Peening. In: *ICSP-1*. Paris: 1981.

[343] Al-Hassani S. An Engineering Approach to Shot Peening Mechanics. In: *ICSP-2*. Chicago: 1984. p. 275-282.

[344] Li J, Li H, Yao M, Wang R. Prediction of Fatigue Strength of Shot-Peened and Then Ground Specimens. In: *ICSP-4*. Tokyo, Japan: 1990. p. 419-426.

[345] Li J, Yao M, Wang R. A New Concept for Fatigue Strength Evaluation of Shot Peened Specimens. In: *ICSP-4*. Tokyo, Japan: 1990. p. 255-262.

[346] Bezyazichny VF, Kozhina T. Calculated Determination of the Parameters of the Surface Layer. In: *ICSP-7*. Warsaw, Poland: 1999. p. 228-234.

[347] Wang T, Platts M, Levers A. A Process Model for Shot Peen Forming. *A.S.N.E.Journal*. 2006; 172159-162.

[348] Iida K. Dent and Affected Layer Produced by Shot Peening. In: *ICSP-2*. Chicago: 1984. p. 283-292.

[349] Chang S, Lawrence Jr. F. An Empirical Model for Weld Fatigue Resistance. In: *ICSP-2*. Chicago: 1984. p. 297-305.

[350] Nakonieczny A. Consideration of Shot Peening in Fatigue Strength Calculation of Machine Parts. In: *ICSP-4*. Tokyo, Japan: 1990. p. 399-410.

[351] Amano A, Sakakibara T, Wakita M. Calculated Determination of Optimum Peening Condition in terms of Collision Energy on Coil Spring by Shot Peening. In: *ICSP-9*. Paris, France: 2005.

[352] Lieurade HP, Castelucci P, Flavenot J, Lu J, Bel G. Efficiency of Fatigue Improvement Techniques as a Function of the Type of Welded Joint. In: *ICSP-5*. Oxford, UK: 1993. p. 359-370.

[353] Palin-Luc T, Flavenot J, Lasserre S, Charles J, Leguernic Y. Fatigue Strength Prediction of a Shot Peened 42CrMo4 Steel Using Multiaxial Fatigue Criterion and Surface Roughness Correction. In: *ICSP-9*. Paris, France: 2005.

[354] Guechichi H, Castex L. Fatigue Limits Prediction of Shot Peened Materials. In: *ICSP-9*. Paris, France: 2005.

[355] Gardiner D, Platts J. Towards Peen Forming Process Optimisation. In: *ICSP-7*. Warsaw, Poland: 1999. p. 235-243.

[356] Korsunsky AM. On the Modelling of Residual Stresses Due to Surface Peening Using Eigenstrain Distributions. *Journal of Strain Analysis for Engineering Design*. 2005; 40(8):817-824.

[357] Satraki M, Evans A, King A, Bruno G, Wither P. Eigenstrain Generated by Shot Peening in Udimet 720LI Inferred By Means of Finite Element and Analytical Models. In: *ICSP-9*. Paris, France: 2005.

[358] Meguid SA, Shagal G, Stranart JC. Finite Element Modelling of Shot-peening Residual Stresses. *Journal of Materials Processing Technology*. 1999; 92-93401-404.

[359] Meguid SA, Shagal G, Stranart JC, Daly J. Three-dimensional Dynamic Finite Element Analysis of Shot-peening Induced Residual Stresses. *Finite Elements in Analysis and Design*. 1999; 31(3):179-191.

[360] Hills DA, Waterhouse RB, Noble B. Analysis of Shot Peening. *Journal of Strain Analysis for Engineering Design*. 1983; 18(2):95-100.

[361] Schiffner K, Droste Gen. Helling C. Simulation of Residual Stresses by Shot Peening. *Computers and Structures*. 1999; 72(1):329-340.

[362] Nakonieczny A, Borkowski P, Wymyslowski P. *Example of the Computer Simulation of Shot Peening Process*. Garmisch-Partenkirchen, Germany: 2002. p. 523-525.

[363] Curtis S, De los Rios ER, Rodopoulos CA, Levers A. Analysis of the Effects of Controlled Shot Peening on Fatigue Damage of High Strength Aluminium Alloys. *International Journal of Fatigue*. 2003; 25(1):59-66.

[364] Baragetti S. Three-dimensional Finite-element Procedures for Shot Peening Residual Stress Field Prediction. *International Journal of Computer Applications in Technology*. 2001; 14(1-3):51-63.

[365] Kyriacou S. Shot Peening Mechanics, a Theoretical Study. In: *ICSP-6*. San Francisco, USA: 1996. p. 505-516.

[366] Hirai N, Tosha K, Rouhaud E. Finite Element Analysis of Shot Peening -on the Form of a Single Dent. In: ICSP-9. Paris, France: 2005.

[367] Al-Hassani S, Kormi K, Webb K. Numerical Simulation of Multiple Shot Impact. In: *ICSP-7*. Warsaw, Poland: 1999. p. 217-227.

[368] Guagliano M, Vergani L, Bandini M, Gili F. An Approach to Relate the Shot Peening Parameters to the Induced Residual Stresses. In: *ICSP-7.* Warsaw, Poland: 1999. p. 274-282.

[369] Guagliano M. Relating Almen Intensity to Residual Stresses Induced by Shot Peening: A Numerical Approach. *Journal of Materials Processing Technology.* 2001; 110(3):277-286.

[370] Li Y, Wang F, Lu J, Xu K. Finite Element Analysis for Shot-peening Models of Single Shot Impact and Multiple Impingement. *Hsi-An Chiao Tung Ta Hsueh/Journal of Xi'an Jiaotong University.* 2007; 41(3):348-352.

[371] Webster GA, Ezeilo AN. Residual Stress Distributions and Their Influence on Fatigue Lifetimes. *International Journal of Fatigue.* 2001; 23(SUPPL. 1):S375-S383.

[372] Han K, Owen D, Peric D. Combined Finite/Discrete Element and Explicit/Implicit Simulations of Peen Forming Processes. *Engineering Computations.* 2002; 19(1):92-118.

[373] Meo M, Vignjevic R. Finite Element Analysis of Residual Stress Induced by Shot Peening Process. *Advances in Engineering Software.* 2003; 34(9):569-575.

[374] Deslaef D, Rouhaud E, Rasouli-Yazdi S. 3D Finite Element Models of Shot Peening Processes. *Materials Science Forum.* 2000; 347241-246.

[375] Helling CDG, Schiffner K. Numerical Simulation of Shot Peening. *Computational and Experimental Methods.* 2001; 6303-311.

[376] Al-Obaid YF. Dynamic Nonlinearity in Shot Peening Mechanics: Shakedown Analysis. In: *ICSP-5.* Oxford, UK: 1993. p. 104-115.

[377] Meguid SA, Shagal G, Stranart JC. 3D FE Analysis of Peening of Strain-Rate Sensitive Materials using Multiple Impingement Model. *International Journal of Impact Engineering.* 2002; 27119-134.

[378] Majzoobi GH, Azizi R, Alavi Nia A. A Three-dimensional Simulation of Shot Peening Process Using Multiple Shot Impacts. *Journal of Materials Processing Technology.* 2005; 164-1651226-1234.

[379] Majzoobi G, Azizi R. A 3D Numerical Study of Shot Peening Process Using Multiple Shot Impacts. In: *ICSP-9.* Paris, France: 2005.

[380] Eibella A, Rami V, Hogirala J. A Sensitivity Analysis of Shot Peening Parameters. In: *Proceedings of the ASME Design Engineering Technical Conference.* 2006.

[381] Kopp R, Schulz J. Optimising the Double-Sided Simultaneous Shot Peen Forming. In: *ICSP-8.* Garmisch-Partenkirchen, Germany: 2002. p. 227-233.

[382] Wang T, Platts J. Finite Element Impact Modelling for Shot Peen Forming. In: *ICSP-8.* Garmisch-Partenkirchen, Germany: 2002. p. 540-546.

[383] Zeng Y. Finite Element Simulation of Shot Peen Forming. In: *ICSP-8.* Garmisch-Partenkirchen, Germany: 2002. p. 554-561.

[384] Schwarzer J, Schulze V, Vöhringer O. Finite Element Simulation of Shot Peening – A Method to Evaluate the Influence of Peening Parameters on Surface Characteristics. In: *ICSP-8.* Garmisch-Partenkirchen, Germany: 2002. p. 507-515.

[385] Han K, Peric D, Owens, D.R.J., Crook, A.J.L. A Combined Finite/discrete Element Simulation of Shot Peening Processes Part I- Study of 2d Interaction Laws. *Engineering Computations.* 2000; 17(5):593-619.

[386] Han K, Peric D. A Combined Finite/discrete Element Simulation of Shot Peening Processes Part II-3D Interaction Laws. *Engineering Computations*. 2000; 17(5):680-702.

[387] Haga S, Harada Y, Tsubakino H. Relation Between Microstructure of Shot-peening Steel and Its Mechanical Characteristics. In: *ICSP-9*. Paris, France: 2005.

[388] Wang JY, Tan YX, Zhu R. Transmission Electron Microscopy Analysis of Shot Peening Strengthened Layer for Some Industrial Alloys. In: *ICSP-4*. Tokyo, Japan: 1990. p. 191-198.

[389] Lu J, Lu K. Generation of Nanostructure by SMAT (surface Mechanical Attrition Treatment): Basic Concept, Different Processes and Applications. In: *ICSP-9*. Paris, France: 2005.

[390] Mano H, Satoru K, Matsumuro A, Imura T. Characterization of Nanocrystalline Surface Layer Induced by Shot Peening and Effect on Their Fatigue Strength. In: *Materials Research Society Symposium Proceedings*. 2005. p. 67-72.

[391] Roland T, Retraint D, Lu K, Lu J. Fatigue Life Improvement Through Nanostructuring of Stainless Steel. In: *ICSP-9*. Paris, France: 2005.

[392] Mano H, Kondo S, Imura T, Matsumuro A. Effect of Nanocrystalline Surface Layer Produced by Shot Peening on Fatigue Strength. N*ippon Kinzoku Gakkaishi/Journal of the Japan Institute of Metals*. 2005; 69(2):213-216.

[393] Xinling M, Wei W, Wei Y. Simulation for Surface Self-nanocrystallization under Shot-Peening. *Acta Mechanica Sinica*. 2003; 19(2):172-180.

[394] Dai K, Shaw L. Comparison Between Shot Peening and Surface Nanocrystallization and Hardening Processes. *Materials Science and Engineering* A. 2007; 463(1-2):46-53.

[395] Liu JL, Umemoto M, Todaka Y, Tsuchiya K. Formation of a Nanocrystalline Surface Layer on Steels by Air Blast Shot Peening. *Journal of Materials Scienc*e. 2007; 42(18):7716-7720.

[396] Eltobgy MS, Ng E, Elbestawi MA. Three-dimensional Elastoplastic Finite Element Model for Residual Stresses in the Shot Peening Process. Proceedings of the Institution of Mechanical Engineers, Part B: *Journal of Engineering Manufacture*. 2004; 218(11):1471-1481.

[397] Fathallah R, Inglebert G, Castex L. Modelling of Shot Peening Residual Stresses and Plastic Deformation Induced in Metallic Parts. In: *ICSP-6*. San Francisco, USA: 1996. p. 464-480.

[398] Fathallah R, Inglebert G, Castex L. Prediction of Plastic Deformation and Residual Stresses Induced in Metallic Parts by Shot Peening. *Materials Science and Technology*. 1998; 14(7):631-639.

[399] Frija M, Hassine T, Fathallah R, Bouraoui C, Dogui A. Finite Element Modelling of Shot Peening Process: Prediction of the Compressive Residual Stresses, the Plastic Deformations and the Surface Integrity. *Materials Science and Engineering* A. 2006; 426(1-2):173-180.

[400] Lillamand I, Barrallier L, Lalanne B, Castex L. Cyclic Modelling of the Mechanical State Produced by Shot-peening. *Fatigue and Fracture of Engineering Materials and Structures*. 2001; 24(2):93-104.

[401] Al-Hassani STS. Shot Peening of Metals- Mechanics and Structures. In: *Shot Peening for Advanced Aerospace Design. Anaheim California* : 1982. p. 1-13.

[402] Slim S, Harm T, Inglebert G, Point N. Prediction of Residual Stresses Due to Shot Peening Using Modelling of the Material Behavior. *Materials and Manufacturing Processes*. 1995; 10(3):579-588.

[403] Guechichi H, Castex L, Frelat J, Inglebert G. Predicting Residual Stresses due to Shot Peening. In: *Proceedings of the 10th Conference on Shot Peening* (CETIM-ITI). Senlis, France: 1986. p. 23-34.

[404] Rouquette S, Rouhaud E, Pron H, François M, Bissieux C. Thermo-elasto-plastic Model for Shot Peening: A Numerical and Experimental Approach. In: *ICSP-9*. Paris, France: 2005.

[405] Schutz, W. Fatigue Life Improvement of High-strength Materials by Shot Peening. In: ICSP-1. Paris: 1981. p. 423-434.

[406] Kopp R. A Statistical Analysis of the Shot Peening Intensity Measurement In: *ICSP-5*. Oxford, UK: 1993.

[407] Vahratian A, Garibay RP. Application of Statistically Capable Shot Peening to Automotive Component Design. In: *ICSP*-5. Oxford, UK: 1993. p. 73-84.

[408] Mason C, I'Anson K. The Practical Use of Statistical Process Control and Computer Monitored Equipment in the Shot Peening Process. In: *ICSP-6*. San Francisco, USA: 1996. p. 328-337.

[409] Simpson R S , Bamesky R. A Statistical Analysis of the Internal Statistical Capability of Several Common Controlled Shot Peening Specifications? In: *ICSP-5*. Oxford, UK: 1993.

[410] Chardin H, Inglebert G, Slim S. Modeling the Influence of the Shot Size Dispersion on the Evolution of the Residual Stress Profiles vs. Time in Shot Peening. In: *ICSP-6*. San Francisco, USA: 1996. p. 348-355.

[411] Kopp R, Wustefeld F, Linnemann W. High Precision Shot Peen Forming. In: *ICSP-5*. Oxford, UK: 1993. p. 91-102.

[412] Abyaneh M. Fundamental Aspects of Shot Peening Coverage Control Part II: Simulation of Single And Multiple Impacting. In: *ICSP-6*. San Francisco, USA: 1996. p. 448-455.

[413] Abyaneh M, Kirk D. Fundamental Aspects of Shot Peening Coverage Control Part III: Coverage Control Versus Fatigue. In: *ICSP-6*. San Francisco, USA: 1996. p. 456-463.

[414] Abyaneh M. Fundamental Aspects of Shot Peening Coverage Control Part I: Formulation of Single and Multiple Impacting. In: *ICSP-6*. San Francisco, USA: 1996. p. 438-447.

[415] Nevarez I, Nelson D, Esterman M, Ishii K. Shot Peening and Robust Design for Fatigue Performance. In: *ICSP-6*. San Francisco, USA: 1996. p. 517-529.

[416] Knotek O, Elsing R. Computer Simulation of Different Surface Topographies of Metals Produced by Blasting Processes. In: *ICSP-3*. Garmisch-Partenkirchen, Germany : 1987. p. 361-368.

[417] Clausen R, Bruder M. Simulation of Spherical Shaped Surfaces, Produced by Peen Forming with Pointed Ball Shooting at Lower Coverages. In: *ICSP-4*. Tokyo, Japan: 1990. p. 615-624.

[418] Bailey P, Thompson R, Lombardo D, Popp H. Full Assurance Shot Peening of Aircraft Gas Turbine Engine Components. In: *ICSP-6*. San Francisco, USA: 1996. p. 320-327.

[419] Tange A, Okada H. Shot Peening and Coverage. In: *ICSP-8*. Garmisch-Partenkirchen, Germany: 2002. p. 516-522.

[420] Wohlfahrt H. Practical Aspects of the Application of Shot Peening to Improve the Fatigue Behaviour of Metals and Structural Components. In: *ICSP-3*. Garmisch-Partenkirchen, Germany : 1987. p. 563-584.

[421] Kiefer B. Shot Peening, Special Application and Procedure. In: *ICSP-3*. Garmisch-Partenkirchen, Germany : 1987. p. 157-162.

[422] Gao Y-, Yao M, Li J-. An Analysis of Residual Stress Fields Caused by Shot Peening. *Metallurgical and Materials Transactions A: Physical Metallurgy and Materials Science*. 2002; 33(6):1775-1778.

[423] Xu J., Zhang D., Shen B., The Fatigue Strength and Fracture Morphology of Leaf Spring Steel After Prestressed Shot Peening. In: *ICSP-1*. Paris: 1981. p. 367-374.

[424] Simpson RS, Probst RL. Topography and Fracture Studies of Surfaces Shot Peened at Different Intensities. In: *ICSP-3*. Garmisch-Partenkirchen, Germany : 1987. p. 261-268.

[425] Wigmore, G., Miles, L. The Use of Shot Peening to Delay Stress Corrosion Crack Initiation in Austenitic 8Mn8Ni4Cr Generator End Ring Steel. In: *ICSP-1*. Paris: 1981. p. 61-70.

[426] Takemoto M, Shinohars T, Shirai M. Control of Stress Corrosion Cracking by Shot Peening. In: *ICSP-1*. Paris: 1981. p. 521-528.

[427] Mukhopadhyay N, Prasad C, Sharma MC. Selection of Optimum Peening Parameters to Improve the Stress Corrosion Cracking Resistance of 18Mn4Cr Generator End Retaining Ring Steel. In: *ICSP-4*. Tokyo, Japan: 1990. p. 543-550.

[428] Speidel MO. Effect of Shot Peening on Stress Corrosion Cracking and Corrosion Fatigue. In: *ICSP-1*. Paris: 1981. p. 625-636.

[429] Rodopoulos CA, Curtis SA, De Los Rios ER, SolisRomero J. Optimisation of the Fatigue Resistance of 2024-T351 Aluminium Alloys by Controlled Shot Peening - Methodology, Results and Analysis. *International Journal of Fatigue*. 2004; 26(8):849-856.

[430] Prevéy PS, Cammett JT, The Effect of Shot Peening Coverage on Residual Stress, Cold Work and Fatigue in a Ni-Cr-Mo Low Alloy Steel. In: *ICSP-8*. Garmisch-Partenkirchen, Germany: 2002.

[431] Mogul J, Diepart CP. Process Controls the Key to Reliability of Shot Peening. *Industrial Heating*. 1995; 62(11):34-36.

[432] Daly J. The Use of the Shot Peen Process as a Design Tool Through the Use of Microprocessors. In: *ICSP-3*. Garmisch-Partenkirchen, Germany : 1987. p. 87-91.

[433] Happ M, Daly J. Computer Monitored Controlled Deflector Peening. In: *ICSP-4*. Tokyo, Japan: 1990. p. 607-614.

[434] Kirk D. Behavior of Peen-Formed Steel Strip on Isochronal Annealing. In: *ICSP-2*. Chicago: 1984. p. 231-236.

[435] Kittel S, Linnemann W, Wustefeld F, Kopp R. Tight tolerance Peen Forming with On-Line Shape Control. In: *ICSP-7*. Warsaw, Poland: 1999. p. 301-307.

[436] Petit-Renaud F. Optimization of the Shot Peening Parameters. In: *ICSP-8*. Garmisch-Partenkirchen, Germany: 2002. p. 119-129.

[437] Seno T, Horiuchi H, Naito T. Effect of Various Peening Parameters on Compressive Residual Stress for Carburized Steel. In: *ICSP-4*. Tokyo, Japan: 1990. p. 505-512.

[438] Nakamura M, Matsushima Y, Hasegawa T, Nakatani Y, Kouka T. Effect of Alloying Elements and Shot Peening on Fatigue Strength of Carburized Steel. In: *ICSP-4*. Tokyo, Japan: 1990. p. 485-494.

[439] Tufft M. Shot Peen Impact on Life part 1: Designed Experiment Using Rene 88Dt. In: *ICSP-7*. Warsaw, Poland: 1999. p. 244-253.

[440] Tufft M. Shot Peen Impact on Life part 2: single Particle Impact Tests Using Production Shot. In: *ICSP-7*. Warsaw, Poland: 1999. p. 254-263.

[441] Lassithiotakis D, Aylott CJ, Shaw BA, Ooi J, Petit-Renaud F. Optimising Shot Peening Parameters using DoE. In: *ICSP-9*. Paris, France: 2005.

[442] Tufft M, Sighurd R. Modeling of Fatigue Behavior due to Shot Peening Conditions. In: *ICSP-8*. Garmisch-Partenkirchen, Germany: 2002. p. 530-539.

[443] Tufft M. Shot Peen Impact on Life part 3: Development of a Fracture Mechanics/Threshold Behaviour Predictive Model. In: *ICSP-7*. Warsaw, Poland: 1999. p. 264-273.

[444] Sharma MC, Mubeen A. Effect of Shot Size on Peening Intensity for Local Peening of Different Thickness Samples. In: *ICSP-2*. Chicago: 1984. p. 62-66.

[445] Happ MB. Dovetail Peening by the Deflector Method. In: *ICSP-3*. Garmisch-Partenkirchen, Germany : 1987. p. 117-123.

[446] Adachi S. Fatigue Strength of Gear Steels Shot Peened in Extremely High Intensity Conditions. In: *ICSP-4*. Tokyo, Japan: 1990. p. 363-372.

[447] Okada H, Tange A, Tango K, Ando K. Effect of Pre-stress on Stress Shot Peening. Nippon Kikai Gakkai Ronbunshu, *A Hen/Transactions of the Japan Society of Mechanical Engineers, Part A*. 2004; 70(4):647-653.

[448] Barrett CF, Todd R. Investigation of the Effects of Elastic Pre-Stressing Technique on Magnitude of Compressive Residual Stress Induced by Sot Peen Forming of Thick Aluminum Plates. In: *ICSP-2*. Chicago: 1984. p. 15-21.

[449] Barret L, Cailliau J. Analysis of Some Particular Cases of Shot-Peening and their Influence on the Field of Residual Stresses. In: *ICSP-3*. Garmisch-Partenkirchen, Germany : 1987. p. 399-406.

[450] Robinson CG, Smart E. The Use of Specialised Shot Peening Techniques for Tapered Leaf Suspension Springs for Road Vehicles. In: *ICSP-2*. Chicago: 1984. p. 79-83.

[451] He J, Wu Y, Yu Y, Flavenot J. Shot Peening on Torsion Fatigue. In: *ICSP-3*. Garmisch-Partenkirchen, Germany : 1987. p. 619-624.

[452] Muller E. The Development of Residual Stresses at Bending Specimens under the Influence of Setting and Stress Peening. In: *ICSP-7*. Warsaw, Poland: 1999. p. 88-95.

[453] Wandell JL. Shot Peening of Engine Components. American Society of Mechanical Engineers, *Internal Combustion Engine Division (Publication) ICE*. 1997; 29101-108.

[454] Le Guernic Y, Eckersley J. Peenstress Software Selects Shot Peening Parameters. In: *ICSP-6*. San Francisco, USA: 1996. p. 481-492.

[455] Lewis Zion H, Steven Johnson W. Parametric Two-dimensional Finite Element Investigation: Shot Peening of High-strength Steel. *AIAA Journal*. 2006; 44(9):1973-1982.

[456] Baragetti S, Guagliano M, Vergani L. Numerical Procedure for Shot Peening Optimization by Means of Non-dimensional Factors. *International Journal of Materials and Product Technology*. 2000; 15(1):91-103.

[457] Baragetti S, Terranova A. Non-dimensional Analysis of Shot Peening by Means of DoE. *International Journal of Materials and Product Technology*. 2000; 15(1):131-141.

[458] P.M.George, Nisha Pillai, Nisha George. Optimization of Shot Peening Parameters Using Taguchi Technique. *Journal of Materials Processing Technology*. 2004; 153-154925-930.

[459] Franz HE, Olbricht A. Optimization of Shot Peening to Improve the Fatigue Strength of Ti-6AI-4V. In: *ICSP-3*. Garmisch-Partenkirchen, Germany : 1987. p. 439-446.

[460] Kopp R, Wustefeld F. Modern Simulation and Optimization of Peen Forming Processes. In: *ICSP-4*. Tokyo, Japan: 1990. p. 561-572.

[461] Altenberger I. Alternative Mechanical Surface Treatments: Microstructures, Residual Stressesand Fatigue Behavior. In: *ICSP-8*. Garmisch-Partenkirchen, Germany: 2002. p. 421-434.

[462] Tange A, Takamura N. Relation Between Shot-Peening Residual Stress Distribution and Fatigue Crack Propagation Life in Spring Steel. In: *ICSP-4*. Tokyo, Japan: 1990. p. 243-254.

[463] Naito K, Ochi T, Takahashi T, Suzuki N. Effect of Shot Peening on the Fatigue Strength of Carburized Steels. In: *ICSP-4*. Tokyo, Japan: 1990. p. 519-526.

[464] Zhang Jian-Zong, Hu Nai-sai , Zhou Hui-jiu. Effect of Shot Peening on Hardened High Carbon Steels and Its Mechanism. In: *ICSP-2*. Chicago: 1984. p. 225-230.

[465] Ogawa K, Yamada H, Saruki K, Yokoi M, Inuzuka M. Influence of Residual Stress on Fatigue Strength of Carburized and Shot Peened Notched Specimens. In: *ICSP-4*. Tokyo, Japan: 1990. p. 445-454.

[466] Gerdes C, Lutjering G. Influence of Shot Peening on Notched Fatigue Strength of Ti-6Al-4V. In: *ICSP-2*. Chicago: 1984. p. 175-180.

[467] Honda T, Ramulu M, Kobayashi AS. Effect of Shot Peening on Fatigue Crack Growth in 7075-T7351. *Journal of ASTM International*. 2005; 2(6):39-52.

[468] Sabelkin V, Martinez SA, Mall S, Sathish S, Blodgett MP. Effects of Shot-peening Intensity on Fretting Fatigue Crack-initiation Behaviour of Ti-6al-4v. *Fatigue and Fracture of Engineering Materials and Structures*. 2005; 28(3):321-332.

[469] Dorr T, Wagner L. Effect of Shot Peening on Residual Life of Fatigue Pre-Damaged 2024 AL. In: *ICSP-6*. San Francisco, USA: 1996. p. 174-183.

[470] Turnbull A, De los Rios ER, Tait R, Laurant C, Boabaid J. Improving the Fatigue Crack Resistance of Waspalloy by Shot Peening. *Fatigue and Fracture of Engineering Materials and Structures*. 1998; 211513-1524.

[471] Ferreira J. Effects of Surface Treatments on the Fatigue of Notched Bend Specimens. *Fatigue and Fracture of Engineering Materials and Structures*. 1996; 19(1):111-117.

[472] Jaensson, B. The Influence on the Fatigue Strength of Aluminium Alloy Parts of the Relationship Between the Surface Residual Stress State and Shot Peening. In: *ICSP-1*. Paris: 1981.

[473] Hanagarth H, Vohringer O, Macherauch E. Fatigue Behaviour of Shot Peened TiA16V4 in the Temperature Range 477 20¡ C :5 T :5 4500 C. In: *ICSP-3*. Garmisch-Partenkirchen, Germany : 1987. p. 477-484.

[474] Gray H, Wagner L, Lutjering G. Influence of Shot Peening Induced Surface Roughness, Residual Macrostresses and Dislocation Density on the Elevated Temperature HCF-Properties of Ti Alloys. In: *ICSP-3*. Garmisch-Partenkirchen, Germany : 1987. p. 447-458.

[475] Kopp R. Some Aspects of Stress Peening of Coil Springs for Vehicle Suspensions. In: *ICSP-5*. Oxford, UK: 1993. p. 341-348.

[476] Feng Z, Zhang J, Chen X. Fatigue Crack Propagation Behaviors in Shot Peened Layer of Melals. In: *ICSP-6*. San Francisco, USA: 1996. p. 432-437.

[477] Honda T, Ramulu M, Kobayashi AS. Fatigue of Shot Peened 7075-T7351 Senb Specimen - a 3-D Analysis. *Fatigue and Fracture of Engineering Materials and Structures*. 2006; 29(6):416-424.

[478] Mitsubayashi M, Miyata T, Aihara H. Prediction of Improvement in Fatigue Strength by Shot Peening and Selection of Most Effective Peening Condition. Nippon Kikai Gakkai Ronbunshu, *A Hen/Transactions of the Japan Society of Mechanical Engineers, Part A*. 1995; 61(586):1172-1178.

[479] Krug T, Lang K, Löhe, D. Residual Stress Sensitivity of Case-Hardened Notched Specimen. In: *ICSP-9*. Paris, France: 2005.

[480] J. K. Li, Yao Mei, Wang Duo, Wang Renzhi. *An Analysis of Stress Concentrations caused by Shot Peening and its Application in predicting Fatigue Strength. Fatigue and Fracture of Engineering Materials and Structures*. 1992; 15(12):1271-1279.

[481] Guagliano M, Vergani L. An Approach for Prediction of Fatigue Strength of Shot Peened Components. *Engineering Fracture Mechanics*. 2004; 71501-512.

[482] De Los Rios ER, Trull M, Levers A. Modelling Fatigue Crack Growth in Shot-peened Components of Al 2024-t351. *Fatigue and Fracture of Engineering Materials and Structures*. 2000; 23(8):709-716.

[483] Romero JS, De Los Rios E, Fam YH, Levers A. Optimisation of the Shot Peening Process in Terms of Fatigue Resistance. In: *ICSP-7*. Warsaw, Poland: 1999. p. 117-126.

[484] Rodopoulos C, Edwards R, Curtis S, Romero JS. Theoretical Analysis of Beneficial and Detrimental Effects of Controlled Shot Peening in High Strength Aluminium Alloys. In: *ICSP-8*. Garmisch-Partenkirchen, Germany: 2002. p. 547-553.

[485] McClung RC. A Literature Survey on the Stability and Significance of Residual Stresses During Fatigue. *Fatigue and Fracture of Engineering Materials and Structures*. 2007; 30(3):173-205.

[486] Zhuang WZ, Halford GR. Investigation of Residual Stress Relaxation Under Cyclic Load. *International Journal of Fatigue*. 2001; 23(SUPPL. 1):S31-S37.

[487] Almer J, Cohen J, Moran B. The Effects of Residual Macrostresses and Microstresses on Fatigue Crack Initiation. *Materials Science and Engineering A*. 2000; 284268-279.

[488] Schulze V. Stability of Surface Changes Induced By Mechanical Surface Treatments. In: *ICSP-9*. Paris, France: 2005.

[489] Altenberger I, Noster U, Scholtes B. High Temperature Fatigue of Mechanically Surface Treated Materials. In: *ICSP-8*. 2002. p. 483-489.

[490] Zhuang W, Wicks B. Mechanical Surface Treatment Technologies for Gas Turbine Engine Components. *Journal of Engineering for Gas Turbines and Power*. 2003; 125(4):1021-1025.

[491] Yonghe Y. The Effect of Shot Peening on Strain-Controlled Fatigue Behaviours. In: *ICSP-4*. Tokyo, Japan: 1990. p. 411-418.

[492] Bignonnet A. Fatigue Strength of Shot-Peened Grade 35 NCD 16 Steel. Variation of Residual Stresses introduced by Shot Peening according to Type of Loading. In: *ICSP-3*. Garmisch-Partenkirchen, Germany : 1987. p. 659-666.

[493] Holzapfel H, Schulze V, Macherauch E, Vohringer O. Relaxation Behavior of Shot Peening Induced Residual Stresses in AISI 4140 due to Quasistatic Uniaxial Loading at Elevated Temperatures. In: *ICSP-6*. San Francisco, USA: 1996. p. 385-396.

[494] Capello E, Davoli P, Filippini M, Foletti S. Relaxation of Residual Stresses Induced by Turning and Shot Peening on Steels. *Journal of Strain Analysis for Engineering Design*. 2004; 39(3):285-290.

[495] Zhang K, Xu X, Wang N. Residual Stress Concentration and its Effect on Notch Fatigue Strength. In: *ICSP-3*. Garmisch-Partenkirchen, Germany : 1987. p. 625-630.

[496] Desvignes M, Gentil B, Khabou T, Castex L. Fatigue Behaviour of Shot Peened Steel. In: *ICSP-3*. Garmisch-Partenkirchen, Germany : 1987. p. 369-376.

[497] Berns H, Weber L. Fatigue Progress in Shot-Peened Surface Layers. In: *ICSP-3*. Garmisch-Partenkirchen, Germany : 1987. p. 647-654.

[498] Bergstrom J, Ericsson T. Relaxation of Shot Peening Induced Comprssive Stress During Fatigue of Notched Steel Samples. In: *ICSP-2*. Chicago: 1984. p. 241-248.

[499] Bignonnet A, Lieurade HP, Picouet L, Castex L. Shot Peening of Welded Joints and Residual Stress Relaxation during Fatigue. In: *ICSP-3*. Garmisch-Partenkirchen, Germany : 1987. p. 585-594.

[500] Hasegawa N, Watanabe Y, Kato Y. Effect of Shot Peening on Fatigue Strength of Carbon Steel at Elevated Temperature. In: *ICSP-5*. Oxford, UK: 1993. p. 134-139.

[501] Hoffmann J, Scholtes B, Vohringer O, Macherauch E. Thermal Relaxation of Shot Peening Residual Stresses in the Differently Heat Treated Plain Carbon Steel Ck 45. In: *ICSP-3*. Garmisch-Partenkirchen, Germany : 1987. p. 239-246.

[502] Cammett J, Sauer C, Arnold T. Observations on Shot Peening Residual Stresses in a Cold Worked 17Cr7Ni Stainless Steel and their Redistribution by Mechanical Loading. In: *ICSP-5*. Oxford, UK: 1993. p. 282-289.

[503] Torres MAS, Voorwald HJC. An Evaluation of Shot Peening, Residual Stress and Stress Relaxation on the Fatigue Life of AISI 4340 Steel. *International Journal of Fatigue*. 2002; 24877-886.

[504] Krauß M, Scholtes B. *Thermal Fatigue of Shot Peened or Hard Turned Hot-Work Steel AISI H11*. Garmisch-Partenkirchen, Germany: 2002. p. 324-330.

[505] Bergstrom J, Ericsson T. X-Ray Microstructure and Residual Stress Analysis of Shot Peened Surface Layers during Fatigue Loading. In: *ICSP-3*. Garmisch-Partenkirchen, Germany : 1987. p. 221-230.

[506] Menig R, Schulze V, Vöhringer O. Residual Stress Relaxation and Fatigue Strength of AISI 4140 under Torsional Loading after Conventional Shot Peening, Stress Peening and Warm Peening. In: *ICSP-8*. Garmisch-Partenkirchen, Germany: 2002. p. 317-323.

[507] Wick A, Schulze V, Vohringer O. Influence of the Shot Peening Temperature on the Relaxation Behaviour of Residual Stresses during Cyclic Bending. In: *ICSP-7*. Warsaw, Poland: 1999. p. 110-116.

[508] Menig R, Schulze V, Vöhringer O. Influence of Optimized Warm Peening on Residual Stress Stabilityand Fatigue Strength of AISI 4140 in Different Material States. In: *ICSP-8*. Garmisch-Partenkirchen, Germany: 2002. p. 311-316.

[509] Iida K, Taniguchi K. Relaxation of Residual Stress Distribution Produced by Shot Peening Under Fatigue Test. In: *ICSP-6*. San Francisco, USA: 1996. p. 397-402.

[510] Schulze V, Vohringer O, Macherauch E. Thermal Relaxation of Shot Peening Induced Residual Stress in a Quenched and Tempered Steel - 42CrMo4. In: *ICSP-5*. Oxford, UK: 1993. p. 264-273.

[511] Holzapfel H, Schulze V, Macherauch E, Vohringer O. Stability and Relaxation Behaviour of Shot Peening Induced Residual Stresses in AIS14140 During Bending Fatigue. In: *ICSP-6*. San Francisco, USA: 1996. p. 413-423.

[512] Hanagarth H, Vohringer O, Macherauch E. Relaxation of Shot Peening Residual Stresses of the Steel 42 CrMo 4 by Tensile or Compressive Deformation. In: *ICSP-4*. Tokyo, Japan: 1990. p. 327-336.

[513] Eleiche AM, Megahed MM, Abd-Allah NM. The Shot-peening Effect on the Hcf Behavior of High-strength Martensitic Steels. *Journal of Materials Processing Technology*. 2001; 113(1-3):502-508.

[514] Hirsch T, Bohringer O, Macherauch E. Bending Fatigue Behavior of Differently Heat Treated and Shot Peened AlCu5Mg2. In: *ICSP-2*. Chicago: 1984. p. 90-101.

[515] Bonnafe JP, Maeder G, Bathias C. Study of Fatigue Damage in Shot Peened Aluminium Alloys using Acoustic Emission and Macro/ Microscopic X-Ray Diffraction Investigations. In: *ICSP-3*. Garmisch-Partenkirchen, Germany : 1987. p. 485-498.

[516] Roth M, Wortman C. Relaxation of Shot Peening Residual Stresses in the 7050-T7451 Aluminium Alloy after Heat Cycles for Adhesive Bonding. In: *ICSP-8*. Garmisch-Partenkirchen, Germany: 2002. p. 208-214.

[517] Boyce B, Chen X, Peters J, Hutchinson J, Ritchie R. Mechanical Relaxation of Localized Residual Stresses Associated with Foreign Object Damage. *Materials Science and Technology*. 2003; A34948-58.

[518] Lee H, Mall S, Sathish S. Investigation into Effects of Re-shot-peening on Fretting Fatigue Behavior of Ti-6Al-4V. *Materials Science and Engineering A*. 2005; 390(1-2):227-232.

[519] Kato Y, Takafuji S, Kiriyama M. Effect of Shot Peening on Fatigue Strength of Ti-6Al-4V Alloy at Elevated Temperatures. *Zairyo/Journal of the Society of Materials Science, Japan*. 1996; 45(1):43-47.

[520] Berger M, Gregory J. Residual Stress Relaxation in Shot Peened Timetal 21s. *Materials Science and Engineering A*. 1999; 263200-204.

[521] Ofsthun, M. When Fatigue Quality Enhancers Do Not Enhance Fatigue Quality. *International Journal of Fatigue*. 2003; 251223-1228.

[522] Cammett J, Prevéy P, Jayaraman N. The Effect of Shot Peening Coverage on Residual Stress, Cold Work and Fatigue in a Nickel-base Superalloy. In: *ICSP-9*. Paris, France: 2005.

[523] Buchanan D, John R, Ashbaugh N. Thermal Residual Stress Relaxation in Powder Metal IN100 Superalloy. *Journal of ASTM International*. 2004 May; XX(X):1-16.

[524] Masmoudi N, Castex L. The effect of Temperature and Time on the Stress Relaxation of Shot Peened Nickel Based Superalloy In100. In: *International Conference on Residual Stresses*. Nancy, France: 1989. p. 710-715.

[525] Khadhraoui M, Cao W, Castex L, Gue´dou JY. Experimental Investigations and Modelling of Relaxation Behaviour of Shot Peening Residual Stresses at High Temperature for Nickel Base Superalloys. *Materials Science and Technology*. 1997; 13(4):360-367.

[526] Cao W, Khadhraoui M, Brenier B, Guedou J, Castex L. Thermomechanical Relaxation of Residual Stresses in shot peened Nickel Base Superalloy. *Materials Science and Technology*. 1994; 10947-954.

[527] Evans A, Kim S, Shackleton J, Bruno G, Preuss M, Withers P. Relaxation of Residual Stress in Shot Peened Udimet 720li Under High Temperature Isothermal Fatigue. *International Journal of Fatigue*. 2005; 271530-1534.

[528] Kim S-, Evans A, Shackleton J, Bruno G, Preuss M, Withers PJ. Stress Relaxation of Shot-peened Udimet 720li Under Solely Elevated-temperature Exposure and Under Isothermal Fatigue. Metallurgical and Materials Transactions A: *Physical Metallurgy and Materials Science*. 2005; 36(11):3041-3053.

[529] Belassel M, Brauss M, Berkley S. Residual Stress Relaxation in Nickel Base Alloys subjected to Low Cycle Fatigue. In: *ICRS-6*. Oxford, UK: 1994. p. 144-151.

[530] Kumar D, Das S, Radhakrishnan VM. Effect of Shot Peening on Fatigue of a Ni-based Superalloy. *Scandinavian Journal of Metallurgy*. 1987; 16(6):253-256.

[531] Kirk D. Effects of Plastic Straining on Residual Stresses induced by Shot-peening. In: *ICSP-3*. Garmisch-Partenkirchen, Germany : 1987. p. 213-220.

[532] Dattoma V, De Giorgi M, Nobile R. Numerical Evaluation of Residual Stress Relaxation by Cyclic Load. *Journal of Strain Analysis*. 2004; 39(6):663-672.

[533] Wang C, Liu Q. *Predictive Models for Small Fatigue Cracks Growing Through Residual Stress Fields*. 2002 Sep;

[534] Lu J, Flavenot J. Use of a Finite Element Method for the Prediction of the Shot-Peened Residual Stress Relaxation during Fatigue. In: *ICSP-3*. Garmisch-Partenkirchen, Germany : 1987. p. 391-398.

[535] Meguid SA, Shagal G, Stranart JC, Liew KM, Ong LS. Relaxation of Peening Residual Stresses Due to Cyclic Thermo-Mechanical Overload. *Journal of Engineering Materials and Technology*. 2005; 127170-178.

[536] Du J-, Zhou J-, Yang C-, Ni M-, Cao X-, Huang S. Simulation Study on Stress Field After Laser-peening and Shot-peening of Sheet Metal. *Zhongguo Jiguang/Chinese Journal of Lasers*. 2007; 34(SUPPL.):98-101.

[537] Hatamleh O, Lyons J, Forman R. Laser Peening and Shot Peening Effects on Fatigue Life and Surface Roughness of Friction Stir Welded 7075-T7351 Aluminum. *Fatigue and Fracture of Engineering Materials and Structures*. 2007; 30(2):115-130.

[538] Hatamleh O, Lyons J, Forman R. Laser and Shot Peening Effects on Fatigue Crack Growth in Friction Stir Welded 7075-T7351 Aluminum Alloy Joints. *International Journal of Fatigue*. 2007; 29(3):421-434.

[539] Rodopoulos CA, Romero JS, Curtis SA, De los Rios ER, Peyre P. Effect of Controlled Shot Peening and Laser Shock Peening on the Fatigue Performance of 2024-T351 Aluminum Alloy. *Journal of Materials Engineering and Performance*. 2003; 12(4):414-419.

[540] Noordhuis J, De Hosson TM. Microstructure and Mechanical Properties of a Laser Treated Al Alloy. *Acta metallurgica et materialia*. 1993; 41(7):1989-1998.

[541] Dehossn JTM, Noordhuis J. Mechanical Properties and Microstructure of Laser Treatment Al-cu-mg Alloys. *Journal De Physique*. 1993; 3(7 pt 2):927-932.

[542] Noordhuis J, De Hosson JTM. Mechanical Properties and Microstructure of Laser Treated Al Alloys. *Materials Science Forum*. 1994; 163-6(pt 1):405-410.

[543] Noordhuis J, De Hosson JTM. Surface Modification by Means of Laser Melting Combined with Shot Peening. a Novel Approach. *Acta metallurgica et materialia*. 1992; 40(12):3317-3324.

[544] Shikun S, Ziwen C. Laser Shock Processing of Titanium Alloy. In: *International Technology and Innovation Conference 2006 (ITIC 2006).* 2006. p. 1749-1752.

[545] J.P. Chu, J.M. Rigsbee , G. Banas, H.E. Elsayed-Ali. Laser-shock Processing Effects on Surface Microstructure and Mechanical Properties of Low Carbon Steel. *Materials Science and Engineering A.* 1999; 260260-268.

[546] Rankin JE, Hill MR, Halpin J, Chen H-, Hackel LA, Harris F. The Effects of Process Variations on Residual Stress Induced by Laser Peening. *Materials Science Forum.* 2002; 404-40795-100.

[547] Rankin JE, Hill MR, Hackel LA. The Effects of Process Variations on Residual Stress in Laser Peened 7049 T73 Aluminum Alloy. *Materials Science and Engineering A.* 2003; 349(1-2):279-291.

[548] Banas G, Lawrence Jr. F, Namiki K, Hatano A. Shot Peening Versus Laser Shock Processing. In: *ICSP-4.* Tokyo, Japan: 1990. p. 95-104.

[549] Peyre P, Merrien P, Lieurade HP, Fabbro R, Bignonnet A. Optimization of the Residual Stresses induced by Laser-shock Surface Treatment and Fatigue Life Improvement of two Cast Aluminium Alloys. In: *ICSP-5.* Oxford, UK: 1993. p. 301-310.

[550] Peyre P, Scherpereel X, Berthe L, Carboni C, Fabbro R, Be´ranger G, et al. Surface Modifications Induced in 316l Steel by Laser Peening and Shot-peening. Influence on Pitting Corrosion Resistance. *Materials Science and Engineering A.* 2000; 280(2):294-302.

[551] Zhang P, Lindermann J, Kiefer A, Leyens, C. Mechanical Surface Treatments on the High-Strength Wrought Magnesium Alloy AZ80. In: *ICSP-9.* Paris, France: 2005.

[552] Zhang P, Lindemann J. Effect of Roller Burnishing on the High Cycle Fatigue Performance of the High-strength Wrought Magnesium Alloy AZ80. *Scripta Materialia.* 2005; 52(10):1011-1015.

[553] Drechsler A, Kiese J, Wagner L. Effects of Shot Peening and Roller-Burnishing on Fatigue Performance of Various Titanium Alloys. In: I*CSP-7.* Warsaw, Poland: 1999. p. 145-152.

[554] Marcin K, Alfred O, Lothar W. Shot peening and Roller Burnsihing to improve fatigue resistance of the (alpha+beta) Titanium alloy. In: *ICSP-8*. Garmisch-Partenkirchen, Germany: 2002. p. 461-467.

[555] Hassan AM, Momani AMS. Further Improvements in Some Properties of Shot Peened Components Using the Burnishing Process. *International Journal of Machine Tools and Manufacture.* 2000; 40(12):1775-1786.

[556] Lindemann J, Grossmann K, Raczek T. Influence of Shot Peening and Deep Rolling on High-Temperature fatigue of the Ni-Superalloy Udimet 720 Li. In: *ICSP-8*. Garmisch-Partenkirchen, Germany: 2002. p. 454-460.

[557] Muller E. Comparison of Shot (Stress) Peening and Rolling in Spring Steel. In: *ICSP-8.* Garmisch-Partenkirchen, Germany: 2002. p. 441-446.

[558] Prevéy, P. Evolution of the Residual Stresses by Stress Rolling. In: *ICSP-9*. Paris, France: 2005.

[559] Prevey PS, Ravindranath RA, Shepard M, Gabb T. Case Studies of Fatigue Life Improvement Using Low Plasticity Burnishing in Gas Turbine Engine Applications. *Journal of Engineering for Gas Turbines and Power*. 2006; 128(4):865-872.

[560] Sharma MC. Roller Pressing or Shot Peening of Fir-tree root of LPT blades of 500 MW steam turbines. In: *ICSP-8*. Garmisch-Partenkirchen, Germany: 2002. p. 490-497.

[561] Prevey PS, Jayaraman N, Ravindranath R. Introduction of Residual Stresses to Enhance Fatigue Performance in the Initial Design. In: *Proceedings of the ASME Turbo Expo 2004*. p. 231-239.

[562] H.Leitner, H.P.Ganser, W.Eichlseder. Surface Treatment by Shot Peening and Deep Rolling - Mechanisms, Modelling, Methods. *Materialpruefung/Materials Testing. 2007*; 49(7-8):408-413.

[563] Juijerm P, Altenberger I, Scholtes B. Fatigue and Residual Stress Relaxation of Deep Rolled Differently Aged Aluminium Alloy AA6110. *Materials Science and Engineering A*. 2006; 426(1-2):4-10.

[564] Wagner L. Mechanical Surface Treatments on Titanium, Aluminum and Magnesium Alloys. *Materials Science and Engineering A*. 1999; 263210-216.

[565] Nikitin I, Altenberger I. Comparison of the Fatigue Behavior and Residual Stress Stability of Laser-shock Peened and Deep Rolled Austenitic Stainless Steel AISI 304 in the Temperature Range 25-600°C. *Materials Science and Engineering A*. 2007; 465(1-2):176-182.

[566] Hull D, Bacon D. *Introduction to Dislocations*. Butterworth-Heinemann; 2005.

[567] Crump S, Burguete R, Sim W-, De Oliveira A, Van Der Veen S, Boselli J, et al. A Concurrent Approach to Manufacturing Induced Part Distortion in Aerospace Components. *Materials Science and Technology*. 2005; 475-86.

[568] Azizi R, Schulze V. Proposing an Applied Information Based Solution in Studying Shot-Peening Process. In: *ICSP-9*. Paris, France: 2005.

[569] Munjiza A, Latham J, John N. 3D Dynamics of Discrete Element Systems Comprising Irregular Discrete Elements— Integration Solution for Finite Rotations in 3D. *International Journal for Numerical Methods in Engineering*. 2003; 56(1):35-55.

[570] Williams, J. R., O'Connor, R. Discrete Element Simulation and the Contact Problem. *Archival of Computational Methods in Engineering*. 1999; 6(4):279-304.

[571] Owen D, Feng Y. Parallelised Finite/discrete Element Simulation of Multi-fracturing Solids and Discrete Systems. *Engineering Computations*. 2001; 18(3/4):557-576.

[572] Ramnath V, Wiggs G. DACE Based Probabilistic Optimization of Mechanical Components. In: *Proceedings of GT2006 Expo: Power for Land,Sea and Air*. Barcelona, Spain: 2006.

In: Computational Materials
Editor: Wilhelm U. Oster

ISBN: 978-1-60456-896-7

Chapter 3

VIBRATIONAL DYNAMICS OF BULK METALLIC GLASSES STUDIED BY PSEUDOPOTENTIAL THEORY

*Aditya M. Vora**

Humanities and Social Science Department, STBS College of Diploma Enginnering, Opp. Spinning Mill, Varachha Road, Surat 395006, South Gujarat, India

ABSTRACT

In this chapter, we discuss a vibrational dynamics of some bulk metallic glasses (BMG) in terms of the phonon eigen frequencies of the localized collective excitations using model potential formalism at room temperature for the first time. The theoretical effective atom model (EAM) with Wills-Harrison (WH) form are used to compute the interatomic pair potential and pair correlation function (PCF) for the glassy systems. The phonon dispersion curves are computed from the three approaches proposed by Hubbard-Beeby (HB), Takeno-Goda (TG) and Bhatia-Singh (BS). Various elastic and thermodynamic properties have been studied from the elastic limit of the dispersion relations. Different types of the local field correction functions are used for the first time in the present investigation to study the screening influence on the aforesaid properties. The pseudo-alloy-atom (PAA) model is applied for the first time instead of Vegard's Law.

Keywords: pseudopotential; pair potential; bulk metallic glasses (BMG); pair correlation function; phonon dispersion curves (PDC); thermodynamic properties; elastic properties.

1. INTRODUCTION

In the long history of materials science and technology, which is embellished with many a fascinating account of the discovery and application of a new ceramic, polymeric or composite materials, the recent emergence of rapidly solidified metals particularly the exotic

* Tel. : +91-2832-256424, E-mail address : voraam@yahoo.com.

non-crystalline ones called metallic glasses, as new engineering materials of tremendous promise which constitute a unique and bright new chapter. The most noteworthy aspect of the fast development of the rapid quenching of metallic melts, which many discerning engineering scientists consider to be the most important breakthrough in materials in materials technology. Such solids have electronic properties normally associated with metals but atomic arrangement is not spatially periodic. They made up of number of components of metals which provide us with physically interesting systems for theoretical investigations. For the understanding of the thermodynamic, transport and other properties of such an interesting metallic glass on a microscopic level, knowledge of their atomic structure and dynamics on the basis of interatomic forces is required. Interatomic forces can be studied either experimentally by inelastic scattering experiments if the energy change of the scattered particle can be directly related to the atomic dynamics of the system or theoretically, also, by pseudopotential calculations [1-32].

In most of the theoretical studies of vibrational dynamics of bulk metallic glasses (BMG) [26-29], the Vegard's law was used to calculate electron-ion interaction from the potential of the pure components. Also in bulk metallic glasses, the translational symmetry is broken, and therefore, the momentum (or quasi-momentum) should not be used to describe the state of the system. The virtual crystal approximation enables us to keep the concept of the momentum only in an approximate way. But, it is well established that pseudo-alloy-atom (PAA) is more meaningful approach to explain such kind of interactions in binary systems [16-28]. In the PAA approach a hypothetical monoatomic crystal is supposed to be composed of pseudo-alloy-atoms, which occupy the lattice sites and from a perfect lattice in the same way as pure metals. In this model the hypothetical crystal made up of PAA is supposed to have the same properties as the actual disordered alloy material and the pseudopotential theory is then applied to studying various properties of alloy systems. The complete miscibility in the glassy alloy systems is considered as a rare case. Therefore, in such binary systems the atomic matrix elements in the pure states are affected by the characteristics of alloys such as lattice distortion effects and charging effects. In the PAA model, such effects are involved implicitly. In addition to this it also takes into account the self-consistent treatment implicitly.

This chapter introduces pseudopotential based theory to address the problems of vibrational dynamics of bulk metallic glasses and their related elastic as well as thermodynamic properties with the method of computing the properties under investigation. Three main theoretical approaches given by Hubbard-Beeby (HB) [33], Takeno-Goda (TG) [34, 35] and Bhatia-Singh (BS) [36, 36] are used in the present study for computing the phonon frequencies of the bulk non-crystalline or glassy alloys. Five local field correction functions viz. Hartree (H) [38], Taylor (T) [39], Ichimaru-Utsumi (IU) [40], Farid *et al.* (F) [41] and Sarkar *et al.* (S) [42] are used for the first time in the present investigation to study the screening influence on the aforesaid properties of bulk metallic glasses (BMG). Besides, the thermodynamic properties such as longitudinal sound velocity υ_L, transverse sound velocity υ_T and Debye temperature θ_D, low temperature specific heat capacity C_V and some elastic properties viz. the isothermal bulk modulus B_T, modulus of rigidity G, Poisson's ratio σ and Young's modulus Y are also calculated from the elastic part of the phonon dispersion curves (PDC).

Permanent Address : Parmeshwari 165, Vijaynagar Area, Hospital Road, Bhuj – Kutch, 370 001, Gujarat, INDIA

2. Theoretical Methodology

In this section, we have addressed about the various theoretical methodology for computing the vibrational properties of some bulk metallic glasses in detail.

2.1. Interatomic Pair Potential

The fundamental ingredient, which goes into the calculation of the vibrational dynamics of bulk metallic glasses (BMG), is the interatomic pair potential [16-32]. In the present study, the interatomic pair potential is to treat in such a way that the binary system as a one component metallic fluid, i.e. the concept of effective atom [16-32]. In this concept a simple binary disordered system $fA + gB + hC + iD$ can be looked upon as an assembly of the effective atom (i.e. one component system). Interatomic interactions between these effective atoms of the alloy are then calculated in the usual way followed for a single component liquid metal instead of taking the alloy as the mixture of individual components AA, AB, AC, AD, BC, BD, BB, CD, CC and DD. This model is known as effective atom model (EAM). In the present study we have considered here all the glasses as a one component fluid for investigating the phonon frequencies and their related elastic and thermodynamic properties.

It is difficult to know the exact pair interactions in terms of interatomic pair potential through simple pseudopotential theories particularly for non-simple transition metal based binary metallic glasses. The reason is that the linear response method used to express the cohesive energy of a simple metal as a sum of a volume term and of pair and higher order interaction cannot be used for transition metal d-bands. This has been achieved by using a simplified tight-binding description and a momentum decomposition of the d-electron density of states that result in d-bonding pair interaction proportional to the width of the d-band and a repulsive interaction arising from the shift of the centre of gravity of the d-band. These pair interaction add to the simple metal like interatomic pair potential mediated by the s-electrons, the total potential being much stronger than in simple metals. For non-simple transition metal based metallic glasses, in the present study Wills-Harrison (WH) [43] approach is used to generate the interatomic pair potential.

The interatomic pair potential in Wills-Harrison (WH) form is written as [43]

$$V(r) = V_S(r) + V_b(r) + V_r(r). \tag{1}$$

The s-electron contribution to the interatomic pair potential $V_S(r)$ is calculated from [16-28],

$$V_s(r) = \left(\frac{Z_S^{\,2} e^2}{r}\right) + \frac{\Omega_O}{\pi^2} \int F(q) \left[\frac{Sin(qr)}{qr}\right] q^2 \, dq. \tag{2}$$

Here, by integrating the partial s-density of states resulting from self-consistent band structure calculation, a value of $Z_S \sim 1.5$ is obtained for the entire 3d and 4d series [16-28, 43], while Ω_O the atomic volume of the one component fluid.

The energy wave number characteristics appearing in the equation (2) is written as [16-28]

$$F(q) = \frac{-\Omega_O\, q^2}{16\pi} \left| W_B(q) \right|^2 \frac{[\varepsilon_H(q) - 1]}{\{1 + [\varepsilon_H(q) - 1][1 - f(q)]\}}. \tag{3}$$

Here, $W_B(q)$ is the bare ion potential $\varepsilon_H(q)$ the modified Hartree dielectric function, which is written as [38]

$$\varepsilon(q) = 1 + (\varepsilon_H(q) - 1)(1 - f(q)). \tag{4}$$

While, $\varepsilon_H(X)$ is the static Hartree dielectric function and the expression of it is given by [38],

$$\varepsilon_H(q) = 1 + \frac{me^2}{2\pi k_F \hbar^2 \eta^2} \left(\frac{1-\eta^2}{2\eta} \ln \left| \frac{1+\eta}{1-\eta} \right| + 1 \right); \eta = \frac{q}{2k_F} \tag{5}$$

here $m, e, \hbar$ are the electronic mass, the electronic charge, the Plank's constant, respectively and $k_F = \left(3\pi^2 Z/\Omega_O\right)^{1/2}$ is the Fermi wave vector, in which Z the valence. While $f(q)$ is the local field correction function. In the present investigation, the local field correction functions due to Hartree (H) [38], Taylor (T) [39], Ichimaru-Utsumi (IU) [40], Farid *et al.* (F) [41] and Sarkar *et al.* (S) [42] are incorporated to see the impact of exchange and correlation effects on the aforesaid properties. The details of all the local field corrections are below.

The Hartree (H) screening function [38] is purely static, and it does not include the exchange and correlation effects. The expression of it is,

$$f(q) = 0. \tag{6}$$

Taylor (T) [39] has introduced an analytical expression for the local field correction function, which satisfies the compressibility sum rule exactly. This is the most commonly used local field correction function and covers the overall features of the various local field correction functions proposed before 1972. According to Taylor (T) [39],

$$f(q) = \frac{q^2}{4k_F^2} \left[1 + \frac{0.1534}{\pi k_F^2} \right]. \tag{7}$$

The Ichimaru-Utsumi (IU) local field correction function [40] is a fitting formula for the dielectric screening function of the degenerate electron liquids at metallic and lower densities, which accurately reproduces the Monte-Carlo results as well as it also, satisfies the self consistency condition in the compressibility sum rule and short range correlations. The fitting formula is

$$f(q)=A_{IU}\left(\frac{q}{k_F}\right)^4+B_{IU}\left(\frac{q}{k_F}\right)^2+C_{IU}+\left[A_{IU}\left(\frac{q}{k_F}\right)^4+\left(B_{IU}+\frac{8A_{IU}}{3}\right)\left(\frac{q}{k_F}\right)^2-C_{IU}\right]\left\{\frac{4-\left(\frac{q}{k_F}\right)^2}{4\left(\frac{q}{k_F}\right)}\ln\left|\frac{2+\left(\frac{q}{k_F}\right)}{2-\left(\frac{q}{k_F}\right)}\right|\right\}. \tag{8}$$

On the basis of Ichimaru-Utsumi (IU) local field correction function [40] local field correction function, Farid *et al.* (F) [41] have given a local field correction function of the form

$$f(q)=A_F\left(\frac{q}{k_F}\right)^4+B_F\left(\frac{q}{k_F}\right)^2+C_F+\left[A_F\left(\frac{q}{k_F}\right)^4+D_F\left(\frac{q}{k_F}\right)^2-C_F\right]\left\{\frac{4-\left(\frac{q}{k_F}\right)^2}{4\left(\frac{q}{k_F}\right)}\ln\left|\frac{2+\left(\frac{q}{k_F}\right)}{2-\left(\frac{q}{k_F}\right)}\right|\right\}. \tag{9}$$

Based on equations (8-9), Sarkar *et al.* (S) [42] have proposed a simple form of local field correction function, which is of the form

$$f(q)=A_S\left\{1-\left(1+B_S\left(\frac{q}{k_F}\right)^4\right)\exp\left(-C_S\left(\frac{q}{k_F}\right)^2\right)\right\}. \tag{10}$$

The parameters $A_{IU}, B_{IU}, C_{IU}, A_F, B_F, C_F, D_F, A_S, B_S$ and C_S are the atomic volume dependent parameters of IU-, F- and S-local field correction functions. The mathematical expressions of these parameters are narrated in the respective papers of the local field correction functions [40-42].

The well recognized model potential $W_B(r)$ [16-28] (in r-space) used in the present computation is of the form,

$$W(r)=\begin{cases}\dfrac{-Ze^2}{r_C^3}\left[2-\exp\left(1-\dfrac{r}{r_C}\right)\right]r^2\ ; & r\le r_C\\[2ex] \dfrac{-Ze^2}{r}\ ; & r\ge r_C\end{cases}. \tag{11}$$

The model potential $W_B(q)$ (in q-space) used in the present computation is of the form [16-28]

$$W_B(q)=\frac{-4\pi e^2 Z}{\Omega_O q^2}\left[\begin{array}{l}\left\{\begin{array}{l}-1+\frac{12}{U^2}+\frac{U^2}{1+U^2}+\frac{6U^2}{(1+U^2)^2}+\frac{18U^2}{(1+U^2)^3}-\frac{6U^4}{(1+U^2)^3}\\+\frac{24U^2}{(1+U^2)^4}-\frac{24U^4}{(1+U^2)^4}\end{array}\right\}\cos(U)\\ \qquad + \\ \left\{\begin{array}{l}\frac{6}{U}-\frac{12}{U^3}+\frac{U}{1+U^2}+\frac{3U}{(1+U^2)^2}-\frac{3U^3}{(1+U^2)^2}+\frac{6U}{(1+U^2)^3}\\-\frac{18U^3}{(1+U^2)^3}+\frac{6U}{(1+U^2)^4}-\frac{36U^3}{(1+U^2)^4}+\frac{6U^5}{(1+U^2)^4}\end{array}\right\}\sin(U)\\ \qquad + \\ 24U^2\exp(1)\left\{\frac{U^2-1}{(1+U^2)^4}\right\}\end{array}\right]. \quad (12)$$

This form has feature of a Coulombic term out side the core and varying cancellation due to a repulsive and an attractive contribution to the potential within the core. Hence it is assumed that the potential within the core should not be zero nor constant but it should very as a function of r. Thus, the model potential has the novel feature of representing varying cancellation of potential within the core over and above its continuity at $r=r_C$ and weak nature [16-28]. Here r_C is the parameter of the model potential of bulk metallic glasses (BMG). The model potential parameter r_C is calculated from the well known formula [16-28] as follows :

$$r_C=\left[\frac{0.51\,r_S}{(Z)^{1/3}}\right]. \quad (13)$$

Here r_S is the Wigner-Seitz radius of the bulk metallic glasses (BMG).

The d-electron contributions to the pair potential are expressed in terms of the number of d-electron Z_d, the d-state radii r_d and the nearest-neighbour coordination number N as follows:

$$V_b(r)=-Z_d\left(1-\frac{Z_d}{10}\right)\left(\frac{12}{N}\right)^{1/2}\left(\frac{28.06}{\pi}\right)\frac{2r_d^{\,3}}{r^5}, \quad (14)$$

and

$$V_r(r) = Z_d\left(\frac{450}{\pi^2}\right)\frac{r_d^6}{r^8}. \tag{15}$$

Here, $V_b(r)$ takes into account the Friedel-model band broadening contribution to the transition metal cohesion and $V_r(r)$ arises from the repulsion of the d-electron muffin-tin orbital on different sites due to their non-orthogonality. Wills-Harrison (WH) [43] have studied the effect of the s- and d-bands. The parameters Z_d, Z_S, r_d and N can be calculated by the following expressions,

$$Z_d = fZ_{dA} + g\,Z_{dB} + hZ_{dC} + iZ_{dD}, \tag{16}$$

$$Z_S = fZ_{SA} + g\,Z_{SB} + hZ_{SC} + iZ_{SD}, \tag{17}$$

$$r_d = fr_{dA} + g\,r_{dB} + hr_{dC} + ir_{dD}, \tag{18}$$

and

$$N = fN_A + g\,N_B + hN_C + iN_D, \tag{19}$$

Where A, B, C and D are denoted the first, second, third and forth pure metallic components of the bulk metallic glasses (BMG) while f, g, h and i the concentration factor of the first, second, third and forth metallic components. Z_d, Z_S and r_d are determined from the band structure data of the pure component available in the literature [38]. The values used in the present study are listed in table 1.

Table 1. Input parameters and constants used in the present computation

Bulk Metallic Glasses	Z	Z_S	Z_d	Ω_0 $(au)^3$	M (amu)	N	ρ_M (gm/cm^3)	r_C (au)	r_d (au)
$Pd_{77.5}Si_{16.5}Cu_6$	3.05	1.50	7.16	103.84	90.91	10.02	9.8073	0.7071	1.14
$Fe_{80}B_{14}Si_6$	3.06	1.50	5.20	77.52	47.88	6.40	6.9190	0.6400	1.21
$Fe_{80}B_{10}Si_{10}$	3.10	1.50	5.20	80.93	48.57	6.40	6.7230	0.6437	1.21
$Fe_{40}Ni_{40}B_{20}$	3.40	1.50	6.00	70.83	47.98	8.00	7.5884	0.5789	1.07
$Ni_{80}B_{10}Si_{20}$	3.90	1.50	6.80	77.77	50.85	9.60	7.3246	0.5450	1.54
$Fe_{60}Ni_{20}B_{10}Si_{10}$	4.30	1.65	6.80	91.19	53.66	9.60	6.5919	0.5385	1.07

2.2. Pair Correlation Function

A quantity which is equally important as the pair potential while studying a disorder system is the pair correlation function (PCF) $g(r)$. It provides the statistical description of the structure of the system under investigation. The complete information of the precise position and momentum of each particle at each instant of time is contained in this function. The function $g(r)$ can be obtained either experimentally by X-ray diffraction and neutron diffraction technique [44, 45] or computed theoretically from the interatomic pair potentials [43]. Instead of using experimentally available $g(r)$, here the pair correlation function for all bulk metallic glasses (BMG) are generated from presently obtained interatomic pair potentials. The function $g(r)$ is presently calculated using the expression [46],

$$g(r) = \exp\left[\frac{-V(r)}{k_B T}\right] - 1. \tag{20}$$

Here k_B is the Boltzmann's constant and T the room temperature of the system under investigation.

2.3. Phonon Dispersion Curves (PDC)

Much progress has also been made recently in lattice dynamics of disordered crystal lattices, impure crystals and mixed crystals or alloys. Inspite of their wide spread excitations in nature, however the vibrational properties of amorphous solids or glassy solids belonging to a different class of disordered solids have received little attention among theoretical solid state physicists. The neutron scattering data from short wavelength phonons in amorphous solids are easily available. Further it has been shown that there is a remarkable similarity between the scattered neutron spectra of amorphous or poly-crystalline solids near melting temperatures. There are attempts to calculate phonon frequencies of binary metallic glasses with various approaches and model potentials [16-32].

In the present study, three main approaches are used to generate the phonon dispersion curves (PDC) of bulk metallic glasses (BMG): (1) Hubbard-Beeby (HB) [33], (2) Takeno-Goda (TG) [34, 35] and (3) Bhatia-Singh (BS) [36, 37]. The brief description of each approach is given in this section.

2.3.1. Hubbard-Beeby (HB) Approach

Hubbard and Beeby (HB) [33] have given the theory in 1969. The theory is based upon a new approximation which may be regarded as either a generalization of the random phase approximation (RPA) or alternatively as a generalizations of the phonon theory of solids. The theory is applied to predict the form of the inelastic coherent neutron scattering from liquids making use of particular model. According to this theory, the liquid differs from the crystalline solid in two principle ways: (a) firstly, the atoms in the liquids do not have periodicity, which they are disordered and (b) secondly, the motions of atoms in the liquids are free than in the solid.

Therefore, the present phonon theory can be given in two steps : (1) a theory of phonons in a disordered but a stationary system of two atoms, i.e., a cold amorphous solid is developed by adopting the phonon problem to discuss the propagation of electrons in a disordered system. The virtue of this approach is that it becomes exact when specialized to the case of a crystalline solid, showing that it has adequately included correlation effects and (2) this stationary theory is then generalized in a straight forward manner for the motion of the disordered atoms.

With the physical argument that the products of the static pair correlation function $g(r)$ and the second derivative of interatomic pair potential $V''(r)$ is peaked near some radius σ, known as the hard sphere diameter, the expressions for longitudinal phonon frequency ω_L and transverse phonon frequency ω_T can be written according to Hubbard and Beeby (HB) as [33],

$$\omega_L^2(q) = \omega_E^2\left[1 - \frac{\sin(q\sigma)}{q\sigma} - \frac{6\cos(q\sigma)}{(q\sigma)^2} + \frac{6\sin(q\sigma)}{(q\sigma)^3}\right], \tag{21}$$

$$\omega_T^2(q) = \omega_E^2\left[1 - \frac{3\cos(q\sigma)}{(q\sigma)^2} + \frac{3\sin(q\sigma)}{(q\sigma)^3}\right], \tag{22}$$

with

$$\omega_E^2 = \left(\frac{4\pi\rho}{3M}\right)\int_0^\infty g(r)V''(r)r^2 dr, \tag{23}$$

is the maximum phonon frequency. M is the atomic mass and ρ the e number density while $V''(r)$ is the second derivative of the interatomic pair potential.

2.3.2. Takeno-Goda (TG) Approach

Takeno and Goda (TG) [34, 35] have proposed the theory in 1971. They have adopted a theory of phonons of amorphous solids which is successfully applied to high frequency phonons i.e. collective modes in simple liquids. The study of collective modes leads to the computation of eigen frequencies of the longitudinal and transverse systems.

Since, many physical quantities are calculated systematically through averaged Green's function. Therefore, in this method, configurationally averaged Green function $\langle G\rangle$ is first calculated instead of the averaged equation of motion. The set of equations is solved by a decoupling procedure. This leads us to a self-consistent treatment for phonon eigen values because the phonon eigen frequencies are determined by effective force constant, which depends upon the correlation function for the displacement of atoms in solids whereas the correlation function of displacement itself depends on the phonon frequencies.

For the computation of frequency spectrum, conditional averaging procedure (to take into account the atomic vibrations to all orders of magnitude) is applied. Green functions leading to a set of hierarchy equations, which are designed to be written in terms of many body correlation functions of atoms in amorphous solids. Hierarchy is truncated employing the lowest order decoupling approximation, called the quasi-crystalline approximation, to give the following secular equation for the phonon dispersion [34, 35].

$$\det\left|\omega^2\,\delta(\alpha\beta)-D_{\alpha\beta}(q)\right|=0. \tag{24}$$

The dynamical matrix $D_{\alpha\beta}(q)$ is written as

$$D_{\alpha\beta}(q)=\left(\frac{\rho}{M}\right)\int dr\, g(r)\nabla_{\alpha}\nabla_{\beta}\langle V(r)\rangle\{1-\exp(-i\,q\cdot r)\}, \tag{25}$$

where ρ is the number density of the ion, M the mass, $g(r)$ the pair correlation function and $\langle V(r)\rangle$ the same interatomic pair potential between the ions, respectively.

For the evaluation of the integral in equation (25) the Cartesian coordinate system with Z-axis in the direction of wave vector q is used and performing the volume integral in spherical polar coordinates, we have

$$D_{ZZ}(q)=\omega_L^2(q)=\left(\frac{4\pi\rho_{eff}}{M_{eff}}\right)\int_0^\infty dr\, g(r)\left[\left\{rV'(r)\left(1-\frac{\sin(qr)}{qr}\right)\right\}+\right.$$
$$\left.\left\{r^2V''(r)-rV'(r)\right\}\left(\frac{1}{3}-\frac{\sin(qr)}{qr}-\frac{2\cos(qr)}{(qr)^2}+\frac{2\sin(qr)}{(qr)^3}\right)\right], \tag{26}$$

and

$$D_{XX}(q)=D_{YY}(q)=\omega_T^2(q)=\left(\frac{4\pi\rho_{eff}}{M_{eff}}\right)\int_0^\infty dr\, g(r)\left[\left\{rV'(r)\left(1-\frac{\sin(qr)}{qr}\right)\right\}+\right.$$
$$\left.\left\{r^2V''(r)-rV'(r)\right\}\left(\frac{1}{3}+\frac{2\cos(qr)}{(qr)^2}+\frac{2\sin(qr)}{(qr)^3}\right)\right]. \tag{27}$$

As we use spherically symmetric interatomic pair potential, all non-diagonal terms vanish.

2.3.3. Bhatia-Singh (BS) Approach

This model of Bhatia and Singh (BS) [36, 37] is a very simple and an adhoc model to generate the PDC of metallic glass. This model is an extension of the model of cubic metals. In this approach, they have considered that a two component metallic glass is virtually a binary metallic alloys. And also this model retained the interatomic interactions in metallic glass effective between the first nearest neighbours giving rise to two model parameters, β and δ. The electron-ion interaction was considered to be same as that used in metals. Such an interaction resulted in another model parameter k_e, the bulk modulus of the electron gas.

The model is based on following two assumptions: (1) the ions interact with a central pair wise potential $V(r)$, which is effective between the nearest neighbours only and (2) the force on an ion due to volume-dependent energies in the metal (kinetic and exchange energies of the conduction electrons, the ground-state energy of the electron, etc.) could be calculated using the Thomas-Fermi method.

Under these assumptions the equations determining $\omega - q$ relations may be written as [36, 37]

$$\rho_a \omega_L^2(q) = \frac{2N}{q^2}(\beta I_0 + \delta I_2) + \frac{k_e k_{TF}^2 q^2 |G(qr_S)|^2}{q^2 + k_{TF}^2 \varepsilon(q)}, \tag{28}$$

and

$$\rho_a \omega_T^2(q) = \frac{2N}{q^2}\left(\beta I_0 + \frac{1}{2}\delta(I_0 - I_2)\right). \tag{29}$$

With

$$\beta = \frac{\rho_a a^2}{2M}\left[\frac{1}{r}\frac{dV(r)}{dr}\right]_{r=a}, \tag{30}$$

$$\delta = \frac{\rho_a a^3}{2M}\left[\frac{d}{dr}\left(\frac{1}{r}\frac{dV(r)}{dr}\right)\right]_{r=a}, \tag{31}$$

The expressions of I_0 and I_2 are, with $x = qa$,

$$I_0 = 1 - \frac{\sin(x)}{x}, \tag{32}$$

and

$$I_2 = \frac{1}{3} - \sin(x)\left[\frac{1}{x} - \frac{2}{x^3}\right] - \frac{2\cos(x)}{x^2}. \tag{33}$$

Where $\rho_a = n_i M$, is the atomic density in which n_i the ion-number density, $V(r)$ the interatomic pair potential, a the nearest neighbours distance of the glassy system.

In equation (28), $|G(qr_S)|^2$ is defined as the shape factor and is considered in order to take into account the cancellation effects which occur between kinetic and potential energies inside the core of the ions making the effective potential weak in the core. Then

$$|G(qr_S)|^2 = \left[\frac{3\{\sin(qr_S) - (qr_S)\cos(qr_S)\}}{(qr_S)^3}\right]^2. \tag{34}$$

Where $r_S = [3/4\pi n_i]^{1/3}$ being the Wigner-Seitz radius of the glassy system. We note that $G(qr_S) \to 1$ for $q \to 0$ and dwindles to zero for large q. Further, to compensate for the fact that electrons are not entirely free, the factor k_e is defined as [36, 37]

$$k_e = \frac{4\pi n_e n_i Z e^2}{k_{TF}^2}. \tag{35}$$

Here k_e is the bulk modulus of the electron gas, in which $n_e = n_i Z$ the electron-number density. k_{TF}^2 is the Thomas – Fermi screening length, given by $k_{TF}^2 = \frac{4k_F}{\pi}$.

According to Bhatia and Singh (BS) [36], they have not contained $\varepsilon(q)$ term in numerator of equation (28). For $q \to 0$, $\varepsilon(q) = 1$. Thus this term has no influence for $q \to 0$ but for higher q, $\varepsilon(q)$ drops rapidly and has a great Influence in the numerical calculation. Recently the approach of Bhatia and Singh (BS) [36] is modified by Shukla and Campanha [37], they have used $\varepsilon(q)$ in numerator of equation (28). Due to this modification, equation (28) can be written as

$$\rho_a \omega_L^2(q) = \frac{2N}{q^2}(\beta I_0 + \delta I_2) + \frac{k_e k_{TF}^2 q^2 \varepsilon(q) |G(qr_S)|^2}{q^2 + k_{TF}^2 \varepsilon(q)}, \tag{36}$$

This modified equation is used in the present computations. The parameters β, δ and k_e are calculated from the equations (28, 29 and 36).

2.4. Thermodynamic and Elastic Properties

In the case of metallic glasses, it is convenient to discuss density fluctuations in q-space and obtain macroscopic properties by taking $q \to 0$ limit (in low frequency region). The introduction of the elastic model is a good example of the same, because the stress can be written most conveniently in q-space. Elastic behaviour of the system is in general given by the response of it to the propagation of the density fluctuations in the wavelength limit. Therefore, the dispersion relations prove to be useful in deriving the elastic as well as thermodynamic properties of bulk metallic glasses (BMG).

The present study include some elastic properties such as the isothermal bulk modulus B_T, modulus of rigidity G, Poisson's ratio σ and Young's modulus Y and some thermodynamic properties such as longitudinal sound velocity υ_L, transverse sound velocity υ_T and Debye temperature θ_D for the bulk metallic glasses (BMG).

In the long wavelength limit of the frequency spectrum, both the frequencies i.e. transverse and longitudinal are proportional to the wave vectors and obey the relationships [16-32],

$$\omega_L \propto q \text{ and } \omega_T \propto q,$$

$$\omega_L = \upsilon_L q \text{ and } \omega_T = \upsilon_T q. \tag{37}$$

Where υ_L and υ_T are the longitudinal and transverse sound velocities in the glass, respectively.

For HB approach the formulations for υ_L and υ_T are given by [33]

$$\upsilon_L(HB) = \omega_E \sqrt{\frac{3\sigma^2}{10}}, \tag{38}$$

and

$$\upsilon_T(HB) = \omega_E \sqrt{\frac{\sigma^2}{10}}. \tag{39}$$

Where ω_E is calculated from the equation (23).

In TG approach the expressions for υ_L and υ_T are written by [34, 35]

$$\upsilon_L(TG) = \left[\left(\frac{4\pi\rho}{30\,M}\right) \int_0^\infty dr\; g(r)\, r^3 \left\{ r V''(r) - 4V'(r) \right\} \right]^{1/2}, \tag{40}$$

and

$$\upsilon_T\,(TG) = \left[\left(\frac{4\pi\rho}{30\,M}\right)\int_0^\infty dr\; g(r)\; r^3\;\{3rV''(r) - 4V'(r)\}\right]^{1/2}. \tag{41}$$

The formulations for υ_L and υ_T in BS approach are as follows [36, 37],

$$\upsilon_L\,(BS) = \left[\frac{N}{\rho_a}\left(\frac{1}{3}\beta + \frac{1}{5}\delta\right) + \frac{k_e}{3}\right]^{1/2}, \tag{42}$$

and

$$\upsilon_T\,(BS) = \left[\frac{N}{\rho_a}\left(\frac{1}{3}\beta + \frac{1}{15}\delta\right)\right]^{1/2}, \tag{43}$$

The isothermal bulk modulus B_T, modulus of rigidity G, Poisson's ratio σ and Young's modulus Y are found using the expressions [16-32],

$$B_T = \rho_M\left(\upsilon_L^2 - \frac{4}{3}\upsilon_T^2\right), \tag{44}$$

$$G = \rho_M\,\upsilon_T^2\,. \tag{45}$$

With ρ_M is the isotropic number density of the bulk metallic glasses (BMG).

$$\sigma = \frac{1 - 2\left(\dfrac{\upsilon_T^2}{\upsilon_L^2}\right)}{2 - 2\left(\dfrac{\upsilon_T^2}{\upsilon_L^2}\right)}, \tag{46}$$

and

$$Y = 2G(\sigma + 1). \tag{47}$$

The Debye temperature is given in terms of both the velocities as [16-28],

$$\theta_D = \frac{\hbar\omega_D}{k_B} = \frac{\hbar}{k_B}2\pi\left[\frac{9\rho_{eff}}{4\pi}\right]^{1/3}\left[\frac{1}{\upsilon_L^3}+\frac{2}{\upsilon_T^3}\right]^{(-1/3)}, \tag{48}$$

here k_B the Boltzmann's constant and ω_D the Debye frequency, respectively.

2.5. Low Temperature Heat Capacity (C_V)

Now a day it is firmly established that for all amorphous solids, the universal temperature behaviour of vibrational contributions to the heat capacity (C_V) differs essentially from that of crystal. In the thermodynamic limit ($N_O \to 0$, $\Omega_0 \to 0$ and $N_O/\Omega_0 =$ constant, with N_O is the number of atoms in the unit cell), one can obtain [47]

$$C_V = \frac{\Omega_0\hbar^2}{k_B T^2}\sum_\lambda \int \frac{d^3q}{(2\pi)^3}\frac{\omega_\lambda^2(q)}{\left[\exp\left(\frac{\hbar\omega_\lambda(q)}{k_B T}\right)-1\right]\left[1-\exp\left(-\frac{\hbar\omega_\lambda(q)}{k_B T}\right)\right]}. \tag{49}$$

Where, T is the temperature of the system, respectively. The basic features of temperature dependence of C_V are determined by the behavior of $\omega_L(q)$ and $\omega_T(q)$.

3. Method of Computation

In the present study on vibrational dynamics of bulk metallic glasses (BMG), first we have to compute the interatomic pair potential of the bulk metallic glasses (BMG) using equation (1). The integrations, in equation (1), for $V(r)$ is carried out using Simpson's $1/3$ rule with step size $10^{-3}\,k_F$ over the range $0 \le q \le 20\,k_F$. Hence, any artificial cut off is avoided in generating the interatomic pair potential of glassy system. Using this interatomic pair potential $V(r)$, the pair correlation function $g(r)$ is generated. In computing PDC, using HB, TG and BS approaches, we need $V'(r)$ and $V''(r)$. In computing ω_L and ω_T, the integration over r is carried out $10^{-1}\,r$. For investigating screening influences on the aforesaid properties, we have discussed the results of HB, TG and BS approaches with various local field correction functions. Thus generated ω_L and ω_T are used to study the thermodynamic and elastic properties of glasses through equations (44) to (49). The law temperature specific heat capacity is generated using ω_L and ω_T. The summation λ is over the longitudinal and transverse branches. The schematic diagram of performed computations of vibrational dynamics of bulk metallic glasses (BMG) is shown in figure 1. The necessary programs to study all these vibrational dynamics of bulk metallic glasses (BMG) are prepared indigenously in the laboratory. While preparing the programs for specific applications, enough care has been taken to ensure maximum accuracy, stability and portability.

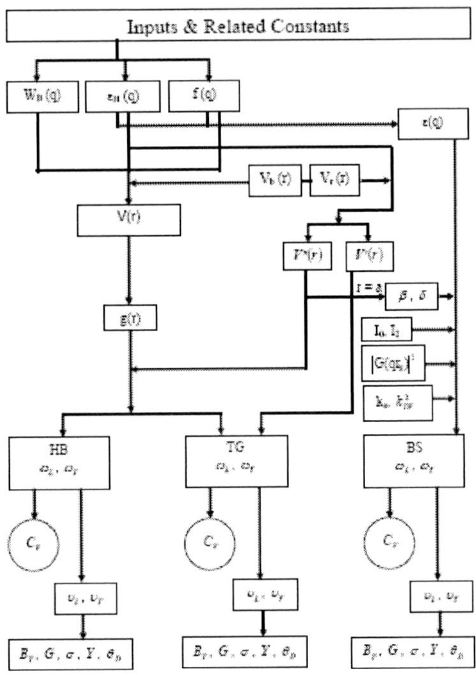

Figure 1. Schematic diagram of the performed computations of vibrational dynamics of bulk metallic glasses.

4. Results and Discussion

After performing the numerical treatment to the bulk metallic glasses, the results which are generated in the present work are discussed and narrated in this section with the references (wherever found in the literature and relevant to present study) for each glass. Here attempt is made to prepare discussion for each glass which contains the graphical representation of interatomic pair potential, pair correlation functions, phonon dispersion curves (PDC), the specific heat C_V, table of thermodynamic and elastic properties and related discussion. When other comparisons of the results are not available for most of the bulk metallic glasses, we have avoided putting here any remarks about present findings.

4.1. $Pd_{77.5}Si_{16.5}Cu_6$ Bulk Metallic Glass

The $Pd_{77.5}Si_{16.5}Cu_6$ is the most important candidate of bulk metallic glasses. The presently computed interatomic pair potentials of this glass are shown in figure 2. It is seen that the inclusion of exchange and correlation effects in the static H-dielectric screening changes the nature of the interatomic pair potential, significantly. The first zero of the interatomic pair potential $V(r = r_0)$ due to all local field correction functions occurs at $r_0 = 3.3$ au. The interatomic pair potential well width and its minimum position $V_{min}(r)$ are also affected by the nature of the screening. The maximum depth in the interatomic pair potential is obtained for S-function. The present results do not show oscillatory behaviour and potential energy remains negative and constant in the large r - region. Thus, the Coulomb repulsive potential part dominates the oscillations due to ion-electron-ion interactions, which show the waving shape of the interatomic pair potential after $r = 10.0$ au. Hence, the interatomic pair potentials converge towards a finite value instead of zero in attractive region.

The pair correlation function (PCF) $g(r)$ computed theoretically through the interatomic pair potential is shown in figure 3. It is found that, the peak positions due to S-function show higher while those due to H-function show lower. The screening effect is also observed in the nature of the PCF. The ratio (r_2/r_1) of the position of the second peak (r_2) to that of the first peak (r_1) is found 1.37, 1.55, 1.52, 1.52 and 1.60 for H-, T-, IU-, F- and S-function, respectively. While the ratio (r_3/r_1) of the position of the third peak (r_3) to that of the first peak (r_1) i.e. the ratio of the third atomic shell radius to the nearest-neighbour distance is found 1.89, 2.13, 2.09, 2.09 and 2.21 for H-, T-, IU-, F- and S-function, respectively. The (r_2/r_1) ratio is close to the c/a ratio in close-packed hexagonal structure i.e. c/a = 1.63, which suggests that the short range order of nearest neighbours is influenced more or less by the atomic arrangement of the crystalline structure. This result is typical of a metallic glass with a large main peak at the nearest-nearest distance followed by smaller peaks corresponding to more distant neighbours. The computed pair correlation function using various local field correction functions are overlapped with each other after $r = 9.4$ au. Therefore, it is very difficult to draw the remarks regarding the disorder is invisible after $r = 9.4$ au in the figure 3, because the experimental data is not available of this glass. Actually, this long range order is normal and it may be due to the waving shape of the interatomic pair potential.

The PDC of $Pd_{77.5}Si_{16.5}Cu_6$ glass has been theoretically investigated by Agarwal *et al.* [29, 30] and by Aziz-Ray [31] using BS approach assuming the force among nearest neighbours as central and volume dependent. The experimental data of PDC of this glass is not available in the literature. Therefore, we have reported vibrational properties of this glass using pseudopotential theory for the first time. The phonon eigen frequencies for longitudinal and transverse phonon modes calculated using HB, TG and BS approaches with the five screening functions are shown in figures 4-6. It can be seen from figures 3-5 that the inclusion of exchange and correlation effect enhances the phonon frequencies in both longitudinal as well as transverse branches. The present results of PDC due to T-, IU- and F-function are lying between those due to H- and S-screening. The first minimum in the longitudinal branch of HB approach is found around at $q \approx 3.2Å^{-1}$ for H-, $q \approx 3.2Å^{-1}$ for T-, IU-, F-function and S-function. While, the first minimum in the longitudinal branch of TG approach is found around at $q \approx 2.4Å^{-1}$ for H-, $q \approx 1.8Å^{-1}$ for T-, $q \approx 3.4Å^{-1}$ for IU- as well as F-function and $q \approx 1.9Å^{-1}$ for S-function. The first minimum in the longitudinal branch of BS approach is

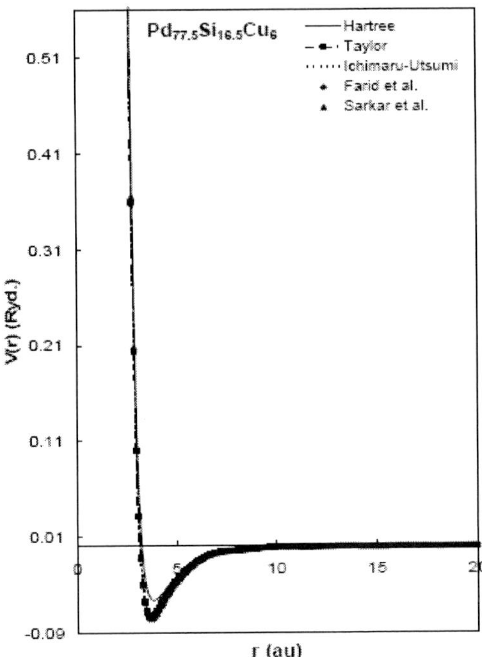

Figure 2. Dependence on screening on pair potentials of $Pd_{77.5}Si_{16.5}Cu_6$ bulk metallic glass.

found around at $q \approx 1.5\text{Å}^{-1}$ for H-, T-, IU-, F- and S-function. Characteristically, the dispersion relations show a minimum near q_p, the wave vector where the static structure factor $S(\mathbf{q})$ of the glass has its first maximum. The first maximum in the longitudinal branch ω_L of HB approach is found around at $q \approx 1.6\text{Å}^{-1}$ for H-, T-, IU- as well as F-function and $q \approx 1.7\text{Å}^{-1}$ for S-function. While, the first maximum in the longitudinal branch of TG approach is found around at $q \approx 2.0\text{Å}^{-1}$ for H-, $q \approx 0.7\text{Å}^{-1}$ for T-, $q \approx 1.1\text{Å}^{-1}$ for IU- as well as F-function and $q \approx 0.8\text{Å}^{-1}$ for S-function. The first maximum in the longitudinal branch of BS approach is found around at $q \approx 0.7\text{Å}^{-1}$ for most of the local field correction functions.

It is also observed from the figures 3-5 that, the oscillations are more prominent in the longitudinal phonon modes as compared to the transverse modes in all three approaches. This shows the existence of collective excitations at larger momentum transfer due to longitudinal phonons only and the instability of the transverse phonons due to the anharmonicity of the atomic vibrations in the metallic systems. Moreover, the present outcomes of both the phonon modes due to HB as well as TG approaches are more enhanced than the BS approach. The first crossover position of ω_L and ω_T in the HB and BS approaches is observed at $q \approx 2.1\text{Å}^{-1}$ and $q \approx 1.5\text{Å}^{-1}$ for most of the local field correction functions, respectively. While, the first crossover position of ω_L and ω_T in the TG approach is observed at $q \approx 2.0\text{Å}^{-1}$ for H-, $q \approx 1.8\text{Å}^{-1}$ for T-, $q \approx 2.5\text{Å}^{-1}$ for IU- as well as F-function and $q \approx 1.4\text{Å}^{-1}$ for S-function.

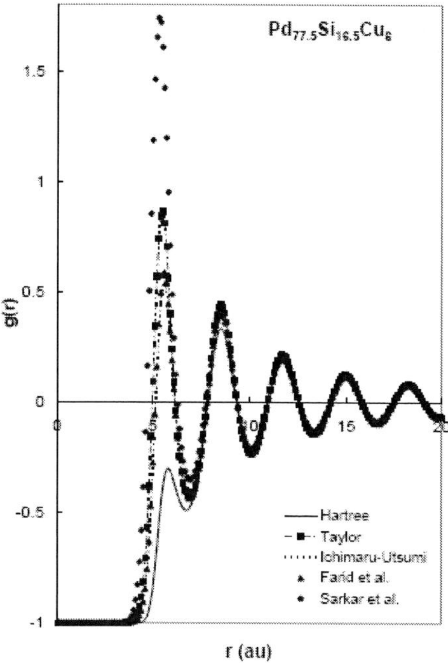

Figure 3. Dependence on screening on pair correlation function of $Pd_{77.5}Si_{16.5}Cu_6$ bulk metallic glass.

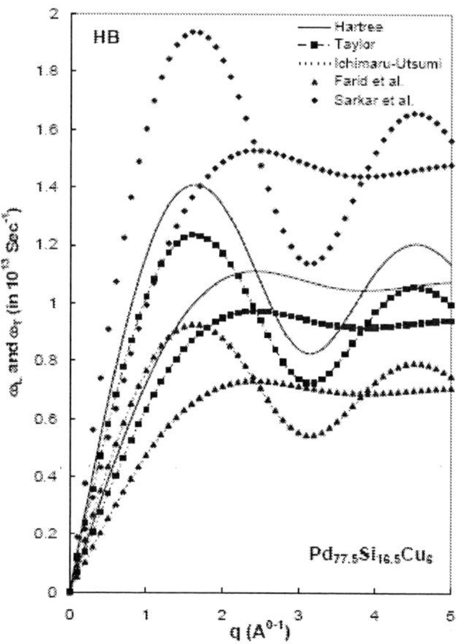

Figure 4. Dependence on screening on phonon dispersion curves of $Pd_{77.5}Si_{16.5}Cu_6$ bulk metallic glass using HB approach.

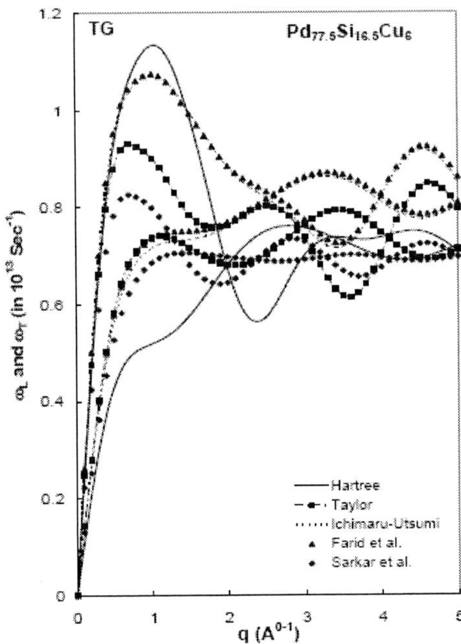

Figure 5. Dependence on screening on phonon dispersion curves of $Pd_{77.5}Si_{16.5}Cu_6$ bulk metallic glass using TG approach.

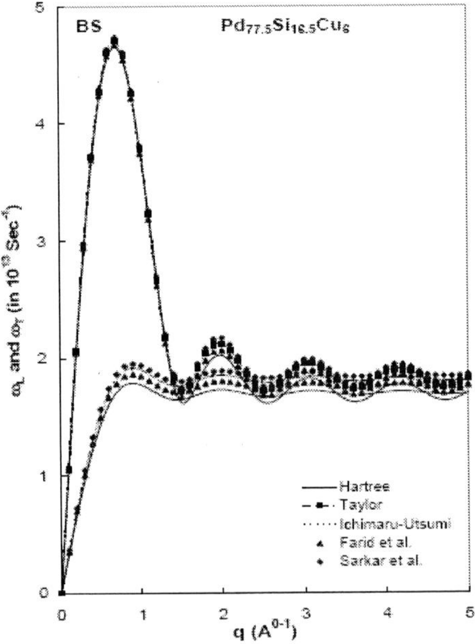

Figure 6. Dependence on screening on phonon dispersion curves of $Pd_{77.5}Si_{16.5}Cu_6$ bulk metallic glass using BS approach.

As shown in figures 7-9, the exchange and correlation functions also affect the anomalous behaviour (i.e. deviation from the T^3 law) which is observed in the specific heat (C_V). The reason behind the anomalous behaviour may be due to the low frequency modes modify the generalized vibrational density of states of the glass with that of the polycrystal. These modes are mainly responsible for the difference in the temperature dependence of the specific heat which departs from the normal behaviour. At low temperature region high bump is observed in HB, TG and BS approaches.

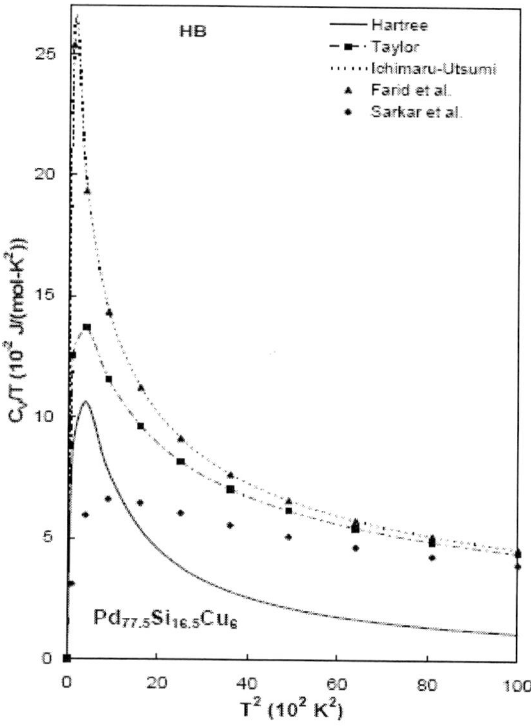

Figure 7. Dependence on screening on low temperature specific heat of $Pd_{77.5}Si_{16.5}Cu_6$ bulk metallic glass using HB approach.

Furthermore, the thermodynamic and elastic properties estimated from the elastic part of the PDC are tabulated in table 2. Among the five screening functions, the results of υ_L and υ_T are influenced more due to S-function. The comparison with other such theoretical results [26] favours the present calculation and suggests that proper choice of dielectric screening is important part for explaining the thermodynamic and elastic properties of $Pd_{77.5}Si_{16.5}Cu_6$ glass.

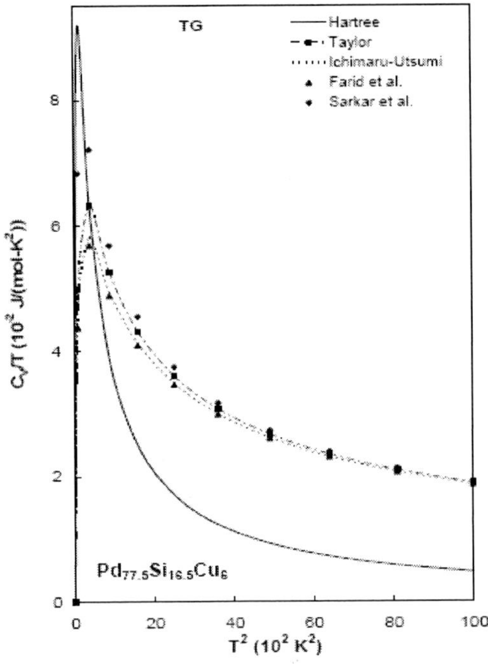

Figure 8. Dependence on screening on low temperature specific heat of $Pd_{77.5}Si_{16.5}Cu_6$ bulk metallic glass using TG approach.

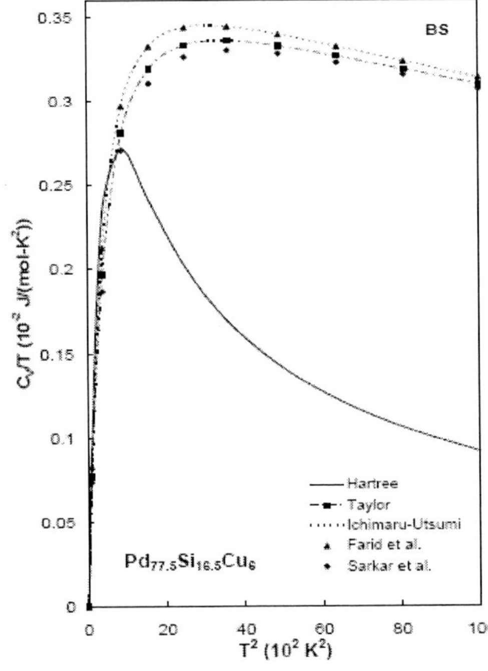

Figure 9. Dependence on screening on low temperature specific heat of $Pd_{77.5}Si_{16.5}Cu_6$ bulk metallic glass using BS approach.

4.2. $Fe_{80}B_{14}Si_6$ Bulk Metallic Glass

The $Fe_{80}B_{14}Si_6$ metallic glass is the most important candidate of transition metal-metalloid group. In our literature survey we have not found any reports on the experimental as well as theoretical work based on pseudopotential theory related to PDC of this glass. But, the structural properties of this glass have been studied by Sulir *et al.* [44]. Therefore, the vibrational properties of this glass are reported for the first time.

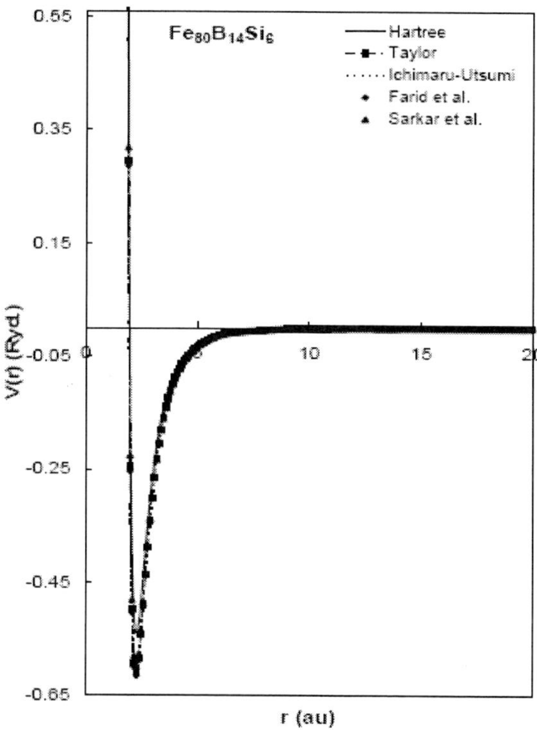

Figure 10. Dependence on screening on pair potentials of $Fe_{80}B_{14}Si_6$ bulk metallic glass.

The presently computed interatomic pair potentials of this glass are shown in figure 10. It is seen that the inclusion of exchange and correlation effects in the static H-dielectric screening changes the nature of the interatomic pair potential, significantly. The first zero of the interatomic pair potential $V(r = r_0)$ due to all local field correction functions occurs at $r_0 = 2.3$ au. The interatomic pair potential well width and its minimum position $V_{min}(r)$ are also affected by the nature of the screening. The maximum depth in the interatomic pair potential is obtained for F-function, while minimum is for H-screening function. The present results do not show oscillatory behaviour and potential energy remains negative in the large r - region. Thus, the Coulomb repulsive potential part dominates the oscillations due to ion-electron-ion interactions, which show the waving shape of the interatomic pair potential after $r = 7.0$ au. Hence, the interatomic pair potentials converge towards a finite value instead of zero in attractive region.

Table 2. Thermodynamic and Elastic properties of $Pd_{77.5}Si_{16.5}Cu_6$ bulk metallic glass

App.	SCR	υ_L x 10^5 cm/s	υ_T x 10^5 cm/s	B_T x 10^{11} dyne/cm^2	G x 10^{11} dyne/cm^2	σ	Y x 10^{11} dyne/cm^2	θ_D (K)
HB	H	1.38	0.80	1.04	0.62	0.25	1.56	105.90
	T	1.21	0.70	0.80	0.48	0.25	1.20	92.87
	IU	0.90	0.52	0.44	0.27	0.25	0.66	69.17
	F	0.91	0.52	0.45	0.27	0.25	0.67	69.75
	S	1.90	1.10	1.97	1.18	0.25	2.95	145.77
TG	H	2.36	1.11	3.82	1.21	0.36	3.29	149.92
	T	2.53	1.46	3.51	2.09	0.25	5.23	194.12
	IU	2.59	1.42	3.94	1.96	0.29	5.05	189.00
	F	2.63	1.45	4.03	2.06	0.28	5.28	193.41
	S	2.26	1.31	2.77	1.69	0.25	4.21	174.38
BS	H	5.80	1.89	28.35	3.52	0.44	10.14	258.25
	T	5.89	2.03	28.70	4.03	0.43	11.54	275.94
	IU	5.85	1.97	28.47	3.82	0.44	10.97	268.90
	F	5.85	1.98	28.45	3.83	0.44	11.00	269.24
	S	5.93	2.06	28.89	4.16	0.43	11.91	280.40
Others [26]		-	-	7.62, 9.06, 13.06, 13.66, 17.26, 18.30	3.39, 3.40	-	8.88, 9.06, 9.38, 9.41, 9.57, 9.60	303.60, 305.69, 309.44, 309.78, 311.67, 312.05

The pair correlation function (PCF) $g(r)$ computed theoretically through the interatomic pair potential is shown in figure 11. It is found that, the peak positions due to S-function show higher while those due to H-function show lower. The screening effect is also observed in the nature of the PCF. The ratio (r_2/r_1) of the position of the second peak (r_2) to that of the first peak (r_1) is found 1.28, 1.51, 1.51, 1.51 and 1.57 for H-, T-, IU-, F- and S-function, respectively. While the ratio (r_3/r_1) of the position of the third peak (r_3) to that of the first peak (r_1) i.e. the ratio of the third atomic shell radius to the nearest-neighbour distance is found 1.77, 2.08, 2.08, 2.08 and 2.16 for H-, T-, IU-, F- and S-function, respectively. The (r_2/r_1) ratio is close 1.63 characteristic for the disordered closed pack crystallographic structures and therefore this component of the second peak might eventually be due to not complete amorphousation of the samples [44]. This suggests that the short range order of nearest neighbours is influenced more or less by the atomic arrangement of the crystalline structure. This result is typical of a metallic glass with a large main peak at the nearest-nearest distance followed by smaller peaks corresponding to more distant neighbours. The computed pair correlation function using various local field correction functions are overlapped with each other after $r = 8.5$ au. Therefore, it is very difficult to draw the remarks regarding the disorder is invisible after $r = 8.5$ au in the figure 11. Actually, this long range order is normal and it may be due to the waving shape of the interatomic pair potential.

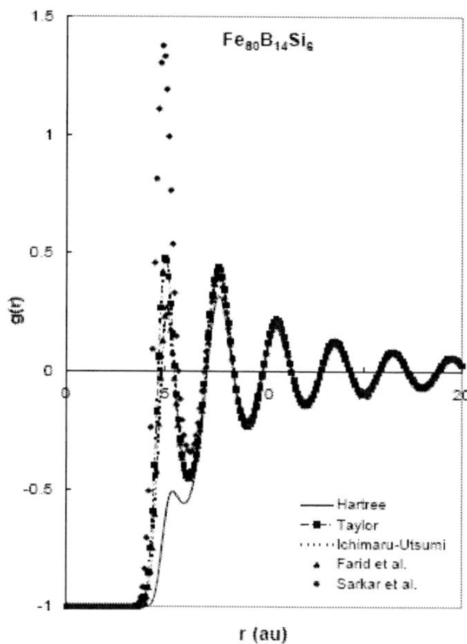

Figure 11. Dependence on screening on pair correlation function of $Fe_{80}B_{14}Si_6$ bulk metallic glass.

The phonon eigen frequencies for longitudinal and transverse phonon modes calculated using HB, TG and BS approaches with the five screening functions are shown in figures 12-14. It can be seen that the inclusion of exchange and correlation effect enhances the phonon frequencies in both longitudinal as well as transverse branches. The present results of the PDC due to H-, T- and IU-function are lying between those due to F- and S-screening in HB approach. While, the computed outcomes of the PDC due to T-, IU- and F-function are lying between those due to H- and S-screening in TG approach. Also in BS approach, the present results of the PDC due to T-, IU- and F-function are lying between those due to H- and S-screening. The first minimum in the longitudinal branch of HB approach is found around at $q \approx 3.5\text{Å}^{-1}$ for H-, T-, IU-, F- and S-function. While, the first minimum in the longitudinal branch of TG approach is found around at $q \approx 2.9\text{Å}^{-1}$ for H-, $q \approx 3.7\text{Å}^{-1}$ for T-, $q \approx 3.2\text{Å}^{-1}$ for IU- as well as F-function and $q \approx 2.2\text{Å}^{-1}$ for S-function. The first minimum in the longitudinal branch of BS approach is found around at $q \approx 1.5\text{Å}^{-1}$ for H-, T-, IU-, F- and S-function. The first maximum in the longitudinal branch ω_L of HB approach is found around at $q \approx 1.8\text{Å}^{-1}$ for H-, T-, IU-, F- and S-function. While, the first maximum in the longitudinal branch of TG approach is found around at $q \approx 1.4\text{Å}^{-1}$ for H-, T-, IU- as well as F-function and $q \approx 0.9\text{Å}^{-1}$ for S-function. The first maximum in the longitudinal branch of BS approach is found around at $q \approx 0.7\text{Å}^{-1}$ for most of the local field correction functions.

It is also observed from the figures 12-14 that, the oscillations are more prominent in the longitudinal phonon modes as compared to the transverse modes in all three approaches. This shows the existence of collective excitations at larger momentum transfer due to longitudinal phonons only and the instability of the transverse phonons due to the anharmonicity of the atomic vibrations in the metallic systems. Moreover, the present outcomes of both the phonon modes due to HB as well as TG approaches are more enhanced than the BS approach. The first crossover position of ω_L and ω_T in the HB and BS approaches is observed at $q \approx 2.7$Å^{-1} and $q \approx 1.5$Å^{-1} for most of the local field correction functions, respectively. While, the first crossover position of ω_L and ω_T in the TG approach is observed at $q \approx 2.2$Å^{-1} for H-, $q \approx 2.6$Å^{-1} for T-, $q \approx 2.5$Å^{-1} for IU- as well as F-function and $q \approx 1.7$Å^{-1} for S-function.

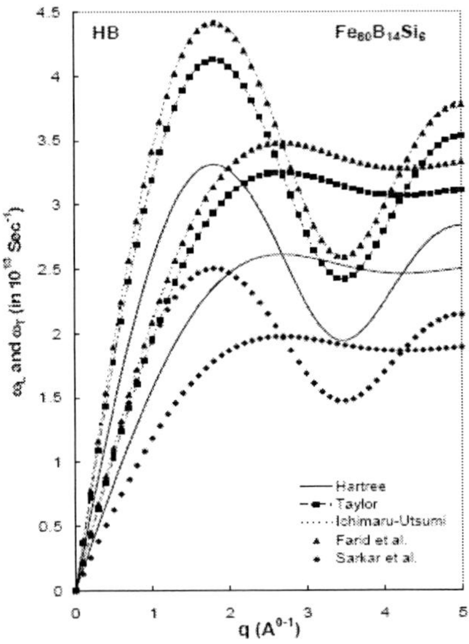

Figure 12. Dependence on screening on phonon dispersion curves of $Fe_{80}B_{14}Si_6$ bulk metallic glass using HB approach.

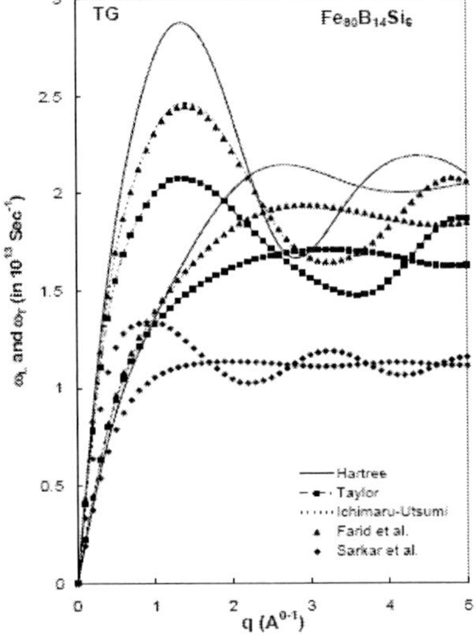

Figure 13. Dependence on screening on phonon dispersion curves of $Fe_{80}B_{14}Si_6$ bulk metallic glass using TG approach.

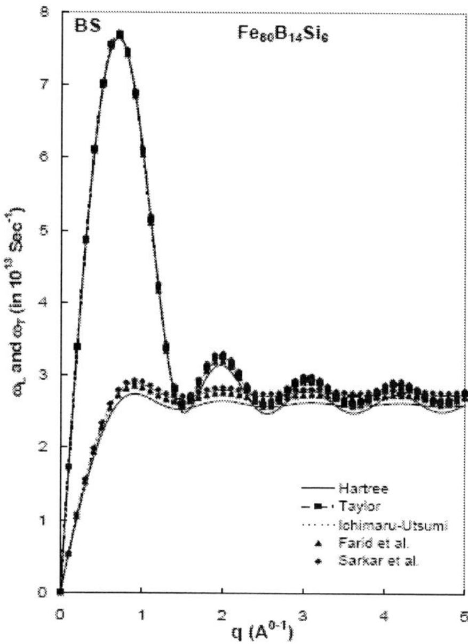

Figure 14. Dependence on screening on phonon dispersion curves of $Fe_{80}B_{14}Si_6$ bulk metallic glass using BS approach.

The $C_V/T \rightarrow T^2$ relation is shown in figures 15-17. For all the three approaches as temperature increases, the initial rise in specific heat is observed at low temperature region and then it decreases. This observation is deviated for S-screening in HB and TG approaches, while those for IU-function in BS, where very high bump in C_V/T for low temperature is absent.

It is noticed from table 3 that the υ_L and υ_T for HB and TG approaches are influenced significantly by various exchange and correlation functions as compare for BS approach. However, very high compressibility is observed for BS approach. As υ_L and υ_T are depend on the nature of screening as well as the method adopted, the other thermodynamic and elastic properties are also reflecting the same behaviour.

4.3. $Fe_{80}B_{10}Si_{10}$ Bulk Metallic Glass

The $Fe_{80}B_{10}Si_{10}$ bulk metallic glass is the most important candidate of transition metal-metalloid group. In our literature survey we have not found any reports on the experimental as well as theoretical work based on pseudopotential theory related to PDC of this glass. But, the structural properties of this glass have been studied by Sulir *et al.* [44]. Therefore, the vibrational properties of this glass are reported for the first time.

The presently computed interatomic pair potentials of this glass are shown in figure 18. It is seen that the inclusion of exchange and correlation effects in the static H-dielectric screening changes the nature of the interatomic pair potential, significantly.

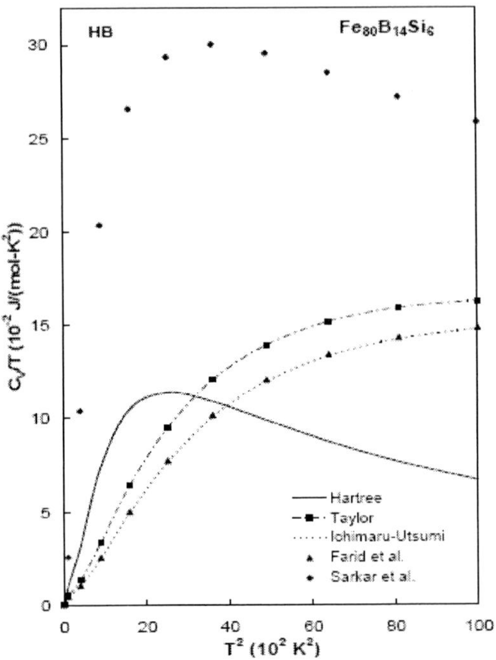

Figure 15. Dependence on screening on low temperature specific heat of $Fe_{80}B_{14}Si_6$ bulk metallic glass using HB approach.

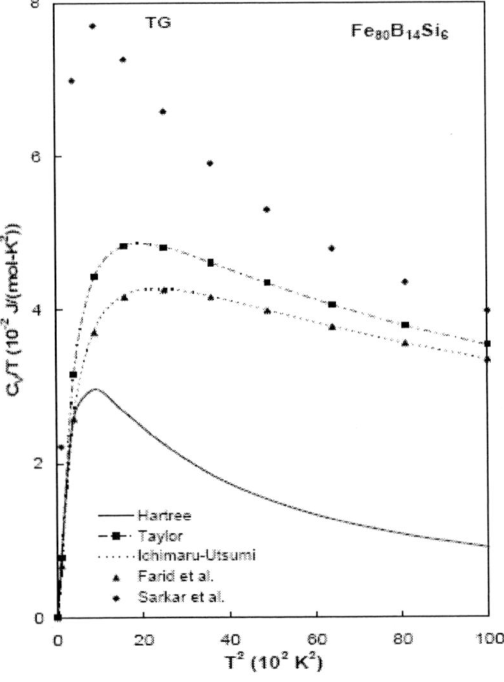

Figure 16. Dependence on screening on low temperature specific heat of $Fe_{80}B_{14}Si_6$ bulk metallic glass using TG approach.

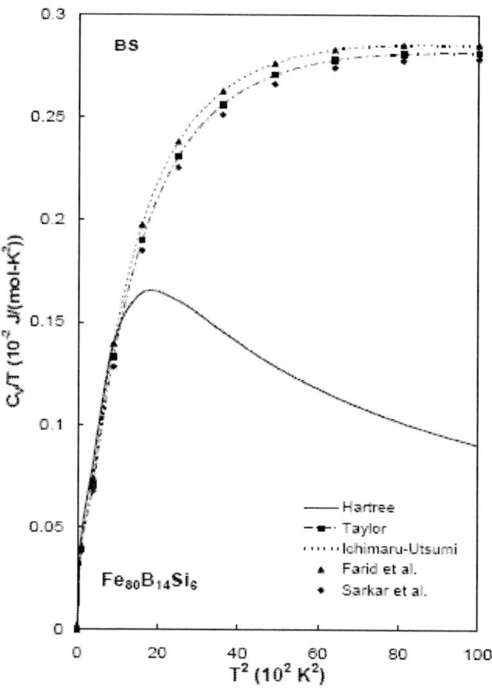

Figure 17. Dependence on screening on low temperature specific heat of $Fe_{80}B_{14}Si_6$ bulk metallic glass using BS approach.

Table 3. Thermodynamic and Elastic properties of $Fe_{80}B_{14}Si_6$bulk metallic glass

App.	SCR	υ_L x 10^5 cm/s	υ_T x 10^5 cm/s	B_T x 10^{11} dyne/cm^2	G x 10^{11} dyne/cm^2	σ	Y x 10^{11} dyne/cm^2	θ_D (K)
HB	H	2.95	1.70	3.33	2.00	0.25	5.00	249.22
	T	3.67	2.12	5.17	3.10	0.25	7.76	310.35
	IU	3.93	2.27	5.95	3.57	0.25	8.93	332.92
	F	3.92	2.26	5.92	3.55	0.25	8.87	331.94
	S	2.23	1.29	1.91	1.14	0.25	2.86	188.46
TG	H	4.28	2.05	8.78	2.92	0.35	7.88	304.79
	T	4.09	2.27	6.79	3.57	0.28	9.11	333.93
	IU	4.26	2.28	7.79	3.58	0.30	9.32	335.62
	F	4.31	2.31	7.90	3.70	0.30	9.60	340.92
	S	3.39	1.94	4.46	2.62	0.25	6.56	285.10
BS	H	8.72	2.61	46.29	4.70	0.45	13.63	392.00
	T	8.81	2.75	46.69	5.24	0.45	15.16	413.86
	IU	8.78	2.71	46.52	5.08	0.45	14.72	407.65
	F	8.78	2.71	46.50	5.09	0.45	14.73	407.85
	S	8.83	2.78	46.84	5.36	0.44	15.49	418.50

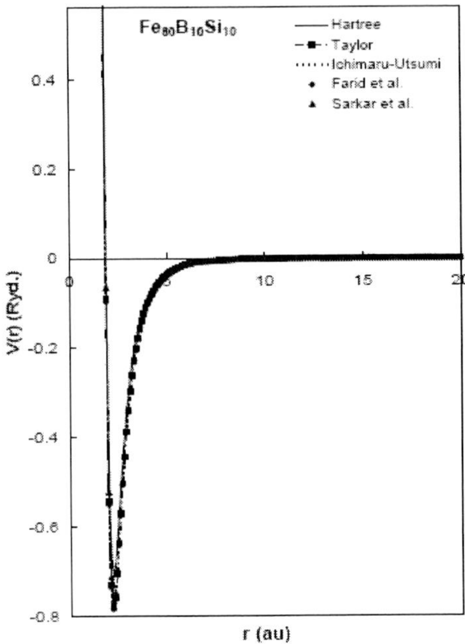

Figure 18. Dependence on screening on pair potentials of $Fe_{80}B_{10}Si_{10}$ bulk metallic glass.

The first zero of the interatomic pair potential $V(r = r_0)$ due to all local field correction functions occurs at $r_0 = 2.2$ au. The interatomic pair potential well width and its minimum position $V_{\min}(r)$ are also affected by the nature of the screening. The maximum depth in the interatomic pair potential is obtained for F-function. The present results do not show oscillatory behaviour and potential energy remains negative in the large r - region. Thus, the Coulomb repulsive potential part dominates the oscillations due to ion-electron-ion interactions, which show the waving shape of the interatomic pair potential after $r = 5.7$ au. Hence, the interatomic pair potentials converge towards a finite value instead of zero in attractive region.

The pair correlation function (PCF) $g(r)$ computed theoretically through the interatomic pair potential is shown in figure 19. It is found that, the peak positions due to S-function show higher while those due to H-function show lower. The screening effect is also observed in the nature of the PCF. The ratio (r_2/r_1) of the position of the second peak (r_2) to that of the first peak (r_1) is found 1.56, 1.53, 1.50, 1.53 and 1.59 for H-, T-, IU-, F- and S-function, respectively.

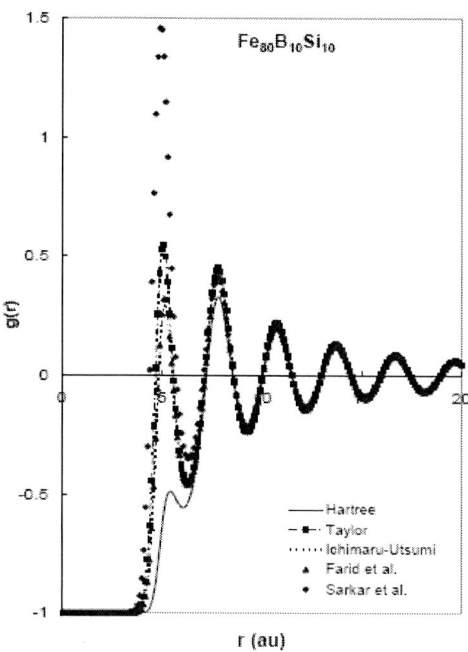

Figure 19. Dependence on screening on pair correlation function of $Fe_{80}B_{10}Si_{10}$ bulk metallic glass.

While the ratio (r_3/r_1) of the position of the third peak (r_3) to that of the first peak (r_1) i.e. the ratio of the third atomic shell radius to the nearest-neighbour distance is found 2.14, 2.10, 2.06, 2.10 and 2.18 for H-, T-, IU-, F- and S-function, respectively. The (r_2/r_1) ratio is close 1.63 characteristic for the disordered closed pack crystallographic structures and therefore this component of the second peak might eventually be due to not complete amorphousation of the samples [44]. This suggests that the short range order of nearest neighbours is influenced more or less by the atomic arrangement of the crystalline structure. This result is typical of a metallic glass with a large main peak at the nearest-nearest distance followed by smaller peaks corresponding to more distant neighbours. The computed pair correlation function using various local field correction functions are overlapped with each other after $r = 8.5$ au. Therefore, it is very difficult to draw the remarks regarding the disorder is invisible after $r = 8.5$ au in the figure 19. In fact, this long range order is normal and it may be due to the waving shape of the interatomic pair potential.

The phonon eigen frequencies for longitudinal and transverse phonon modes calculated using HB, TG and BS approaches with the five screening functions are shown in figures 20-22. It can be seen from figures 20-21 that the inclusion of exchange and correlation effect enhances the phonon frequencies in both longitudinal as well as transverse branches. The present results of the PDC due to H-, T- and F-function are lying between those due to IU- and S-screening in HB approach. While, the computed outcomes of the PDC due to T-, IU- and F-function are lying between those due to H- and S-screening in TG approach. Also in BS approach, the present results of the PDC due to T-, IU- and F-function are lying between those due to H- and S-screening. The first minimum in the longitudinal branch of HB approach is found around at $q \approx 3.5 Å^{-1}$ for H-, T-, IU-, F- and S-function. While, the first minimum in the longitudinal branch of TG approach is found around at $q \approx 2.9Å^{-1}$ for H-, $q \approx 3.6Å^{-1}$ for T-, $q \approx 3.3Å^{-1}$ for IU- as well as F-function and $q \approx 2.2Å^{-1}$ for S-function. The first minimum in the longitudinal branch of BS approach is found around at $q \approx 1.5Å^{-1}$ for H-, T-, IU-, F- and S-function. The first maximum in the longitudinal branch ω_L of HB approach is found around at $q \approx 1.8Å^{-1}$ for H-, T-, IU-, F- and S-function. While, the first maximum in the longitudinal branch of TG approach is found around at $q \approx 1.4Å^{-1}$ for H-, T-, IU- as well as F-function and $q \approx 0.9Å^{-1}$ for S-function. The first maximum in the longitudinal branch of BS approach is found around at $q \approx 0.7Å^{-1}$ for most of the local field correction functions.

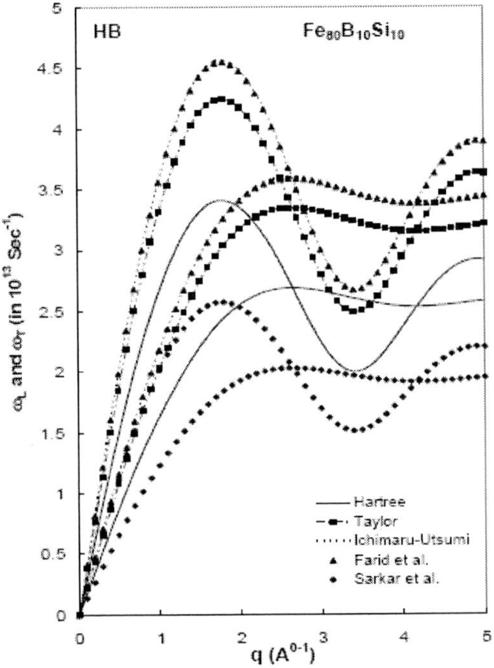

Figure 20. Dependence on screening on phonon dispersion curves of $Fe_{80}B_{10}Si_{10}$ bulk metallic glass using HB approach.

It is also observed from the figures 20-22 that, the oscillations are more prominent in the longitudinal phonon modes as compared to the transverse modes in all three approaches. This shows the existence of collective excitations at larger momentum transfer due to longitudinal phonons only and the instability of the transverse phonons due to the anharmonicity of the atomic vibrations in the metallic systems. Moreover, the present outcomes of both the phonon modes due to HB as well as TG approaches are more enhanced than the BS approach. The first crossover position of ω_L and ω_T in the HB and BS approaches is observed at $q \approx 2.7Å^{-1}$ and $q \approx 1.5Å^{-1}$ for most of the local field correction functions, respectively. While, the first crossover position of ω_L and ω_T in the TG approach is observed at $q \approx 2.3Å^{-1}$ for H-, T-, IU- as well as F-function and $q \approx 1.7Å^{-1}$ for S-function.

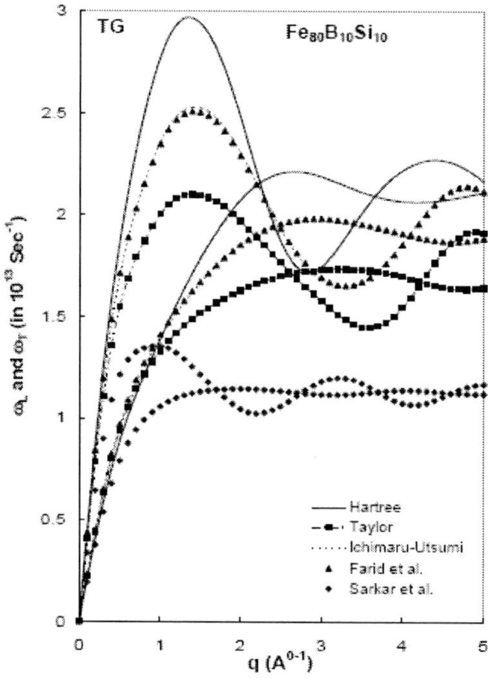

Figure 21. Dependence on screening on phonon dispersion curves of $Fe_{80}B_{10}Si_{10}$ bulk metallic glass using TG approach.

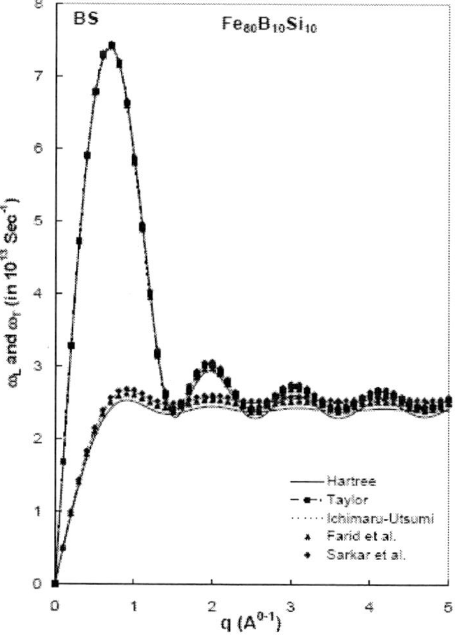

Figure 22. Dependence on screening on phonon dispersion curves of $Fe_{80}B_{10}Si_{10}$ bulk metallic glass using BS approach.

The temperature dependency of the specific heat C_V shows anomalous nature is highlighted in figures 23-25. It is apparent from the nature that C_V is more sensitive to screening. The initial rise on the C_V/T values is observed for low temperature and then further increase of temperature give convergent value.

The calculated thermodynamic and elastic properties of this glass are shown in table 4. Here it is found that the incorporation of exchange and correlation effects in the static H-dielectric function enhance the longitudinal and transverse sound velocities in HB and TG approaches while in BS approach suppression on the velocities is observed. For this glass, also, we do not have any comparison for B_T, G, σ, Y and θ_D hence we avoid to put any remarks.

Table 4. Thermodynamic and Elastic properties of $Fe_{80}B_{10}Si_{10}$bulk metallic glass

App.	SCR	υ_L x 10^5 cm/s	υ_T x 10^5 cm/s	B_T x 10^{11} dyne/cm^2	G x 10^{11} dyne/cm^2	σ	Y x 10^{11} dyne/cm^2	θ_D (K)
HB	H	3.08	1.78	3.53	2.12	0.25	5.30	256.56
	T	3.83	2.21	5.48	3.29	0.25	8.22	319.42
	IU	4.12	2.38	6.33	3.80	0.25	9.50	343.44
	F	4.10	2.37	6.29	3.77	0.25	9.44	342.33
	S	2.32	1.34	2.01	1.21	0.25	3.02	193.62
TG	H	4.35	2.08	8.85	2.90	0.35	7.84	303.95
	T	4.09	2.28	6.60	3.50	0.28	8.91	330.40
	IU	4.30	2.29	7.71	3.53	0.30	9.19	333.12
	F	4.34	2.33	7.80	3.64	0.30	9.46	338.28
	S	3.41	1.95	4.40	2.56	0.26	6.43	281.93
BS	H	8.54	2.44	43.74	3.99	0.46	11.61	361.42
	T	8.62	2.56	44.05	4.42	0.45	12.82	380.07
	IU	8.59	2.53	43.91	4.29	0.45	12.46	374.60
	F	8.59	2.53	43.90	4.29	0.45	12.48	374.92
	S	8.64	2.59	44.17	4.50	0.45	13.07	383.87

4.4. $Fe_{40}Ni_{40}B_{20}$ Bulk Metallic Glass

The $Fe_{40}Ni_{40}B_{20}$ bulk metallic glass is the most important candidate of transition metal-metalloid group. The PDC of this glass has been theoretically studied by Gupta *et al.* [32] by HB approach using experimental structure factor. The experimental data of PDC of this glass is not available in the literature. Therefore, we have reported vibrational properties of this glass using pseudopotential theory for the first time with more advanced screening functions.

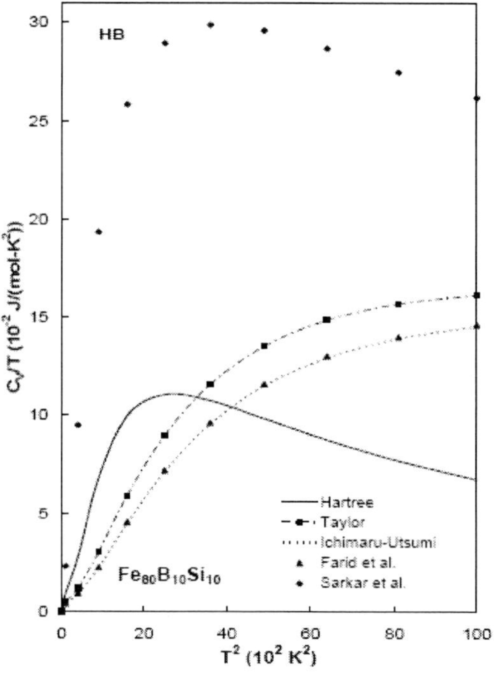

Figure 23. Dependence on screening on low temperature specific heat of $Fe_{80}B_{10}Si_{10}$ bulk metallic glass using HB approach.

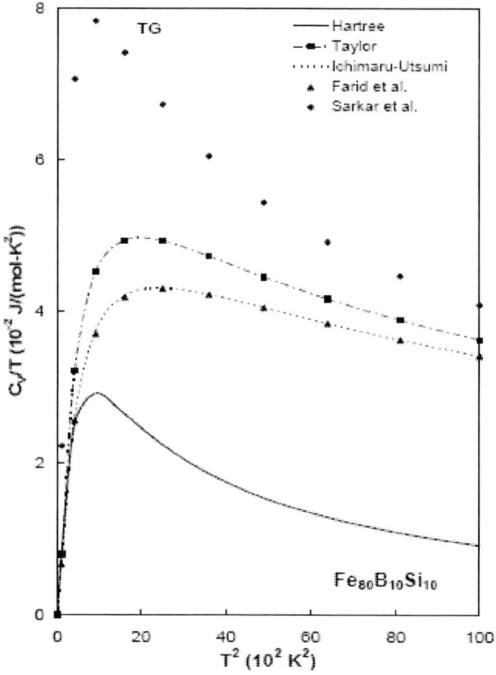

Figure 24. Dependence on screening on low temperature specific heat of $Fe_{80}B_{10}Si_{10}$ bulk metallic glass using TG approach.

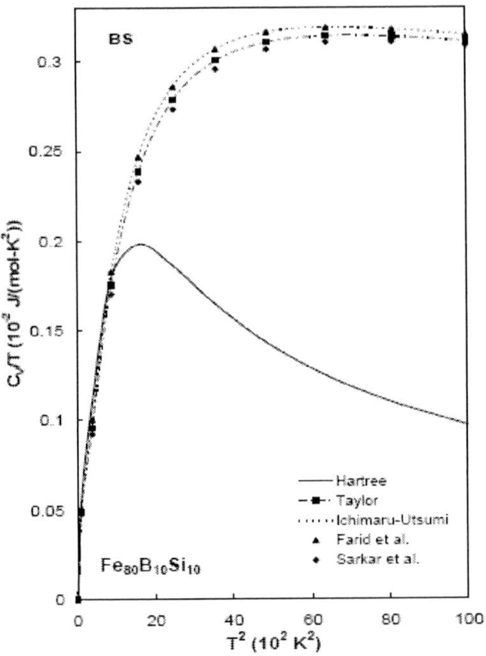

Figure 25. Dependence on screening on low temperature specific heat of $Fe_{80}B_{10}Si_{10}$ bulk metallic glass using BS approach.

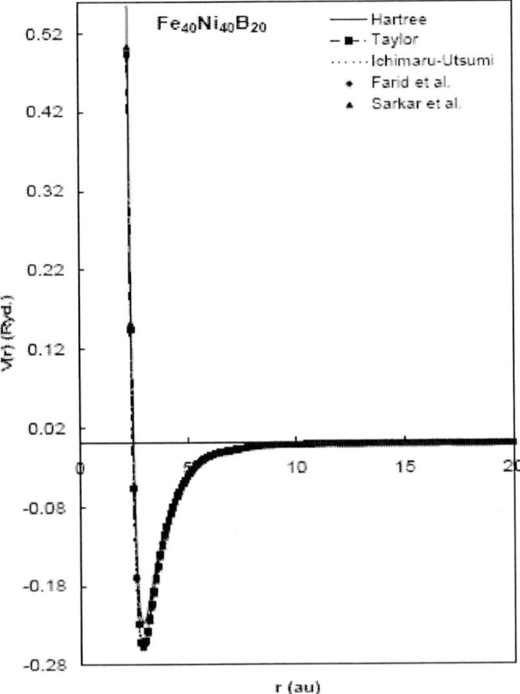

Figure 26. Dependence on screening on pair potentials of $Fe_{40}Ni_{40}B_{20}$ bulk metallic glass.

The computed interatomic pair potentials $V(r)$ of this glass are shown in figure 26. It is seen that the inclusion of exchange and correlation effects in the static H-dielectric screening changes the nature of the interatomic pair potential, significantly. The first zero of the interatomic pair potential $V(r = r_0)$ due to all local field correction functions occurs at $r_0 = 2.9$ au, which is very close to the r_P value at which the pair correlation function $g(r)$ shows its first peak [32]. The interatomic pair potential well width and its minimum position $V_{\min}(r)$ are also affected by the nature of the screening. The maximum depth in the interatomic pair potential is obtained for F-function, while minimum is for H-screening function. The present results do not show oscillatory behaviour and potential energy remains negative in the large r - region. Thus, the Coulomb repulsive potential part dominates the oscillations due to ion-electron-ion interactions, which show the waving shape of the interatomic pair potential after $r = 8.7$ au. Hence, the interatomic pair potentials converge towards a finite value instead of zero in attractive region.

The computed pair correlation function (PCF) $g(r)$ of this glass is shown in figure 27. It is found that, the peak positions due to S-function show higher while those due to H-function show lower. The screening effect is also observed in the nature of the PCF. The ratio (r_2/r_1) of the position of the second peak (r_2) to that of the first peak (r_1) is found 1.29, 1.53, 1.50, 1.50 and 1.57 for H-, T-, IU-, F- and S-function, respectively. While the ratio (r_3/r_1) of the position of the third peak (r_3) to that of the first peak (r_1) i.e. the ratio of the third atomic shell radius to the nearest-neighbour distance is found 1.77, 2.11, 2.08, 2.08 and 2.15 for H-, T-, IU-, F- and S-function, respectively. The (r_2/r_1) ratio is close to the c/a ratio in close-packed hexagonal structure i.e. c/a = 1.63, which suggests that the short range order of nearest neighbours is influenced more or less by the atomic arrangement of the crystalline structure. This suggests that the short range order of nearest neighbours is influenced more or less by the atomic arrangement of the crystalline structure. This result is typical of a metallic glass with a large main peak at the nearest-nearest distance followed by smaller peaks corresponding to more distant neighbours. The computed pair correlation function using various local field correction functions are overlapped with each other after $r = 7.9$. Therefore, it is very difficult to draw the remarks regarding the disorder is invisible after this point. Actually, this long range order is normal and it may be due to the waving shape of the interatomic pair potential.

The screening dependence of the phonon eigen frequencies for longitudinal and transverse phonon modes calculated using three approaches are shown in figures 28-30. It is observed that the inclusion of exchange and correlation effect enhances the phonon frequencies in both longitudinal as well as transverse branches. The present results of the PDC due to H-, T- and F-function are lying between those due to IU- and S-screening in HB approach. While, the computed outcomes of the PDC due to T-, IU- and F-function are lying between those due to H- and S-screening in TG approach. Also in BS approach, the present results of the PDC due to T-, IU- and F-function are lying between those due to H- and S-screening. The first minimum in the longitudinal branch of HB approach is found around at $q \approx 3.6\text{Å}^{-1}$ for H-, T-, IU-, F- and S-function. While, the first minimum in the longitudinal branch of TG approach is found around at $q \approx 2.8\text{Å}^{-1}$ for H-, $q \approx 2.5\text{Å}^{-1}$ for T-, $q \approx 2.7\text{Å}^{-1}$ for

IU- as well as F-function and $q \approx 2.4\text{Å}^{-1}$ for S-function. The first minimum in the longitudinal branch of BS approach is found around at $q \approx 1.4\text{Å}^{-1}$ for H-, T-, IU-, F- and S-function. The first maximum in the longitudinal branch ω_L of HB approach is found around at $q \approx 1.8\text{Å}^{-1}$ for H-, T-, IU-, F- and S-function. While, the first maximum in the longitudinal branch of TG approach is found around at $q \approx 1.3\text{Å}^{-1}$ for H-, T-, IU- as well as F-function and $q \approx 0.7\text{Å}^{-1}$ for S-function. The first maximum in the longitudinal branch of BS approach is found around at $q \approx 0.7\text{Å}^{-1}$ for most of the local field correction functions.

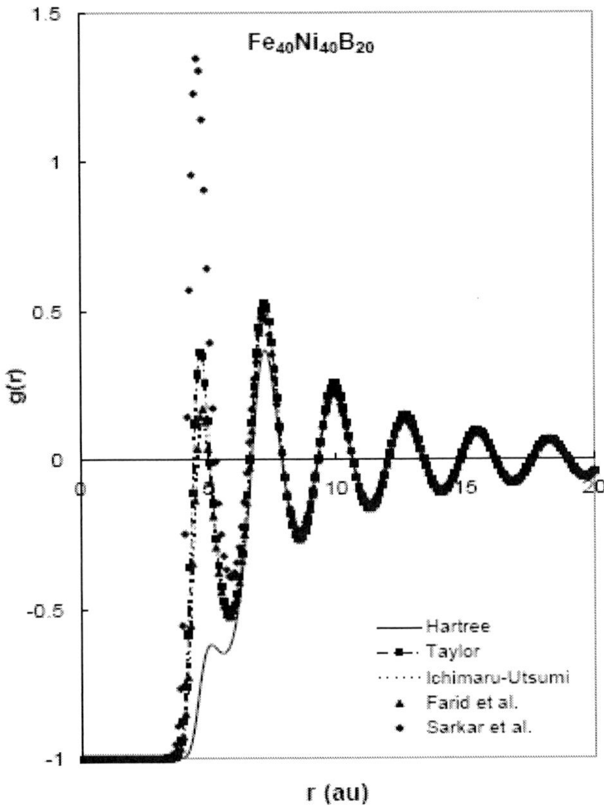

Figure 27. Dependence on screening on pair correlation function of $Fe_{40}Ni_{40}B_{20}$ bulk metallic glass.

It is also observed from the figures 28-30 that, the oscillations are more prominent in the longitudinal phonon modes as compared to the transverse modes in all three approaches. This shows the existence of collective excitations at larger momentum transfer due to longitudinal phonons only and the instability of the transverse phonons due to the anharmonicity of the atomic vibrations in the metallic systems. Moreover, the present outcomes of both the phonon modes due to HB as well as TG approaches are more enhanced than the BS approach. The first crossover position of ω_L and ω_T in the HB and BS approaches is observed at $q \approx 2.8\text{Å}^{-1}$ and $q \approx 1.4\text{Å}^{-1}$ for most of the local field correction functions, respectively. While, the first crossover position of ω_L and ω_T in the TG approach is observed at $q \approx 2.2\text{Å}^{-1}$ for H-, T-,

IU- as well as F-function and $q \approx 1.8Å^{-1}$ for S-function. The specific heat C_V for various temperatures is shown in figures 31-33. It is found that behaviour is more sensitive to local field correction functions. The three approaches used in the present study generate different results for low temperature region.

Furthermore, the thermodynamic and elastic properties estimated from the elastic limit of the PDC are tabulated in table 5. It is seen that the screening theory plays an important role in the prediction of the thermodynamic and elastic properties of this glass. As phonon modes obtained for BS approach is very high compared to HB and TG approaches, the thermodynamic and elastic properties obtained for BS approach is also higher in magnitude. The presently computed results of the thermodynamic and elastic properties are found to be in qualitative agreement with the available theoretical or experimental data [29] in the literature.

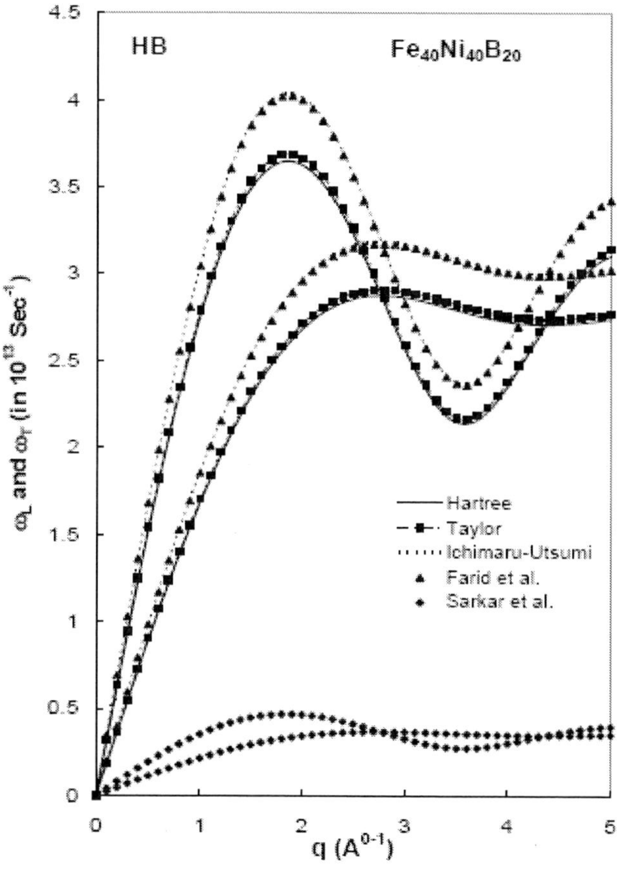

Figure 28. Dependence on screening on phonon dispersion curves of $Fe_{40}Ni_{40}B_{20}$ bulk metallic glass using HB approach.

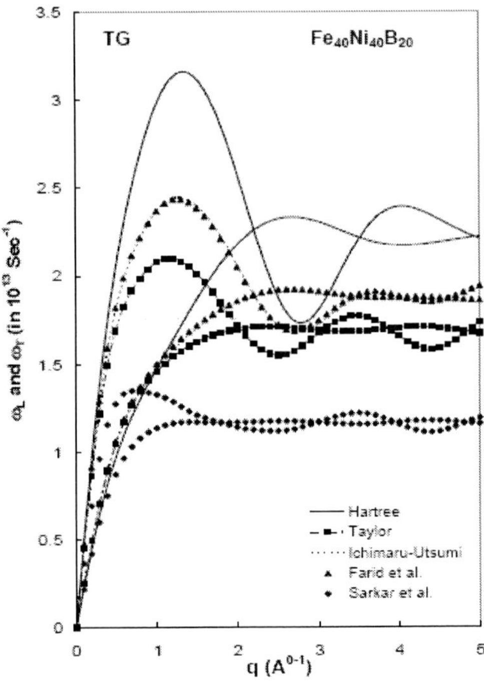

Figure 29. Dependence on screening on phonon dispersion curves of $Fe_{40}Ni_{40}B_{20}$ bulk metallic glass using TG approach.

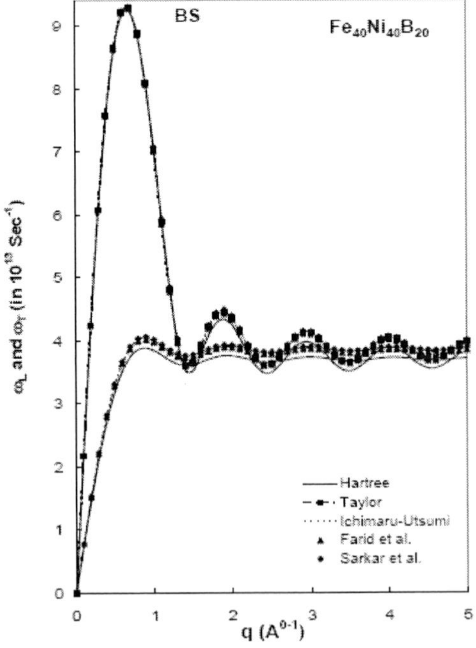

Figure 30. Dependence on screening on phonon dispersion curves of $Fe_{40}Ni_{40}B_{20}$ bulk metallic glass using BS approach.

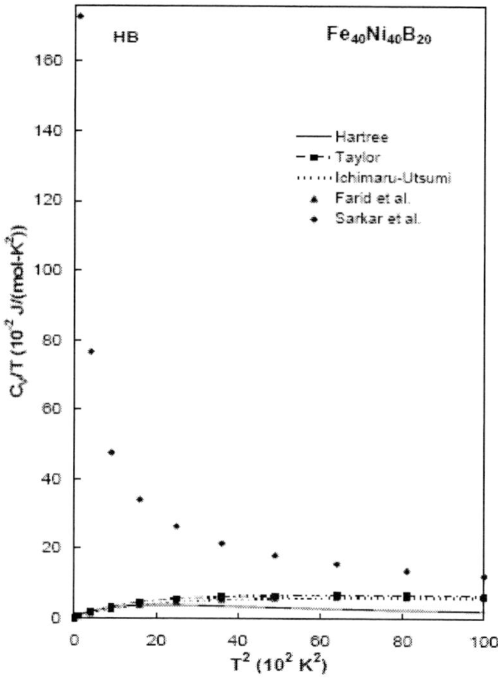

Figure 31. Dependence on screening on low temperature specific heat of $Fe_{40}Ni_{40}B_{20}$ bulk metallic glass using HB approach.

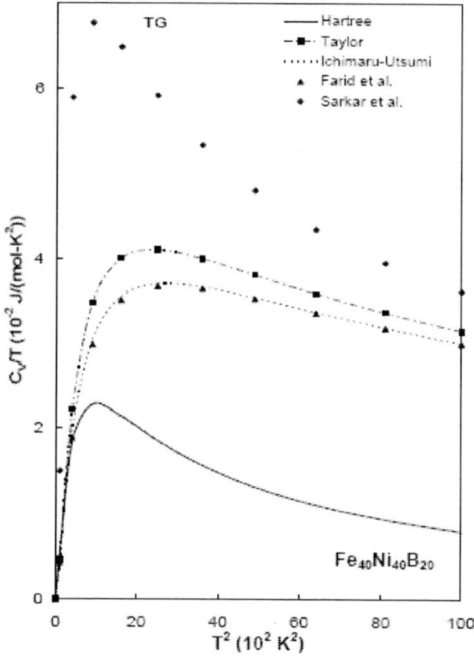

Figure 32. Dependence on screening on low temperature specific heat of $Fe_{40}Ni_{40}B_{20}$ bulk metallic glass using TG approach.

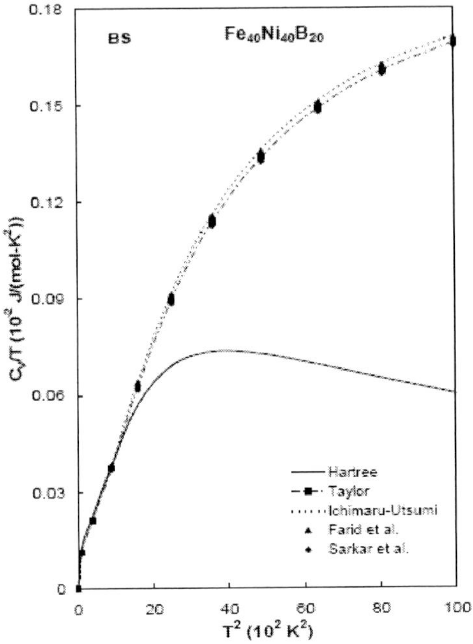

Figure 33. Dependence on screening on low temperature specific heat of $Fe_{40}Ni_{40}B_{20}$ bulk metallic glass using BS approach.

Table 5. Thermodynamic and Elastic properties of $Fe_{40}Ni_{40}B_{20}$ bulk metallic glass

App.	SCR	υ_L x 10^5 cm/s	υ_T x 10^5 cm/s	B_T x 10^{11} dyne/cm^2	G x 10^{11} dyne/cm^2	σ	Y x 10^{11} dyne/cm^2	θ_D (K)
HB	H	3.15	1.82	4.18	2.51	0.25	6.27	274.49
	T	3.18	1.84	4.27	2.56	0.25	6.41	277.59
	IU	3.48	2.01	5.12	3.07	0.25	7.67	303.76
	F	3.47	2.00	5.08	3.05	0.25	7.62	302.61
	S	0.40	0.23	0.07	0.04	0.25	0.10	35.25
TG	H	4.78	2.29	12.04	3.97	0.35	10.73	349.90
	T	4.53	2.54	9.03	4.91	0.27	12.46	384.99
	IU	4.67	2.52	10.09	4.83	0.29	12.48	382.87
	F	4.72	2.57	10.26	4.99	0.29	12.89	389.33
	S	3.69	2.16	5.62	3.55	0.24	8.80	326.21
BS	H	10.25	3.57	66.81	9.65	0.43	27.63	551.63
	T	10.36	3.71	67.44	10.45	0.43	29.82	573.62
	IU	10.33	3.68	67.23	10.28	0.43	29.35	568.93
	F	10.33	3.69	67.25	10.32	0.43	29.45	570.00
	S	10.37	3.73	67.55	10.54	0.43	30.04	575.83
Others [32]		3.96	2.34	0.61	0.393	0.233	1.132	353.45
Expt. [32]		4.47	-	1.773	-	0.341	1.60	350.00

4.5. $Ni_{80}B_{10}Si_{20}$ Bulk Metallic Glass

The $Ni_{80}B_{10}Si_{20}$ metallic glass is the most important candidate of transition metal-metalloid group. In our literature survey we have not found any reports on the experimental as well as theoretical work based on pseudopotential theory related to PDC of this glass. But, the structural properties of this glass have been studied by Sulir and Pyka [45]. Therefore, the vibrational properties of this glass are reported for the first time.

The computed interatomic pair potentials $V(r)$ of this glass are shown in figure 34. It is seen that the inclusion of exchange and correlation effects in the static H-dielectric screening changes the nature of the interatomic pair potential, significantly. The first zero of the interatomic pair potential $V(r = r_0)$ due to all local field correction functions occurs at $r_0 = 2.5$ au, which is very close to the r_P value at which the pair correlation function $g(r)$ shows its first peak. The interatomic pair potential well width and its minimum position $V_{\min}(r)$ are also affected by the nature of the screening. The maximum depth in the interatomic pair potential is obtained for F-function, while minimum is for H-screening function. The present results do not show oscillatory behaviour and potential energy remains negative in the large r - region. Thus, the Coulomb repulsive potential part dominates the oscillations due to ion-electron-ion interactions, which show the waving shape of the interatomic pair potential after $r = 8.0$ au. Hence, the interatomic pair potentials converge towards a finite value instead of zero in attractive region.

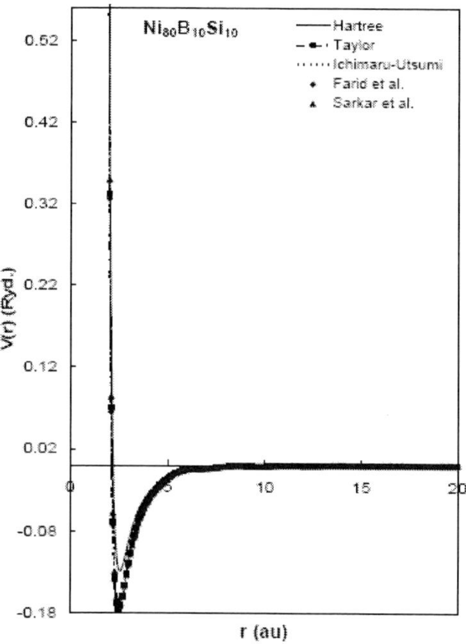

Figure 34. Dependence on screening on pair potentials of $Ni_{80}B_{10}Si_{20}$ bulk metallic glass.

The pair correlation function (PCF) $g(r)$ computed theoretically through the interatomic pair potential is shown in figure 35. It is found that, the peak positions due to S-function show higher while those due to H-function show lower. The screening effect is also observed in the nature of the PCF. The ratio (r_2/r_1) of the position of the second peak (r_2) to that of the first peak (r_1) is found 1.29, 1.54, 1.51, 1.51 and 1.54 for H-, T-, IU-, F- and S-function, respectively. While the ratio (r_3/r_1) of the position of the third peak (r_3) to that of the first peak (r_1) i.e. the ratio of the third atomic shell radius to the nearest-neighbour distance is found 1.78, 2.13, 2.09, 2.09 and 2.13 for H-, T-, IU-, F- and S-function, respectively. The (r_2/r_1) ratio is close 1.63 characteristic for the disordered bcc type crystallographic structures and therefore this component of the second peak might eventually be due to not complete amorphousation of the samples [45]. This suggests that the short range order of nearest neighbours is influenced more or less by the atomic arrangement of the crystalline structure. This result is typical of a metallic glass with a large main peak at the nearest-nearest distance followed by smaller peaks corresponding to more distant neighbours. The computed pair correlation function using various local field correction functions are overlapped with each other after $r = 7.8$ au. Therefore, it is very difficult to draw the remarks regarding the disorder is invisible after this point in the figure 35. In fact, this long range order is normal and it may be due to the waving shape of the interatomic pair potential.

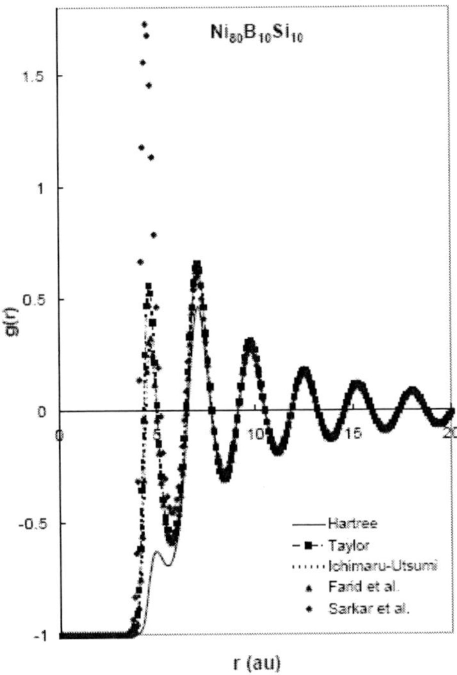

Figure 35. Dependence on screening on pair correlation function of $Ni_{80}B_{10}Si_{20}$ bulk metallic glass.

The screening dependence of the phonon eigen frequencies for longitudinal and transverse phonon modes calculated using three approaches are shown in figures 36-38. It is observed that the inclusion of exchange and correlation effect enhances the phonon frequencies in both longitudinal as well as transverse branches. The present results of the PDC due to H-, T- and F-function are lying between those due to IU- and S-screening in HB approach. While, the computed outcomes of the PDC due to T-, IU- and F-function are lying between those due to H- and S-screening in TG approach. Also in BS approach, the present results of the PDC due to T-, IU- and F-function are lying between those due to H- and S-screening.

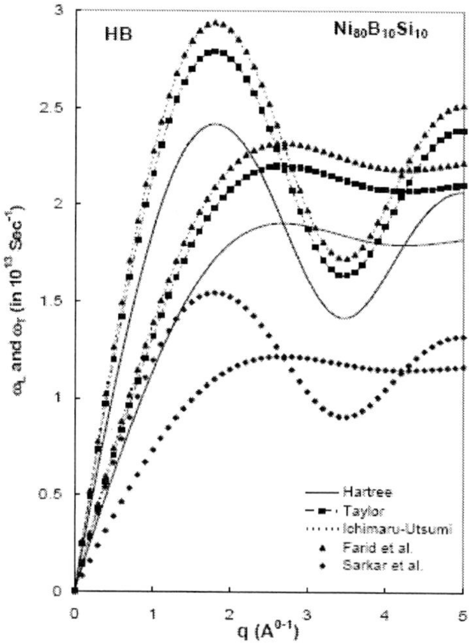

Figure 36. Dependence on screening on phonon dispersion curves of $Ni_{80}B_{10}Si_{20}$ bulk metallic glass using HB approach.

The first minimum in the longitudinal branch of HB approach is found around at $q \approx 3.5Å^{-1}$ for H-, T-, IU-, F- and S-function. While, the first minimum in the longitudinal branch of TG approach is found around at $q \approx 2.7Å^{-1}$ for H-, $q \approx 2.3Å^{-1}$ for T-, $q \approx 2.5Å^{-1}$ for IU- as well as F-function and $q \approx 1.9Å^{-1}$ for S-function. The first minimum in the longitudinal branch of BS approach is found around at $q \approx 1.4Å^{-1}$ for H-, T-, IU-, F- and S-function. The first maximum in the longitudinal branch ω_L of HB approach is found around at $q \approx 1.8Å^{-1}$ for H-, T-, IU-, F- and S-function. While, the first maximum in the longitudinal branch of TG approach is found around at $q \approx 1.2Å^{-1}$ for H-, T-, IU- as well as F-function and $q \approx 0.7Å^{-1}$ for S-function. The first maximum in the longitudinal branch of BS approach is found around at $q \approx 0.6Å^{-1}$ for most of the local field correction functions.

It is also observed from the figures 36-38 that, the oscillations are more prominent in the longitudinal phonon modes as compared to the transverse modes in all three approaches. This shows the existence of collective excitations at larger momentum transfer due to longitudinal phonons only and the instability of the transverse phonons due to the anharmonicity of the atomic vibrations in the metallic systems. Moreover, the present outcomes of both the phonon modes due to HB as well as TG approaches are more enhanced than the BS approach. The first crossover position of ω_L and ω_T in the HB and BS approaches is observed at $q \approx 2.7$Å^{-}

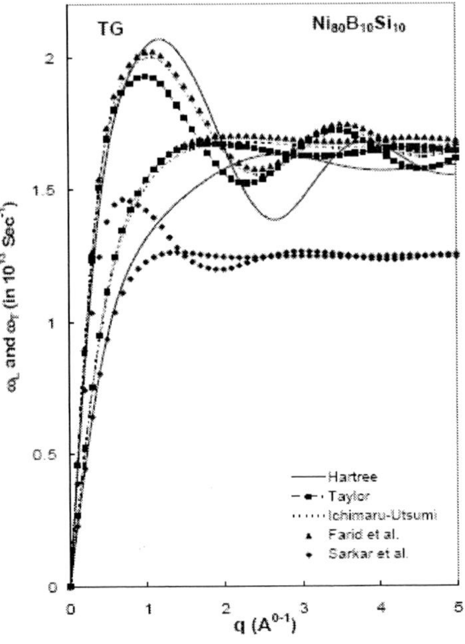

Figure 37. Dependence on screening on phonon dispersion curves of $Ni_{80}B_{10}Si_{20}$ bulk metallic glass using TG approach.

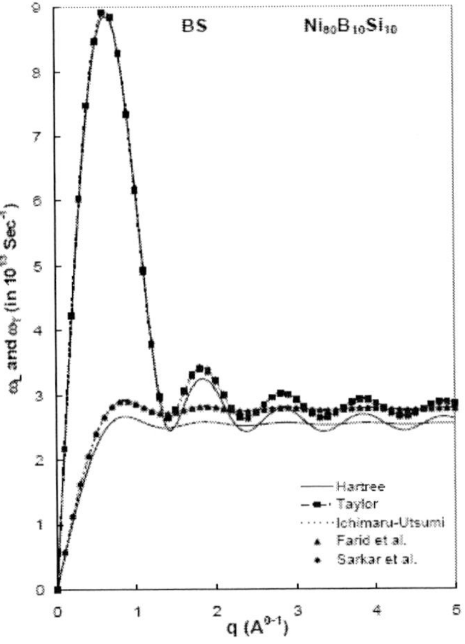

Figure 38. Dependence on screening on phonon dispersion curves of $Ni_{80}B_{10}Si_{20}$ bulk metallic glass using BS approach.

[1] and $q \approx 1.4\text{Å}^{-1}$ for most of the local field correction functions, respectively. While, the first crossover position of ω_L and ω_T in the TG approach is observed at $q \approx 2.1\text{Å}^{-1}$ for H-, $q \approx 1.7$ for T-, $q \approx 1.9$ for IU- as well as F-function and $q \approx 1.4\text{Å}^{-1}$ for S-function.

As shown in figures 39-41, the exchange and correlation functions also affect the anomalous behaviour (i.e. deviation from the T^3 law) which is observed in the specific heat (C_V). The reason behind the anomalous behaviour may be due to the low frequency modes modify the generalized vibrational density of states of the glass with that of the polycrystal. These modes are mainly responsible for the difference in the temperature dependence of the specific heat which departs from the normal behaviour. At low temperature region high bump is observed in HB, TG and BS approaches.

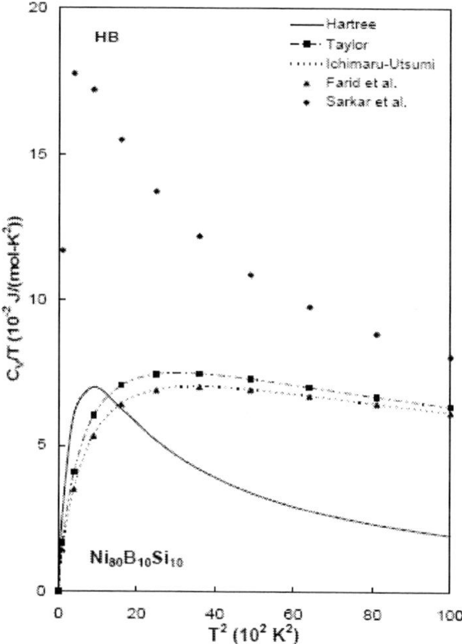

Figure 39. Dependence on screening on low temperature specific heat of $Ni_{80}B_{10}Si_{20}$ bulk metallic glass using HB approach.

The thermodynamic and elastic properties estimated from the elastic limit of the PDC are narrated in table 6. It is seen that the results due to HB and TG approaches are very low in comparison with BS approach. The effect of various local field correction functions is also distinguishable on the thermodynamic and elastic properties of $Ni_{80}B_{10}Si_{20}$ glass.

Table 6. Thermodynamic and Elastic properties of $Ni_{80}B_{10}Si_{20}$bulk metallic glass

App.	SCR	υ_L x 10^5 cm/s	υ_T x 10^5 cm/s	B_T x 10^{11} dyne/cm^2	G x 10^{11} dyne/cm^2	σ	Y x 10^{11} dyne/cm^2	θ_D (K)
HB	H	2.15	1.24	1.88	1.13	0.25	2.82	181.81
	T	2.48	1.43	2.51	1.51	0.25	3.77	209.96
	IU	2.61	1.51	2.77	1.66	0.25	4.15	220.48
	F	2.62	1.51	2.79	1.67	0.25	4.18	221.14
	S	1.38	0.79	0.77	0.46	0.25	1.15	116.25
TG	H	4.26	2.32	8.05	3.93	0.29	10.14	340.80
	T	4.68	2.71	8.88	5.39	0.25	13.45	397.07
	IU	4.67	2.67	9.02	5.22	0.26	13.12	391.06
	F	4.75	2.72	9.29	5.41	0.26	13.60	398.24
	S	3.97	2.32	6.31	3.93	0.24	9.77	338.98
BS	H	10.06	2.49	68.12	4.54	0.47	13.32	374.77
	T	10.19	2.71	68.91	5.39	0.46	15.77	408.33
	IU	10.17	2.68	68.74	5.28	0.46	15.44	404.00
	F	10.18	2.70	68.83	5.35	0.46	15.66	406.88
	S	10.19	2.71	68.87	5.37	0.46	15.69	407.31

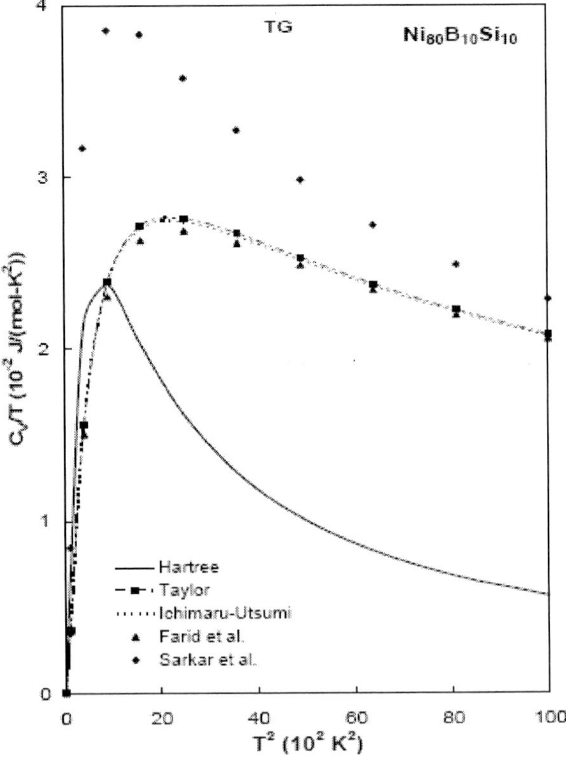

Figure 40. Dependence on screening on low temperature specific heat of $Ni_{80}B_{10}Si_{20}$ bulk metallic glass using TG approach.

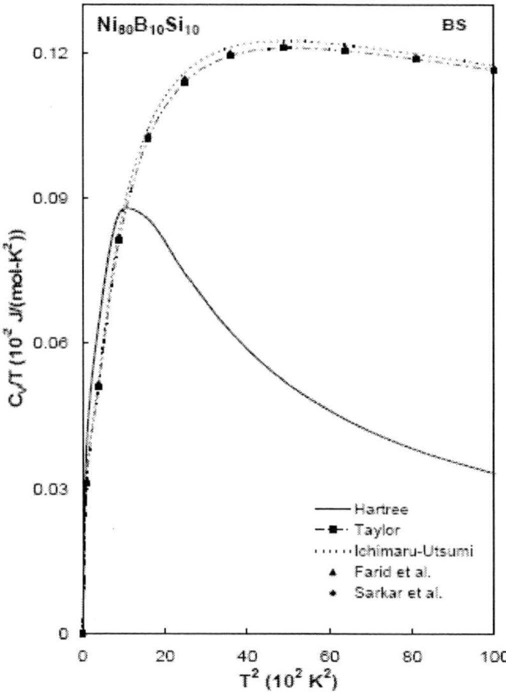

Figure 41. Dependence on screening on low temperature specific heat of $Ni_{80}B_{10}Si_{20}$ bulk metallic glass using BS approach.

4.6. $Fe_{60}Ni_{20}B_{10}Si_{10}$ Bulk Metallic Glass

The $Fe_{60}Ni_{20}B_{10}Si_{10}$ metallic glass is the most important candidate of transition metal-metalloid group. In our literature survey we have not found any reports on the experimental as well as theoretical work based on pseudopotential theory related to PDC of this glass. But, the structural properties of this glass have been studied by Sulir and Pyka [45]. Therefore, the vibrational properties of this glass are reported for the first time.

The computed interatomic pair potentials $V(r)$ of this glass are shown in figure 42. It is seen that the inclusion of exchange and correlation effects in the static H-dielectric screening changes the nature of the interatomic pair potential, significantly. The first zero of the interatomic pair potential $V(r = r_0)$ due to all local field correction functions occurs at $r_0 = 2.6$ au. The interatomic pair potential well width and its minimum position $V_{min}(r)$ are also affected by the nature of the screening. The maximum depth in the interatomic pair potential is obtained for F-function, while minimum is for H-screening function. The present results do not show oscillatory behaviour and potential energy remains negative in the large r - region. Thus, the Coulomb repulsive potential part dominates the oscillations due to ion-electron-ion interactions, which show the waving shape of the interatomic pair potential after $r = 7.8$ au. Hence, the interatomic pair potentials converge towards a finite value instead of zero in attractive region.

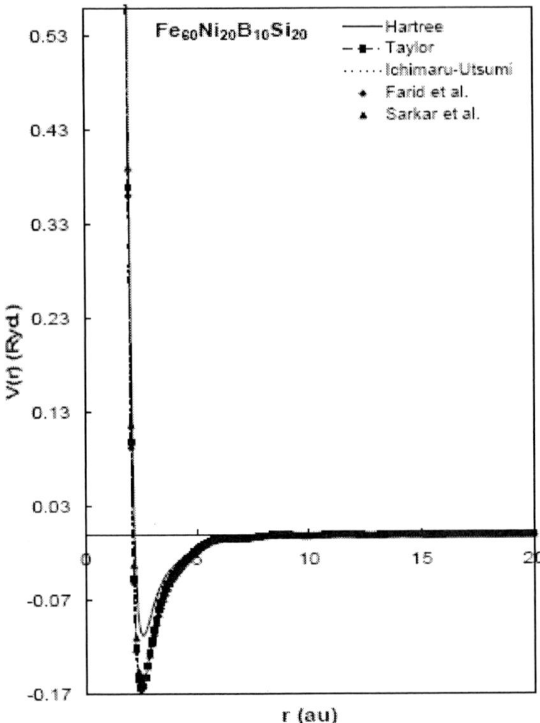

Figure 42. Dependence on screening on pair potentials of $Fe_{60}Ni_{20}B_{10}Si_{10}$ bulk metallic glass.

The pair correlation function (PCF) $g(r)$ computed theoretically through the interatomic pair potential is shown in figure 43. It is found that, the peak positions due to S-function show higher while those due to H-function show lower. The screening effect is also observed in the nature of the PCF. The ratio (r_2/r_1) of the position of the second peak (r_2) to that of the first peak (r_1) is found 1.28, 1.53, 1.53, 1.53 and 1.57 for H-, T-, IU-, F- and S-function, respectively. While the ratio (r_3/r_1) of the position of the third peak (r_3) to that of the first peak (r_1) i.e. the ratio of the third atomic shell radius to the nearest-neighbour distance is found 1.75, 2.13, 2.13, 2.13 and 2.17 for H-, T-, IU-, F- and S-function, respectively. The (r_2/r_1) ratio is close 1.63 characteristic for the disordered bcc type crystallographic structures and therefore this component of the second peak might eventually be due to not complete amorphousation of the samples [45]. This suggests that the short range order of nearest neighbours is influenced more or less by the atomic arrangement of the crystalline structure. This result is typical of a metallic glass with a large main peak at the nearest-nearest distance followed by smaller peaks corresponding to more distant neighbours. The computed pair correlation function using various local field correction functions are overlapped with each other after $r = 7.9$ au. Therefore, it is very difficult to draw the remarks regarding the disorder is invisible after this point in the figure 43. In fact, this long range order is normal and it may be due to the waving shape of the interatomic pair potential.

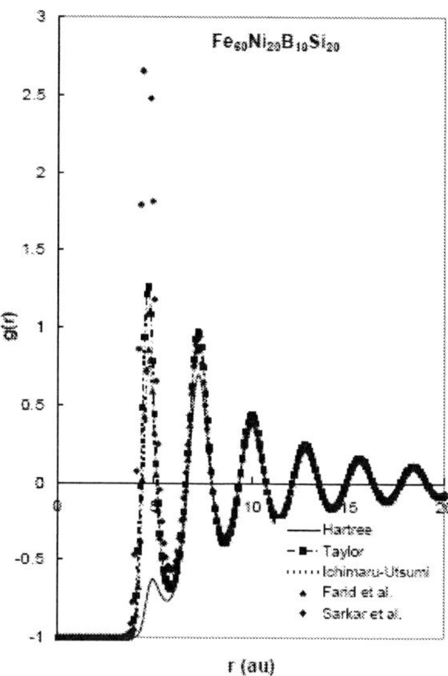

Figure 43. Dependence on screening on pair correlation function of $Fe_{60}Ni_{20}B_{10}Si_{10}$ bulk metallic glass.

The screening dependence of the phonon eigen frequencies for longitudinal and transverse phonon modes calculated using three approaches are shown in figures 44-46. It is observed that the inclusion of exchange and correlation effect enhances the phonon frequencies in both longitudinal as well as transverse branches. The present results of the PDC due to H-, T- and F-function are lying between those due to IU- and S-screening in HB approach. While, the computed outcomes of the PDC due to T-, IU- and F-function are lying between those due to H- and S-screening in TG approach. Also in BS approach, the present results of the PDC due to T-, IU- and F-function are lying between those due to H- and S-screening.

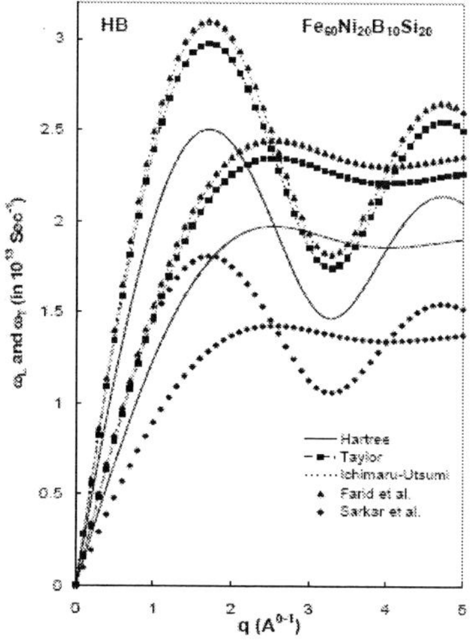

Figure 44. Dependence on screening on phonon dispersion curves of $Fe_{60}Ni_{20}B_{10}Si_{10}$ bulk metallic glass using HB approach.

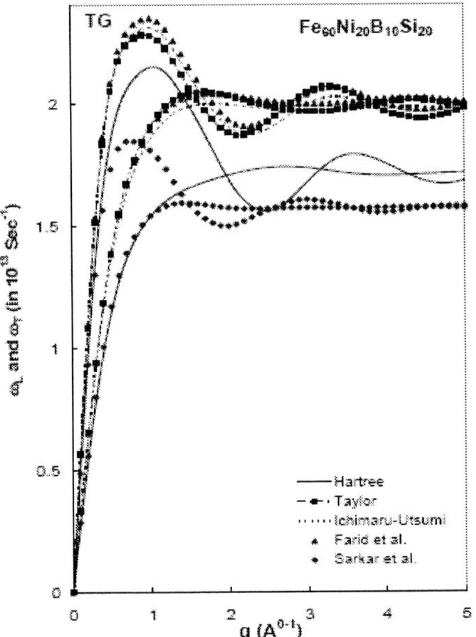

Figure 45. Dependence on screening on phonon dispersion curves of $Fe_{60}Ni_{20}B_{10}Si_{10}$ bulk metallic glass using TG approach.

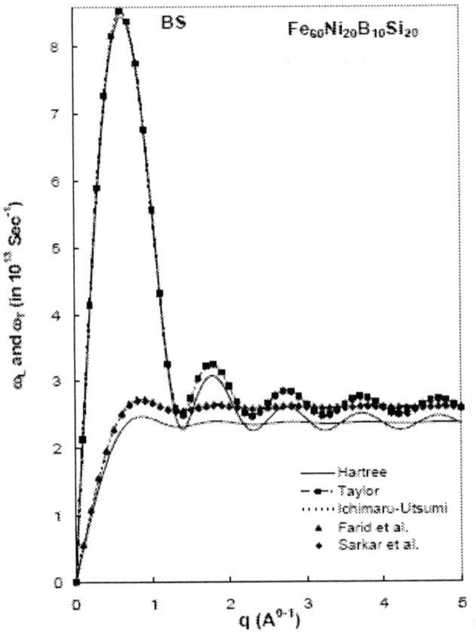

Figure 46. Dependence on screening on phonon dispersion curves of $Fe_{60}Ni_{20}B_{10}Si_{10}$ bulk metallic glass using BS approach.

The first minimum in the longitudinal branch of HB approach is found around at $q \approx 3.3Å^{-1}$ for H-, T-, IU-, F- and S-function. While, the first minimum in the longitudinal branch of TG approach is found around at $q \approx 2.5Å^{-1}$ for H-, $q \approx 2.1Å^{-1}$ for T-, $q \approx 2.3Å^{-1}$ for IU- as well as F-function and $q \approx 1.9Å^{-1}$ for S-function. The first minimum in the longitudinal branch of BS approach is found around at $q \approx 1.4\text{Å}^{-1}$ for H-, T-, IU-, F- and S-function. The first maximum in the longitudinal branch ω_L of HB approach is found around at $q \approx 1.7\text{Å}^{-1}$ for H-, T-, IU-, F- and S-function. While, the first maximum in the longitudinal branch of TG approach is found around at $q \approx 1.0\text{Å}^{-1}$ for H-, T-, IU- as well as F-function and $q \approx 0.8\text{Å}^{-1}$ for S-function. The first maximum in the longitudinal branch of BS approach is found around at $q \approx 0.6\text{Å}^{-1}$ for most of the local field correction functions.

It is also observed from the figures 44-46 that, the oscillations are more prominent in the longitudinal phonon modes as compared to the transverse modes in all three approaches. This shows the existence of collective excitations at larger momentum transfer due to longitudinal phonons only and the instability of the transverse phonons due to the anharmonicity of the atomic vibrations in the metallic systems. Moreover, the present outcomes of both the phonon modes due to HB as well as TG approaches are more enhanced than the BS approach. The first crossover position of ω_L and ω_T in the HB and BS approaches is observed at $q \approx 2.5\text{Å}^{-1}$ and $q \approx 1.3\text{Å}^{-1}$ for most of the local field correction functions, respectively. While, the first crossover position of ω_L and ω_T in the TG approach is observed at $q \approx 2.0\text{Å}^{-1}$ for H-, $q \approx 1.5$ for T-, $q \approx 1.8$ for IU- as well as F-function and $q \approx 1.5\text{Å}^{-1}$ for S-function.

The $C_V/T \rightarrow T^2$ relation is shown in figures 47-49. For all the three approaches as temperature increases, the initial rise in specific heat is observed at low temperature region and then it decreases. This observation is deviated for S-screening in HB and TG approaches, while those for IU-function in BS, where very high bump in C_V/T for low temperature is absent.

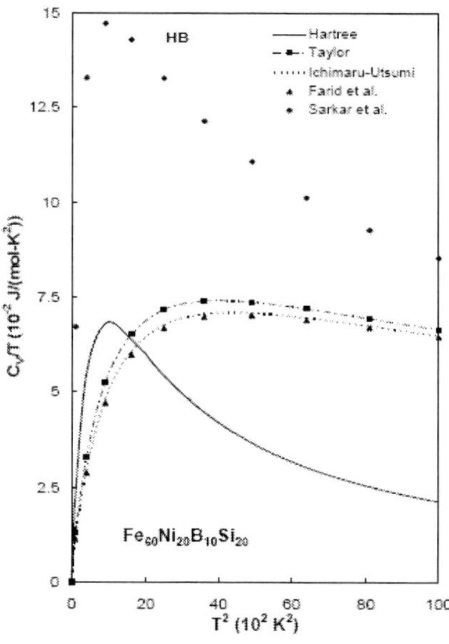

Figure 47. Dependence on screening on low temperature specific heat of $Fe_{60}Ni_{20}B_{10}Si_{10}$ bulk metallic glass using HB approach.

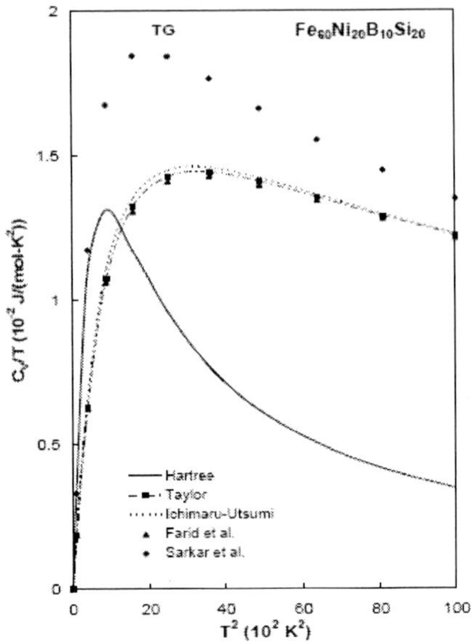

Figure 48. Dependence on screening on low temperature specific heat of $Fe_{60}Ni_{20}B_{10}Si_{10}$ bulk metallic glass using TG approach.

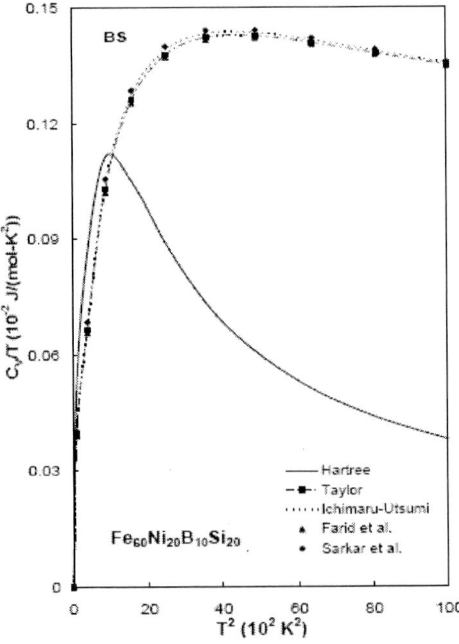

Figure 49. Dependence on screening on low temperature specific heat of $Fe_{60}Ni_{20}B_{10}Si_{10}$ bulk metallic glass using BS approach.

From the elastic limit of the PDC, the calculated υ_L, υ_T, B_T, G, σ, Y and θ_D are tabulated in table 7. As other comparison for these properties is not available it is difficult to draw any remarks at this stage. But among the three approaches adopted, the outcome due to HB and TG approaches are lower than those due to BS approach.

Table 7. Thermodynamic and Elastic properties of $Fe_{60}Ni_{20}B_{10}Si_{10b}$bulk metallic glass

App.	SCR	υ_L x 10^5 cm/s	υ_T x 10^5 cm/s	B_T x 10^{11} dyne/cm^2	G x 10^{11} dyne/cm^2	σ	Y x 10^{11} dyne/cm^2	θ_D (K)
HB	H	2.35	1.36	2.02	1.21	0.25	3.03	188.27
	T	2.79	1.61	2.86	1.71	0.25	4.28	223.83
	IU	2.89	1.67	3.06	1.83	0.25	4.59	231.60
	F	2.91	1.68	3.10	1.86	0.25	4.66	233.36
	S	1.70	0.98	1.06	0.63	0.25	1.58	136.09
TG	H	5.04	2.82	9.76	5.23	0.27	13.31	392.04
	T	5.76	3.38	11.85	7.51	0.24	18.61	468.08
	IU	5.71	3.31	11.87	7.22	0.25	18.00	459.20
	F	5.81	3.37	12.27	7.50	0.25	18.69	468.03
	S	4.97	2.89	8.93	5.51	0.24	13.72	401.22
BS	H	10.07	2.39	61.88	3.76	0.47	11.06	341.21
	T	10.22	2.64	62.76	4.58	0.46	13.42	376.37
	IU	10.21	2.62	62.65	4.53	0.46	13.27	374.24
	F	10.23	2.65	62.78	4.62	0.46	13.52	377.80
	S	10.20	2.61	62.60	4.49	0.47	13.16	372.65

5. Concluding Remarks

All the interatomic pair potentials show the combined effect of the s- and d-electrons. It is also noticed that when volume Ω_O of the bulk glassy alloys decreases (Ω_O is more for $Pd_{77.5}Si_{16.5}Cu_6$ glass), the interatomic pair potential depth deepens. All the interatomic pair potentials show the combined effect of the s- and d-electrons. Bretonnet and Derouiche [48] are observed that the repulsive part of $V(r)$ is drawn lower and its attractive part is deeper due to the d-electron effect. When we go from $Pd_{77.5}Si_{16.5}Cu_6 \rightarrow Fe_{60}Ni_{20}B_{10}Si_{10}$, the net number of d-electron r_d decreases, hence the $V(r)$ is shifted towards the lower r-values. Therefore, the present results are supported the d-electron effect as noted by Bretonnet and Derouiche [48].

Here in transverse branch, the frequencies increase with the wave vector q and then saturates at $\approx q = 2.0\text{Å}^{-1}$, which supports the well known Thorpe model [49] in which, it describes a glass like a solid containing finite liquid cluster. The transverse phonons are absorbed for frequencies larger than the smallest eigen frequencies of the largest cluster.

From the overall picture of the present study it is noticed that, the proposed model potential is successfully applicable to study the vibrational dynamics of some bulk metallic glass (BMG). The influences of various local field correction functions are also observed in the present study. The experimental or theoretical data of most of the bulk metallic glass (BMG) are not available in the literature but the present study is very useful to form a set of theoretical data of particular bulk metallic glass.

In all three approaches, it is very difficult to judge which approaches is best for computations of vibrational dynamics of bulk metallic glasses (BMG), because each approximation has its own identity. The HB approach is simplest and older one, which generating consistent results of the vibrational data of these bulk metallic glass (BMG), because the HB approaches needs minimum number of parameters. While TG approach is developed upon the quasi-crystalline approximation in which effective force constant depends upon the correlation function for the displacement of atoms and correlation function of displacement itself depends on the phonon frequencies. The BS approach is retained the interatomic interactions effective between the first nearest neighbours only hence, the disorderness of the atoms in the formation of metallic glasses is more which show deviation in magnitude of the PDC and their related properties. From the present study we are concluded that, all three approaches are suitable for studying the vibrational dynamics of the amorphous materials. Hence, successful application of the model potential with three approaches is observed from the present study.

The dielectric function plays an important role in the evaluation of potential due to the screening of the electron gas. For this purpose in the present investigations the local filed correction function due to H, T, IU, F and S are used. Reason for selecting these functions is that H-function does not include exchange and correlation effect and represents only static dielectric function, while T-function cover the overall features of the various local field correction functions proposed before 1972. While, IU, F and S functions are recent one among the existing functions and not exploited rigorously in such study. This helps us to study the relative effects of exchange and correlation in the aforesaid properties. Hence, the five different local field correction functions show variations up to an order of magnitude in the vibrational properties.

6. Conclusions

Finally, it is concluded that the PDC generated form three approaches with five local field correction functions reproduce all broad characteristics of dispersion curves. The well recognized model potential with more advanced IU, F and S-local field correction functions generate consistent results. The experimentally or theoretically observed data of most of the bulk metallic glasses (BMG) are not available in the literature. Therefore, it is difficult to draw any special remarks. However, the present study is very useful to provide important information regarding the particular glass. Also, the present computation confirms the applicability of the model potential in the aforesaid properties and supports the present approach of PAA. Such study on phonon dynamics of other binary liquid alloys and bulk metallic glasses (BMG) is in progress.

References

[1] Inoue A.; Hashimoto K. Amorphous and Nanocrystalline Materials: Preparations, Properties and Applications; *Springer-Verlag*: Berlin, 2001.

[2] Kovalenko N. P.; Krasny Y. P.; Krey U.; *Physics of Amorphous Metals*, Wiley-VCH: Berlin, 2001.

[3] Güntherodt H.–J.; Beck H. Glassy Metals-III: Amorphization Techniques, Catalysis, Electronic and Ionic Structure; *Springer-Verlag*: Berlin, 1994.

[4] Güntherodt H.–J.; Beck H. Glassy Metals-II: Atomic Structure and Dynamics, Electronic Structure, Magnetic Properties; *Springer-Verlag*: Berlin, 1983.

[5] Güntherodt H.–J.; Beck H. Glassy Metals-I: Ionic Structure, Electronic Transport and Crystallization; *Springer-Verlag*: Berlin, 1981.

[6] Rowson H. Glasses and their Applications; The Institute of Metals: London, 1991.

[7] Elliot S. R. Physics of Amorphous Materials; Longman Scientific and Technical: London, 1990.

[8] Cusack N. E. The Physics of Structurally Disordered Solids; Institute of Physics: Adam Hilger, England, 1987.

[9] Anantharaman T. R. Metallic Glasses: Productions, Properties and Applications; Trans Tech: Switzerland, 1984.

[10] Zallen R. The Physics of Amorphous Solids; John Wiley and Sons: New York, 1983.

[11] Luborsky F. E. Amorphous Metallic Alloys; Butterworths: London, 1983.

[12] Waseda Y. The Structure of Non-Crystalline Materials; McGraw-Hill Int. Book Com: New York, 1980.

[13] Mitra S. S. Physics of Structurally Disordered Solids; Plenum Press: New York, 1976.

[14] Doremus R. H. Glass Science; John Wiley and Sons: New York, 1973.

[15] J. Hafner, Acta Mater. 2000, Vol. 48, pp 71-92.

[16] Vora Aditya M. Ph. D. Thesis; Sardar Patel University: INDIA, 2004.

[17] Gajjar P. N.; Vora A. M.; Jani A. R. Proc. 9th Asia Pacific Phys. Conf., The Gioi Publ.: Vietnam, 2006, pp 429-433.

[18] Vora A. M.; Patel M. H.; Gajjar P. N.; Jani A. R., Solid State Phys. 2003, 46, 315-316.

[19] Vora Aditya M. Chinese Phys. Lett. 2006, Vol. 23, pp 1872-1875.

[20] Vora Aditya M. *J. Non-Cryst. Sol.* 2006, Vol. 52, 3217-3223.

[21] Vora Aditya M. *J. Mater. Sci.* 2007, Vol. 43, 935-940.

[22] Vora Aditya M. *Acta Phys. Polo*. A. 2007, Vol. 111, 859-871.

[23] Vora Aditya M. Front. *Mater. Sci.* China 2007, Vol. 1, 366-378.

[24] Vora Aditya M. FIZIKA A. 2007, Vol. 16, pp 187-206.

[25] Vora Aditya M. *Romanian J. Phys*. 2008, Vol. 53, pp 517-533.

[26] Vora Aditya M. Inter. J. Thero. Phys., Group. Therory and Nonlin. Opt. 2008, Vol. 13, pp 31-50.

[27] Vora Aditya M. *Glass Phys. Chem.* 2008, Vol. 34, pp 671-682.

[28] Vora Aditya M. *Turkish J. Phys.* 2008, Vol. 32, pp 323-330.

[29] Agarwal P. C.; Aziz K. A.; Kachhava C. M. *Acta Phys*. Hung. 1992, Vol. 72, pp 183-192.

[30] Agarwal P. C.; Aziz K. A.; Kachhava C. M. *Phys. Stat. Sol.* (b). 1993, Vol. 178, pp 303-310.

[31] Aziz K. A.; Ray A. K. *J. Mater. Sci.: Mater. Elec*. 1997, Vol. 8, pp 7-8.

[32] Gupta A.; Bhandari D.; Jain K. C.; Saxena N. S. Phys. Stat. Sol. (b). 1996, Vol. 195, pp 367-374.

[33] Hubbard J.; Beeby J. L. *J. Phys. C: Solid State Phys*. 1969, Vol. 2, pp 556-571.

[34] Takeno S.; Goda M. Prog. Thero. Phys. 1971, Vol. 45, pp 331-352.

[35] Takeno S.; Goda M. Prog. Thero. Phys. 1972, Vol. 47, pp 790-806.
[36] Bhatia A. B.; Singh R. N. *Phys. Rev* B. 1985, Vol. 31, pp 4751-4758.
[37] Shukla M. M.; Campanha *J. R. Acta Phys. Pol.* A. 1998, Vol. 94, pp 655-660.
[38] Harrison W. A. Elementary Electronic Structure; World Scientific: Singapore, 1999.
[39] Taylor R. *J. Phys. F: Met. Phys*. 1978, Vol. 8, pp 1699-1702.
[40] Ichimaru S.; Utsumi K. Phys. Rev. B. 1981, Vol. 24, pp 7385-7388.
[41] Farid B.; Heine V.; Engel G.; Robertson I. *J. Phys. Rev.* B. 1993, Vol. 48, pp 11602-11621.
[42] Sarkar A.; Sen D. S.; Haldar S.; Roy D. Mod. Phys. Lett. B. 1998, Vol. 12, pp 639-648.
[43] Wills J. M.; Harrison W. A. *Phys. Rev* B. 1983, Vol. 28, pp 4363-4373.
[44] Sulir L. D.; Pyka M.; Kozlowski A. *Acta Phys. Pol.* A. 1983, Vol. 63, pp 435-443.
[45] Sulir L. D.; Pyka M. *Acta Phys. Pol.* A. 1983, Vol. 63, pp 747-754.
[46] Faber T. E. Introduction to the Theory of Liquid Metals; Cambridge Uni. Press: London, 1972.
[47] Kovalenko N. P.; Krasny Y. P. Physics B. 1990, Vol. 162, pp 115-121.
[48] Bretonnet J. L.; Derouiche A. *Phys. Rev.* B. 1990, Vol. 43, pp 8924- 8929.
[49] Thorpe M. F. *J. Non-Cryst. Sol.* 1983, Vol. 57, 355-370.

In: Computational Materials
Editor: Wilhelm U. Oster

ISBN: 978-1-60456-896-7

Chapter 4

A COMPUTATIONAL STUDY OF THE PHONON DYNAMICS OF SOME COMPLEX OXIDES

Prafulla K. Jha[1 1] ***and Mina Talati***[2]

[1] Computational Condensed Matter Physics and Nanophononics Laboratory, Department of Physics, Bhavnagar University, Bhavnagar-364022, India

[2] Laboratoire de Physique de la Matière Condensée et Nanostructures (Laboratoire PMCN) Université Lyon 1; CNRS, UMR 5586 Bat. Léon Brillouin Domaine Scientifique de la Doua F-69622 Villeurbanne cedex;. France

ABSTRACT

Understanding the mechanism of strong correlation effects of spin, charge and lattice responsible for the fascinating properties such as large Resistivity changes, huge volume changes, high T_c superconductivity, strong thermoelectric response, gigantic non linear optical effect and finally colossal magnetoresistance (CMR) in the transition metal oxides is an important challenge. One of the central issues of concern in many manganites is to understand the occurrence of colossal magnetoresistance around the insulator (I) to metal (M) transition along with the paramagnetic to ferromagnetic transition at same temperature. Double exchange interaction and electron-phonon interaction triggered by Jahn-Teller (JT) distortions of the MnO_6 octahedral have been proposed. In this situation, correlated information on the phonon properties is essential for a detailed understanding of the lattice distortion in these technological and fundamentally important compounds. The present chapter focuses on recent advances in understanding the phonon properties at ambient condition and far from the ambient to correlate the JT distortions and magnetoresistance in manganites. LaMnO3, which undergoes phase transition at high temperature, is considered in its cubic and rhombohedral phase. The chapter also considers the study of the effect of Sr-doping at La-site. The difference in structural symmetry of cubic and rhombohedral manganites is manifested in their phonon spectra and Phonon density of states. In addition, to understand the effect of pressure and temperature, 30% Sr-doped $LaMnO_3$ *i.e.* $La_{0.7}Sr_{0.3}MnO_3$ (LSMO) is considered at

[1] email : prafullaj@yahoo.com, pkj@bhavuni.edu.

different applied pressure and temperature. The phonon properties are also reported for the $NaCoO_2$ compound in its two different geometry positions.

Keywords : *Transition Metal Oxides, Manganites, Colossal Magnetorsistance, Jahn-Teller distortion, Phonons, Cobalt Oxides.*

1. Introduction

1.1. Transition Metal Oxides

The discovery [1] of high-temperature superconductivity in the cuprates renewed interest in the study of transition metal oxides (TMO). The transition metal oxides form a large, rich and still not well understood class of materials in fundamental physics as well as in technological applications have been attracting intense attention from the condensed matter community [2]. This fascinating class of inorganic solids exhibits a wide variety of exotic and imperfectly understood structures, properties, and phenomena [3, 4], the origin of which is not yet clear and remains an important problem in solid state physics. Its difficulty arises from the fact that the valence electrons in these materials have a strong mutual interaction. The approximation that is often made to solve the Hamiltonian of a solid containing $\sim 10^{23}$ electrons is that these electrons are non-interacting, or that their interaction can be treated in some average way. The material properties resulting from strong interactions can be controlled in various ways: temperature, mechanical pressure, chemical composition, oxygen concentration, external magnetic field, and electric field providing wide practical applications. Exploiting these properties in novel electronic devices remains an active area of research. In this context, transition metal oxides represent a veritable playground for condensed matter physicists and materials scientists.

Physical systems, which are understood well, correspond to ensembles of free particles. For example, semiconductors and most metals can be described as having non-interacting electrons. This simple approach is valid because the interaction (Coulomb) energy of electrons is much smaller than their kinetic energy. Another example is alkali atoms, that Bose condense at low temperatures. Alkali atoms can be treated as non-interacting bosons because their scattering length (*i.e.* the length at which they interact with each other) is much smaller than the average distance between the particles. However, there are important systems for which interactions between the electrons are not weak. In such systems, the on-site Coulomb repulsion energy (the cost of putting two electrons on the same lattice site) between electrons (U) is comparable to or larger than the bandwidth W. Therefore, the electrons no longer remain independent but become "correlated" and their strong correlation play a major role in determining the properties of such "strongly-correlated electron systems (SCES)". The study of these strongly-correlated electron systems is the foremost area of research in contemporary condensed-matter physics. A characteristic feature of the physics in transition-metal oxides is that the charge, spin, and lattice degrees of freedom are strongly coupled and the key ingredient controlling these strong couplings is the orbital degree of freedom (ODF) [5, 6](cf. Figure 1).

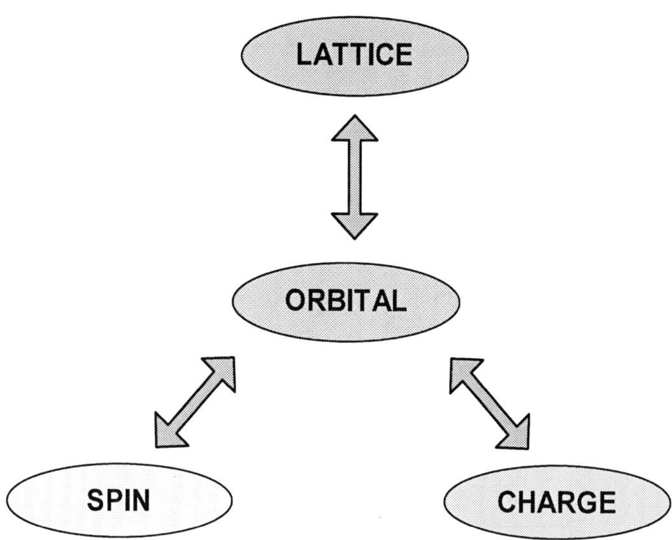

Figure 1. Schematic diagram of strong interplay among spin, charge and lattice degrees of freedom existing in transition metal oxides (TMO). The orbital degree of freedom plays crucial roles in mediating these coupling.

A recent discovered example of transition metal oxides is vanadates (V_2O_3), which undergoes a metal-insulator transition with a seven orders of magnitude change in the electrical conductivity [7]. Several other examples of systems belonging to this class are conventional superconductors, high-T_C superconductors[1], the "colossal"[8] magnetoresistance [9] manganese oxides [10], heavy fermions, Quantum Hall systems, one dimensional electron systems, insulating state of bosonic atoms in a periodic potential (optical lattices) and magnetic systems exhibiting variety of ordering patterns due to the interplay of their spin, charge and lattice degrees of freedom.

A class of Mott insulators, namely transition metal oxides is a part of strongly correlated systems where the localized outer *d*-electrons of the transition metal ions are largely responsible for the interesting electronic and magnetic properties. The on-site Coulomb energy is so huge in such compounds that it localizes the *d*-electrons on each lattice site leading to vanishing electrical conductivity [2]. In other words, the movement of electrons is hindered by the Coulomb repulsion leading to the formation of a conductivity gap. The description of insulating state by strong Coulomb repulsion is now commonly known as Mott-Hubbard theory [3]. Hund's rules for the partially filled *d*-levels provide for the large magnetic moment. The coupling of these moments through both direct and indirect exchange mechanisms produce various long-range magnetic order (depending on the sign of the exchange and the details of the coupling) including ferromagnetic, antiferromagnetic, and ferrimagnetic. In addition to magnetic ordering, TMOs exhibit various other phase separations: charge ordering, orbital ordering, and ferroelectric ordering.

1.1.1.Specific TMO Compounds

Transition metal oxides, existing in a vast number display a wealth of interrelated properties and yield a rich field of fundamental research and technological applications. Versatile properties of transition metal oxides are at least, enough to engage the condensed matter and materials science community for quite some time. This chapter focuses on vibrational dynamics of a specific class of such systems, displaying magnetic and strong correlation effects, which includes the manganites and cobaltates, intrinsic phase of water bound recently discovered superconducting compound ($Na_xCoO_2{\cdot}y\,H_2O$) [11].

Next to the high -T_C cuprates, the manganites represent possibly the next well-known TMO system owing to the discovery of colossal magnetoresistance. Manganites ($RMnO_3$) form different crystal structures depending on the size of the rare-earth ion *R*. Figure 2 presents Shannon ionic radii [11] for the lanthanide series (circles). The horizontal line approximately separates the orthorhombic and hexagonal phases.

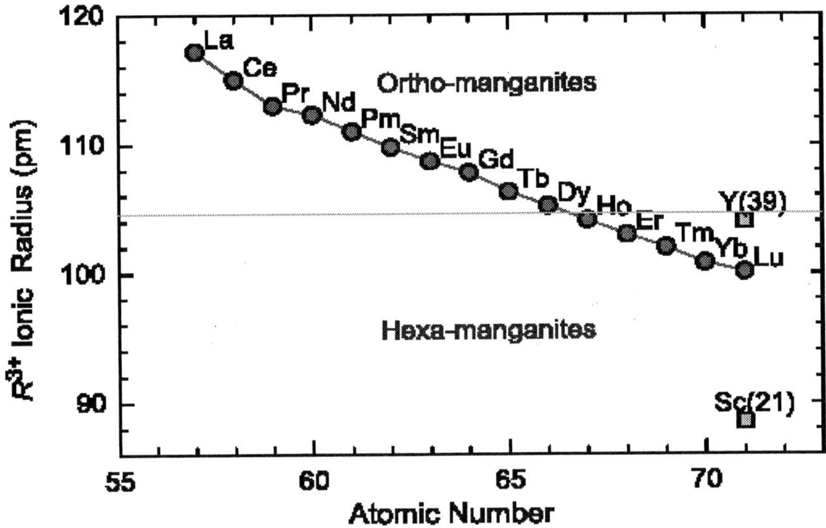

Figure 2. Ionic radii of R3+ ions. Smaller ionic radii form a hexagonal crystal structure.

For ionic radius larger than approximately 105 pm, the manganites stabilize in the orthorhombic phase. For smaller radii, the hexagonal crystal is realized. Yttrium and scandium based manganites also crystallize in hexagonal lattice other than lanthanides. Moreover, the synthesis conditions or the application of mechanical stress decides the exact phase of the ions lying near the separatrix (*e.g.* Y or Ho) between two crystal structures. The remainder of this chapter will be dedicated to introducing the structural, electronic, and vibrational properties of manganites.

1.1.2.Strong Correlation Effects on Condensed Matter Systems

Understanding the strong correlation effects is an important challenge for some of the fascinating properties of condensed matter systems such as large resistivity changes, huge volume changes, high T_C superconductivity, strong thermoelectric response, gigantic non-linear optical effect and finally, colossal magnetoresistance effect. At this point, there is no clear understanding of the mechanism, which leads to a rich variety of behaviors in this class of compounds. However, the active role of outer atomic electrons can not be overlooked in the origin of such enormous variety of properties. Two limiting descriptions of the outer atomic electrons in solids consist of the localized or ligand-field, appropriate for insulators, and the delocalized band theory, appropriate for metals. For tightly bound electrons and weak interatomic interactions, localized picture is most effective. This corresponds to a small electron bandwidth W relative to the electron-electron Coulomb energy U, *i.e.*, $W << U$. On the other hand, band theory is applicable for appreciable overlap of neighboring orbitals, in which case $W >> U$. Typically, the *d* electrons in transition metal oxides exhibit narrow bands overlapping with relatively broad *s-p* bands and neither limit applies. In this intermediate case, $W \sim U$ and the behavior lies between local and band theory. Theoretical models describing this strong correlation regime attempt to explain the interactions responsible for the tendency of electrons to localize more than predicted from noninteracting pictures. Electron-electron (el-el), electron-phonon (el-ph) and spin-phonon coupling are examples of these interactions. The strong correlation regime leads to both localized and itinerant electron behavior. Intra-atomic exchange (Hund's rule splitting) and electron-phonon interactions favor localized electrons [3]. At finite temperatures el-el and el-ph interactions become important. For el-ph interactions, Frohlich's coupling constant λ characterizes the strength of the interaction. Weak coupling gives rise to large polarons with itinerant transport, while strong coupling produces small polarons that tend to localize and display activated hopping transport.

One of the best studied and simplest models including strong correlation effects is that due to Hubbard [12]. This single band model describes the splitting of bands into upper and lower Hubbard bands. As W increases the upper and lower bands come together and the energy gap disappears. The Hubbard model is given [12] by

$$H_{Hub} = -\sum_{i,\delta} t_\delta \left(d^\dagger_{i\alpha} d_{i+\delta\,\alpha} + H.c. \right) + U \sum_i n_{i\uparrow} n_{i\downarrow} \tag{1}$$

where i denotes a lattice site, δ is the displacement vector between sites, α is the spin, and $n_{i\uparrow}\left(n_{i\uparrow}\right)$ denotes the electron density on site i with spin up (down). For $n = 1$ electron per site and finite U, the ground state is insulating, known as a Mott Insulator. With increased hole-doping, $x = 1 - n$, the electron-electron interaction tendency decreases. This results in a metal-insulator transition (Mott transition) at some critical doping concentration. The Mott-Hubbard model is appropriate for magnetic insulators, *e.g.* 3*d* TMOs with a small electronic bandwidth W^3. A single band model, the Hubbard model is relevant only for t and U small compared to the energy gap between other bands and may not apply for transitions to other levels, for example O 2*p* to Mn 3*d*.

In addition to mutual interaction, outer electrons interact with the core spins localized on the transition metal ions. Coupling of the itinerant carriers with the core spins are important for magnetic TMO (*e.g.* the hole-doped orthorhombic manganites). The Kondo lattice model adds a term to the Hamiltonian that includes carrier interactions with the core spins

$$H_{Hund} = -J_H \sum_{\substack{i,a \\ \alpha,\beta}} S_i \cdot \sigma_{\alpha\beta}\, d^\dagger_{i\alpha} d_{i\beta} \tag{2}$$

where J_H denotes the Hund's coupling, S_i denotes the core spin on site i, and $\sigma_{\alpha\beta}$ denotes the carrier spin. In the limit where the coupling energy $J_H S_C$ is large compared to W, the carrier spins align with the core spins.

For the relatively narrow bands of the transition metal oxides, the tight binding model plus interactions well-describes the calculated band structure. The tight-binding Hamiltonian has the form

$$H_{Hub} = -\sum_{\substack{i,\delta \\ ab}} t_{\delta}^{ab} \left(e^{i\frac{e}{c}A\cdot\delta} d_{ia\alpha}^{\dagger} d_{i+\delta b\alpha} + H.c. \right) + H_{int} \tag{3}$$

where i represents lattice sites, δ denotes a displacement connecting two lattice sites, a and b represent electron orbitals on a given site, and H_{int} represents additional interaction effects. Interactions may take the form of Hund's exchange coupling, el-el interactions, el-ph interactions, etc. Hopping matrix elements t_{δ}^{ab} are estimated from fits to band structure calculations. The introduction of the Peierls phase term $e^{i\frac{e}{c}A\cdot\delta}$ affords consideration of the interaction with electromagnetic radiation relevant to optical conductivity studies. The tight binding parameterization, valid only for the nearest neighbor hopping, facilitates calculations over the full Hamiltonian approach.

1.1.3. Practical Applications

Transition metal oxides offer promising applications in the emerging field of magnetoelectronics or "spintronics", the new approach to electronics, which makes use of the spin of the carriers. Possible "spintronic" devices offer the possibility of combining information processing with non-volatile information storage. Potential device applications include [13](i) spin valves, (ii) nonvolatile memory, (iii) spin-based field-effect transistors (FET), (iv) spin-based light emitting diodes (LED), and (v) Spin resonant tunneling diodes (RTD). The success of these devices depends on understanding spin interactions in solid state systems, including the role of dimensionality, defects, and band structure [13]. One of the most pursued applications is the implementation of "non-volatile" random access memories (RAM). In the existing semiconductor technology, a power failure wipes out the RAM because transistors need an applied voltage to work. A spin valve, which acts as a logic gate, would not spend so much energy and would not lose the information when switched off. Moreover, the response time for such devices could be as small as nanoseconds.

The rapid realization of discovery in product of the spin-polarized transport employed in giant magnetoresistance (GMR) materials significantly advanced magnetic storage capacity. GMR materials consist of alternating layers between magnetic and nonmagnetic thin films. An external field polarizes spins in the magnetic layers and minimizes spin-dependent scattering. As a result, large resistive changes occur for small changes in applied field, making GMR materials sensitive magnetic read heads. The GMR effect, initially reported [14] in 1988, first found commercial application in magnetic field sensors in 1994 and in

HDD read heads in 1997. The successful application of the GMR effect fuels interest in related materials with even larger magnetoresistance. These aptly-named "colossal magnetoresistance" (CMR) materials have attracted a great deal of attention, and generated a wealth of unexpected condensed matter phenomena. However, the CMR materials have not yet enjoyed the applied success of their merely giant cousins.

The application of GMR materials in magnetic read heads produced immediate benefits to information technology. A broader goal involves integration of both spin and charge functionality not currently existing in either ferromagnets or semiconductors alone [15]. Such functionality includes demonstrating control of the magnetic ordering by the application of an electric field and vice versa. Several groups have demonstrated this extra degree of freedom in TMO systems. Ohno *et al.*[16] have demonstrated electric field control of ferromagnetism in (In,Mn)As heterostructures. Cheong *et al.* [17] have demonstrated similar effects in the multiferroic $TbMn_2O_5$.

1.2. Colossal Magnetoresistance and Manganese Oxide Perovskites: Fundamental Features

1.2.1. Colossal Magnetoresistance and Applications

The mixed-valence manganese oxide perovskites (here onwards, *manganites*) have been studied for more than five decades but are still considered modern materials because of their wide potential for technological applications. The remarkable phenomenon from which the hole-doped manganites derive their name is the observed magnetoresistance. It is a property of magnetic materials, which is being crucial for a rapid development of new technologies. Other compounds such as double perovskites, manganese oxide pyrochlores and europium hexaborides, among others, have presented the same striking property.

Magnetoresistance (MR) is defined as the change in the electrical resistance produced by the application of an external magnetic field. It is usually given as a percentage and expressed as

$$MR = 100 \times \frac{\rho(H) - \rho(H = 0)}{\rho(H = 0)} \qquad (4)$$

where $\rho(H)$ and $\rho(H = 0)$ designate resistivity in the zero field and in the applied field, respectively. With this definition, magnetoresistance has a maximum value of 100%. The decrease in resistivity in field [$\rho(H) < \rho(H = 0)$] leads to a negative magnetoresistance. Figure 3 (a) shows % MR as a function of applied field. Usually, the magnitude of the magnetic fields necessary to get this large magnetoresistance ranges in the order of several Teslas.

The magnetoresistance effect is common for all metals and is usually quite low, $Fe_{0.2}Ni_{0.8}$ permalloy for example, has MR ~ 3%. In metallic multilayer thin films, magnetoresistance can be significantly enhanced, at least at low temperatures, and MR ratios from 5 to 150% are achieved [8].

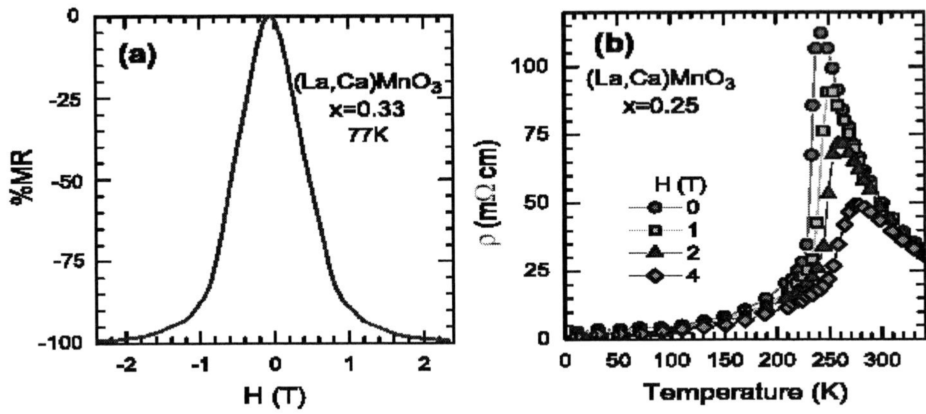

Figure 3. Magnetic-field dependent resistivity in (La, Ca) MnO3 for x ≈ 0.3. (a) Magnetoresistance as a function of applied magnetic field for x = 0.33 after ref. [8]. (b) Resistivity versus temperature for x = 0.25 at different applied magnetic field ref. [18].

This phenomenon is called giant magnetoresistance, GMR [14] and it involves the interfaces of the artificial layers in the thin film. Jin and coworkers [8] observed ~ 1300 % MR near room temperature and ~ 127000 % MR near 77 K, in thin films of $La_{0.67}Ca_{0.33}MnO_3$. The MR observed in single crystalline $Na_{0.5}Pb_{0.5}MnO_3$ [9] and thin films of $La_{0.67}Ca_{0.33}MnO_3$ [8] is an intrinsic effect of the material, not induced by the interfaces, like GMR. This type of magnetoresistance is called colossal magnetoresistance (CMR).

Magnetoresistance could also be caused by extrinsic properties. It arises in polycrystalline samples and artificially created barriers. In these cases, low magnetic fields (~ 100 *Oe*) can produce a decrease of the resistance in a wide range of temperature below T_C. The CMR effect is temperature dependent and is so large due to a metal-insulator transition, which takes place as the magnetic field forces the ordering of the spins. Figure 3 (b) shows the resistivity as a function of temperature for several different externally applied magnetic fields [18]. As the applied field increases, the resistivity drops dramatically and the resistivity peak shifts to higher temperatures. Potential applications of the CMR effect in mixed-valence manganites include magnetic sensors, magnetoresistive read heads, magnetoresistive random access memory (MRAM), and bolometers etc. (cf. Figure 4).

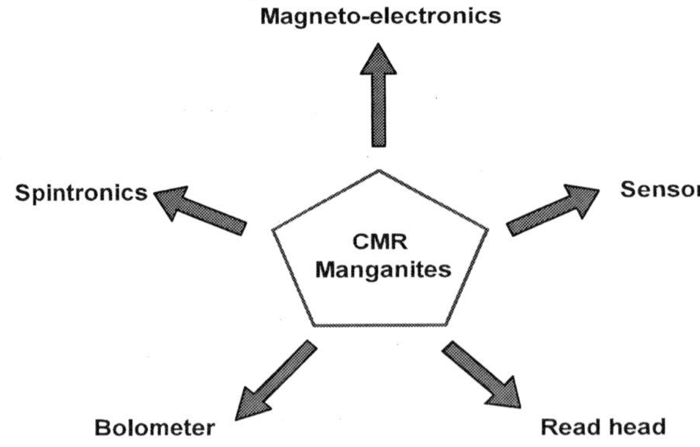

Figure 4. Schematic diagram displaying applications of CMR materials.

Read heads for magnetic recording devices are one of the most straightforward applications of the magnetoresistance. A material with a large magnetoresistance is a very good magnetic sensor. The magnetic fields emanating from the magnetic media change the magnetization of the read head and its resistivity changes correspondingly. The larger the magnetoresistance, the smaller the change in magnetization that is required for the transport through the read head to change sensitively. GMR multilayer structures have already enabled the construction of better "read" heads for magnetic hard disk drives [19], which have allowed storage densities to be increased. The storage density before GMR (*i.e.* 1996) was 1 Gigabyte per inch squared (Gb/in^2). In the year 2000 the density achieved thanks to GMR was 20 Gb/in^2. CMR, as direct inheritor of GMR, could further improve these numbers. In addition to colossal magnetoresistance, the ferromagnetic mixed-valence manganites may also exhibit a very large magnetocaloric effect, and they seem to be promising candidates as working substances in magnetic refrigeration technology. Manganites play an important role as electrodes in solid oxide fuel cells and they may also be used for catalysis and in oxygen sensors.

1.2.2. Manganese Oxide Perovskites: Description and Crystal Structures

1.2.2.1. ABO_3 Structure

Stoichiometric perovskites oxides of the empirical formula ABO_3 form a large family of oxides. The ideal perovskite structure is cubic, and many perovskites adopt this structure at high temperature, but are distorted at lower temperatures. The structure may be conceived as a close-packed array formed of O^{2-} anions and A cations with B cations located at the octahedral interstitial sites. The BO_6 octahedra make contact to each other by their vertices and form a three-dimensional network. Figure 5 (a) represents a unit cell of prototype cubic perovskite, ABO_3 with A ions situated at the corners, the B ions in the center and the oxygen ions at the centers of the faces. $SrTiO_3$ is one such prototype cubic perovskite. Most of the ABO_3 perovskites exhibit a distortion of the cubic cell, except $SrTiO_3$ which is an exception in the group of $ATiO_3$ titanates. Even the mineral perovskite $CaTiO_3$ is not cubic after which the "perovskite" name was coined. Stoichiometric ABO_3 perovskites can be classified into three categories according to the valence of the A and B elements: $A^{I}B^{V}O_3$ (A = Na, Ag, K and more rarely Rb, Tl, Cs; B = Nb, Ta, I), $A^{II}B^{IV}O_3$ (A = Ba, Sr, Ca, Pb and more rarely Cd; B = Ti, Sn, Zr, Hf, Mn, Mo, Th, Fe, Ce, Pr, U), $A^{III}B^{III}O_3$ (A = Ln, Bi, Y; B = Fe, Cr, Co, Mn, Ti, V, Al, Sc, Ga, In, Rh). Many of the ferroelectric oxides belong to the first two series and are characterized by a small distortion of cubic cell. In the third family, most of the compounds exhibit the same kind of the distortion and are orthorhombic; these perovskite are the $GdFeO_3$ type perovskite.

1.2.2.2. $R_{1-x}A_xMnO_3$ Structure

Manganites are three dimensional perovskites and have the general formula, $R_{1-x}A_xMnO_3$, where *R* stands for the trivalent rare earth element such as La, Pr, Nd, Sm, Eu, Gd, Ho, Tb, Y, and etc. or for Bi^{3+} and A for the divalent alkaline earth ions such as Sr, Ca and Ba or for Pb^{2+}. '*x*' is the hole doping concentration. The (*R*, A) site (so-called perovskite A-site) can in most cases form homogeneous solid solution. It is quite clear from the general formula of manganites that the B-sites are occupied by Mn ions.

(a)

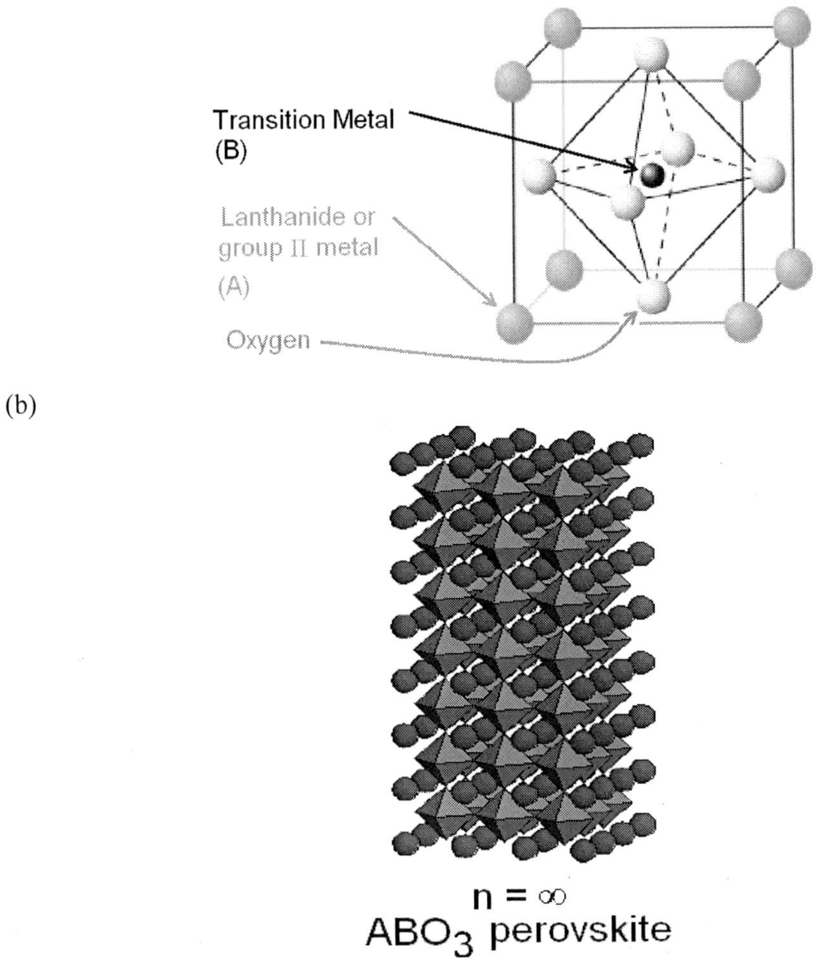

Figure 5. (a) Unit cell structure of "classic" cubic perovskite, ABO_3 (b) Schematic crystal structure of Ruddlesden-Popper phase of the ideal cubic perovskite.

Other important structure derived from the perovskite structure is the Ruddelsden-Popper series or the so-called layered perovskite structure. Figure 5(b) illustrates the Ruddelsden-Popper phase of the ideal cubic perovskite structure. The layered materials have the general formula $(R_{1-x}A_x)_{n+1}Mn_nO_{3n+1}$, where n is the number of corner-shared MnO_6 octahedral sheets forming the layer. Three dimensional perovskite correspond to $n = \infty$, double layered to $n = 2$ and single layered to $n = 1$ [20].

The stability of the perovskite-based structures depends strongly on the A-site and B-site ions. If there is a size mismatch between the A-site and B-site ions and the space in the lattice where they reside, the perovskite structure will become distorted forming the rhombohedral and orthorhombic lattices (so called $GdFeO_3$ - type). Another possible origin in the lattice distortion is the deformation of the MnO_6 octahedron arising from the Jahn-Teller effect that is inherent to the high spin ($S = 2$) Mn^{3+} with double degeneracy of the e_g orbital. In these distorted perovskites, the MnO_6 octahedra show alternating buckling. Such a lattice distortion of the perovskite in the form of ABO_3 is governed by a tolerance factor f [21], which is defined as

$f = (r_A + r_O) / (r_B + r_O)$ (5)

Here, r_A and r_B are the mean radii of the ions occupying the A-site and B-site, respectively, and r_O is the ionic radius of the oxygen. The tolerance factor f measure, by definition, the lattice-mismatching of the sequential AO and BO_2 places. For the ideal cubic perovskite $f = 1$. As r_A or equivalently f decreases, the lattice structure transforms to the rhombohedral ($0.96 < f < 1$) and then to the orthorhombic structure ($f < 0.96$), in which the B-O-B bond angle is bent and deviated from 180°. Doped manganites have other kinds of distorted perovskite structures depending on the nature of the dopant, the degree of doping, temperature, pressure etc. which leads to their distinct physical properties. In the present chapter, the vibrational properties are at main focus and hence the effect of external parameters such as doping concentration, pressure, temperature etc. on the vibrational properties of manganites, particularly lanthanum manganites, is discussed in the subsequent sections.

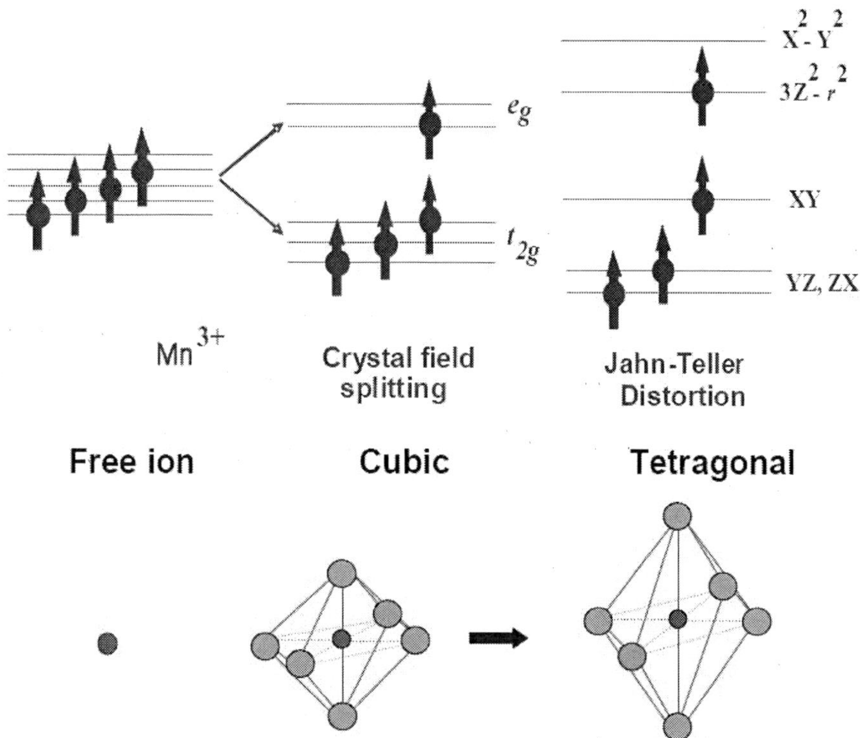

Figure 6. Crystal field splitting of Mn d-levels resulting from cubic and tetragonal crystal fields.

1.2.3. Electronic Structure and Double Exchange (DE) Model

The manganese (Mn) ion having an incomplete *d*-shell (Mn: [Ar] $3d^5 4s^2$) is surrounded by the oxygen octahedron in the CMR manganites. These *d*-electrons comprise the key players determining the electronic properties of the manganites. The number of electrons per Mn is $4 - x$. Hund's rule constrains the pairing of unpaired electrons spin in the outer *d*-shell in order to minimize the energy. Thus, only five *d*-levels with majority parallel spins are accessible. In isolated atoms, the five *d*-orbitals are degenerate but they split in to a lower t_{2g}

triplet (d_{xy}, d_{yz} and d_{zx}) and an upper e_g doublet ($d_{x^2-y^2}$ and $d_{3z^2-r^2}$) in the cubic crystal environment.

Figure 6 shows an energy level diagram of crystal field splitting of the Mn *d*-levels. The upper-lying e_g orbitals are higher in energy than t_{2g} orbitals because the lobes of maximal probability of t_{2g} orbitals point away from both the apical and the equatorial oxygen ions and those of e_g orbitals direct towards oxygens. Mn^{4+} has three electrons in the outer *d*-shell that can be considered as localized in the three t_{2g} levels giving a total spin $S = 3/2$ (core spin). The two e_g levels remain empty. On the other hand, Mn^{3+} has an extra electron that fills one of the e_g levels ($S = 2$). Mn^{3+} is known to be a strong Jahn-Teller ion in octahedral sites *i.e.* it can benefit energetically from a distortion of the oxygen octahedron [22]. Tetragonal distortions, arising from cation disorder and the Jahn -Teller effect, introduce additional crystal field splitting of the degenerate e_g levels by energy E_{JT} and partially lift the degeneracy of the t_{2g} levels as seen in Figure 7. e_g levels are the active ones for conductivity whose hybridization with oxygen *p*- levels [23] constitutes the conduction band and bandwidth depends upon the extent of the overlap of e_g orbitals of Mn and *p*-levels of oxygen. The antiparallel spin levels are very high in energy, which implies that only majority parallel spins can conduct. For this reason, manganites are called half-metals. Tilting of octahedra can be measured with the distortion of the Mn-O-Mn bond angle θ which is 180° for cubic symmetry and can range from 150° to 180° for particular compositions. This distortion affects the conduction band which appears as the hybridization of *p*-level of the oxygen and e_g levels of Mn. The orbitals overlap decreases with the distortion and the relation between the bandwidth W and θ has been estimated as $W \sim \cos^2\theta$ [24]. Also, when the bond-angle decreases, the residual resistivity value ρ_0 increases by orders of magnitude. Bandwidth is closely related to the magnitude of the critical temperature T_C which could be increased by chemical pressure, namely, by choosing the right cations A and B (or, equivalently, by the application of external pressure). Two parameters are important here, the mean value of the cation sizes $\langle r_{A/B}\rangle$ and its variance $\sigma^2 = \langle r^2_{A/B}\rangle - \langle r_{A/B}\rangle^2$ [25]. The main result is that a decrease of the cation disorder implies an increase of the critical temperature. The exchange coupling associated with the electron transfer between Mn^{3+} and Mn^{4+} site is called the double exchange. This double-exchange model (DE) first proposed by Zener [20] explains the metal-insulator phase transition in terms of the Mn *d*-electrons, namely the strong Hund's coupling between the three electrons localized in the t_{2g} orbitals and the 1-*x* electrons in the e_g orbitals.

In this model, the e_g electrons hop from Mn^{3+} site to adjacent Mn^{4+} site via the interstitial O^{2-} ion as seen in Figure 7, providing the alignment of the core spins. Parallel alignment of the core spins maximizes the hopping probability between Mn sites for the electrons residing on the e_g orbitals, producing metallic conduction for $T < T_C$. Thus, ferromagnetic order favours hopping in the double exchange picture [26]. The effective hopping interaction (neglecting the Berry phase of Anderson- Hasegawa [26] is given by,

$$t_{ij} = t_{ij}^0 \cos\left(\theta_{ij}/2\right) \quad (6)$$

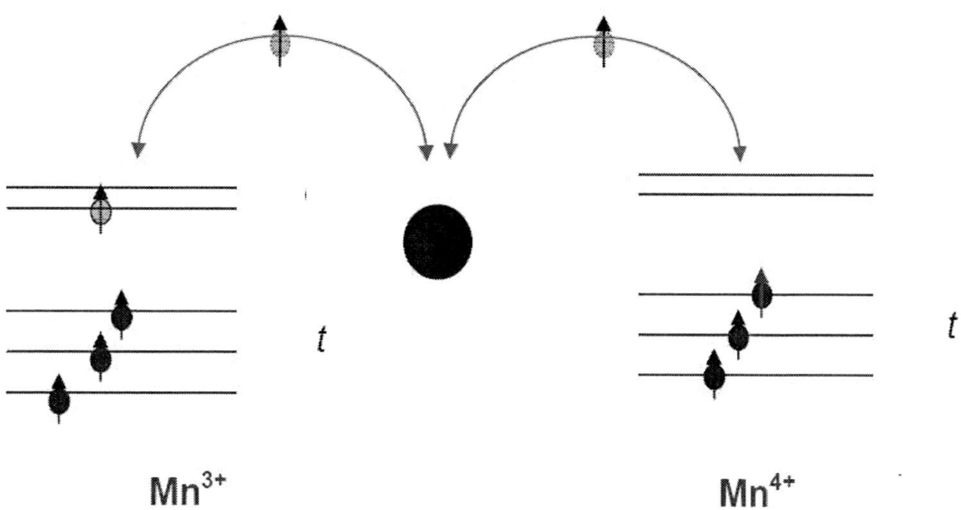

Figure 7. Schematic diagram of double exchange coupling associated with the electron transfer between Mn3+ and Mn4+ site of eg orbitals via interstitial O^{2-} ion.

where, t^0_{ij} is the spin-independent hopping matrix element and θ_{ij} is the angle between neighboring spins. The hopping parameter given by eq. (6) qualitatively describes the observed transport behaviour. The increase of electron vacancies with hole doping affords hopping of e_g electrons from Mn^{3+} to Mn^{4+} sites. At low temperatures, ferromagnetic alignment of the Mn spins maximizes the kinetic energy of the carriers and stabilizes the ferromagnetic state [5]. As temperature increases above T_C, the spins order randomly and the effective hopping reduces. Mainly, double-exchange underlies the complex behaviour of manganites. This model has been deeply studied and, in certain regimes, qualitative agreement with experiments has been found. In particular, magnetic properties are better described than electric properties. But some manganites, such as hole doped $La_{1-x}Sr_xMnO_3$ show metallic behavior for the whole range of the temperatures [27] While the recent investigations bring out the essential role of DE, they also suggest the competing role of the lattice and the electron-lattice interactions to understand the physical properties of these systems. In understanding the novel transport properties, consideration of the overall lattice symmetry as well as local lattice distortions is crucial on equal footing.

1.2.4. Jahn-Teller Small Polaron

In the doping range $0.2 < x < 0.5$, $R_{1-x}A_xMnO_3$ is a ferromagnetic metal at low temperature T and a poorly conducting paramagnet at high temperature; the paramagnetic-ferromagnetic transition occurs at an x dependent transition temperature $T_C(x) \sim 300$ K and is accompanied by a large drop in the resistivity [28, 29]. The colossal magnetoresistance, which has stimulated the recent interest in these materials, is observed for temperatures near $T_C(x)$[8]. The double exchange [20,26,30] has been widely regarded [31, 32, 9, 33] as the only significant physics in the regime $0.2 < x < 0.5$. However, the double exchange alone cannot account for the very large resistivity above T_C [34] or for the sharp drop in resistivity just below T_C. This inadequacy of DE model led to the new physics involving a strong electron-phonon coupling due in part to a Jahn-Teller (JT) [35, 36] splitting of the Mn e_g states. The cubic-tetragonal phase transition observed for $0 \leq x \leq 0.2$ is known to be due to a frozen-in (static) Jahn-Teller distortion with long-range order while it is dynamic for $x > 0.2$ [35, 36]. Jahn-Teller theorem states that a system having a degenerate ground state will spontaneously deform to lower its symmetry unless the degeneracy is simply spin degeneracy [37].This theorem applies to Mn^{3+} breaking the degeneracy of the two e_g levels as shown in Figure 7. The octahedra deformations (Q_1, Q_2, and Q_3 modes) imply local changes in the Mn-Mn distances and are associated to the splitting of the e_g levels when Mn^{3+} is involved. Q_1 is the breathing distortion that occurs due to the different sizes of Mn^{4+} and Mn^{3+}. Q_2 and Q_3 are the two Jahn-Teller modes of distortion of the oxygen octahedra associated to the splitting of the e_g levels of the Mn^{3+} [35]. The interaction between Jahn-Teller distortion modes and orbitals is called cooperative Jahn-Teller. Therefore, the carrier travels in the paramagnetic state of the doped manganites with an associated lattice distortion forming lattice polarons, which is a self-trapping effect.

Experimental results, which support the presence of strong electron-lattice interaction effects in the manganites includes (i) shifts in the IR phonon frequencies related to the Mn-O bonds in $La_{0.7}Ca_{0.3}MnO_3$ near T_C [38],(ii) anomalies in the local structure of the MnO_6 octahedron of $La_{1-x}Ca_xMnO_3$, near T_C obtained from neutron scattering studies [39], (iii) magnetic-field driven structural phase transformation in $La_{0.83}Sr_{0.17}MnO_3$ [40], and (iv) disappearance of small polaronic behaviour below the insulator-metal transition in $La_{0.7}Ca_{0.3}MnO_3$ observed in the thermopower measurements [41]. Early optical studies on $Nd_{0.7}Sr_{0.3}MnO_3$ report evidence of a small polaron signature in the optical conductivity[42].

1.2.5. Inhomogeneities, Charge and Orbital Ordering in Manganites

The basic interactions in the manganite perovskites allow three phases: a ferromagnetic metal, a charge/orbital ordered antiferromagnetic insulator and a paramagnetic polaronic liquid. Metallicity is obtained by introducing holes by doping in antiferromagnetic, insulating $LaMnO_3$ manganite. In the doping region $0.2 < x < 0.50$, $La_{1-x}Ca_xMnO_3$ is both metallic and ferromagnetic, as the interactions are dominated by double exchange. However, with $0.10 < x < 0.20$, $La_{1-x}Ca_xMnO_3$ has a ferromagnetic insulating ground state. The origin of the coexistence of ferromagnetism with insulating behaviour is not clear, but it might stem from a delicate balance of charge localization by orbital ordering (OO), due to the Jahn-Teller (JT) effect, and ferromagnetic interaction between Mn^{3+}-Mn^{4+}. The Ca-doped phase diagram is somewhat less complex, as there is no orthorhombic-rhombohedral structural transition, where as the situation is complicated for Sr-doped manganites because the number of phases is more due to the rhombohedral structure at $x > 0.18$ and pronounced charge ordering at 'x' ~ 0.125 [43]. Therefore, the phase diagram of Sr doped manganites has been explored in great detail and a JT related structural phase transition is observed above the magnetic ordering temperature, $T > T_C$ at 'x' ~ 0.12. Below T_C, a transition to CO or OO is observed, where the cooperative JT distortion is significantly reduced [44,45]. However, as a matter of fact, the parent compound $LaMnO_3$ presents ferromagnetic layers anti-ferromagnetically coupled to neighbouring layers while, $CaMnO_3$ does not present any ferromagnetic interactions. Moreover, CMR appears in the hole doped regime while the electron doped regime is dominated by insulating phases. The present chapter presents the vibrational properties of A-site doped $LaMnO_3$.

A large number of different phases have been reported in manganites. The most obvious example of *orbital ordering* is the case of the parent compound $LaMnO_3$ which is a typical A-type antiferromagnetic insulator (AFM-I). (For the definition of various magnetic structures like ferromagnetic (FM), A-, C-, and G-type AF states, refer to Figure 8.) It is believed that the A-AF ordering of $LaMnO_3$ is accompanied with orbital ordering (OO), which is stabilized by cooperative JT distortion. In this orbital ordered state, two kinds of e_g orbitals, $3x^2$-r^2 and $3y^2$-r^2, are alternately occupied at neighbouring Mn^{3+} sites in the *ab*-plane. Orbital ordering is a generic name for 'static cooperative Jahn-Teller distortion' and it results into the splitting of Mn - e_g orbitals.

Manganites are typically considered to be disordered compounds, where Mn^{3+} and Mn^{4+} cations are randomly distributed in the lattice. However, under certain chemical and temperature conditions, and especially when Mn^{3+} and Mn^{4+} cations are present in equal amounts, these cations order coherently over long distances to form a charge-ordered (CO) lattice. In manganites, charge ordering would be expected to be favored when 'x' = 0.5, due to the presence of equal proportions of the Mn^{3+} and Mn^{4+} states. However, it is found in various compositions in the range $0.3 < x < 0.75$, depending on the *Ln* and A ions [46]

The localization of charges renders the material insulating and also promotes antiferromagnetism. The charge ordering and double exchange are competing phenomena. Moreover, it is found that the CO-state can readily be "melted": the application of magnetic field [47], or pressure [48], or exposure to x-ray photons [49] can destroy the CO-state in favour of a ferromagnetic metal state. At low temperatures and low doping concentrations where the double exchange is not dominant, the electrons are localized with the spins antiferromagnetically coupled due to the super-exchange process.

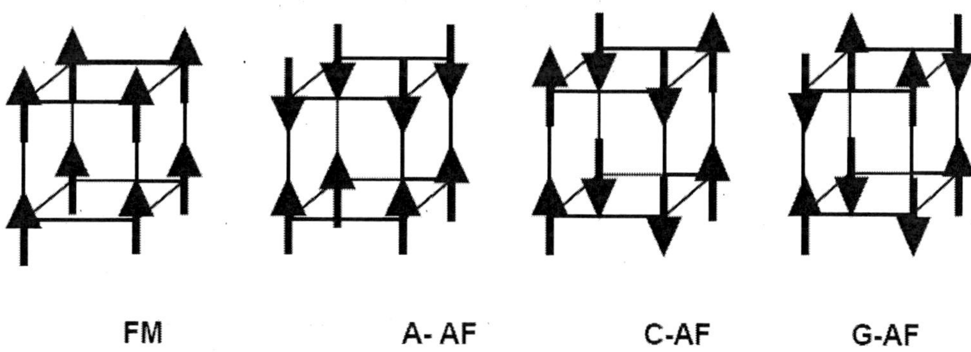

Figure 8. Various magnetic structures are schematically shown. From left to right, FM, A- type antiferromagnetic (A-AF) and so on.

For larger doping concentration, the double exchange can stabilize at low temperatures a spin ordered phase (ferromagnetic state), while the dominant effect of the electron-phonon interaction stabilizes a charged ordered phase.

1.2.6. Phonons and its Importance in Manganese Oxide Perovskites

The energy of a lattice vibration is quantized, and the quantum of energy is called a phonon. Phonons obey Bose-Einstein statistics and represent the simplest elementary excitation in a solid. Elastic waves in crystals are made up of phonons, and thermal vibrations in crystals are thermally excited phonons. The energy of an elastic mode with angular frequency ω is $\varepsilon = \left(n + \frac{1}{2}\right)\hbar\omega$, when the mode is excited to quantum number n, the mode is occupied by n phonons. A phonon with wave vector $\boldsymbol{K}$ will interact with other particles such as electrons, photons and neutrons, as if it had a momentum $\hbar\boldsymbol{K}$, but a phonon does not carry any physical momentum. The reason for this is that a phonon coordinate involves relative coordinates of the atom. The physical momentum of a crystal is

$$\boldsymbol{p} = m\frac{d}{dt}\sum \boldsymbol{u}_S \tag{7}$$

This turns out to be zero, with exception for $K = 0$, which represents uniform translation of the crystal, and such a translation does carry momentum.

Phonons play a critical role in phenomena such as superconductivity and many types of phase transitions, and are the basis for the acoustic, thermal, elastic, and infrared properties of solids [50]. Most undoped or weakly doped manganites show long range (Jahn–Teller type) distortions from the ideal perovskite structure, and changes in the magnetic or orbital ordering patterns are usually accompanied by changes in the crystal structure. As discussed in section 1, colossal magnetoresistance (CMR) occurs in many of the manganites for $0.2 < x < 0.5$ around the insulator (I) to metal (M) transition at the temperature $T_S \sim T_C$, where the system undergoes a paramagnetic to ferromagnetic (FM) transition. A detailed understanding of these phenomena has not yet been achieved, but key roles of the double-exchange interaction, the basic mechanism responsible for electrical transport below critical temperature, T_C and of the crystal lattice, through the electron-phonon coupling triggered by Jahn-Teller distortions of the MnO_6 octahedra, have been proposed [51]. In this scenario, correlated information on lattice dynamics, as that provided by Raman and infrared (IR) phonon studies, and on other vibrational and transport properties can be essential for a detailed understanding of the role of lattice distortions in these technological and fundamentally important compounds.

1.2.7. Other CMR Materials

The discovery of CMR in manganites stirred up the intense interest and a lot of effort has been done (and still is) to find out other materials with a large magnetoresistance and half-metallic character. The main examples of these materials are manganese oxide pyrochlores [52], europium hexaborides (EuB_6) [53], Cr-based chalcogenide spinels [54] and double perovskites [55]. Double perovskites and chromium dioxides are, in fact, double exchange systems but for rest of them neither double-exchange nor Jahn-Teller distortions apply.

$Tl_2Mn_2O_7$ is the parent compound of manganese oxides pyrochlores. These pyrochlores, although they are manganese oxides, present a great deal of differences with manganites. Pyrochlores, when first reported, seemed a very fascinating alternative to manganites because magnetic and electric properties were not strongly interdependent. Therefore, to increase magnetoresistance without causing a great decrease of T_C was possible. However, a complete elucidation of their behaviour still remains elusive. Europium hexaboride (EuB_6) is a ferromagnetic semimetal, which shows a very large magnetoresistance but is completely different from the pyrochlores. A spin-flip Raman scattering experiment [56] shows the existence of magnetic polarons that could be the clue to understand these materials. Double perovskite oxides have the form $A_2B'B''O_6$ where A is an alkaline or rare earth ion and the transition metal sites are occupied alternatively by cations B' and B". Sr_2FeMoO_6 is an example of double perovskite oxides [53]. These materials are half-metallic and possess large T_C overcoming above 400 K (Ca_2FeReO_6 has T_C = 540 K), which makes them very interesting. Single crystals of these materials do not show a significant magnetoresistance while polycrystals show [57]. In fact, their low field magnetoresistance remains much larger than in manganite polycrystals upon heating [54, 58].

In a nutshell, a wide range of half-metals is now in the stage and colossal magnetoresistance is not exclusive of manganites. There may be some basic explanation for these materials beyond the various different interactions, which characterize each of them.

1.2.8. Outline of this Chapter

The present chapter describes the results of some of our own investigation on the vibrational properties of some colossal magnetoresistance materials, especially $LaMnO_3$ and manganites with composition $La_{0.7}A_{0.3}MnO_3$ (A: divalent Sr, Ba, Pb) at ambient as well as different external parameters such as pressure, temperature and doping concentration by using lattice dynamical model theories.

Section I includes the general discussion on transition metal oxides, CMR materials as particular class of TMOs and their physical properties in addition to the most striking phenomenon, namely colossal magnetoresistance and related potential applications. Double exchange model as well as importance of Jahn Teller small polarons and phonons in CMR materials, in particular lanthanum manganites are also described in this section.

Section II, which is devoted to the some preliminaries of the lattice vibrations and present model theory and methodology used for the computation of vibrational properties of above-mentioned materials. The theoretical description related to the computed phonon properties is discussed at length.

In *Section III,* the results of the investigation of phonons and related properties for undoped cubic and rhombohedral $LaMnO_3$ and rhombohedral Sr-doped $LaMnO_3$ systems are discussed. In this section, the results of the theoretical investigations by using lattice dynamical simulation method based on the rigid ion model with physically significant parameters, on the vibrational properties of undoped $LaMnO_3$ in its cubic and rhombohedral phases are reported to understand the role of phonons in such systems. Moreover, the results on lattice dynamical properties of 30% Sr-doped $LaMnO_3$ *i.e.* $La_{0.7}Sr_{0.3}MnO_3$ compound in its rhombohedral structure are also reported at a length. Substitution of La^{3+} by Sr^{2+} changes the radius of La-site, which leads to the changes in various phonon properties. The red shift of Raman active A_{1g} phonon mode is observed for 30% concentration of Sr^{2+} with corresponding changes in phonon dispersion curves, phonon density of states and lattice specific heat..

Section IV describes the computed report of temperature effect on vibrational properties of $La_{0.7}Sr_{0.3}MnO_3$. In order to observe the temperature dependence of the phonon properties, the calculations are carried out at ambient and at very low temperatures. The results reveal that the Raman active A_{1g}, E_g and infra red A_{2u} modes are sensitive to the temperature variation. The difference in phonon density of states of these systems at different temperatures is revealed with appearance of merged or split peaks. The effect of temperature on specific heat and Debye temperature are also discussed.

In *Section V,* the results of lattice dynamical properties of $La_{1-x}Sr_xMnO_3$ (for 'x' = 0.3) in its rhombohedral structure at various external pressures are discussed to understand the role of pressure on phonon modes and phonon related properties. Because of the subtle balance and complicated interaction among the charge, spin, and lattice structure the pressure affect the transport properties and cause changes in magnetic and/or structural properties. It is observed that the A_{1g} mode behaves anomalously while the some IR and Raman modes show linear increase. A pronounced shift in the positions and shape of the peaks in the phonon DOS is observed at high pressure. To see the effect of internal pressure determined by the doping of different divalent atoms and correlate this with the external pressure, vibrational properties of $LaMnO_3$ with Ba and Pb doping are also reported.

In *Section VI,* the results of lattice dynamical properties of intrinsic phase of cobalt oxide superconductor, $NaCoO_2$ are reported. Calculated Raman, IR phonon modes, phonon dispersion curves and generalized phonon density of states have enabled an atomic level understanding of the phonon density of states. The effect of Na site occupancy on the phonon spectra are discussed and found significant differences. The lattice specific heat at constant volume and Debye temperature are also reported. The estimated Debye temperature reveals the importance of phonons in the transport properties in the intrinsic insulating phase of the $NaCoO_2$ compound at room temperature. A summery is given at the end in *Section VII.*

2. Preliminaries, Present Model and Method of Computations

2.1. Cohesive Energy and Lattice Vibrations

The study of the lattice vibrations is of considerable interest because several physical properties of crystal like specific heat, thermal expansion, thermal conductivity, phase transitions, and its interaction with photons, neutrons and x-rays are related to the vibrations of atoms in solids. The collective motions of atoms in solids form traveling waves called lattice vibrations which are quantized in terms of "PHONONS''. In order to understand the physical properties of solids it is of interest to study the energy-wavelength relation (phonon dispersion curves) of thermal motions the atoms determined by the interatomic interactions. This chapter covers discussion on the dynamical theory and underlying physics of the lattice dynamical model used for the study of lattice dynamical properties of perovskite oxides considered in the present thesis. Besides these, a subsection is devoted for the brief discussion about the some experimental techniques, most suitable for the study of phonons.

2.1.1. Cohesive Energy

In order for a crystal to hold together, there exist attractive interactions. It is an electrostatic attraction between positive and negative charges, which is common for all solids. As the atoms come close together, their closed electron shells start to overlap, for which electrons have to be excited to higher states. The excitation to higher energy states costs energy and leads to a repulsive interaction between the atoms. The repulsive interaction dominates for short distances between atoms, while the attractive interaction dominates at large distances. The actual atomic spacing in a crystal is defined by the equilibrium where the potential energy exhibits a minimum. The cohesive energy is the energy that must be added to the crystal to separate it to neutral free atoms at rest, at infinite separation. For ionic crystals the term lattice energy is used instead; the definition is similar except that the energy is defined for relatively free ions at rest at infinite separation. The importance of cohesive energy is that it is the ground-state energy of solid, sign of which determines whether the solids will be stable or not. Its generalization to nonzero temperatures, the Helmholtz free energy, a function of volume and temperature, contains all equilibrium thermodynamic information about the solid. Depending on the distribution of the outer electrons with respect to the ions, different binding types can occur. There are four types of interactions, which can be treated by simple models and give a good approximation to atomic distances and cohesive energies: (i) Ionic bonding (ii) van der Waals bonding (iii) Metallic bonding (iv) Covalent bonding. Crystalline solids are therefore, classified into four principal types: (a) Ionic crystals (b) van der Waals crystals (c) Metals and (d) Valency crystals.

2.2. Theoretical Formalism and Method of Computations for Lattice Vibrations

The formalism of lattice dynamics is based on the Born–Oppenheimer or adiabatic approximation. Ions are about 10^3 –10^5 times heavier than electrons and move much slower than the electrons. The electrons contribute an additional effective potential for the nuclear motions and the lattice vibrations are associated only with nuclear motions. However, a thorough discussion of lattice dynamics in the harmonic approximation can be found in several books and monographs on this subject [59-60], we present here a short account of the theory mainly to clarify the notations and a basis for our model. The crystal potential energy is a function of the instantaneous positions of the atoms. If $\mathbf{u}\begin{pmatrix} l \\ k \end{pmatrix}$ is the displacement of the kth atom (k = 1, 2,. . . ,n) in the lth primitive cell (l = 1,2,. . . ,N) about its equilibrium position $\mathbf{r}\begin{pmatrix} l \\ k \end{pmatrix}$, for small displacement of the atoms, the crystal potential energy ϕ can be expanded using the Taylor expansion as follows:

$$\phi = \phi_0 + \sum_{lk\alpha} \left.\frac{\partial \phi}{\partial u_\alpha\begin{pmatrix} l \\ k \end{pmatrix}}\right|_0 u_\alpha\begin{pmatrix} l \\ k \end{pmatrix} + \frac{1}{2}\sum_{lk\alpha}\sum_{l'k'\beta} \left.\frac{\partial^2 \phi}{\partial u_\alpha\begin{pmatrix} l \\ k \end{pmatrix}\partial u_\beta\begin{pmatrix} l' \\ k' \end{pmatrix}}\right|_0 u_\alpha\begin{pmatrix} l \\ k \end{pmatrix} u_\beta\begin{pmatrix} l' \\ k' \end{pmatrix} + \ldots\ldots, \tag{8}$$

where, suffices α and β denote Cartesian coordinates. The cubic and higher order terms are neglected in the harmonic expansion. In the equilibrium configuration, the force on any atom must be zero. Therefore, $\dfrac{\partial \phi}{\partial u_\alpha \begin{pmatrix} l \\ k \end{pmatrix}} = 0$ for every α, k, l. The tensor force constant between atoms (lk) and ($l'k'$) can also be defined as:

$$\phi_{\alpha\beta}\begin{pmatrix} l & l' \\ k & k' \end{pmatrix} = \frac{\partial^2 \phi}{\partial u_\alpha \begin{pmatrix} l \\ k \end{pmatrix} \partial u_\beta \begin{pmatrix} l' \\ k' \end{pmatrix}} \tag{9}$$

Thus,

$$\phi = \phi_0 + \frac{1}{2} \sum_{lk\alpha} \sum_{l'k'\beta} \phi_{\alpha\beta}\begin{pmatrix} l & l' \\ k & k' \end{pmatrix} u_\alpha \begin{pmatrix} l \\ k \end{pmatrix} u_\beta \begin{pmatrix} l' \\ k' \end{pmatrix} \tag{10}$$

The equation of motion for the displacements of the $(lk)^{th}$ atom becomes

$$m_k \ddot{\boldsymbol{u}}_\alpha \begin{pmatrix} l \\ k \end{pmatrix} = - \sum_{l'k'\beta} \phi_{\alpha\beta}\begin{pmatrix} l & l' \\ k & k' \end{pmatrix} \boldsymbol{u}_\beta \begin{pmatrix} l' \\ k' \end{pmatrix} \tag{11}$$

where m_k is the mass of the $(lk)^{th}$ atom. The crystal periodicity suggests that the solutions of eq. (11) must be such that the displacements of atoms in different unit cells must be same apart from phase factor. The equations of motion (11) are solved by assuming wave like solutions of the type

$$\boldsymbol{u}_\alpha \begin{pmatrix} l \\ k \end{pmatrix} = U_\alpha (k \mid \boldsymbol{q}) \exp\left\{ i \left(\boldsymbol{q} \cdot \boldsymbol{r} \begin{pmatrix} l \\ k \end{pmatrix} - \omega(\boldsymbol{q}) t \right) \right\} \tag{12}$$

Here, $U_\alpha(k \mid \boldsymbol{q})$ is the amplitude of the wave, $\boldsymbol{q}$ is the wave vector and $\omega(\boldsymbol{q})$, the angular frequency associated with the wave. Substituting (12) in (11)

$$m_k \omega(\boldsymbol{q}) U_\alpha (k \mid \boldsymbol{q}) = \sum_{k'\beta} D_{\alpha\beta} \begin{pmatrix} \boldsymbol{q} \\ kk' \end{pmatrix} U_\beta (k' \mid \boldsymbol{q}) \tag{13}$$

where $\boldsymbol{r}\begin{pmatrix} l \\ k \end{pmatrix}$ is the position co-ordinate of the $(lk)^{th}$ atom.

The dynamical matrix $D_{\alpha\beta}\begin{pmatrix} \boldsymbol{q} \\ kk' \end{pmatrix}$ is given by

$$D_{\alpha\beta}\begin{pmatrix} \boldsymbol{q} \\ kk' \end{pmatrix} = \sum_{l'} \phi_{\alpha\beta}\begin{pmatrix} l & l' \\ k & k' \end{pmatrix} \exp\left[i\left\{ \boldsymbol{q} \cdot \left(\boldsymbol{r}\begin{pmatrix} l' \\ k' \end{pmatrix} - \boldsymbol{r}\begin{pmatrix} l \\ k \end{pmatrix} \right) \right\} \right] \tag{14}$$

The dimension of the dynamical matrix is $3n$. Thus for a periodic crystal we have a set of $3n$ equations. The frequencies of the normal modes and eigenvectors are determined by diagonalizing the dynamical matrix:

$$\det\left| m_k \omega^2(\boldsymbol{q}) \delta_{kk'} \delta_{\alpha\beta} - D_{\alpha\beta}\begin{pmatrix} \boldsymbol{q} \\ kk' \end{pmatrix} \right| = 0 \tag{15}$$

Lattice dynamics studies of solids are usually carried out by a lattice dynamical model based on the interatomic interaction and formation of a dynamical matrix. The interatomic interactions are selected based on the type of bonding present in any solids. We describe below the brief description to calculate the different phonon properties, many time even useful to test the success of the selected interatomic potentials by comparing with available experimental data.

2.2.1. Group Theoretical Analysis

Group theoretical symmetry analysis enables a classification of phonon modes belonging to various representations enabling their direct comparisons with observed Raman and infrared scattering data. In the experiments due to the selection rules (governed by the symmetry of the system and the scattering geometry in experiments), only phonon modes belonging to certain group theoretical representations are active in typical Raman, infrared and inelastic neutron scattering experiments which enable their mode assignment. The general theoretical scheme is based on irreducible representations described in detail in refs. [60]. The procedure essentially involves construction of symmetry adapted vectors, required for block diagonalizing the dynamical matrix, which yields the phonon modes belonging to a given representation.

2.2.2. Phonon Frequencies and Phonon Dispersion Relation

The solutions of eq. (15) yield the $3n$ eigenvalues of $\omega_j^2(\boldsymbol{q})$ (j = 1, 2... $3n$). The dynamical matrix is Hermitian, therefore its eigenvalues $\omega_j^2(\boldsymbol{q})$ are real, and eigenvectors $\xi_j(\boldsymbol{q})$ are orthonormal. However, for the stability of the lattice $\omega_j^2(\boldsymbol{q})$ must be positive. The components of the eigenvectors $\xi_j(\boldsymbol{q})$ determine the pattern of displacement of the atoms in a particular mode of vibration. These yield the normal modes of vibrations characteristic of the crystal. The variation of $\omega_j^2(\boldsymbol{q})$, ($j$ = 1, 2...$3n$) with wave vector $\boldsymbol{q}$ gives the phonon dispersion relation or phonon band structure, which is of interest to understand the physical properties of solids. Though, some of these modes are degenerate because of symmetry, in general they are distinct. The form of dispersion relation depends on the crystal structure as well as on the nature of the interatomic forces. A cyclic crystal always has three zero frequency modes at $\boldsymbol{q} = 0$, which correspond to lateral translational of the crystal along three mutually perpendicular directions. These three branches are referred to as acoustic branches. The remaining ($3n$ - 3) branches have finite frequencies at $\boldsymbol{q} = 0$, which are labeled as optic branches. These phonons can propagate in the lattice of a single crystal as a wave and exhibit dispersion depending on their wavelength or equivalently their wave vector in the Brillouin zone.

2.2.3. Total and Partial Phonon Density of States and Anisotropic Thermal Expansion

From the phonon band structure $\omega_j(\boldsymbol{q})$, it is straight forward to find the corresponding density of states as it gives the information of phonons in whole Brillouin zone (BZ). In order to obtain the information about the whole phonon spectrum, Brillouin-zone scanning is necessary. Such a scanning consists in $D_{\alpha\beta}\begin{pmatrix} \boldsymbol{q} \\ kk' \end{pmatrix}$-matrix diagonalization over the three dimensional net of wave vector $\boldsymbol{q} = \left(\frac{\boldsymbol{a}^*}{n_1}, \frac{\boldsymbol{b}^*}{n_2}, \frac{\boldsymbol{c}^*}{n_3} \right)$, at n_1, n_2, n_3 = - N,…., N. In total, this includes $N_i = (2N + 1)^3$ points in Brillouin zone. The phonon density of states (DOS) is determined by summation over all the phonon states and is defined by

$$g(\omega) = D' \int_{BZ} \sum_j \delta\left(\omega - \omega_j(\boldsymbol{q})\right) dq = D' \int_{BZ} \sum_{jp} \delta\left(\omega - \omega_j(\boldsymbol{q})\right) dq_p \tag{16}$$

where, BZ corresponds to the Brillouin zone, D' is a normalization constant such that $\int g(\omega) d\omega = 1$; that is, $g(\omega)d\omega$ is the fraction of phonons which have energies within a range from ω to ω + dω. 'p' is the mesh index characterizing '$\boldsymbol{q}$' in the discretized irreducible Brillouin zone and d$\boldsymbol{q}_p$ provides the weighting factor corresponding to the volume of pth mesh in $\boldsymbol{q}$-space. Partial atomic density of state (PDOS) shows the contribution of different atoms to phonon density of states (DOS) and therefore, it essentially helps in understanding the atomic level contribution to the total phonon DOS. It is defined as

$$g(\omega) = D' \sum_{jp} \delta\left(\omega - \omega_j(\boldsymbol{q})\right) \frac{\left|\xi_j(\boldsymbol{q})\right|^2}{\sum_{jp} \left|\xi_j(\boldsymbol{q})\right|^2} \tag{17}$$

For a solid at a temperature T, the mean number of phonons with energy $\hbar\omega_j(\boldsymbol{q})$ is given by the Bose-Einstein distribution $n_{jq}(T) = \left[\exp \frac{\left(\hbar\omega(\boldsymbol{q})\right)}{K_B T} - 1 \right]^{-1}$. The mean square displacement of a single quantum mechanical harmonic oscillator, $\langle u^2 \rangle = \left(\frac{\hbar}{m\omega}\right)\left(n+\frac{1}{2}\right)$ can easily be generalized to that of a single atom in the direction i as

$$\left\langle u_{ki}^{2} \right\rangle = \left[\frac{V}{2\pi^{3}}\right] \cdot \left(\frac{\hbar}{m_k}\right) \sum \int_{BZ} \left|\varepsilon_{jki}(\boldsymbol{q})\right|^{2} \cdot \left[\frac{\left\{n_{jq}(T) + \frac{1}{2}\right\}}{\omega_j(\boldsymbol{q})}\right] \cdot d\boldsymbol{q} \tag{18}$$

It can be seen from the above expression that light atoms vibrating at low frequencies exhibit large zero point motions. The off-diagonal elements $\left\langle \bar{\boldsymbol{u}}_{ki} \bar{\boldsymbol{u}}_{kj} \right\rangle$ can be calculated in a similar way. The thermal and zero point motion of the atoms are often described using the matrix of anisotropic temperature factors $\bar{B}$. For an atom $\boldsymbol{k}$, it is defined by

$$\bar{B}_{ij}\left(\bar{k}\right) = 8\pi^{2}\left\langle \bar{u}_{ki} \bar{u}_{kj} \right\rangle \tag{19}$$

2.2.4. Thermodynamic Functions and Specific Heat

The theory of lattice dynamics described above allows us to determine the phonon frequencies in the harmonic approximation. Anharmonic effects are relatively small at low temperature in most crystals and become more important at high temperatures. The thermodynamic properties of a crystal may be calculated in the quasiharmonic approximations [1-3]. In quasiharmonic approximation the vibrations of atoms at any finite temperature are in principle, assumed to be harmonic about their mean positions appropriate to that temperature. The free energy, in three dimensional case, is a function of temperature T and the volume V and the equation of state is given by

$$P = -\left(\frac{\partial F}{\partial V}\right)_T \tag{20}$$

where, $F^{P}(T) = \frac{K_B T}{N_t} \sum_{jp} \ln\left(2\sinh\left(\beta_{jp}\right)\right).$ (21)

Here, $\beta_{jp} = \frac{\hbar\omega_j\left(q_p\right)}{2K_B T}$. The quantity $\hbar\omega_j\left(q_p\right)$ is the phonon energy and, $\hbar = h/2\pi$ with h and $\omega_j(\boldsymbol{q}_p)$ as Planck constants and phonon branch at $\boldsymbol{q}_p$ respectively. K_B is Boltzmann's constants and T is temperature. The internal energy U and the entropy S are given by the relations

$$U = F - T\left(\frac{\partial F}{\partial T}\right)_V = F + TS\,. \tag{22}$$

Here, $S(T) = \frac{1}{N_t}\sum_{jp} \beta_{jp} \coth\left(\beta_{jp}\right) - \ln\left(2\sinh\left(\beta_{jp}\right)\right).$ (23)

The Heat capacity is defined by $C(T) = \frac{1}{N}\sum_{jp}\left[\frac{\beta_{jp}}{2\sinh(\beta_{jp})}\right]^2.$ (24)

The relations (eq. (21), (23), and (24)) concern with the specific values per unit cell. However, the specific heat $C_V(T)$ can also be expressed as

$$C_V(T) = \frac{dE}{dT} = \frac{\hbar^2}{K_B T}\int_0^{\omega_D} \frac{\omega^2 exp\left(\hbar\omega/K_B T\right)}{\left(exp\left(\hbar\omega/K_B T\right) - 1\right)^2} g(\omega)\, d\omega \tag{25}$$

where $g(\omega)$ is the phonon density of states. Since the experimental measurements usually provides C_P value, which differs from C_V value at elevated temperature, the following corrections is required: $C_P = C_V + B\beta^2 V_m T$, where B is the bulk modulus expressed as $B = -V(dP/dV)_{V0}$, β is the thermal expansion coefficient and V_m is the molar volume at equilibrium.

2.2.5. Atomic Thermal Parameters and Pair Distribution Function

Atomic thermal parameters (ATP), or atomic thermal amplitudes, are determined by the expression:

$$a_{i,\alpha\beta}(T) = \frac{1}{N_t}\sum_{jp} \frac{h}{2\omega_j(\mathbf{q})} coth\left(\beta_{jp}\right) e_{i\alpha}(\mathbf{q}) e_{i\beta}(\mathbf{q}) \tag{26}$$

Pair distribution function (PDF)-the probability distribution of the interatomic distances-within the quasiharmonic approximation is determined by the expression:

$$f(R,T) = \sum_j \frac{1}{\sqrt{\pi\sigma_j}} exp\left(-\frac{\left(R - R_j\right)^2}{\sigma_j}\right) \tag{27}$$

Here, R_j is the temperature-dependent distance between atoms i and j which is related to the static equilibrium interatomic distance R_{j0} by the relation: $R_j^2(T) = R_{j0}^2 + \sigma_j$. Where σ_j is the corresponding dispersion determined by the relation:

$\sigma_j = \langle u_i u_i \rangle + \langle u_j u_j \rangle - 2\langle u_i u_j \rangle$. The isotropic atomic displacement correlation is defined by the relation:

$$\left\langle \bar{u}_i \bar{u}_j \right\rangle = \frac{1}{N_t} \sum_{jp} \frac{h}{2\omega_j(q)} coth\left(\beta_{jp}\right)\left(e_{ix}e_{jx} + e_{iy}e_{jy} + e_{iz}e_{jz}\right) \tag{28}$$

2.2.6. Infrared and Raman intensity

It is always necessary to calculate the infrared and Raman modes at zone centre of Brillouin zone to compare with the experimental measured Raman and infrared (IR) data. Infrared intensity is proportional to the square of the oscillator strength, $\xi_j^I = \frac{dP}{dQ_j}$, which is also associated with atomic displacements (*i.e.* eigen vectors) as $\xi_j^I = \sum_i P_i \xi_{ij}$. Here, *P* is the polarization per unit cell. The corresponding contribution of the mode *j* to the static dielectric constant is $\Delta\varepsilon_j = \frac{4\pi}{\lambda_j v_c}\left(\xi_j^I\right)^2$, where v_c is the unit cell volume. Other important characteristics of the IR spectrum are longitudinal (LO) frequencies, which are determined using LST relation:

$$\varepsilon(\omega) = \varepsilon^{\infty} + \frac{4\pi}{v_C} \sum_j \frac{\left(\xi_j^I\right)^2}{\omega_j^2 - \omega^2} \tag{29}$$

The Raman scattering intensity, which is proportional to the square of the derivative $\zeta_j = \frac{d\varepsilon^{\infty}}{dQ_j}$, is also associated with atomic displacements as $\zeta_j = \sum_i \varepsilon_i^{\infty} \xi_{ij}$. Here,

$\varepsilon_i^{\varepsilon} = \frac{d\varepsilon_i^{\varepsilon}}{dX_i}$ is the atomic derivatives of the dielectric constant.

2.2.7. Elastic Constants

The determination of the elastic constants of crystals has been of importance due to its use for the study of many properties mainly the sound velocities [61]. For any direction in a single crystal three types of sound wave may be propagated: one longitudinal and two transverse waves. These three corresponding velocities are the roots of a cubic equation called as Christoffel's equation, whose coefficients are the functions of the elastic constants and the direction cosines for the direction of propagation of the sound. In the case of a cubic crystal there are only three independent elastic constants while for other structures the number of elastic constants is more. The elastic constant expression can be written as

$$C_{\mu\nu} = \frac{1}{v_c}\left(V_{\mu\nu}^{uu} - \sum_j \lambda_j Q_{j\mu} Q_{j\nu}\right), \tag{30}$$

where atomic relaxation along the normal co-ordinate is defined as

$$Q_{j\mu} = \frac{dQ_j}{du_\mu} = -\frac{1}{\lambda_j} V_{j\mu} \tag{31}$$

under u_μ and $V_{j\mu} = \dfrac{d^2V}{dQ_j du_\mu} = \sum_i V_{i\mu}^{xu}\, \xi_{ij}$. (32)

2.3. Interatomic Potential and Parameters Determination

As it is widely accepted fact that the dominant interactions in the perovskite like compounds are ionic type, a model of rigid-ion or shell model type are the obvious choices. In the rigid ion model [62], the ions are treated as point charges, which are not rigid during vibrations in reality. The electric field set up by the displacements of the ions is modified by their electronic polarizability, which in turn modifies the force on them and affects the phonon frequencies. This may be described by a shell model [63], in which each ion is regarded as composed of a rigid or non-polarizable core and charged shell with effective charges *X(K)* and *Y(K)*, respectively. In general the shell model is the extension of the rigid ion model. In the present chapter, the lattice dynamics of the perovskite oxides is carried out in the frame work of the shell model. The model is not only appropriate for the perovskite-like oxides such as high T_C superconductors and manganites because in accordance with their predominately ionicity, the interatomic interactions are represented as sums of long-range Coulomb interactions and short-range interactions but also has been quite successful in predicting the phonon properties of the perovskite oxides. In addition, it considers the ionic polarizability of the ions. In the present shell model every ion is represented by a shell coupled to a core through a harmonic force constants *K*, and the short-range interactions between ions are represented by pair potentials of the Born-Mayor form and the total potential along with the Coulomb interactions are written as

$$U_{ij}\left(r_{ij}\right) = \frac{Z_i Z_j e^2}{r_{ij}} + b_{ij} \exp\left(-\frac{r_{ij}}{\rho_{ij}}\right) \tag{33}$$

where the first and second term represent the long-range Coulomb potential and Born-Mayer repulsion energies, respectively. Here, r_{ij} is the inter atomic distance between atoms *i* and *j*, Z_i and Z_j are the effective charges of the respective atoms, b_{ij} and ρ_{ij} are the short range potential parameters (hardness and strength, respectively) for each pair of atoms usually found by fitting to experimental data. As a first step in the present model calculation for $LaMnO_3$ in cubic and rhombohedral phase and $La_{0.7}Sr_{0.3}MnO_3$ in rhombohedral phase, we have transferred the O-O short-range (SR) interactions from [64], which were successfully used for the modeling of many oxides [65-67]. The cation-anion SR potentials parameters are obtained by ensuring that the total stress and forces for the given structure vanishes. In the present study, in principle no fitting of the experimental data on phonon frequencies is done to obtain the parameters. However, it is ensured that the physically significant parameters are obtained which give nearly vanishing forces on all the atoms and right magnitude of the eigen frequencies in the harmonic approximation. For the right magnitude of the eigen frequencies guidance has been taken from the experimental A_{1g} phonon mode. The electronic polarization of the lattice is included by the shell model, in which an ion is represented by a massless shell of charge Y and a core of charge X which are coupled by a harmonic spring constant K. The free ionic polarizability is expressed by

$$\alpha = (Ye)^2 / K \tag{34}$$

where Y is the dimensionless shell charges and e is the absolute value of the electron charge. To calculate the phonon properties the software LADY for lattice dynamical simulation is used [68] which have been quite successfully used recently by us [67]. In the present case, only the oxygen ions are considered to be polarizable and the short-range interactions are restricted to only nearest neighbor shell-shell interaction.

2.4. Experimental Techniques: Measurement of Phonons

Experimental studies of lattice vibrations include use of techniques like Raman spectroscopy, infrared absorption, inelastic neutron scattering, inelastic X-ray scattering, etc. Unlike Raman and infrared studies, which probe only the long wavelength excitations in one-phonon scattering, inelastic neutron and X-ray scattering can directly probe the phonons in the entire Brillouin zone. The most powerful technique currently used for studying lattice vibrations is inelastic neutron scattering. While inelastic neutron scattering is widely used for such measurements, inelastic X-ray scattering has also been used [69-72] at intense synchrotron sources for the study of phonons in a few materials. Despite the wealth of information which has been obtained from this technique, it does have some limitations such as the need for expensive equipment, a relatively low resolution, and the fact some materials cannot be investigated because they have a low scattering cross section or high absorption cross section for neutrons. Infrared absorption and Raman and Brillouin light scattering provide complimentary techniques for investigating lattice vibrations. These methods have higher resolution than neutron scattering but first order phonon processes are limited to the center of Brillouin zone by momentum conservation. Second order processes are not subject to this restriction but it is usually quite difficult to unfold the combined density of phonon states. These techniques are also limited by rigorous selection rules.

2.4.1. General Theory of Scattering

Scattering experiments, for instance, are very important for studies in material science and condensed matter physics, since they allow a unique insight into the arrangement of the atoms through the observation of the electron distributions and their fluctuations in space and time. A general scattering experiment is shown schematically in Figure 9.

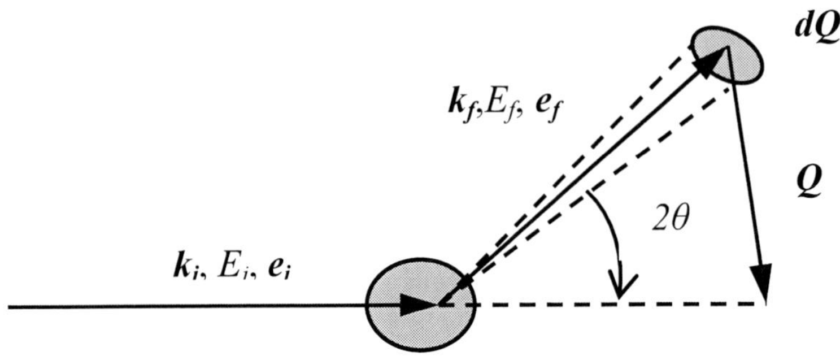

Figure 9. Scattering Geometry.

This schematic arrangement is valid for all probes such as neutrons, electron beams, and electromagnetic radiation. The incident beam of well defined wave vector $\boldsymbol{k}_i$, energy E_i and polarization unit vector $\boldsymbol{e}_i$ is scattered into the solid-angle element d**Ω** under the scattering angle 2θ. The scattered beam is completely defined by the new wavevector $\boldsymbol{k}_f$, the energy E_f and the polarization unit vector $\boldsymbol{e}_f$. The scattered intensity is described by the double-differential cross section $\mathrm{d}^2\sigma/(\mathrm{d}\boldsymbol{\Omega}\mathrm{d}\omega_f)$. It is given by the removal rate of particles out of the incident beam as the result of being scattered into a solid angle d**Ω** with a frequency range of $\mathrm{d}\omega_f$. The scattered beam is usually distributed over a range of energies E_f. There can be beam contributions that have been scattered elastically with no change of energy and other contributions that have changed energy due to inelastic scattering. Therefore, the scattering process contains information on energy and momentum transfers by

$$E = \hbar\omega \equiv E_i - E_f \text{ and } \hbar\boldsymbol{Q} \equiv \hbar\left(\boldsymbol{k}_i - \boldsymbol{k}_f\right) \tag{35}$$

Here, a brief discussion on scattering of a probe is restricted to the transferred energy smaller than the photon energy ($E << E_i$). In this case, the momentum transfer $\hbar\boldsymbol{Q}$ is simply connected with the scattering angle θ by

$$\hbar\boldsymbol{Q} \equiv 2\hbar\, k_i \sin\theta \tag{36}$$

The double-differential scattering cross section, which is expressed as [71],

$$\frac{\mathrm{d}^2}{\mathrm{d}\boldsymbol{\Omega}\ \mathrm{d}\omega_f} = \left(\frac{\mathrm{d}\sigma}{\mathrm{d}\boldsymbol{\Omega}}\right)_0 \cdot S(\boldsymbol{Q},\omega) \tag{37}$$

This involves two contributions: (1) the coupling of the beam to the scattering system characterized by the intrinsic cross section $\left(\frac{\mathrm{d}\sigma}{\mathrm{d}\boldsymbol{\Omega}}\right)_0$ and (2) the scattering function $S(\boldsymbol{Q}, \omega)$

which expresses the properties of the sample in the absence of the perturbing probe. The sophisticated form of scattering function can be given by

$$S(\boldsymbol{Q},\omega)=\frac{1}{2\pi}\int e^{-i\omega t}\left\langle \gamma_i \left| \sum_{j,j'} e^{-i\boldsymbol{Q}\boldsymbol{r}_j(t)} e^{i\boldsymbol{Q}\boldsymbol{r}_j(0)} \right| \gamma_i \right\rangle \mathrm{d}t \tag{38}$$

which describes the correlations of the scattering phases of the particles at positions $\boldsymbol{r}_j(t)$ at different times t. The phase of the scattering amplitude is described by $e^{i\boldsymbol{Q}\boldsymbol{r}_j}$. In classical limit, it represents essentially the Fourier transform in time of the density correlation function and gives information on the particle fluctuations in the scattering system in the same state, γ_i, at different times. The dependence of the scattering phases on the term $\boldsymbol{Q}\boldsymbol{r}_j$ can be used for a classification of the scattering process [72]. Thereby, the inverse of the transferred momentum $\hbar\boldsymbol{Q}$ has to be compared with a characteristic length ζ_{char} of the scattering system describing the spatial inhomogeniety. This can be an interparticle distance or a screening length.

For $Q\zeta_{char} \ll 1$, there exists interference between the scattering amplitudes from many particles of the system. Consequently, mainly the collective behaviour of the particles will be detectable. Therefore, collective motions of the scattering system like phonons, magnons or plasmons can be observed if, in addition, the transferred energy is in the characteristic frequency range of these motions.

For $Q\zeta_{char} \gg 1$, the interference of the scattering amplitudes is negligible and the scattering contributions of different particles are independent. Therefore, single-particle properties are observed, like, for example, Compton scattering [72] in the case of photon interaction with an electron system, if the photon energy is large compared with the binding energy of the electron.

In the intermediate ranges $Q\zeta_{char} \approx 1$, both collective and single-particle properties are visible. The scattering function (eq. (38)) is often transformed to representations that are more suitable to describe important physical properties of a particular system. The discussion on the general scattering theory by a probe is based on the review of Burkel [73].

2.4.2. Neutron Scattering

Neutron scattering in solids is of considerable interest to solid state and reactor physicists and slow neutron scattering by solids has developed into a powerful tool for investigating details of lattice vibrations. The average energy of neutrons that have reached thermal equilibrium with the atoms of any moderating material at temperature T is nearly $(3/2)K_BT$, where K_B is the Boltzmann constant. Corresponding to room temperature ($T \sim 300$ K) this energy is ~ 0.04 eV. Such neutrons represent an excellent probe for two reasons: (i) the de Broglie wavelength of such neutrons ($= h/(3m_0K_BT)^{1/2}$, where m_0 is the mass of neutron) is of the same order as the interatomic distance in crystals ($\sim 10^{-8}$ cm) so that, like X-rays, they can be used for diffraction studies, and (ii) the energy of such neutrons being of the same order as the thermal energy of the atoms in a solid, on collision the relative change in energy of the neutron can be large and readily measured. A study of the energy of scattered neutrons thus provides a direct method of studying lattice dynamics.

Thermal neutrons have velocities of the order of 3 x 10^5 cm/sec. The time that these neutrons take to cover the distance 3 Å is ≈ 10^{-13} sec, which is of the same order as the characteristic time of atomic vibrations. Hence, they can notice the atomic motions in their passage through a crystal and provide a method for studying a lattice dynamics. If the nuclei have zero spin and no isotope, the scattering of neutron waves interfere with the scattering from others. The interference part of the scattering is called *coherent* scattering and both elastic and inelastic scattering can give rise to it. If the nuclei composing the lattice have a spin or exist in more than one isotopic state, because of their random distribution, the different nuclei scatter independently and part of both elastic and inelastic scattering is *incoherent.*

2.4.2.2. Inelastic Neutron Scattering

Scattering of a neutron, which involves a change in its initial energy as a result of emission, or absorption of one or more phonons is called *inelastic scattering.* However, it is only the one-phonon scattering which gives us information about the frequency distribution function and the dispersion relation. Inelastic coherent scattering leads directly to the information of phonon dispersion relations, while a study of inelastic incoherent scattering determines directly the frequency distribution function of the scatterer. Hence, by measuring the energy distribution of neutrons, which are incoherently scattered through a certain angle, the frequency spectrum (phonon density of states) of scatterer can be determined. The intensity of inelastic coherent neutron scattering is proportional to the space and time Fourier Transforms of the time-dependent pair correlation function, $G(r,t)$ = probability of finding a particle at position 'r' at time 't' when there is a particle at $r = 0$ and $t = 0$. For inelastic incoherent scattering, the intensity is proportional to the space and time Fourier Transforms of the self-correlation function, $G_s(r,t)$ *i.e.* the probability of finding a particle at position 'r' at time 't' when the same particle was at $r = 0$ at $t = 0$.

2.4.2.3. Elastic Neutron Scattering

By elastic scattering are implied those scattering processes, such as Bragg reflections, in which neutron energy remains unaltered. Elastic coherent scattering studies lead to information about the structure of the crystal, including information about the magnetic state of the crystal. Elastic incoherent scattering does not give any direct useful information, though its temperature variation gives information about the validity of any model for the frequency distribution function. The intensity of elastic, coherent neutron scattering is proportional to the spatial Fourier Transform of the Pair Correlation Function, $G(r)$ *i.e.* the probability of finding a particle at position 'r' if there is simultaneously a particle at $r = 0$.

2.4.3. Raman Scattering

Raman spectroscopy is one of the most powerful, versatile and fascinating tools for the investigation of matter. It is the measurement of the wavelength and intensity of inelastically scattered light from molecules. It is possible for the incident photons to interact with the molecules in such a way that the energy is either gained or lost so that the scattered photons are shifted in frequency. Such inelastic scattering is called Raman scattering. Raman scattering can occur with a change in vibrational, rotational or electronic energy of a molecule. The difference in energy between the incident photon and the Raman scattered photon is equal to the energy of a vibration of the scattering molecule. A plot of intensity of scattered light versus energy difference is a Raman spectrum. Raman scattering occurs due to the change in polarizability during the molecular vibrations. The change is described by the polarizability derivative $\frac{\partial \alpha}{\partial \boldsymbol{Q}}$, where $\boldsymbol{Q}$ is the normal coordinate of the vibration. The selection rule for a Raman-active vibration emphasizes that there be a change in polarizability during the vibration and it is given as,

$$\frac{\partial \alpha}{\partial \boldsymbol{Q}} \neq 0 \tag{39}$$

For polarizable molecules, the incident photon energy can excite vibrational modes of the molecules, yielding scattered photons, which are diminished in energy by the amount of the vibrational transition energies. From group theory, it is straightforward to show that if a molecule has a center of symmetry, vibrations, which are Raman active, will be silent in the infrared, and vice versa. Scattering intensity is proportional to the square of the induced dipole moment *i.e.*

$$I_{Raman} \propto \left(\frac{\partial \alpha}{\partial Q}\right)_0^2 \tag{40}$$

To a good approximation Raman scattering occurs from zero-wavevector phonons. However, to the extent that the phonon wavevector differs from zero, phonon selection rules will deviate from the zero-wavevector rules and will depend on the angle between the direction of propagation of the incident and scattered light. For "optical phonons," which have zero dispersion at the zone center, any direction dependence in the Raman shift is quite small. On the other hand, for the "acoustic phonons," which have a linear dispersion near the zone center, the angular dependence of the Raman shift is more pronounced.

A spectral analysis of the scattered light under these circumstances reveals spectral satellite lines below the Rayleigh scattering peak at the incident frequency. Such lines are called "Stokes lines". If there is significant excitation of vibrational excited states of the scattering molecules, then it is also possible to observe scattering at frequencies above the incident frequency as the vibrational energy is added to the incident photon energy. These lines, generally weaker, are called “anti-Stokes lines”.

Numerically, the energy difference between the initial and final vibrational levels, $\bar{\nu}$, or Raman shift in wave numbers (cm^{-1}), is calculated by,

$$\bar{\nu} = \frac{1}{\lambda_{incident}} - \frac{1}{\lambda_{scattered}} \tag{41}$$

in which $\lambda_{incident}$ and $\lambda_{scattered}$ are the wavelengths (in cm) of the incident and Raman scattered photons, respectively. The Stokes and anti-Stokes spectra contain the same frequency information. The ratio of anti-Stokes to Stokes intensity at any vibrational frequency is a measure of temperature. The energy of a vibrational mode depends on molecular structure and environment. Atomic mass, bond order, molecular substituents, molecular geometry, and hydrogen bonding all affect the vibrational force constant, which, in turn dictates the vibrational energy. Vibrational Raman spectroscopy is an extraordinarily versatile probe into a wide range of phenomena ranging across disciplines from physical biochemistry to materials science.

2.4.4. Infrared Scattering

The vibrational spectrum of a molecule is considered to be a unique physical property and is characteristic of the molecule. As such, the infrared spectrum can be used as a fingerprint for identification by the comparison of the spectrum from an ''unknown'' with previously recorded reference spectra.

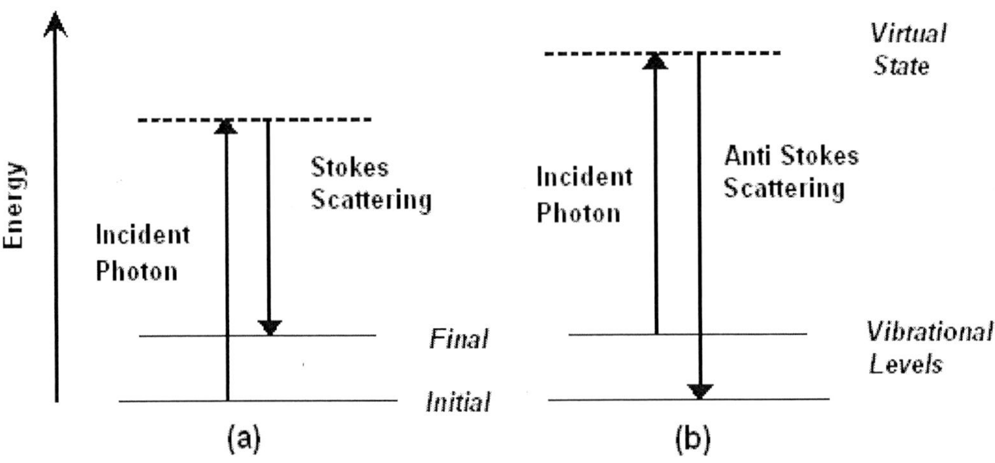

Figure 10. Energy level diagram for Raman scattering: (a) Stokes Raman scattering (b) Anti-Stokes Raman scattering.

For a molecule to absorb infrared radiation (IR), the vibrations or rotations within a molecule must cause a net change in the dipole moment of the molecule. The intensity of an infrared absorption band I_{IR} depends on the change of the dipole moment μ during this vibration:

$$I_{IR} \propto \left(\frac{\partial \mu}{\partial Q}\right)_0^2 \tag{42}$$

The alternating electrical field of the radiation interacts with fluctuations in the dipole moment of the molecule. If the frequency of the radiation matches the vibrational frequency of the molecule, the radiation will be absorbed, causing a change in the amplitude of molecular vibration. The positions of atoms in a molecule are not fixed; they are subject to a number of different vibrations with two main categories of *stretching* and *bending*. For simple systems, the atoms can be considered as point masses, linked by a 'spring' having a force constant k and following Hooke's Law. Using this simple approximation, the equation shown below can be utilized to approximate the characteristic stretching frequency (in cm^{-1}) of two atoms of masses m_1 and m_2, linked by a bond with a force constant k:

$$\nu = \frac{1}{2\pi c}\sqrt{\frac{k}{\mu}} \tag{43}$$

where, $\mu = m_1 m_2/(m_1+m_2)$ (termed the "reduced mass"), and c is the velocity of light. In order to be IR active, vibration must cause a change in the dipole moment of the molecule. In general, the larger the dipole change, the stronger is the intensity of the band in an IR spectrum. One selection rule that influences the intensity of infrared absorptions is that a change in dipole moment should occur for a vibration to absorb infrared energy. Molecular asymmetry is a requirement for the excitation by infrared radiation and fully symmetric molecules do not display absorbance in this region unless asymmetric stretching or bending transitions are possible.

2.4.5. Brillouin Scattering

Raman scattering from low-energy acoustic phonons is known as Brillouin Scattering. Brillouin scattering is a powerful and promising probe to study the surface and bulk acoustic phonons as well as magnetic excitations in opaque solids and the elasticity of materials at extreme conditions [74]. Brillouin light scattering is generally referred to as inelastic scattering of an incident optical wave field by thermally excited elastic waves in a sample. From a strictly classical point of view, the compression of the medium will change the index of refraction and therefore lead to some reflection or scattering at any point where the index changes. From a quantum point of view, the process can be considered as interaction of light photons with acoustic or vibrational quanta (phonons). Brillouin spectroscopy is an experimental method of performing such velocity measurements on small samples of high-pressure phases. The Brillouin spectrum of light scattered from thermal phonons contains, in its shift, the phase velocity of sound and, in its line width, the acoustic absorption. Brillouin scattering manifests as extra phonons, at low energy. The essential difference between Raman and Brillouin scattering is the sensitivity of the "Brillouin shift" to the relative angle of scattering. Brillouin scattering is a nondestructive light scattering technique, which allows for extracting all necessary information from exceptionally small samples. The properties obtained by Brillouin techniques include, but are not limited to, a full set of single crystal elastic moduli, aggregate bulk and shear moduli, and density as a function of pressure. Due to the requirements of such small samples, Brillouin scattering experiments can be readily combined with the Diamond Anvil Cell (DAC) or high-temperature cells to obtain high-pressure and/or high-temperature data.

2.4.6. X-ray Scattering

X-rays are tools with a very wide field of application. Traditionally, experimental determination of lattice dynamics is the domain of inelastic neutron scattering, but the restrictions on sample size imposed by the technique relegated the achievable information to low or at most moderate pressures (~10 GPa). Characterizing the effect of pressure on the propagation of elastic wave is instead singularly important of understanding elasticity, mechanical stability of solids, material strength, inter-atomic interactions, and phase transition mechanism. The elastic properties and the sound wave anisotropy of hexagonal metals at high pressure are experimentally investigated by Inelastic X-ray Scattering (IXS). This technique allows the collection of the phonon dispersion curve and is particularly well suited for extreme conditions. X-rays, originally a tool for structural investigations and imaging purposes, are nowadays successfully applied in many ways in an enormously broad field: materials science, biology and medicine among others. Another advantage is the accessible range in energy and momentum transfer, particularly advantageous in the study of noncrystalline matter. Furthermore, elements having too large absorption or incoherent cross sections for neutrons can be investigated.

3. Vibrational Properties of Lanthanum Manganites: Effect of Structure and Doping

3.1. Introduction

Perovskite manganites are a class of materials that have attracted enormous attention in recent years, mainly because of the colossal magnetoresistance (CMR) effect [5], which is discussed at a length in section 1. At high temperatures, these materials are paramagnetic insulators (PM), while at low temperatures, they exhibit a number of phases including ferromagnetic metallic (FM) and antiferromagnetic insulating (AFM). Understanding the paramagnetic insulating (PI) phase, which has properties that remain largely unexplained, is a central issue to unraveling the mystery of CMR effect. Recently, evidence of temperature [9, 10], pressure [9], magnetic field [10, 75], doping [76, 77] dependence of structural, magnetic, vibrational and electronic properties due to the strong interplay of spin, charge and lattice degree of freedom has been reported.

$LaMnO_3$ is a prototype of ABO_3 perovskite investigated in detail. Synthesis conditions [78, 79] play an important role in determining the well-defined phases of undoped $LaMnO_3$ crystals. At room temperature, undoped $LaMnO_3$ behaves as paramagnetic insulator (PI) and can adopt either orthorhombic $(Pbnm)$ or rhombohedral $(R\bar{3}c)$ phase. $LaMnO_3$ however, remains paramagnetic insulator (PI) in either structural phase at room temperature. At the ambient pressure, the most known phase of orthorhombic $LaMnO_3$ undergoes structural phase transition at about 750 K [80] and becomes cubic $(Pm\bar{3}m)$, with the disappearance of orbital ordering. The $LaMnO_3$ in these structures exhibit completely different properties such as Curie temperature (T_C) and magnetoresistance (MR). This is believed to be the result of the different structure and changes (distortions) of Mn-O-Mn bond angles [81]. While distortion is maximum in orthorhombic structure, it is minimum in cubic structure through rhombohedral phase. This change in distortion affects the relative strength of the electron-phonon interaction, which is responsible for the lattice distortion in paramagnetic insulting (PI) phase. To explain the colossal magnetoresistance in manganites, Zener [20] proposed a mechanism better known as the double-exchange (DE) mechanism. Recently Millis *et al.* [82] have proposed existence of a strong electron-phonon coupling, which localizes e_g electrons as polarons due to slowly fluctuating local Jahn-Teller distortion (dynamical Jahn-Teller distortion). The Jahn-Teller distortion favors the insulating state. This is partly removed below T_C and is accompanied by the changes in the structure of these systems. The electron-phonon correlation in these systems results into the formation of the vibronic ground state. However, the nature of the ground states of manganites is still under question.

It has been observed that these lattice effects of doped and undoped $LaMnO_3$ in rhombohedral symmetry are well displayed in their infrared (IR) and Raman spectra of phonons [83-86]. The Raman investigations in doped $LaMnO_3$ with $\left(R\bar{3}c\right)$ space symmetry reveal that A_{1g} -like phonon mode is sensitive to both value and kind of dopant [83]. Despite several experimental investigations performed by Raman spectroscopy, there is a controversy on the interpretation of the Raman spectra even for stoichiometric $RMnO_3$ [87-88]. Rhombohedral $LaMnO_3$ have been investigated recently by means of lattice dynamical calculations [89], however the studies were focused to the calculations of some selected zone-centre phonon modes only and there was an emphasis for detailed investigations. Due to the inversion symmetry at all lattice sites of a perfect cubic crystal, the Raman investigations are forbidden in the perfect cubic manganites. Cubic $LaMnO_3$ is, however investigated theoretically by Kovaleva *et al.* [90] only, for the evaluation of electronic and ionic polarization energies associated with holes localized at Mn^{3+} cation and O^{2-} anion. Reichardt and Braden [91] have investigated the phonon branches in rhombohedral $La_{0.8}Sr_{0.2}MnO_3$ and $La_{0.7}Sr_{0.3}MnO_3$ at 15 K by means of inelastic neutron scattering and shell model calculations by including screening in cubic phase, and observed the anomalous features. In this calculation, they have not only fitted the zone centre phonons but also the calculation was done for rhombohedral systems by considering the cubic phase.

Among a number of perovskite type manganese oxides with various combinations of rare earth and alkaline earth ions, $La_{1-x}Sr_xMnO_3$ (LSMO) is considered as prototypical and reference material. Depending on both doping and temperature, $La_{1-x}Sr_xMnO_3$ compounds exhibit a number of magnetic phases (para and ferromagnetic insulators and metals, canted antiferromagnetic insulators) as well as some structural transitions related to cubic, rhombohedral, tetragonal, orthorhombic and monoclinic crystal symmetries [81, 92]. For $La_{1-x}Sr_xMnO_3$ the ferromagnetic state appears below T_C and above critical composition 'x_c' = 0.17 (compositional I-M phase boundary) at which the structural transformations occur from orthorhombic phase to rhombohedral phase [5, 92]. This suggests that ferromagnetic metallic state prefers the rhombohedral structure which in $La_{1-x}Sr_xMnO_3$ develops up to $T_C = 370$ K at the doping level of 'x' = 0.3-0.5 [5, 88, 93]. Neutron diffraction measurements performed in $La_{1-x}Sr_xMnO_3$ for various doping concentration show that the $La_{1-x}Sr_xMnO_3$ prefer orthorhombic phase for $x < 0.14$ and rhombohedral phase for $x \geq 0.16$. Magnetic field induced large negative magneto resistance ($\approx$ 0.95) has been observed around ferromagnetic phase transition $(T_C)^5$ that corroborates the fact that the electronic and magnetic properties critically depend on 'x' in the hole doped Mn oxides. The strong coupling of the vibrational and electronic systems in the $La_{1-x}A_xMnO_3$ (A = Sr, Ba, Pb etc.) lattice was found to play an important role in the mechanism of the colossal magnetoresistance [82]. In particular, the Jahn-Teller (JT) effect was found to affect both the magnetic and transport properties of $La_{1-x}Sr_xMnO_3$ systems [82]. While the double-exchange model describes the ferromagnetic-metallic state of $La_{1-x}Sr_xMnO_3$ systems, a mechanism based on the strong electron-phonon coupling was shown to be important in the paramagnetic insulating phase. The JT effect mainly responsible for the distortion of the cubic perovskite structure is therefore expected to leave footprints in the phonon spectra [93]. There is also large consensus on the existence of lattice polarons observed through the neutron scattering [94, 95] and isotope shift [36] experiments. A clear demonstration of this effect in manganites is the frustration of the insulator to metallic transition, only by the isotope substitution of O^{16} by O^{18} in a $La_{0.33}Nd_{0.33}Ca_{0.33}MnO_3$ compound [37].

In terms of the total amount of work devoted so far to the $La_{1-x}Sr_xMnO_3$ materials, little work appears to have been done on the lattice dynamical properties of these systems particularly by using theoretical models inspite of the importance of phonons in these compounds. The situation is slight better in case of experimental measurements[83, 84, 85, 98]. Some attempts have been made to understand the vibrational and electronic excitations, phase transitions, lattice transformations etc. of these compounds by using infrared spectroscopy and Raman spectroscopy[83, 98, 99]. The Raman investigations in the $La_{1-x}Sr_xMnO_3$ reveal that the A_{1g}-like phonon mode, which arises due to rotation of MnO_6 octrahedra, is sensitive to both value and kind of dopant [83]. $La_{0.7}Sr_{03}MnO_3$ with rhombohedral structure is the most appropriate system to study the rhombohedral distortion, which is measured by softening of A_{1g} phonon modes. Raman experiments performed on single crystal $La_{1-x}Sr_xMnO_3$ [99] as a function of temperature show an anomalous softening of the in-phase stretching mode of oxygen cage and out of phase bending modes.

Therefore, it seems important to perform a systematic and detailed investigation of phonon properties of undoped and doped $LaMnO_3$ to understand the effect of structure and doping. This will not only lead to know the role of phonons in these compounds but also to understand the interatomic forces in manganites. In this section, we present the results [100] of our own systematic and detailed investigation on the phonon properties of cubic and rhombohedral phase of undoped $LaMnO_3$ and rhombohedral $La_{0.7}Sr_{0.3}MnO_3$ is presented for ambient conditions by using a lattice dynamical theory and methodology discussed below. The model parameters obtained so far the undoped and doped $LaMnO_3$ systems are presented in Table I and II respectively. As mentioned in section 2, the O-O interactions have been taken from ref. [64] and other interactions are obtained by ensuring the stability of the structure and neutrality of charge. The difference in the value of strength parameter for two different structure of $LaMnO_3$ is obvious as it is chosen by ensuring all frequencies in the whole Brillouin zone to be positive and stable structure.

Table I. Model Parameters for $LaMnO_3$ in cubic and rhombohedral phases ref [100]

	Short - range interactions				Effective charge		
	LaMnO3 $(Pm\bar{3}m)$		LaMnO3 $(R\bar{3}c)$			LaMnO3 $(Pm\bar{3}m)$	LaMnO3 $(R\bar{3}c)$
Interactions	b_{ij} (eV)	ρ_{ij} (Å)	b_{ij} (eV)	ρ_{ij} (Å)	Ion	Z(e)	Z(e)
La – O	1876	0.3404	1514	0.3153	La	2.0	4.0
Mn – O	281	0.1930	1514	0.3453	Mn	4.0	2.0
O – O	22764	0.3243	22764	0.4165	O	-2.0	-2.0

The short range O–O potential is taken from Ref. [64].

Table II. Model Parameters for $La_{0.7}Sr_{0.3}MnO_3$ in rhombohedral phase

	Short - range interactions		Effective Charge	
	b_{ij} (eV)	ρ_{ij} (Å)	Ion	Z(e)
La/Sr–O	1350	0.1960	La/Sr	2.85
Mn–O	1850	0.2160	Mn	2.85
O–O	22764	0.1090	O	-1.90

Shell Model[a]

Ion	Y(e)	K (eV- Å^2)
O	-2.869	74.92

[a]Y and K refer to the shell charge and harmonic spring constant respectively.

3.2. Zone centre phonons

$LaMnO_3$ is an antiferromagnetic insulator below 135 K and undergoes a structural phase transition at a $T_S \sim 750$ K and its high-temperature phase is believed to be cubic. The low-temperature phase is approximately tetragonal, with one lattice constant about 0.15 Å shorter than the other two [101]. Several other distortions such as, small-amplitude (~ 0.01Å) also occur at temperatures less than or equal to T_S [102] and the structure at room temperature is orthorhombic. Stoichiometric $LaMnO_3$ with orthorhombic structure is Jahn-Teller distorted due to a mismatch between La-O and Mn-O bond lengths. Synthesis condition is an essential parameter in determining the structure of undoped $LaMnO_3$ at room temperature, which could be either orthorhombic or rhombohedral. The rhombohedral phase is of special interest as it is typical for some $R_{1-x}A_xMnO_3$ and some specific features associated with the CMR such as variations of Jahn-Teller distortions (electron-phonon coupling) with the Mn^{4+}/Mn^{3+} ratio, temperature as well as pressure, are expected to be reflected in the phonon spectra of this phase. The composition of compound is also of importance because the presence of a small amount of Mn^{4+} ions disturbs the system so much, that the structure and magnetic order can be different from those corresponding to the stoichiometric compound. It is an observed fact that an increase in Mn^{3+} content prefers orthorhombic phase and that in Mn^{4+} favors rhombohedral phase[5, 92]. A gradual increase in Sr concentration 'x' in $La_{1-x}Sr_xMnO_3$ (LSMO) brings about the structural phase transition from orthorhombic to rhombohedral phase. For 'x' = 0.3, LSMO possesses well defined rhombohedral phase. Both the rhombohedral $LaMnO_3$ [89] and $La_{0.7}Sr_{0.3}MnO_3$ [77] have the symmetry of the space group $(R\bar{3}c)$ $(D_{3d}^6, z=2)$ and belongs to the family of rotationally distorted perovskites. The $(R\bar{3}c)$ structure can be obtained from the simple-cubic perovskite $(Pm\bar{3}m)$ by the rotation of the adjacent MnO_6 octahedra in opposite directions around the $[111]_c$ (cubic) directions.

The factor group analysis of the zone centre vibrational modes in the present case has been done by using the correlation table method. Results of this analysis are presented in Table III (a) and III (b) for rhombohedral $LaMnO_3$ and $La_{0.7}Sr_{0.3}MnO_3$ respectively and in Table III (c) for cubic $LaMnO_3$. For rhombohedral structure, out of total 30 Γ-point modes, the A_{1g} + 4 E_g modes are Raman active, the 3 A_{2u} + 5 E_u are only IR active and the remaining 2 A_{1u} +3 A_{2g} modes are silent modes. La atoms participate in four Γ – point phonon modes ($A_{2g} + A_{2u} + E_g + E_u$). While Mn atoms contribute to four ($A_{1u} + A_{2u} + 2\ E_u$) Γ – point phonon modes, O atoms take part in twelve ($A_{1g} + A_{1u} + A_{2g} + 2\ A_{2u} + 3\ E_g + 3\ E_u$) Γ – point phonon modes in the rhombohedral lattice. In $(R\bar{3}c)$ $LaMnO_3$, La and Mn atoms occupy 2*a* (1/4, 1/4, 1/4) and 2*b* (0, 0, 0) sites and O atoms possess 6*e* $\left(x, \bar{x} + \frac{1}{2}, \frac{1}{4}\right)$ atomic positions with x = 0.447[79].

In $La_{0.7}Sr_{0.3}MnO_3$, La/Sr and Mn atoms occupy 6*a* (0, 0, 0.25) and 6*b* (0, 0, 0) sites and O atoms prefer 18*e* (x, 0, 0.25) sites with x = 0.4577 [79]. For better understanding of the relationship between the frequency and vibrational patterns of the modes investigated, it is instructive to recall the correlation between the modes in distorted $(R\bar{3}c)$ and undistorted $(Pm\bar{3}m)$ structures. Additionally, the triply degenerated modes of the cubic structure split into pairs of non-degenerated and doubly degenerated modes in the rhombohedral structure. The ideal perovskite $LaMnO_3$ [103] of cubic symmetry (space group: O_h^1, $(Pm\bar{3}m)$) presents 15 normal modes of vibrations, out of which three of F_{1u} irreducible representations correspond to three IR active optical modes. The other F_{1u} mode corresponds to a triply degenerate acoustic mode. While, IR active modes are allowed for the cubic symmetry, the first order Raman modes are forbidden due to the inversion symmetry at all lattice site of a perfect cubic structure. In the cubic symmetry, F_{2u} is a silent mode, which is neither Raman nor IR active [105]. In lattice, while La atoms are positioned at 1*a* (0, 0, 0), Mn atoms occupy 1*b* (1/2, 1/2, 1/2) site and O atoms take positions at 3*c* [(1/2, 1/2, 0), (1/2, 0, 1/2), (0,

Table III(a). Factor group analysis and selection rules for the zone-centre vibrational modes of the rhombohedral $LaMnO_3$ $(R\bar{3}c)$, z = 2

Atom	Number of equivalent positions (Wyckoff notations)	Site symmetry	Irreducible representations of modes
La	2(*a*)	D_3 (32)	$A_{2g} + A_{2u} + E_g + E_u$
Mn	2(*b*)	S_6 (-3.)	$A_{1u} + A_{2u} + 2\ E_u$
O	6(*e*)	C_2 (.2)	$A_{1g} + 2\ A_{2g} + 3\ E_g + A_{1u} + 2\ A_{2u} + 3\ E_u$

$\Gamma_{Raman} = A_{1g} + 4\ E_g$

$\Gamma_{IR} = 3\ A_{2u} + 5\ E_u$

$\Gamma_{acoustic} = A_{2u} + E_u$

$\Gamma_{Silent} = 3\ A_{2g} + 2\ A_{1u}$

Raman Selection Rules:

A_{1g} : $\alpha_{xx} + \alpha_{yy}$, α_{zz}

E_g : ($\alpha_{xx} - \alpha_{yy}$, α_{xy}), (α_{xz},α_{yz})

Table III (b). Factor group analysis and selection rules for the zone-centre vibrational modes of the rhombohedral $La_{0.7}Sr_{0.3}MnO_3$ $(R\bar{3}c)$, z = 2

Atom	Number of equivalent positions (Wyckoff notations)	Site symmetry	Irreducible representations of modes
La/Sr	6(a)	D_3 (32)	$A_{2g} + A_{2u} + E_g + E_u$
Mn	6(b)	S_6 (-3.)	$A_{1u} + A_{2u} + 2\ E_u$
O	18(e)	C_2 (.2)	$A_{1g} + 2\ A_{2g} + 3\ E_g + A_{1u} + 2\ A_{2u} + 3\ E_u$

$\Gamma_{Raman} = A_{1g} + 4\ E_g$

$\Gamma_{IR} = 3A_{2u} + 5\ E_u$

$\Gamma_{acoustic} = A_{2u} + E_u$

$\Gamma_{Silent} = 3\ A_{2g} + 2\ A_{1u}$

Raman Selection Rules:

$A_{1g}: \alpha_{xx} + \alpha_{yy}, \alpha_{zz}$

$E_g: (\alpha_{xx} - \alpha_{yy}, \alpha_{xy}), (\alpha_{xz}, \alpha_{yz})$

Table III (c). Factor group analysis and selection rules for the zone-centre vibrational modes of the cubic $LaMnO_3$ $(Pm\bar{3}m)$, z = 1

Atom	Number of equivalent positions (Wyckoff notations)	Site symmetry	Irreducible representations of modes
La	1(*a*)	T_h (m-3m)	F_{1u}
Mn	1(*b*)	T_h (m-3m)	F_{1u}
O	3(*c*)	O_h (4/mm.m)	$F_{1u} + F_{2u}$

$\Gamma_{IR} = 3\ F_{1u} + 1\ F_{2u}$.

1/2, 1/2)] sites, respectively with lattice constant a = 3.934 Å [45]. The calculated zone centre phonon frequencies for rhombohedral $LaMnO_3$, $La_{0.7}Sr_{0.3}MnO_3$, and cubic $LaMnO_3$ are listed in Table IV and compared with available experimental data [98, 99, 106].

Table IV reveals that there is in general good agreement with the available experimental data. Based on the lattice dynamical calculations and similarity in Raman spectra of rhombohedral and orthorhombic $LaMnO_3$, Abrashev *et al.*[22] assigned pure low frequency La vibrations in hexagonal $(001)_h$ plane to E_g mode. Peculiar phonon mode due to the rotation of the oxygen octahedra around the hexagonal $[001]_h$ direction is assigned to A_{1g} mode, pure mid frequency oxygen bending vibrations and high frequency out-of-phase stretching vibrations to two E_g modes. However, only A_{1g} mode is of prime importance allowed for the $(R\bar{3}c)$ phase since this mode involves the atomic motions that cause the rhombohedral distortion (static rotational displacement of the oxygen octahedra around the hexagonal $[001]_h$ direction *i.e.* it is a "soft" mode and its position could be used as a measure of the degree of

Table IV. Calculated Raman frequencies for cubic and rhombohedral $LaMnO_3$ and $La_{0.7}Sr_{0.3}MnO_3$ ref. [100].

Phonon Frequencies in cm^{-1}				
Rhombohedral			Cubic	
D_{3d}^{6} Modes	$LaMnO_3$	$La_{0.7}Sr_{0.3}MnO_3$	O_h^1 Modes	$LaMnO_3$
Raman-active Modes				
A_{1g}	195.38, 236[a], 249[b]	209.03, 199[c], 180[d]		
E_g	-	28.91, 42[c]		
E_g	241.6,179[a], 163[b]	173.49		
E_g	374.6, 329[a]. 468[b]	408.40		
E_g	624.5,515[a], 646[b]	447.83		
IR-active Modes				
A_{2u}	-	66.90	F_{1u}	165.76
A_{2u}	-	561.62		271.77
A_{2u}	642.4,645[b] (LO)	647.70, 641[e], 580[f],576[g]		586.23
E_u	-	43.51	T_{2u}	423.93
E_u	-	72.22		
E_u	-	112.57		
E_u	207.2	179.69		
E_u	478.3	416.67		
Silent Modes				
A_{2g}	-	343.88		
A_{2g}	186.14, 441[b]	478.54		
A_{2g}	528.40, 716[b]	538.25		
A_{1u}	-	174.18		
A_{1u}	639.95	549.71		

[a,b]Experimental and Calculated Raman data for rhombohedral $LaMnO_3$ [89].
[c]Experimental Raman data at 10 K [99].
[d]Experimental Raman data at 300 K for $La_{0.67}Sr_{0.33}MnO_3$ [98].
[e,f]Calculated and experimental IR data at 405 K respectively [91].
[g]Inelastic neutron scattering data [106].

the distortion. The A_{1g} mode frequency should soften to zero with increasing temperature approaching the temperature of structural phase transition to cubic ($Pm\bar{3}m$) phase *i.e.* this mode is one of the two "soft" modes. The large softening of this mode also occurs with the increase of Sr content, which can be explained by the rhombohedral distortion. This fact is supported by the reports of Podobedov *et al.*[83]. Moreover, the frequency of A_{1g} mode should also mainly correlate with angle α of the rhombohedral distortion connected with the 'x' parameter of oxygen-site positions and it is defined by $x = \frac{1}{2}\left(1 \pm \frac{1}{\sqrt{3}} \tan\alpha\right)$.

Indeed, angle $\alpha \approx 11^\circ$ and $\approx 9^\circ$ are obtained using the values of x = 0.447 for $LaMnO_3$ and x = 0.4577 for $La_{0.7}Sr_{0.3}MnO_3$, respectively which shows linear relation with A_{1g} mode frequency. In orthorhombic distortion, there are two types of static rotational displacements for the oxygen octahedra, which results in the existence of three Raman-active "soft" modes with orthorhombic A_g, B_{1g}, and B_{2g} symmetries, respectively. The Raman active A_g mode corresponding to an in-phase rotation of the MnO_6 octahedra around the *b*-axis in lower symmetry (distorted) orthorhombic structure of these compounds is closely related with the A_{1g} mode in higher symmetry rhombohedral $La_{0.7}Sr_{0.3}MnO_3$ and found to be the most sensitive to the kind and value of the dopant [83]. According to the symmetry, the following correlations hold between the long-wavelength IR active 3 A_{2u} and 5 E_u modes of the rhombohedrally distorted $(R\bar{3}c)$ perovskites and phonons of the cubic perovskite structure compounds:

(1) Three A_{2u} and three out of five E_u modes make pairs and originate from the three zone centre triply degenerate IR-active F_{1u} modes of the ideal cubic perovskite. These three pairs correspond to (a) the low-frequency vibrations of Mn-O sublattice against La atoms, (b) the middle-frequency bending vibrations of Mn and two apical O atoms against the other four oxygen of the octahedron, (c) the high-frequency stretching vibrations of O and Mn atoms against rigid oxygen octahedron.
(2) A triply degenerate inactive (silent) F_{2u} (Γ) mode at the Γ point of the cubic Brillouin zone (torsional oxygen vibrations of the oxygen octahedra) contributes to in making pairs of the remaining two E_u modes in the rhombohedral phase with two inactive A_{1u} modes.

The vibrational patterns exhibiting rhombohedral distortions in $LaMnO_3$ and $La_{0.7}Sr_{0.3}MnO_3$ for some phonon modes, which have been selected on the basis of main atomic motions are presented in Figure 11 (a) and (c), while for cubic $LaMnO_3$ the vibrational pattern of zone centre phonon modes are presented in Figure 11(b).

3.3. Phonon Band Structure

In order to have essence of phonon vibrations in manganites, the phonon spectra of cubic $(Pm\bar{3}m)$ and rhombohedral $(R\bar{3}c)$ undoped $LaMnO_3$ and Srdoped $LaMnO_3$ are calculated in the present study. Phonon dispersion curves (PDC) of undoped $LaMnO_3$ in its two different structures are presented in Figure 12 (a). This figure also includes the inelastic neutrons scattering (INS) data obtained at 15 K for $(R\bar{3}c)$ $La_{0.7}Sr_{0.3}MnO_3$ to have an idea of the success of present calculation. The difference observed in the present calculation and INS data is quite obvious due to difference in temperature. However, the present results predict some of the modes quite satisfactorily within the limitations. The phonon dispersion curve is presented only in one direction $(\bar{q}\bar{q}\bar{q})$ of the Brillouin zone for $(R\bar{3}c)$ structured $LaMnO_3$ (cf. Figure 12 (a (ii))), as it is sufficient to note the differences and understand the role of phonons. The number of phonon modes at zone centre is more in rhombohedral, which is obvious due to more number of atoms in the unit cell. It is well observed from this figure, that the degeneracy of the phonon mode is removed in cubic $LaMnO_3$ at points other than zone centre.

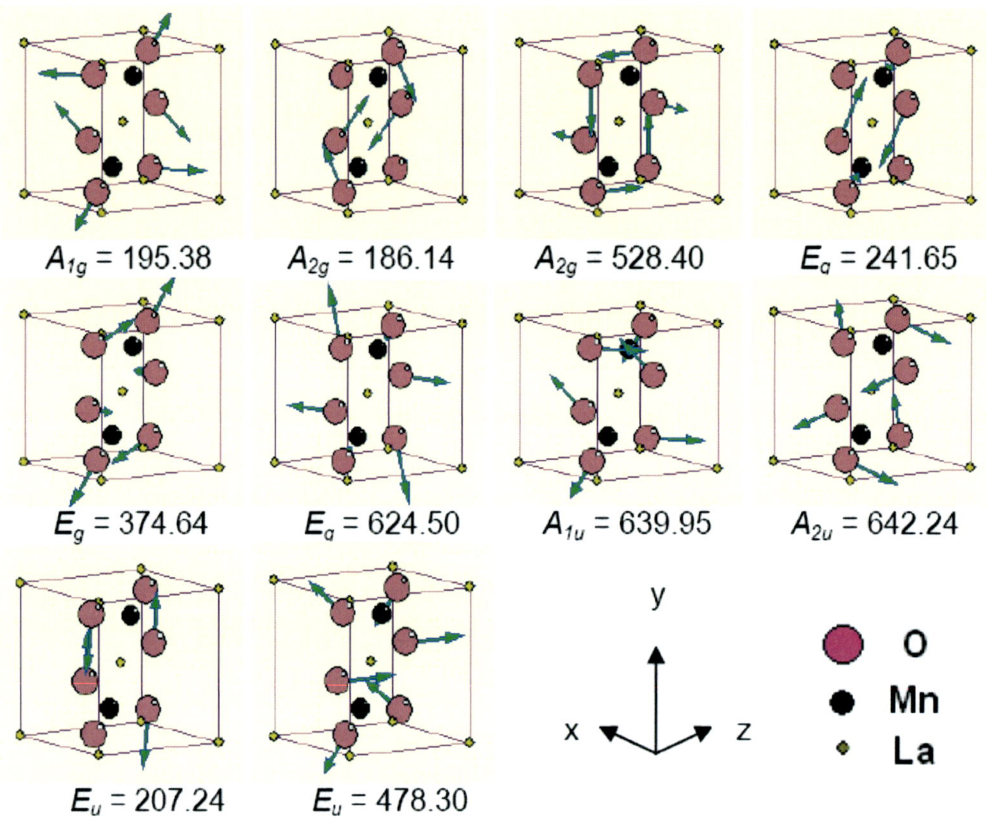

Figure 11(a). The vibrational pattern of zone centre phonon modes of rhombohedral $LaMnO_3$.

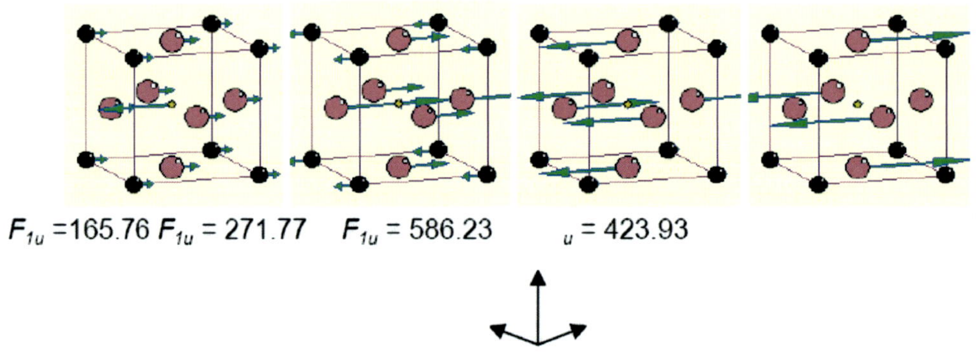

Figure 11(b). The vibrational pattern of zone centre phonon modes of cubic $LaMnO_3$.

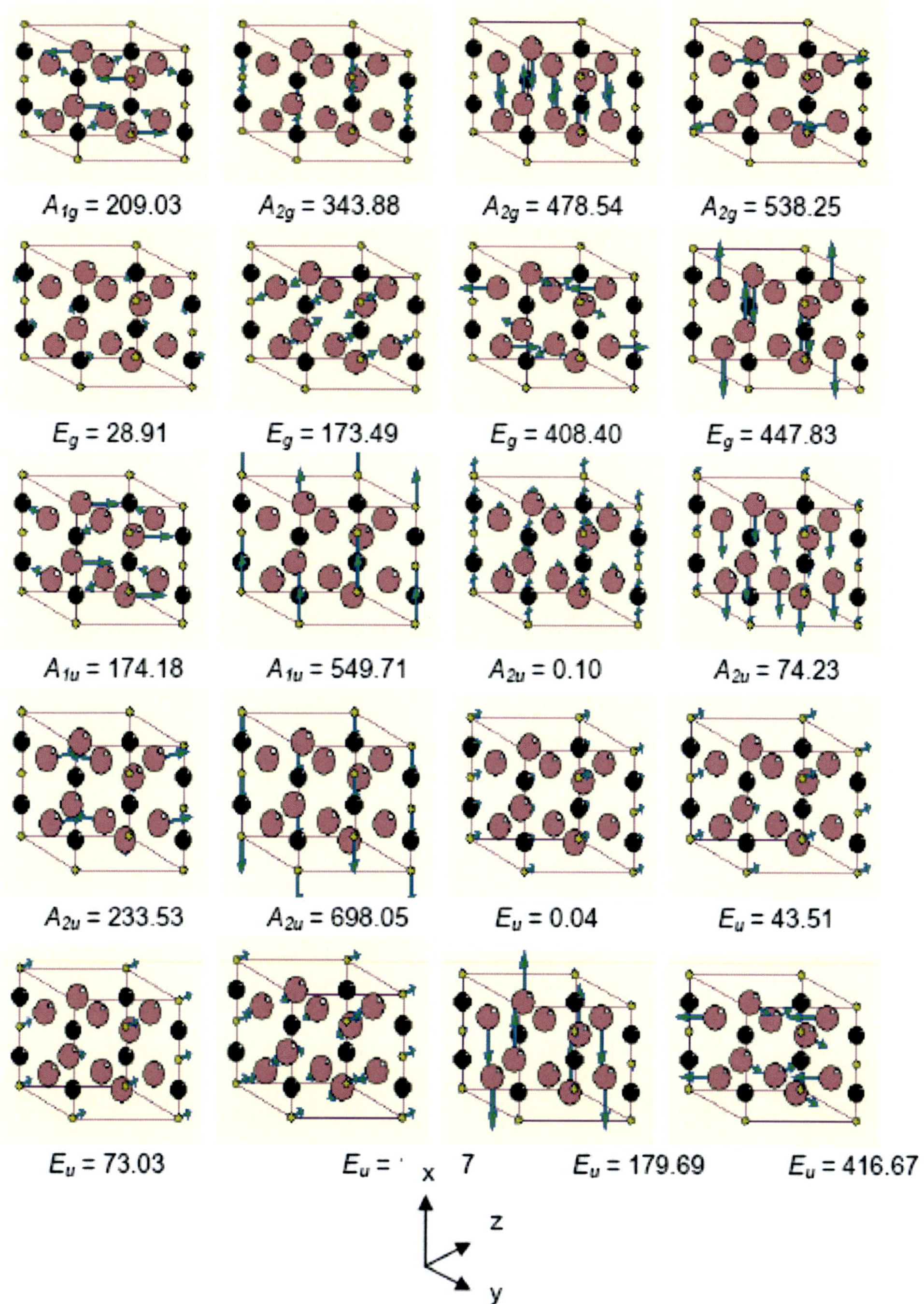

Figure 11(c). The vibrational pattern of zone centre phonon modes of rhombohedral $La_{1-x}Sr_xMnO_3$ for x = 0.3 ref [100].

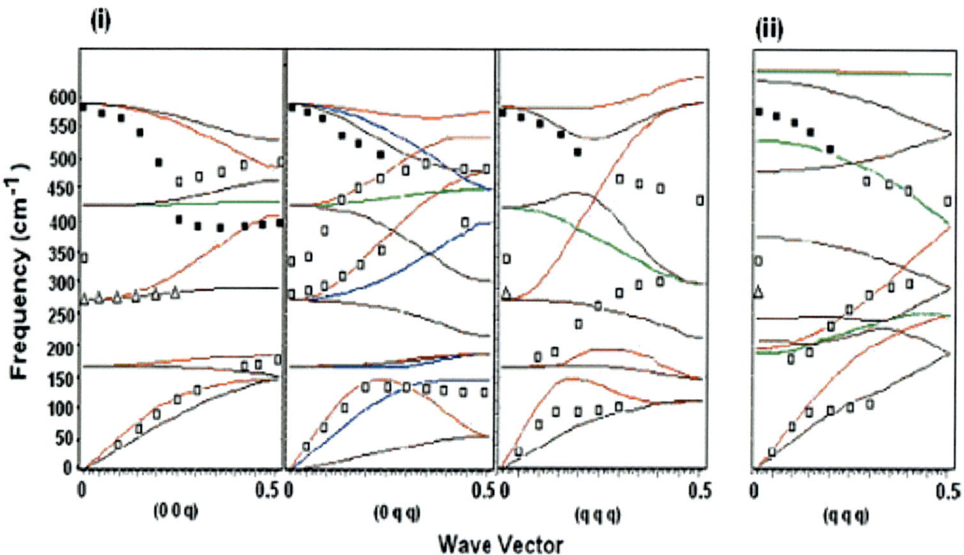

Figure 12(a). Phonon dispersion curves for $LaMnO_3$ in (i) cubic and (ii) rhombohedral phases ref [100]. Open and solid symbols are the inelastic neutron scattering data for $La_{0.7}Sr_{0.3}MnO_3$ in rhombohedral phase at 15K ref [91].

The main features of phonon dispersion curves of cubic $LaMnO_3$ are as follows:

(1) Degeneracy of all IR active phonon modes of cubic $LaMnO_3$ is revealed in $(\overline{0}\,\overline{q}\,\overline{q})$ direction.
(2) Phonon modes at 424 cm^{-1} show completely different behaviour in $(\overline{q}\,\overline{q}\,\overline{q})$ direction.
(3) Phonon modes at 272 cm^{-1} are more dispersive in $(\overline{q}\,\overline{q}\,\overline{q})$ direction with disappearance of one of their degenerate branches. It can be seen from the phonon dispersion curves in $(\overline{q}\,\overline{0}\,\overline{0})$ direction of Brillouin zone (BZ) for rhombohedral phase that there is a noticeable difference in the character of phonon dispersion curves.

Approximately, the phonon spectra of manganites can be separated into external (~ 185 cm^{-1}), bending (~ 350 cm^{-1}) and stretching (~ 550 cm^{-1}) modes with respect to cubic $(Pm\overline{3}m)$ symmetry. Depending on ion size and doping concentration, these triply degenerate modes split into pairs of non-degenerate (A) and doubly degenerate (E) modes, and, moreover, they become broader and overlap [107,108]. Furthermore, due to the larger unit cell additional modes emerge. Figure 12 (b) presents the phonon dispersion curves of LSMO for $x = 0.3$ in high symmetry directions of the Brillouin zone (BZ) and reveals distinct features of all phonon modes of Sr doped LSMO. The tilt of the octahedra results in a doubling of the cubic unit cell, and 8 ($5E_u$ and $3A_{2u}$) modes are IR active. Figure 12 (b) also includes the measured (open symbols) and calculated phonon modes of $La_{0.7}Sr_{0.3}MnO_3$ obtained by Reichardt and Braden [91]. The phonon branches are mainly classified into three categories for $La_{0.7}Sr_{0.3}MnO_3$:

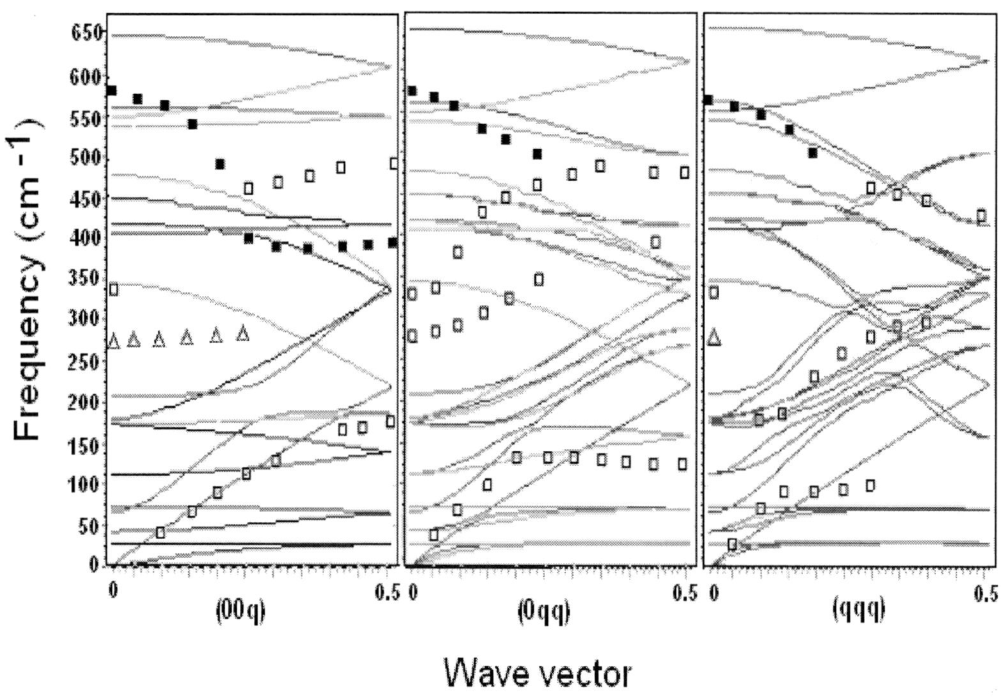

Figure 12(b). Phonon dispersion for rhombohedral $La_{1-x}Sr_xMnO_3$ for x = 0.3 ref. [100]. Open and Solid symbols are the inelastic neutron scattering data at 15 K ref. [91]

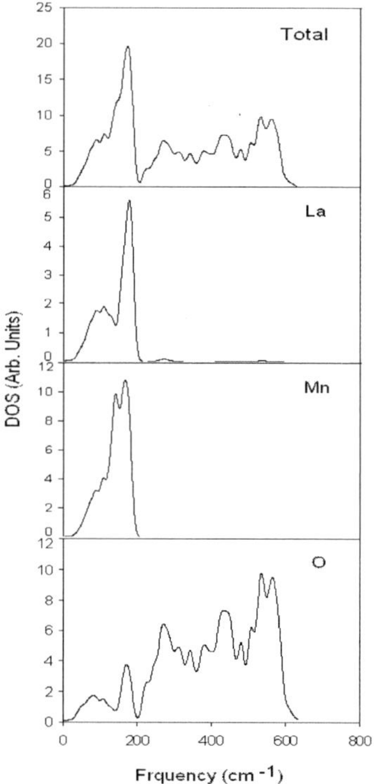

Figure 13. Total and partial phonon density of states of cubic $LaMnO_3$ ref. [100].

(1) External longitudinal optical phonon mode corresponding to the vibrations of La/Sr ions, which is also the lowest frequency branch at Brillouin zone centre (Γ-point) in phonon dispersion curves of $La_{0.7}Sr_{0.3}MnO_3$.
(2) Mn-O-Mn bending modes associated with the excitation of Jahn-Teller phonon modes, which contribute to mid-frequency range of phonon dispersion curves.
(3) Stretching or breathing mode of MnO_6 octahedra, which is a JT phonon mode with Mn-O bond character contributing to the highest frequency region of phonon dispersion curves.

3.4. Total and Partial Phonon Density of States

The phonon density of states (DOS) is obtained by histogram sampling of frequencies over bins of 2 cm^{-1} and then smoothened by Gaussians of full width at half maxima (FWHM) of 5 cm^{-1}. The knowledge of partial phonon density of states enables us to determine the mean square displacements of various atoms, which in turn leads to the determination of the vibrational amplitudes of individual atoms. This also provides information about a particular atom moving in specific directions as a function of phonon energy, which is very useful for the interpretation of the inelastic neutron scattering data. To understand the origin of peaks in the total phonon density of states (DOS) of cubic $LaMnO_3$, it is compared with its partial DOS displayed in Figure 13. The prominent features are discussed as follows:

(1) Lower frequency states (0-200 cm^{-1}) are occupied by mainly La and Mn atoms with small contribution from oxygen vibrations. Sharp peak in DOS in this region is positioned about ~ 200 cm^{-1} with small shoulders around 100 cm^{-1}. Contributions of La and O atoms are seen as small hump centered about 100 cm^{-1}.
(2) In mid frequency region (from 200 to 400 cm^{-1}) O-vibrations mainly dominate however, there appears a small but significant contribution of La atoms.
(3) The frequency region (400-600) cm^{-1} is mainly dominated by O-vibrations with negligible occupancy of La atoms. Overall features of cubic $LaMnO_3$ reveal that mid and high frequency range in DOS mimics the maximum contribution from oxygen atoms.

In order to understand variation of phonon density of states upon structural changes, the DOS of both rhombohedral and cubic phases of $LaMnO_3$ are compared. It is clearly seen from Figure 14 that DOS of both phases span the entire frequency region, with apparent red shifts of peaks in cubic phase, which could be due to its higher symmetry phase. However, some significant features are as follows:

Lower frequency region (0-200 cm^{-1}) is well-defined in cubic phase which is however, mixed with mid frequency region in rhombohedral structure. Peak about ~ 200 cm^{-1} in cubic phase is shifted to lower wave numbers with a well-observed smaller peak at 100 cm^{-1} corresponding to significant La and Mn – vibrations.

In the mid frequency region (200-400 cm^{-1}) of a rhombohedral phase, a broad hump with smaller peaks and shoulders corresponding to Mn and O vibrations shift to lower wave numbers in cubic phase.

Two well-defined sharp peaks arising to prominent vibrations of oxygen atoms seem to merge in a broader peak of cubic phase with observable red shift in frequency region (400-600 cm^{-1}). The gross features of the calculated DOS for $(R\bar{3}c)$ structure are similar to the calculated partial density of states (PDOS) and Raman spectra obtained by Iliev *et al.*[109].

The compositional changes also affect the phonon density of states significantly in addition to structural changes, which are studied by comparing DOS of rhombohedral undoped $LaMnO_3$ and 30 % Sr-doped $LaMnO_3$. Sr-doping essentially modified the DOS of rhombohedral $LaMnO_3$ as observed from Figure 5 with the following prominent features:

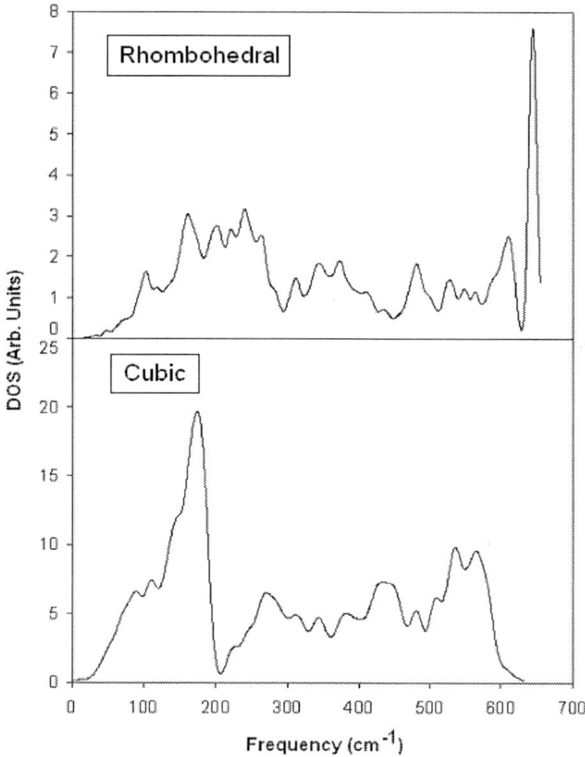

Figure 14. Total density of states (DOS) of rhombohedral and cubic LaMnO3.

(1) In lower frequency region, two well-defined peaks at ~ 100 cm^{-1} and ~ 200 cm^{-1} in cubic phase shifts towards low wave numbers and appear to be distinct from Sr-doped $LaMnO_3$. The origin of these peaks is lying in the significant vibrations of La/Sr, Mn and oxygen.In particular, the contribution of Mn appears to be prominent as a sharp peak in lower frequency region for 'x' = 0.3, which is clearly seen from its partial density of states as well (cf. Figure 16). The features of partial density of states are more or less similar to those for cubic $LaMnO_3$ except that the contribution of Mn atoms in high frequency region in the Sr-doped systems is noticeable.

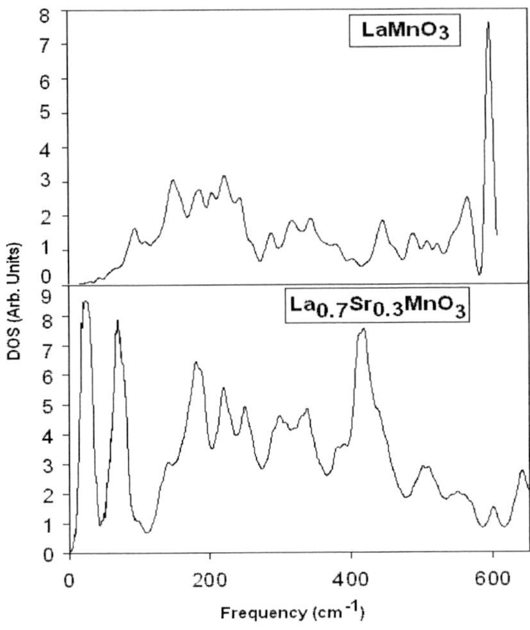

Figure 15. Total phonon density of states of rhombohedral $LaMnO_3$ and $La_{0.7}Sr_{0.3}MnO_3$ ref. [100].

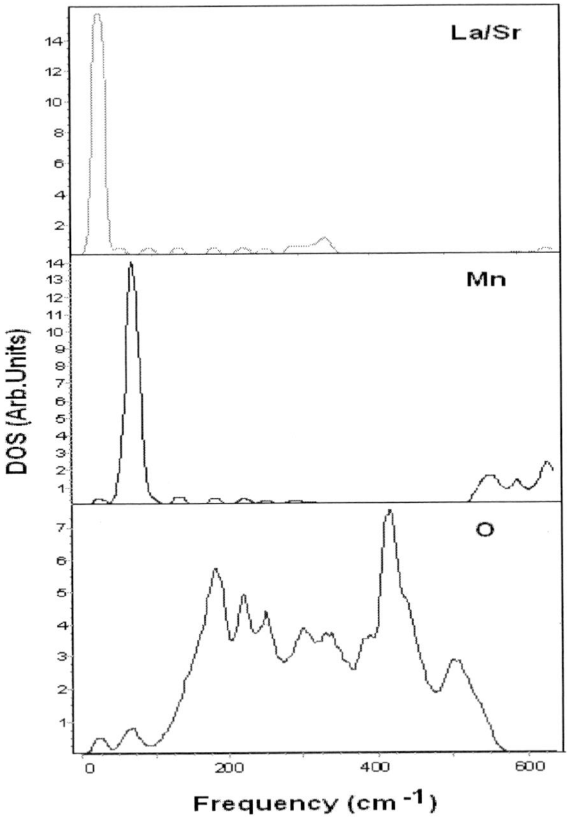

Figure 16. Partial phonon density of states of rhombohedral $La_{0.7}Sr_{0.3}MnO_3$ ref [100].

(2)Sharp peaks about 200 cm^{-1} evolve as broad peaks and almost whole intermediate frequency region is gradually spanned with some new peaks which is clearly observed for 'x' = 0.3. This could be due to increased Mn-O vibrations with Sr doping suggesting the onset of highly metallic state.

(3)The higher frequency sharp peaks arising mainly due to oxygen vibrations merge to some smaller but broader peaks for 'x' = 0.3. In absence of any existing experimental density of states of LSMO systems, the gross features of calculated DOS are observed to be in good accordance with their experimental Raman investigations [83, 88, 99,110] and are similar to the smeared PDOS of rhombohedral $LaMnO_3$ investigated by Iliev *et al.*[107]. Besides $La_{0.7}Sr_{0.3}MnO_3$ being common compound under consideration for later section also, the discussion on total phonon density of states and partial phonon density of states for 'x' = 0.3 are exclusive to the present section only.

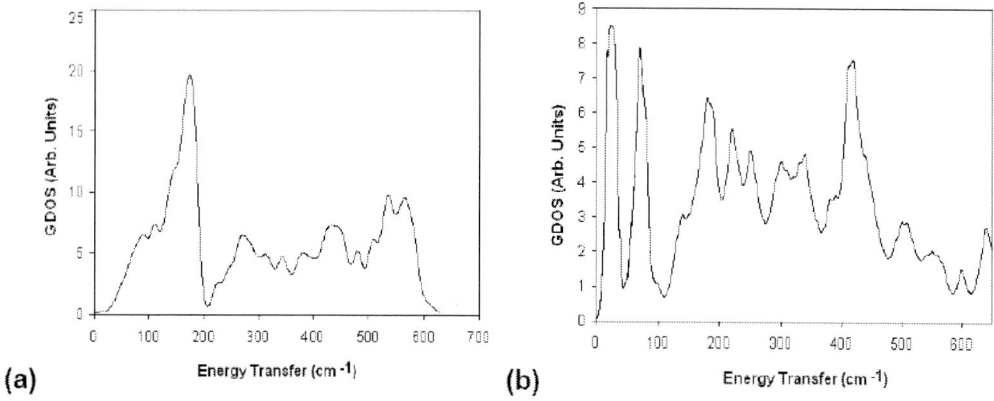

Figure 17. Neutron weighted phonon density of states (GDOS) of (a) Cubic $LaMnO_3$ and (b) Rhombohedral $La_{0.7}Sr_{0.3}MnO_3$ ref. [100].

3.5. Neutron Weighted Phonon Density of States

The Neutron weighted generalized phonon density of states (GDOS) is bare total phonon density of states (DOS) phonon density of states (DOS) divided by (σ_i / M_i), where σ_i and M_i are the scattering cross section and mass of the ith atom. The distinct features of generalized phonon density of states (GDOS) of cubic $LaMnO_3$ and rhombohedral $La_{1-x}Sr_xMnO_3$ systems for 'x' = 0.3 calculated in the present study are presented in Figure 17 (a) and (b) respectively. The generalized phonon density of states (GDOS) can be directly compared with the experimental neutron scattering spectra whenever they are available. However, at present these results could not be compared with any experimental data due to nonavailability. The description of peak in GDOS in these figures is similar to the DOS presented in Figure 15.

3.6. Thermodynamic Properties: Specific Heat and Debye Temperature

The thermo dynamical properties such as lattice specific heat at constant volume (C_V) and Debye temperature (θ_D) gives an idea about critical parameters below which lattice remains stiff. In other words, these parameters decide stiffness limit up to which lattice sustains a particular phase. The thermodynamical parameters are evaluated for both the structures of $LaMnO_3$ and Sr-doped rhombohedral $LaMnO_3$ in the present study by using the methodology presented in section 2 and calculated phonon density of states in the present section discussed above. Figure 18 demonstrates the variation of the lattice specific heat of cubic and rhombohedral $LaMnO_3$ and $La_{0.7}Sr_{0.3}MnO_3$ with temperature. There appear no anomalies in these quantities however, a clear difference in the specific heat at lower temperature is observed, which is due to the difference in the behaviour of higher frequency phonon modes. In the absence of any experimental or theoretical investigations, the results of thermodynamic properties particularly the lattice specific heat, are compared with those of other La-based manganites [111-113] and are found to agree satisfactorily.

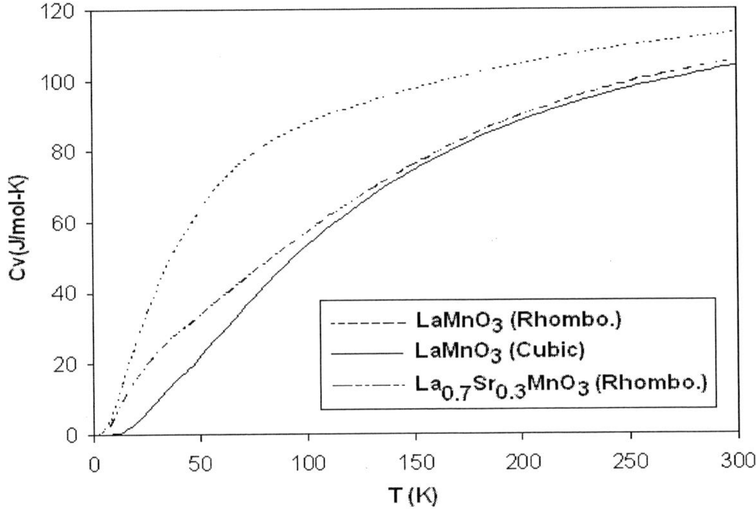

Figure 18. Specific heat (CV) as a function of temperature, T (K) for cubic and rhombohedral $LaMnO_3$.

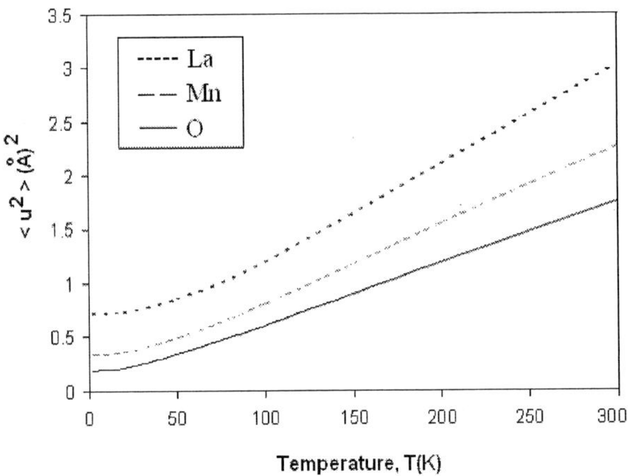

Figure 19. Atomic rms displacement in unit cell of $La_{0.7}Sr_{0.3}MnO_3$ for all atoms as a function of temperature.

The values calculated at room temperature for cubic $LaMnO_3$, rhombohedral $LaMnO_3$ and $La_{0.7}Sr_{0.3}MnO_3$ are 103.96, 112.98 and 105.47 J/mol-K, respectively which are reasonably close to the experimental value of CP (125 J/mol-K) for $La_{0.75}Sr_{0.25}MnO_3$ [110]. The calculated Debye temperatures (θ_D) of rhombohedral and cubic $LaMnO_3$ and $La_{0.7}Sr_{0.3}MnO_3$ are 449, 470 and 491.76 K respectively, at room temperature. The present results on Debye temperature are slightly higher than the reported values for polycrystalline $La_{0.7}Sr_{0.3}MnO_3$ [38]. Nevertheless, they are very much in the range of Debye temperatures observed for the La-based manganites for which values between 360 K and 532 K have been reported [114-115]. It can be concluded that the cubic $LaMnO_3$ with highest symmetry has lowest specific heat and Debye temperature and the increase in Sr concentration is responsible for the lattice softening at high temperatures (above 100 K).

3.7. Isothermal Parameters

Figure 19 presents the temperature dependence of the atomic *rms* displacement for all atoms in unit cell of $La_{0.7}Sr_{0.3}MnO_3$. It is seen from the Figure 6 that the temperature dependence of the *rms* displacement for all atoms are almost linear which is on the line of the theory of the dynamical Jahn-Teller effect [13]. The temperature dependence of the *rms* displacements in the case of other considered systems in the present study are similar.

4. Vibrational Study of Sr-Doped Lanthanum Manganites: Effect of Temperature

4.1. Introduction

Transition metal oxides form crystals that have phases with exotic properties such as high temperature superconductivity, ferromagnetism, ferroelectricity, and charge and orbital ordering. Mixed valent perovskite manganites of the type $R_{1-x}A_xMnO_3$ (R = rare earth element; A = divalent element) are one such class of transition metal oxides. They have recently been the subject of scientific investigations due to the exhibition of colossal negative magnetoresistance effects [116], charge and spin ordering effects as a function of Mn^{3+}/Mn^{4+} ratio [117] and similarity of many issues with the problems of High T_C superconductivity in the cuprates [118]. The most extensively studied derivatives of the above general class of manganites are the compounds with the formula $La_{1-x}A_xMnO_3$ (A = Ca, Sr, Pb, Ba etc.). Sr-doped lanthanum manganites are one of the widely investigated systems with general formula $La_{1-x}Sr_xMnO_3$, or in short LSMO for tolerance factor 't' ($0 < t < 1$), all 'x' ($0 < x < 1$), and temperature [86, 88, 89, 99, 119-120]. Especially, systems with doping up to 'x' = 0.3 have received maximum attention due to its interesting phase diagram and physical properties [74, 88, 91,121]. However, there are only few reports on the investigation of phonons in LSMO [91]. While both end compounds of LSMO *i.e.* $LaMnO_3$ and $SrMnO_3$ are antiferromagnetic insulators at low temperatures and possess orthorhombic structure ($Pnma$) [122], the hole doping by substitution of the La with Sr gradually leads to the changes in their lattice (*i.e.* orthorhombic to rhombohedral) and electronic (*i.e.* antiferromagnetic insulating to ferromagnetic metallic) structures [76, 123]. The electronic and structural phase diagram clearly shows the persistence of antiferromagnetic phase of LSMO up to 'x' = 0.15 and appearance of ferromagnetic phase below T_C. Further doping of Sr content 'x', increases the ferromagnetism in this compound up to the value of 'x' = 0.3, which finally saturates after this value of 'x'. $La_{1-x}Sr_xMnO_3$ with doping level 'x' = 0.3 *i.e.* $La_{0.7}Sr_{0.3}MnO_3$ compound is found to possess rhombohedral symmetry at all temperatures [76, 123]. There is a report by Takeda *et al.* [124] which demonstrates that the rhombohedral modifications appear for 'x' $\geq$ 0.2 in the case of $La_{1-x}Sr_xMnO_3$. The coexistence of FM ordering and the metallic behaviour have been traditionally explained within a framework of the double exchange (DE) model [123]. However, a recent calculation by Millis *et al.* [34,82] indicated that the DE model alone cannot explain resistivity and magnetoresistance effects quantitatively and emphasized the importance of Jahn-Teller splitting of Mn^{3+} e_g levels. The Jahn-Teller (JT) effect based on strong electron-lattice interaction, leading to formation of polarons (local lattice distortion) was found to affect both the magnetic and transport properties of $La_{1-x}Sr_xMnO_3$ systems [34]. This effect is mainly responsible for the distortion of the cubic perovskite structure and its influence is therefore, expected to be displayed in the phonon spectra [84]. Recently, Kim *et al.* [19] observed the frequency shifts of internal phonon modes such as bending and stretching modes, near metal-insulator (MI) transition of polycrystalline $La_{0.7}Ca_{0.3}MnO_3$ and explained the shifts of internal modes in terms of polaron (local lattice distortion) model as the external phonon modes are not affected by local lattice distortion. In addition to the local lattice distortion, the external parameters such as temperature, pressure, doping etc. also play a key role in the appearance of different phonon modes in the phonon spectra due to the changes in Mn-O bond lengths and Mn-O-Mn bond angles. The Raman investigation reveals the sensitivity of the A_{1g} -like phonon mode in the rhombohedral $La_{1-x}Sr_xMnO_3$ to the external parameters [88]. Reichardt and Braden [91] have investigated the phonon branches in rhombohedral $La_{0.8}Sr_{0.2}MnO_3$ and $La_{0.7}Sr_{0.3}MnO_3$ at 15 K by using inelastic neutron scattering. They also performed the shell model calculations for these compounds, but in the cubic phase. The model parameters for this cubic phase calculations were obtained from a fit to the phonon data in rhombohedral unit cell. Granado *et al.* [99] performed the Raman experiments on single crystal $La_{1-x}Sr_xMnO_3$ as a function of temperature and observed an anomalous softening of the in-phase stretching mode of oxygen cage and out of phase bending modes. Okimoto *et al.* [126] measured optical spectra of a single crystal $La_{1-x}Sr_xMnO_3$, but its phonon spectra were not studied in detail. Despite the importance of phonons and role of external parameters on the phase diagrams of these compounds, detailed and systematic investigation on phonons under external parameters is lacking. However, there are some reported experimental [84, 88, 99, 126-127] and theoretical [89, 91] studies. Experimental investigations so far performed mainly focused on the Raman and IR studies in doped manganites.

In the present section we present our own results of a detailed and systematic study on phonon properties as a function of temperature is reported, which may provide valuable information on the importance of phonons in these materials [130]. The investigation of phonon properties include the calculation of phonon dispersion curves (PDC), total phonon density of states (DOS), partial phonon density of states (PDOS), Debye Waller factors, specific heat, and Debye temperature, which have been performed by the methodology and model discussed in section 2. The short-range parameters (La/Sr-O and Mn-O) at 1.6 K have been slightly tuned to produce the positive Eigen frequencies in the entire Brillouin zone (BZ) as with the same set of interactions some frequencies are driven negative. The parameters so obtained are listed in Table V. The crystallographic data for rhombohedral $La_{0.7}Sr_{0.3}MnO_3$ have been taken from ref. [76,123].

4.2. Zone Centre Phonon Modes

The number of phonon modes of LSMO at zone centre is less in rhombohedral phase than in orthorhombic structure due to increased symmetry. The general features of irreducible representation of $(R\,\overline{3}c)$ LSMO are discussed in section 3. All zone centre phonon frequencies are calculated for rhombohedral $La_{0.7}Sr_{0.3}MnO_3$ at 1.6 K and 300 K. However, the phonon frequencies corresponding to A_{1g}, E_g and A_{2u} modes are only presented in Table II as these modes essentially reflect the internal structure of such complex perovskites and pronounced temperature effect. With change in temperature, the changes in bond strength and bond lengths are expected, and can be seen as hardening or softening of the phonon modes.

Table V. Model Parameters for rhombohedral $La_{0.7}Sr_{0.3}MnO_3$ at 1.6 K and room temperature ref. [130]

Short - range interactions for $La_{0.7}Sr_{0.3}MnO_3$ for	T = 1.6 K		T = 300 K	
Interactions	b_{ij} (eV)	ρ_{ij} (Å)	b_{ij} (eV)	ρ_{ij} (Å)
La/Sr – O	1500.0	0.3066	1500.0	0.3066
Mn – O	2310.0	0.3154	3030.0	0.3354
O – O	22764.0	0.4550	22764.0	0.4550

The O-O interaction is taken from ref. [64].

Table VI. Calculated Raman frequencies for rhombohedral $La_{1-x}Sr_xMnO_3$ at 1.6K and room temperature

Phonon frequencies of $La_{0.7}Sr_{0.3}MnO_3$ in cm^{-1}			
D_{3d}^{6} Modes	at RT	at 1.6 K	Experimental
A_{1g}	209.03	188.28	199[a], ~180[b]
E_g	28.91	28.89	42[a]
E_g	173.49	158.60	-
E_g	408.40	386.03	426[a]
A_{2u}	647.70	675.81	641[c], 580[d], 576[e]

[a] Experimental Raman data at 10 K [99].
[b] Experimental Raman data at 300 K for $La_{0.67}Sr_{0.33}MnO_3$ [129].
[c,d] Calculated and experimental IR data at 405 K respectively [106].
[e] Inelastic neutron scattering data [91].

Table VI reveals that the A_{1g} mode, which corresponds to the rotation of the MnO_6 octahedra around the hexagonal *c*-axis [99], softens upon lowering the temperature. This softening of A_{1g} phonon mode may be due to the gradual ceasing of the in-phase rotation of MnO_6 octahedra with the decrease in temperature. While A_{1g} mode involves the motion of the oxygen ions, E_g mode arises due to both oxygen site and La (Sr) site vibrations. The A_{2u} mode observed at 580 cm^{-1} for polycrystalline $La_{0.7}Ca_{0.3}MnO_3$ [131, 132], sensitive to Mn-O bond length, is slightly overestimated at both temperatures in the present study similar to the previous calculation of Abrashev *et al.*[89]; its hardening clearly follows qualitatively upon reduction of temperature. The similar feature of stretch mode in $La_{0.7}Ca_{0.3}MnO_3$ and $La_{0.67}Ca_{0.33}MnO_3$ has been observed from optical reflectivity and infra red measurements [115, 132]. One of the possible reasons for hardening of this mode is the decrease in the Mn-O bond lengths due to decrease in temperature [133]. The present calculated phonon frequencies of Raman active A_{1g} phonon mode at 1.6 K are close to the 10 K experimental Raman data obtained by Granado *et al.* [99], which shows hardening at elevated temperature. The phonon mode frequencies of IR active mode, particularly, A_{2u} mode lies close to the experimental IR frequencies reported at 405 K [133].

4.3. Phonon Dispersion Curves

The rhombohedral ($R\bar{3}c$) phase of the $La_{0.7}Sr_{0.3}MnO_3$ (LSMO) has 10 atoms in the primitive cell and thus 30 phonon modes at each wave vector. The calculated temperature dependence of the phonon dispersion curves (PDC) in the rhombohedral phase of LSMO is shown in Figure 20 ((a) and (b)). It reveals the temperature dependent shift of all phonon modes. However, to understand the behavior of phonon modes in this compound, the general features of the phonon dispersion curves common at both temperatures are briefly discussed. It can be seen from the figure that the phonon frequencies are stable in whole Brillouin zone and more or less there is a considerable dispersion for most of the phonon branches.

The phonon branches are classified into three categories: (1) External longitudinal optical phonon mode corresponding to the vibrations of La/Sr ions, which is also the lowest optical frequency branch at Brillouin zone, centre (Γ-point) in phonon dispersion curves of $La_{0.7}Sr_{0.3}MnO_3$. (2) Mn-O-Mn bending modes associated with the excitation of Jahn-Teller phonon modes which contribute to the mid-frequency range of the phonon dispersion curves and (3) stretching or breathing mode of MnO_6 octahedra which is a JT phonon mode with Mn-O bond character contributing to the highest frequency region of phonon dispersion curves.

To have an idea about the success of the present approach in predicting the phonon dispersion curves of LSMO, the experimental points obtained from the inelastic neutron scattering at 15 K [91] are also included in Figure 20 ((a) and (b)). It is seen from the figure 20 that the general features of INS data are better produced by the present calculation in the [111] direction than the [100] and [110] directions of the BZ. It is observed from the present figure that the calculated $\sim$ 580 cm^{-1} phonon mode at both temperatures 300 K and 1.6 K softens going from zone centre to the zone boundary in the [111] direction similar to the INS data. Softening of this mode reflects a tendency to a JT instability which basically causes the lattice distortion. There is a decrease or damping of this mode at zone centre due to the

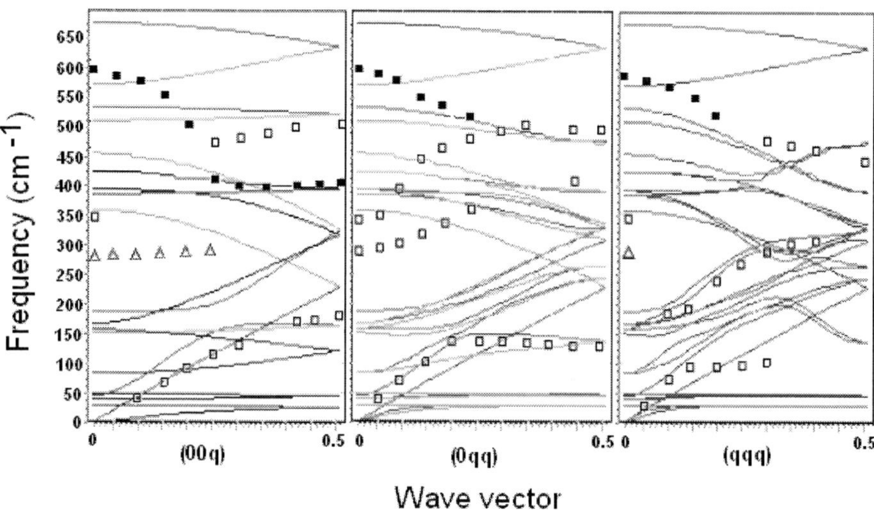

Figure 20(a). Phonon dispersion curves of rhombohedral $La_{0.7}Sr_{0.3}MnO_3$ at 1.6K. Open and filled symbols represent experimental INS data [11].

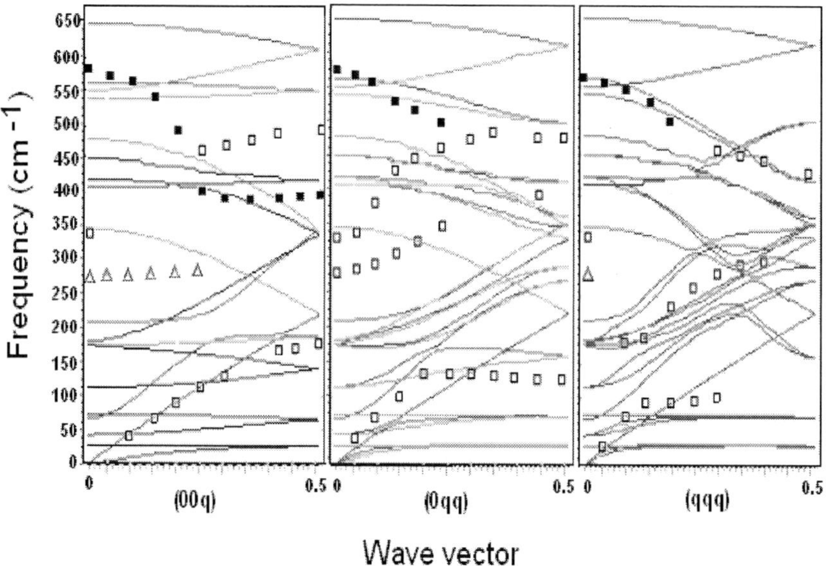

Figure 20 (b). Phonon dispersion curves of rhombohedral $La_{0.7}Sr_{0.3}MnO_3$ at room temperature (300K). Open and filled symbols represent experimental INS data [91].

reduction of the temperature. The hardening of the phonon mode at ~ 175 cm^{-1} is well observed in [111] direction of BZ at both temperature. This mode is due to the vibration of La/Sratoms against the MnO_6 octahedra and is an external longitudinal optical phonon mode. The phonon mode at ~ 350 cm^{-1} are due to the linear breathing character and could be produced well at zone centre, while the mode at ~ 300 cm^{-1} arising due to bending could not be produced so well in the present calculation. These both modes are JT modes. While the mode at ~ 350 cm^{-1} shows the softening with the decrease in temperature, the mode 175 cm^{-1} is unaffected. It can be observed that the highest phonon mode exhibits a remarkable increase of mode frequency with the decrease in temperature.

4.4. Phonon Density of States

The phonon density of states (PDOS) at two temperatures are presented in Figure 21. With increase of temperature the phonon spectra shifted to higher frequencies. However, a significant result is the significant difference between the spectra at two temperatures. It is observed from the Figure 21 that there exists a gap in PDOS at 1.6 K in higher frequency region, which disappears with the increase in temperature.

The main features of phonon density of states at both considered temperatures are given below:

(1) Low frequency peaks shift faster towards high wave numbers as compared to the mid and high frequency peaks. Moreover, in lower frequency region, two separated peaks at 1.6 K temperature seem to split into two distinct separate peaks at 300 K.
(2) While the peaks at 220 cm^{-1} and 275 cm^{-1} seem to diminish at low temperature, the shoulders at 422 cm^{-1} and 510 cm^{-1} become intense at 1.6 K.
(3) Shoulder around 425 cm^{-1} (for 1.6 K data) disappears at 300 K.
(4) Frequency gap seen at 1.6 K in density of states (DOS) around 550-570 cm^{-1} disappears forming a smaller hump in this region at 300K.
(5) The broad peak with few shoulders observed at 1.6 K mainly due to low lying dispersionless optical phonons and acoustical phonons is divided in two peaks at 300 K on both sides of this peak. Figure 2 also reveals that the oxygen vibrations are sensitive to the temperature and hence reduce significantly from 300 K to 1.6 K. This is manifested in reduction of maximum frequency from 625 cm^{-1} to 548.09cm^{-1}. Although the present phonon density of states could not be compared with any existing experimental density of states of LSMO, they are in general in good agreement with the experimental Raman spectrum of LSMO [99] and smeared PDOS of rhombohedral $LaMnO_3$ [134].

4.5 .Specific Heat and Debye Temperature

The specific heat at constant volume (C_V) and Debye temperature (θ_D) for $La_{0.7}Sr_{0.3}MnO_3$ at both considered temperatures 1.6 K and 300 K are reported in figure. 22, which have been calculated by using the total phonon density of states calculated in the previous section and the expressions presented in section 2. The values of C_V for $La_{0.7}Sr_{0.3}MnO_3$ are 105.47 and 106.68 J/mol-K respectively. It can be seen that there is no significant difference between the values of C_V for both considered values; however, a slight reduction is observed for the C_V at 300 K. Since the present results could not be compared with the experimental data however their values are reasonably close to the experimental value of C_P (125 J/mol-K) for $La_{0.75}Sr_{0.25}MnO_3$ [111] and give confidence to the present model calculation. Figure 22 presents the variation of calculated specific heat at constant volume for LSMO at 1.6 and 300 K.

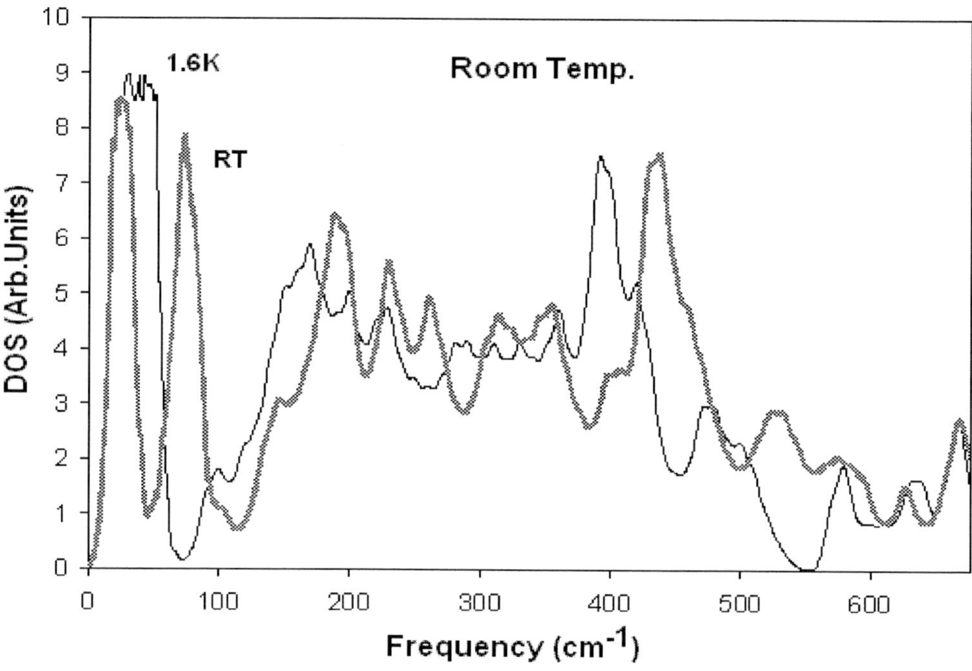

Figure 21. Total Phonon density of states (PDOS) of rhombohedral $La_{0.7}Sr_{0.3}MnO_3$ at 1.6 and 300 K.

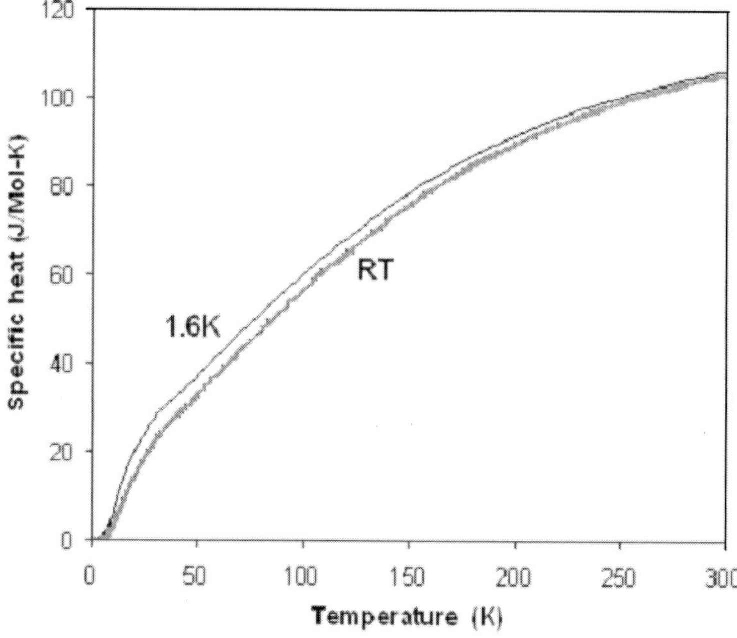

Figure 22. Specific heat as a function of temperature for $La_{0.7}Sr_{0.3}MnO_3$ calculated at 1.6 and 300 K.

The Debye temperature (θ_D) which gives an idea about the critical temperature up to which lattice remains stiff, has also been evaluated by using the present calculated specific heat data. The calculated Debye temperature for $La_{0.7}Sr_{0.3}MnO_3$ at 1.6 K and 300 K are 495.05 K and 491.76 K respectively. The present results on Debye temperature are slightly higher than the reported values for polycrystalline $La_{0.7}Sr_{0.3}MnO_3$ [111]. However, these values are very much in the range for the La-based manganites for which the values of θ_D lie between 360 K and 532 K [95,114]. The decrease in Debye temperature with the increase in temperature indicates the lattice softening.

4.6. Effective Grüneisen Parameter

The lattice excitations play a key role in the discussion of two generic ground states, ferromagnetism, and charge-order seen in perovskite manganites. This is proportional to exp (-2W), where W being the Debye-Waller factor is expressed as $2W(\boldsymbol{q}) = \left\langle (\boldsymbol{q} \cdot \boldsymbol{u})^2 \right\rangle$ with the atomic displacement is $\left\langle u(T)^2 \right\rangle$. The temperature dependence of the atomic displacement is correlated with the anharmonic lattice distortion via the effective Grüneisen parameter γ_{eff} for rhombohedral $La_{0.7}Sr_{0.3}MnO_3$ systems. The effective Grüneisen parameter, which describes the anharmonic coupling of the external strain to the thermal phonon modes, is the Debye specific heat and can be expressed by

$$\frac{W(T2)}{W(T1)} = \left[\frac{V(T2)}{V(T1)}\right]^{2\gamma_{eff}} \tag{44}$$

where, V represents volume. While the typical value of Grüneisen parameter is in the range of 2-3 for most of the solids, its value for rhombohedral $La_{0.7}Sr_{0.3}MnO_3$ is found to be 69.84 in the present case. This indicates the presence of anharmonic lattice modes in the manganite perovskites. In other words, if the observed anharmonicity in W is only due to volume expansion, one would expect the effective Grüneisen parameter γ_{eff} , computed from W and V to be less than 3. Instead, it is found to be 69.84 for T between 1.6 and 300 K, suggesting that the volume expansion is too small to account for the change in W. Die *et al.*[135] reported that γ_{eff} varies from 85 (260-40K) to 25 (600 and 260 K) for $La_{0.65}Ca_{0.35}MnO_3$ in the coherent neutron elastic scattering experiment which very much supports the present high value of γ_{eff}. Also, the high value of γ_{eff} confirms the contribution from anharmonic phonon modes arising due to the distortion present in the lattice. Similar results are also reported by Radaelli *et al.*[136].

5. VIBRATIONAL STUDY OF SR-DOPED LANTHANUM MANGANITES: EFFECT OF PRESSURE

5.1. Introduction

The phase diagram of $R_{1-x}A_xMnO_3$ (R = rare earth La, Pr, Nd, Dy; A= alkaline earth ions Sr, Ca, Ba, Pb) manganites is rich and complex and variables such as pressure [75], temperature [9], magnetic field [9-10] and A-site average ionic radius (chemical pressure) [5] determine a wide range of ground states in these systems. Since the physical properties of these states are often sensitive to even a small change in intrinsic and external conditions, there appears number of colossal effects. The colossal magnetoresistance (CMR) effect in these perovskite manganites is one of the best known examples of such an effect [138].

Table VII. Model Parameters for rhombohedral $La_{0.7}Sr_{0.3}MnO_3$ at all considered pressure

Short-range interactions at ambient pressure for $La_{0.7}Sr_{0.3}MnO_3$, $La_{0.7}Ba_{0.3}MnO_3$, $La_{0.7}Pb_{0.3}MnO_3$

Interactions	b_{ij} (eV)	ρ_{ij} (Å)
La/Sr-O	1500.0	0.3066
La/Ba-O	1555.0	0.2978
La/Pb	1575.0	0.2921
Mn-O	3030.0	0.3354
O-O	2100.0	0.4550

Shell Model [a]

Ion	Y (e)	K (eV- Å^2)
O^{2-}	-2.86902	74.92

[a] Y and K refer to the shell charge and harmonic spring constant, respectively.

It is observed that the application of a magnetic field induces a transition from a paramagnetic insulating (PI) to a ferromagnetic metallic (FM) phase. The large difference between the resistivity of these two phases lies at the heart of the CMR effect, which can be qualitatively explained by the double exchange (DE) model, first proposed by Zener [125]. But the DE model could not predict measured resistivity quantitatively and a large difference is observed in predicted and measured values. Millis and co-workers [82] argued that the DE alone could not explain the resistivity in these systems and the Jahn-Teller (JT) type lattice distortions of the MnO_6 octahedra should be considered owing to JT distortion. This JT distortion results into the splitting of Mn^{3+} e_g orbital and therefore the energy of the occupied orbital reduces and localizes the states.

In the cubic perovskite ABO_3, due to the radius mismatch of the A and the B-site atoms, structural distortions is induced. The chemical substitution (doping elements and concentrations 'x') at A-site changes the number of electrons in $3d$ band of Mn, lattice parameters, Mn-O bond length and Mn-O-Mn bond angle [2, 138-143]. The changes in the doping levels also result in changes in the MnO_6 octahedra and therefore the local distortion is affected. The similar effects can also be achieved by the externally applied hydrostatic pressure. The application of pressure also results in the stabilization of rhombohedral phase of $La_{1-x}Sr_xMnO_3$ ('x' = 0.12-0.18) [144-145] and $La_{0.8}Ba_{0.2}MnO_3$ [146] systems. The application of pressure reduces the lattice constants, increases the Mn-O-Mn bond angles and the unit cell becomes more cubic, and reduces the local distortion of the MnO_6 octahedra [147-149]. In addition, an increase in T_C is observed as the pressure is increased [75, 149]. It is an established fact that the JT distortion plays an important role and hence influences many properties in the manganites.

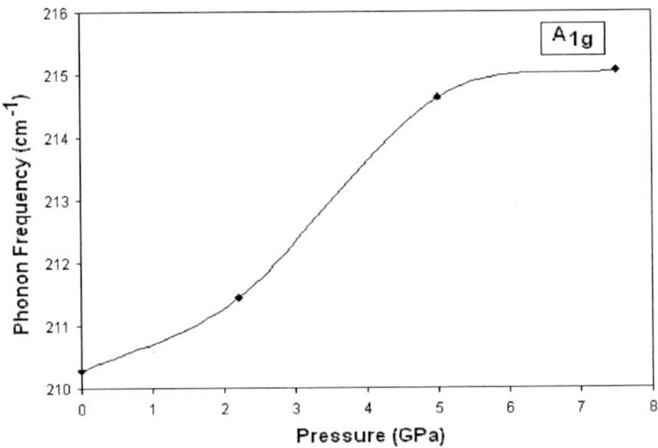

Figure 23(a). Pressure variation of zone centre A_{1g} phonon modes.

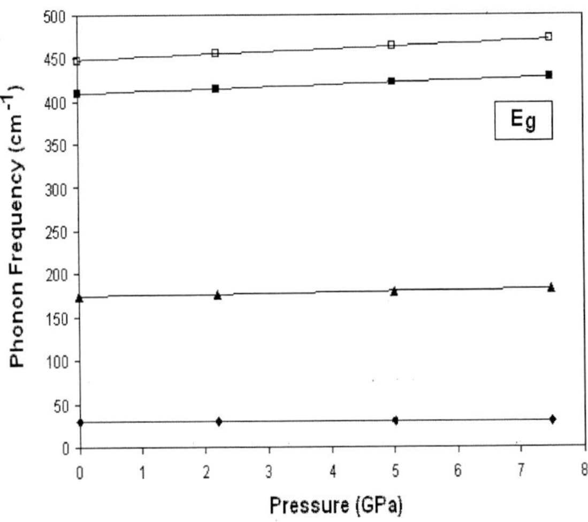

Figure 23(b). Pressure variation of zone centre E_g phonon modes.

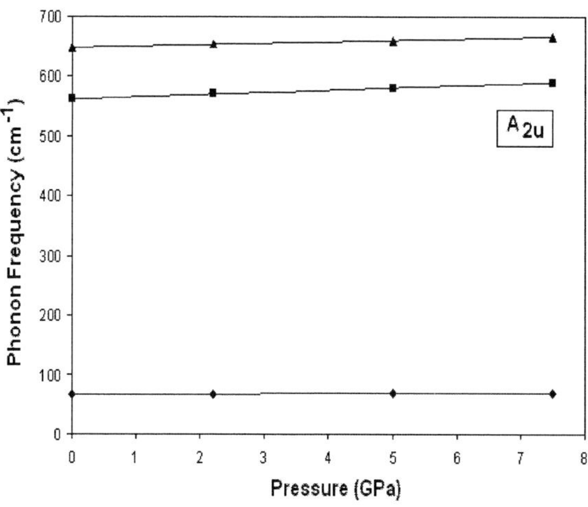

Figure 23(c). Pressure variation of zone centre A_{2u} phonon modes.

The application of pressure both chemical and external reduces the volume around the A ion in the cage of MnO_6 octahedra which results into the symmetrisation of the surrounding structure and reduction in JT distortion [148-150]. The coherent and incoherent distortions affected by the high pressure and doping are abruptly reduced when crossing into the ferromagnetic (FM) phase [88]. The electron-phonon (el-ph) interaction is affected by the pressure through the modification of the frequency of the octahedral bending and stretching modes [38,51].

In recent times, there have been some speculations that for pressures above 2 GPa, the behavior of manganites may be different from that observed in the low-pressure measurements [154, 155]. However, Hwang *et al.* [75] found that the effect of pressure could be mapped onto the average radius of the A-site atoms with a conversion factor of 0.00375 Å/GPa in the pressure range below ~ 2 GPa. The Raman scattering study of orthorhombic $La_{0.75}Sr_{0.25}MnO_3$ by Congeduti *et al.*[148] showed that the pressure above 7.5 GPa induces a new phase other than the predicted metallic phase. The lattice compression due to the application of pressure results in abrupt change in phonon frequency and strong phonon broadening suggesting increase in electron-phonon interaction. A long-range static/dynamic Jahn-Teller distortion and more distorted MnO_6 octahedra are observed by Meneghini *et al.* [91].

To the best of our knowledge, the high pressure effects on the CMR and its related properties including phonon properties of the optimally doped manganites with the rhombohedral crystal structure of $\left(R\bar{3}c\right)$ symmetry are not studied so far. *The present section* reports the results of our calculation on the complete phonon properties of rhombohedral $La_{0.7}Sr_{0.3}MnO_3$ under high pressure [156]. The rhombohedral $La_{0.7}Sr_{0.3}MnO_3$ (LSMO) compound exhibits the metallic like temperature behavior of resistivity and transforms to the ferromagnetic (FM) state at $T_C \sim 370$ K [157]. The $dT_C/dP \approx 5$ $KGPa^{-1}$ for LSMO [158] is much smaller in comparison to that obtained in the same pressure range for the other manganites with a close chemical content but in the orthorhombic phase [159]. The results on the investigation of the effect of A-site average atomic radius ($\langle r_A \rangle$) on some of the selected phonon modes and phonon spectra are also reported to correlate the effect of two different kinds of pressure. To see the effect of ($\langle r_A \rangle$), Ba and Pb doped $LaMnO_3$ systems are selected with Sr doped $LaMnO_3$ as reference system. It will be seen in what follows that the effect of these two pressures is different and there is clear difference in the phonon properties for low and high $\langle r_A \rangle$. The lattice dynamical investigations are done by using the lattice dynamical simulation method discussed in section 2. For the pressure dependent study, the shell model is used considering only oxygen atom to be polarizable. This model has been quite successfully used in recent time by us [156, 160, 161]. Here, same set of model parameters are used to determine the phonon properties of $La_{0.7}Sr_{0.3}MnO_3$ at all considered pressures but a variation of 3 to 5% in the case of A/A'-O short-range interactions for Ba and Pb doped systems are allowed. However, it is ensured that the parameters yield stable structure and positive frequencies in whole BZ. The parameters so obtained are listed in Table VII.

5.2. Zone Centre Phonon Modes

The rhombohedral $(R\bar{3}c)$ $La_{0.7}Sr_{0.3}MnO_3$ (LSMO) has five Raman active (A_{1g} + 4 E_g) and eight infrared (3 A_{2u} + 5 E_u) active phonon modes in rhombohedral LSMO manganites. The structure parameters of $La_{0.7}Sr_{0.3}MnO_3$ at selected pressures and ambient temperature for the calculation of phonon properties in the present study are used from the ref. [147]. Since, in the present section our main aim is to study the effect of pressure on the phonon properties of $La_{0.7}Sr_{0.3}MnO_3$ and find the most affected phonon modes with the application of pressure and therefore responsible for any unusual behavior particularly related to JT distortion (electron-phonon interactions) and MnO_6 octahedra. For this reason, present study reports the zone-centre phonon frequencies, phonon dispersion curves, phonon density of states and thermal properties of LSMO at pressure up to 7.5 GPa. They are seen to match reasonably well with the available data [64, 99,129,161]. Since the A_{1g} and E_g modes are directly related with the MnO_6 octahedra resulting due to the JT distortion in these compounds [28], the discussion on the effect of pressure is limited to these phonon modes only. The A_{1g} mode is due to the rotation of MnO_6 octahedra around the hexagonal *c-axis*, while the E_g modes are due to the bending of MnO_6 octahedra [99]. The Raman active A_g mode corresponding to an in-phase rotation of the MnO_6 octahedra around the *b-axis* in lower symmetry (distorted) orthorhombic structure of these compounds is closely related with the A_{1g} mode in higher symmetry rhombohedral $La_{0.7}Sr_{0.3}MnO_3$ and found to be the most sensitive to the kind and value of the dopant [28]. Figure 23((a), (b) and (c)) present pressure variation of some selected zone centre phonon modes. It is seen from this figure that both the (IR and Raman) frequencies of modes involving the vibrations of La atom do not show any change with pressure while the modes (E_g and A_{2u}) involving oxygen atom vibrations (high frequency) and related to the MnO_6 octahedra show linear pressure induced hardening.

The low frequency phonons are ascribed to pure A-atom vibrations, which actually do not depend on the octahedral distortion. The pressure variation of the Raman active A_{1g} mode shows an unusual couple of slope changes in the considered pressure range, which may be ascribed to some abrupt change of the JT distortion. The $d\omega/dP$ is not too large for this mode but it saturates for pressures above 5.0GPa. The pressure behavior of the stretching and rotational modes (A_{2u} and E_g respectively)[162] with $d\omega/dP$ of about 4 cm^{-1}/GPa suggests that the Mn-O-Mn angle is close to the ideal 180° value of the cubic structure and therefore the system would be of more metallic character [147]. The rotational frequency of MnO_6 octahedra is not changing fast indicating that the octahedra are not severely distorted by the application of pressure. Therefore, it can be concluded that the hardening of phonon frequencies is consistent with the increase in T_C with pressure [147].

The similar increase in the stretching mode is observed earlier for $La_{0.75}Ca_{0.25}MnO_3$ [148, 149]. The concentration and temperature dependent Raman investigation performed by Bjornsson *et al.*[161] shows the appearance of new phonon modes at about 230 cm^{-1} and 420 cm^{-1} in rhombohedral phase of $La_{1-x}Sr_xMnO_3$ ('x' = 0.2). With increase in doping concentration the mode at ~230 cm^{-1} (A_{1g}) shifts towards lower frequencies at low temperature. Such a mode does not appear for 'x' = 0.1. The pressure induced blue shift in A_{1g} phonon mode frequency as observed from Table VII could be due to a modification of the local symmetry at room temperature.

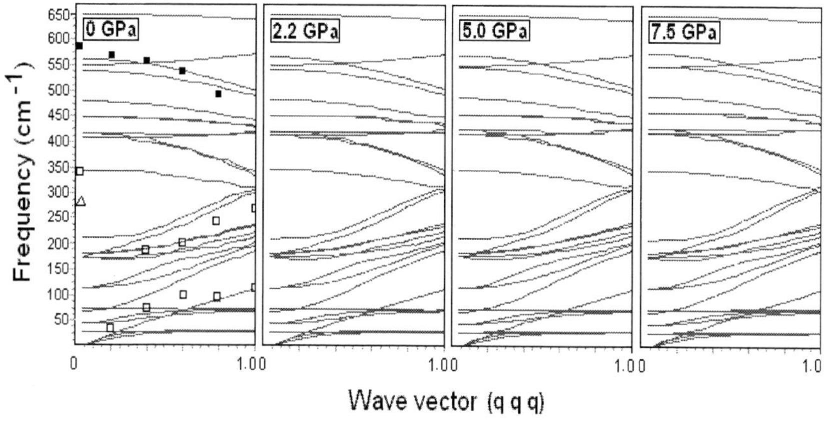

Figure 24. Pressure variation of Phonon dispersion curves (PDC) of $La_{0.7}Sr_{0.3}MnO_3$. Open and filled symbols represent INS data at ambient pressure [91].

5.3. Phonon Dispersion Curves

The phonon band structure is known as the dispersion relation *i.e.* the relation between the phonon mode frequency and wave vector. There are in all 30 vibrational degrees of freedom of atoms in the primitive cell of rhombohedral $La_{0.75}Ca_{0.25}MnO_3$ giving rise to 30 vibrational modes at the Brillouin zone centre (q = 0), which are well distributed among the various irreducible representation. Figure 24 presents the phonon dispersion curves (PDC) of $La_{0.7}Sr_{0.3}MnO_3$ in high symmetry direction of the Brillouin zone (BZ) at four different pressures. The ambient pressure phonon dispersion curves of $La_{0.7}Sr_{0.3}MnO_3$ presented in Figure 24 can be utilized to quantitatively explain the behavior of modes and their origin based on some conjectures drawn by using a quite crude classification [152]. Based on this classification the modes at high (500 -700 cm^{-1}), intermediate (200-500 cm^{-1}) and low (below 200 cm^{-1}) frequencies can be ascribed to Mn-O bond stretching, tilting or rotation of the octahedra and vibrations of the heavy rare earth La and Sr atoms, respectively. Figure 24 also includes the experimental data obtained from the inelastic neutron scattering (INS) experiments at ambient pressure for rhombohedral $La_{0.7}Sr_{0.3}MnO_3$ [91]. It is seen from the

Figure 24 that the phonon modes of the intermediate and higher frequency ranges are hardened with the application of pressure. This reflects that the frequencies of the modes related to the Mn-O bond and tilting or rotation of octahedra increase with the pressure, which may be due to the increase of strain in Mn-O bond. From Figure 24, it is clear that a distinct gap between 200 and 350 cm^{-1} at zone centre increases with the pressure.

5.4. Phonon Density of States

In order to investigate the phonon properties, the understanding of phonon density of states is vital, as it requires the computation of phonon modes in the entire BZ. In addition, the phonon density of states presents an overall view of the range and extent of various phonon modes in the lattice and calculated by using the expression presented section 2. This also tells us how particular atom moves in particular directions as a function of phonon energy, which is very useful for the interpretation of the inelastic neutron scattering data.

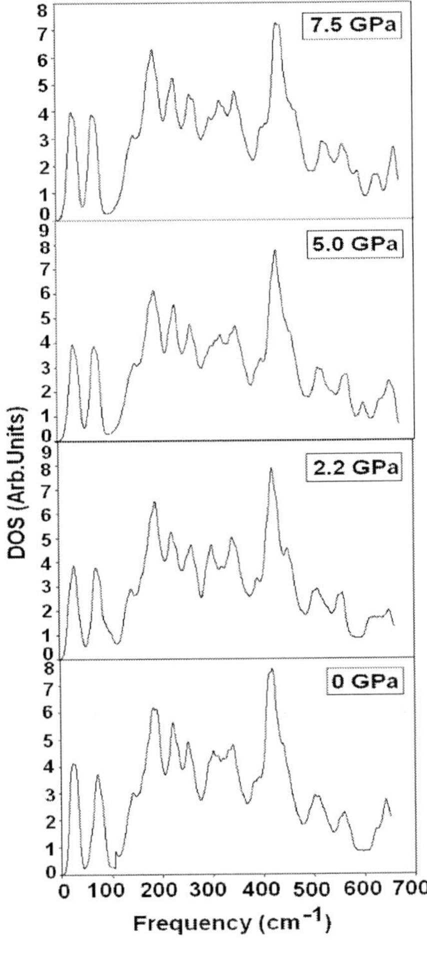

Figure 25. Pressure variation of the total phonon density of states of $La_{0.7}Sr_{0.3}MnO_3$.

To understand the origin of peaks in the total phonon density of states (DOS), the ambient condition spectra of total DOS along with the partial DOS displayed in figure 15 and 16 of section 3.4 are examined. From the partial and total phonon DOS some conjectures can be drawn exploiting a crude classification based on the atomic contributions. Hence, the total DOS can be classified into three regions. Figure 25 presents the total phonon DOS at different pressures and reveals that there is a pronounced shift in the peak positions and change in their shapes as the pressure is increased. Most prominent change observed is that the peaks after 600 cm^{-1}, 400 cm^{-1} and 300 cm^{-1} evolve and further sharpens with the increase in pressure. The sharpening of the peaks can be attributed to the reduction of broadening of phonon peaks due to the shortage of charge introduced by the pressure, which reduces the JT distortion in MnO_6 octahedra and decreases to some extent the lattice disorder. As a matter of fact, both electron-phonon interaction and structural disorder cause an increase of phonon life time thus closing the peak profiles [148, 149]. The shift of the peak positions in the DOS can be attributed to the variation in the percentage contribution of individual atoms with the pressure.

5.5. Anharmonicity and Jahn-Teller Distortion

For a solid at a temperature *T*, the mean number of phonons with energy is given by the Bose-Einstein distribution. The mean square displacement of a single quantum mechanical harmonic oscillator $\langle u^2 \rangle = \left(\frac{\hbar}{m\omega}\right)\left(n+\frac{1}{2}\right)$, can easily be generalized to that of a single atom in the direction *i* as

$$\left\langle u_{ki}^2 \right\rangle = \left[\frac{V}{2\pi^3}\right] \cdot \left(\frac{\hbar}{m_k}\right) \sum \int \left|\varepsilon_{jki}\left(\bar{q}\right)\right|^2 \cdot \left[\frac{\left\{n_{jq}(T)+\frac{1}{2}\right\}}{\omega_j\left(\bar{q}\right)}\right] d\bar{q} \qquad (45)$$

It can be seen from the above expression that light atoms vibrating at low frequencies exhibit large zero point motions. The off-diagonal elements $\left\langle \bar{u}_{ki} \bar{u}_{kj} \right\rangle$ can be calculated in similar way. The thermal and zero point motion of the atoms are often described using the matrix of anisotropic temperature factors. For an atom *k*, it is defined by eq. (19). The pressure dependent anisotropic temperature factors at 300 K calculated by using eq. (19) are presented in Table VIII.

The lattice excitations play a key role in the discussion of two generic ground states, ferromagnetism and charge order observed in perovskite manganites. The lattice excitation is proportional to exp (-2*W*), where *W* being the Debye-Waller factor and expressed as $2W\left(\bar{q}\right) = \left\langle \left(\bar{q} \cdot \bar{u}\right)^2 \right\rangle$ with the atomic displacement. Normally the temperature dependence of the atomic displacement $\left\langle u(T)^2 \right\rangle$ is corrected with the lattice distortion via the effective

Grüneisen parameter γ_{eff} for any system. Since the pressure also causes the distortion like the temperature, pressure dependence of the atomic displacement at fixed temperature can be correlated with the lattice distortion and the effective Grüneisen parameter γ_{eff}. The effective Grüneisen parameter can be expressed as [135]

$$\frac{W(P2)}{W(P1)} = \left[\frac{V(P2)}{V(P1)}\right]^{2\gamma_{eff}} . \tag{46}$$

where $V(P_i)$: $i = 1, 2$ represents the volume at pressure P_i. It is observed that the typical value of effective Grüneisen parameter is in the range of 2-3 for most of the ordered solids [142]. In the present case, the effective Grüneisen parameter has been calculated for three different regions between 0 to 7.5 GPa. The value of γ_{eff} is 2.78, 1.19 and 0.87 for 0-2.2, 2.2-5 and 5-7.5 GPa, respectively.

Table VIII. Vibrational amplitudes of individual atoms of $La_{0.7}Sr_{0.3}MnO_3$ at 0, 2.2, 5.0 and 7.5 GPa at room temperature

Atoms	$8\pi^2<u^2>/3$ (Å)			
	0 GPa	2.2 GPa	5.0 GPa	7.5 GPa
La/Sr	3.2680	3.2063	3.0889	3.0301
Mn	1.5516	1.5117	1.4513	1.4126
O	0.6954	0.6777	0.6542	0.6368

The variation in γ_{eff} is so dramatic that its value drops rapidly from the quite highvalue of 2.78 in the range of 0-2.2 GPa to 0.87 in the range of 5-7.5 GPa.

These values of γ_{eff} reflect that the γ_{eff} *i.e* anharmonicity decreases for the range of pressure going from lower to higher which is in agreement with the idea of a pressure-induced reduction of the JT distortion [75].

5.6. Pressure Dependent Specific Heat and Debye Temperature

The pressure dependent lattice specific heat at constant volume of $La_{0.7}Sr_{0.3}MnO_3$ has been calculated by using the phonon density of states and can be expressed as

$$C_V(T) = K_B \int \left(\frac{h\omega}{2\pi K_B T}\right)^2 \left[\frac{\exp\left(\frac{h\omega}{2\pi K_B T}\right)}{\left(\exp\left(\frac{h\omega}{2\pi K_B T}\right)-1\right)^2}\right] g(\omega)d\omega \tag{47}$$

where, $g(\omega)$ is the total phonon DOS. The values of room temperature specific heat at constant volume for the considered manganite are 104.6, 105.07, 104.55, 104.1 J/mol-K at 0, 2.2, 5.0 and 7.5 GPa respectively. It can be seen that there is no significant difference between the values of C_V for all considered pressure; however, a slight increase in C_V at 2.2 GPa indicates lattice expansion under pressure which further reduces upon increase in pressure. Since the present results could not be compared with the experimental data, however, their values are reasonably close to the experimental value of C_P (125 J/mol-K) for $La_{0.75}Sr_{0.25}MnO_3$ [111] and give confidence to the present model calculation.

5.7.Effect of Internal Pressure and Correlation between Internal and External Pressures

It is a known fact that the phase diagram of manganites depends very much on several variables such as pressure, applied magnetic field, doping concentration 'x', temperature, and A-site average atomic radius ($<r_A>$)[9-10, 75, 137, 138]. Temperature dependent diffraction measurements show that the JT distortion of MnO_6 octahedra in the insulating state reduces at IM transition [136]. The Mn^{3+} to Mn^{4+} ratio changes due to the effect of external parameters and results into a change in the Mn-O bond length and Mn-O-Mn bond angle and hence in the perovskite structure. Increase in $<r_A>$ has been found responsible for the reduction of the octahedral distortion, enhancement of the metallic character and increase in T_C [75, 163] similar to the effect of external pressure [75, 149]. Since IM transition can be related with the narrowing of octahedra and frequency hardening of the octahedral bending and stretching modes, it will be of interest to see the effect of internal pressure determined by $<rA>$ on the phonon modes of manganites and find if there is any correlation between the internal (chemical) and external (applied) pressure effect on phonons. It is expected that the increase in average atomic dimension at the A-site will enhance the pressure (internal) which may reduce the free volume around the A site and finally the IM transition. In this case, the octahedra become distorted and the Mn-O-Mn angle tends to 180° [75]. Therefore, investigation to the phonon properties of$La_{0.7}A'_{0.3}MnO_3$ (A': Sr, Pb, and Ba) at ambient pressure is extended not only to see the behavior of phonons in Ba and Pb doped $LaMnO_3$ in addition to Sr doped systems reported above but also to see the effect of internal pressure determined by the average atomic radius, $<r_A>$. Since, the present study mainly focuses on the pressure dependent phonon behavior in $La_{0.7}Sr_{0.3}MnO_3$, the detailed $<r_A>$ dependent phonon properties are not reported here. However, an effort is made to analyze the effect of internal and external pressures on the phonon modes particularly related to the MnO_6 octahedra and phonon spectrum (DOS). The Ba and Pb doped systems are chosen along with the Sr doped $LaMnO_3$ due to their similar rhombohedral crystallographic structure at 'x' = 0.3 doping concentration. The crystallographic structure parameters of the $La_{0.7}Ba_{0.3}MnO_3$ and $La_{0.7}Pb_{0.3}MnO_3$ for the calculation of phonon properties are used from the ref. [136,164].

5.7.1. Zone Centre Phonon Modes

Table IX presents the A_{1g} and E_g Raman active phonon modes of $La_{0.7}Sr_{0.3}MnO_3$, $La_{0.7}Ba_{0.3}MnO_3$ and $La_{0.7}Pb_{0.3}MnO_3$. The results for Pb and Ba doped systems could not be compared with any measured data, while for the Sr doped system, it is already discussed above. As mentioned above, A_{1g} and E_g modes are directly related with the octahedron, the discussion is restricted to these phonon modes only. Table IX reveals that the A_{1g} mode arising due to the rotation of MnO_6 octahedra shows different behavior in the present case than the one observed under external pressure. The frequency of the A_{1g} mode increases with the increase of $\langle r_A \rangle$ similar to the effect of external pressure, but a dramatic increase is observed for Pb-doped $LaMnO_3$ which has the highest $\langle r_A \rangle$ of 1.399 Å [165] in these considered manganite systems. The average A-site atomic radii for Sr and Ba-doped $LaMnO_3$ are 1.24 Å and 1.29 Å, respectively [136]. This increase of the frequency of A_{1g} mode for the Pb-doped system is quite large in comparison to the external pressure even at maximum considered pressure of 7.5 GPa for Sr-doped compound. The hardening of A_{1g} phonon mode with increasing internal pressure determined by $\langle r_A \rangle$ indicates that there is a reduction in the el-ph interaction and hence in the distortion of the octahedral with the increase of internal pressure similar to the external pressure. The similar trend is observed for the E_g modes except for the higher frequency mode involving oxygen atom vibration and Mn-O bond length.

Table IX. A1g and Eg phonon modes in $La_{1-x}A'_xMnO_3$ (A′ = Sr, Ba and Pb; x = 0.3)

Raman Active Phonon Frequencies (cm^{-1})			
Modes	$La_{0.7}Sr_{0.3}MnO_3$	$La_{0.7}Ba_{0.3}MnO_3$	$La_{0.7}Pb_{0.3}MnO_3$
A_{1g}	210.78	216.08	264.10
	29.08	52.15	83.80
	174.25	188.05	257.50
E_g	408.72	464.05	449.60
	447.57	506.78	489.32

A remarkable feature is observed for the Pb-doped $LaMnO_3$ compound that the frequency of higher E_g mode involving oxygen atom vibrations and Mn-O bond length decreases inspite of the increase of $\langle r_A \rangle$, a trend opposite to the Ba-doped system. While the small increase of $\langle r_A \rangle$ *i.e.* going from Sr to Ba-doped system shows more or less same trend of increase of frequency similar to the effect of applied pressure, the frequencies for Pb-doped system where there is a large increase in $\langle r_A \rangle$, are different. In the present considered manganites case, a distinct increase in the frequency of A_{1g} and two E_g phonon modes as well as a significant decrease in the higher frequency E_g mode of Pb-doped system are observed with reference to Ba-doped $LaMnO_3$. Although these calculated frequencies or the trend could not be compared with any theoretical or experimental data, a conclusion can be drawn for the different trend in frequencies based on changes in octahedra of rhombohedral $(R\bar{3}c)$ Sr, Ba and Pb-doped $LaMnO_3$ caused due to two different kinds of pressure. An analogy can be seen from the temperature and concentration dependent Raman measurement on $La_{1-x}Sr_xMnO_3$ performed by Bjornsson *et al.*[162] which shows the appearance of new A_{1g} Raman active phonon mode with the increase of Sr concentration from 'x' = 0.1 to 'x' = 0.2 *i.e.* increasing the internal pressure. The reason for the frequency of A_{1g} and two E_g modes closure in Sr and Ba-doped and relatively significant difference for A_{1g} mode and decrease of E_g mode frequency in Pb-doped system may be correlated with the low and high $\langle r_A \rangle$. At low $\langle r_A \rangle$ both the increase of the Mn-O-Mn bond angle and the decrease of Mn -O bond length contribute to the A_{1g} and E_g modes, while at high $<r_A>$, the two structural parameters have opposite effects[111]. This may

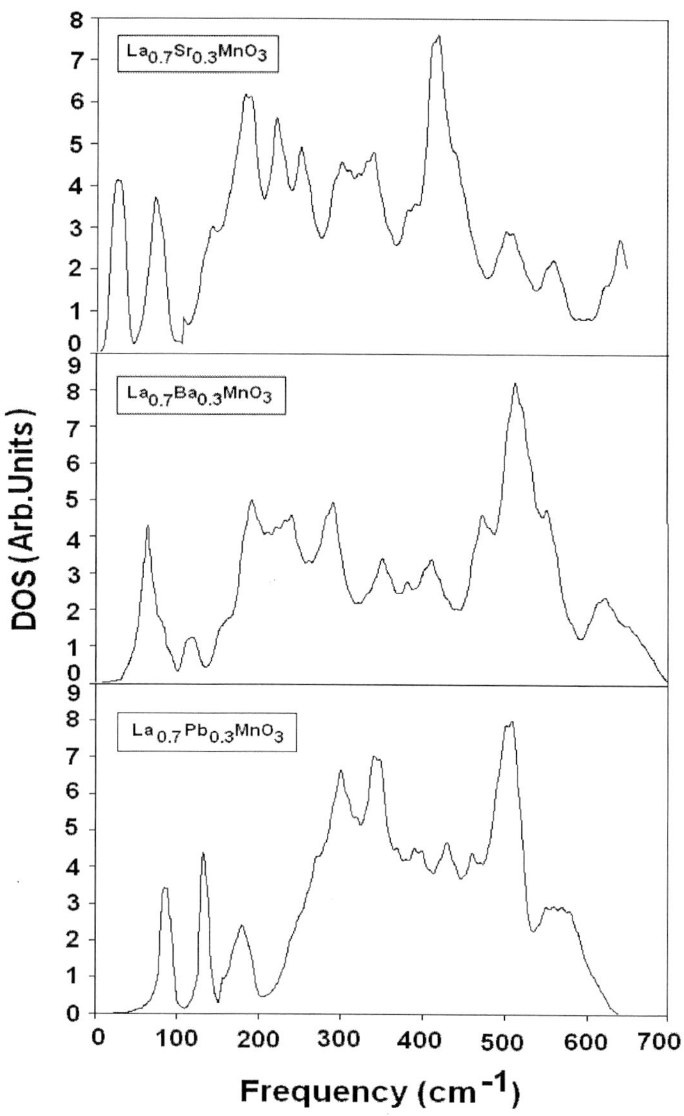

Figure 26. Total phonon density of states of $La_{1-x}A'_xMnO_3$ (A′ = Sr, Ba and Pb; x = 0.3).

be the reason for the increase in A_{1g} and E_g frequencies of Ba-doped system and significant increase in A_{1g} and decrease in the higher frequency E_g mode of Pb-doped system in comparison to Ba-doped $LaMnO_3$ as these modes arise due to the Mn-O-Mn bond angle and Mn-O bond length, respectively. It is also observed that A_{1g} phonon mode frequency is expected to increase both as functions of increasing $<r_A>$ and applied pressure. However, the rate of increase of frequency for low $<r_A>$ and applied pressure is similar and slow but for high $<r_A>$ it is faster. The application of external pressure on these compounds produces quite different structural effects than the chemical or internal pressure. The external pressure compresses all bond lengths and a slight increase of the Mn-O-Mn bond angle while the internal pressure enhances Mn-O-Mn bond angle but suppresses Mn-O bond length. The anomalous change in the frequency of A_{1g} mode for $La_{0.7}Pb_{0.3}MnO_3$ may be due to some abrupt reduction of JT distortion and electron-phonon interactions.

5.7.2. Phonon Density of States

In Figure 26, the phonon density of states (phonon spectrum) of Sr, Ba and Pb-doped $LaMnO_3$ compounds for the same concentration (x) are presented to see the effect of A- Site average atomic radius $<r_A>$ on the phonon spectrum. Figure 26 reveals that there are prominent shifts of the peak positions and change in their shapes on increasing $<r_A>$. There is a gradual shift of last peak in Sr doped system to lower wave numbers going from Sr to Pb *i.e.* increasing $<r_A>$. Also, the three small peaks at ~ 500 cm^{-1} in Sr-doped system gets converted into two with sharpness in the case of Ba-doped system and to a very sharp peak at 500 cm^{-1} and one broad and small peak centered around 550 cm^{-1} in the case of Pb-doped system. The prominent peak at about 450 cm^{-1} observed in Sr-doped system almost disappears in Ba and Sr-doped $LaMnO_3$. A noticeable difference in the middle region of the spectra is clearly visible. The peaks in the spectrum from 200 to 275 cm^{-1} in $La_{0.7}Sr_{0.3}MnO_3$ start diminishing in Ba-doped system and finally disappear in Pb-doped system. A shoulder in the case of Sr-doped system evolves in a well defined peak below 200 cm^{-1} in the case of Pb-doped system. The two phonon peaks observed below 100 cm^{-1} in La-Pb and La-Sr system convert in to one very prominent and sharp peak in Ba-doped system due to almost same mass for La and Ba atoms. As far as the comparison between the phonon spectra under the influence of two different modes of pressure is concerned, it is seen from figs. 25 and 26 that the effect ofinternal pressure determined by average A- site atomic radius is more prominent and spectra changes significantly.

6. Phonon Dynamics of Intrinsic Phase of Cobalt Oxide Superconductor: NaCoO2

6.1. Introduction

In recent years cobalt oxide systems have been studied intensively because of their wide range of unique magnetic [166, 167] and thermoelectric [168] properties as well as for possible analogies to colossal magentoresistive manganite materials or high transition temperature superconducting cuprate oxides. Thereafter, the recent discovery of superconductivity in water-intercalated Na_xCoO_2 compound ($Na_xCoO_2 \cdot yH_2O$) which is a breakthrough in the search for new layered transition metal oxide superconductors immediately spurred tremendous round of intense interest in this system [169-170]. Although the superconducting transition temperature ($T_C \approx 5$ K), is much lower than T_C's in cuprate superconductor, both system share many common features. Similar to cuprates the Co- based superconductor represents a strongly correlated and anisotropic type II system [171-174]. As far as superconducting mechanism in this compound is concerned, a large number of experimental investigations suggest that the superconductivity in this material is unconventional [171,175-176] To understand the mechanism of superconductivity in this material some theoretical models such as resonating valence bond [177-180] and spin triplet superconductivity [181-184] have been proposed. But, these theoretical investigations do not lead to any firm conclusions as far as mechanism of superconductivity in this material is concerned. However, some experiments suggest the presence of strong electron-phonon coupling in this compound [185-187]. Crystal structures of these compounds involve CoO_2 layers formed from edge sharing CoO_6 octahedra, alternating with Na cations in two partially occupied sites – Na(1) and Na(2). The physical properties in these materials strongly depend on the sodium concentration and for 'x' = 0.5 the system undergoes a transition to the insulating state at ~ 53K. The observed insulating transition is said to be associated to a charge ordering in the CoO_2 planes, which further induces an ordering of the sodium cations. Therefore, it is a right juncture to investigate the phonon properties of $Na_xCoO_2 \cdot y\ H_2O$ and its parent compound. A detailed understanding of the lattice vibrations is in view of the possible role of phonons in the superconductivity and in the understanding of the physical nature of the interatomic forces in these compounds. It is also expected that the detailed investigation of the phonons in cobaltate may lead to drawsome conclusion on the role of phonons in manganite in particular and complex oxides in general.

As far as the investigations of detailed and systematic phonon properties for these compounds are concerned very few attempts have been made so far. While there are some Raman [188-189], infrared [185, 190] and neutron scattering [191, 192] experiments on the investigations of the phonon properties of these compounds very scant attention has been so far paid toward the theoretical investigations [192, 193]. Very recently Li *et al.* [193] have performed the lattice dynamical calculations for parent $NaCoO_2$ compound by using first principles method. In this study they report the phonon frequencies throughout the Brillouin zone and phonon density of states. However, the results of the first principles calculation of Li *et al.*[193] are in general good agreement with the experimentally investigated Raman and infrared phonon modes but failed in predicting the experimental phonon density of states (DOS) obtained by using inelastic neutron scattering [192]. While the measured spectra reports phonon density up to 100 meV (≈ 800 cm^{-1}), the phonon DOS calculated by using first principles report the phonon density only up to 80 meV (≈ 600 cm^{-1}). Also, there is contradiction on the sensitive phonon modes to the Na site occupancy with the shell model results of Lemmens *et al.* [189]. The results of the shell model calculation of Lemmens *et al.* [189] are only reported for Raman modes and hence a detailed and systematic lattice dynamical calculation for $NaCoO_2$ compound is necessary. It is indeed essential to find out that if the present simple model based on interatomic interaction with more realistic and physical parameters is sufficient to describe the vibrational properties of the phonon properties of intrinsic insulating phase of the cobalt oxide superconductor $NaCoO_2$. *The present section* reports a systematic and detailed investigation by us on the phonon properties of the intrinsic insulating phase of the cobalt oxide superconductor $NaCoO_2$ is described [160]. The methodology to obtain these phonon properties are discussed in section 2. The model parameters for the shell model lattice dynamical computaion applied in $NaCoO_2$ compound are presented in Table X. To understand effect of different site occupancy of the sodium atoms on the phonon properties the calculations are done in two geometries discussed later. In the two different geometries of the $NaCoO_2$ compound the sodium (Na) atoms occupy two different Wyckoff positions. These two positions are independent of each other and not occupied simultaneously.

Table X. Model Parameters of the Potential for $NaCoO_2$.

Short - range interactions			Effective charge	
Interactions	b_{ij} (e*V*)	ρ_{ij} (Å)	ion	*Z*(*e*)
Na – O	1025.4	0.3734	Na	1.67
Co – O	1341.4	0.3214	Co	2.13
O – O	22764.00	0.1490	O	-1.90

Shell Model[a]

Ion	*Y*(e)	*K* (e*V*- Å^2)
O^{2-}	-2. 86902	74.92

[a] Y and K refer to the shell charge and harmonic spring constant respectively.

6.2. Zone Center Phonon Modes and Atomic Positions

For the present calculations of the phonon properties, the intrinsic $NaCoO_2$ is considered having a hexagonal structure (space group # 194, $P6_3/mmc$) with lattice constants of a = 2.82 Å, c = 10.92 Å and the structural parameters Co at 2*a* (0,0,0); O at 4*f* (1/3,2/3,z); Na1 at 2*d* (2/3,1/3,14) and Na2 at 2*b* (0,0,1/4) [188]. Na1 and Na2 represent the positions of Na atoms in two different geometries. A symmetry analysis taking into account the $P6_3/mmc$ point group for $NaCoO_2$ leads to the following zone center phonon modes

$$\Gamma\,(P6_3/mmc) = A_{1g} + 2\,B_{1g} + E_{1g} + 2\,E_{2g} + 3\,A_{2u} + 2\,B_{2u} + 3\,E_{1u} + 2\,E_{2u} \qquad (48)$$

where, the $A_{1g}+E_{1g}+2\,E_{2g}$ modes are Raman active, the $3\,A_{2u}+3\,E_{1u}$ modes are infrared active. Each of the E_{1g}, E_{2g} or E_{1u} modes is doubly degenerate. The B_{1g}, B_{2u} and E_{2u} are silent modes.

The Raman active modes A_{1g} and E_{1g} involve vibrations from oxygen atoms only while E_{2g} modes are related to Na and oxygen. Due to full point group symmetry the Co sites do not contribute to the Raman scattering. The vibrational pattern of zone center phonon modes are presented in Figure 27 and calculated frequencies of the phonon modes at zone centre in two geometries are presented in Table II along with the other available calculated and experimental values. As can be seen from the Table XI that the present calculation is quite successful in reproducing the phonon modes at zone center except few discrepancies in comparison to the available experimentally measured values by using Raman scattering [188, 189[and infrared [185, 190] and calculated values of Li *et al.*[193] by using first principles method.

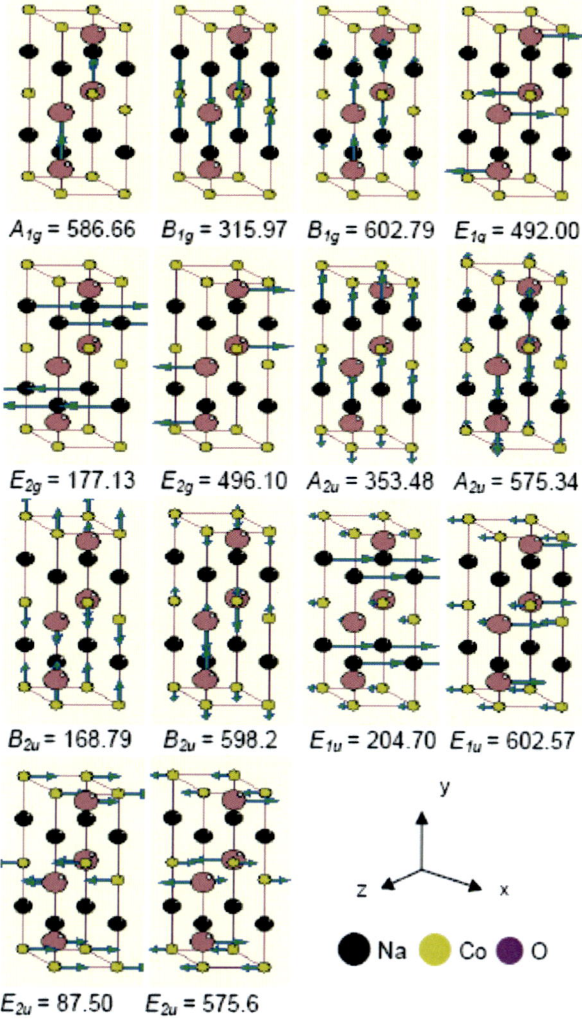

Figure 27. Vibrational pattern of some selected zone center phonon modes.

Table XI. Zone center optical phonon modes of $NaCoO_2$. All frequencies are in cm^{-1}

Modes	Present Calculation		Others
Raman active modes			
	Geometry A	Geometry B	
A_{1g}	586.66	597.92	608.0[a], 604.6[b], 574[c], 598[d], 582[e], 588[f]
E_{1g}	492.00	504.12	477.1[a], 482.1[b], 458[c], 480[d], 469[e]
E_{2g}	177.13	194.99	172.9[a], 185.7[b]
E_{2g}	496.10	515.10	483.7[a], 489.8[b], 494[c]
Infrared active modes			
E_{1u}	204.70	218.12	201.2[a], 216.5[b]
E_{1u}	602.57	610.00	586.7[a], 590.0[b], 570[g], (505,530,560,575)[h]
A_{2u}	353.48	333.94	397.5[a], 337.0[b]
A_{2u}	575.34	566.11	569.8[a], 566.5[b]

Silent modes			
B_{1g}	315.97	302.60	351.7[a], 309.9[b]
B_{1g}	602.79	613.40	622.4[a], 616.0[b]
B_{2u}	168.79	160.03	197.0[a], 172.1[b]
B_{2u}	598.20	610.23	616.2[a], 610.0[b]
E_{2u}	87.50	92.51	88.0[a], 95.2[b]
E_{2u}	575.6	585.10	582.8[a], 585.6[b]

[a]Ist principles calculation (Ref. 193) of $NaCoO_2$ in their geometry A
[b]Ist principles calculation (Ref. 193) of $NaCoO_2$ in their geometry B
[c]Raman frequency (Ref. 188) of $Na_{0.7}CoO_2$
[d]Raman frequency (Ref. 189) of $Na_xCoO_2 \cdot y H_2O$
[e]Raman frequency (Ref. 189) of $Na_{0.3}CoO_2 . 1.3H_2O$
[f]Raman frequency (Ref. 189) of $Na_{0.7}CoO_2$ Single Crystal
[g]Infrared frequency (Ref. 185) of $Na_{0.57}CoO_2$
[h]Infrared frequency (Ref. 185) of $Na_{0.7}CoO_2$

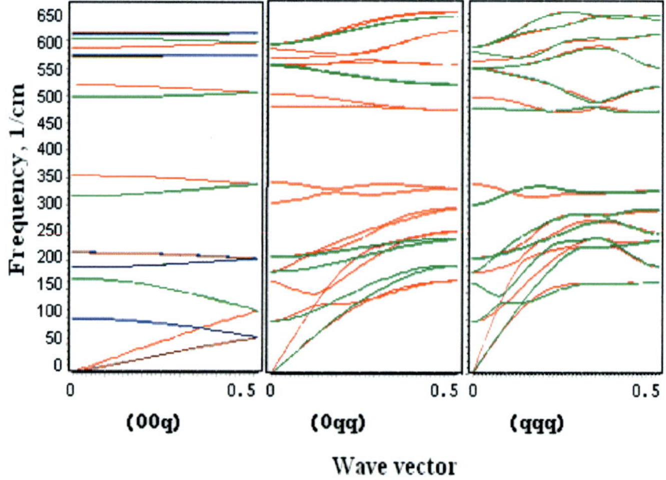

Figure 28. Phonon dispersion curves of $NaCoO_2$ in considered geometry A.

6.3. Phonon Dispersion Curves

Figure 28 presents the calculated phonon dispersion curves of $NaCoO_2$. It can be seen from the present figure that the phonon modes are separated in two frequency groups similar to the first principles calculation [193]. The lower frequency group, which is up to about 370 cm^{-1} is due to the acoustic phonons while the higher frequency group with frequencies between 490 cm^{-1} and 650 cm^{-1} is due to optical phonons. These regions have been named as soft and hard phonons by the Li *et al.* [28]. It can be seen from the phonon dispersion curves that the phonon modes are more dispersive in $(\overline{qq0})$ and $(\overline{qqq})$ directions of the Brillouin zone. It has been shown by Jorgensen *et al.* [194] and Huang *et al.* [31] from their structural

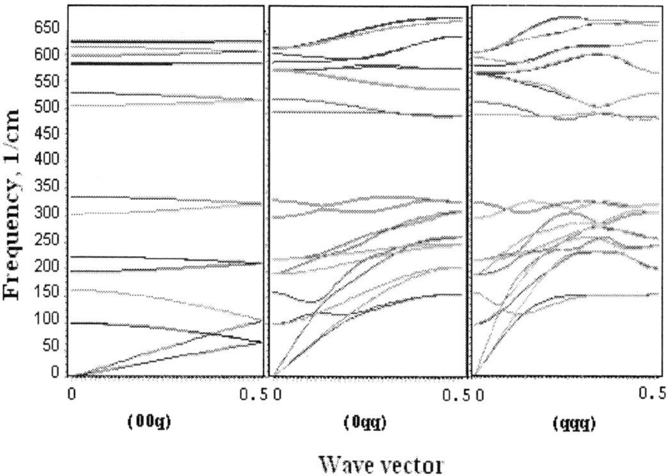

Figure 29. Phonon dispersion curves of $NaCoO_2$ in considered geometry B.

studies that the Na atoms occupy two different sites 2*d* and 2*b*. To see the effect of different site occupancy by Na atoms, we also considered two geometries for the $NaCoO_2$ similar to the consideration of Li *et al.* [193] in their first principles calculation. In first case (geometry A) the sodium atoms occupy 2*b* site while in second case (geometry B) the sodium atoms occupy the 2*d* site for which the PDCs are presented in Figure 28 and 29 respectively and discussed above. These figures reveal that there is not any significant difference in the gross features of the phonon dispersion curves in these two geometries except that the behavior of some phonon modes particularly involves the Na atoms vibration. In addition, some phonon modes giving significant different frequencies near the zone center are noticed for the two geometries, which is obvious due to the different atomic positions of the Na atoms.

6.4. Phonon Density of States

In order to investigate the phonon properties, the understanding of phonon density of states is vital, as it requires the computation of phonon modes in the entire Brillouin zone. The calculated total phonon density of states (DOS) and its partial components corresponding to the different atoms are presented in Figure 30.

The latter are used to calculate the vibrational amplitudes of the different atoms as presented in Table XII while the former are used to calculate the specific heat and Debye temperature. To understand the origin of peaks in the DOS, we examine the spectra of partial DOS displayed in Figure 30 along with the total DOS. From a comparison of partial and total phonon DOS, it is clear that the total DOS has three different regions. The first region is below 200 cm^{-1}, where the peaks are mainly due to Co atoms with a small contribution from oxygen; the second region is between 200 and 370 cm^{-1}, where the contributions are mainly due to sodium atoms with some contribution from Co and O atoms, while the third and final region above 475 cm^{-1} is only due to Oxygen atoms. The contributions to frequency ranges result from significant difference in atomic weight.

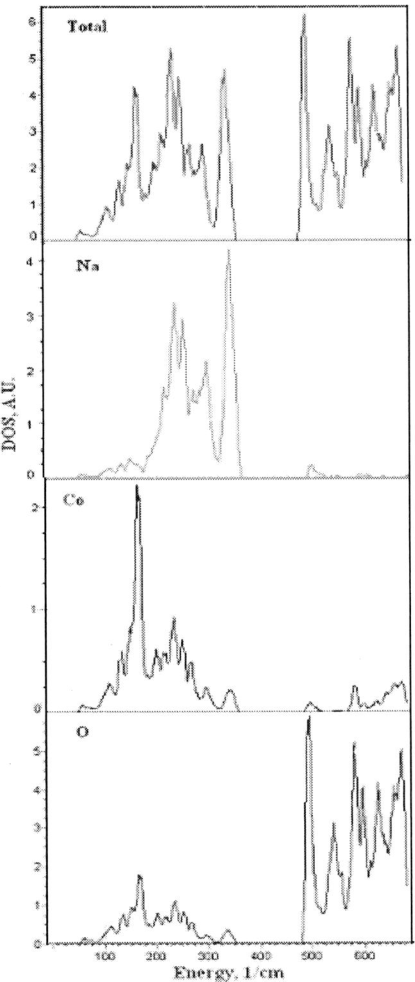

Figure 30: Total and partial phonon density of $NaCoO_2$ compound.

The neutron weighted phonon density of states is the quantity measured by the experiments and therefore its calculation is vital to test the success of the model calculation. Also, this calculation is important as there was a disagreement between the first principles calculated DOS and experimentally obtained neutron scattering spectra [189]. Neutron weighted phonon density of states (generalized phonon density ofstates) is also calculated to compare with the experimentally measured GDOS to understand the success of the present predictions. The calculated GDOS for both geometries are presented in Figure 31. The GDOS is bare DOS weighted with σi/Mi (scattering cross section over mass) and therefore the gross features in GDOS is similar to the total phonon DOS. This quantity for the atoms Na, Co and O is 0.3626, 0.1575 and 0.0423 barn/amu, respectively. It is seen from the Figure 31((a) and (b)) that there are two phonon groups similar to the experimental GDOS [189] and first principles DOS [193].

The present calculation similar to the first principles calculations [193] predicts low frequency DOS comparable to the experimental GDOS [189]. However, it seems that the present calculation is better in reproducing the higher frequency side of the spectra and compares well with the experimental DOS [189] in comparison to the first principal calculation [193] as far as range of DOS is concerned. The present calculation could also not predict the DOS up to 800 cm^{-1}as observed in the experimental spectra obtained from inelastic neutron scattering. As far as failure of these calculations in predicting the complete neutron scattering data [24], which show that the DOS goes zero at 100 meV, is concerned, it could be because of the scattering for greater than equal to 100 meV in INS measurement is due to the multiphonon scattering. There are some noticeable significant changes in the lower frequency side of the phonon density of states in two different geometries. It can be seen from the figures. 31(a) and 31(b) that many of the peaks in lower frequency side of the spectra which are prominent in geometry A (Figure 31(a)) either disappear or become weak in the case of geometry B (Figure 31(b)). In addition, many peaks convert into the shoulders around the main peaks. However, no significant change in the higher frequency side of the spectra is observed.

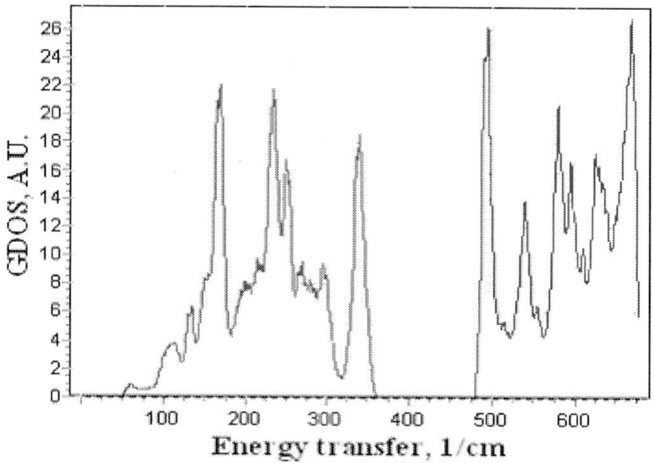

Figure 31(a). Generalized Phonon density of States of $NaCoO_2$ Geometry A.

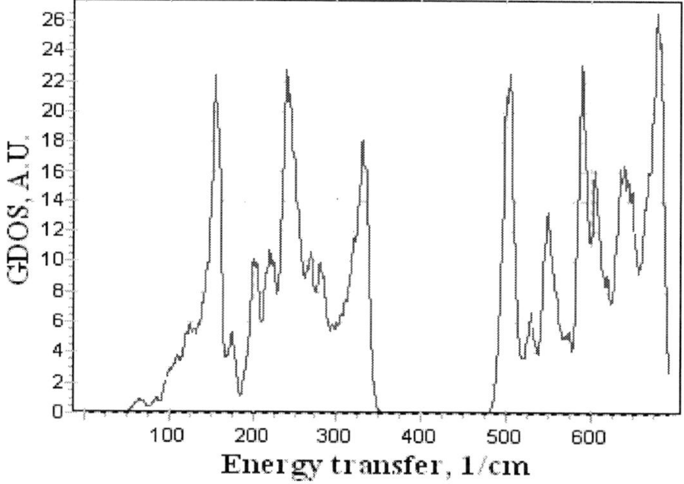

Figure 31(b). Generalized Phonon density of States of $NaCoO_2$ Geometry B.

6.5. Specific Heat and Debye Temperature

The temperature dependent lattice specific heat at constant volume of $NaCoO_2$ has been calculated by using the phonon density of states and expression presented in section 2. In Figure 32, the heat capacity at constant volume with temperature is presented. To the best of our knowledge, this is the first reported specific heat data for this compound. The Debye temperature at room temperature has been calculated for $NaCoO_2$ from the present specific heat data. Our estimated value of Debye temperature (θ_D) is 793 K, which could not be compared with any experimental or theoretical data due to non-availability. However, the Debye temperature of 793 K at room temperature suggests that the transport properties in insulating phase of cobalt oxide superconductor are dominated by electron-phonon interaction.

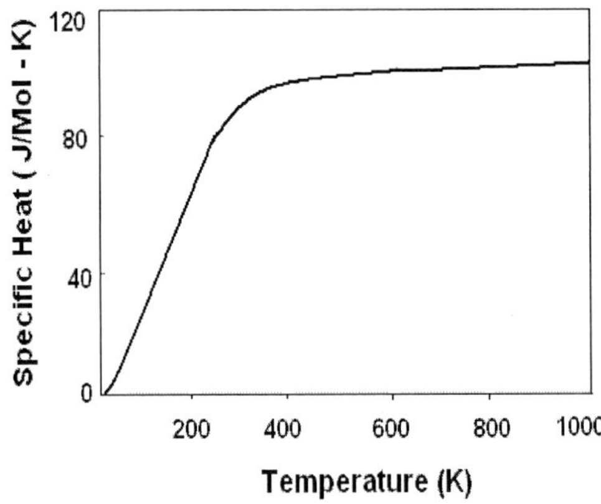

Figure 32. The lattice Specific heat at constant volume of $NaCoO_2$.

6.6. Isotropic Thermal Parameters

The computed partial phonon density of states is used to calculate the vibrational amplitudes of different species of $NaCoO_2$ compound and is presented in Table XII. This table reveals that the vibrational amplitudes increase with the increase in temperature similar to the experimental data [24]. The vibrational amplitudes do not behave anomalously in the studied temperature range.

Table XII. Vibrational amplitudes of the various atoms in $NaCoO_2$.

Atoms	$8\pi^2 < u^2 >/3$ (Å)				
	2K	50 K	100 K	200 K	300 K
Na	0.2191	0.2244	0.2446	0.3325	0.4464
Co	0.1056	0.1123	0.1360	0.2111	0.2965
O	0.1812	0.1869	0.2037	0.2575	0.3251

7. SUMMARY

Manganites being complex systems are interesting from underlying physics point of view in addition to their promising technological applications. The present chapter reports the results on the investigation of vibrational properties of manganites under far from equilibrium conditions and an attempt is made to complement these properties of lanthanum manganites to understand the CMR physics. In these systems, an important issue is the influence of phononic and electronic degrees of freedom on phenomenon like CMR. High pressure and temperature tunes the interplay between lattice and electronic degrees of freedom in manganites. Internal pressure induced by increasing $<r_A>$ (*i.e.* average ionic radius of A-site ion) also brings a delicate balance in a similar fashion. The double exchange mechanism introduced to understand ferromagnetism in manganites, is not sufficient alone to account for their magnetotransport properties (sections 1 and 3). Some features exhibited by doped manganites are difficult to understand without taking into account the Jahn-Teller coupling between the electrons and phonons. This view is further supported by the close relationship between the structural and electronic phase diagram (section 1).

As a matter of fact, $LaMnO_3$, which undergoes phase transition at high temperature, is considered in its cubic and rhombohedral phase. It is also considered to study the effect Sr-doping at A-site. In addition, to understand the effect of pressure and temperature, 30 % Sr-doped $LaMnO_3$ *i.e.* $La_{0.7}Sr_{0.3}MnO_3$ (LSMO) is considered at different applied pressure and temperature. Due to decreased symmetry of rhombohedral $LaMnO_3$ and more number of atoms in rhombohedral unit cell as compared to cubic $LaMnO_3$, the number of phonon modes at zone centre of rhombohedral $LaMnO_3$ is more. The difference in structural symmetry of cubic and rhombohedral manganites is manifested in their phonon spectra and density of states. Structure phase transition from orthorhombic-rhombohedral-cubic gradually removes buckling of MnO_6 octahedral and leads to reduction in electron-phonon interaction accompanied by achieving higher symmetry. Besides structural symmetry, A-site ion radius, doping, temperature and pressure also cause the modification in phonon dispersion curves, phonon density of states and phonon branches, splitting the phonon modes into a pair of non-degenerate (A) and doubly degenerate (E) modes with broadening and overlapping. Doping by Sr atoms at La- site changes A- site ionic radius ($<r_A>$) which untilts octahedral framework and the far away oxygen atoms in first co-ordination shell start to feel an attractive interaction as they come closer to A-site. This increased attractive interaction is transmitted to Mn-O network as an effective internal pressure, which is accompanied by the decrease in Jahn-Teller distortion resulting into the increase in T_C and apparently metallicity of manganites increases (ferromagnetism). In vibrational spectra, this is complemented with the softening of A_{1g} phonon modes originating due to MnO_6 rotation and high frequency stretching mode (section 3). In insulating orthorhombic phase of parent $LaMnO_3$, this mode is higher than that in the Sr-doped rhombohedral $LaMnO_3$. The correlation of the frequency of A_{1g} mode with angle α of the rhombohedral distortion indicates that the doping at A-site of undoped $LaMnO_3$ reduces the Jahn-Teller distortion and manifested in softening of the A_{1g} mode following the linear relation. Substitution of Sr atoms at A-site regularizes the crystal structure maintaining the tilting of the octahedra. The application of internal pressure introducing a divalent atom at A-site thereby reduces the electron-phonon coupling and system achieves character that is more metallic. In addition, calculation of the thermodynamic parameters exhibit that the cubic $LaMnO_3$ with the highest symmetry has the lowest specific heat and Debye temperature among undoped cubic $LaMnO_3$ and rhombohedral $LaMnO_3$, and rhombohedral Sr-doped $LaMnO_3$. The doping at La-site by Sr atom softens the lattice at high temperatures (above 100 K), which is reflected in softening of A_{1g} phonon modes in rhombohedral structure.

The use of temperature and pressure as thermodynamic variables provides with a simple but powerful means by which interaction within the system can be modified without changing the doping level. Sr-doped $LaMnO_3$ in rhombohedral phase for 'x' = 0.3 is considered to see the effects of temperature and pressure on phonon properties. Softening of the stretching mode in the phonon dispersion curve of $La_{0.7}Sr_{0.3}MnO_3$ at specific temperature reflects the instability of phonon modes causing Jahn Teller lattice distortion (section 4). The A_{1g} phonon mode corresponding to the rotations of MnO_6 octahedra also undergoes softening while the stretching mode of MnO_6 octahedra, which is a JT phonon mode with Mn-O bond character contributing to the highest frequency region, hardens upon reduction in temperature. The lattice softening as a result of temperature variation is manifested in decrease in Debye temperature (θ_D) at 300 K. The temperature dependence of the atomic displacement is correlated with the anharmonic lattice distortion via the effective Grüneisen parameter γ_{eff} for rhombohedral $La_{0.7}Sr_{0.3}MnO_3$ systems. The presence of anharmonic lattice modes is also indicated by anomalously high value of Grüneisen parameter.

The application of external pressure always increases the Curie temperature (T_C) widening the Mn-O-Mn bond angle and compressing Mn-O and A-O bond lengths. It also reduces the local distortion of the MnO_6 octahedra and structure of systems becomes more cubic with reduction in electron-phonon coupling through the modification of the frequency of the octahedral bending and stretching modes. The A_{1g} mode is the most sensitive to pressure variation, which seems to be related with the changes in JT distortion. Phonon dispersion curves reveal that both the (IR and Raman) frequencies of modes involving the vibrations of La atom which actually do not depend on the octahedral distortion, are almost pressure independent while the stretching and rotational modes (E_g and A_{2u}) involving oxygen atom vibrations (high frequency) and related to the MnO_6 octahedra show linear pressure induced hardening (section 5). The pressure behaviour of stretching and rotational modes (A_{2u} and E_g) also suggests that the Mn-O-Mn angle is close to the ideal 180° value of the cubic structure and therefore the system would be of more metallic character. The rotational frequency of MnO_6 octahedra is not changing fast which indicates that the octahedra are not severely distorted by the application of pressure. Therefore, it can be concluded that the hardening of phonon frequencies is consistent with the increase in T_C with pressure. The phonon peaks in the higher energy side of the phonon density of states show decrease in peak width with the increase of pressure and hence there appears reduction in electron-phonon interaction and lattice disorder responsible for the JT distortion. The effective mode Grüneisen parameter depicts that pressure induces the reduction of JT distortion, which is signature of the rhombohedral structure.

The phonon properties investigated for the $NaCoO_2$ compound in its two different geometry positions are significantly different for the sodium atoms. The in-plane mode is found to occur at lower frequency and is insensitive to Na positions, while the out-of-plane mode is sensitive to the sodium position, and gives rise to the higher frequency modes with Na(2) position and lower with Na(1) positions. A temperature dependent investigations of phonons can lead to some understanding of insulator-Metal transition and charger ordering, which is not discussed here.

ACKNOWLEDGEMENT

This research was supported by the Department of Science and Technology, Ministry of Science and Technology and Council of Scientific and Industrial Reseasrch of India. A portion of this research was perfirmed at Brazilian Centre for Physical Research, Rio de Janeiro, Brazil. It is a pleasure to acknowledge the very fruitful collaborations with Orif. A. Troper and Prof. I C. da Cunha Lima . We also thank Prof. S.P. Sanyal for support and encouragement..

REFERENCES

[1] J. G. Bednorz, and K. A. Muller, *Z. Phys*. B 64, 189 (1986).
[2] M. Imada, A. Fujimori, and Y. Tokura, *Rev. Mod. Phys*. 70, 1039 (1998).
[3] C. N. R. Rao, and B. Raveau, *Transition Metal Oxides* (VCH Publishers, New York: 1995).
[4] A. J. Millis, *Nature* 392, 147 (1998).
[5] Y. Tokura, editor, *Colossal Magnetoresistive Oxides* (Gordon and Breach Science Publishers, The Netherlands: 2000).
[6] Y. Tokura, and N. Nagaosa, *Science* 288, 462 (2000).
[7] N. F. Mott, *Metal-Insulator Transitions* (Taylor and Francis, London: 1990).
[8] S. Jin, T. H. Tiefel, M. McCormack, R. A. Fastnacht, R. Ramesh, and L. H. Chen, *Science* 264,413 (1994).
[9] R. M. Kusters, J. Singleton, D. A. Keen, R. McGreevy, and W. Hayes, *Physica* B 155, 362 (1989).
[10] G. H. Jonker, and J. H. van Santen, *Physica* B 16, 337 (1950).
[11] K. Takada, H. Sakurai, E. Takayama-Muromachi, F. Izami, R. A Dilanian and T. Sasaki, *Nature* (London) 422, 53 (2003)
[12] A. J. Millis, J. Electron Spectrosc. Relat. Phenom. 114, 669 (2001); G. A. Prinz, *Science* 282, 1660 (1998).
[13] S. A. Wolf, D. D. Awschalom, R. A. Buhrman, J. M. Daughton, S. von Moln'ar, M. L. Roukes, A. Y. Chtchelkanova, and D. M. Treger, *Science* 294, 1488 (2001).
[14] M. N. Baibich, J. M. Broto, A. Fert, F. Nguyen Van Dau, F. Petroff, P. Eitenne, G. Creuzet, A. Friederich, and J. Chazelas, *Phys. Rev. Lett*. 61, 2472 (1988).
[15] D. D. Awschalom, and R. K. Kawakami, *Nature* 408, 923 (2000).
[16] H. Ohno, D. Chiba, F. Matsukura, T. Omiya, E. Abe, T. Dietl, Y. Ohno, and K. Ohtani, *Nature* 408, 944 (2000).
[17] N. Hur, S. Park, P. A. Sharma, J. S. Ahn, S. Guha, and S -W. Cheong, *Nature* 429, 392 (2004).
[18] P. Schiffer, A. P. Ramirez, W. Bao, and S -W. Cheong, *Phys. Rev. Lett*. 75, 3336(1995).
[19] *Wall Street Journal* (1997), 10 November, page B8.

[20] C. Zener, *Phys. Rev*. 81, 440 (1951).

[21] For a review, see J. B. Goodenough and J. M. Longon, *Magnetic and Other Properties of Oxides and Related Compounds*, edited by K. - H. Hellwege, and O. Madelung, Landolt –Bornstein, New Series, Group III, Vol. 4, Pt. a (Springer-Verlage, Berlin: 1970).

[22] K. Yosida, *Theory of Magnetism* (Springer-Verlag, Berlin, Heidelberg: 1996).

[23] W. E. Pickett, and D. J. Singh, *Phys. Rev.* B 53, 1146 (1996).

[24] J. L. García-Muñoz, J. Fontcuberta, M. Suaaidi, and X. Obradors, *J. Phys.: Condens. Matter* 8, L787 (1996).

[25] L. M. Rodríguez-Martínez, and J. P. Attfield, *Phys. Rev*. B 63, 024424 (2001).

[26] P. W. Anderson, and H. Hasegawa, Phys. Rev. 100, 675 (1955).

[27] Y. Lyanda-Geller, S. H. Chun, M. B. Salamon, P. M. Goldbart, P. D. Han, Y. Tomioka, A. Asamitsu, and Y. Tokura, *Phys. Rev.*B 63, 184426 (2001).

[28] E. O. Wollan, and W. C. Koehler, *Phys. Rev*. 100, 545 (1955).

[29] G. Matsumoto, *J. Phys. Soc. Jpn*. 29, 613 (1970).

[30] P -G. de Gennes, *Phys. Rev.* 118, 141 (1960).

[31] C. W. Searle, and S. T. Wang, Can. *J. Phys.* 48, 2023 (1970).

[32] K. Kubo, and N. Ohata, *J. Phys. Soc. Jpn.* 33, 21 (1972)

[33] N. Furukawa, J. Phys. Soc. Jpn. 63, 3214 (1994); 64, 2734 (1995); 64, 2754 (1995); *ibid.* 64, 3164 (1995).

[34] A. J. Millis, P. B. Littlewood, and B. I. Shraiman, *Phys. Rev. Lett.* 74, 5144 (1995).

[35] J. Kanamori, *J. Appl. Phys*. 31, 14S (1960).

[36] J. B. A. A. Elemans, B. van Laar, K. R. van Der Veen, and B. O. Loopstra, *J. Solid State Chem*. 3, 238 (1971).

[37] W. A. Harrison, *Solid State Theory* (Dover Publications Inc, New York: 1979).

[38] K. H. Kim, J. Y. Gu, H. S. Choi, G. W. Park, and T. W. Noh, *Phys. Rev. Lett*. 77, 1877 (1996).

[39] S. J. L. Billinge, R. G. DiFrancesco, G. H. Kwei, J. J. Neumeier, and J. D. Thompson, *Phys. Rev. Lett*. 77, 715 (1996).

[40] Asamitsu, Y. Moritomo, Y. Tomioka, T. Arima, and Y. Tokura, *Nature* (London) 373, 407 (1995).

[41] M. Jaime, M. B. Salamon, K. Pettit, M. Rubinstein, R. E. Treece, J. S. Horwitz, and D. B. Chrisey, *Appl. Phys. Lett*. 68, 1576 (1996).

[42] S. G. Kaplan, M. Quijada, H. D. Drew, D. B. Tanner, G. C. Xiong, R. Ramesh, C. Kwon, and T. Venkatesan, *Phys. Rev. Lett*. 77, 2081 (1996).

[43] Y. Yamada, O. Hino, S. Nohdo, and R. Kanao, T. Inami, and S. Katano, *Phys. Rev. Lett.* 77, 904 (1996).

[44] D. N. Argyriou, J. F. Mitchell, C. D. Potter, D. G. Hinks, J. D. Jorgensen, and S. D. Bader, *Phys. Rev. Lett*. 76, 3826 (1996).

[45] Y. Endoh, K. Hirota, S. Ishihara, S. Okamoto, Y. Murakami, A. Nishizawa, T. Fukuda, H. Kimura, H. Nojiri, K. Kaneko, and S. Maekawa, *Phys. Rev. Lett*. 82, 4328 (1999).

[46] C. N. R. Rao, A. Arulraj, A. K. Cheetham, and B. Raveau, *J. Phys. Condens. Matter* 12, R83 (2000).

[47] H. Kuwahara, Y. Tomioka, A. M. Asamistu, Y. Moritomo and Y. Tokura, *Science* 270, 961 (1995).

[48] Y. Moritomo, H. Kuwahara, Y. Tomioka, and Y. Tokura, *Phys. Rev.* B 55, 7549 (1997).

[49] P. G. Radaelli, D. E. Cox, M. Marezio, and S -W. Cheong, *Phys. Rev.* B 55, 3015 (1997).

[50] P. K. Jha and S. P. Sanyal, Physica C 330, 39 (2000); Physica C 322, 110 (1999); *Physica* B 262, 322 (1999).

[51] T. S. Bush. J. D. Gale, C. R. A. Catlow, and P. D. Battle, *Mater. Chem.* 4, 831 (1994); J. D. Gale, *Phil. Mag.* B 73, 3, 1996.

[52] Y. Shimakawa, Y. Kubo, and T. Manako, *Nature* (London) 379, 53 (1996).

[53] S. Süllow, I. Prasad, S. Bogdanovich, M. C. Aronson, J. L. Sarrao, and Z. Fisk, Conference contribution MMM'99, San Jose; cond-mat/9912390 (1999).

[54] A. P. Ramirez, R. J. Cava, and J. Krajewski, *Nature* (London) 386, 156 (1997).

[55] K. -I. Kobayashi, T. Kimura, H. Sawada, K. Terakura, and Y. Tokura, *Nature* (London) 395, 677 (1998).

[56] P. Nyhus, S. Yoon, M. Kauffman, S. L. Cooper, Z. Fisk, and J. Sarrao, *Phys. Rev.* B 56, 2717 (1997).

[57] Y. Tomioka, T. Okuda, Y. Okimoto, R. Kumai, K -I. Kobayashi, and Y. Tokura, *Phys. Rev.* B 61, 422 (2000).

[58] T. H. Kim, M. Uehara, S -W. Cheong, and S. Lee, *Appl. Phys. Lett.* 74, 1737 (1999).

[59] M. Born, and K. Huang, *Dynamical Theory of Crystal Lattices*, (Oxford University Press, London: 1954);P. Brüsch, *Phonons: Theory and Experiments I*, (Springer-Verlag, Berlin, Heidelberg, New York: 1982).

[60] A.A. Maradudin, and S. H. Vosko, Rev. Mod. Phys. 40, 1 (1968); G. Venktaraman, L. Feldkamp, and V. C. Sahni, *Dynamics of Perfect Crystals*, (MIT Press, Cambridge: 1975).

[61] H. B. Huntington, The Elastic Constants of Crystals, *Solid State Physics*, 7, 213 (Academic Press, New York, N.Y: 1958).

[62] E. W. Kellerman, Phil. Trans. Roy. Soc. (Lond.) A 230, 513 (1940); *Proc. Roy. Soc.* (Lond.) A 178, 17 (1941).

[63] W. Cochran, Proc. R. Soc. London, Ser. A 253, 260 (1959); *Phys. Rev. Lett.* 2, 495 (1959).

[64] G. V. Lewis, and C. R. A. Catlow, J. Phys. C: Solid State Physics 18, 1149 (1985).

[65] T. S. Bush. J. Gale, C. R. A. Catlow, and P. D. Battle, *Mater. Chem.* 4, 831 (1994); J. Gale, *Phil. Mag.* B 73, 3 (1996).

[66] S. L. Chaplot, *Phys. Rev.* B 37, 7435 (1988); 45, 4885 (1992).

[67] P. K. Jha, A. Troper, I. C. da Cumha Lima, M. Talati, and S. P. Sanyal, Physica B 366, 153 (2005) ; P. K. Jha, and S. P. Sanyal, Physica C 265, 259 (1996); P. K. Jha, *Phys. Rev.* B 72, 214502 (2005).

[68] M. B. Smirnov, and V. Yu Kazimirov, *Preprint JINR E* 14,159 (2001).

[69] R. Lubbers, H. F. Grunstendel, A. I. Chumakov, and G. Wortmann, *Science* 287, 1250 (2000).

[70] M. Schwoerer-Böhning, A. T. Macrander, and D. A. Arms, *Phys. Rev. Lett.* 80, 5572, (1998).

[71] Shukla, M. Calandra, M. d'Astuto, M. Lazzeri, F. Mauri, C. Bellin, M. Krisch, J. Karpinski, S. M. Kazakov, J. Jun, D. Daghero, and K. Parlinski, *Phys. Rev. Lett.* 90, 95506 (2003).

[72] A. Q. R. Baron, H. Uchiyama, Y. Tanaka, S. Tsutsui, D. Ishikawa, S. Lee, R. Heid, K - P. Bohnen, S. Tajima, and T. Ishikawa, *Phys. Rev. Lett.* 92, 197004 (2004).

[73] E. Burkel, *Rep. Prog. Phys*. 63, 171 (2000).

[74] P. Murugavel, C. Narayana, A. R. Raju, A. K. Sood, and C. N. R. Rao, *Europhys. Lett.* 52, 461 (2000).

[75] H. Y. Hwang, T. T. M. Palstra, S -W. Cheong, and B. Batlogg, *Phys. Rev*. B 52, 15046 (1995).

[76] Urushibara, Y. Moritomo, T. Arima, A. Asamitsu, G. Kido, and Y. Tokura, *Phys. Rev.* B 51, 14103 (1995).

[77] P. G. Radaelli, G. Iannone, M. Marezio, H. Y. Hwang, S -W. Cheong, J. D. Jorgensen, and D. N. Argyriou, *Phys. Rev.* B 56, 8265 (1997).

[78] P. Norby, I. G. Krogh Andersen, and E. Krogh Andersen, *J. Solid State Chem.* 119, 191 (1995).

[79] Q. Huang, A. Santoro, J. W. Lynn, R. W. Erwin, J. A. Borchers, J. L. Peng, and R. L. Green, *Phys. Rev.* B 55, 14987 (1997).

[80] L. Pinsard-Gaudart, J. Rodríguez-Carvajal, A. Daoud-Aladine, I. Goncharenko, M. Medrade, R. I. Smith, and A. Revcolevschi, *Phys. Rev.* B 64, 064426 (2001).

[81] H. Y. Hwang, S -W. Cheong, P. G. Radaelli, M. Marezio, and B. Batlogg, *Phys. Rev. Lett.* 75, 914 (1995).

[82] J. Millis, B. I. Shraiman, and R. Mueller, Phys. Rev. Lett. 77, 175 (1996); A. J. Millis, *Phys. Rev.* B 53, 8434 (1996).

[83] V. B. Podobedov, A. Weber, D. B. Romero, J. P. Rice, and H. D. Drew, Sol. State. Commun. 105, 589 (1998); V. B. Podobedov, A. Weber, D. B. Romero, J. P. Rice, and H. D. Drew, *Bulletin Am. Phys. Soc.* 42, 341 (1997).

[84] K. H. Kim, J. Y. Gu, H. S. Choi, G. W. Park, and T. W. Noh, *Phys. Rev. Lett.* 77, 1877 (1996).

[85] Y. Okimoto, T. Katsufuji, T. Ishikawa, T. Arima, and Y. Tokura, *Phys. Rev.* B 55, 4206 (1997).

[86] M. N. Iliev, M. V. Abrashev, V. N. Popov, V. G. Ivanov, H -G. Lee, W. H. Mc Carrol, M. Greenblatt, R. L. Meng, and C. W. Chu, *Physica* C 341, 2257 (2000).

[87] M. N. Iliev, M. V. Abrashev, H -G. Lee, V. N. Popov, Y. Y. Sun, C. Thomsen, R. L. Meng, and C. W. Chu, *Phys. Rev.* 57, 2872 (1998).

[88] V. B. Podobedov, A. Weber, D. B. Romero, J. P. Rice, and H. D. Drew, *Phy. Rev.* B 58, 43 (1998).

[89] M. V. Abrashev, A. P. Litvinchuk, M. N. Iliev, R. L. Meng, V. N. Popov, V. G. Ivanov, and R. A. Chakalov, C. Thomsen, *Phys. Rev.* B 59, 4146 (1999).

[90] N. N. Kovaleva, J. L. Gavartin, A. L. Shluger, A. V. Boris, and A. M. Stoneham, *Physica* B 312, 734 (2002).

[91] W. Reichardt, and M. Braden, *Physica* B 263, 416 (1999).

[92] R. Mahendiran, S. K. Tiwary, A. K. Raychaudhuri, T. V. Ramakrishnan, R. Mahesh, N. Rangavittal, C. N. R. Rao, *Phys. Rev*. B 53, 3348 (1996).

[93] Urushibara, Y. Moritomo, T. Arima, A. Asamitsu, G. Kido, and Y. Tokura, Phys. Rev. B 51, 14103 (1995).

[94] S. J. L. Billinge, R. G. DiFrancesco, G. H. Kwei, J. J. Neumeier, and J. D. Thompson, *Phys. Rev. Lett.* 77, 715 (1996).

[95] P. Dai, J. Zhang, H. A. Mook, S -H. Liou, P. A. Dowben, and E. W. Plummer, *Phys. Rev.* B 54, R3694 (1996).

[96] G. Zhao, K. Conder, H. Keller, and K. A. Muller, *Nature (*London) 381, 676 (1996).

[97] M. R. Ibarra, Guo-meng Zhao, J. M. De Teresa, B. García-Landa, Z. Arnold, C. Marquina, P. A. Algarabel, H. Keller, C. Ritter, *Phys. Rev.* B 57, 7446 (1998).

[98] L. Martín-Carrón, A. de Andrés, M. J. Martínez-Lope, M. T. Casais, and J. A. Alonso, *Phys. Rev.* B 66,174303 (2002).

[99] E. Granodo, N. O. Moreno, A. Garćia, J. A. Sanjurjo, C. Rettori, I. Torriani, S. B. Oseroff, J. J. Neumeier, K. J. McClellan, S -W. Cheong, and Y. Tokura, *Phys. Rev.* B 58, 11435 (1998).

[100] M. Talati and P. K. Jha, *Computational Mat. Sci.* 37, 64 (2006); M. Talati, Ph.D. thesis, The M.S. University of Baroda, (2006).

[101] E. O. Wollan, and W. C. Koehler, *Phys. Rev.* 100, 545 (1955).

[102] J. B. A. A. Elemans, B. van Laar, K. R. van der Veen, and B. O. Loopstra, *J. Solid State Chem.* 3, 328 (1971).

[103] J. Rodríquez-Carvajal, M. Hennion, F. Moussa, A. H. Moudden, L. Pinsard, and A. Revcolevschi, *Phys. Rev.* B 57, R3189 (1998).

[104] B. D. Silverman, and G. F. Koster, Z. Phys. 163, 158 (1961); J. T. Last, *Phys. Rev.* 105, 1740 (1957).

[105] I. V. Solovyev, and K. Terakura, *Phys. Rev. Lett.* 82, 2959 (1999).

[106] B. F. Woodfield, M. L. Wilson, and J. M. Byers, *Phys. Rev. Lett.* 78, 3201 (1997).

[107] A. Paolone, P. Roy, A. Pimenov, A. Loidl, O. K. Melnikov, and A. Y. Shapiro, *Phys. Rev.* B 61, 11 255 (2000).

[108] F. Mayr, Ph.D. thesis, Universität Augsburg (2002).

[109] M. N. Iliev, M. V. Abrashev, V. N. Popov, and V. G. Hadjiev, *Phys. Rev.* B 67, 212301 (2003).

[110] G. De Marzi, H. J. Trodahl, J. Bok, A. Cantarero, and F. Sapina ,cond-mat/0305323 (2003).

[111] Kim, B. L. Zink, F. Hellman, and J. M. D. Coey, *Phys. Rev.* B 65, 214424 (2002).

[112] S. N. Bai, Y. Y. Chen, Y. D. Yao, L. H. Chen, S. H. Lin, and Y. Liou, Chin. *J. Phys.* 34, 798 (1996).

[113] L. Ghivelder, I. Abrego Castillo, M. A. Gusmão, J. A. Alonso, and L. F. Cohen, *Phys. Rev.* B 60, 12184 (1999).

[114] J. J. Hamilton, E. L. Keatley, H. L. Ju, A. K. Raychaudhuri, V. N. Smolyaninova, and R. L. Greene, Phys. Rev. B 54, 14926 (1996); J. Tanaka, and T. Mitsuhashi, *J. Phys. Soc. Jpn.* 53, 24 (1984).

[115] M. Premila, T. N. Sairam, and C. S. Sundar, *Current Science*, 88, 102 (2005).

[116] R. von Helmolt, J. Wecker, B. Holzapfel, L. Schultz, and K. Samwer, Phys. Rev. Lett. 71, 2331 (1993); S. Jin, M. McCormack, T. H. Tiefel, and R. Ramesh, *J. Appl. Phys.* 76, 6929 (1994).

[117] H. Yoshizawa, H. Kawano, Y. Tomioka, and Y. Tokura, Phys. Rev. B 52, 13145 (1995); J. B. Goodenough, *Phys. Rev.* 164, 785 (1967).

[118] D. I. Khomskii and G. A. Sawatzky, Solid State Commun. 102, 87 (1997); J. M. Tranquada, B. J. Sternlieb, J. D. Axe, Y. Nakamura, and S. Uchida, *Nature* (London) 375, 561 (1995).

[119] E. Liarokapis, Th. Leventouri, D. Lampakis, D. Palles, J. J. Neumeier, and D. H. Goodwin, *Phys. Rev.* B 60, 12758 (1999).

[120] J. Mira, J. Rivas, L. E. Hueso, F. Rivadulla, M. A. Lõpez Quintela, M. A. Señarís Rodríguez, and C. A. Ramos, *Phys. Rev.* B 65, 024418 (2001).

[121] D. P. Kozlenko, I. N. Goncharenko, B. N. Savenko, and V. I. Voronin, *J. Phys.: Condens. Matter* 16, 6755 (2004).
[122] M. Valkeapää, Ph.D Thesis, Göteborg University, Göteborg, Sweden (2004).
[123] Kawano, R. Kajimoto, M. Kubota, and H. Yoshizawa, *Phys. Rev.* B 53, R14709 (1996); *ibid.* 2202 (1996).
[124] Y. Takeda, S. Nakai, R. Kanno, N. Imanishi, G. Q. Shen, and O. Yamamoto, *Mat. Res. Bull.* 26, 153 (1991).
[125] C. Zener, Phys. Rev. 82, 403 (1951); P -G. de *Gennes, ibid.* 118, 141 (1960).
[126] Y. Okimoto, T. Katsufuji, T. Ishikawa, A. Urushibara, T. Arima, and Y. Tokura, *Phys. Rev. Lett* 75, 109 (1995).
[127] R. Gupta, A. K. Sood, R. Mahesh, and C. N. R. Rao, *Phys. Rev.* B 54, 14899 (1996).
[128] V. B. Podobedov, A. Weber, D. B. Romero, J. P. Rice, and H. D. Drew, *Bull. Am. Phys. Soc.* 42, 341 (1997).
[129] L. Martín-Carrón, A. de Andrés, M. J. Martínez-Lope, M. T. Casais, and J. A. Alonso, *Phys. Rev.* B 66, 174303 (2002).
[130] M. Talati and P. K. Jha, *Int. J. of Modern Phys.* B (Submitted).
[131] J. J. Hamilton, E. L. Keatley, H. L. Ju, A. K. Raychaudhuri, V. N. Smolyaninova, and R. L. Greene, Phys. Rev. B 54, 14926 (1996); J. Tanaka, and T. Mitsuhashi, *J. Phys. Soc. Jpn.* 53, 24 (1984).
[132] G. V. Lewis, and C. R. A. Catlow, J. Phys. C: Solid State Phys. 18, 1149 (1985).
[133] Ch. Hartinger, F. Mayr, A. Loidl, and T. Kopp, *Phys. Rev.* B 71, 184421 (2005).
[134] M. N. Iliev, M. V. Abrashev, V. N. Popov, and V. G. Hadjiev, *Phys. Rev.* B 67, 212301 (2003).
[135] P. Dai, J. Zhang, H. A. Mook, S -H. Liou, P. A. Dowben, and E. W. Plummer, *Phys. Rev.* B 54, R3694 (1996).
[136] P. G. Radaelli, M. Marezio, H. Y. Hwang, S -W. Cheong, and B. Batlogg, *Phys. Rev.* B 54, 8992 (1996).
[137] R. von Helmolt, J. Wecker, B. Holzapfel, L. Schultz, and K. Samwer, *Phys. Rev. Lett.* 71, 2331 (1993).
[138] M. B. Salamon, and M. Jaime, *Rev. Mod. Phys.* 73, 583 (2001).
[139] 139 Y. Tokura, and Y. Tomioka, *J. Magn. Magn. Mater.* 200, 1 (1999).
[140] J. M. D. Coey, M. Viret, and S. von Moln'ar, *Adv. Phys.* 48, 167 (1999).
[141] T. Venkatesan, M. Rajeswari, Z -W. Dong, S. B. Ogale, and R. Ramesh, *Philos. Trans. R. Soc. London*, Ser. A 356, 1661 (1998).
[142] A. P. Ramirez, *J. Phys.:Condens. Matter* 9, 8171 (1997).
[143] C. N. R. Rao, and A. K. Cheetham, *Adv. Mater.* 9, 1009 (1997).
[144] M. Itoh, K. Nishi, J. D. Yu, and Y. Inaguma, *Phys. Rev.* B 55, 14408 (1997).
[145] Y. Moritomo, A. Asamitsu, and Y. Tokura, *Phys. Rev.* B 56, 12190 (1997).
[146] V. Laukhin, B. Martinez, J. Fontcuberta, and Y. M. Mukovskii, *Phys. Rev.* B 63, 214417 (2001).
[147] D. P. Kozlenko, I. N. Goncharenko, B. N. Savenko, and V. I. Voronin, *J. Phys.: Condens. Matter* 16, 6755 (2004).
[148] A. Congeduti, P. Postorino, E. Caramagno, M. Nardone, A. Kumar and, D. D. Sarma, *Phys. Rev. Lett.* 86, 1251 (2001).
[149] V. Laukhin, J. Fontcuberta, J. I. Garcia-Muñoz, and X. Obradors, *Phys. Rev.* B 56, R10009 (1997).

[150] Y. Moritoma, A. Asamitsu, H. Kuwahara, and Y. Tokura, *Nature* (London) 380, 141 (1996).

[151] J. D. Lee, and B. I. Min, Phys. Rev. B 55, 12 454 (1997); U. Yu, B. I. Min, and J. D. Lee, *Phys. Rev*. B 61, 84 (2000).

[152] S. Yoon, H. L. Liu, G. Schollerer, and S. L. Cooper, P. D. Han, D. A. Payne, S -W. Cheong, and Z. Fisk, Phys. Rev. B 58, 2795 (1998); M. V. Abrashev, V. G. Ivanov, M. N. Iliev, R. A. Chakalov, R. I. Chakalovo, and C. Thomsen, *Phys. Stat. Sol.* (b) 215, 631 (1999).

[153] X. Wang, Y. W. Pan, and Q. L. Cui, *J. Solid State Chem*. 160, 307 (2001); X. Wang, Q. Cui, Y. Pan, G. Zou, and J. Liu, Gaoya Wuli Xuebao 15, 60 (2001).

[154] L. Pinsard-Gaudart, J. Rodríguez-Carvajal, A. Daoud-Aladine, I. Goncharenko, M. Medarde, R. I. Smith, and A. Revcolevschi, *Phys. Rev*. B 64, 064426 (2001).

[155] D. Kozlenko, and B. N. Savenko, *J. Phys. Condens. Matter* 16, 9031 (2004).

[156] M. Talati and P. K. Jha, Phys. Rev. B 74, 134406 (2006); *Journal of Molecular Structure,* 838, 227 (2007).

[157] Y. Moritomo, A. Asamitsu, and Y. Tokura, *Phys. Rev*. B 51, 16491(1995).

[158] P. Postorino, A. Congeduti, P. Dore, A. Sacchetti, F. Gorelli, L. Ulivi, A. Kumar, and D. D. Sarma, *Phys. Rev. Lett.* 91, 175501 (2003).

[159] J. J. Neumeier, M. F. Hundley, J. D. Thompson, and R. H. Heffner, *Phys. Rev.* B 52, R7006 (1995).

[160] P. K. Jha, A. Troper, I C. da Cunha Lima, M. Talati, and S. P. Sanyal, *Physica* B 366, 153 (2005).

[161] P. K. Jha, *Phys. Rev*. B 72, 214502 (2005).

[162] P. Björnsson, M. Rübhausen. J. Bäckström, M. Käll, S. Eriksson, J. Eriksen, and L. Börjesson, *Phys. Rev.* B 61, 1193 (2000).

[163] J. Fontcuberta, V. Laukhin, and X. Obradors, *Appl. Phys. Lett.* 72, 2607 (1998).

[164] J. Gutiérrez, A. Peña, J. M. Barandiarǎn, J. L. Pizarro, T. Hernǎndez, L. Lezama, M. Insausti,and T. Rojo, *Phys. Rev.* B 61, 9028 (2000).

[165] N. Ghosh, S. Elizabeth, H. L. Bhat, U. K. Rössler, K. Nenkov, S. Rössler, K. Dörr and K. H. Müller, *Phys. Rev.* B 70, 184436 (2004).

[166] T. Motohashi, R. Ueda, E. Naujalis, T. Tojo, I. Terasaki, T. Atake, M. Karppinen, and H. Yamauchi, *Phys. Rev.* B 67, 064406 (2003).

[167] J. Sugiyama, H. Itahara, J. H. Brewer, E. J. Ansaldo, T. Motohashi, M. Karppinen, and H. Yamauchi, *Phys. Rev.* B 67, 214420 (2003).168

[168] Terasaki, Y. Sasago, and K. Uchinokura, *Phys. Rev.* B 56, R12685 (1997).169.

[169] K. Takada, H. Sakurai, E. Takayama-Muromachi, F. Izumi, R. A. Dilanian, and T. Sasaki, *Nature* 422, 53 (2003). 170

[170] R. E. Schaak, T. Klimczuk, M. L. Foo, and R. J. Cava, *Nature* 424, 527 (2003).

[171] H. Sakurai, K. Takada, S. Yoshii, T. Sasaki, K. Kindo, and E. Takayama-Muromachi, *Phys. Rev.* B 68, 132507 (2003).

[172] G. Cao, C. Feng, Y. Xu, W. Lu, J. Shen, M. Fong, and Z. Xu, *J. Phys. : Condens. Matter* 15, L519 (2003).

[173] R. Jin, B. C. Sales, P. Khalifah, and D. Mandrus, *Phys. Rev. Lett.* 91, 217001 (2003).

[174] B. Lorenz, J. Cmaidalka, R. L. Meng, and C. W. Chu, *Phys. Rev.* B 68, 132504 (2003).

[175] K. Ishida, Y. Ihara, Y. Maeno, C. Michioka, M. Kato, K. Yoshimura, K. Takada, T. Sasaki, H. Sakurai, and E. Takayama-Muromachi, Cond.-mat/0308506 (2003).

[176] W. Higemoto, K. Ohishi, A. Kodo, R. Kadono, K. Ishida, K. Takada, H. Sakurai, E. Takayama-Muromachi, and T. Sasaki, cond-mat/0310324 (2003).
[177] G. Baskaran, *Phys. Rev. Lett.* 91, 097003 (2003).
[178] B. Kumar and B. S. Shastry, *Phys. Rev.* B 68, 104508 (2003).
[179] C. Honerkamp, *Phys. Rev.* B 68, 104510 (2003).
[180] Q -H. Wang, D -H. Lee, and P. A. Lee, *Phys. Rev.* B 69, 092504 (2004).
[181] A. Tanaka, and X. Hu, *Phys. Rev. Lett.* 91, 257006 (2003).
[182] D. J. Singh, *Phys. Rev.* B 61, 13397 (2003); 68, 020503 (2003).
[183] H. Ikeda, Y. Nisikawa, and K. Yamada, *J. Phys. Soc. Jpn.* 73, 17 (2004).
[184] M. M. Maska, M. Mierzejewski, B. Adrzejewski, M. L. Foo, R. J. Cava, and T. Klimczuk, *Phys. Rev.* B 70, 144516 (2004).
[185] S. Lupi, M. Ortolani, and P. Calvani, *Phys. Rev.* B 69, R180506 (2004).
[186] C. J. Milne, D. N. Argyriou, A. Chemseddine, N. Aliouane, J. Veira, S. Landsgesell and D. Alber, cond-mat/0401273 (2004).; M. L. Foo, Y. Wang, S. Watauchi, H. W. Zandbergen, T. He, R. J. Cava, and N. P. Ong, *Phys. Rev. Lett.* 92, 247001 (2004)
[187] Y. Kobayashi, M. Yokoi, and M. Sato, *J. Phys. Soc. Jpn.* 72, 2161(2003).
[188] M. N. Iliev, A. P. Litvinchuk, R. L. Meng, Y. Y. Sun, J. Cmaidalka, and C. W. Chu, *Physica* C 402, 239 (2004).
[189] P. Lemmens, V. Gnezdilov, N. N. Kovaleva, K. Y. Choi, H. Sakurai, E. Takayama-Muromachi, K. Takada, T. Sasaki, F. C. Chou, D. P. Chen, C. T. Lin, and B. Keimer, *J. Phys. : Condens. Matter*, 16, S857 (2004).
[190] N. L. Wang, P. Zheng, D. Wu, Y. C. Ma, T. Xiang, R. Y. Jin, and D. Mandrus, cond-mat/0312630 (2004).
[191] A.T. Boothroyd, R. Coldea, and D. A. Tennant, D. Prabhakaran, L. M. Helme, and C. D. Frost, cond-mat/0312589 (2004).
[192] J. W. Lynn, Q. Huang, C. M. Brown, V. L. Miller, M. L. Foo, R. E. Schaak, C. Y. Jones, E. A. Mackey, and R. J. Cava, *Phys. Rev.* B 68, 214516 (2003).
[193] Z. Li, J. Yang, J. G. Hov, and Q. Zhu, *Phys. Rev.* B 70, 144518 (2004).
[194] J. D. Jorgensen, M. Avdeev, D. G. Hinks, J. C Burley, and S. Short, *Physical Rev.* B 68, 214 517 (2004).

In: Computational Materials
Editor: Wilhelm U. Oster

ISBN: 978-1-60456-896-7

Chapter 5

DIFFUSION MECHANISMS NEAR TILT GRAIN BOUNDARIES IN NI, CU, AL AND NI_3AL

***G.M. Poletaev*[1 1], *M.D. Starostenkov*[1] *and S.V. Dmitriev*[2 2]**
[1] Altay State Technical University, Russia
[2] Institute for Metals Superplasticity Problems RAS, Russia

With the use of the molecular dynamics technique the diffusion mechanisms along tilt grain boundaries <111> and <100> are investigated in pure metals Ni, Cu, Al and in Ni3Al intermetallide. The following three basic mechanisms of grain boundary diffusion were revealed: migration of atoms along the cores of grain boundary dislocations, cyclic mechanism near the core, and the formation of the chains of atoms displaced from the core of one dislocation to the core of the other one. The density of steps at grain boundary dislocations strongly affects the probability of the realization of all three mechanisms. Temperature and misorientation angle also were found to be important factors. Main peculiarity of grain boundary diffusion in Ni3Al intermetallide is related to the fact that it occurs mainly by the displacements of Ni atoms over their sublattices in L12 superstructure. As a result, the short-range order is almost preserved. At high temperature Al atoms start to participate in the migration process. Although their displacements are small in comparison to the displacements of Ni atoms, they cause an essential reduction of a superstructural order in the alloy.

INTRODUCTION

Metals are normally used in polycrystalline form and grain boundaries constitute an integral part of their structure. Grain boundaries influence greatly many physical and mechanical properties of metals such as plasticity, diffusion, creep, processes of recrystallization, failure, melting, and so on. In spite of the existence of many works devoted

[1] Lenin St. 46, Barnaul 656038, Russia, genphys@mail.ru.
[2] Khalturina St. 39, Ufa 450001, Russia.

to grain boundaries, there still exist a number of problems related to grain boundary structure and structural changes near grain boundaries under heat treatment and deformation.

Kinetic properties and especially the diffusion mobility of atoms are the most developed parts of grain boundary physics having the longest history and many important application. From the one hand, diffusion is closely related to the structure of grain boundaries and it is a universal tool in the study of structure and the mechanisms of atomic displacements near the grain boundaries. From the other hand, diffusion determines kinetics of more complex processes, e.g., the ones related to the relaxation of stress in the plastically deformed materials.

The two main problems in the study of the diffusion in materials containing grain boundaries are (i) to determine the coefficients of grain boundary diffusion and (ii) to identify the main diffusion mechanisms. The main difficulty in solving the first problem is separation of contributions from the grain boundary diffusion and the bulk diffusion. The second problem is difficult to solve because one cannot observe experimentally atomic displacements in dynamics; only their initial and final positions can be seen.

The history of the research of diffusion over grain boundaries can be divided into three stages. At the first stage, diffusion process was studied by measuring the penetration depth of diffusing atoms as the function of grain size of a polycrystalline material. The second stage is characterized by the research of diffusion along grain boundaries in polycrystals. At this stage, diffusion along grain boundaries was estimated by averaging over all grain boundaries of different types. That is why, they characterized not individual grain boundary with its peculiarities, but averaged influence of all grain boundaries on the kinetics of diffusion processes in polycrystal materials. During the third period, covering last three decades, diffusion experiments have been carried out for bicrystals with single grain boundary with particular crystallographic, geometric, and chemical properties.

The first experiments devoted to diffusion over grain boundaries in metals have been done in 1922. Studying self-diffusion of radioactive isotopes of lead, Khiveshi and Obrucheva found that diffusion velocity in a lead foil with a small value of grain size is considerably larger than that in coarse-grained lead molding [1]. Geiss and Van-Limpt obtained analogous results studying interdiffusion of molybdenum and tungsten [1]. Bulgakov and his co-workers have established the first trends in the grain boundary diffusion at the beginning of thirties. Measuring diffusion coefficients D of zinc in a polycrystal copper with small grains, it was found that linear dependence between lnD and reverse temperature T^{-1} was violated in a broad interval of temperatures (i.e. Arrhenius law is violated). Experimental points deviated from the straight line lnD (T^{-1}) at the temperatures less than $(0.6\text{-}0.7)\cdot T_{melt}$, where T_{melt} is melting temperature [2]. This deviation is explained by the peculiarity of diffusion contribution over grain boundaries into a general diffusion flux: the coefficient of boundary diffusion is bigger than that of the bulk one, but the activation energy is smaller, because it leads to the deviation from the Arrhenius law at relatively low temperatures. However, the mechanism of the appearance of such deviation has not been explained yet. At present, it is explained by the contribution from grain boundary migration which becomes noticeable for temperature about $0.6\text{-}0.7\cdot T_{melt}$ [1].

It was very difficult to determine grain boundary diffusion flux in comparison with the bulk diffusion flux relying on the Fick's law. In this connection, it was necessary to develop mathematical apparatus allowing for a correct separation of two fluxes, along the boundaries and in the bulk. In 1951, Fisher [3] suggested the model describing main physical features of the process of grain boundary diffusion correctly: substances pass the diffusion ahead over the boundary and then they were withdrawn from the boundary into the volume. In Fisher model [3], concentration profile had the form of "diffusion wedge". Fisher studied diffusion in a bicrystal. In any case, grains in his model were so big that diffusion flows from neighboring grains did not cross. The boundary in the model was presented by a uniform isotropic plate of width δ (δ was equal to 5Å in the calculations) located perpendicular to the surface between two semi-infinite grains. The obtained equations are rather awkward but they allow to determine main diffusion parameters under certain assumptions.

At present, exact solutions of Fisher's model are available and their asymptotic expansions make it possible to reproduce various experimental conditions, research methods, and the way of the treatment of experimental results properly. In all similar experiments, the product of diffusion coefficient D and boundary width δ is calculated. That is why main difficulty is a correct determination of the second value. To define diffusion characteristics, concentration curves are usually found using the method of local x-ray analysis, curves of layer activity (by radioactive isotopes) [4]. Recently, the methods having high space resolution – mass spectroscopy of secondary ions, auger spectroscopy and x-ray photoelectron spectroscopy has become very popular in the research of narrow diffusion zones [4].

The research of diffusion over single grain boundaries is of great interest because it allows finding the connection of diffusion processes and boundaries structure. In early experiments, made in fifties of the last century, diffusion anisotropy was studied. The penetration of migrating atoms in parallel and perpendicular directions to the axis of boundary tilt, and also boundary penetration or the coefficient of boundary diffusion in the dependence on misorientation angle of neighboring grains were analyzed.

Hoffman and Turnbull [5] studied self-diffusion of silver along and across of symmetrical boundaries with tilt axis $\langle 111 \rangle$. It was found that diffusion anisotropy decreased smoothly with the increase of misorientation angle, but it was observed up to maximum angle 45°. Analogous measurings were made by Kuling and Smoluchowski [4, 6]. They studied zinc diffusion along and across tilt boundaries $\langle 111 \rangle$ in copper by autoradiography method. The biggest anisotropy was seen at small misorientation angles. Lange and Jurish measured diffusion velocity over tilt boundaries $\langle 111 \rangle$ in Al bicrystal. They also found that anisotropy took place at all angles, but it decreased with the increase of the angle [4]. All those results indicate on crystal structures of boundaries, otherwise diffusion would be isotropic one.

Experiments devoted to the research of orientation dependence of penetration depth showed that the depth, and consequently, the coefficient of boundary diffusion depended on misorientation angle of neighboring grains [7-13]. For the first time, such studies were made by Achter and Smoluchowski [4]. Studying diffusion of copper and silver, they have not found advantageous penetration over symmetric boundaries at misorientation angles $\theta<10°$ and $\theta>70°$. In the interval from 10° to 70° the depth of penetration along the boundary exceeded the depth of volume diffusion. The diffusivity increased with the increase of θ, reached maximum at 45°, then it decreased to zero at 70°. Smoluchowski suggested the model called "condensation of dislocations" in a large-angle boundary at the increase of misorientation angle θ on the basis of those results [6].

Upergrov and Sinnot observed monotonous increase of depth penetration with the increase of misorientation angle studying Ni self-diffusion over symmetric boundaries with tilt axis ⟨111⟩. The maximum increase in the area 45° and the following decrease to volume level at the angles of about 90° were seen [13]. Later, additional minimums at some misorientation angle were found by Heinson, Smoluchowski, Arkharov, Jukava and others [14]. The presence of minimums on orientation dependence of diffusion permeability is important evidence in the favor of structural models, lattices of coinciding knots and others.

We can find many experimental data concerning not only polycrystal materials but also bicrystals with attested grain boundary. All these data testifies to a rapid diffusion over intergrain boundary: diffusion coefficients in volume and boundary differ in several orders. Anisotropy of grain boundary diffusion is also very important result that testifies to crystal structure of grain boundaries.

The considered results of the determination of diffusion coefficients and activation energy give quantitative description of grain boundary processes and show the connection between diffusion parameters and boundaries structure clearly, but they don't allow solving the problem on the leading mechanisms of grain boundary diffusion and their difference from the mechanisms acting in the volume of grains.

Studying the structure of grain boundaries in the limits of dislocation model, Turnbull and Hoffman suggested the "pipe" diffusion mechanism [4]. It can be seen from their model that the lattice between dislocations is deformed, but it is relatively perfect, and its diffusion permeability is nearly the same as in a perfect lattice. Dislocation cores (pipes) are disordered very much and they are characterized by higher diffusion coefficient. Thus, the authors presented an intergrain boundary not in a form of place of uniform width with constant diffusion coefficient (as for example, in Fisher's pioneer model [3]) but in a form of "tubes" set with definite squares of cross section, located in one plane at some distance from each other. The distance can be calculated if we know the angle of grains misorientation. The model of Turnbull and Hoffman was proved experimentally for low-angle grain boundaries in Ag [4].

The models explaining the peculiarities of diffusion processes in dislocation core were suggested by Loze, Vever and others [4]. They introduced the notions "vacancy and interstitial atoms in dislocation cores". To their opinion, vacancy - interstitial atom pairs appear in the process of thermal fluctuations in dislocation cores. When interstitial atom appeared in the cores recombines with "another's" vacancy (not with the vacancy appeared in the cores simultaneously), the resulting diffusion flow directed along the cores is observed in dislocation cores. When interstitial atom recombines with "native" vacancy and when the complex vacancy-interstitial atom does not take part in a general movement, diffusion flow is not seen. The authors point out that diffusion stipulated by the movement of vacancies or by interstitial atoms can prevail in the dependence on the properties of metal.

More difficult problem is the determination of diffusion mechanism between dislocation cores. High values of diffusion coefficient are also observed along the given direction [4, 14]. Smoluchowski and Li tried to explain high diffusion permeability in the direction perpendicular to dislocations "pipes" by the unification of dislocation cores at large angles of grains misorientation [4]. But their model contradicts experimentally observed anisotropy of boundary diffusion at any angles of grains misorientation.

In recent works by Sorensen, Mishin, Suzuki, Farkas and others [15-20], based on computer simulations, the research of diffusion mechanisms over grain boundaries is made mainly by the evaluation of activation energy of atomic jumps in different directions. These papers mainly deal with special and symmetric grain boundaries. They have relatively low diffusion permeability because of their high structural ordering (density of coincident sites). It is very difficult to simulate diffusion in such boundaries by molecular dynamics – it requires high temperature and long duration of the experiment. But the authors of works [15-20] did not try to simulate diffusion over grain boundaries in dynamics. The papers are devoted to the search of activation energy of single atomic jumps in different directions by vacancy and interstitial mechanisms in the area of the considered grain boundaries. The authors made suppositions concerning the mechanism of grain boundary diffusion using the obtained data. One of the questions extensively discussed in frame of these works is about the dominating mechanism, vacancy or interstitial, for the grain boundary diffusion. Different authors give different answers to this question. For example, Liu and Plimpton [20] studied energies of vacancy migration and interstitial atom migration in special boundaries Σ5(310)[001] and Σ13(320)[001] in Ag by molecular static (for medium temperatures) and molecular dynamics (for high temperatures). They have concluded that the interstitial diffusion mechanism takes place at moderate temperatures, while vacancy diffusion mechanism dominates at high temperatures. However, this result was not approved in the other works, such as papers by Sorensen, Mishin, Suzuki [15, 16], who spoke about the domination of vacancy diffusion.

Suzuki and Mishin [16] considered that migration of a vacancy or an interstitial atom along grain boundary often had long-period character, i.e. it included several atomic jumps taking place “without stop” simultaneously. The authors did not find the cause of the appearance of such cooperative atomic motion, they only related it with the instability of point defects having interstitial position in the chain. In papers [15-20], main mechanisms of grain boundary diffusion were not described, only energetical characteristics of single atomic jumps were described in details. Thus, any explanation for the anisotropy of grain boundary diffusion and for the deviation from the Arrhenius dependence at high temperatures were not offered.

In the works [21-24] the diffusion along non-equilibrium grain boundaries has been analyzed numerically. With the help of molecular quasi-static approach, the vacancy formation and activation energies have been computed for a set of typical non-equilibrium grain boundaries under tensile stress. In some cases dislocations and disclination dipoles were introduced in the grain boundaries. Tilt as well as general-type grain boundaries have been analyzed. It has been demonstrated that internal stresses due to various defects introduced in the grain boundary structure can increase the diffusivity along grain boundaries by three to four orders of magnitude.

It can be seen from the above presented review that the microscopic theory of diffusion processes near grain boundaries still has a lack of understanding of the grain boundary diffusion mechanisms at an atomic level. Mathematical models of diffusion processes have, as a rule, a number of assumptions and contain empirical constants. Often they are based on presumed micromechanisms of diffusion. Besides, diffusion mechanisms taking place along and across the dislocation cores, diffusion from the grain boundary inside the grains, the role of grain boundary migration in diffusion process, and the reason of the deviation from the Arrhenius law for grain boundary diffusion have not been explained. All above mentioned problems require further molecular dynamics studies of the grain boundary structure and dynamics at an atomic level.

Diffusion processes in alloys are far more complicated in comparison to that in pure metals. This is so because the variety of point defects and the number of mechanisms of their migration in allows is greater than in pure metals. There are special classes of alloys, e.g., ordered alloys and intermetallides, that have a number of distinct properties. One of the most intriguing and very important for applications property is positive temperature dependence of yield stress. Such behavior is typical, e.g., for Ni_3Al intermetallide [25]. Besides, Ni-Al system is the basis of different superalloys. Physical and mechanical properties of intermetallides depend on the type of superstructure and on the degree of atomic ordering.

The objective of the present work is the investigation of the diffusion mechanisms along tilt grain boundaries in the pure metals Ni, Cu, Al and in the Ni_3Al intermetallide by the method of molecular dynamics.

THE COMPUTER MODEL

Tilt boundaries with misorientation axes <111> and <100> were chosen in our study of diffusion along grain boundaries. Such choice is stipulated by the prevalence of the planes of grain boundaries with small indices in metals (in the given case {110} and {100}) [25-29]. We consider tilt boundaries with relatively simple structure, rather than the boundaries of mixed type, for the sake of simplicity.

Many researches, for example, Hofmann, Wolf, MacLaren, Vitek, and others [30-38] consider that the use of many-body potentials for the description of interatomic interactions does not lead to qualitative change of the results of computer simulation and new data in comparison with calculations made upon using pair potentials. The errors appearing during the obtaining of different characteristics (especially energetical) by both types of potentials are not very different that is stipulated by two reasons. The first reason is that energetic characteristics of metal used for the choice of potentials parameters have errors themselves not less (sometimes bigger – several dozens, even hundreds percents) than the difference provided by the use of different potentials. The second cause: the technique of parameterization of many-body potentials has a drawback connected with the absence of reliable criterion evaluating the contribution of many-body component for the calculation of potentials parameters. Cauchy pressure $\frac{1}{2}(C_{12}-C_{44})$ is usually such criterion (all other fitting characteristics can be described by pair component). But it is impossible to fit elastic modules ideally and Cauchy pressure has, as a rule, big errors. Besides, this criterion corresponds to an ideal crystal lattice. That is why, the contribution of many-body component in case of non-ideal structure, for example, near a defect can not be evaluated for sure. In this connection many-body component can have overestimated or underestimated contribution. In the first case, many-body potentials have the behavior similar to the behavior of pair potentials. In the second case, the structure of simulated metals tends to the increase of coordination number (local density) sometimes to the detriment of long order of crystal because of an excessive contribution of many-body component, as in paper [39].

To study main regularities of the dynamics of atomic structure, and diffusion mechanisms, pair potentials can be used. To simulate diffusion in the calculated blocks containing large number of atoms, highly efficient calculations are necessary for the conduction of long-term computer experiments. In this case, pair potentials have an essential advantage, because they require less expenses of computer time in comparison with many-body potentials.

To describe interatomic interactions, pair central Morse potentials were used in the work:

$$\varphi(r) = D\beta e^{-\alpha r}\left(\beta e^{-\alpha r} - 2\right), \quad (1)$$

where α, β, D are the potential parameters; r is the distance between two atoms. The parameters of Morse potentials for pure metals were determined taking into account five coordination spheres over the lattice parameter, bulk modulus, sublimation energy. Besides, the approbation of the potentials was made on temperature coefficient of linear expansion, velocities of longitudinal and transversal elastic waves. The obtained values agree with experimental data satisfactory.

The radius of the potentials acting included five coordination spheres. Time integration step in the equations of motion was equal to 0.01 ps (10^{-14} s).

The temperature in molecular-dynamic model was given by initial velocities of atoms in correspondence with Maxwell's distribution. Initial velocities were given to be equal in absolute value but having arbitrary directions. The sum kinetic energy corresponded to the given temperature, but the total impulse of the calculated cell was equal to zero:

$$\left|\mathbf{v}_i\right| = v_{sq}\sqrt{2} = \sqrt{\frac{2\xi kT}{m_i}}, \quad \sum_{i=1}^{N} m_i \mathbf{v}_i = 0 , \quad (2)$$

where k is the Boltzman constant, T is temperature, v_{sq} is the root-mean-square velocity of atoms, ξ is model dimensionality, m_i – mass of i-th atom, N – number of atoms in the computational cell. Initial distribution of velocities in the process of the computer experiment approaches to Maxwell distribution rapidly (during several iterations) [40, 41].

The present work makes several demands to molecular-dynamic models of grain boundaries. Firstly, the calculated block should have one boundary with definite crystallographic parameters at least in the beginning of the experiments. Secondly, the calculated block should be as big as possible. It influences on the certainty of the results, their statistics, the accuracy of measuring of different parameters connected with grain boundary. Big sizes of the calculated block are necessary to observe the processes and phenomena covering nanometer scale (for example, cooperative displacements of atoms and shears). Besides, big sizes of the calculated block decrease the influence of the conditions applied on the block boundary. The third demand to the model is keeping of grain boundary during all the experiment in a central area of the calculated block to avoid it's migration outside the block. The dynamics of the structure is directed to decrease of free energy. Every molecular-dynamic experiment would finish with the disappearance of intergrain boundary from the calculated block leading to the decrease of energy if the third condition is not satisfied.

The stage of the preparation of initial structures of the calculated blocks containing different grain boundaries preceded main molecular-dynamic experiments. The preparation included the construction of the calculated block with grain boundary having the given parameters, dynamic relaxation of the block structure with its reduction into the state with minimum energy and the following cooling.

Tilt grain boundary was made in the middle of the calculated block by the turn of two crystals (two parts of the block) to the angles of misorientation θ and β around the axes <111> or <100>. Angle β defined the orientation of one grain relatively the boundary, θ -

the angle of grain misorientation (Figure 1). The obtained calculated block was cut at the edges so to have the form of parallelepiped did not have cavities at the edges. The atoms located outside the boundary of the parallelepiped were removed. Then the atoms located outside the line of intergrain boundary in the area of the other grain were removed if they were nearer than $0.7 \cdot r_1$, where r_1 – radius of the first coordination sphere. The given minimum interatomic distance at the conjugation of grains was the most optimal, intergrain boundaries in that case had the minimum energy. It is not a critical condition in the method of molecular dynamics, because the method allows to make the most realistic dynamical relaxation (at the big number of atoms contained in the calculated block) leading the system to the most energetically profitable conjunction of grains.

Primary dynamical relaxation was made at the initial temperature 0 K during 10-20 ps after the procedure of removing of unnecessary atoms. In this connection, atoms displaced in the positions corresponding to the minimum of energy that was accompanied by insignificant heating of the calculated block. Rapid cooling of the calculated block to 0 K was made after the stabilization of temperature. Then, the calculated block was subjected to insignificant overall deformation of tension or pressure to the reaching of energetical minimum to avoid surplus internal tensions in grains (interatomic distances were changed, as a rule, in less than 0.1%). Dynamic relaxation and the following cooling were made once again. The obtained calculated blocks were used in main computer experiments as start ones.

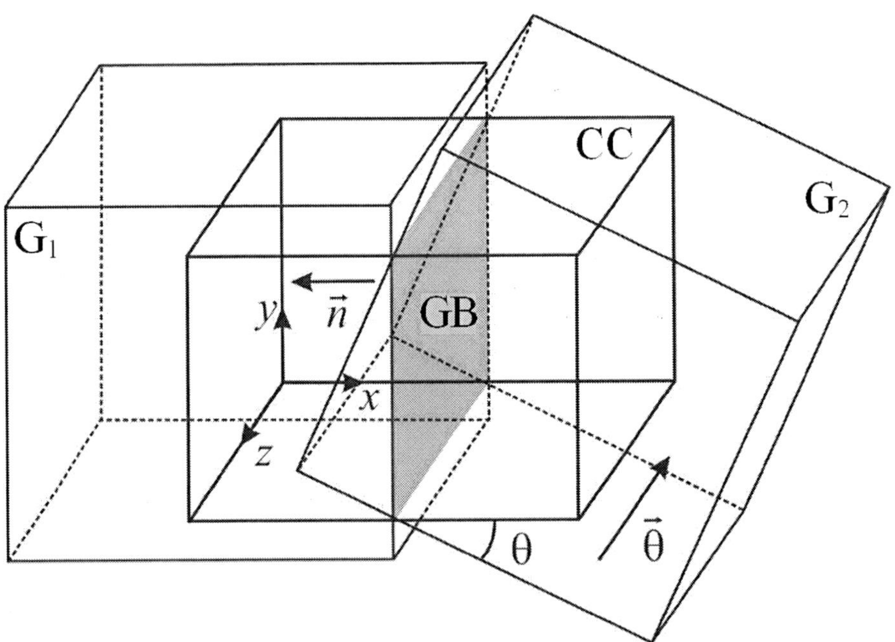

Figure 1. Schematic of the construction of the three-dimensional computational cell containing a tilt grain boundary. G1 and G2 denote the grains, CC denotes the computational cell, GB is grain boundary, $\vec{\theta}$ is the vector of grain misorientation, $\vec{n}$ is the unit vector normal to the GB.

The number of atoms in three-dimensional calculated blocks ranged from 20000 to 50000. The following coordinate system was introduced (Figure 1): axis X was directed perpendicular to the plane of intergrain boundary inside a grain; axis Y along the grain boundary and perpendicular to the axis of tilt; axis Z – along the axis of tilt. Rigid conditions were applied along axes X and Y, and along Z – periodical conditions. Thus, the structure of the sides of the calculated block parallel to the axis of grains tilt was fixed. That is why, intergrain boundary was not able to migrate outside the limits of the block in the process of main molecular-dynamic experiment. But rigid boundary conditions were able to influence on the dynamics inside the block because of the fixation of structure near the sides of the calculated block. The sizes of the calculated block were chosen to be maximum possible (with the length of the block sides of 10 nm order) to avoid the influence of rigid conditions.

The duration of molecular-dynamic experiments was equal to 0.2-0.3 ns in the process of determination of diffusion coefficients and the temperature of the calculated block was stable. The cooling of the block to 0 K was made to avoid thermal displacements of atoms in the final stage of the experiment. Then, diffusion coefficients were calculated by intergrain boundary along the directions X, Y and Z. Diffusion coefficients were determined for small cells with the sizes 5x5x5 Å by the formulas

$$D_{xj} = \frac{\sum_{i=1}^{N_j}(x_{0i} - x_i)^2}{2Nt};\ D_{yj} = \frac{\sum_{i=1}^{N_j}(y_{0i} - y_i)^2}{2Nt};\ D_{zj} = \frac{\sum_{i=1}^{N_j}(z_{0i} - z_i)^2}{2Nt}, \tag{3}$$

where D_{xj}, D_{yj}, D_{zj} – the diffusion coefficients along axes X, Y and Z in i-th cell, N_j – the number of atoms in j-th cell; x_{0i}, y_{0i} and z_{0i} – the coordinates of an initial position of i-th atom; x_i, y_i and z_i – coordinates of i-th atom at the moment of time t. The position of the cell was changed to 1 Å along X (perpendicular to the plane of an intergrain boundary), when two other coordinates were stable. Then one of coordinates Y or Z of the cell was changed to 1 Å, coordinate X of the cell was changed again. Thus, the whole calculated block was studied. The maximum values of the diffusion coefficients D_{xj}, D_{yj} and D_{zj} were calculated at every change of coordinate X of the cell. Then, the arithmetic means of the values D_x, D_y, D_z for all displacements along X were determined. The average values of the maximum diffusion coefficients corresponded to diffusion over a grain boundary.

The study of boundaries structure was made after the construction of the calculated blocks and the dynamical relaxation. The visualizators of potential energy and local stresses were used for a pictorial representation of the structure. In this connection, atoms were plotted in different grades of grey color in correspondence with the module of the change of bond energy in comparison with the bond energy of an ideal crystal. The atoms which bond energies were less than 0.5% in comparison with an ideal crystal were not shown.

RESULTS

The boundary structures were studied after the construction of calculated blocks containing grain boundaries. Dislocations in low-angle tilt boundary <111> θ=9 are shown in the section XY in Figure 2 (a). Dislocations of the given types of boundaries are united in pairs and they represent top dislocations. Their Burger's vector is equal to ½<110>. Similar structure of low-angle grains has an experimental confirmation. Figure 2 (c) shows the photo obtained by electronic microscope [42] low-angle grain boundary θ=9° in Cu. Top (60-

degree) grain boundary dislocations which are similar to the dislocations obtained in the model can be seen in the photo.

More complex form of top dislocations is typical for boundaries <100> (Figure 2 b). Top dislocations consist of three partial dislocations with total Burger's vector ½<110> in the case of small angles of misorientation ($\theta \leq 7^{o}$). Top dislocations include four dislocations at more high angles of misorientation, Burger's vector has the value 1<100> (Figure 2 b). Such dislocations are more stable to external influences than dislocations in boundary <111>.

The probability of the appearance of diffusion, as it was found, was stipulated by the presence of steps on dislocations. Figure 3 shows the position of dislocation cores in metals with low-angle tilt boundaries <111> and <100> using the visualizator of potential energy. As it can be seen, the distance between steps in boundaries <111> is not equal, it is determined by a crystallographic packing of such boundaries. It is worth to note that the density of steps on dislocations in the boundary <100> is higher than in the boundaries <111>.

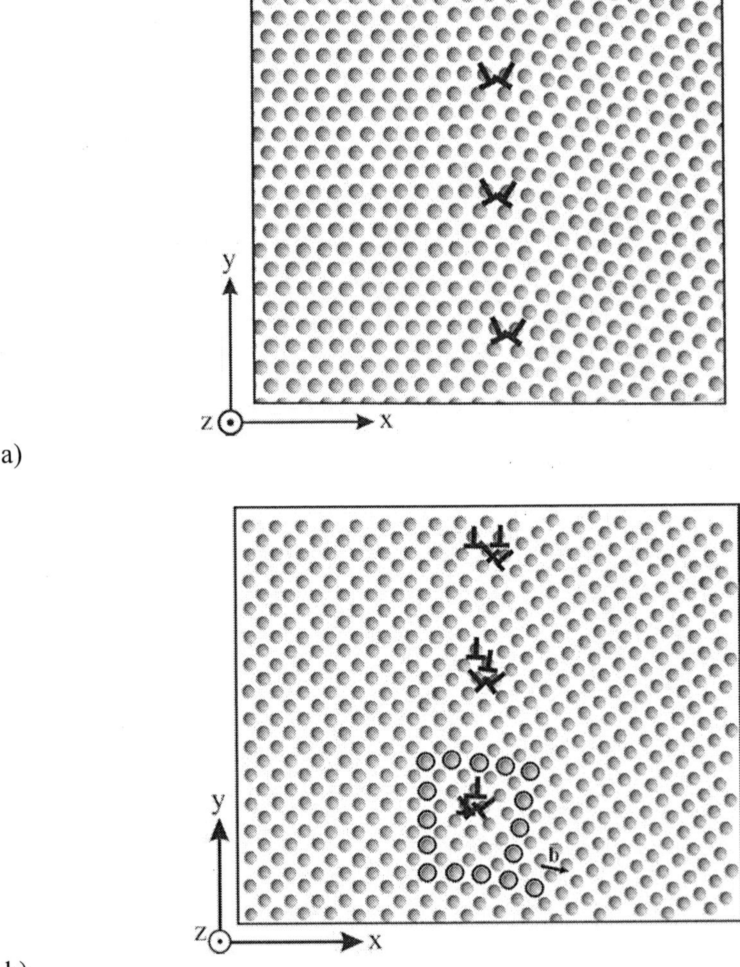

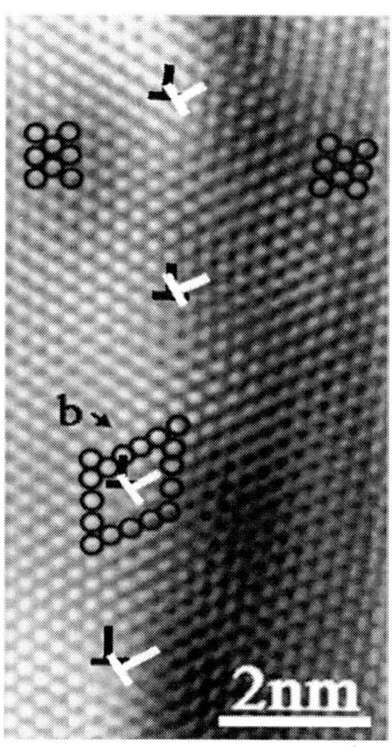

c)

Figure 2. Grain boundary dislocations in Cu projected on XY plane: a) the <111> θ=9o tilt boundary; b) <100> θ=12° boundary; c) the high-resolution electron microscopy image of the <111> θ=9°boundary reported in [42] (not all of the grain boundary dislocations are indicated in this image). $\vec{b}$ is the Burger's vector.

The structure of grain boundary dislocations could change as the result of thermoactivation. The reconstructions of dislocations structure connected mainly with the decrease of steps density were observed for the most dislocations in low-angle boundaries <111> at high temperatures. Dislocations in boundaries <100> appeared to be stable and changed their structure considerably more rare. Partial unification of dislocation cores, i.e. the areas of stresses in some places along the cores of two neighboring dislocations were observed at the increase of the angle of grains misorientation.

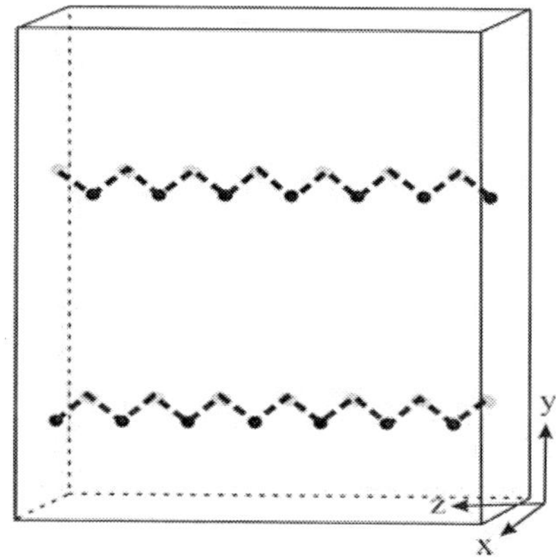

a)

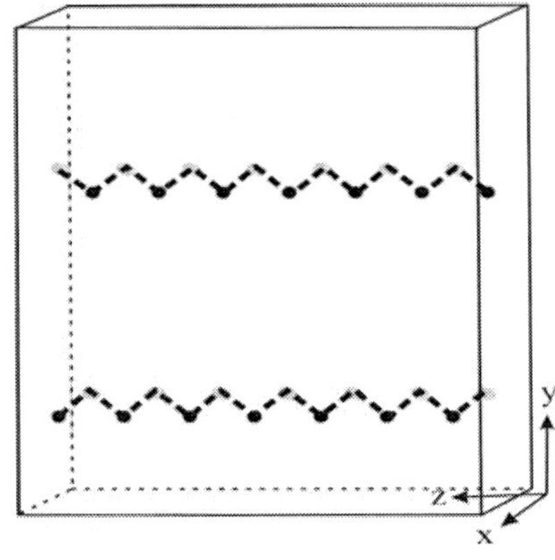

b)

Figure 3. The steps on grain boundary dislocations in boundaries <111> (a) and <100> (b). The positions of dislocation cores marked by the dotted line. The figure shows only the atoms having the values of potential energy which are nearly equal to maximum values.

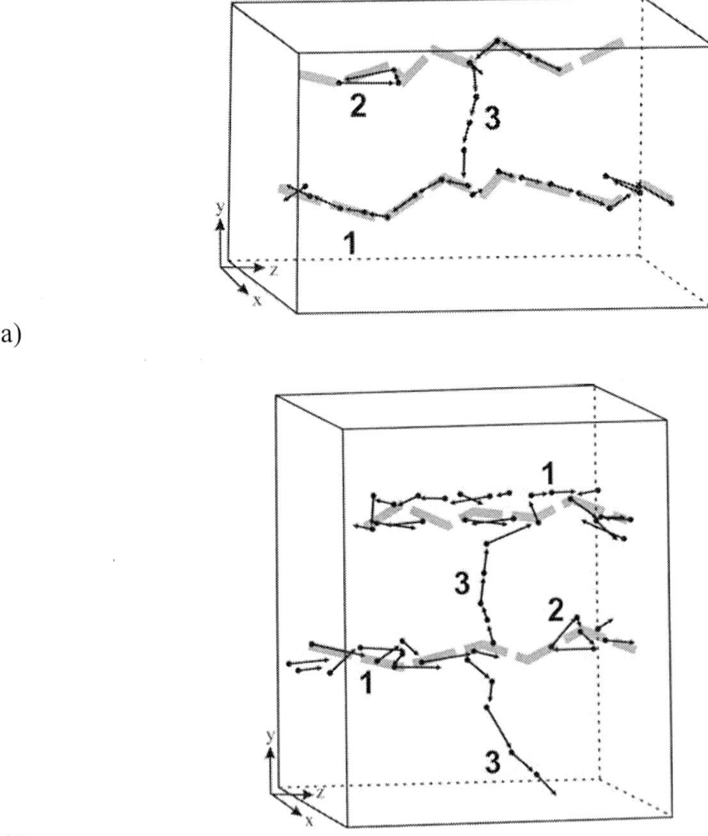

a)

b)

1) migration of atoms along the cores of grain boundary dislocation; 2) cyclic mechanism near the cores of dislocation; 3) the formation of the chain of the displaced atoms from one cores to the other one. Positions of dislocation cores are shown by heavy grey lines.

Figure 4. The examples of diffusion mechanisms over grain boundaries of tilt <111> (a) and <100> (b).

To determine diffusion mechanisms, molecular-dynamic experiments were made. They were being carried out during 100-300 ps at different temperatures and different angles of grain misorientation. The regulations of atomic migration were studied during the experiments. Three main mechanisms of grain boundary diffusion in metals Ni, Cu, Al and intermetallide Ni_3Al with grain boundary diffusion of tilt <111> and <100> were found (Figure 4): migration of atoms along the cores of grain boundary dislocations (1), cyclic mechanism near the core (2), the formation of the formation of the chain of the displaced atoms from one cores to the other one (3).

The migration of atoms along the cores of dislocations in boundaries <111> (mechanism 1, or "tube" mechanism) was observed even at small temperatures and short duration of the experiment. The chain of displaced atoms along the cores of dislocation was formed in that case. In this connection, elementary act was in successive displacement of atoms from one step of dislocation to the other one. "Tube" mechanism was observed at all angles of misorientation and in the biggest range of temperatures being the most probable mechanism.

Cyclic mechanism (2) was also initiated near the cores of dislocations. The mechanism was the consequence of the above mentioned mechanism (1). The number of the displaced atoms in cyclic mechanism could be equal from three to several dozens. The probability of the appearance of cyclic mechanisms including a big number of atoms was increased with the increase of temperature. Cyclic mechanism took place at big temperatures (as a rule, bigger than $0.6\text{-}0.7 \cdot T_{melt}$, were T_{melt} – melting temperature).

Diffusion mechanism consisting in migration of atoms from one dislocation to the other one (3) was observed mainly at high temperatures and big angles of misorientation, i.e. density of dislocations. In the simpliest case, the mechanism represented the chain of the displaced atoms from the step of one dislocation to the step of the other one. The probability of the realization of the mechanism was increased with the increase of misorientation angle θ and temperature. The cores of grain boundary dislocations during the transition of low-angle boundaries to large-angle ones were applied, and the mechanism took place in combination with the first mechanism and had nearly the same probability of realization.

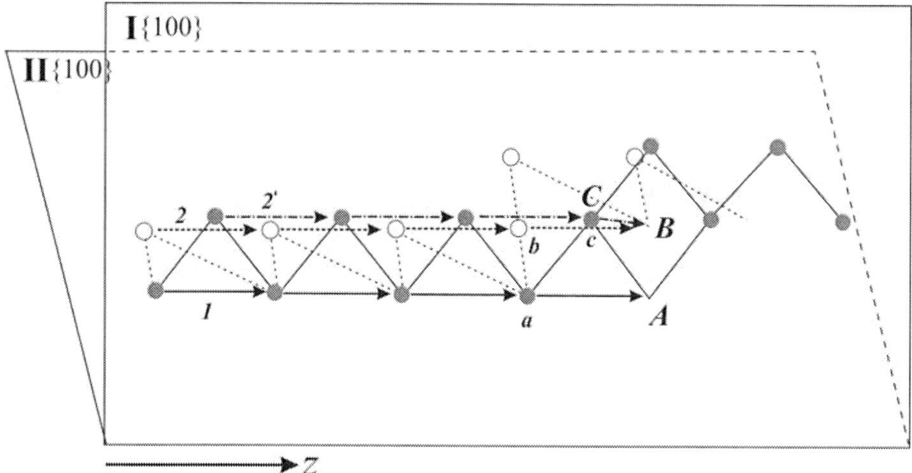

Figure 5. The scheme of the formation of the chains of displaced atoms along dislocations in boundaries <100>. Atoms located in *I* extraplane of grain boundary dislocation are shown by dark circles, in *II* – by light circles.

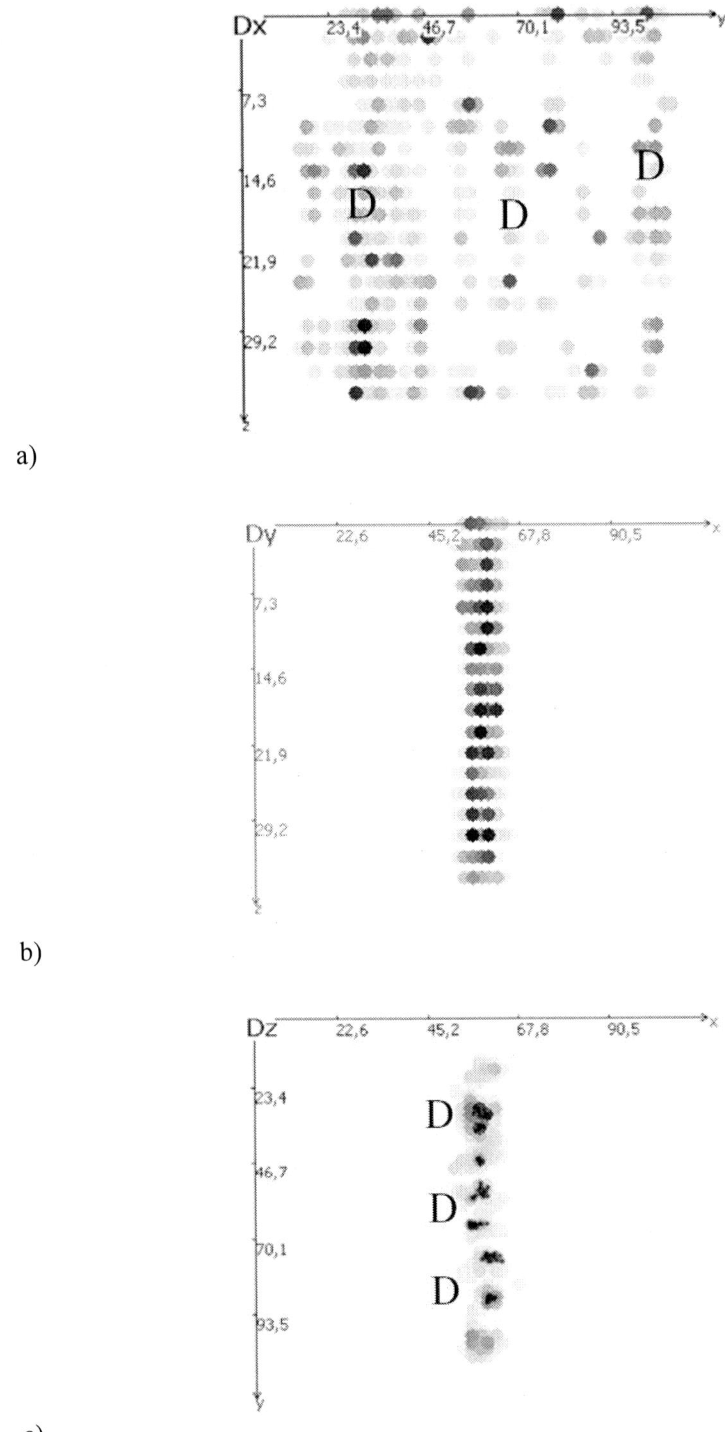

Figure 6. The distribution of diffusion intensity in Ni with grain boundary <111> θ=15° projected on the planes (a) *YZ*, (b) *XZ*, and (c) *XY* observed during 200 ps at temperature $0.7 \cdot T_{melt}$. Dark areas correspond to higher values of diffusion coefficients. The positions of dislocations are marked by the letter *D*.

The considered mechanisms were the basic in diffusion processes over grain boundaries of tilt <111>. It should be noted that atoms displaced mainly along closely packed directions during the realization of all three mechanisms. The initiation of mechanisms took place, as a rule, on the steps of dislocations. The migration of atoms led to the change of the steps positions. The directions where the contributions of the considered mechanisms are essential can be defined: the first mechanism – mainly in the direction of axis *X*, the second and the third mechanisms – mainly in the direction *Y*, less – along axis *X* and the least – along *Z*.

In case of tilt boundaries <100> (Figure 4 b), the character of mechanisms proceeding was different because of the peculiarities of the structure of those boundaries. Elementary diffusion jumps in mechanism (2) and (3) also took place mainly along closely-packed directions. In this connection, the total length of the chains of displaced atoms between neighboring dislocations in boundaries <100> was bigger than in boundaries <111>, i.e. relatively big energy was required for their activation. From the other hand, more high density of steps on dislocations promoted more intensive diffusion along the cores of dislocations.

The character of diffusion proceeding over boundaries <100> had some differences in comparison with diffusion over boundaries <111>. Two parallel chains of displaced atoms, beginning and finishing at the steps of dislocations, formed, as a rule, during the migration of atoms along a dislocation in boundary <100>. To explain the appearance of parallel chains, the scheme is shown in Figure 5. The atoms located at the edges of semiplanes of partial dislocations and the steps connected with them are shown on the scheme. The figure depicts two extraplanes of partial dislocations *I* and *II*. Atom *a* displaces in direction *1* because of thermal fluctuations and completes the step of dislocation in plane *I* in point *A*. In plane *II*, atom *b* displaces in direction *2* and completes the same step displacing in knot *B*. The step is completed simultaneously in two extraplanes of partial dislocations. The formation of the second completing chain is also possible in plane *I*, if atom *c* is displaced in point *B* over trajectory 2'. The first chain of displaced atoms caused by thermal fluctuations destroys the structure of dislocation and decreases the energy of defect. The second chain of displaced atoms corrects the appeared structural imperfections of a dislocation caused by the first chain that is promoted to the decrease of the system energy. The appearance of the second chain of displaced atoms appears after the appearance of the first chain after ~0,02 ps. Diffusion jumps of atoms take place mainly along the plane of the boundary in directions <100> during the realization of the given mechanism.

Cyclic mechanisms in boundaries <100> (2 in Figure 4 b) appear at high temperatures (of order $0.7 \div 0{,}8 \cdot T_{melt}$). They were the consequence of mechanisms (1) or (3). The mechanism consisting of the formation of the chains of displaced atoms from one dislocation to the other one (3) is realized in boundaries <100> more rare than in boundaries <111> at other equal conditions.

To study distribution of diffusion intensity in the calculated block, special visualizator is used. The diffusion coefficients were obtained in small cells of 5x5x5 Å size in directions *X*, *Y*, *Z*. The projection of the calculated block to the plane which was perpendicular to this direction is colored in different shades of grey color in correspondence with the values of diffusion coefficients in cells for the given directions. The step of the change of cells coordinates is equal to 0.02 Å. Figure 6 shows the example of distribution of diffusion intensity in Ni with boundary <111> θ=15°. The projection data depict qualitative picture of the distribution of maximum displacements of atoms in definite direction (because $D \sim \Delta^2$, where Δ – displacement of atoms). That is why, the distribution of atomic displacements and distribution of diffusion intensity can be identical. Figure 6 (a) shows the distribution of diffusion intensity in the direction which is perpendicular to the plane of grain boundary. The biggest values of diffusion coefficients are observed near grain boundary dislocations (marked by letter *D*). Diffusion intensity over grain boundary increase with the increase of temperature and misorientation angle θ. It leads to the washing of dark sections corresponding to maximum diffusion. Figure 6 (b) depicts diffusion distribution along the plane of intergrain boundary in the direction perpendicular to the cores of dislocations. It is seen that the distribution of diffusion intensity is localized in the area of grain boundary. This area expands with the increase of temperature that is connected with the increase of intensity of atomic displacements near grain boundary. The distribution of diffusion intensity in projection perpendicular to dislocations cores (Figure 6 c) has maximums in the area of cores (the examples are marked by letter *D*) and minimums between them. The maxima nearly merge at the increase of misorientation angle θ and temperature.

Figure 7 presents the example of atomic displacements near low-angle grain boundary in Cu during grain boundary diffusion. Diffusion by the "pipe" mechanism (i.e. mechanism 1 in Figure 4) prevails for the given misorientation angle and temperature. It is clearly seen in the figure that diffusion takes place mainly near dislocations cores. It agrees with the distribution of diffusion intensity presented in Figure 6. Besides, the chains of displaced atoms which are far enough from dislocation cores can be observed. The marked chains return to the same cores by the closed trajectory because the distance between cores is big for the given misorientation angle. Probably these chains lead to the realization of mechanism (2) at the increase of misorientation angle θ.

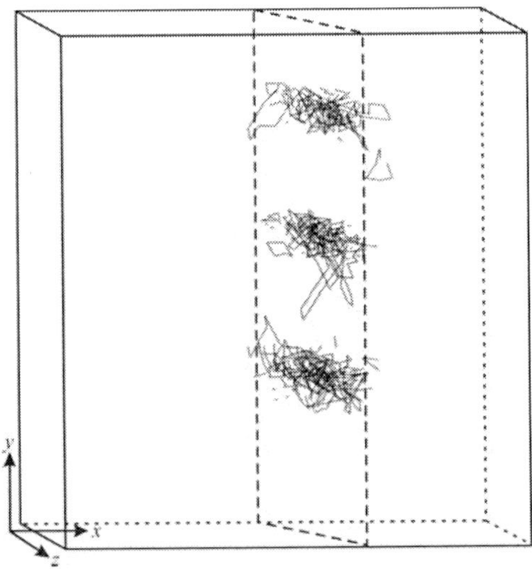

Figure 7. Example of atomic displacements near <111> θ=7° low-angle grain boundary in Cu within 150 ps simulation run at the temperature of $0.6 \cdot T_{melt}$.

Using the approach described above, the values of grain boundary diffusion coefficients along *X*, *Y*, and *Z* axes were obtained. The most intensive diffusion was observed along the *Z* axis, i.e., along the dislocation cores, for all considered temperatures and misorientation angles. Diffusion in the direction of *Y* axis, i.e., between the dislocation cores took place with a smaller intensity. The diffusion coefficient D_y was measured to be several times smaller than D_z for the boundaries with moderate values of θ. The least intensive diffusion was observed in the directions of X axis (diffusion from the boundary inside the grain). On the other hand, at high temperatures, diffusion in the directions X and Y is nearly same. This qualitative change is explained by the increase of diffusion intensity from the boundary inside the grain mainly because of the contribution from the cyclic diffusion mechanism and boundary migration.

The values of the coefficients of grain boundary diffusion along axes X, Y, Z were obtained for grain boundaries under study. The width of boundaries was equal to 5 Å. The most intensive diffusion was noticed along axis Z, i.e. along dislocations cores for all considered temperature and the angles of grain misorientations. Diffusion in the direction of axis Y, i.e. between dislocations took place less intensive. The least intensive diffusion was observed in the direction of axis X (from the boundary inside the grain). Diffusion intensity increased along all directions with the increase of misorientation angle θ. The increase of diffusion coefficients with the increase of θ was connected with the increase of the density of grain boundary dislocations. For large-angle boundaries, the values of diffusion along Y were close to the values of diffusion coefficient along Z.

The dependence graphs $\ln D_b$ on T^1 (D_b - the average coefficient of grain boundary diffusion) for tilt grain boundaries <111> and <100> in Cu with different misorientation angles are presented in Figure 8. As it can be seen from the figure s, the graphs have the curvature that agrees with experimental data [1, 2, 13, 14]. The position of the curvature point depends on θ and it is equal to nearly $0.7 \cdot T_{melt}$ for boundaries <111> and $0.8 \cdot T_{melt}$ for boundaries <100> at θ=16°. The presence of curvature on the dependences can be explained by the presentation of diffusion coefficient as the sum of classical equations of Arrhenius corresponding to the single mechanisms. Usually the fracture is connected with the contribution of boundaries migration into diffusion. Having reached temperature order 0.6-$0.7 \cdot T_{melt}$, migration is essentially intensified [1]. But, it has been found lately that the mechanism of boundaries migration is often connected with cyclic mechanism of atomic migration [43-45]. Thus, the fracture on the dependences can be explained by the inclusion of cyclic mechanism (the second mechanism in the paper) into the general process of diffusion.

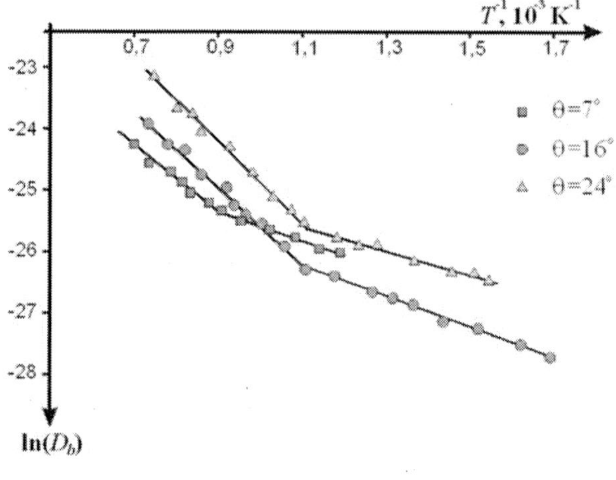

a)

Figure 8 (Continued on next page.)

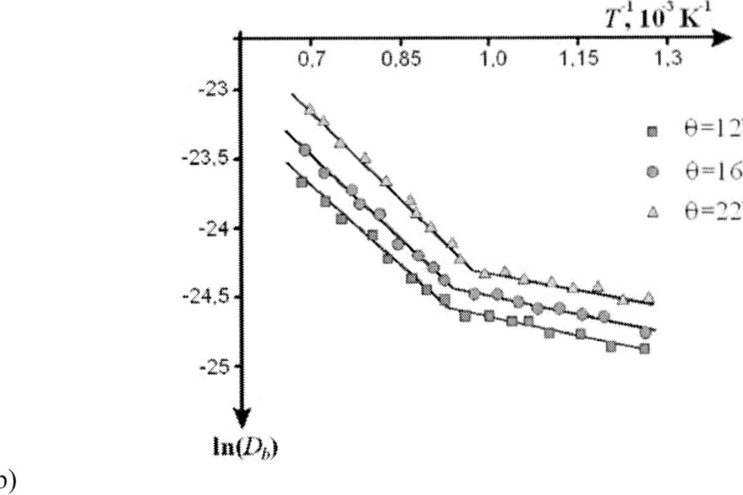

b)

Figure 8. Dependence of lnDb on T-1 for different misorientation angles of grains in Cu with the boundaries (a) <111> and (b) <100>.

The activation energy of grain boundary diffusion for the two temperature ranges was estimated from the graphs presented in Figure 8. The values obtained for Cu are shown in Table 1 for different misorientation angles θ.

For grain boundaries <111> and <100> at $T<(0.7\div0.8)T_{\text{melt}}$, migration of atoms along dislocations cores is main diffusion mechanism. The contribution of cyclic mechanisms and the mechanism of atomic migration between neighboring dislocations appears to be considerable at $T>(0.7\div0.8)T_{\text{melt}}$ along axes X and Y. The studies showed that the essential contribution was made by the mechanisms (2) and (3) at the temperature which depended on misorientation angle θ. It was mainly connected with the decrease of diffusion intensity between dislocations with the growth of θ. In this connection, the curvature point at the dependence $\ln D_b$ on T^{-1}displaced to low temperatures.

The values of the diffusion activation energy for the two temperature ranges in Ni, Cu and Al crystals having a tilt boundary <111> or <100> θ=16° are shown in Table 2. The values obtained in the model agree very well with the previously reported data. The diffusion activation energies at high temperatures, laying in the range of 1.13÷1.36 eV, have been presented in [46, 47] for Ni polycrystal. For Ni with grain boundaries <100> θ=20°, activation energy at high temperatures is equal to 1.11 eV, according to [47].

Table 1. The diffusion activation energy for the two temperature ranges in Cu for the tilt boundaries <111> and <100> with three different misorientation angles

Diffusion over tilt boundaries <111>			Diffusion over tilt boundaries <100>		
θ	The diffusion activation energy, eV		θ	The diffusion activation energy, eV	
	$T>0.7 \cdot T_{melt}$	$T<0.7 \cdot T_{melt}$		$T>0.8 \cdot T_{melt}$	$T<0.8 \cdot T_{melt}$
7°	0.68	0.21	12°	0.62	0.25
14°	0.65	0.20	14°	0.59	0.24
24°	0.60	0.22	22°	0.55	0.26

Table 2. Activation energy of grain boundary diffusion for the two temperature ranges in Ni, Al and Cu with tilt boundaries <111> and <100> θ=16°

Metal	The diffusion activation energy, eV			
	Tilt boundary <111>		Tilt boundary <100>	
	T>0.7·Tmelt	T<0,7·Tmelt	T>0,8·Tmelt	T<0,8·Tmelt
Ni	1.44	0.43	1.26	0.32
Cu	0.64	0.21	0.56	0.28
Al	0.57	0.17	0.38	0.23

Diffusion in the Ni_3Al intermetallide had some peculiarities besides three mechanisms mentioned earlier earlier. Migration of Ni atoms course of computer experiments. Al atoms were practically immobile. That is why the superstructural order was not noticeably reduced during grain boundary diffusion at relatively low temperatures. However, with the increase of temperature, the second and the third diffusion mechanisms come into play and Al atoms began to take part in the diffusion. As a result, the grain boundary diffusion at high temperatures was accompanied by a decrease of the superstructural order.

Figure 9 shows the dependences of ln$D$ on $1/T$ for the tilt grain boundaries <111> and <100> with misorientation angle θ=16°. It can be seen from the graphs that Ni atoms have higher diffusion mobility than Al atoms within the whole investigated temperature range. Besides, a change of slope of the dependence can be clearly seen at the temperature of 1150-1200 K (about 0.6-0.7·T_{melt}).

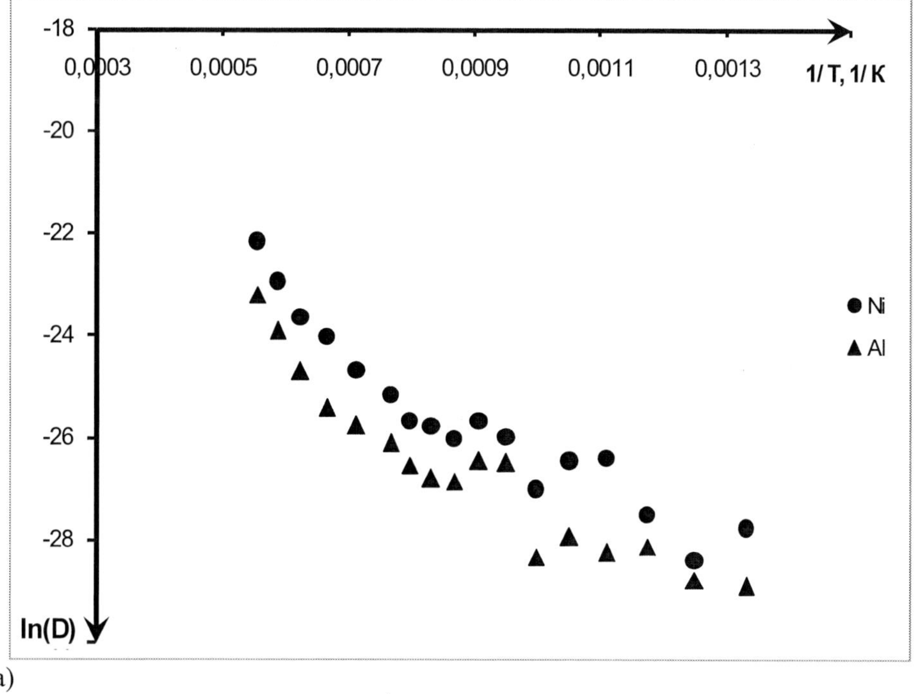

a)

Figure 9. (Continued.)

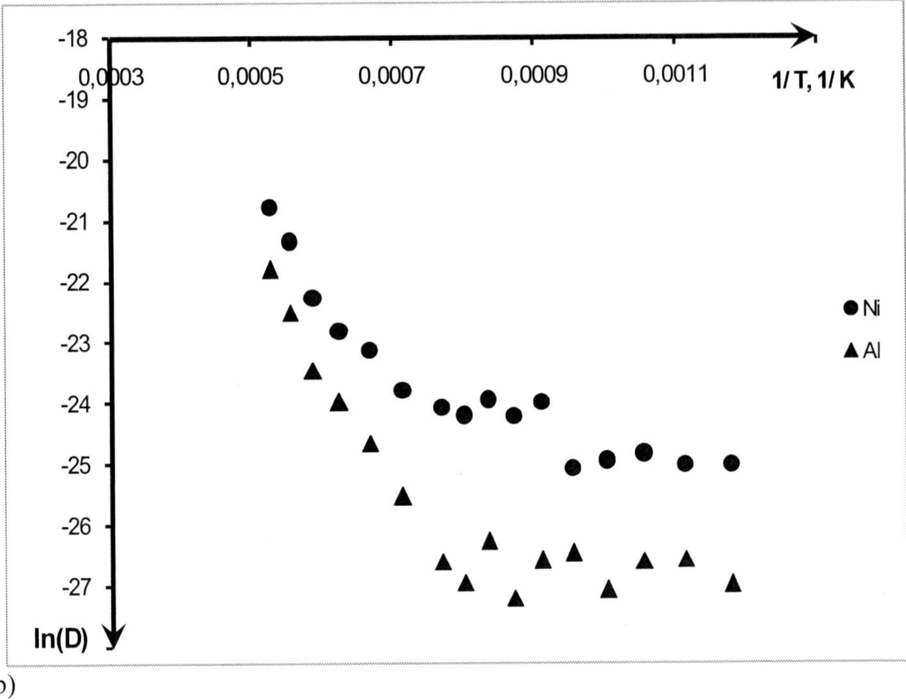

b)

Figure 9. lnD as the function of 1/T for the tilt grain boundaries (a) <111> and (b) <100> with misorientation angle of θ=16°

CONCLUSIONS

Our molecular-dynamic simulations showed that the diffusion along tilt grain boundaries in fcc metals and Ni_3Al intermetallide takes place mainly by the following three mechanisms: migration of atoms along the cores of grain boundary dislocations, cyclic mechanism near the cores, and the formation of the chains of displaced atoms from the core of one dislocation to the core of another one. The probability of observe the diffusion by the third mechanism essentially increases with increase in the misorientation angle. The density of steps on grain boundary dislocations influences greatly the probability of realization of all three diffusion mechanisms along grain boundaries. The chains of displaced atoms at the realization of the third mechanism start and end, as a rule, on dislocations steps. The deviation from the Arrhenius law for the diffusion along grain boundaries was explained through the presence of several diffusion mechanisms having different activation energies. The plots of ln*D* as the functions of $1/T$ have a change-of-slope point with the abscissa dependent on the misorientation angle θ. With increase in θ, the change-of-slope point shifts to lower temperatures. This can be explained by the decrease of the diffusion activation energy for the third diffusion mechanism with increase in θ.

In Ni_3Al intermetallide, the grain boundary diffusion takes place mainly by the displacements of Ni atoms over their sublattices in $L1_2$ superstructure. As a result, the short-range order was not seriously affected by the diffusion. However, at higher temperatures, Al atoms are also involved in the migration process. Their displacements from the original positions are small but they cause noticeable reduction of the superstructure ordering in the alloy.

REFERENCES

[1] Kaybyshev O. A., and Valiev R. Z. (1987) Grain boundaries and properties of metals. Moscow: *Metallurgiya.*

[2] Kopetzky Ch. V., Orlov A. N., and Fionova L. K. (1987) Grain boundaries in pure metals. Moscow: *Nauka.*

[3] Fisher J. C. (1951) Calculation of Penetration Curves of Surface and Grain Boundary Diffusion. *J. Appl. Phys.* 22, 74-80.

[4] Glater G., and Chalmers B. (1975) Large-angle grain boundaries. Moscow: *Metallurgizdat.*

[5] Turnbull D., and Hoffman R. The effect of relative crystal and boundary orientations on grain boundary diffusion rates // *Acta Met.* - 1954. - V.2. - P. 419-425.

[6] Achter M. R., and Smoluchowski R. (1951) Anisotropy of Diffusion in Grain Boundaries. *Phys. Rev.* 83. 163-170.

[7] Gupta D., Kampbell D., and Kho P. 1982 *Diffusion over grain boundaries.* In: *Thin films, mutual diffusion and reactions.* (163-249) Moscow: *Mir*,. - P.

[8] Lubashevsky I. A., Alatortzev V. L. (1988) Peculiarities of spatial distribution of diffused atoms in regular polycrystals. *Fizika metallov i metallovedenie,* 65, №5, 858-867.

[9] Kondratiev V. V., and Trahtenberg I. Sh. (1986) Grain boundary diffusion of atoms in the model of structurally inhomogeneous boundaries. *Fizika metallov i metallovedenie,* 62, №3, 434-441.

[10] Bokshtein S. Z., Bolberova E.V., Kishkin S. T., and Razumovsky I. M. (1981) Diffusion characteristics of grain boundaries of eutectic alloys with the directed structures. *Fizika metallov i metallovedenie,* 51, №1, 101-107.

[11] Bokshtein S. Z., Bolberova E. V., Kishkin S. T., Kostyukova E. P., Mishin Yu. M., and Razumovsky I. M. (1984) The peculiarities of diffusion in grain boundaries of Ni alloys obtained by the method of directed crystallization, *Fizika metallov i metallovedenie*, 58, №1, 189-191.

[12] Aleshin A. N., Bokshtein B. S., and Shvindlerman L. S. (1982) The research of diffusion over individual grain boundaries in metal, *Poverhnost,* №6, 1-12.

[13] Fedorov G. B., and Smirnov E. A. (1978) Diffusion in a reactor materials. Moscow: *Atomisdat.*

[14] Bokshtein B. S., Kopetzky Ch. V., and Shvindlerman L. S. (1986) Thermodynamics and kinetics of grain boundaries in metals. Moscow: *Metallurgia.*

[15] Sorensen M. R., Mishin Y., and Voter A. F. (2000) Diffusion mechanisms in Cu grain boundaries. *Phys. Rev.* B.62, 3658-3673.

[16] Suzuki A., and Mishin Y. (2003) Atomistic modeling of point defects and diffusion in copper grain boundary, *Interface Science*, №11, 131-148.

[17] Suzuki A., and Mishin Y. (2004) Diffusion mechanisms in grain boundaries. *Journal of Metastable and Nanocrystalline Materials*, 19, 1-23.

[18] Suzuki A., and Mishin Y. (2005) Atomic mechanisms of grain boundary diffusion: Low versus high temperatures. *Journal of Materials Science*, 40, 3155-3162.

[19] Farkas D. (2000) Atomistic theory and computer simulation of grain boundary structure and diffusion. *Journal of Physics: Condensed Matter*, №12, R497-R516.

[20] Liu C. L., and Plimpton S. J. (1995) Molecular-statics and molecular-dynamics study of diffusion along [001] tilt grain boundaries in Ag. *Phys. Rev.* B51, 4523-4529.

[21] Murzaev R. T., and Nazarov A. A. (2005) Vacancy formation energy in [001] tilt gain boundaries in nickel: computer simulation. *Phys. Metals Metallogr.*, V. 100, №3, 228-234.

[22] Murzaev R. T., and Nazarov A. A. (2006) Activation energy for vacancy migration in [001] tilt boundaries in nickel. *Phys. Metals Metallogr.*, V. 101, №1, 86-92.

[23] Murzaev R. T., and Nazarov A. A. (2006) Energies of formation and activation for migration of grain-boundary vacancies in a nickel bicrystal containing a disclination. *Phys. Metals Metallogr.*, V. 102, №2, 198-204.

[24] Murzaev R. T., and Nazarov A. A. (2006) Vacancy formation and migration energies in nonequilibrium grain boundaries. *TMS Letters,* V. 3, №2, 29.

[25] Grinberg B. A., and Ivanov M. A. (2002) Ni_3Al and TiAl intermetallides: microstructure, deformation behavior. Ekaterinburg: UrO RAS.

[26] Yu Pan, and B. L. Adams. (1994) On the CSL grain boundary distributions in polycrystals. *Scripta Met.* V. 30, №8, 1055-1060.

[27] Randle V. (1994) Asymmetric tilt boundaries in polycrystalline nickel. *Acta Cryst.* A50, №5, 588-595.

[28] Lim L. C., and Raj R. (1984) On the distribution of Σ for grain boundaries in polycrystalline nickel prepared by strain annealing technique. *Acta Met.,* V.32, №8, 1177-1181.

[29] Gertzman V. Yu., Danilenko V. N., and Valiev R. Z. (1990) Distribution of grain boundaries on misorientation in nihrom. *Metallophysica,* V.12, №3, 120-121.

[30] Hofmann D., and Finnis M. W. (1994) Theoretical and experimental analysis of near Σ=3(211) boundaries in silver. *Acta Met.,* V.42, №10, 3555-3567.

[31] MacLaren J. M., Crampin S., Vvedensky D. and D., Eberhart M. E. (1989) Mechanical stability and charge densities near stacking faults. *Phys. Rev. Lett.,* V.63, №23, 2586-2589.

[32] Wolf D. (1989) Correlation between the energy and structure of grain boundaries in bcc metals. I. Symmetrical boundaries on the (110) and (100) planes. *Phil. Mag.* B., V.59, №6, 667-680.

[33] Wolf D. (1990) Structure-energy correlation for grain boundaries in fcc metals. III. Symmetrical tilt boundaries. *Acta Met.,* V.38, №5, 781-790.

[34] Plimpton S. J. and Wolf E. D. (1990) Effect of interatomic potential on simulated grain boundary and bulk diffusion: A molecular-dynamic study. *Phys. Rev.* B41, №5, 2712-2721.

[35] De Hasson J. Th. M., and Vitek V. (1990) Atomic structure of (111) twist grain boundaries in fcc metals. *Phil. Mag.* A61, №2, 305-327.

[36] Vitek V., and Chen S. P. (1991) Modeling of grain boundary structures and properties in intermetallic compounds. *Scripta Met.,* V.32, №6, 1237-1242.

[37] Alber I., Bassani J.L., Khantha M., Vitek V., and Wang G.J. (1992) Grain boundaries as heterogeneous systems: atomic and continuum elastic properties. *Phil. Trans. Roy. Soc. London* A, V. 339, №1655, 555-586.

[38] Holian B.L., and Ravelo R. (1995) Fracture simulations using large-scale molecular dynamics. *Phys. Rev.*B51, №17, 11275-11288.

[39] Osetsky Yu. N., Serra A., Victoria M., Golubov S. I., and Priego V. (1999) Vacancy loops and stacking-fault tetrahedra in copper I. Structure and properties studied by pair and many-body potentials. *Phil. Mag.* A79, №9, 2259-2283.

[40] Chirkov A. G., Ponomarev A. G., and Chudinov V. G. (2004) Dynamical properties of Ni, Cu, Fe in a condensed state (molecular dynamics method). *Zhurnal tehnicheskoy fiziki*, V. 74, №2, 62-65.

[41] Poletaev G. M., Starostenkov D. M., Demyanov B. F., Starostenkov M. D., and Krasnov V. Yu. (2006) Dynamic collective atomic displacements in metals. *Fundam. Probl. Sovrem. Materialoved.*, №4, 130-134.

[42] Huang J.Y., Zhu Y.T., Jiang H., and Lowe T.C. (2001) Microstructures and dislocation configurations in nanostructured Cu processed by repetitive corrugation and straightening. *Acta Mater.*, V. 49, 1497-1505.

[43] Rakitin R. Yu., Poletaev G. M., Aksenov M. S., and Starostenkov M. D. (2005) Molecular-dynamic research of diffusion over grain boundaries in two-dimensional metals. *Fundam. Probl. Sovrem. Materialoved.*, №2, 5-8.

[44] Rakitin R.Yu., Poletaev G.M., Aksenov M.S., and Starostenkov M.D. (2005) The mechanisms of structural transformation near grain boundaries in fcc metals in the conditions of deformation. *Fundam. Probl. Sovrem. Materialoved.*, V.2, №3, 46-50.

[45] Suzuki A., and Mishin Y. (2005) Atomic mechanisms of grain boundary motion. *Materials Science Forum*, V.502, 157-162.

[46] Larikov L. N., and Isaychev V. I. (1987) Diffusion in metals and alloys. Kiev: *Naukova dumka.*

[47] Smitls K. J. (1980) Metalls: Moscow: *Metallurgiya.*

In: Computational Materials
Editor: Wilhelm U. Oster

ISBN: 978-1-60456-896-7

Chapter 6

COMPUTATIONAL MATERIALS: III-V SEMICONDUCTOR CLUSTERS

Hamidreza Simchi*
Semiconductor Component Inustry
P.O.Box: 19575-199, Tehran, Iran

SUMMARY

Semiconductor materials have been playing an important role in daily life. Among them, the narrow gap III-V semiconductor materials InAs, GaSb, InSb and their alloys are particularly interesting and useful materials since they offer the promise of being able to access the 2-10 microns wavelength region and should provide the next generation of LEDs lasers and photo detectors for applications such as sensors, molecular spectroscopy and thermal imaging. On the other hand the discovery of fullerenes and nanotubes of carbon have led to a wide spread interest to understand their properties, applications and development of unconventional forms (i.e Clusters) of materials. Therefore it is expected, that, the clusters of III-V semiconductors have been widely studied. One of the main interests in semiconductor clusters is the variation of band gap with size which affects the photoluminescence properties. Here we review the progress in our understanding of the structure, and electronic spectra of InAs, GaAs, InSb and CdSe clusters. First principles approaches based on Hartree-Fock and Density Functional Theory have become central to such studies and a brief overview of them is given.

Keywords: Clusters, electronic properties, Raman spectra, III-V semiconductors, Hartree-Fock and density Functional Theory, relativistic effects, symmetry

* E-Mail: yamahdiadrecny@yahoo.com

1. INTRODUCTION

The progress made in physics and technology of semiconductors depends mainly on two families of materials: the group IV elements and the III-V compounds. The first report of the formation of III-V compounds was published in 1910 by Thiel and Koelsch [1]. They synthesized a compound of indium and phosphorus and reached the conclusion that its formula is very probably InP. The fact that one of the III-V compounds, InSb is a semiconductor akin to Ge and α-Sn, was reported in 1950 by Blum, et al [2]. The particular features that attracted interest were the low effective mass, high electron mobility, and the ionic component in the crystal binding. After the first exploratory research, nearly all of the work was limited to InSb, partly because this was an almost ideal vehicle for such studies, but also because it was easy to grow InSb single crystals. The technology of the other compounds fell far behind this, for the purification was very difficult and there was no great incentive to work in this field.

The invention of the semiconductor laser and the discovery of the Gunn effects have caused a marked change in this situation. There is now an increased prospect of industrial application of these materials, and we may expect a growing interest in GaAs, GaP, and compounds or alloys of compounds which can emit visible or microwave radiation. Now a days, III-V compound semiconductors are widely used as substrates for optical devices such as LED's and laser diodes and for electronic devices such as FET's, HEMT's, HBT's and IC's. These applications are becoming key elements in an advanced information society. A complete overview of the technologies necessary to grow bulk signal-crystal substrates, and grow hetero-and homoepitaxial III-V films using molecular beam epitaxy (MBE) or metal-organic chemical vapor deposition (MOCVD) is provided in reference [3].

On the other hand the discovery of fullerenes and nanotubes of carbon have led to a wide spread interest to understand their properties, applications and development of unconventional forms (i.e Clusters) of materials. Like crystals a molecule can be represented as a system of electrons traveling in the field of ions. Hence the approach where the crystal is simulated by a certain "large molecule" fragment with corresponding boundary conditions is quite valid. Such a large molecule has been called a cluster while the approach itself is called the cluster approach. The cluster approach utilizes a variety of techniques from quantum chemistry employed to describe the molecular electronic properties. Of course clusters are not suitable for describing physical phenomena associated with long-range order due to their finite size. At the same time phenomena associated with short-range order can be successfully analyzed by means of the cluster approach. For example, point defects in crystals [4], localized excitations, self-localized quasiparticles, chemisorption [5], etc., have been successfully analyzed within the framework of the cluster approach.

Here we review the progress in our understanding of the structure, and electronic spectra of InAs, GaAs, InSb and CdSe clusters. First principles approaches based on Hartree-Fock and Density Functional Theory have become central to such studies and a brief overview of them is given.

2. Computer Simulations Using First Principles Approaches

2.1. Self-Consistent Field (Scf) Approximation

In considering the motion of electrons in condensed matter we are dealing with the problem of describing the motion of an enormous number of electrons and nuclei (~ 10^{23}) obeying the laws of quantum mechanics. An exact solution of this problem is impossible and hence it is necessary to rely on a wide range of simplifying approximation.

Consider a system of N interacting electrons which move in an external potential, set up, for example, by the positively charged nuclei of the system. We can divide the Hamiltonian H of the electronic system into two parts:

$$H = \sum_{i=1}^{N} h(i) + \frac{1}{2}\sum_{i \neq j} v(i,j) \qquad (0.0.1)$$

The single-electron operator

$$h(i) = -\frac{1}{2m}\nabla_i^2 + v(i) \qquad (0.0.2)$$

represents the sum of the kinetic energy $-(1/2m)\nabla_i^2$ of the electron and its energy in the external potential V(i). Should the external potential be attributable to nuclei of charges $-Z_\lambda e^2$, we would have $v(i) = -\sum_\lambda Z_\lambda e^2 / r_{\lambda i}$, where $r_{\lambda i}$ is the distance between nucleus λ and electron i. The term $v(i,j) = e^2 / r_{ij}$ describes the Coulomb repulsion two electrons i and j which are a distance $r_{ij} = |r_i - r_j|$ apart.

Next the Hamiltonian is written in second quantized form, for which we introduce electron field operators $\psi(\vec{r})$. They satisfy the following anti-commutation relations:

$$[\psi_\sigma^+(\vec{r}), \psi_{\sigma'}(\vec{r}')]_+ = \delta_{\sigma\sigma'}\delta(\vec{r} - \vec{r}')$$
$$[\psi_\sigma(\vec{r}), \psi_{\sigma'}(\vec{r}')]_+ = [\psi_\sigma^+(\vec{r}), \psi_{\sigma'}^+(\vec{r}')]_+ = 0 \qquad (0.0.3)$$

We chose $[\,,\,]_+$ instead of $[\,,\,]_-$, since Hamiltonian must be positive definite.
If we write the Lagrangian and Hamiltonian density as

$$L = -\psi_\sigma^+(\vec{r})(-\frac{\nabla^2}{2m} + V(\vec{r}))\psi_\sigma(\vec{r}) + i\frac{\partial\psi_\sigma^+}{\partial t}\psi_\sigma \qquad (0.0.4)$$

$$H = \Pi_{\psi_\sigma^+}\frac{\partial\psi_\sigma^+}{\partial t} - L \qquad (0.0.5)$$

Where

$$\Pi_{\psi_\sigma^+} = \frac{\partial L}{\partial(\partial\psi_\sigma^+ / \partial t)} \tag{0.0.6}$$

The Hamiltonian will be

$$H = \sum_\sigma \int d^3r(Hamiltonian - density) = \sum_\sigma \int d^3r \psi_\sigma^+(\vec{r})(-\frac{\nabla^2}{2m} + V(\vec{r}))\psi_\sigma(\vec{r}) \tag{0.0.7}$$

Therefore in terms of them, the Hamiltonian H takes the form

$$H = \sum_\sigma \int d^3r \psi_\sigma^+(\vec{r})(-\frac{1}{2m}\nabla^2 + V(\vec{r}))\psi_\sigma(\vec{r}) + \frac{e^2}{2}\sum_{\sigma\sigma'}\int d^3r d^3r' \psi_\sigma^+(\vec{r})\psi_\sigma(\vec{r})\frac{1}{|\vec{r}-\vec{r}'|}\psi_{\sigma'}^+(\vec{r}')\psi_{\sigma'}(\vec{r}') \tag{0.0.8}$$

In most cases one is trying to find approximate eigenstates of H within a given set of L basis functions $f_i(\vec{r})$. If $a_{j\sigma}^+, a_{j\sigma}$ are electron creation and annihilation operators, an approximation for $\psi_\sigma(\vec{r})$ can be written as:

$$\psi_\sigma(\vec{r}) = \sum_{i=1}^{L} a_{i\sigma} f_i(\vec{r}) \tag{0.0.9}$$

The functions $f_i(\vec{r})$ are generally not orthogonal to each other. Their overlap matrix is defined by:

$$S_{ij} = \int d^3r f_i^*(\vec{r}) f_j(\vec{r}) \tag{0.0.10}$$

From (2.1.4) and (2.1.6), it follows that

$$\begin{aligned} \left[a_{i\sigma}^+, a_{j\sigma'}\right]_+ &= S_{ji}^{-1}\delta_{\sigma\sigma'} \\ \left[a_{i\sigma}^+, a_{j\sigma'}^+\right]_+ &= \left[a_{i\sigma}, a_{j\sigma'}\right]_+ = 0 \end{aligned} \tag{0.0.11}$$

Within the space spanned by the function $f_i(\vec{r})$, the Hamiltonian (2.1.4) becomes

$$H = \sum_{ij\sigma} t_{ij} a_{i\sigma}^+ a_{j\sigma} + \frac{1}{2}\sum_{\substack{ijkl \\ \sigma\sigma'}} V_{ijkl} a_{i\sigma}^+ a_{k\sigma'}^+ a_{l\sigma'} a_{j\sigma} \tag{0.0.12}$$

Where the matrices t_{ij} and V_{ijkl} are

$$t_{ij} = \int d^3 r f_i^*(\vec{r})(-\frac{1}{2m}\nabla^2 + V(\vec{r}))f_j(\vec{r}) \tag{0.0.13}$$

$$V_{ijkl} = e^2 \int d^3 r d^3 r' f_i^*(\vec{r}) f_j(\vec{r}) \frac{1}{|\vec{r} - \vec{r}'|} f_k^*(\vec{r}') f_l(\vec{r}')$$

In the following discussion, a Hamiltonian of the form (2.1.12) will be used frequently.

Finding the eigenfunctions and eigenvalues of an electronic system described by the Hamiltonian (2.1.12) is impossible without drastic simplifying assumptions. Hartree [6], Fock [7], and Slater [8] treated the electrons as being independent of each other and introduced the idea of the Self-Consistent Field (SCF). The latter is the interaction fields an electron experiences when we take a spatial average over the positions of all the other electrons. Within the independent-electron approximation, we can assume the wave function φ_1 can be exist at r_1, r_2,…r_N, the wave function φ_2 can be exist at r_1, r_2,…r_N and so on i.e we have a charge distribution. In this condition the ground state wave function is of the form

$$\Phi(\vec{r}_1\,\sigma_1,\ldots,\vec{r}_N\,\sigma_N) = \frac{1}{\sqrt{N}!}\begin{vmatrix} \varphi_1(\vec{r}_1\,\sigma_1)\ldots\ldots\ldots\varphi_N(\vec{r}_N\,\sigma_N) \\ \ldots\ldots\ldots\ldots\ldots\ldots\ldots\ldots \\ \ldots\ldots\ldots\ldots\ldots\ldots\ldots\ldots \\ \varphi_1(\vec{r}_N\,\sigma_N)\ldots\ldots\ldots\varphi_N(\vec{r}_N\,\sigma_N) \end{vmatrix} \tag{0.0.14}$$

Without loss of generality, the functions $\varphi_\nu(\vec{r}\,\sigma)$ can be assumed to be orthogonal to each other.On the other hand, we want not only the expectation value for the ground-state energy, $E_0 = \prec \Phi|H|\Phi \succ$, is minimized by variation of $\varphi_\mu(\vec{r}\,\sigma)$ and $\varphi_\mu^*(\vec{r}\,\sigma)$ but also the constraint $\prec \varphi_\mu|\varphi_\mu \succ = 1$ is satisfied. These conditions are taken into account by introducing Lagrange parameters ε_μ when doing the variation. The requirement is therefore

$$\delta(E_0 - \sum_\mu \varepsilon_\mu \prec \varphi_\mu|\varphi_\mu \succ) = 0 \tag{0.0.15}$$

But

$$E_0 = \sum_{\nu=1}^{N} \prec \varphi_\nu|h|\varphi_\nu \succ + \frac{1}{2}\sum_{\mu\nu}^{N}[\prec \varphi_\mu\varphi_\nu|v|\varphi_\mu\,\varphi_\nu \succ - \prec \varphi_\mu\varphi_\nu|v|\varphi_\nu\,\varphi_\mu \succ] \tag{0.0.16}$$

Where

$$\prec \varphi_\nu|h|\varphi_\nu \succ = \int d^3 r \chi_\mu^*(\vec{r}) h(\vec{r}) \chi_\nu(\vec{r}) \tag{0.0.17}$$

$$\prec \varphi_\mu\varphi_\tau|v|\varphi_\nu\varphi_\rho \succ = \delta_{\sigma_\mu\sigma_\nu}\delta_{\sigma_\tau\sigma_\rho}\int d^3 r d^3 r' \chi_\mu^*(\vec{r})\chi_\nu(\vec{r}) v(\vec{r},\vec{r}')\chi_\tau^*(\vec{r}')\chi_\rho(\vec{r}') \tag{0.0.18}$$

and $\varphi_\mu(\vec{r}) = \chi_\mu(\vec{r})\sigma$ are a product of a spatial orbital $\chi_\mu(\vec{r})$ and a two-component spinor σ. The latter equal equals $\alpha = \begin{pmatrix}1\\0\end{pmatrix}$ for spin-up and $\beta = \begin{pmatrix}0\\1\end{pmatrix}$ for spin-down electrons with respect to a preferred axis.

By substituting (2.1.16) in (2.1.15), it can be shown

$$F\left|\varphi_\mu\right\succ = \varepsilon_\mu \left|\varphi_\mu\right\succ \tag{0.0.19}$$

Where F is the Fock operator with matrix element

$$f_{\nu\mu} = \prec\varphi_\nu\left|h\right|\varphi_\mu\succ + \sum_\tau^N (\prec\varphi_\nu\varphi_\tau\left|v\right|\varphi_\mu\varphi_\tau\succ - \prec\varphi_\nu\varphi_\tau\left|v\right|\varphi_\tau\varphi_\mu\succ) \tag{0.0.20}$$

Equation (2.1.19) constitutes the well-known Hartree-Fock (HF) equations.

The eigenvalues ε_μ are obtained from the diagonal form of the Fock matrix and the total energy E is not simply the sum over ε_μ but, instead, given by

$$E_0 = \sum_\mu^N \varepsilon_\mu - \frac{1}{2}\sum_{\mu\nu}^N (\prec\varphi_\mu\varphi_\nu\left|v\right|\varphi_\mu\varphi_\nu\succ - \prec\varphi_\mu\varphi_\nu\left|v\right|\varphi_\nu\varphi_\mu\succ) \tag{0.0.21}$$

Solving the HF equation exactly is not possible except in trivial cases such as that of the homogeneous electron gas. For inhomogeneous systems calculations are done with a set of basis functions $f_i(\vec{r})$, in which case the Hamiltonian takes the form of (2.1.12). In order to derive the Fock operator here, the determinant (2.1.14) is written in second quantized form as

$$\left|\Phi\right\succ = \prod_{\mu\sigma} c_{\mu\sigma}^+ \left|0\right\succ \tag{0.0.22}$$

Where $c_{\mu\sigma}^+$ is the creation operator for an electron in spin orbital $\varphi_\mu(\vec{r}\,\sigma)$ and $\left|0\right\succ$ the vacuum state.

Because the $\varphi_\mu(\vec{r}\,\sigma)$ can be assumed to be orthogonal, the $c_{\mu\sigma}^+$ satisfy the usual fermionic anticommutation relations

$$\begin{aligned}&[c_{\mu\nu}^+, c_{\nu\sigma'}]_+ = \delta_{\mu\nu}\delta_{\sigma\sigma'}\\&[c_{\mu\sigma}^+, c_{\nu\sigma'}^+] = [c_{\mu\sigma}, c_{\nu\sigma'}]_+ = 0\end{aligned} \tag{0.0.23}$$

The wave function (2.2.21) is denoted by $\left|\Phi_{SCF}\right\succ$ in the text below, the corresponding expectation value of the Hamiltonian (2.1.12) is given by

$$E_0 = \sum_{ij}^{L}\sum_{\sigma} t_{ij} \prec \Phi_{SCF}\left|a_{i\sigma}^{+}a_{j\sigma}\right|\Phi_{SCF} \succ + \frac{1}{2}\sum_{ijkl}\sum_{\sigma\sigma'} V_{ijkl} \prec \Phi_{SCF}\left|a_{i\sigma}^{+}a_{k\sigma'}^{+}a_{l\sigma'}a_{j\sigma}\right|\Phi_{SCF} \succ \quad (0.0.24)$$

The occupied orbital of the ground state $|\Phi_{SCF} \succ$ are expanded in terms of the basis functions $f_i(\vec{r})$ as

$$c_{\mu\sigma}^{+} = \sum_{n=1}^{L} d_{\mu n} \hat{a}_{n\sigma}^{+} \quad (0.0.25)$$

and the bond-order matrix, P_{ij} is defined as

$$P_{ij} = \sum_{\sigma} \prec a_{i\sigma}^{+}a_{j\sigma} \succ = 2\sum_{\nu}^{occ} d_{\nu i}^{*}d_{\nu j} \quad (0.0.26)$$

Now if we repeat the condition (2.1.15), we will find

$$\sum_{j=1}^{L}(f_{ij} - \varepsilon_{\mu}S_{ij})d_{ij} = 0 \quad (0.0.27)$$

Where the f_{ij} are given by

$$f_{ij} = t_{ij} + \sum_{kl}(V_{ijkl} - \frac{1}{2}V_{ilkj})P_{kl} \quad (0.0.28)$$

These are the Self-Consistent Field (SCF) equations for a finite basis set [9]. We can satisfy (2.1.26) if we introduce the Fock operator

$$F = \sum_{ij\sigma} f_{ij}(a_{i\sigma}^{+}a_{j\sigma} - \prec a_{i\sigma}^{+}a_{j\sigma} \succ) \quad (0.0.29)$$

and require that the following self-consistent relation hold

$$\begin{aligned} Fc_{\mu\sigma}|\Phi_{SCF} \succ &= -\varepsilon_{\mu}c_{\mu\sigma}|\Phi_{SCF} \succ \\ Fc_{i\sigma}^{+}|\Phi_{SCF} \succ &= \varepsilon_{i}c_{i\sigma}^{+}|\Phi_{SCF} \succ \end{aligned} \quad (0.0.30)$$

depending on whether the coefficient for occupied $(d_{\mu n})$ or unoccupied (d_{in}) orbital are considered. The corresponding spin orbital $\varphi_{\mu}(\vec{r}\,\sigma)$ and $\varphi_{i}(\vec{r}\,\sigma)$ are the canonical molecular orbital (MOs). It can be checked that $F|\Phi_{SCF} \succ = 0$.

The solution of the SCF equation must fulfill symmetry requirements which restrict their form. For example, in addition to being a solution of (2.1.26) , a molecular orbital must be an eigenfunction of the different symmetry operations with which the Hamiltonian commutes, i.e., it must transform according to an irreducible representation of the point group of the molecule.

In addition to the symmetry requirements, there are also equivalence restrictions. An equivalence restriction is, for example, that the orbital function of $4\sigma_{\uparrow}$ is the same as that of $4\sigma_{\downarrow}$ .The equivalence restrictions are consequences of conservation laws.

Unrestricted Hartree-Fock wave functions break both symmetry and equivalence requirements. With these wave functions one often obtain lower energies than when working with restricted Hartree-Fock functions. The reason is that breaking a symmetry implies that electronic correlations are partially taken into account.

2.2. Density Functional Theory

The density functional approach expresses ground-state properties- such as total energy, equilibrium positions and moments- in terms of the electronic density $\rho(\vec{r})$ or spin density $\rho_\sigma(\vec{r})$ and provides a scheme for calculating them. The method avoids the problem of calculating the ground-state wave function.

2.2.1. Thomas-Fermi Method

Consider a charge distribution of electron, $\rho(\vec{r})$. If $V(\vec{r})$ is a given external electrostatic potential, the energy of an electronic density $\rho(\vec{r})$ in this potential will be equal to $\int d^3rV(\vec{r})\rho(\vec{r})$. Also for a classical charge distribution the repulsion energy is equal to $\frac{e^2}{2}\int d^3rd^3r'\frac{\rho(r)\rho(\vec{r'})}{|\vec{r}-\vec{r'}|}$ and the kinetic energy density of homogenous system is equal to $\frac{3}{10m}(3\Pi^2)^{2/3}\int d^3r\rho^{5/3}(\vec{r})$. The crucial assumption of the Thomas-Fermi method is the form of the ground-state energy functional $E_{TF}[\rho,V]$, which one writes as [10, 11]

$$E_{TF}[\rho,V]=\int d^3rV(\vec{r})\rho(\vec{r})+\frac{e^2}{2}\int d^3rd^3r'\frac{\rho(\vec{r})\rho(\vec{r'})}{|\vec{r}-\vec{r'}|}+\frac{3}{10m}(3\Pi^2)^{2/3}\int d^3r\rho^{5/3}(\vec{r}) \qquad (2.2.1.1)$$

Now we have two conditions

$$\delta E_{TF}[\rho,V]=0 \text{ i.e. } E_{TF}[\rho,V] \text{ is minimum} \qquad (2.2.1.2)$$

$$\delta\int d^3r\rho(\vec{r})=0 \text{ i.e. total electron number remains constant}$$

By using Lagrange parameter μ, the requirement is therefore

$$\delta(E_{TF}[\rho,V]-\mu\int d^3r\rho(\vec{r}))=0 \qquad (2.2.1.3)$$

By substituting (2.2.1) in (2.2.3), one finds

$$V(\vec{r}) + e^2 \int d^3r' \frac{\rho(\vec{r}')}{\left|\vec{r} - \vec{r}'\right|} + \frac{1}{2m}(3\Pi^2)^{2/3} - \mu = 0 \tag{2.2.1.4}$$

By introducing the effective potential, $V_c(\vec{r})$, as

$$V_c(\vec{r}) = e^2 \int d^3r' \frac{\rho(\vec{r}')}{\left|\vec{r} - \vec{r}'\right|} \tag{2.2.1.5}$$

$$\nabla^2 V_c(\vec{r}) = -4\Pi e^2 \rho(\vec{r})$$

One can writes

$$\rho(\vec{r}) = \frac{(2m)^{3/2}}{3\Pi^2}[\mu - V(\vec{r}) - V_c(\vec{r})]^{3/2} \tag{2.2.1.6}$$

Equations (2.2.1.5) and (2.2.1.6) yield a differential equation for $V_c(\vec{r})$. Here we neglected the exchange and correlation effects.

2.2.2. Hohenberg-Kohn-Sham Theory

There are two important theorems, called Hohenberg and Kohn theorems [12]

a)- The ground-state energy, E, of a many-electron system in the presence of an external potential $V(\vec{r})$ is a functional of the electronic density $\rho(\vec{r})$, and one can writes

$$E[\rho, V] = \int d^3r V(\vec{r})\rho(\vec{r}) + F(\rho) \tag{2.2.2.1}$$

Where $F(\rho)$ is an unknown, but universal functional of density $\rho(\vec{r})$ only, and does not depend on $V(\vec{r})$.

b)- The $E[\rho, V]$ is minimized by the ground state density.

Now if one assumes

$$F[\rho] = \frac{e^2}{2} \int d^3r d^3r' \frac{\rho(\vec{r})\rho(\vec{r}')}{\left|\vec{r} - \vec{r}'\right|} + T_0[\rho] + E_{xc}[\rho] \tag{2.2.2.2}$$

where the first term is repulsive energy, the second term is kinetic energy and the third term is exchange and correlation energy.

By using (2.2.1.2) one can find

$$\int d^3r\delta\rho(\vec{r})\left\{V(\vec{r})+e^2\int d^3r'\frac{\rho(\vec{r}')}{\left|\vec{r}-\vec{r}'\right|}+\frac{\delta T_0[\rho]}{\delta\rho(\vec{r})}+\frac{\delta E_{xc}[\rho]}{\delta\rho(\vec{r})}\right\}=0 \tag{2.2.2.3}$$

Now if one define the effective potential as

$$V_{eff}(\vec{r})=V(\vec{r})+e^2\int d^3r'\frac{\rho(\vec{r}')}{\left|\vec{r}-\vec{r}'\right|}+V_{xc}(\vec{r}) \tag{2.2.2.4}$$

$$V_{xc}(\vec{r})=\frac{\delta E_{xc}[\rho]}{\delta\rho(\vec{r})}$$

instead of solving many body equation, he/she can solve the following single electron Schrödinger equation

$$[-\frac{1}{2m}\nabla^2+V_{eff}(\vec{r})]\chi_\mu(\vec{r})=\varepsilon_\mu\chi_\mu(\vec{r}) \tag{2.2.2.5}$$

$$\rho(\vec{r})=2\sum_\mu^{N/2}\left|\chi_\mu(\vec{r})\right|^2$$

The sum is over the eigenfuctions with the lowest eigenvalues. This equation is called Kohn-Sham equation.

A comment is in order on the physical significance of the eigenvalues ε_μ of (2.2.2.5). We have here

$$2\sum_\mu^{N/2}\varepsilon_\mu=T_0[\rho]+\int d^3rV_{eff}(\vec{r})\rho(\vec{r}) \tag{2.2.2.6}$$

From (2.2.2.1) and (2.2.2.2) it follows that the total energy is given by

$$E[\rho,V]=2\sum_\mu^{N/2}\varepsilon_\mu-\frac{e^2}{2}\int d^3rd^3r'\frac{\rho(\vec{r})\rho(\vec{r}')}{\left|\vec{r}-\vec{r}'\right|}+E_{xc}[\rho]-\int d^3rv_{xc}(\vec{r})\rho(\vec{r}) \tag{2.2.2.7}$$

This relation can be compared with the corresponding one (2.1.21) for independent electrons. The real eigenvalues ε_μ do not describe electronic excitation energies, generally understood to be complex quantities due to finite lifetimes of excitations. However, it turns out that for infinite systems with extended states the energy of the highest occupied level is equal to the chemical potential.

2.2.3. Local Density Approximation (LDA)

The Local Density Approximation (LDA) consists of replacing exchange correlation energy $E_{xc}[\rho]$ by

$$E_{xc}[\rho] = \int d^3r \rho(\vec{r})\varepsilon_{xc}(\rho(\vec{r})) \tag{2.2.3.1}$$

where $\varepsilon_{xc}(\rho(\vec{r}))$ is the exchange and correlation energy per electron of a homogenous electron gas of density ρ, considered to be known. Therefore

$$V_{xc}(\vec{r}) = \frac{d[\rho(\vec{r})\varepsilon_{xc}(\rho(\vec{r}))}{d\rho(\vec{r})} \tag{2.2.3.2}$$

and $V_{eff}(\vec{r})$ depends only on $\rho(\vec{r})$ and the Kohn-Sham equation take simpler form than Hartree-Fock equation.

2.2.4. Local Spin Density Approximation (LSDA)

We can add the effect of spin to LDA, if we write $\rho = \rho_\uparrow + \rho_\downarrow$, where $\uparrow$ stands for spin up and $\downarrow$ stands for spin down. The exchange correlation energy is $\varepsilon_{xc}(\rho \uparrow, \rho \downarrow)$ and the (2X2) matrix equation, reduces to two coupled equations written as

$$(-\frac{1}{2m}\nabla^2 - \mu_B \vec{\sigma}.H(\vec{r}) + V_\sigma^{eff}(\vec{r}))\phi_{\mu\sigma}(\vec{r}) = \varepsilon_{\mu\sigma}\phi_{\mu\sigma}(\vec{r}) \tag{2.2.4.1}$$

The Zeeman term has been included, and $H(\vec{r})$ is an applied uniform magnetic field. The spin dependent effective single particle potential is given by

$$V_\sigma^{eff}(\vec{r}) = V(\vec{r}) + e^2 \int d^3r' \frac{\rho(\vec{r}')}{|\vec{r} - \vec{r}'|} + V_\sigma^{xc}(\vec{r}) \tag{2.2.4.2}$$

Where

$$V_\sigma^{xc}(\vec{r}) = \frac{d}{d\rho_\sigma(\vec{r})}\{[\rho_\uparrow(\vec{r}) + \rho_\downarrow(\vec{r})]\varepsilon_{xc}(\rho_\uparrow(\vec{r}), \rho_\downarrow(\vec{r})\} \tag{2.2.4.3}$$

The spin densities are obtained from the functions $\phi_{\mu\sigma}(\vec{r})$ through

$$\rho_\sigma(\vec{r}) = \sum_\mu^{occ} |\phi_{\mu\sigma}(\vec{r})|^2 \tag{2.2.4.4}$$

The sum is over all occupied orbitals with spin σ.

2.3. Electron Correlations in Semiconductors

Electron correlations are very important in the study of semiconductors. Correlation effects appear even more dramatic when excited states are considered. The energy gaps in semiconductors calculated within the SCF or HF approximation come out much too large and this is easy to understand in terms of the correlation picture.

When an electron is excited from a valance into a conduction band, the added particle polarizes the bonds in its neighborhood because the system is locally no longer charge neutral. The polarized bonds form a polarization cloud which moves with the extra electron (hole) and together they form a quasiparticle. It costs much less energy to create an electron-hole pair with a polarization cloud than without. The generation of such polarization cloud is a correlation effect, which is not taken into account in the independent electron approximation.

In LDA to density functional theory, one can not distinguish the correlation holes around the added and the remaining electrons. The density dependent exchange correlation potential remains unchanged when an extra electron is added to the infinite system. The so-called energy gap problem finds its origin here. One can circumvent it by applying the local ansatz to the computation of energy bands in semiconductors.

2.3.1. Ground State Correlation

Correlation energy calculation for covalent semiconductors become very simple, and in fact can be performed analytically, when the bond-orbital approximation (BOA) is made. In the discussion below, a diamond-like structure is considered.

Our starting point is a Hamiltonian of the form

$$H = \sum_{ij\sigma} t_{ij} a_{i\sigma}^{+} a_{j\sigma} + \frac{1}{2} \sum_{\substack{ijkl \\ \sigma\sigma'}} V_{ijkl} a_{i\sigma}^{+} a_{k\sigma'}^{+} a_{l\sigma'} a_{j\sigma} \tag{2.3.1.1}$$

where

$$V_{ijkl} = \int d^3r d^3r' f_i^*(\vec{r}) f_j(\vec{r}) \frac{e^2}{\left|\vec{r} - \vec{r}'\right|} f_k^*(\vec{r}') f_l(\vec{r}') \tag{2.3.1.2}$$

and $f_j(\vec{r})$ are the set of L basis function which within them, the approximate eigenstate of H is found. The atomic sp^3 hybrid functions take place of the basis function $f_j(\vec{r})$. The t_{ij} are parameters which can be determined from more sophisticated band-structure calculations.

The $a_{i\sigma}^{+}$ are creation operators for electrons in orthognalized, tetrahedral atomic hybrids. Hence, they fulfill the relations

$$[a_{i\sigma}, a_{j\sigma'}^{+}] = \delta_{ij} \delta_{\sigma\sigma'} \tag{2.3.1.2}$$

In the BOA all calculations are carried out in terms of bonding and anti-bonding wavefunctions with their corresponding creation operators

$$B_{I\sigma}^{+} = \frac{1}{\sqrt{2}}(a_{I1\sigma}^{+} + a_{I2\sigma}^{+}) \quad (2.3.1.3)$$

$$A_{I\sigma}^{+} = \frac{1}{\sqrt{2}}(a_{I1\sigma}^{+} - a_{I2\sigma}^{+})$$

Here I is a bond index. The two hybrids forming bond I are indexed by the subscripts 1 and 2, respectively. The SCF ground state is assumed to be the form

$$\left|\Phi_{BOA} \succ = \prod_{I\sigma} B_{I\sigma}^{+} \left|0\right. \succ \quad (2.3.1.4)$$

i.e. the one particle density matrix $p_{ij}^{\sigma\sigma'}$ is simply

$$p_{ij}^{\sigma\sigma'} = \prec a_{i\sigma}^{+} a_{j\sigma'} \succ = \begin{cases} \delta_{\sigma\sigma'}/2 \rightarrow for i.j \in I \\ 0 \rightarrow Otherwise. \end{cases} \quad (2.3.1.5)$$

There are two electrons in each bond.

The Hilbert space within which we calculate the correlated ground state $|\psi_0 \succ$ is spanned by the state $|\Phi_{BOA} \succ$, $A_{I\sigma}^{+} B_{I\sigma} |\Phi_{BOA} \succ$ and $A_{I\sigma}^{+} A_{J\sigma'}^{+} B_{J\sigma'} B_{I\sigma} |\Phi_{BOA} \succ$.

For correlated ground state wavefunction, one can make the ansatz

$$|\psi_0 \succ = e^{S} |\Phi_{BOA} \succ$$

$$|\psi_0 \succ = (1 + \sum_{i\mu} \alpha_{\mu}^{i} c_i^{+} c_{\mu} + \sum_{\substack{i \prec j \\ \mu \prec \theta}} \alpha_{\mu\nu}^{ij} c_i^{+} c_j^{+} c_{\nu} c_{\mu} + \ldots) |\Phi_{BOA} \succ \quad (2.3.1.6)$$

$$|\psi_0 \succ = (1 + \sum_{i\mu} \alpha_{\mu}^{i} \omega_{\mu}^{i} + \sum_{\substack{i \prec j \\ \mu \prec \nu}} \alpha_{\mu\nu}^{ij} \omega_{\mu\nu}^{ij} + \ldots) |\Phi_{BOA} \succ$$

Here $\omega_{\mu}^{i} = c_i^{+} c_{\mu}, \omega_{\mu\nu}^{ij} = c_i^{+} c_j^{+} c_{\nu} c_{\mu}$, and so on. The c_i, c_i^{+} are defined in (2.1.25). We obtain the $\alpha_{\mu}^{i}, \alpha_{\mu\nu}^{ij}, etc.$ diagonalizing H, within a Hilbert space of a given dimension. Alternatively, one may consider the coefficient as variational parameters fixed by minimizing the energy $E = \prec \psi_0 |H| \psi_0 \succ / \prec \psi_0 | \psi_0 \succ$.

The ground state energy is equal to

$$E_0 = \prec H \succ + \sum_{i\mu} \prec H\omega_{\mu}^{i} \succ \alpha_{\mu}^{i} + \sum_{\substack{i \prec j \\ \mu \prec \nu}} \prec H\omega_{\mu\nu}^{ij} \succ \alpha_{\mu\nu}^{ij}$$

Or

$$E_0 = \prec H \succ + \sum_{\mu \prec \nu} E_{\mu\nu},$$
$$E_{\mu\nu} = \sum_{i \prec j} \prec H\omega_{\mu\nu}^{ij} \succ \alpha_{\mu\nu}^{ij} \tag{2.3.1.7}$$

The electrons are annihilated and created in localized molecular (as well as local) orbitals, therefore $\omega_{\mu}^{i} = c_i^{+} c_{\mu}, \omega_{\mu\nu}^{ij} = c_i^{+} c_j^{+} c_{\nu} c_{\mu}$, etc. and c_i, c_i^{+} refer to localized molecular oebitals.

2.3.2. Excited States

Electron correlations in semiconductors with electrons in excited states show features which are absent in the ground state, a fact intimately connected with the so-called energy-gap problem. While calculations done within LDA to the density functional method always seem to give too small energy gaps when applied to semiconductors, the form of different bands agrees quite well with experiment.

Our starting point is again the Hamiltonian (2.3.1.1) with a SCF part

$$H_{SCF} = \sum_{ij\sigma} f_{ij} (a_{i\sigma}^{+} a_{j\sigma} - \prec a_{i\sigma}^{+} a_{j\sigma} \succ) + E_{SCF} \tag{2.3.2.1}$$

It will turn out to be useful to separate the Fock matrix f_{ij} into two parts

$$f_{ij} = f_{ij}^{H} + f_{ij}^{x} \tag{2.3.2.2}$$

Where

$$f_{ij}^{H} = t_{ij} + \sum_{kl\sigma'} V_{ijkl} \prec a_{k\sigma'}^{+} a_{l\sigma'} \succ \tag{2.3.2.3}$$

is the Hartree part and

$$f_{ij}^{x} = -\frac{1}{2} \sum_{kl\sigma'} V_{ijkl} \prec a_{k\sigma'}^{+} a_{l\sigma'} \succ \tag{2.3.2.4}$$

is the exchange part.

By using (2.3.1.2), the notation

$$\left. \begin{aligned} U &= V_{iiii} \\ J_1 &= V_{iijj} \\ K_1 &= V_{ijji} \end{aligned} \right\} i \neq j, i, j \in I \tag{2.3.2.5}$$

is introduced. Here I is a bond index.

The SCF eigenstate of the (N+1)-electron system is written in the form

$$|\Phi_{k c\sigma} \succ = c^{+}_{k c\sigma} |\Phi_{SCF} \succ \tag{2.3.2.6}$$

where $c^{+}_{kc\sigma}$ creates the extra electron in conduction band c with momentum **k** and spin σ. The corresponding SCF energy is

$$[H_{SCF}, c^{+}_{kc\sigma}]_{-} = \varepsilon^{SCF}_{c\sigma}(k) c^{+}_{kc\sigma} \tag{2.3.2.7}$$

We apply the BOA in order to evaluate the effect of exchange on the energy bands. Now the exchange can be evaluated analytically because the one particle density matrix is of the simple form of (2.3.1.5). The SCF conduction and valanced bands are obtained from

$$\begin{aligned} [H_{SCF}, c^{+}_{kc\sigma}]_{-} &= \varepsilon^{SCF}_{c}(k) c^{+}_{kc\sigma} \\ [H_{SCF}, c_{k\nu\sigma}]_{-} &= -\varepsilon^{SCF}_{\nu}(k) c_{k\nu\sigma} \end{aligned} \tag{2.3.2.8}$$

One can writes

$$\begin{aligned} \varepsilon^{SCF}_{c}(k) &= (f^{H}_{ii} - \frac{U}{2} - \frac{3}{2}V_{1331} - \ldots) - (f^{H}_{12} - \frac{J_1}{2} - \frac{1}{2}K_1 - \ldots) + \varepsilon^{\sim SCF}_{c}(k) \text{ , c=1,...4} \\ \varepsilon^{SCF}_{\nu}(k) &= (f^{H}_{ii} - \frac{U}{2} - \frac{3}{2}V_{1331} - \ldots) - (f^{H}_{12} - \frac{J_1}{2} - \frac{1}{2}K_1 - \ldots) + \varepsilon^{\sim SCF}_{\nu}(k) \text{ , c=1,...4} \end{aligned} \tag{2.3.2.9}$$

The $\varepsilon^{\sim SCF}(k)$ differ fro the $\varepsilon^{\sim H}(k)$ by the inclusion of exchange contributions. The energies U and J_1, the dominant interaction matrix elements, are defined by (2.3.2.5).

The effect of correlation can be calculated by making use of the projection method. Let $|\psi^{N}_{0} \succ$ denote the ground state of an elemental semiconductor with energy $E^{(N)}_{0}$. When an electron with momentum **k** and spin σ is put into a conduction band c, the energy of the exact (N+1) electron eigenstate is $E^{(N+1)}_{0}(k,c)$. The excitation energy is defined by

$$\varepsilon_c(k) = E^{(N+1)}_{0}(k,c) - E^{(N)}_{0} \tag{2.3.2.10}$$

and is contained in the one particle correlation function

$$R_{\sigma}(k,c,\tau) = \prec \psi_0 | c_{kc\sigma}\, e^{-\tau[H - E^{(N)}_{0}]} c^{+}_{kc\sigma} |\psi^{(N)}_{0} \succ \tag{2.3.2.11}$$

Within the quasiparticle approximation this expression reduces to

$$R_{\sigma}(k,c,\tau) = \prec \psi^{(N)}_{0} | c_{kc\sigma}\, c^{+}_{kc\sigma} |\psi^{(N)}_{0} \succ e^{-\tau\varepsilon_c(k)} \tag{2.3.2.12}$$

where $\varepsilon_c(k)$ is the quasiparticle energy. It shows up as a pole the Laplace transform $R_{\sigma}(k,c,z)$, i.e.

$$R_{\sigma}(k,c,z) = (c_{kc\sigma}\Omega \left| \frac{1}{z-(L_{SCF}+H_{res})} c^{+}_{kc\sigma}\Omega \right) \tag{2.3.2.13}$$

The operator Ω describes the exact ground state $|\psi_0^N\succ$. The L_{SCF} is Liouville operator belonging to H_{SCF} and defined through $L_0A = [H_{SCF}, A]_-$ where A is arbitrary operator. We assumed $H = H_{SCF} + H_{res}$, the eigenstates and eigenvalues of H_{SCF} are known and the effect of H_{res} on the ground state energy is relatively small.

The excitation energies are given by the poles of $R_{\sigma}(k,c,z)$ [13].

2.4. Relativistic Effective Core Potential (RECP)

One of the most fundamental assumptions in chemistry is that low lying core electrons are relatively inert, and are not perturbed by a molecular environment. This assumption is supported by the chemical similarity of elements in the same column of the periodic table. Most of the important chemical properties of atoms and molecules are determined by the interaction of their valence electrons with the valence electrons of other atomic or molecular species. For molecules containing heavy atoms, large computational savings may be realized if a particular mathematical form for the atomic orbitals is assumed, *a priori*, so that the total number of parameters which must be optimized in, say, the computation of a Hartree-Fock wavefunction may be reduced. This approximation on its own is known as the frozen core approximation. A more extensive savings may be achieved if, instead of simply assuming a particular form for the core orbitals, all terms describing the interaction of the electrons in these orbitals with each other and those outside the core region are simply replaced by a scalar ``effective potential". This provides the additional benefit of reducing the basis set size requirements, since no basis functions are now needed to explicitly describe the core orbitals. This has the effect of drastically reducing the number of two-electron quantities, such as electron repulsion integrals, which must be considered. In correlated electronic structure calculations, the molecular orbitals which correspond to atomic core orbitals are often excluded from the active orbital space. Even when such orbitals are included in the active space, the resultant contributions are typically small for most chemical properties. The use of effective core orbitals in these methods ideally should not significantly affect the quality of the resulting molecular property predictions. Because correlated methods scale higher than Hartree-Fock with respect to basis set size, the decreased basis set size resulting from the use of effective potentials will introduce in an even more dramatic computational savings over all-electron methods.

By using (2.1.8), the Fock operator is defined as

$$\hat{F} = -\frac{\nabla_r^2}{2} - \frac{Z}{r} + \frac{l(l+1)}{2r^2} + \sum_{j=1}^{n}(\hat{J}_i - \hat{K}_j) \tag{2.4.1}$$

where the action of the operator $\hat{J}_j$ and $\hat{K}_j$ is defined by

$$\hat{J}_j \phi_i(\vec{r}_1) = (\int dr_2 \phi_j^*(\vec{r}_2) \frac{1}{r_{12}} \phi_j(\vec{r}_2)) \phi_i(\vec{r}_1)$$
$$\hat{K}_j \phi_i(\vec{r}_1) = (\int dr_2 \phi_j^*(\vec{r}_2) \frac{1}{r_{12}} \phi_i(\vec{r}_2)) \phi_j(\vec{r}_1) \qquad (2.4.2)$$

The ϕ_i are the single particle eigenfunctions of the Fock operator.

If we now divide the orbitals into a group of N_c core orbitals and V_v valence orbitals, we may re-express the fock operator as

$$\hat{F} = -\frac{\nabla_r^2}{2} - \frac{Z}{r} + \frac{l(l+1)}{2r^2} + \hat{V}^{core} + \hat{V}^{val} \qquad (2.4.3)$$

Where $\hat{V}^{core} = \sum_a^{N_c} (\hat{J}_a - \hat{K}_a)$ and $\hat{V}^{val} = \sum_i^{N_v} (\hat{J}_i - \hat{K}_i)$.

For the molecular case, one can expand the Fock operator, if he/she replaces $\hat{V}^{core}$ by an effective core potential (ECP), $\hat{V}^{eff}$, a local potential, and the nuclear charge, Z, by

$Z_{eff} = Z - N_c$, where N_c is the number of core electrons. The Fock equation will take the form

$$(-\frac{\nabla_r^2}{2} - \frac{Z_{eff}}{r} + \frac{l(l+1)}{2r^2} + \hat{V}^{eff} + \hat{V}^{val}) \phi_i = \varepsilon_l \phi_i \qquad (2.4.4)$$

The form of this effective potential is typically derived from numerical Hartree-Fock atomic solution, in accordance with the frozen core approximation.

If $\hat{I}$ is angular momentum operator of a valance atomic orbital, and we replace $\{\phi_i\}$ by approximate pseudo-orbitals, $\{\chi_i\}$, which are nodeless, we can rewrite (2.2.4) as

$$(-\frac{\nabla_r^2}{2} - \frac{Z}{r} + \frac{l(l+1)}{2r^2} + V_I^{eff} + V_{val}^I) \chi_i^I = \varepsilon_i^I \chi_i^I \qquad (2.4.5)$$

By solving (2.4.5) for V_I^{eff} one find

$$V_I^{eff} = \varepsilon_i^I + \frac{Z_{eff}}{r} - \frac{l(l+1)}{2r^2} + \frac{(\frac{1}{2}\nabla_r^2 - V_{val}^I)\chi_i^I}{\chi_i^I} \qquad (2.4.6)$$

The total V_{eff} for an atom, then, is written

$$V_{eff} = \sum_I V_I^{eff} \sum_{m=-I}^{I} |lm \succ \prec lm| \qquad (2.4.7)$$

So that the orthonormality of the spherical harmonics, $|lm \succ$, pairs each atomic wavefunction with proper V_I^{eff} term.

Relativistic effective core potential (RECP) may be obtained in a manner directly analogous to the non-relativistic EDP's. The starting point for RECP's is the atomic Dirac-Coulomb-Fock equation. The valence solution, $\{\phi_i\}$, are four-component spinors, but the large radial component generally accounts for over 99% of the electronic density. Therefore the re-normalized large component are used as the starting points for the relativistic pseudo-orbitals, χ_i. Once $\{\chi_i\}$ have been determined, the RECP's are obtained. The new form of (2.4.5) is

$$(-\frac{\nabla_r^2}{2} - \frac{Z_{eff}}{r} + \frac{l(l+1)}{2r^2} + \hat{V}^{eff}_{I,j} + \hat{V}^{I}_{val})\chi_i^{I,j} = \varepsilon_i^{I,j}\chi_i^{I,j} \tag{2.4.6}$$

where $\hat{j} = \hat{l} + \hat{s}$ is total angular momentum operator.

Since the RECP's are typically used within the framework of non-relativistic quantum chemistry, where single particle states are eigenfunction of $\hat{l}$ and $\hat{s}$ rather than $\hat{j}$, it is necessary to construct a V^{eff} which is only $\hat{I}$ dependent. This may be accomplished by statistically averaging over all of the appreciate j-dependent RECP's which are associated with a particular I-value to give what are known as I-averaged RECP's (AREP).

$$V_I^{AREP} = (2l+1)^{-1}[lV^{REP}_{l,j=l-\frac{1}{2}} + (l+1)V^{REP}_{l,j=l+\frac{1}{2}}] \tag{2.4.7}$$

The total V^{AREP} may be given by

$$V^{AREP} = \sum_I V_I^{AREP} \sum_{m=-l}^{I} |lm \succ \prec lm| \tag{2.4.8}$$

which is completely analogous to (2.4.7).

For atoms beyond the third row of the periodic table i.e for these very large nuclei, electrons near the nucleus are treated in an approximate way, via ECPs. This treatment includes some relativistic effects, which are important in these atoms.

2.5. Symmetry of Clusters

The space group of many of the simpler crystal structures are, referred to as point space groups [14]. Every element of a point space group may be written as the product of an element of a translation group with an element of a point group, every element of the latter being either a rotation or the product of a rotation and an inversion.

In clusters we deal with the point group only. The zinc blende structure has the point group T_d^2, and the face-centered cubic structure has the point space group O_h^5. The space group for the diamond structure, O_h^7, is not a point space group [14].

The lines and points of symmetry in the Brillouin zone of the zinc blende structure are shown in figure 1.

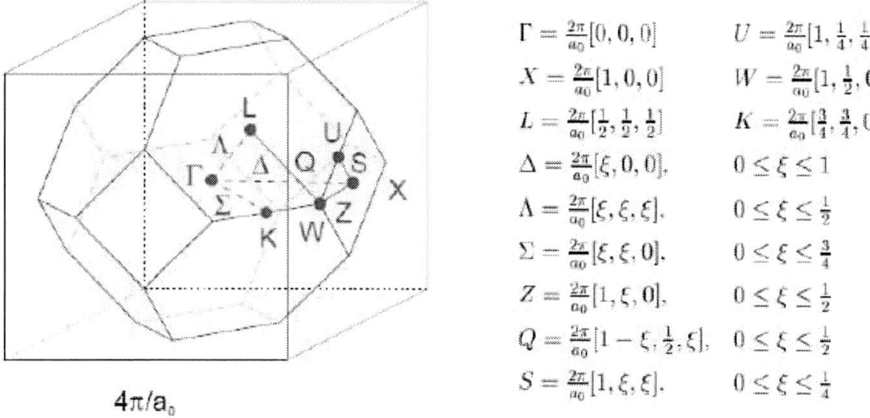

Figure 1. Lines and Points of symmetry in the Brillouin zone of the zinc blende structure.

As an example, rotation by 90^0 about the axis Δ followed by an inversion, followed by the time reversal symmetry operation (replacing $\vec{k}$ by $-\vec{k}$) will convert a wave function belonging to Δ_3 into one belonging to Δ_4, while leaving $\vec{k}$ and Hamiltonian invariant. Thus Δ_3 and Δ_4 stick together [15].

Also one can define the action integral as

$$S = \int L(\phi, \partial_\mu \phi, x_\mu) d^4 x \qquad (2.5.1)$$

where $L(\phi, \partial_\mu \phi, x_\mu)$ is Largrangian density function.

If Under the transformation of space-time and internal space i.e

$$\begin{aligned} x^\mu &\to x'^\mu \\ \phi(x) &\to \phi'(x') \end{aligned} \qquad (2.5.2)$$

the action S is invariant, it can be shown that, there is a current density as follows

$$j_i^\nu = -L(X_i)^\nu_\mu + \frac{\partial L}{\partial U^A_{,\nu}}[U^A_{,\lambda}(X_i)^\lambda_\mu x^\mu - (X_i)^A_B U^B] \qquad (2.5.3)$$

Where

$$\begin{aligned} x'^\mu &= (X_i)^\nu_\mu x^\mu \\ U^A &= (X_i)^A_B U^B \end{aligned} \qquad (2.5.4)$$

such that $\partial_\nu j^\nu = 0$. It is called Neother's theorem [16].

We can divide the j_i^ν to two parts as

$$j_i^\nu = [-L(X_i)_\mu^\nu x^\mu + \frac{\partial L}{\partial U^A_{,\nu}} U^A_{,\lambda}(X_i)_\mu^\lambda x^\mu] - [\frac{\partial L}{\partial U^A_{,\nu}}(X_i)^A_B U^B] \quad (2.5.5)$$

where first part is referred to space-time transformation and second part is referred to internal space transformation. By usage of the first part, one can find the conservation law of angular momentum or energy-momentum tensor and by usage of second part, he/she can find the conservation law of spin (i.e. internal angular momentum)[16]. Therefore the symmetry of cluster will appear itself in conservation laws and $\vec{k}$, and Hamiltonian invariant.

3. InSb Clusters

It is important to select an appropriate basis set and cluster to treat a particular problem. The clusters are terminated with hydrogen, with the X-H bonds oriented along the bulk bond direction. The cluster itself takes two main forms. The first is atom-centered surrounded by concentric shells of bulk atoms. This cluster therefore possesses a tetrahedral symmetry. For the III-V's, this means that there is an excess of either the group-III or group-V atom type, and consequentially the neutral cluster would have the incorrect number of bonding electrons. For all of the bonding orbital to be filled, the neutral charge state is modeled by charging the cluster by the difference in the number of, group-III and group-V atoms. For example, a Ga centered cluster $Ga_{19}As_{16}H_{36}$ would be charged 3e. The second type of cluster is bond centered, which avoids the charging problem encountered in the atom centered III-V cluster as these clusters contain the same number of group-III and group-V atoms (by symmetry). This type of cluster is referred to as stoichiometric, and possesses trigonal symmetry. One disadvantage of using stoichiometric cluster for III-V compound is that they possess an inherent dipole. Consequentially the central bond is longer than the six equivalent back bonds. However, for most III-V materials all bonds are reproduced to within 3% of the experimental value [17].

We considered $In_4Sb_4H_{18}$ cluster with zinc-blend structure of InSb and used Guassian 2003 Software[18] for doing simulation. The figure 2 shows the cluster. The cohesive energy and band gap of InSb were calculated by Unrestricted Hartee-Fock (UHF) and Density Functional Theory (DFT) method under UHF/STO-3G and B3LYP/STO-3G codes. The results were shown in table 1.

Table 1. The cohesive energy and band gap of InSb, non-relativistic and relativistic cases

	UHF		DFT	
	Cohesive energy (a.u)	Band gap (ev)	Cohesive energy(a.u)	Band gap (ev)
Non-relativistic case	0.3165	4.3224	0.2952	0.9184
Relativistic case	0.1266	1.7049	0.1288	0.6509

It is noted that, the band gap was estimated by the difference in the energies of the highest occupied and lowest unoccupied Khon-Sham eigenvalues. For doing the DFT and UHF calculations we must use suitable basis sets. Basis sets for atoms beyond the third row of the periodic table are handled somewhat differently. For these very large nuclei, electrons near the nucleus are treated in an approximate way, via effective core potential (ECPs). This treatment includes some relativistic effects, which are important in these atoms. The LANL2DZ basis set is the best known of these which was used by us for simulation[18]. The result of simulation after considering the relativistic effects is shown in table 1. By comparing the results in table 1, it is concluded that by adding relativistic effects we could got better result [19].

Also we know the unit cells of the ternary alloys are defined by Vegard's law[20] as follows:

$$a_A B_X C_{1-x} = (x)a_{AB} + (1-x)a_{AC}$$
$$a_{A_{1-x}} B_x C = (1-x)a_{AC} + (x)a_{BC} \quad (3.1)$$

where $a_{ABX}C_{1-X}$, $a_{A_{1-X}}B_XC$, a_{AC}, a_{BC} ,and a_{AB} are the lattice constant of the alloys and compounds respectively and InSb has zinc belende structure. If we assumed the Vegard's law is valid for nano clusters and they have zinc belende structure the We can consider $In_{10}Sb_4H_{23}$ cluster in zinc belende structure i.e its lattice constant equal to 6.4598 A. Since the tetragonal symmetry of InBi suggested that the grown InSbBi might have a tetragonal rather than zinc blende structure[21,22], we kept the dimensions in X-Y plane fix, and changed the dimension in Z direction (called h) after substituting Bi atoms instead of Sb atoms in the cluster. By using B3LYP/LANL2DZ Code, we calculated band gap, cohesive energy, nuclear repulsion energy and symmetry of

cluster in density functional method based on the relativistic bases.

The band gap was estimated by the difference in the energies of the highest occupied and lowest unoccupied Khon-Sham eigenvalues.

The result is shown in table 2. As the results show, when h/4=1.62495, band gap of $In_{10}Sb_3BiH_{23}$ and $In_{10}Sb_2Bi_2H_{23}$ are smaller than $In_{10}Sb_4H_{23}$ (see item 3, 8 and 9). It means that, after adding the Bi to InSb, the structure was changed from zinc blende to tetragonal and band gap was decreased although the symmetry was not changed [23].

4. INAS CLUSTERS

We considered $In_{19}Sb_{16}H_{36}$ and $In_{19}As_{16}H_{36}$ clusters in zinc belende structure i.e its lattice constants are equal to 6.4598 A and 6.058 A respectively.

After adding the As atoms to InSb or adding Sb atoms to InAs the lattice constant can be calculated by Vegard's law[20]. We considered x value equal to the number of guest atoms to the number of host atoms and used the calculated lattice constant by Vegard's law in the simulations. When the number of guest atoms be equal to the number of host atoms, we put the average lattice constant in the simulation.

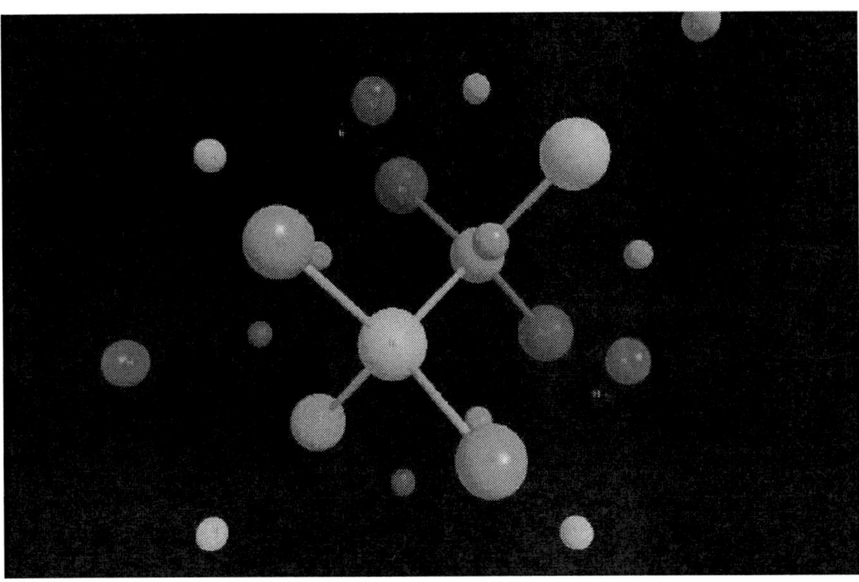

Figure 2. The $In_4Sb_4H_{18}$ Cluster.

Table 2. The simulation results of $In_{10}Sb_4H_{23}$ before and after adding Bi element

Item	Formula	Band Gap, eV	Cohesive energy, a.u	Nuclear Repulsion Energy, Hart.	Full point group	Largest Abelian group	Largest concise Abelian group	Lattice constant in Z direction, h/4, A
1	$In_{10}Sb_4H_{23}$	0.571	-52.69	251.0587	C1, NOP=1	C1, NOP=1	C1, NOP=1	1.61495
2	$In_{10}Sb_3BiH_{23}$	0.557	-52.75	251.0587	C1, NOP=1	C1, NOP=1	C1, NOP=1	1.61495
3	$In_{10}Sb_3BiH_{23}$	0.375	-52.70	250.5582	C1, NOP=1	C1, NOP=1	C1, NOP=1	1.62495
4	$In_{10}Sb_3BiH_{23}$	0.507	-52.73	250.05925	C1, NOP=1	C1, NOP=1	C1, NOP=1	1.63495
5	$In_{10}Sb_3BiH_{23}$	0.535	-52.74	250.3085	C1, NOP=1	C1, NOP=1	C1, NOP=1	1.62995
6	$In_{10}Sb_3BiH_{23}$	0.572	-52.73	250.8082	C1, NOP=1	C1, NOP=1	C1, NOP=1	1.61995
7	$In_{10}Sb_3BiH_{23}$	0.406	-52.71	250.5557	C1, NOP=1	C1, NOP=1	C1, NOP=1	1.625
8	$In_{10}Sb_3BiH_{23}$	0.327	-52.70	250.5607	C1, NOP=1	C1, NOP=1	C1, NOP=1	1.6249

Finally by using B3LYP/LANL2DZ Code, we calculated band gap, cohesive energy, nuclear repulsion energy and symmetry of cluster in density functional method based on the relativistic bases. The results are shown in tables 3 and 4.

As the table 3 shows, when the symmetry of the cluster, $In_{19}Sb_{16}H_{36}$, after adding the As element approaches to the symmetry of the cluster before adding the guest element, the band gap reduction is seen. It means that, not only the element but also the final symmetry of the cluster is important (see item 9 of table 3).

As table 4 shows, when the symmetry of the cluster, $In_{19}As_{16}H_{36}$, after adding the Sb element approaches to the symmetry of the cluster before adding the guest element, the band gap reduction is seen. It means that, not only the element but also the final symmetry of the cluster is important (see items 5 and 9 of table 4).

Table 3. The simulation results of $In_{19}Sb_{16}H_{36}$ before and after adding As element

Item	Formula	Band Gap , eV	Cohesive energy, a.u	Nuclear Repulsion Energy, Hart.	Full point group	Largest Abelian group	Largest concise Abelian group	Lattice constant , a/4, A
1	$In_{19}Sb_{16}H_{36}$	0.387	-141.616	1191.2338	TD, NOP=24	D2, NOP=4	D2, NOP=4	1.61495
2	$In_{19}Sb_{15}AsH_{36}$	0.525	-142.356	1196.0745	C1, NOP=1	C1, NOP=1	C1, NOP=1	1.6082
3	$In_{19}Sb_{14}As_2H_{36}$	0.502	-143.091	1201.85086	C1, NOP=1	C1, NOP=1	C1, NOP=1	1.6006
4	$In_{19}Sb_{13}As_3H_{36}$	0.499	-143.802	1208.5963	C3V, NOP=6	CS, NOP=2	CS, NOP=2	1.59175
5	$In_{19}Sb_{12}As_4H_{36}$	0.548	-144.542	1216.2795	C1, NOP=1	C1, NOP=1	C1, NOP=1	1.58145
6	$In_{19}Sb_{11}As_5H_{36}$	0.489	-145.264	1225.880	C1, NOP=1	C1, NOP=1	C1, NOP=1	1.5693
7	$In_{19}Sb_{10}As_6H_{36}$	0.717	-146.001	1236.9331	C1, NOP=1	C1, NOP=1	C1, NOP=1	1.5546
8	$In_{19}Sb_9As_7H_{36}$	0.638	-146.714	1251.7066	C1, NOP=1	C1, NOP=1	C1, NOP=1	1.5368
9	$In_{19}Sb_8As_8H_{36}$	0.257	-147.428	1229.4901	CS, NOP=2	CS, NOP=2	CS, NOP=2	1.5647

Table 4. The simulation results of $In_{19}As_{16}H_{36}$ before and after adding Sb element

Item	Formula	Band Gap , eV	Cohesive energy, a.u	Nuclear Repulsion Energy, Hart.	Full point group	Largest Abelian group	Largest concise Abelian group	Lattice constant , a/4, A
1	$In_{19}As_{16}H_{36}$	0.969	-153.270	1270.2430	TD, NOP=24	D2, NOP=4	D2, NOP=4	1.5145
2	$In_{19}As_{15}SbH_{36}$	0.488	-152.512	1264.43	C1, NOP=1	C1, NOP=1	C1, NOP=1	1.5212
3	$In_{19}As_{14}Sb_2H_{36}$	0.405	-151.738	1258.3615	C2V, NOP=4	C2V, NOP=4	C2V, NOP=4	1.5288
4	$In_{19}As_{13}Sb_3H_{36}$	0.834	-151.009	1251.1597	C3V, NOP=6	CS, NOP=2	CS, NOP=2	1.5376
5	$In_{19}As_{12}Sb_4H_{36}$	0.229	-150.214	1242.7539	TD, NOP=24	D2, NOP=4	D2, NOP=4	1.548
6	$In_{19}As_{11}Sb_5H_{36}$	0.545	-149.518	1233.0628	C1, NOP=1	C1, NOP=1	C1, NOP=1	1.5602
7	$In_{19}As_{10}Sb_6H_{36}$	0.558	-148.789	1222.0399	C1, NOP=1	C1, NOP=1	C1, NOP=1	1.57477
8	$In_{19}As_9Sb_7H_{36}$	0.509	-148.079	1207.9512	C3V, NOP=6	CS, NOP=2	CS, NOP=2	1.5926
9	$In_{19}As_8Sb_8H_{36}$	0.381	-147.323	1229.4901	CS, NOP=2	CS, NOP=2	CS, NOP=2	1.5647

As the items 3 and 8of table 2 show, when x value is equal to 1/3, the minimum band gap was seen. Also as the item 5 of the table 4 shows again when x value is equal to 4/12, the minimum band gap was seen. In the other words both of them at same value of x show minimum band gap[23].

5. GaAs Clusters

We considered $Ga_4As_4H_{18}$, $Ga_{19}As_{28}H_{24}$, $Ga_4As_3BiH_{18}$ and $Ga_{19}As_{27}BiH_{24}$ clusters and calculated: cohesive energy, band gap by unrestricted Hartee-Fock (UHF) and density functional theory (DFT) methods.

The result of simulation of energy is shown in table 1. We considered the clusters have exactly the zinc bland structure of GaAs.

Table 5. Cohesive energy and Band gap of $Ga_4As_4H_{18}$ and $Ga_{19}As_{28}H_{24}$

	Cohesive energy (a.u)		Band gap (ev)	
	UHF	DFT	UHF	DFT
$Ga_4As_4H_{18}$	0.7369	0.8699	2.2198	1.1246
$Ga_{19}As_{28}H_{24}$	1.3573	1.5931	5.5417	0.5366

The effect of band gap narrowing after adding Bi to $Ga_4As_3BiH_{18}$ and $Ga_{19}As_{27}BiH_{24}$ clusters and the amount of calculated band gap of $Ga_4As_3BiH_{18}$ show is shown in table 6. the LANAL2DZ basis was used.

Table 6. Cohesive energy and Band gap of Ga4As_3BiH_{18} and $Ga_{19}As_{27}BiH_{24}$

	Cohesive energy (a.u)		Band gap (ev)	
	UHF	DFT	UHF	DFT
$Ga_4As_3BiH_{18}$	0.1812	0.2086	4.4818	1.0962
$Ga_{19}As_{27}BiH_{24}$	0.7732	0.7247	2.6712	0.4753

The results of table 5 and table 6 show by inserting Bi instead of As in the both clusters the band gap decrease. As the result of UHF simulation on $Ga_4As_3BiH_{18}$ shows the band gap reach to ~1 ev after inserting Bi.

The ~1 ev value for band gap of GaAsBi alloys was seen theoretically and experimentally[24,25], but although the cluster size (i.e $Ga_4As_3BiH_{18}$) is small and the inclusion of the displacements of the nearest neighbor atoms did not considered, the result of UHF/STO-3G code shows bulk value.

6. CdSe Cluster

It was shown, in contrast to bulk CdSe, the nanosized samples is observed to retain the wurtzite structure well above the bulk phase transition pressure of 3 GPa upstroke and at pressure above 6 GPa, the sample begins to convert to the rock salt structure [26].

Also it was shown, the Alkoxysilane-stabilized CdSe nanocrystals exhibited cubic zinc blend structure [27,28].

In this reason we considered the zinc blend and wurtzite CdSe clusters for ab initio calculations.

We considered $CdSe_4H_{12}$ and $Cd_4Se_4H_{16}$ cluster in zinc-blend structure with lattice constant equal to 4.7 A. The clusters are shown in figure 3 and 4.

Also we considered $Cd_9Se_6H_8$ and $Cd_{15}Se_{10}H_{13}$ clusters in wurtzite structure with lattice constant equal to a=3.43816A and C=5.60168A. The clusters are shown in figure 5 and 6. The calculation results are shown in table 7, table 8, table 9 and table 10 [29].

Figure 3. Zinc-blend Structure of $CdSe_4$ cluster.

Figure 4. Zinc-blend structure of Cd_4Se_4 cluster.

Figure 5. Wurtzite Structure of Cd_9Se_6 clusters.

Figure 6. Wurtzite Structure of $Cd_{20}Se_{13}$ cluster.

Table 7. The cohesive energy, band gap, of zinc-blend structure, Hartree-Fock

Cluster Zinc-Blende	Cohesive energy, a.u	Band gap, eV
$CdSe_4H_{12}$	0.408	6.50
$Cd_4Se_4H_{16}$	0.234	5.36

Table 8. The cohesive energy, band gap, of zinc-blend structure, Density Functional Theory

Cluster Zinc-Blende	Cohésive energy, a.u	Band gap, eV
CdSe4H12	0.387	2.86
Cd4Se4H16	0.243	1.59

Table 9. The cohesive energy, band gap, of wurtzite structure, Hartree-Fock

Cluster Wurtzite	Cohésive energy, a.u	Band gap, eV
$Cd_9Se_6H_8$	0.092	5.31
$Cd_{15}Se_{10}H_{13}$	0.007	3.56

Table 10. The cohesive energy, band gap, of wurtzite structure, Density Functional Theory

Cluster Wurtzite	Cohésive energy, a.u	Band gap, eV
$Cd_9Se_6H_8$	0.12	1.18
$Cd_{15}Se_{10}H_{13}$	0.036	1.37

7. References

[1] A. Thiel and H. Koelsch, Z. Anorg. *Chem.* 66, 288 (1910)
[2] A. I. Blum, N. P. Mokrovski, and A. R. Regel, Seventh All-Union Conference on the Properties of Semiconductors (Kiev, USSR, 1950), *Izv. Akad. Nauk* SSSR Ser. Fiz. 16, 139 (1952)
[3] Handbook of Compound Semiconductors, edited by Paul H. Holloway and Gary E. McGuire, 1995 by Noyes Publications
[4] V. V. Nemoshkalenko and Yu. N. Kucherenko, Computational Physics Techniques in Solid State Theory, Nonideal Crystals, *Naukova Dumka Press*, Kiev, 1985
[5] Theory of Chemisorption, ed ited by G. Smith (Mir Press, Moscow, 1983)(Russian trans.)
[6] D. R. Hartree: Proc. Cambridge Philos. Soc. 24, 89 (1928)
[7] V. Fock: Z. Phys. 61, 126 (1930)
[8] J. C. Slater: *Phys. Rev.* 35, 210 (1930)
[9] C. C. J. Roothaan: *Rev. Mod. Phys.* 23, 69 (1951)
[10] L. H. Thomas: Proc. Cambridge Philos. Soc. 23, 542 (1927)
[11] E. Fermi: *Z. Phys.* 48, 73 (1928)
[12] M. Levy: *Proc. Natl. Acad. Sci.* (USA) 76, 6062 (1979)
[13] Peter Fulde, "Electron Correlations in Molecules and Solids", Third Enlarged Edition, pp 207-215, *Springer (*1995)
[14] W. H. Zachariasen, "Theory of X-ray diffraction in crystals", Chap. 2, John Wiely and Sons, Inc. New York, (1945)
[15] R. H. Parmenter, *Phys. Rev.* Vol. 100, No. 2, pp 573-579, Oct./15/1955
[16] Lewis H. Ryder, "Quantum Field Theory", pp 87-89, Cambridge University Press (1987)
[17] J.P. Goss, PhD., thesis in Physics, Tan. 1997.
[18] J. B. Foresman, A. Frisc, "Exploring Chemistry with Electronic Methods", Gaussian Inc., 1996
[19] H. Simchi, Comp. Matt., Scie., 40, pp 557-561, 2007
[20] L. Vegard, *Z. Phys.* 5, 17, 1921
[21] J.J Lee, and M. Razeghi, SPIE, Vol. 3287, 0277-786X, 1998
[22] V. G. Deybuk and Ya. I. Vikluk, Phys. and Techno. of Semicon. V.36, No.10, 2002 (in Russian language)
[23] H.Simchi, "The theoretical study on nano InSb clusters after adding Bi, As and Al elements", *submitted to optic and quantum electronic journal*
[24] T. Miyamoto, K. Takeuchi, F. Koyama, and K. Iga, *IEEE Photonics Technol.* Lett. 9, 1448, 1997.
[25] S. Tixier, and others, *App. Phys. Letters*, 86 112113, 2005.
[26] .H. Tolbert and A.P. Alivisatos, "Rock salt structural transformation" *J. hem.Phys.* 102, 4642, 1995
[27] A. L. Rogach et. al., "Synthesis and characterization of a size series of extremely small Thiol-stabilized CdSe nanocrystalas" *J. Phys. Chem.* B, 103, 3065-3069, 1999
[28] V. Ptatschek et. al. *J. Phys. Chem.* B, 101, 8898, 1997

[29] H. Simchi " Binding Energy, Band Gap,Infrared Intensities and absorption spectra of nano CdSe clusters" *submitted to Com. Matt. Sci Journal.*

In: Computational Materials
Editor: Wilhelm U. Oster

ISBN: 978-1-60456-896-7

Chapter 7

COMPUTATIONAL PROGNOSES OF DAMAGE GROWTH AND FAILURE IN SOME STEEL ELEMENTS OF A LIQUEFIED NATURAL GAS TERMINAL COVERING THE TEMPERATURE RANGE OF THE DUCTILE-TO-BRITTLE TRANSITION REGION

J. Jackiewicz[1]
The Department of Mechanical Engineering
at the University of Technology and Life Sciences, Prof. Kaliskiego 7,
Bydgoszcz, PL 85-796, Poland

ABSTRACT

The paper discusses a method for numerical simulation of the transition from ductile to brittle cleavage behavior in some steel elements of a liquefied natural gas terminal. To determine an onset of material failure, a coupled mechanistic model is developed for the effect of preceding ductile damage on cleavage fracture. The aim of the research is to improve our understanding of transferable and applicable, formulated fracture criterion when the cleavage crack initiation is preceded by ductile crack growth giving a large scatter to values of fracture toughness. For grains and interfaces, results of micromechanical simulations refer to macroscopic predictions, which can be obtained by means of a version of the Gurson type model incorporating effects of a microvoid aspect ratio.

An additional motivation of the presented work is further development of combined method of contour elements because the standard finite element method has some limitations in several applications. In the paper implemented, new numerical formulations, that are built-up on analytical integration methods, are briefly described and implemented.

The validity of the developed fracture model is checked by comparisons of results of standardized experimental tests of fracture mechanics with results of micromechanical simulations obtained by means of the combined method of contour elements.

[1] phone: +48-52-340-8252; fax: +48-52-340-8250; e-mail: Jacek.Jackiewicz@mail.utp.edu.pl.

Index Terms—Contour subdomain method, Ductile to brittle transition, Mesomechanics behavior of the polycrystalline microstructure, Weibull statistics, Methods of elastic-plastic analysis, Liquefied natural gas, Ferritic and duplex steels.

1. INTRODUCTION

As a result of the worldwide energy crisis, much emphasis is currently being placed on expanding the production and storage facilities for energy producing fuels, such as liquefied natural gas. Liquefied natural gas or LNG is natural gas that has been converted to liquid form. The process of natural gas liquefaction involves removal of certain components (such as dust, helium, or impurities that could cause difficulty downstream, e.g. water, and heavy hydrocarbons) and then condensation into a liquid by cooling it to approximately -163^{0}C at close to atmospheric pressure (maximum transport pressure set around 25kPa). Liquefaction reduces the natural gas volume by approximately 600 times, making it more economical to transport or storage because much of the world's natural gas resources are generally located far from the main consumption areas. Nowadays liquid natural gas is stored at LNG terminals on a huge scale because of its high energy density and the cleanliness of its combustion exhaust gas. The LNG terminal consists of storage and regasification facilities as shown in figure 1. The regasification facilities are used to convert the LNG, stored in specially made storage tanks, from the liquefied phase to the gaseous phase, which is moved to the final destination through the natural gas pipeline system.

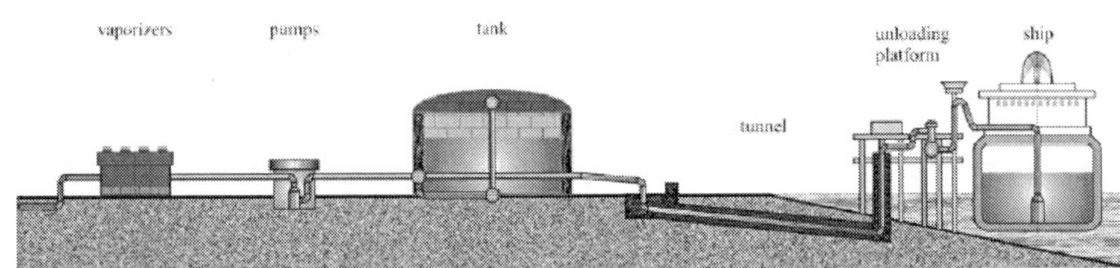

Figure 1. LNG storage concept.

The LNG storage tank is in fact a tank within a tank that is filled with super insulation under high vacuum. The inner tank, in contact with the LNG, is made of 9% nickel steel suitable for cryogenic service and structural loading of the LNG. The outer tank is generally made of carbon steel. The maximum working pressure for these tanks is normally 2MPa or lower, with product stored at 345-830kPa. The LNG fueling pump is a single or multistage centrifugal pump submerged in the storage tank or in a separate vacuum-insulated sump allowing on-demand LNG fueling.

As a liquid, LNG is not explosive. If LNG leaks from a storage tank, due to damage in the tank wall or piping, the LNG will receive heat flux from the floor of the safety dike and vigorously vaporize above it, and a flammable gas (main component methane) would diffuse in the atmosphere. The LNG vapor will only explode if it is concentrated in an enclosed space within the flammable range of 5%-15% when mixed with air. However, when enough LNG is spilled on water at a very fast rate, a rapid phase transition occurs. Heat is transferred from the water to the LNG, causing the LNG to instantly convert from its liquid phase to its gaseous phase. A large amount of energy is released during this rapid transition between phases and a physical explosion can occur. While there is no combustion, this physical explosion can be hazardous to any nearby person or buildings.

The problems we face today, regarding serious disaster protection for LNG facilities, take on greater significance when we consider the large value of the LNG terminal installations, the limited alternative supplies of LNG, and the very serious effects of interrupted service. A wide use of cryogenic tanks for storing and transporting LNG requires special purpose methods of assessment of strength and durability of these vessels because any one of LNG storage tanks might be destroyed in seconds, but effects of this are usually remembered for years.

Traditionally, LNG tanks are designed with due regard for pressure, weight, transporting and structural loads, etc. A distinctive feature of cryogenic ranks is generation of considerable temperature stresses under unsteady conditions of filling (cooling) and discharging (warming) which affect the strength and durability of the structure. The major difficulty encountered in correct determination of temperature stresses in elements of cryogenic tanks arises from a lack of enough reliable data on temperature fields produced on cooling and warming. In the strict sense analytical numerical solutions of the problem of unsteady heat exchange have been obtained only in a certain approximation due to extreme complexity of mathematical modeling of physical phenomena of heat transfer in a boiling cryoproduct. However, experimental tests of small models do not guarantee adequacy as we pass on to a full-scale object because temperature fields are dependent on scale factor.

In view of these reasons, for calculation of strength of LNG tanks submitted to a certain number of stress cycles it is convenient to use numerical simulations, according to which the amplitude of temperature stresses in a tank can be assessed. In a first approximation, it was assumed that in the filling process the temperature of the tank wall close to the phase interface changes irregularly (from the ambient temperature to the temperature of the liquid product). Such a temperature distribution determines the maximum amplitude of the temperature stresses and, consequently, can indicate the minimum durability of the tank.

9% nickel steel is still seldom used above 40mm thickness, and some of the difficulties lie in the fact that some design codes do not consider such large wall thickness in their calculations. To fully reassure potential users and designers on the availability of such materials and their intrinsic safety so that it would become possible to design and build large storage tanks, a new coupled micromechanical numerical model based on the local criterion approach is proposed. Taking into consideration thermal contraction and brittleness risk at the temperature range of the ductile-to-brittle transition region, the coupled micromechanical numerical model can be used to evaluating the possibility of damage growth and an onset of material failure in a LNG tank wall or piping.

In the paper a calibration procedure of material parameters for the coupled micromechanical numerical model embracing the fracture behavior of a pure ductile mechanism that competes with a pure cleavage mechanism is described. Using the combined fracture model with the previously calibrated material parameters the probabilities of structural failures of some steel elements of the LNG storage tank can be determined.

2. Materials for LNG Tank Construction

The materials used in the vessels which keep the gas at liquefaction temperatures need to remain ductile and crack resistant with a high level of safety (see Ref. [1]). The material also needs to have high strength in order to reduce the wall thickness of the container and it must permit welding without any risk of brittle fracture. The materials used for the various working temperatures are listed in table 1.

In the case of LEG/LNG, large land-based tanks, at temperatures of below -105°C only 5-9% nickel steels have the optimum properties. However, large LNG carriers have the spherical tanks made of aluminum for to reduce weight. When the temperatures are even lower it is necessary to use austenitic stainless steels or aluminum alloys.

Table 1. Liquefaction temperatures of gases and used types of parent materials.

Gas	Liquefaction temperature (°C)	Type of parent material used
Ammonia	-33.4	Carbon steel
Propane (LPG)	-45.5 to -42.1	Fine grain Al-killed steel
Propylene	-47.7	2.25% Ni steel
Carbon disulphide	-50.2	
Carbon dioxide	-78.5	3.5% Ni steel
Acetylene	-84	
Ethane	-88.4	
Ethylene (LEG)	-103.8	5-9% Ni steel
Krypton	-151	
Methane (LNG)	-163	
Oxygen	-182.9	
Nitrogen	-195.8	Austenitic stainless steel
Neon	-246.1	
Heavy Hydrogen	-249.6	Al alloys
Hydrogen	-252.8	
Helium	-268.9	
Absolute zero	-273.18	

Table 2. Typical properties for cryogenic steels

	ASTM 304L EN 10088-1 1.4305	ASTM A553Gr1 EN 10028-4 X8Ni9	EN10028-4 12Ni19	ASTM A203GrE EN10028-4 12Ni14
C max. %	0.03	0.13	0.12	0.15
Mn max. %	2.0	0.9	0.8	0.8
Cr %	16.5			
Ni %	9.5	9	5	3.5
Yield strength, R_p (MPa) min.	190	585	390	345
Ultimate strength, R_m (MPa) min.	490	690	530-710	490-640
Charpy V-notch impact energy (Joule)	>60 (-196°C)	>70 (-196°C)	> 34 (-120°C)	> 27 (-100°C)

5-9 % nickel steels

The 9% nickel steels provide a combination of properties at a reasonable price. The excellent low temperature impact properties are the result of a fine grained structure of tough nickel-ferrite. Small amounts of stable austenite formed by tempering improve impact resistance after heat treatment. The cryogenic nickel steels are usually quenched (to form martensite) and tempered (to modify the martensite and produce carbides) in a very narrow temperature range to optimize the microstructure and thereby its properties.

Stainless steels

Stainless steels for cryogenic applications down to liquid helium temperatures are well established and play an important role in LNG ships and in piping systems and associated equipment. The standard grade of wrought stainless steel for general cryogenic applications is type 304L, but various additional grades have been specified, including 304LN and 316LN. An important feature of the austenitic stainless steels is their very good weldability and corrosion resistance.

The 9%Ni steel is being used for cryogenic applications, such as LNG tankers. Its nil-ductility-temperature is reported to be below the boiling point of liquid nitrogen. However, more precise measurements of low-temperature toughness have been difficult to obtain. Attempts to measure K_{IC} at -196^0C have been complicated by excess plasticity, but values of K_Q corresponding to extension of a fatigue pre-crack and based on 5% secant offset are in the range 110-140 MPam$^{1/2}$ (see Ref. [2]).

The plane strain K_{IC} fracture toughness test involves the loading to failure of pre-cracked notched specimens in tension or three-point bending. The test is unusual in that the calculation of a valid toughness value can only be determined after the test has been completed, via examination of the load-displacement plot and measurement of the fatigue pre-crack crack length. The provisional fracture toughness K_Q is first calculated from the following equation:

$$K_Q = \frac{P_Q}{B\sqrt{W}} f(a/W), \tag{1}$$

where P_Q is the load corresponding to a defined increment of crack length, B is the specimen thickness, W the width, and the function $f(a/W)$ is a geometry dependent factor that relates the compliance of the specimen to the ratio of crack length and width. Only when specific validity criteria are satisfied, can the provisional fracture toughness K_Q be quoted as valid plane strain fracture toughness K_{IC}.

Fracture-toughness properties of materials are important from two viewpoints; propagation resistance and initiation resistance. If a fracture should get started, propagation resistance should keep it as short as possible. Namely, when a break occurs in a pipeline, pressure is not released from everywhere in the line at once; it is exhausted from behind a decompression front which propagates through the pressurizing medium at the acoustic velocity. If the fracture is faster than the decompression velocity, the fracture cannot unload (the nominal stress at the propagating fracture front remains the same as the original nominal stress before the break) and the fracture tends to continue to propagate. On the other hand, if the fracture is slower than the decompression velocity, it can unload and therefore would be expected to arrest.

Steels in use for gas-transmission service have transition temperatures below which they fail in a cleavage mode with speeds higher than the decompression velocity, and above which they fail in a shear mode with speeds lower than the decompression velocity. Charpy V-notch impact tests and the Battelle drop-weight tear test are roughly capable of identifying this transition temperature (see figures 2 and 3).

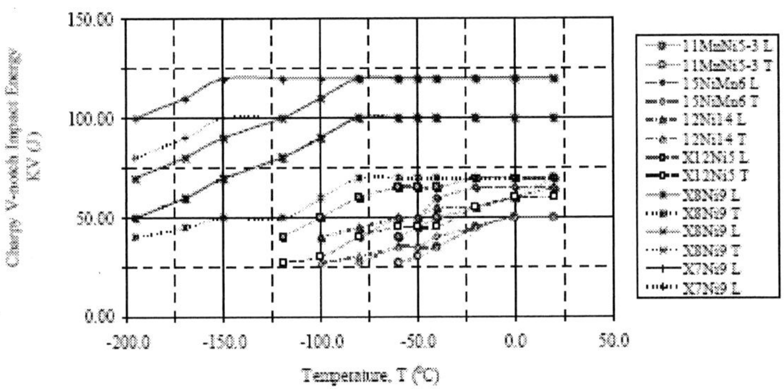

Figure 2. Charpy V-notch impact test results from Ref. [3].

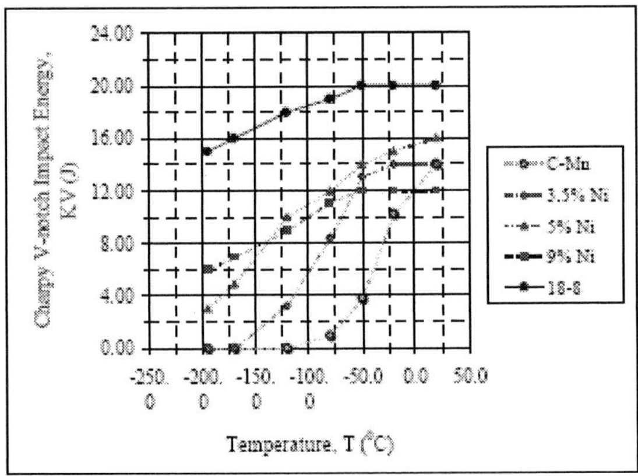

Figure 3. Charpy V-notch impact test results from Ref. [4].

It should be noted that materials which do not have transition temperatures and which, therefore, always fail in relatively slow-speed shear, are usually more expensive. Metals that have a face-centered-cubic crystal lattice such as aluminum show no loss of ductility at low temperatures. Members of the body-centered-cubic (bcc) class such as iron fail with limited ductility below a certain transition-temperature range. Austenitic stainless steels have a face-centered-cubic (fcc) lattice and, therefore, have no ductile-brittle transition temperature. In other ferrous alloys, the transition temperature is lowered by reducing carbon content or by increasing nickel content.

A major problem in the construction of an LNG pipeline is the loss of ductility exhibited by some materials at low temperatures. The use of Invar pipe without expansion joints shows several practical advantages over the other materials. Figure 4 shows a proposed design of Invar pipe with foamed-in-place insulation. The invar alloy, $Fe_{65}Ni_{35}$, is well known as the material which has a very small thermal expansion coefficient around room temperature ($<2\times10^{-6}\,K^{-1}$ compared to most metallic materials which have a thermal expansion coefficient of $10\div20\times10^{-6}\,K^{-1}$, namely, compared with stainless steel, Invar alloy has a coefficient of linear expansion of approximately one-tenth) and is widely used in industrial applications. In addition to the thermal expansion anomaly, Fe-rich fcc Fe-Ni alloys show many other anomalous properties, such as large negative pressure effects on the magnetization and on the Curie temperature (the transition temperature between ferromagnetic and paramagnetic phases), a large forced volume magnetostriction (the volume expansion induced by an applied magnetic field), and an anomalous temperature dependence of the elastic constants.

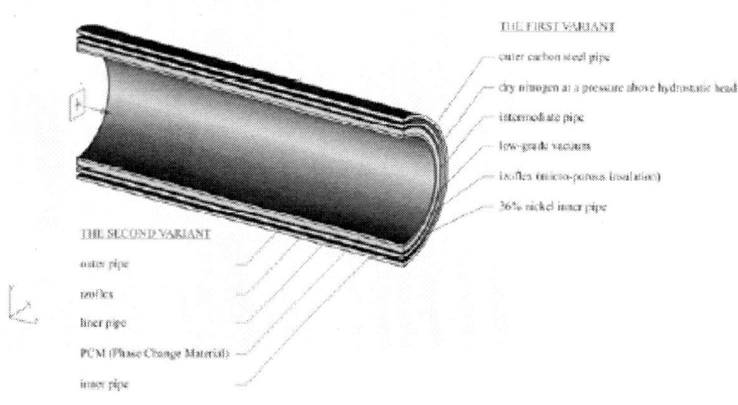

Figure 4. Proposed designs of Invar pipe with foamed-in-place insulation.

Applying Invar alloy as material for LNG pipes makes it possible to build an LNG piping system solely made of straight pipes, enabling significant reduction in construction cost. In particular, when an Invar alloy LNG piping is installed in a tunnel, reduction of the diameter of the tunnel will lower the tunnel construction cost (see figure 5).

Steels are microstructurally complex and thus versatile. Some steels are much stronger than mild steel, but also tough.

Nickel (Ni) is an essential alloying element in some steels. Ni is added to increase solid-solution strength and hardness as well as to increase hardenability. Since Ni does not form carbide in steel it is outstanding for lowering the transition temperature of steels.

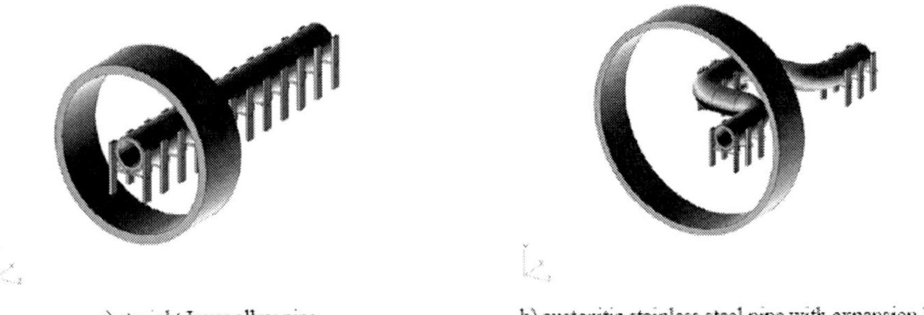

Figure 5. LNG piping installed in a tunnel.

The ferritic stainless steels are basically chromium steels with chromium ranging between 10.5 and 27%. These alloys deliberately lack high nickel contents, because nickel renders the steels austenitic (as previously mentioned). The ferritic stainless steels are the lower-cost stainless steels; because they contain less expensive alloying elements, and they do not contain nickel (nickel is more expensive than chromium). The microscopic fracture mechanisms of crack initiation and propagation are found to change in polycrystalline ferritic steels from brittle at lower temperatures to ductile at higher temperatures. This results in the ductile-to-brittle transition curve, where a parameter such as Charpy impact energy or fracture toughness is plotted against temperature as shown in figure 3. At higher temperatures corresponding to the upper shelf of this transition curve, crack initiation and propagation is a result of plasticity and void formation usually at second phase inclusions or carbide precipitates. It is the coalescence of these voids that effects crack propagation. By comparison at the lower temperatures corresponding to the lower shelf, brittle fracture is considered to occur predominantly by transgranular cleavage on {1 0 0} planes before extensive plastic flow intervenes.

Duplex stainless steels have a mixed microstructure of ferrite and austenite. There is only one standard type of duplex stainless steel, which contains 23 to 28% Cr, 2.5 to 5.0% Ni, and 1.0 to 2.0% Mo. The duplex stainless steels have corrosion resistance similar to an austenitic stainless steel but possess higher tensile and yield strengths and improved resistance to stress-corrosion cracking.

Austenitic stainless steels have a microstructure of austenite at room temperature. Thus, they are nonmagnetic. Austenitic stainless steel (such as the popular type 304) has been called 18/8 stainless steel, because it contains nominally 18% Cr and 8% Ni. All the austenitic stainless steels are essentially chromium-nickel alloys. The chromium varies between 15 and 24% and the nickel between 3 and 22%. The corrosion resistance of the austenitic stainless steels is superior to other types of stainless steel.

In order to achieve the fracture-toughness requirements and lower basic-material costs the chemical component of nickel in steel elements of the liquefied natural gas terminal should be optimized.

3. Fracture Mechanisms for Ferritic and Duplex Steels

Cleavage fracture is characterized by crack propagation with very little plastic deformation and occurring in a crystallographic fashion along planes of low indices, i.e. high atomic density. This behavior would appear to be an intrinsic characteristic of iron but it has been shown that highly purified iron (i.e., containing minimal concentrations of carbon, oxygen and nitrogen) is very ductile even at extremely low temperatures. As the carbon and nitrogen content of the iron is increased, the ductile to brittle transition (DBT) takes place at increasing temperatures.

The nucleation and the propagation of a cleavage crack must be distinguished. Nucleation occurs when a critical value of the effective shear stress is reached corresponding to a critical grouping of dislocations, which can create a crack nucleus. In contrast, propagation of a crack depends on the magnitude of the local tensile stress, which must reach a critical level. This critical stress does not appear to be temperature dependent. At low temperatures the yield stress is higher, so the crack propagates when the plastic zone ahead of the crack is small, whereas at higher temperatures a larger plastic zone is required to achieve the critical local tensile stress in the consequence of smaller values of the yield stress.

Detailed microfractographic observations (see Ref. [5]) show that different types of cleavage initiation sites can be distinguished. They are classified as follows: (i) small, approximately spherical inclusions (diameter approximately 2-3 μm), often composed of MnS and/or oxide, (ii) large, elongated inclusions, mainly composed of MnS, in form of a stringer (length in the order of 100 μm) or clusters of large inclusions, and (iii) any other particular features that are encountered in the cleavage initiation area, and that played an uncertain role in cleavage initiation, namely grain boundaries, clusters of large cleavage facets (indicated abnormally large grains), ductile/cleavage interfaces, carbides inside a grain and stretch zones.

In the tensile test plastic deformation involves shearing slip along crystal planes within the crystals, but in the presence of tension of equal magnitude in each principal direction, shearing stresses are absent, plastic deformation is prevented and a brittle fracture occurs as soon as the cohesive strength of the material is exceeded. If the margin between cohesive strength and plastic yield strength is small, a brittle fracture may occur in a material ordinarily considered highly ductile.

Changes in ductility and typical associated mechanisms of fracture for bcc materials are illustrated in figure 6. As shown in this figure there is a marked increase in tensile strain to fracture, and also in the work of fracture, at about 0.018–0.25 of the melting point (T_m). Similar changes occur in other measures of ductility such Charpy values (compare with figure 3). Exceptions are certain fcc metals and alloys (Al, Cu, Ni, Pb) that do not normally cleave. As such, there is no transition in values, which gradually rise with temperature. The increase in ductility over the ''transition temperature range'' is followed by a gradual drop beyond approximately $0.35 \cdot T_m$. It is believed that it happens due to the continuing fall in the Pierls–Nabarro stress which opposes dislocation movement, coupled with the emergence of cross-

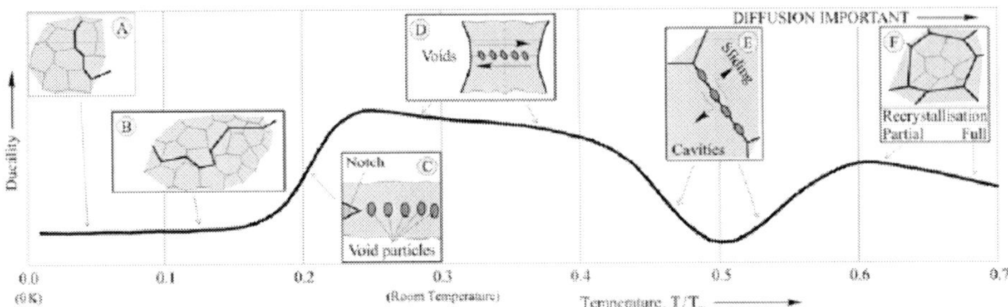

Figure 6. Changes in ductility and typical associated mechanisms of fracture for bcc materials: (A) low-temperature intergranular cracks, (B) twinning or slip leading to cleavage, (C) shear fracture at particles, (D) low energy shear at particles, (E) cavities along grain faces and (F) re-crystallization suppresses cavitation.

slip (as opposed to Frank–Read sources) as a dislocation generator as the temperature is raised (see Ref. [5]). In the opinion of Komarovsky and Astakhov [6], the cause is in dilations–compressions reactions. At high temperatures, the influence of grain boundaries become significant. Below approximately $0.45{\cdot}T_m$, grain boundaries act principally as barriers inhibiting cleavage and causing dislocation pile-ups. At higher temperatures, the regions of intense deformation, which are contained within the grains at lower temperatures, now shift to the grain boundaries themselves. Voids are nucleated and cracks then develop on the grain boundaries. Shear stresses on the boundaries cause relative sliding of the grains, and voids are reduced in regions of stress concentrations (position E in figure 6). Therefore, around this temperature region can be referred as the ductility valley.

At temperatures $(0.5 \div 0.6){\cdot}T_m$, recovery and re-crystallization processes set in (recovery relates to a re-distribution of dislocation sources so that dislocation movement is easier, and in re-crystallization, the energy of dislocations generated during prior deformation is used to nucleate and grow new grains, thus effecting an annealed structure over a long time). The net effect is increased ductility, causing a bump on the ductility curve as shown in figure 6.

The classical fracture mechanics using K_{Ic} or J_c criteria cannot describe an unstable fracture behavior of a low-alloy steel because in a temperature range of the DBT the steel fails through a mixture of dimpled ductile failure and remnant transgranular cleavage. Even if the cleavage initiation and propagation in steels containing isolated carbides are qualitatively well understood no one from the well known damage concepts can explain the sharp upturn of the stress intensity at fracture in the transition temperature regime. Therefore, estimating the behavior of materials in the temperature range of the DBT has been a problem of interest for the last several years in the solid mechanics community.

Goldthorpe and Wiesner [8] developed a coupled mechanistic model for numerical simulation of the DBT in a structural, low-alloy steel, which combined the Beremin model [9] for cleavage fracture with the Gurson ductile damage model [10]. However, whilst attempts have been made to couple cleavage and ductile fracture models with some success (see Ref. [11]); these methods are not yet well established.

The present work consists in further development of the coupled mechanistic model of preceding ductile damage on cleavage fracture. Moreover, the motivation of this work is to establish a computational method, which implies an opportunity to evaluate the material behavior at the mesoscale, and implement of a new combined method of contour elements because the standard finite element method (FEM) has its limitations. A locking phenomenon (caused by too small values of element displacements for the fast achievement of the convergence of the numerical solution) and numerical instabilities (such as so-called hourglass modes) among others belong to the known limitations of the FEM.

4. Coupled Mechanistic Model of Ductile Damage and Cleavage Fracture

4.1. Model of Cleavage Fracture

The simplest brittle fracture criterion states that fracture is initiated in a brittle solid when the greatest tensile principal stress, $\sigma_{MAX(1)}$, reaches a critical magnitude for tensile strength, σ_{TS}:

$$\text{fracture} \quad when \quad \sigma_{MAX(1)} \geq \sigma_{TS} . \tag{2}$$

However, criterion (2) is too crude for many applications. The tensile strength of the brittle solid usually shows considerable statistical scatter because the likelihood of failure is determined by the probability of finding a critical microcrack in a highly stressed region of the material. The model for cleavage fracture is based on the following assumptions:

- microcracks nucleate at some brittle particles under the action of plastic flow,
- these microcracks propagate unstably into the steel matrix if a local stress exceeds the critical stress value,
- cleavage obeys the weakest-link principle;

which were stated by Beremin [9].

Weibull statistics refers to a technique used to predict the probability of failure of the brittle material. The approach is to test a large number of specimens with identical size and shape under uniform tensile stress, and determine the survival probability as a function of stress. The function of the survival probability, $P_{S(V_0)}(\sigma_1)$, is fit by the Weibull distribution:

$$P_{S(V_0)}(\sigma_1) = e^{-(\sigma_1/\sigma_u)^m} , \tag{3}$$

where σ_u and m are material constants, and V_0 is the corresponding volume of the specimen. The exponent, m, is independent of the specimen volume. The parameter, σ_u, is the stress at which the probability of survival is e^{-1} (about 36.8%) and vary with the specimen volume. For given σ_u, m and V_0, the survival probability of a volume, V, of the material subjected to uniform uniaxial stress, σ_1, follows as

$$P_{S(V)}(\sigma_1) = e^{-\frac{V}{V_0}\left(\frac{\sigma_1}{\sigma_u}\right)^m} . \tag{4}$$

Note that the volume, V, can be thought of as containing $n = V/V_0$ corresponding specimen volumes. The probability that they all survive is $\left[P_{S(V_0)}(\sigma_1)\right]^n$. As the root of the Weibull method of failure prediction is a weakest link consideration, then survival of the structure depends on simultaneous non-failure of all of the constituent elements of its material subjected to nonuniform uniaxial stresses, σ_{1_i}. Hence, the total survival probability of the structure, P_{ST}, may be calculated by:

$$P_{ST} = \prod_{i=1}^{n} \exp\left[-\frac{V_i}{V_0}\left(\frac{\sigma_{1_i}}{\sigma_u}\right)^m\right], \tag{5}$$

where i refers to each elemental volume V_i at stress σ_{1_i}. Beremin [9] introduced a Weibull stress defined in the case of the structure submitted to an inhomogeneous stress field:

$$\sigma_W{}^m = \int_{V_p} \sigma_{MAX(1)}{}^m \cdot \frac{dV}{V_0}, \tag{6}$$

where V_p is the yield volume and dV is the volume submitted to the homogeneous maximum principal stress, $\sigma_{MAX(1)}$. Therefore, the fracture probability of the entire structure follows a two parameter distribution function

$$P_f = 1 - \exp\left[-\left(\frac{\sigma_W}{\sigma_u}\right)^m\right]. \tag{7}$$

4.2. Model of Ductile Damage

Ductile fracture in tension occurs by the nucleation, growth and coalescence of voids in the material. Therefore, to model ductile damage sophisticated approaches are used, in which the void volume fraction is an explicit state variable. The Gurson constitutive model is an example. The yield function of the Gurson-type model (GM), defined in Ref. [10], can be rewritten as

$$\tilde{\Phi}_{GM}\left(\mathbf{S},\sigma_M\left(\kappa^{pl}\right),f^*\right)\equiv\frac{S_{eq}}{D_{nom}^{GM}\left(S_h,\sigma_M\left(\kappa^{pl}\right),f^*\right)}-\sigma_M\left(\kappa^{pl}\right)=0, \tag{8}$$

where $S_{eq}\equiv\sqrt{\frac{3}{2}\mathbf{S}^{(dev)}:\mathbf{S}^{(dev)}}$ is the scalar equivalent of the stress tensor: $\mathbf{S}$, $\sigma_M\left(\kappa^{pl}\right)$ is the rational stress/strain curve with the cumulated plastic strain: κ^{pl}, which governs the expansion of the yield surface, and

$$D_{nom}^{GM}\left(S_h,\sigma_M\left(\kappa^{pl}\right),f^*\right)\equiv\sqrt{1+\left(q_{GM}\,f^*\right)^2-2\,q_{GM}\,f^*\cosh\left(\frac{3S_h}{2\sigma_M\left(\kappa^{pl}\right)}\right)} \tag{9}$$

is the parameter that determines the material deterioration (see Ref. [12]). In Eq. (9) q_{GM} is the material factor, which controls the rate of damage accumulation (for spherical voids the theoretical value of $q_{GM}=4/e=1.47$ was derived by Perrin and Leblond [13]). The GM is a multi-scale model of micro and macro scales (macro Piola-Kirchhoff – micro Cauchy true stress). Note that the effective stress:

$$S_{eff}\equiv S_{eq}/D_{nom}^{GM}\left(S_h,\sigma_M\left(\kappa^{pl}\right),f^*\right) \tag{10}$$

has a singularity when $D_{nom}^{GM}\left(S_h,\sigma_M\left(\kappa^{pl}\right),f^*\right)$ tends to zero for chosen values of its parameters as shown in figure 7.

According to the micromechanical approach ductile fracture is considered as a continuous damage process composed of three stages: (a) nucleation: is mainly due to deformation incompatibilities between metallic matrix and non metallic inclusions (such as carbides or sulfides) giving rise to the formation of voids; (b) cavity growth: is corresponding to the cavities growth under loading of the porous media; and (c) coalescence: is the final stage where shearing is occurring between existing voids and gives the final fracture or crack advance.

The data from chemical analysis and/or quantitative metallography should be used to estimate the initial void volume fraction f_0. It is recommended that f_0 should relate to the volume fraction of the critical particles in the ductile fracture process. For cases where manganese sulphide inclusions dominate the fracture process, Franklin's formula [14] may be used to estimate f_0 as follows:

$$f_0=\frac{0.054\sqrt{d_x\,d_y}}{d_z}\left(\mathrm{S}(\%)-\frac{0.001}{\mathrm{Mn}(\%)}\right) \tag{11}$$

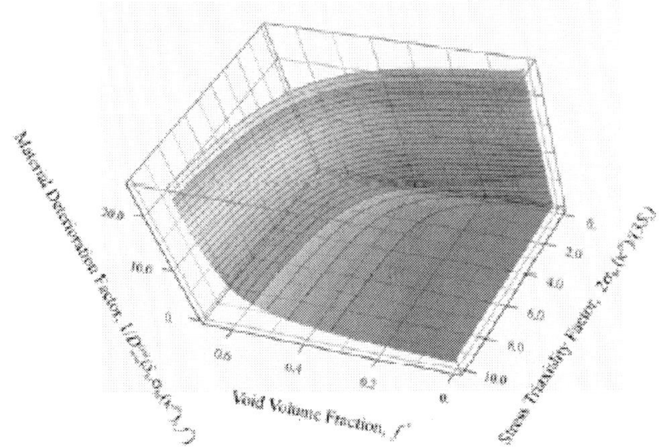

Figure 7. Material deterioration factor as a function of the void volume fraction and the stress triaxiality factor.

where d_x, d_y and d_z are the average inclusion diameters in the x, y and z directions, respectively, and the chemical compositions of Manganese (Mn) and Sulphur (S) are measured in weight %. Equation (11) is based on the observation that the size of the voids formed at MnS inclusions depends on the two dimensions d_x and d_y perpendicular to the direction of load z.

The simplest model for coalescence is to assume a constant critical value of the void volume fraction, f_C, and to simulate the coalescence process by accelerating the rate of increase of the void volume fraction after f_C is attained. The void coalescence process can be simulated by the function (introduced by Tvergaard and Needleman [15]) of the parameter $f^*(f)$, which is related to the void volume fraction f, by:

$$f^*(f) = \begin{cases} f & \text{when} \quad f < f_C \\ f_C + K_{inst}\left(f - f_C\right) & \text{when} \quad f \geq f_C \end{cases} \tag{12}$$

where K_{inst} is an adjustment parameter (i.e., the damage acceleration slope) that varies between 3 and 8 and that can be fitted on void cell calculations or on experimental data.

A simple failure criterion that accounts for the substantial reduction in strength caused by the presence of tensile hydrostatic stress, S_h, is given by

$$\text{failure} \quad when \quad q_M \leq q_{M_{cr}} \quad \left(for\, S_h > 0\right). \tag{13}$$

According to criterion (13) the material fails when the quotient of stress multiaxiality, $q_M = S_{eqv}/S_h$, is close to the critical magnitude, $q_{M_{cr}}$. The multiaxiality quotient was first defined by Clausmeyer [16] and considers the interaction between shear as well as tension initiating shear and cleavage fracture. The quantification of the transition from shear fracture to cleavage fracture is one of the major questions in the safety analysis of components. The physical interpretation of the meaning of q_M is given in figure 8 (which is, in fact, the magnified view of figure 7). Note that q_M allows to determine the substantial reduction in strength caused by the presence of tensile hydrostatic stress, S_h.

More accurate predictions can be obtained if in Definition (9) the material factor q_{GM} is replaced by a parameter q^*_{GM}, which will be calibrated as a function of f^* and q_M:

$$q^*_{GM} = \begin{cases} 0.5\, q_{GM} & \text{if} \quad q_M \le q_{GM} \Big/ \left(\sqrt{3} \ln\left(1 \Big/ \left(q_{GM} f^* \right) \right) \right); \\ q_{GM} & \text{otherwise}; \end{cases} \tag{14}$$

and an additional condition for f^* that depends both on f and q_M will be introduced as follows:

$$\text{if} \quad \left(f < f_C \ \text{ and } \ q_M \le q_{GM} \Big/ \left(\sqrt{3} \ln\left(1 \Big/ \left(q_{GM} f^* \right) \right) \right) \right) \quad \text{then} \quad f^*\left(f, q_M \right) = f_C . \tag{15}$$

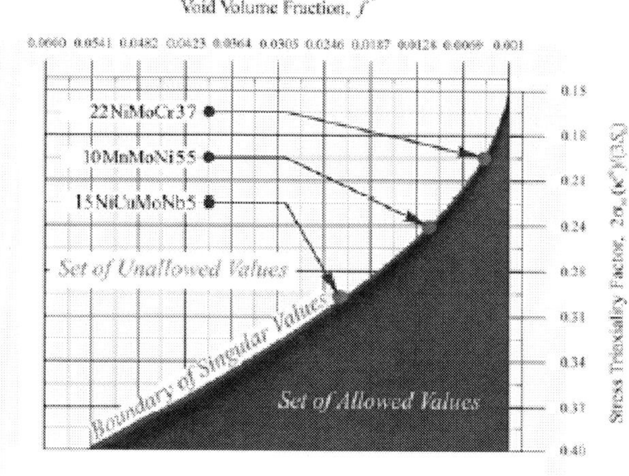

Figure 8. Magnified view of figure 7 (Material deterioration factor as a function of the void volume fraction and the stress triaxiality factor).

5. Methods of Elastic-Plastic Analysis

For an elasto-plastic simulation the material stiffness is continually varying and therefore the incremental stress/strain relationship should be established instantaneously. By analogy with the decomposition of the total strain increment: $\delta\varepsilon_{ij} = \delta\varepsilon^e_{ij} + \delta\varepsilon^a_{ij}$ into elastic components, $\delta\varepsilon_{ij}^{e}$, and inelastic components, $\delta\varepsilon_{ij}^{a}$, the present study concerns the decomposition of the total pseudo-stress increment, $\delta\widetilde{\sigma}_{ij}$, (see figure 9) into elastic and inelastic components:

$$\delta\widetilde{\sigma}_{ij} = \delta\sigma_{ij}^{e} + \delta\widetilde{\sigma}_{ij}^{a} \quad \textit{for} \quad \text{2D problems}: \left(i, j = 1, 2 \right), \tag{16}$$

where

$$\delta\sigma_{ij}^{e} = 2\mu\, \delta\varepsilon_{ij}^{e} + \frac{2\mu\nu}{1-2\nu}\, \delta\varepsilon_{kk}^{e}\, \delta_{ij} \quad \left(k = 1, 2\right) \tag{17a}$$

and

$$\delta\widetilde{\sigma}_{ij}^{a} = 2\mu\, \delta\varepsilon_{ij}^{a} + \frac{2\mu\nu}{1-2\nu}\, \delta\varepsilon_{kk}^{a}\, \delta_{ij}\,. \tag{17b}$$

In Eqs (17) μ is the shear modulus of elasticity, ν is the Poisson's ratio and δ_{ij} is the Kronecker delta function. Note that the true stress increment is always related by the elastic modulus to the increment of elastic strain, from hence $\delta\boldsymbol{\sigma} = \delta\boldsymbol{\sigma}^{e} = \delta\widetilde{\boldsymbol{\sigma}} - \delta\boldsymbol{\sigma}^{a}$. In the standard version of the flow theory of plasticity with isotropic hardening or softening, the yield function has the form

$$f\left(\boldsymbol{\sigma}, \kappa\right) = F\left(\boldsymbol{\sigma}\right) - \sigma_{Y}\left(\kappa_{p}\right), \tag{18}$$

where $\boldsymbol{\sigma}$ is the stress tensor, $F\left(\boldsymbol{\sigma}\right)$ is the equivalent stress and $\sigma_{Y}\left(\kappa_{p}\right)$ is the current yield stress. Fundamental equations of associated elastoplasticity ($\mathbf{m} = \mathbf{n}$) include also the elastic stress/strain law:

$$\boldsymbol{\sigma} = \mathbf{D}^{e} : \left(\boldsymbol{\varepsilon} - \boldsymbol{\varepsilon}^{a}\right), \tag{19}$$

the flow rule:

$$\delta\boldsymbol{\varepsilon}^{a} = \delta\lambda\, \mathbf{m}\,, \tag{20}$$

and the loading-unloading conditions:

$$\delta\lambda \geq 0, \quad f\left(\boldsymbol{\sigma}, \kappa_{p}\right) \leq 0 \quad \textit{and} \quad \delta\lambda\, f\left(\boldsymbol{\sigma}, \kappa_{p}\right) = 0\,. \tag{21}$$

Here, $\mathbf{D}^e$ is the matrix of the elastic material stiffness, $\delta\lambda$ is the increment of the plastic multiplier and $\mathbf{m} \equiv \dfrac{\partial G}{\partial \boldsymbol{\sigma}}$ is the gradient of plastic potential function G, which defines the direction of plastic flow. Hence, the consistency condition involves

$$\delta f = \frac{\partial f}{\partial \boldsymbol{\sigma}} : \delta\boldsymbol{\sigma} + \frac{\partial f}{\partial q}\frac{\partial q}{\partial \lambda} \cdot \delta\lambda = \mathbf{n} : \delta\boldsymbol{\sigma} + H \cdot \delta\lambda = \tilde{\mathbf{n}} : \delta\boldsymbol{\varepsilon} - \tilde{\mathbf{H}} \cdot \delta\lambda \leq 0, \left(\mathbf{n} = \frac{\partial f}{\partial \boldsymbol{\sigma}};\ H = \frac{\partial f}{\partial q_{\text{int}}}\frac{\partial q_{\text{int}}}{\partial \lambda} \right) \quad (22)$$

whereby the array $\tilde{\mathbf{n}}$ encloses the gradients of the loading surfaces in the strain space, and the matrix $\tilde{\mathbf{H}}$ contains the hardening parameters. It can be shown that $\tilde{\mathbf{n}} = \mathbf{n} : \mathbf{D}^e$ and $\tilde{\mathbf{H}} = \mathbf{n} : \mathbf{D}^e : \mathbf{m} - H$. In condition (22) q_{int} is the internal variable.

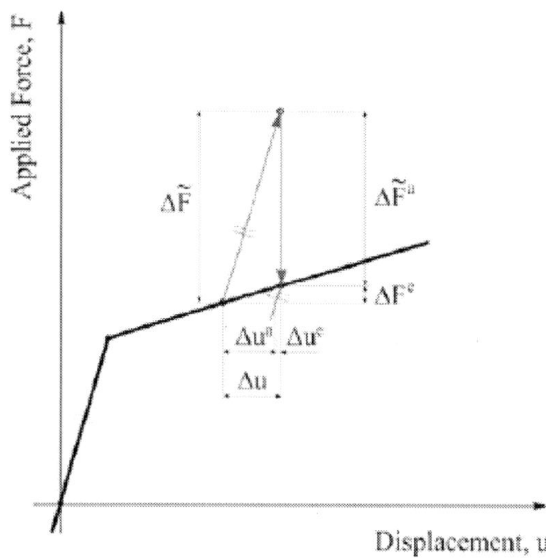

Figure 9. Geometrical representation of the initial load method.

6. Combined Method of Contour Elements with a Nonlocal Characterization Function

Many attempts to characterize a degradation of material stiffness by local inelastic continuum theories (such as plasticity or continuum damage mechanics) have been unsatisfactory because the phenomena of failure zone growth (e.g., crack band and/or shear zone propagation, crystal fault etc.) have a discontinuous nature. Therefore, a realistic prediction of failure caused by progressive damage requires an application of non-standard continuum theories, such as: (a) nonlocal concept, (b) grade n-continua, (c) Cosserat continua and (d) contour-subdomain element technique, which can be classified as a discrete-continuum method. In this paper the contour-subdomain element technique (see Refs [17] and

[18]) is developed to model some plasticity problems coupled to damage in the DBT without uncertainties concerning the distribution of the stress and strain fields close to the crack tip. The proper identification of these crack-tip singular fields is very important to determine Weibull modulus of the brittle fracture model.

Equilibrium in the presence of body forces (per unit volume), $\mathbf{b}$, can be written as

$$\sigma^e_{ij,i} + b_j = 0\,. \tag{23}$$

An alternative form of equilibrium equation (23) is given by

$$\widetilde{\sigma}_{ij,i} + \widetilde{b}_j = 0\,, \tag{24}$$

where $\widetilde{b}_j$ are pseudo-body forces given by

$$\widetilde{b}_i = b_i - \widetilde{\sigma}^a_{ji,j}\,. \tag{25}$$

Accordingly, the pseudo-tractions (i.e., the pseudo-stress vectors), $\widetilde{t}_i$, are related to the outward normal vector, n_j, by

$$\widetilde{t}_i = \widetilde{\sigma}_{ij} n_j = \left(\sigma^e_{ij} + \widetilde{\sigma}^a_{ij}\right) n_j\,. \tag{26}$$

Let Ω denote a finite and simply connected, 2D domain with a regular boundary $\partial(\Omega)$. The direct boundary integral equation for elastoplastic problems can be derived from the Bettis's reciprocal work theorem for two self-equilibrated states of strains $\boldsymbol{\varepsilon} \mid \boldsymbol{\varepsilon}^*$ and pseudo-stresses $\widetilde{\boldsymbol{\sigma}} \mid \widetilde{\boldsymbol{\sigma}}^*$ as in the elastic case

$$\int_\Omega \widetilde{\sigma}_{ij}\, u^*_{i,j}\, d\Omega = \int_\Omega \widetilde{\sigma}^*_{ij}\, u_{i,j}\, d\Omega\,. \tag{27}$$

To achieve the reciprocity relationship in the elastoplastic problem, the elastic fundamental solution and the total strain increment deformed by the total pseudo-stress increment for the desired solution are used. Applying the divergence theorem with some algebraic manipulations, after integrating by parts on the left-hand side of Eq. (27) and then on the right-hand side of the same equation, the following expressions:

$$\int_\Omega \widetilde{\sigma}_{ij} u^*_{i,j}\, d\Omega \;=\; \int_\Omega \left(\widetilde{\sigma}_{ij} u^*_i\right)_{,j} d\Omega - \int_\Omega \widetilde{\sigma}_{ij,j} u^*_i\, d\Omega \;=\; \int_{\partial(\Omega)} u^*_i \widetilde{t}_i\, d\Gamma + \int_\Omega u^*_i \widetilde{b}_i\, d\Omega \tag{28a}$$

and

$$\int_{\Omega} \tilde{\sigma}_{ij}^{*} u_{i,j}\, d\Omega = \int_{\Omega} \left(\tilde{\sigma}_{ij}^{*} u_{i}\right)_{,j} d\Omega - \int_{\Omega} \tilde{\sigma}_{ij,j}^{*} u_{i}\, d\Omega = \int_{\partial(\Omega)} \tilde{t}_{i}^{*} u_{i}\, d\Gamma + \int_{\Omega} \delta(\mathbf{r}-\mathbf{r}') e_{i}(\mathbf{r}') u_{i}\, d\Omega(\mathbf{r}') =$$

$$= \int_{\partial(\Omega)} \tilde{t}_{i}^{*} u_{i}\, d\Gamma + u_{i}(\mathbf{r}) \quad . \tag{28b}$$

can be obtained. When the governing solution of elasticity theory is used, fields of $\mathbf{u}^{*}$, $\mathbf{t}^{*}$ and $\tilde{\mathbf{b}}^{*}$ can be expressed as

$$u_{i}^{*}(\mathbf{r}) = U_{ij}(\mathbf{r},\mathbf{r}') e_{j}(\mathbf{r}') \quad \left(\text{displacement field}\right), \tag{29a}$$

$$t_{i}^{*}(\mathbf{r}) = T_{ij}(\mathbf{r},\mathbf{r}') e_{j}(\mathbf{r}') \quad \left(\text{traction field}\right), \tag{29b}$$

$$\tilde{b}_{i}^{*}(\mathbf{r}) = \delta(\mathbf{r}-\mathbf{r}') e_{i}(\mathbf{r}') \quad \left(\text{body force field}\right) \tag{29c}$$

for the i-direction at any field point $\mathbf{r} = (x, y)$ due to the unit force e_{j} in the j-direction applied at the load point $\mathbf{r}' = (x', y')$. In Eqs (24) $U_{ij}(\mathbf{r},\mathbf{r}')$ and $T_{ij}(\mathbf{r},\mathbf{r}')$ are the fundamental solutions for linear elastic problems (see Ref. [17]). In Eq. (29c) $\delta(\mathbf{r}-\mathbf{r}')$ is the Dirac delta function.

For 2D problems an inelastic formulation of the integral equation, in which traction and body forces are fictitious (depend on the inelastic strains) but the displacements are the actual ones, is derived in an incremental form as

$$\left(\mathbf{R}' \in \partial(\Omega)\right): \qquad \delta u_{i}(\mathbf{r}) + \int_{\partial(\Omega)} T_{ij}(\mathbf{r},\mathbf{R}') \delta u_{j}(\mathbf{R}') d\Gamma(\mathbf{R}') =$$

$$= \int_{\partial(\Omega)} U_{ij}(\mathbf{r},\mathbf{R}') \delta\tilde{t}_{j}(\mathbf{R}') d\Gamma(\mathbf{R}') + \int_{\Omega} U_{ij}(\mathbf{r},\mathbf{r}') \delta\tilde{b}_{j}(\mathbf{r}') d\Omega(\mathbf{r}') \quad . \tag{30}$$

In Eq. (30) the displacements, pseudo-tractions and pseudo-body forces are respectively determined on the boundary $\partial(\Omega)$ and in the domain Ω of a quadrilateral contour-subdomain element. The solution domain is discretized in subdomains, as in the FEM or the finite volume method (FVM), e.g. each subdomain may be surrounded by contour boundary elements.

As can be seen in figure 10, the quadrilateral contour-subdomain element consists of four contour segments and one internal cell. On each contour segment, $\Delta\Gamma_{(\beta)}$, increments of displacement components, $\delta u_{\xi(\beta)}^{(\alpha)}$ and $\delta u_{\eta(\beta)}^{(\alpha)}$, are approximated by the linear interpolation function. However, increments of pseudo-traction components, $\delta\tilde{t}_{\xi(\beta)}^{(\alpha)}$ and $\delta\tilde{t}_{\eta(\beta)}^{(\alpha)}$, are constant along $\Delta\Gamma_{(\beta)}$. Therefore, integrations of integrand functions can be performed exactly. Exact integration is generally faster than numerical integration for a level of reasonable numerical accuracy. The transformation of integration results from the local co-ordinates system $\left(\xi_{(\beta)}^{(\alpha)}, \eta_{(\beta)}^{(\alpha)}\right)$ to the global one (x, y) is straightforward. Increments of inelastic parts of the strain tensor derivatives, $\delta\varepsilon_{rs,s}^{a(C\in\Omega)}$, may be constant and allocated at the center of the combined element or variable over the element to improve the continuity conditions on interfaces of a body partitioned into the element subdomains. If the assumed increments of inelastic parts of the strain tensor are constant inside the contour-subdomain element, $\delta\varepsilon^{a(C\in\Omega)} = const$ (in the case without implementing of the internal cell), the modeling of the elastic-plastic interface falling inside this element is not possible. Therefore, higher order approximations of the inelastic strain distribution are recommended to settle by means of contour segments and internal cells of the enhanced element.

For a given source point, $P_{(\alpha)}$, the boundary form of the integral equation (30) can be discretized into N contour segments and M cells as follows:

$$\delta u_{i(\alpha)} + \chi_c \sum_{\beta=1}^{N}\sum_{s=1}^{2}\left(H_{ij(\beta)}^{F(\alpha)}\,\delta u_{j(\beta)}^{F} + H_{ij(\beta)}^{L(\alpha)}\,\delta u_{j(\beta)}^{L}\right) = \frac{\chi_c}{2\mu}\sum_{\beta=1}^{N}\sum_{s=1}^{2} G_{ij(\beta)}^{(\alpha)}\,\delta\tilde{t}_{j(\beta)} + \frac{\chi_c}{2\mu}\sum_{\gamma=1}^{M}\sum_{s=1}^{2} B_{ij(\gamma)}^{(\alpha)}\,\delta\tilde{b}_{j(\gamma)}$$
$$(\alpha, \beta = 1, 2, \ldots, N; \quad \gamma = 1, 2, \ldots, M) \qquad (31)$$

where

$$\delta u_{i(\alpha)} = \tfrac{1}{2}\left(\delta u_{i(\alpha)}^{F} + \delta u_{i(\alpha)}^{L}\right),$$
$$\delta u_{i(\beta)}^{L} = \delta u_{i(\beta+1)}^{F} \quad for \quad \beta = 1, 2, \ldots, N-1,$$
$$\delta u_{j(1)}^{F} = \delta u_{j(N)}^{L} \quad and \quad \chi_C = -1/\left[4\pi(1-\nu)\right].$$

Due to proper shape functions for the displacement field, strain field and stress field along each contour segment a special treatment, used to circumvent the well-known corner problem for the boundary element method (BEM), is not required. For a given increment of loading, elements of the matrices: $\mathbf{H}_{(\beta)}^{F(\alpha)}$, $\mathbf{H}_{(\beta)}^{L(\alpha)}$, $\mathbf{G}_{(\beta)}^{(\alpha)}$ and $\mathbf{B}_{(\gamma)}^{(\alpha)}$ in Eq. (31) are assessed by integrating the fundamental solutions analytically (see Ref. [17]) without necessity to use numerical integration over each contour segment and each internal cell.

Nonlocal theories are of integral and gradient type and, as a common feature, include one or several intrinsic length scales. The nonlocal theories of linear elasticity were proposed by several authors [19, 20 and 21]. For nonlinear problems the nonlocal theory was developed by Eringen [22].

For a solid with an interface Γ_{int}, the total energy is conserved in the sense that the total energy content within an arbitrary domain D in the solid can only change if energy flows into (or out of) the volume through its boundary. Therefore, the internal energy balance equation is given by

$$\rho\,\dot{e} = \sigma_{ij}\dot{\varepsilon}_{ij} + R^{\bullet} \tag{32}$$

with the following mathematical restrain:

$$\int\limits_{D-\Gamma_{\text{int}}} R^{\bullet}\, dV = 0\,, \tag{33}$$

where ρ is the mass density of the body, e is the internal energy density (per unit mass), the product $\sigma_{ij}\dot{\varepsilon}_{ij}$ represents the internal power supply density induced by deformation and $R^{\bullet}$ is the nonlocal energy residual. In terms of basic laws of thermodynamics and energy balance equation (32) the nonlocal approach may be established.

Considering thermomechanical processes, where the only sources of energy are mechanical work and heat, the principle of conservation of energy states that the rate of change of total energy is equal to the work done by the body forces and surface tractions plus the heat energy delivered to the body by the heat flux and other sources of heat. The internal energy per unit volume is denoted by $\rho\, w^{\text{int}}$ where w^{int} is the internal energy per unit mass. The heat flux per unit area is denoted by a vector $\mathbf{q}$, in units of power per area, and the heat source per unit volume is denoted by $\rho\, s$. The conservation of energy requires that the rate of change of the total energy in the body, which includes both internal energy and kinetic energy, equals the power of the applied forces and the energy added to the body by heat conduction and any heat sources. Hence, the partial differential equation of energy conservation becomes

$$\rho\frac{Dw^{\text{int}}}{Dt} = \mathbf{D}:\boldsymbol{\sigma} - \nabla\cdot\mathbf{q} + \rho\, s\,, \tag{34}$$

where $D\bullet/Dt$ is the material time derivative.

According to the nonlocal plasticity theory of Eringen [23], the stress is computed by averaging the local stress that would be obtained from the local model. This is equivalent to rewriting stress/strain law (19) in the form

$$\langle\boldsymbol{\sigma}\rangle = \left\langle \mathbf{D}^{e} : \left(\boldsymbol{\varepsilon} - \boldsymbol{\varepsilon}^{a}\right)\right\rangle, \tag{35}$$

where the pointed brackets $\langle\ \rangle$ denote the averaging operator. Thus, the nonlocal approach of Eringen can be characterized as averaging of the stress. To analyze the nonlocal mechanical behavior of the polycrystalline metals the elastic parts of the pseudo-tractions, $\sigma_{ij}^{e}\, n_{j}$, are regularized by

$$\left\langle \sigma_{ij}^{e}(\mathbf{r})\right\rangle = \int\limits_{V} \alpha\left(\left\|\mathbf{r}' - \mathbf{r}\right\|_{GD}\right)\sigma_{ij}^{e}(\mathbf{r}')\, dV(\mathbf{r}'), \tag{36}$$

where $\sigma_{ij}^{e}(\mathbf{r}')$ are classical stress components and $\alpha\left(\left\|\mathbf{r}'-\mathbf{r}\right\|_{GD}\right)$ is a nonlocal characterization function. Hence, stress values of the elastic stress components at mid-nodes of the contour segments may be obtained from the modified form of Somigliana's identity (30), which embraces the nonlocal and standard matrices of the kernel $\mathbf{U}$.

The 2D nonlocal characterization function, which is used for assessment of the nonlocal stress components, can be expressed as a form of the Gauss distribution function:

$$\alpha\left(\left\|\mathbf{r}'-\mathbf{r}\right\|_{GD}\right)=\frac{1}{\pi}\left(\frac{k}{l_s}\right)^2 \exp\left[-\left(\frac{k}{l_s}\right)^2\left(\left\|\mathbf{r}'-\mathbf{r}\right\|_{GD}\right)^2\right], \tag{37}$$

where k is a constant and l_s is the characteristic length. The l_s may be selected according to the range and sensitivity of the physical phenomena. For instance, for the perfect crystals, l_s may be taken as the lattice parameter. For granular materials, l_s may be considered to be the average granular distance and for fiber composites, the fiber distance, etc.

Here, the concept of geodetical distance, $\left\|\mathbf{r}'-\mathbf{r}\right\|_{GD}$, suggested by Polizzotto et al. [24], is applied as the length of the shortest path joining $\mathbf{r}$ with $\mathbf{r}'$ without intersecting the boundary surface. In the vicinity of the boundary of a finite body (what is typical for the boundary element analysis), it is assumed that the averaging is performed only on the part of the domain of influence that lies within the solid. Therefore, in this case the formula of averaging (36) is replaced by the more enhanced form:

$$\left\langle\sigma_{ij}(\mathbf{r})\right\rangle=\left[1-\gamma(\mathbf{r})\right]\sigma_{ij}(\mathbf{r})+\int_V \alpha\left(\left\|\mathbf{r}'-\mathbf{r}\right\|_{GD}\right)\sigma_{ij}(\mathbf{r}')\,dV(\mathbf{r}'), \tag{38}$$

where $\gamma(\mathbf{r})=\int_V \alpha\left(\left\|\mathbf{r}'-\mathbf{r}\right\|_{GD}\right)dV(\mathbf{r}')$. The integration volume, V, should exclude the already failed volume, where critical damage has been met in previous computational increments.

7. Application Examples

In order to illustrate the method of assessing computational prognoses of damage growth and failure in some steel elements covering the temperature range of the ductile-to-brittle transition region as an example material *the ferritic reactor pressure vessel steel* (German designation 22 NiMoCr 3 7, which is similar to an ASTM A508 steel) was investigated. For this steel the transition temperature range varies with microstructure and strain rate, but typically is 50^0C to 100^0C wide, centered on -50^0C to 0^0C. As temperature falls within this range, the fracture toughness drops from more than 100 MPa $\mathrm{m}^{1/2}$ to less than 40 MPa $\mathrm{m}^{1/2}$, and the fracture mode changes from rupture to cleavage.

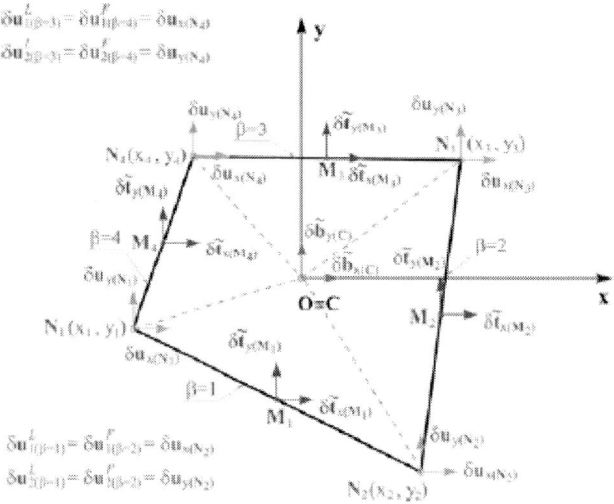

Figure 10. Quadrilateral contour-subdomain element.

Two versions of numerical modeling were carried out for the steel, 22 NiMoCr 3 7: (i) identification of cleavage parameters at the low temperature -150^{0}C using notched tensile specimens (NT) and compact tension specimens (CT), (ii) examination of the DBT by means of CT specimens.

Most of the experiments were carried out on the CT specimens, which were tested to failure. The precracks were put into the 1T CT specimens of 25 mm thickness (B) and 50 mm width (W). True stress vs. true plastic strain curves, made for the steel 22 NiMoCr 3 7 (see Ref. [25]), are shown in figure 11.

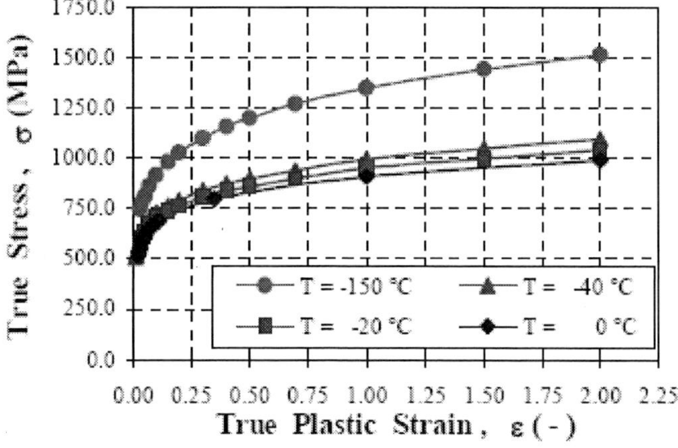

Figure 11. True stress vs. true plastic strain curves for the steel, 22NiMoCr37.

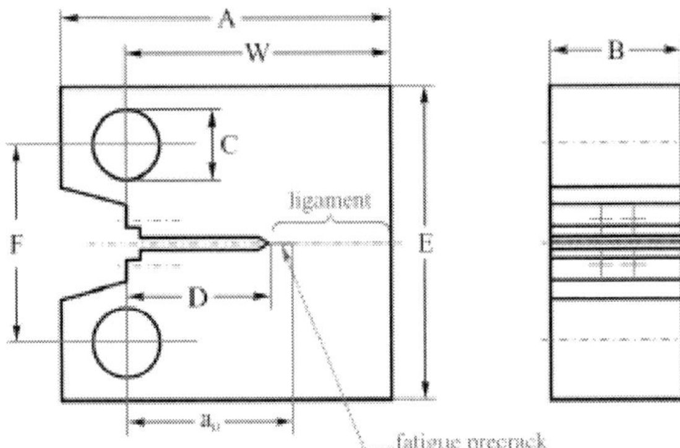

Figure 12. Geometry of a tension specimen of the C(T) type.

7.1. Identification of Cleavage Parameters at the Low Temperature –150^{0}C

In order to calculate the stress states at fracture, the contour subdomain method is implemented. The Weibull stress is computed according to the Beremin [9] model with the imposed characteristic volume $V_0 = (100\text{m}m)^3$, which is large enough to contain defects but sufficiently small to be considered as uniformly loaded. In order to obtain a meaningful statistical cleavage fracture characterization, the complete set of 30 1T CT specimens was tested experimentally (see figure 12 and Ref. [25]).

The statistical variation in the failure stress is possible to predict if Weibull modulus are assessed. Twice taking the logarithm of Eq. (7) yields a linear equation

$$\ln\left[\ln\left(\frac{1}{1-P_f}\right)\right] = m\ln(\sigma_W) - m\ln(\sigma_u) \tag{39}$$

with a slope of m and an intercept of $-m\ln(\sigma_u)$. To estimate m by using Eq. (39), probabilities have to be assigned to all experimental data. Since true probabilities are unknown, P_f has to be estimated. Several studies have been conducted to determine which probability estimator performs better. All probability estimators were found to give biased results, i.e., the average of the estimated values, $\widehat{m}$, is not the same as the true value, m_{true}. For sample sizes $n > 20$, the probability estimator with the least bias is:

$$\widehat{P}_{f_{i_{rank}}} = \frac{i_{rank} - 0.5}{n}, \tag{40}$$

where i_{rank} is the rank of each data point ($i_{rank} = 1,2,\ldots,n$). Recently, Song et al. [26] demonstrated that better estimates of the Weibull modulus can be made when the probability estimator is written in the form:

$$\widehat{P}_{f_{i_{rank}}} = \frac{i_{rank} - \alpha_S}{n + \beta_S}, \tag{41}$$

where α_S and β_S are empirical values that changes with n (e.g., $\alpha_S = 0.66$ and $\beta_S = 0.99$ for $n = 30$).

Last squares regression of $\ln\left[\ln\left(1 - \widehat{P}_{f_{i_{rank}}}\right)^{-1}\right]$ vs. $\ln \sigma_{W\,i_{rank}}$ is used to determine the Weibull parameters, $\widehat{m}$ and $\widehat{\sigma}_u$, and the correction factors, $k_m = \widehat{m}/m$ and $k_{\sigma_u} = \widehat{\sigma}_u/\sigma_u$. If one sample of n specimens is tested to estimate the population parameters, then the confidence intervals for m and σ_u may be calculated as follows:

$$m = \widehat{m}\exp\left(-M_m \pm Z\frac{S_m}{\sqrt{n}}\right), \tag{42}$$

$$\sigma_u = \widehat{\sigma}_u \exp\left(-M_{\sigma_u} \pm Z\frac{S_{\sigma_u}}{\sqrt{n}}\right), \tag{43}$$

where Z is the two-tailed test statistic for a normal distribution ($Z = 1.96$ for the 95% confidence interval).

The mean values of $\ln(k_m)$ and $m\ln(k_{\sigma_u})$ are dependent on sample size and are well fitted by the following relations:

$$M_m = -0.0008n - 0.0448 + 9.3573\frac{1}{n} - 26.7284\frac{1}{n^2}, \tag{44}$$

$$m\,M_{\sigma_u} = -0.0075n + 3.8638 - 9.7062\frac{1}{n} + 15.5686\frac{1}{n^2}, \tag{45}$$

where M_m is the mean of the $\ln(k_m)$ values and M_{σ_u} is the mean of the $\ln(k_{\sigma_u})$ values. The standard deviations of $\ln(k_m)$ and $m\ln(k_{\sigma_u})$ are also dependent on sample size and are well fitted by the following relations:

$$S_m = -0.0003n + 0.0936 + 2.5199\frac{1}{n} - 5.3329\frac{1}{n^2}, \quad (46)$$

$$m\,S_{\sigma_u} = -0.0015n + 0.3447 + 8.1178\frac{1}{n} - 29.1186\frac{1}{n^2}, \quad (47)$$

where S_m is the mean of the $\ln(k_m)$ values and S_{σ_u} is the mean of the $\ln(k_{\sigma_u})$ values. Note that $m \approx \widehat{m}/\exp(M_m)$ and $\sigma_u \approx \widehat{\sigma}_u/\exp(M_{\sigma_u})$.

The value obtained for the Weibull exponent m is 22 with the 95% confidence interval: $[20.75,\ 23.315]$. It was found that the critical cleavage stress σ_u is 2528 MPa with the 95% confidence interval: $[2506.0,\ 2550.2]$. Note that at the low-temperatures failure initiates not from the main crack, but at microcracks associated with brittle particles near the main crack tip.

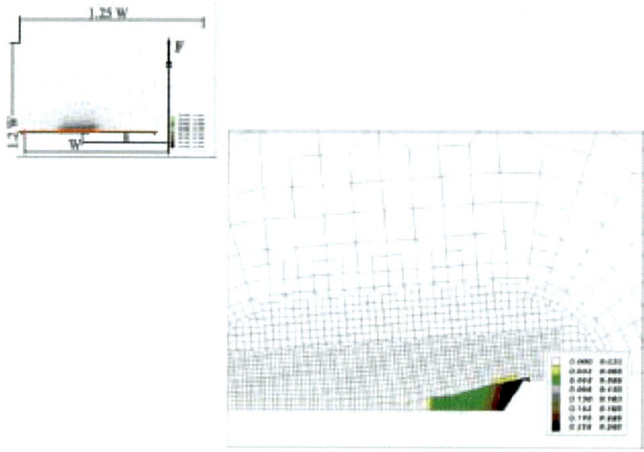

Figure 13. Mesh used for modeling the 1T CT specimen with the porosity distribution, f_{EGM}, in the vicinity of the crack tip (T = 0^{0}C, V$_{LL}$ = 5.8mm, 1852 elements).

7.2. Examination of the DBT

The aim of the numerical simulations by means of the contour subdomain method was to describe the influence of ductile crack extension prior to cleavage failure on fracture behavior of the low-alloy ferritic steel in the DBT. In the DBT an approach to predict cleavage fracture toughness is based on the condition that the main crack propagates in a brittle fashion when the local stress close to one of the brittle particles, which are recognized as microcracks, achieves the cohesive strength of the steel. However, the ductile crack advance causes an increase of the loaded volume susceptible to contain the microcracks, namely potential cleavage initiation sites. Their spatial distribution, variations in size and orientation ahead of the main crack tip are described probabilistically. Therefore, for cleavage fracture prediction, ductile fracture has to be well represented.

The material constants in the constitutive equation are chosen as - the Young's modulus: $E = 210$ GPa and the Poisson`s ratio: $\nu = 0.3$. The material constants for the EGM are set as the following - the initial void volume fraction: $f_0 \approx 9.8 \times 10^{-4}$, the parameter for improving the agreement between experimental and numerical results: $q_{GM} = 1.5$, the damage acceleration slope: $K_{ins} = 4.0$ and the characteristic length for the nonlocal description of the material: $l_C = 0.25\ mm$. The l_C denotes the average spacing between primary directions and allows to improve the continuity of discrete solutions, which are found by the numerical methods. Two material parameters are assigned to identification, namely the critical microvoid volume fraction, f_C, and the critical value of the stress multiaxiality quotient, $q_{M_{cr}}$.

To estimate the unknown parameters the combination of standardized tests and micromechanical simulations was used. The final estimated values of f_C and $q_{M_{cr}}$ are 4.08×10^{-3} and 0.17, respectively. These values allow to predict ductile damage in the complex structures, which are made of the steel, 22 NiMoCr 3 7, properly.

The compact tension specimen was modeled by means of the contour subdomain elements. An example of the 2D model, which consists of 1852 elements, is shown in figure 13.

Results and predictions from coupled micromechanistic modeling, which comprises the simulation of ductile damage parallel to the post processing calculation of Weibull stresses, are presented in figures 14-20. As depicted in figure 20 maximal values of the Weibull stresses, which were calculated, can show the trend to brittle failure during the ductile to brittle transition and precisely evaluate the fracture probability.

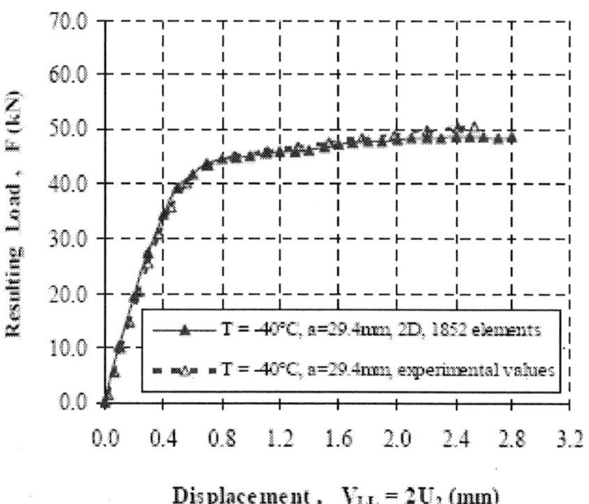

Figure 14. Diagram of the reaction force F as a function of the load line displacement V_{LL} (T = -40^{0}C, 1852 elements).

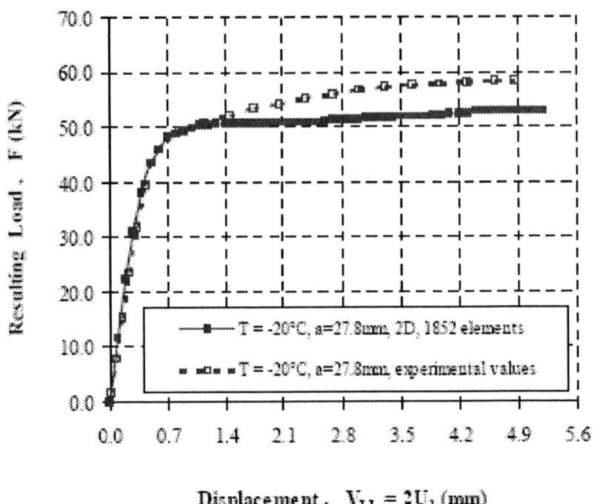

Figure 15. Diagram of the reaction force F as a function of the load line displacement V_{LL} (T = -20^{0}C, 1852 elements).

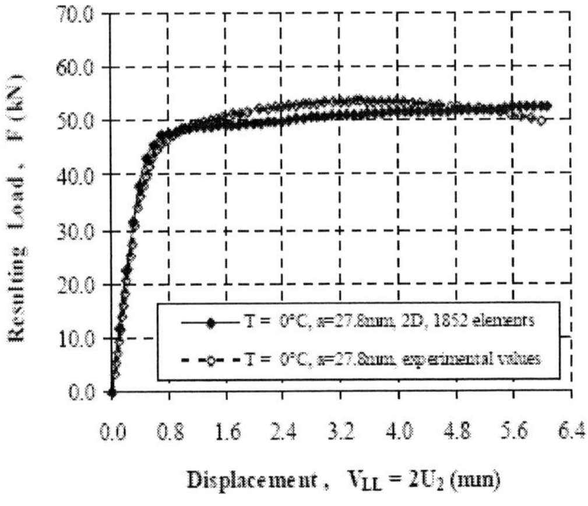

Figure 16. Diagram of the reaction force F as a function of the load line displacement V_{LL} (T = 0^{0}C, 1852 elements).

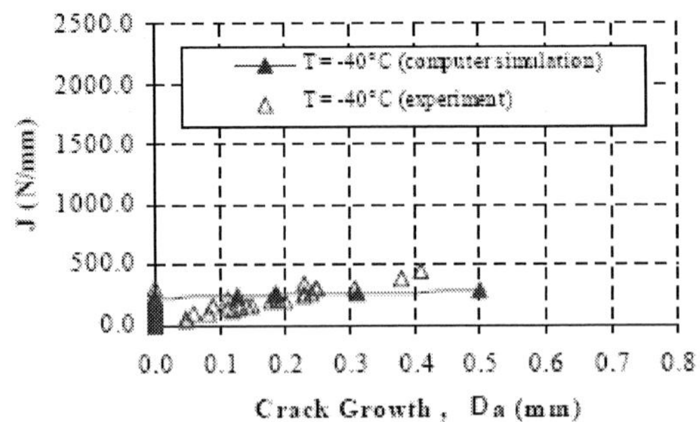

Figure 17. J_R curve at -40^{0}C (values of the J-integral introduced by Rice and Cherepanov vs. the crack growth).

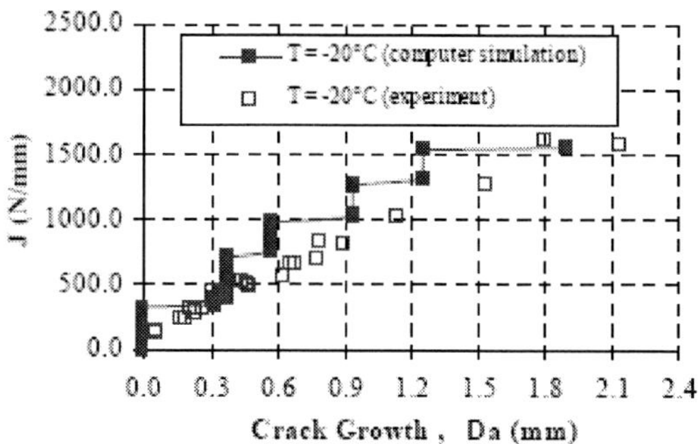

Figure 18. J_R curve at -20^{0}C values of the J-integral introduced by Rice and Cherepanov vs. the crack growth).

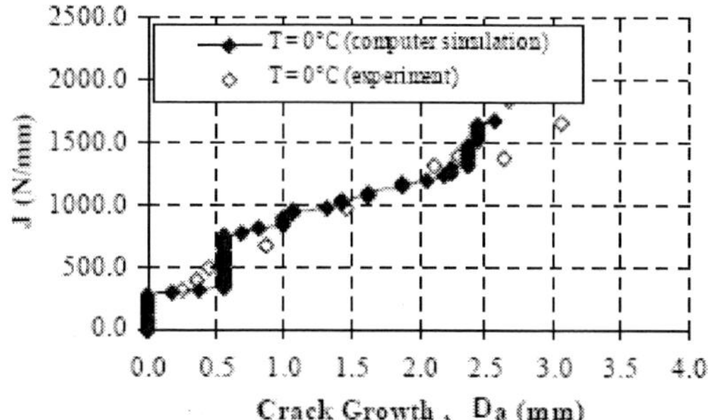

Figure 19. J_R curve at 0^{0}C values of the J-integral introduced by Rice and Cherepanov vs. the crack growth).

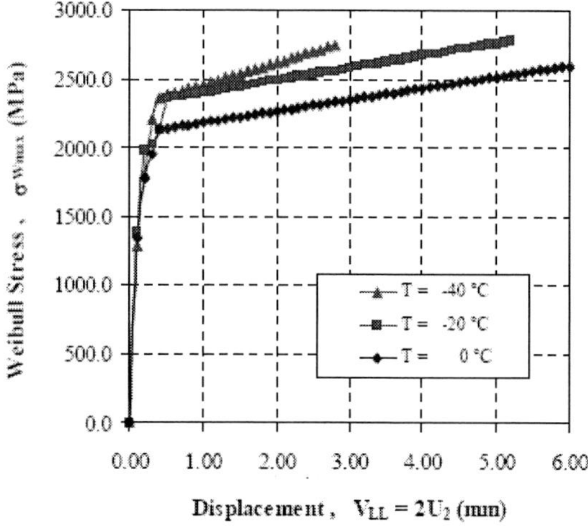

Figure 20. Maximal values of Weibull stresses, σ_{Wmax}, vs. load line displacement, V_{LL}.

8. CONCLUSIONS

The new contour subdomain method was investigated. It can be concluded that the contour subdomain method can be applied to simulate the ductile to brittle transition in the polycrystalline microstructure of ferritic or duplex steels.

In spite of a variety of potential cleavage initiation sites in the material, the relatively simple model of the cleavage fracture founded by Beremin can be useful to evaluate the fracture probability for structural elements of the liquefied natural gas terminal made of ferritic or duplex steels.

For the proper simulation of material damage the mesh size is very important. When the mesh size is too coarse to capture the localization zone the modeling of the crack extension cannot be simulated properly as well as the assessment of values of Weibull stresses.

The proposed method of assessing computational prognoses of damage growth and failure in some steel elements of the liquefied natural gas terminal can be used to reduce costs of completion of nickel LNG storage tanks.

REFERENCES

[1] "Welding liquid natural gas tanks and vessels in 5% and 9% nickel steels," Report published by *ESAB AB*, Box 8004, 402 77 Göteborg, Sweden.

[2] J. Dainora, A.R. Duffy and T.J. Atterbury, "Materials of Construction for Use in an LNG Pipeline," PRCI Report published by Technical Toolboxes, Inc., April 1968, Houston, Texas, *Pipeline Research Council International Catalog* No. L40000e.

[3] European standard EN 10028-4.

[4] S. Butnicki, "Weldability and brittleness of steel," WNT 1991, Warsaw (in polish).

[5] A.G. Atkins, Y.W. Mai, "Elastic and Plastic Fracture: Metals, Polymers. Ceramics, Composites, Biological Materials," John Wiley and Sons, New York, 1985.

[6] A.A. Komarovsky, V.P. Astakhov, "Physics of Strength and Fracture Control: Fundamentals of the Adaptation of Engineering Materials and Structures," CRC Press, Boca Raton, 2002.

[7] G. Z. Wang, Y. G. Liu, J. H. Chen, "Investigation of cleavage fracture initiation in notched specimens of a C-Mn steel with carbides and inclusions," *Materials Science and Engineering*, vol. A369, pp. 181–191, 2004.

[8] M. R. Goldthorpe, C. S. Wiesner, "Micromechanical prediction of fracture toughness for pressure vessel steel using a coupled model," *ASTM STP* 1332, 1998.

[9] F. M. Beremin, "A local criterion for cleavage fracture of a nuclear pressure vessel steel," *Met. Trans.* 14A, pp. 2277-2287, 1983.

[10] A. L. Gurson, "Continuum theory of ductile rupture by void nucleation and growth, Part I: Yield criteria and flow rules for porous ductile media," *J. Eng. Mater. Technolog.* 99, pp. 2-15, 1977.

[11] A. H. Sherry, D. Beardsmore, D. P. G. Lidbury, I. C. Howard, M. A. Sheikh, "Remnant life assessment using the local approach, A prediction of the outcome of the NESC experiment," HSE Seminar on Remnant Life Prediction, Institution of Mechanical Engineers, 26 Nov. 1997.

[12] J. Jackiewicz, "Application of the evolution strategy for assessment of deformation and fracture response of ferritic steel during its manufacturing," Taylor and Francis, Inc., *Materials and Manufacturing Processes* 20-3, pp. 523-542, 2005.

[13] G. Perrin, J.-B. Leblond, Analytical study of a hollow sphere made of plastic porous material and subjected to hydrostatic tension – application to some problems in ductile fracture of metals, Int. J. Plasticity 6 (1990) 677–699.

[14] A.G. Franklin, Comparison between a quantitative microscope and chemical methods for assessment of non-metallic inclusions, J. Iron Steel Institute 207 (1969) 181-186.

[15] V. Tvergaard, A. Needleman, "Analysis of the cup-cone fracture in a round tensile bar," *Acta Metall.* 32, pp. 157-169, 1984.

[16] H. Clausmeyer, K. Kussmaul, E. Roos, "Influence of stress state on the failure behaviour of cracked components made of steel," *J. Appl. Mech.* 44, pp. 77–92, 1991.

[17] J. Jackiewicz, "Boundary integral equations with the divergence free property for elastostatics problems," Advances in Computational and Experimental Engineering and Sciences, *Proceedings of the 2004 International Conference on Computational and Experimental Engineering and Sciences*, Eds: S. N. Atluri and A. J. B. Tadeu, Tech Science Press, pp. 224-229, 2004.

[18] J. Jackiewicz, "Combined method of contour elements with the use of analytical-only integration scheme for arbitrary quadrilateral element shape," *Proceedings of 16th International Conference on Computer Methods in Mechanics CMM-2005*, Czestochowa, Poland, June 21-24, 2005.

[19] A.C. Eringen, D.G.B. Edelen, "On non-local elasticity," *Int. J. Eng. Sci.* 10, pp. 233–248, 1972.

[20] I.A. Kunin, *Elastic Media with Microstructure*, Springer, New York, 1982.

[21] K.C. Valanis, "Gradient field theory of material instabilities," *Arch. Mech.* 52, pp. 817–825, 2000.

[22] A.C. Eringen, "Theory of non-local elasticity and some applications," *Res. Mech.* 21 (1987) 313–342.

[23] A.C. Eringen, "On nonlocal plasticity," *Int. J. Eng. Sci.* 19, pp. 1461–1474, 1981.

[24] C. Polizzotto, P. Fuschi, A.A. Pisano, "A strain-difference-based nonlocal elasticity model," *Int. J. Solids Struc.* 41, pp. 2383-2401, 2004.

[25] C. Poussard, C. Sainte Catherine, ESIS TC 8, Numerical Round Robin on Micro-Mechanical Models, Specification of Phase III for the Simulation of the Brittle to Ductile Transition Curve, CEA Saclay, SEMI/LCMI/RT/02-027/A, March 2003.

[26] L. Song, D. Wu, Y. Li, "Optimal probability estimators for determining Weibull parameters," J. Mater. Sci. Lett. 22, pp. 1651-1653, 2003.

In: Computational Materials
Editor: Wilhelm U. Oster

ISBN: 978-1-60456-896-7

Chapter 8

POROSITY AND MECHANICAL PROPERTIES OF CEMENT MORTARS BASED ON MICROSTRUCTURAL INVESTIGATION

Ali Ugur Ozturk and Bulent Baradan
Department of Civil Engineering, Faculty of Engineering Dokuz Eylul University, 35160, Izmir, Turkey

ABSTRACT

Microstructure and mechanical behaviour of cement based mortars has become more important issue due to the recent developments in microstructural investigations by using computational analysis techniques. Microstructural investigations have a considerable capacity to define the inner structural formation of cementitious materials. A relationship between pore structures of cement mortars and their mechanical properties can be determined by studies based on image processing and analysis.

In the scope of this investigation, the effects of retarders have been investigated by implementation these methods as a case study. Standard cement mortars were prepared by incorporation of retarders with various ratios in order to obtain different pore formations. Microstructures of different cement mortars were investigated by using optical microscope. Micrographs of polished sections of cement mortar samples were taken to determine the area ratio values of pore formations. The pore area ratio values represent total pore area amount in a polished section. The development of pore area ratio values for 1, 2, 7 and 28-day old cement mortars have been determined. The test results indicate that mechanical properties of cement mortars increase as the pore area ratio values decrease. The relationship between pore structures of cement mortars and their mechanical properties has been established by various mathematical models. The chapter shows that image analysis techniques have a remarkable potential to define the relationship between pore area ratios obtained by microstructural investigations and mechanical properties.

INTRODUCTION

Engineers and materials scientists have been involved extensively with the characterization of construction materials. Characterization is necessary in order to understand the behavior of a material under different conditions. Within the past two decades, technological advantages in understanding of materials have indirectly led to improvement of the properties of various materials. Cement and concrete are two good examples of such materials need improved properties as a result of innovations in the construction technology.

Knowledge of the microstructural evolution of cementitious materials at early age is helpful for forecasting their performance. Microstructural studies and numerical simulations become increasingly important to understand the formation of the microstructure of cementitious materials [1]. Image analysis of micrographs of cementitious materials are performed to quantify the microstructure of cement pastes for determination of porosity, pore structure and phases such as undifferentiated hydration products and anhydrous cement content [2-9].

Cements are the most widely used materials in different types of construction. Products of the reaction between cements and water serve as a binder for aggregates and other materials such as fibers. Cement characteristics such as composition and fineness influence fresh and hardened concrete properties. Concrete gains strength gradually as a result of a chemical reaction (hydration) between cement and water; for a specific concrete mixture, strength at any age is related to the degree of hydration. Hydration process is affected by many parameters such as the properties of cement, the environmental conditions (temperature, humidity and etc.). Since the rate of hydration is a function of temperature, the strength development of a given concrete depends on its time-temperature history, assuming that sufficient moisture is available for hydration [10]. During hydration process, cement paste microstructure formation changes by time especially at early ages. Early age characteristics can be defined as rheologic behavior, heat evolution, setting, and strength development. Hydration products also affect hardened concrete properties, as well as the durability of reinforced concrete structures.

Especially, as a result of some problems in concrete technology; the production of high performance concretes which are more durable against physical, biological and chemical attacks has increased considerably. Chemical and mineral admixtures are the main ingredients of high performance concrete besides proper types of cement.

Chemical admixtures improve the specified properties of concrete with some side effects. The most important criteria are; knowledge of their effects on cementitious materials, determining the optimum usage limits, limiting the side effects and improve the efficiency mechanism. Chemical admixtures can lead to a delay or acceleration on the formation of phases such as undifferentiated hydrates, unhydrated cement, calcium hydroxide and pore structure. Retarders, one of these admixtures, which can retard the hydration process of cement in order to extend the casting period especially in hot weather conditions. Retarders may cause a decrease in heat evolution during the hydration of cementitious materials. Retarders like other chemical admixtures affect the early microstructure formation during hydration process.

The pore size levels have significant effects on the durability and mechanical properties of concrete. The diameter size of concrete can be approximately classified into micro (<1 μm), meso (between 1 μm and 1 cm) and macro (greater than 1 cm) l groups. Performance of concrete largely depends on pore structure and hydration degree of a cementitious material besides many factors [11]. However, pore structure is more effective than the hydration degree. A cement mortar may have a higher hydration degree, however if it has a porous structure, it is obvious that this cement mortar will have lower mechanical properties. The strength of concrete is affected by the type, volume and distribution of all voids in concrete (entrapped air, capillary pores, gel pores, cracks and entrained air) and a number of functions based on various researches including the following, have been proposed to express this strength-porosity relationship [12,13] (Eqs. (1) - (4)):

$$\sigma_c = \sigma_{co}(1-n)^m \quad (1)$$

$$\sigma_c = \sigma_{co}\exp(-m.n) \quad (2)$$

$$\sigma_c = \sigma_{co}\ln(n/n_o) \quad (3)$$

$$\sigma_c = \sigma_{co}-m.n \quad (4)$$

where σ_c is the strength of the porous material, σ_{c0} theoretical strength of the material at zero porosity, n the porosity, n_0 the critical porosity corresponding to zero strength, and m is an empirical constant.

Porosity and pore structure and mechanical condition of pore shells have a marked influence on the mechanical properties [1]. Generally, compressive strength increases linearly with the density. However, many other factors such as; pore size distribution, microcracks, interface, etc., so are also important factors that determine mechanical properties of cementitious materials [14]. To determine the effects of pore structure and hydration degree on compressive strength by incorporation of admixtures, it is necessary to implement current experimental technologies and to analyze their results.

Evidence has been presented that mercury intrusion porosimetry (MIP) measurements of pore size distribution systematically misallocates the sizes of almost all of the volume of pores in hydrated cementitious materials, it assigns them to sizes smaller than the threshold diameter regardless of their actual sizes. It appears that the failure of the MIP with cement systems is intrinsic rather than accidental, and derives from the lack of direct accessibility of most of the pore volume (including air voids) to the mercury surrounding the specimen [15].

Recently to investigate polished sections of specimens by microscopy techniques has become more important. There are several advantages of this technique. Although cement paste has a reasonably heterogeneous structure, this technique has a great capacity on determining the relationship between microstructure and mechanical properties by quantitative measures. Microscopy applications improve the ability to characterize the microstructure of cement mortar and concrete, and helps in investigating effects of admixtures, determining durability problems and service life [16].

LeChatelier [17] first examined the thin sections of clinker by using a conventional microscope. Reflected microscopy provides a good view to observe and quantify microstructure features [18-19]. Since the debut in the early of 1980, SEM micrographs under backscattered electron (BE) mode have shown a great potential to investigate cementitious materials [20-21]. Generally, fractured specimens are used for microscopy applications. Minimum requirements to prepare specimen and to identify the topographical images easily, make fractured specimens more attractive. However, imaging process of fractured specimens must be applied more carefully. Since, during preparing specimens, cracks in cementitious material may propagate through the weakest zone and a fractured specimen may not represent the whole specimen [22]. The roughness of fractured specimen affects the image processing inversely [23–25]. To get the benefits of microscopy imaging and image processing flat specimens are necessary.

Flat sections of cementitious materials are prepared by application of some specimen preparation techniques. Specimens covered by epoxy or polyester are polished by diamond paste of micro sprays in order to obtain smooth polished sections.

In this chapter, polished sections of cement mortars were prepared at different ages to investigate the effects of pore structure formation within time on the mechanical properties of cement mortars. The relationship between microstructure and mechanical property is presented by different mathematical models and image analysis results.

EXPERIMENTAL

In order to establish a relationship between pore structure and mechanical properties, experimental studies have been programmed into two stages. In the first stage of experimental studies, mechanical properties of cement mortars (compressive strength and flexural strength) were determined. Cement mortars of different sizes were prepared by incorporation of a naphthalene sulphonate based admixture with different dosages. This type (ASTM C-494 Type G) of chemical admixtures has a retardation effect on the hydration of cement mortar and causes different pore structure formations. The properties of admixture and cement (type CEM I 42.5 R) used in this research are presented in table 1 and 2, respectively. The mix designs of cement mortars are given in table 3. The first dosage (1%) is selected by the manufacturer's recommendations. The second dosage (2%) is a maximum overdosage.

Table 1. Properties of Chemical Admixture

Type	Base	Density	Solid Phase
ASTM-C 494 Type G	Naphthalene Sulphonate	1,16-1,19 kg/lt	30-40%

Table 2. Properties of OPC (Cement Type : CEMI 42,5 R)

Results of Chemical Analysis				
Property	Unit	Value	Test Method	Standard Criteria (TS EN 197-1)
(Cl^-)	(%)	0,01	TS EN 196-2	≤ 0,1%
(SO_3)	(%)	2,82	TS EN 196-2	≤ 4,0%
LOI	(%)	3,63	TS EN 196-2	≤ 5,0%
Results of Physical Analysis				
Property	Unit	Value	Test Method	Standard Criteria (TS EN 197-1)
Initial Setting Time	min	155	TS EN 196-3	≥ 60
Final Setting Time	min	230	TS EN 196-3	
Specific Surface (Blaine)	cm^2/g	3640	TS EN 196-6	
Volume Expansion	mm	0,5	TS EN 196-3	≤ 10
Results of Compressive Strength Test				
Property	Unit	Value	Test Method	Standard Criteria (TS EN 197-1)
(2 day) Strength	N/mm^2	27,60	TS EN 196-1	≥ 20,0
(28 day) Strength	N/mm^2	50,40	TS EN 196-1	≥ 42,5

Table 3. Mix Design of Cement Mortars

Mix Name	Amount by Weight in Mix (g)			
	Aggregate	Cement	Water	Admixture
Control	1350	450	225	
1 % NSB	1350	450	223,43	4,5
2 % NSB	1350	450	221,85	9,0

Compressive and flexural strength properties of cement mortars at the ages of 1, 2, 7 and 28 days were found on the 50 mm cube and 40x40x160 mm prism samples, respectively.

Microstructural investigation is the topic of the second part of experimental studies. Microstructural studies require careful expert practice. Better micrographs and results can only be obtained by careful execution of microstructural investigation procedures. Quantification of phases by performing image analysis requires, in the first stage, good specimen preparation and a proper imaging technique to obtain representative images. An objective phase-segmentation algorithm should be precise and reproducible. For a precise microstructural investigation; preparation of samples, taking of micrographs of prepared sections, image processing and analysis are the processes that also need careful exertion.

To minimize the effects of cutting specimens, some of the mixtures were cast into plastic molds. After 1, 2, 7 and 28 days, hydration processes of the mixes were delayed by submerging specimens into alcohol ispropylique for five days. Before microstructure studies the specimens were covered by a polyester film, and then the surface of each specimen was polished. Each specimen was sanded by 600 and 1200 sandpapers. After sanding, each specimen was polished by 0.25, 1, 3 and 9 µm diamond paste for 120 s [24].

A high-quality original image is a prerequisite for accurate segmentation of phases and subsequent quantification steps. For optimum performance, the electron microscope operating configuration (accelerating voltage, beam current and working distance) must be set depending on the particular contrast produced by the specimen/detector system [26]. Furthermore micrographs of polished sections of specimens at different ages were taken with the same magnification, contrast and brightness level. The micrographs of different cement mortars at various ages with 50X and 100X magnifications are shown in figure 1-12. Segmentation analysis was conducted in order to determine the pore area ratio values of phases.

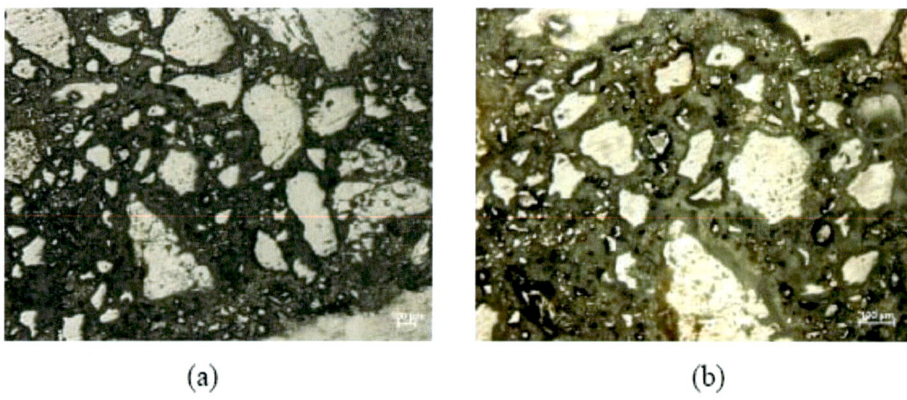

(a) (b)

Figure 1. Micrographs of control specimen at 1 day; (a) Magnification 50X, (b) Magnification 100X .

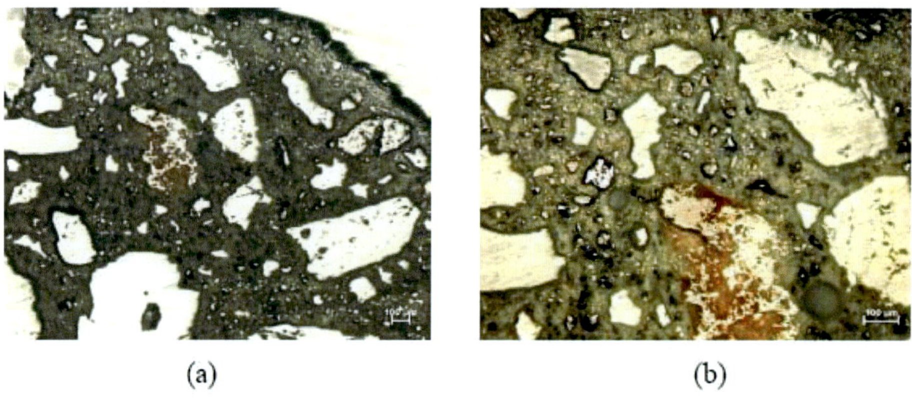

(a) (b)

Figure 2. Micrographs of control specimen at 2 day; (a) Magnification 50X, (b) Magnification 100X .

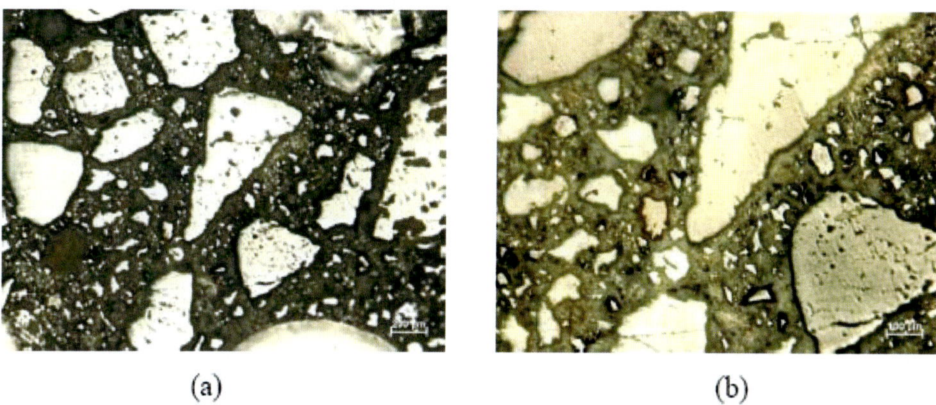

(a) (b)

Figure 3. Micrographs of control specimen at 7 day; (a) Magnification 50X, (b) Magnification 100X.

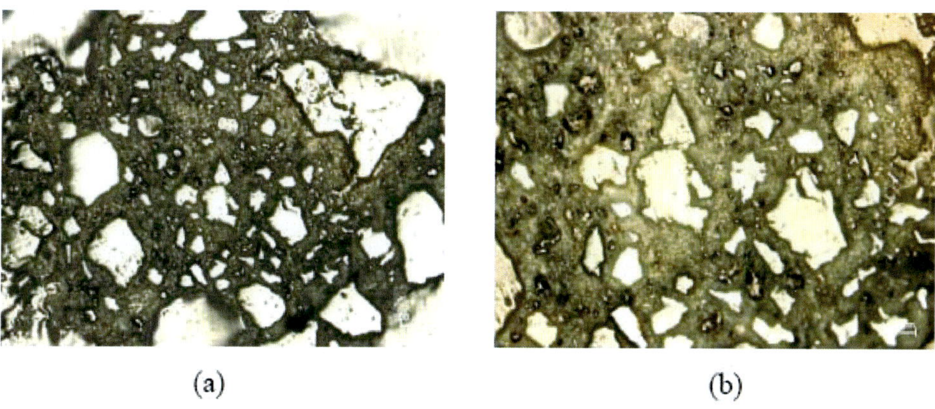

(a) (b)

Figure 4. Micrographs of control specimen at 28 day; (a) Magnification 50X, (b) Magnification 100X.

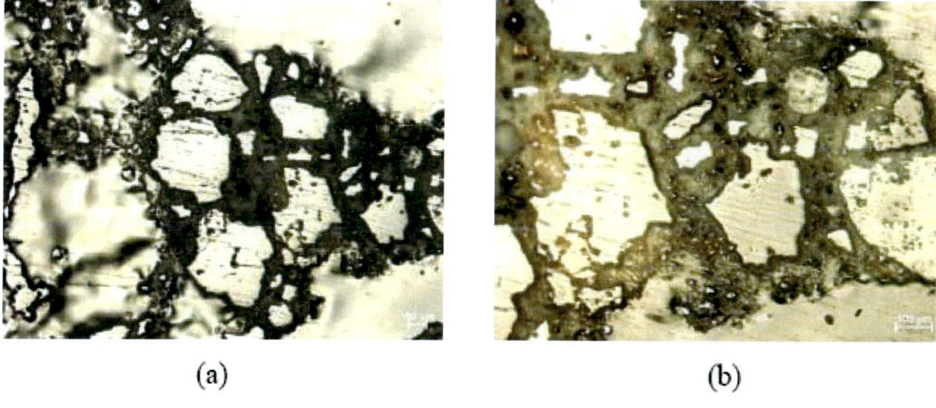

(a) (b)

Figure 5. Micrographs of NSB 1.0 % specimen at 1 day (a) Magnification 50X, (b) Magnification 100X.

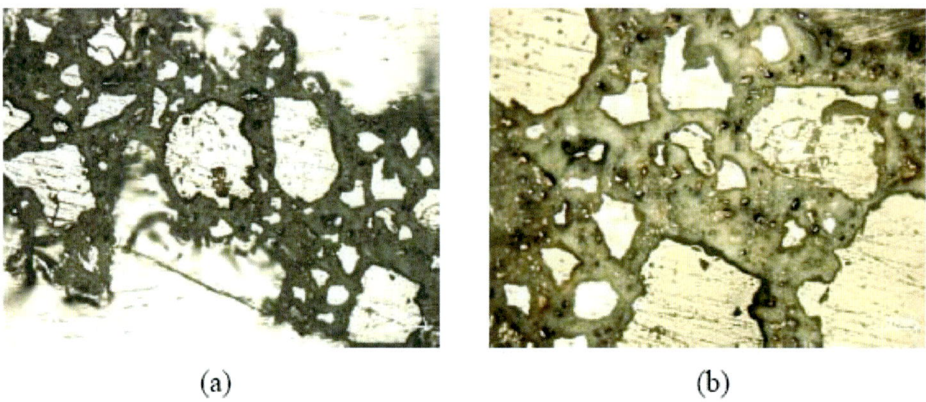

Figure 6. Micrographs of NSB 1.0 % specimen at 2 day (a) Magnification 50X, (b) Magnification 100X.

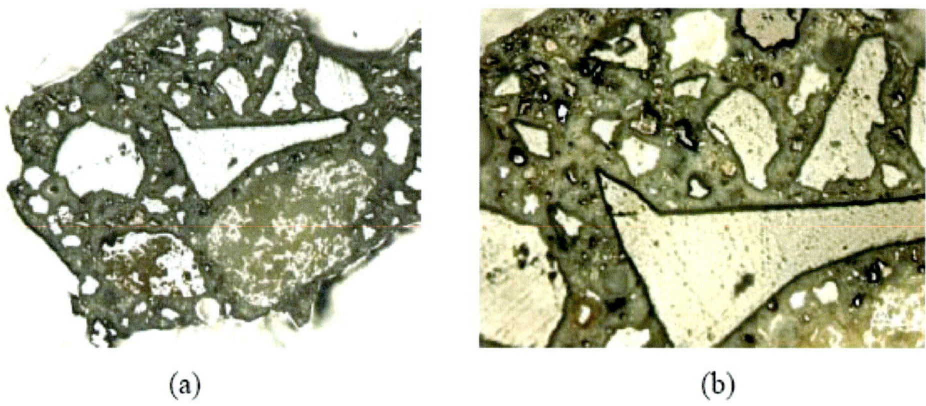

Figure 7. Micrographs of NSB 1.0 % specimen at 7 day (a) Magnification 50X, (b) Magnification 100X.

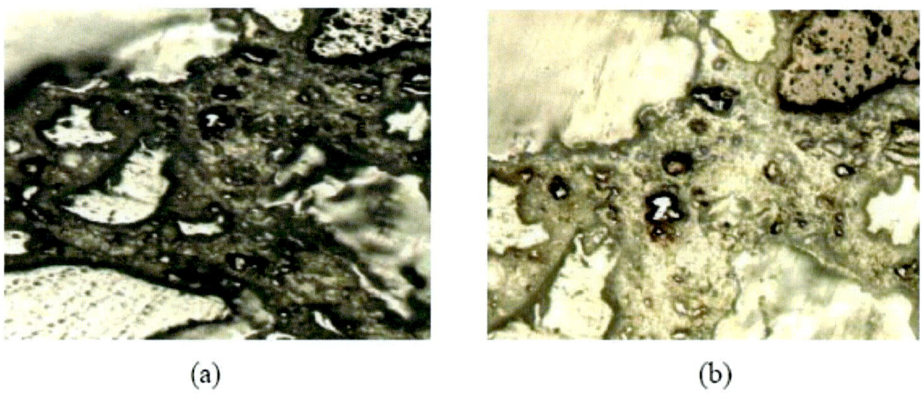

Figure 8. Micrographs of NSB 1.0 % specimen at 28 day (a) Magnification 50X, (b) Magnification 100X.

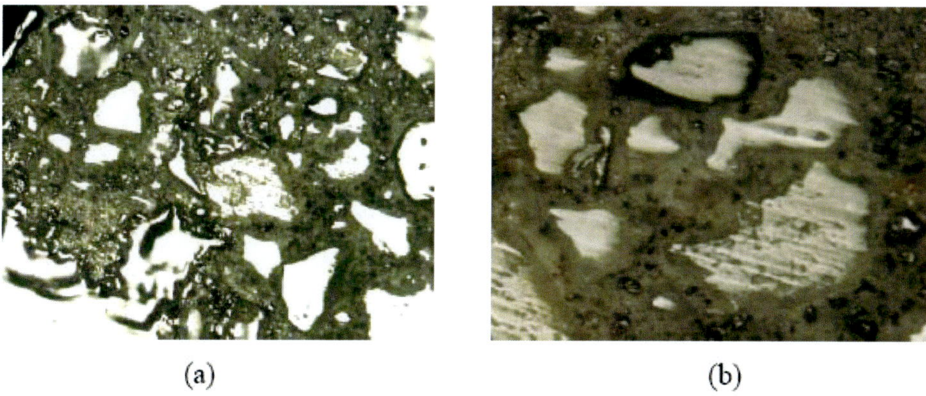

(a) (b)

Figure 9. Micrographs of NSB 1.5 % specimen at 1 day (a) Magnification 50X, (b) Magnification 100X.

(a) (b)

Figure 10. Micrographs of NSB 1.5 % specimen at 2 day (a) Magnification 50X, (b) Magnification 100X.

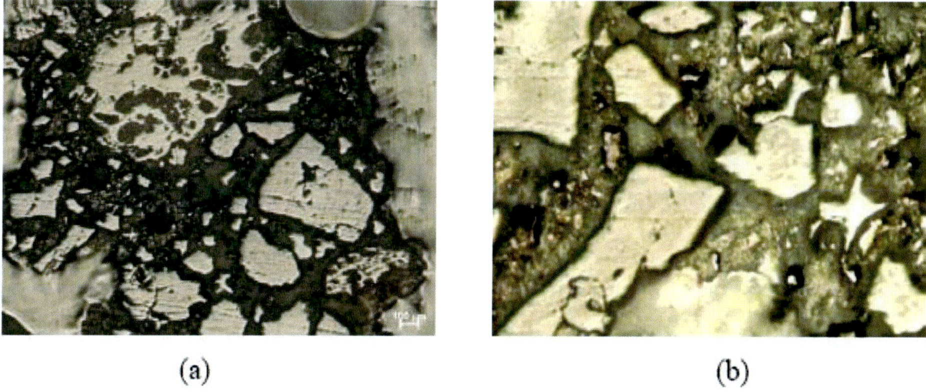

(a) (b)

Figure 11. Micrographs of NSB 1.5 % specimen at 7 day (a) Magnification 50X, (b) Magnification 100X.

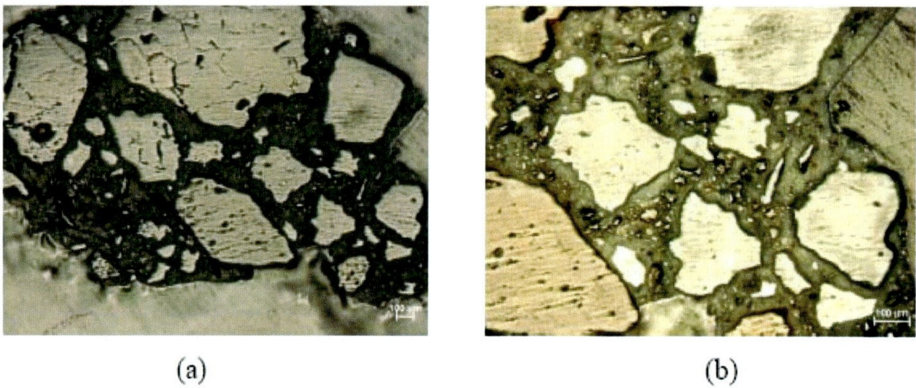

(a) (b)

Figure 12. Micrographs of NSB 1.5 % specimen at 28 day (a) Magnification 50X, (b) Magnification 100X.

The phases in polished sections of cementitious materials should be segmented with accuracy and consistency. Accurate analyses lead to meaningful quantitative data that can be used for comparative studies and to characterize a relationship between microstructure and mechanical behavior. In a micrograph, phases have different brightness intensities. The phases in a polished section can be segmented by their grey level thresholds in the micrograph. The grey level values of phases compose separate peaks in the grey level histogram with their heights proportional to the relative fractions of each phase.

The brightness of each pixel does not necessarily represent a single phase alone. A pixel does not indicate a single phase since the size of a pixel in a digital image is finite and also because of sampling volume effects [26]. This may create a problem for an accurate segmentation.

The lower threshold level for pores can be set to zero (black pixels) and the segmentation process is then reduced to only determining the upper threshold level. However, problems in defining the exact boundary between pores and the surrounding hydrated paste still exist. Grey values of pore formation can be easily separated from the hydration products especially from calcium silicate hydrate gel. To avoid these types of failures in the image analysis segmentation process of pore structure was accomplished by using overflow method [26].

This method depends on the change in segmented area values by grey values in a micrograph of a cement-based material. The segmented area values and their threshold grey values for the control specimen at the age of 1 day are plotted in figure 13 for an example. The critical point where the area segmented starts to overflow can provide a good estimate for the pore threshold level. When the total segmented area is plotted versus the threshold level, the critical overflow point corresponds to the inflection of the cumulative curve, which can be estimated from the intersection between the two linear segments as shown in the figure. The grey value at this intersection can be used as the upper threshold level for porosity [26].

Segmentation analysis has been accomplished in each micrograph of polished sections in order to obtain a relationship between microstructure and mechanical properties of cement mortars. The critical point which indicates the overflow in the segmented area – grey value graph has been determined and plotted for each specimen.

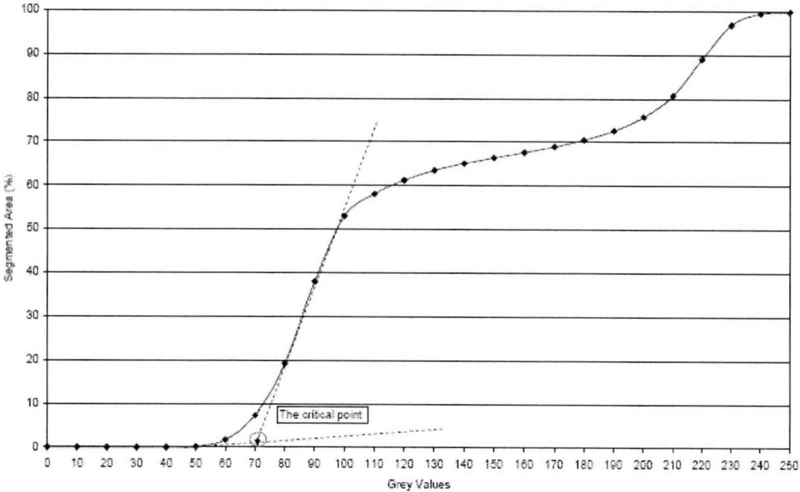

Figure 13. Segmented area - grey value relationship for control specimen at 1 day.

RESULTS

The mechanical properties of cement mortars have been found on 1, 2, 7 and 28 days old specimens. These compressive strength and flexural strength development of cement mortars with time are presented in figures 14 and 15. The compressive strength values were determined on 50 mm cube specimens whereas flexural strength values were found on 40X40X160 mm prisms by mid-point loading. Also the flow values of each mixture are presented in figure 16.

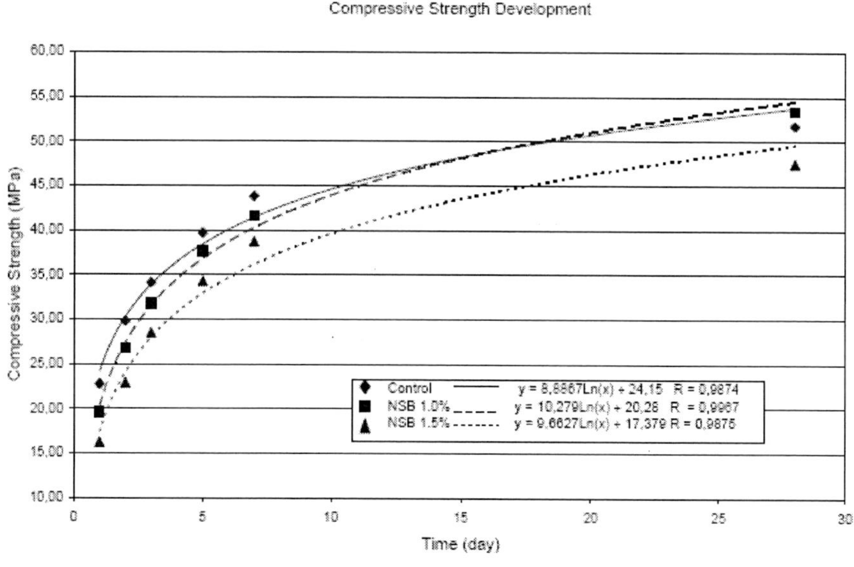

Figure 14. Compressive strength development of cement mortars.

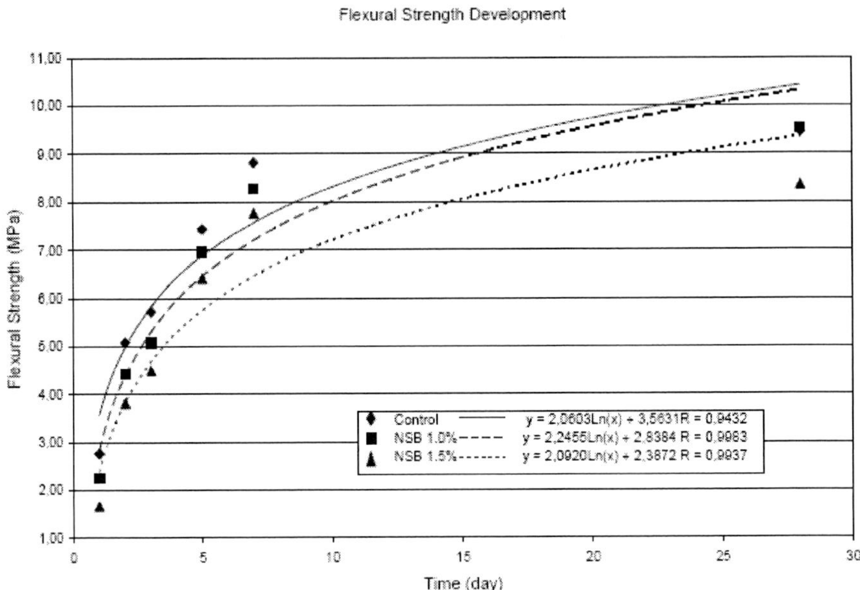

Figure 15. Flexural strength developments of cement mortars.

As can be observed from figures 14 and 15, the lowest 28 day mechanical properties were found on the mixtures with a 1.5 % chemical admixture content. Optimum dosage of chemical admixture in terms of mechanical properties and flowability seems to be 1.0 %.

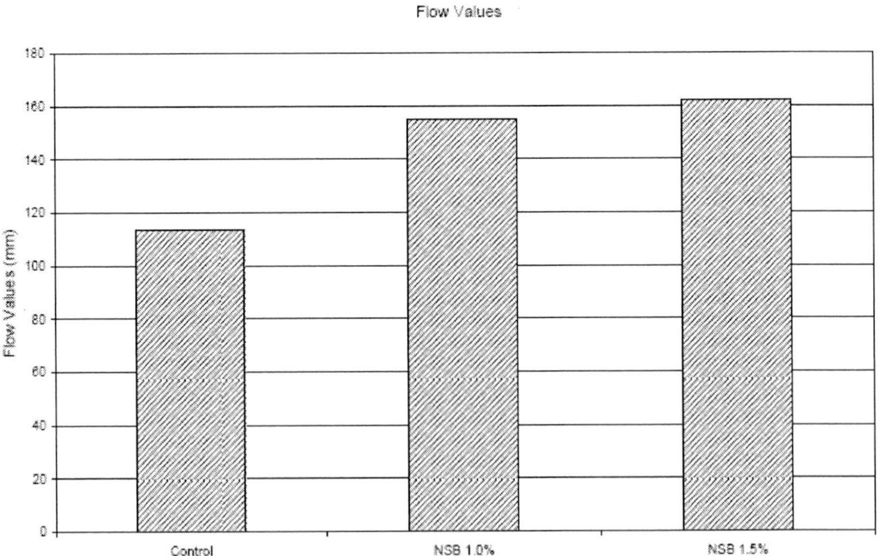

Figure 16. Flow values of cement mortars.

As mentioned before micrographs of cement mortars were taken in order to investigate the effects of retarders on microstructure formation of cement mortars. Micrographs were taken from 3 different areas of polished sections in order to maintain homogeneity of image analysis. Also image processing procedures and segmentation analyses were performed on the images of 1, 2, 7 and 28 days old specimens to determine the pore area ratio values of selected polished sections. The decrease of pore area ratio within time for each specimen is presented in figure 17. Incorporation retarders decreases the rate of hydration process of cement mortars, also results in a more porous texture in the early stage of hydration compared to control specimens.

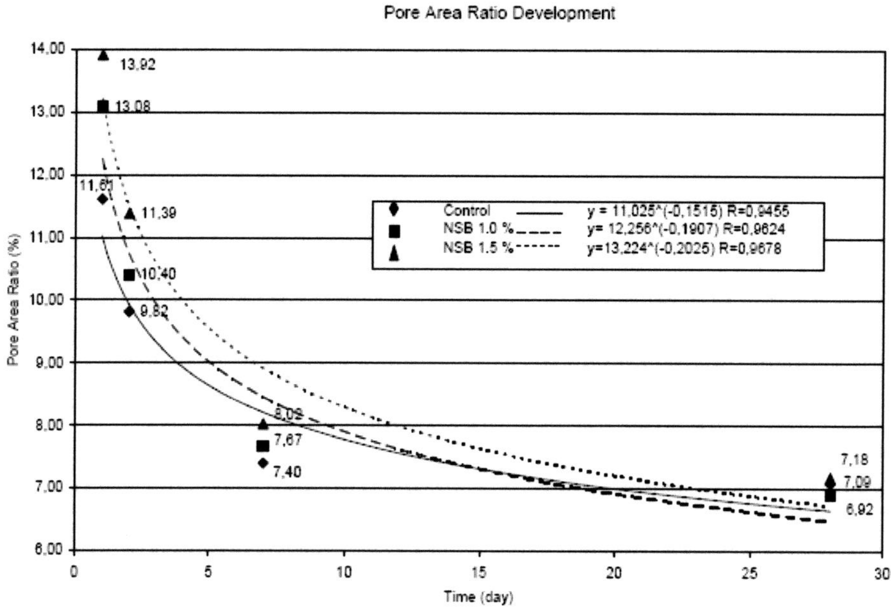

Figure 17. Pore area ratio development.

The segmented figures with pore area ratio values of micrographs of cement mortars are presented in figure 18. The red color in the segmented figures indicates the pore formation of polished sections.

However, in 28 days there is a negligible difference in pore area ratios between control specimens and specimens with retarders. It may be concluded that; mixtures with retarders are more sensitive to chemical, biological and physical attacks, also needs a proper curing process especially in the early stages of hydration. There is a strong correlation ($R > 0.99$) with the compressive strength values and pore area ratios for all mixtures. This relationship is in accordance with the related mathematical models in the literature. The results of image analysis may be used to predict the development of mechanical properties of cement mortars and concrete with time. Similar correlation has been observed between flexural strength and pore area ratio values for all mixtures (figure 20). Pore area ratio values representing the total porosity in a polished section closely were related to mechanical properties of cement mortars. Various mathematical models introduced above have been used to establish a relationship between pore structure and mechanical properties of cement mortars [12, 13]. All models representing the porosity-strength relationship in the literature show a similar trend to

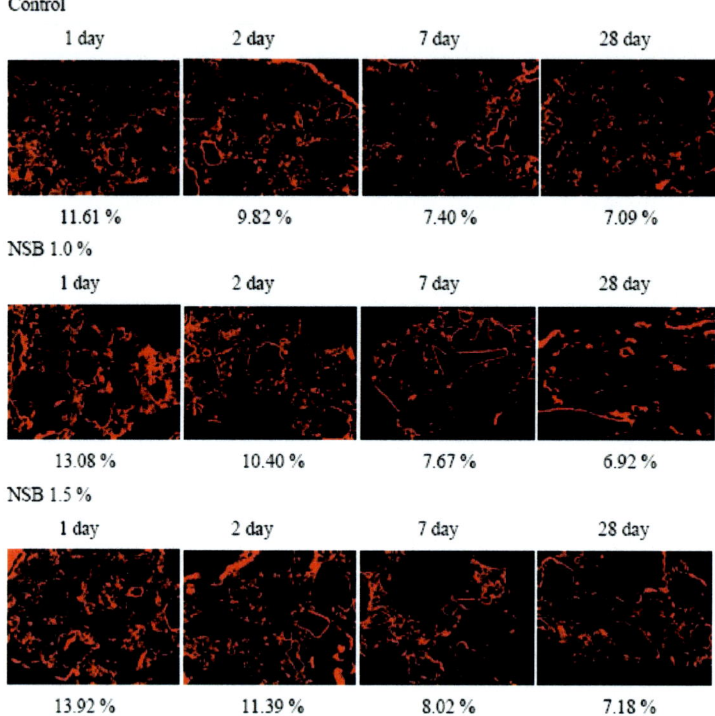

Figure 18. Segmented micrographs of cement mortars.

each other. The relationship established by using image analysis results indicating the pore structure and strength values defining the mechanical behaviour, has a strong correlation values. They are also in conformity with the other mathematical models suggested by other researchers.

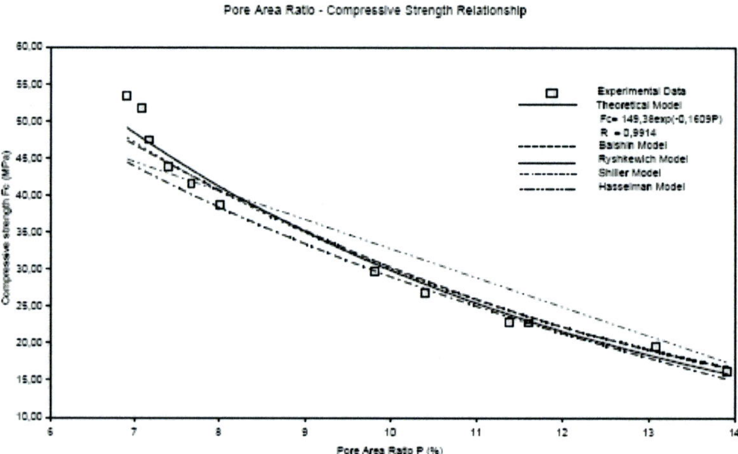

Figure 19. The relationship between pore area ratio and compressive strength.

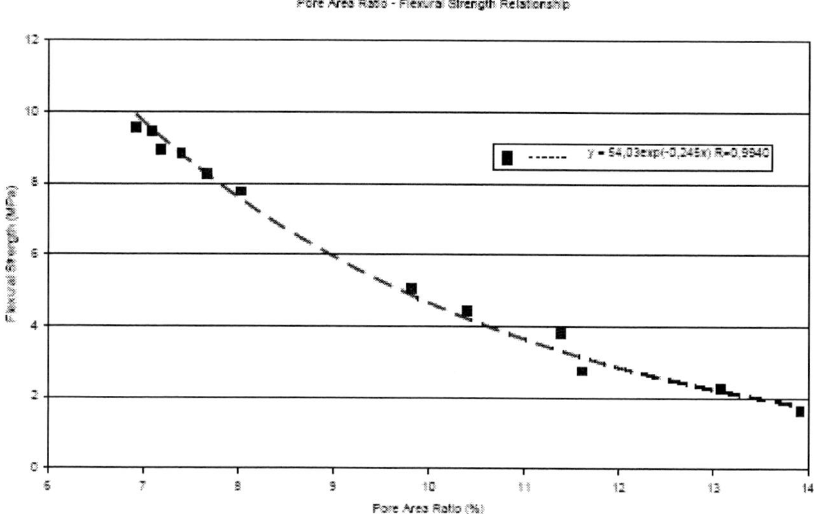

Figure 20. Flexural Strength – Pore structure relationship

CONCLUSIONS

In this chapter, the effects of pore structure being a part of microstructural formation on the mechanical properties of cement mortars have been investigated. The relationship between pore structure and mechanical properties of cementitious materials has been established for various mixtures. The effects of retarders on the formation of pore area ratio have also been investigated.

Cement mortars prepared by incorporation of chemical admixtures (retarders) with different ratios were prepared. Different ratios of admixtures result in various microstructural formations at different ages. Compressive and flexural strength developments of cement mortars were determined for different ages besides microstructure investigations.

The retarding effect of admixtures in the mechanical properties development of specimens prepared with the naphthalene sulphonate based chemical admixtures can be clearly observed in early ages. An excess amount of chemical admixture results in a decrease in the strength properties of cementitious materials. This result may be due to its more porous structure. Mixes with a 1% addition of chemical admixture have higher strength values than those of control specimens at 28 days. Flow value of each mixture was also determined. Cement mortar with 1.5% addition of chemical admixture has the highest flow value.

At early ages, cement mortars with chemical admixtures showed higher pore area ratio values due to the retardation effect of chemical admixtures. Pore area ratio values of all cement mortars decreased with time due to the process of hydration process. Cement mortars with 1.0 % addition of chemical admixture have the lowest pore area ratio values at the 28th day. This case indicates that retarders have significant effects on the early microstructure formation of cement mortars. However, at the later ages the inner structure becomes more solid and homogenous. Indeed, pore area ratio values of cement mortars have a strong correlation with mechanical properties. As can be seen from figure 19 and 20, values for all strength tests increase as pore area ratio values decrease according to the progress of hydration reaction of cement mortars.

The relationship between microstructure and mechanical behavior of cement mortars was established by using different mathematical models. The test results indicate that the mathematical models given in the literature give strong correlations for the relationship between pore area ratio values obtained by image analysis and their mechanical properties. The relationship obtained by image analysis results indicates exponential behavior.

REFERENCES

[1] G.Ye, K. Van Breugel, A.L.A. Faraaij. Experimental study and numerical simulation on the formation of microstructure in cementitious materials at early age. *Cem. Concr. Res.* 2003; 33: 233-239.

[2] K.L. Scrivener, H.H. Patel, P.L. Pratt, L.J. Parrott, Analysis of phases in cement paste using backscattered electron images, in: L.J. Struble, P.W. Brown (Eds.), Microstructural Development During Hydration of Cement, Mater. *Res. Soc. Symp.* Proc. 1987; 85: 67-76.

[3] H. Zhao, D. Darwin, Quantitative backscattered electron analysis for cement paste, *Cem. Concr. Res*. 1992; 22: 695 -706.

[4] K.O. Kjeilsen, R.J. Detwiler, O.E. Gjùrv, Backscattered electron image analysis of cement paste specimens: Specimen preparation and analytical methods, *Cem. Concr. Res.* 1991; 21: 388-390.

[5] D.A. Lange, H.M. Jennings, S.P. Shah, Image analysis techniques for characterization of pore structure of cement based materials, *Cem.Concr. Res*. 1994; 24: 841-853.

[6] Y. Wang, S. Diamond, An approach to quantitative image analysis for cement pastes, Mater. *Res. Soc. Symp. Proc*. 1995; 370: 23 - 32.

[7] D. Darwin, M.N. Abou-Zeid, Application of automated image analysis to the study of cement paste microstructure, *Mat. Res. Soc. Symp. Proc*. 1995; 370: 3-12.

[8] S. Diamond, M.E. Leeman, Pore size distributions in hardened cement paste by SEM image analysis, Mater. *Res. Soc. Symp. Proc*. 1995; 370: 217 - 226.

[9] M. Barrioulet, R. Saada, E. Ringot, A quantitative structural study of fresh cement paste by image analysis: Part 1. Image processing, *Cem. Concr. Res.* 1991; 21: 835-843.

[10] H. A. Razak, H.S. Wong. Strength estimation model for high-strength concrete incorporating metakaolin and silica fume. *Cem. Concr. Res*. 2005; 35: 688-695.

[11] E.P. Kearsley, P.J. Wainwright. The effect of porosity on the strength of foamed concrete. *Cem. Concr. Res.* 2002; 32: 233-239.

[12] M. Rößler, I. Odler. Investigations on the relationship between porosity, structure and strength of hydrated Portland cement paste: I. Effect of porosity. *Cem. Concr. Res.* 1985; 15: 320-330.

[13] G. Fagerlund. Strength and porosity of concrete, Proc. Int. Rilem Symp. Pore Structure, Praque 1973 D51-D73 (Part 2).

[14] R.F. Feldmann, J.J. Beaudin. Microstructure and strength of hydrated cement *Cem. Concr. Res*. 6 (1976) 398–400.

[15] S. Diamond. Mercury porosimetry an appropriate method for the measurement of pore size distributions in cement-based materials. *Cem. Concr. Res*. 2000; 30: 1517-1525.

[16] J. Skalny, J. Gebauer, I. Odler. Scanning Electron Microscopy in Concrete Petrography. Materials Science of Concrete Special Volume: Calcium Hydroxide in Concrete, American Ceramic Society. November 1–3, 2000, Anna Maria Island, Florida, pp. 59–72, 2001.

[17] Standard test method for quantitive determination of phases in Portland cement clinker by microscopical point-count procedure, ASTM C 1356. Annual book of *ASTM* standards, vol. 4.01, 2001.

[18] D.H. Campbell. Microscopical examination and interpretation of Portland cement and clinker. 2nd ed. Portland cement association; 1999 201 p.

[19] F. Hofmänner. Microstructure of Portland cement clinker. Holderbank, Switzerland: Holderbank management and consulting ltd. ; 1973 48 p.

[20] K.L. Scrivener, P.L. Pratt. Characterization of Portland cement hydration by electron optical techniques. In: Electron Microscopy of Materials. *Proc. Mater. Res. Soc. Symp.*, 31, 1983, pp. 351–356.

[21] K.L. Scrivener, P.L. Pratt. Backscattered electron images of polished cement sections in the scanning electron microscope. In: Proc. Sixth Int. Conf. Cement Microscopy, Albuquerque, 1984, pp. 145–55.

[22] R.J. Detwiller, L.J. Powers, U.H. Jakobsen, W.U. Ahmed, K.L. Scrivener, K.O. Kjellsen, *Concr. Int.* (2001) 51–58.

[23] J.L. Goldstein, D.E. Newbury, P. Echlin, D.C. Joy, C. Fiori, E. Lifshin, Scanning Electron Microscopy and X-ray Microanalysis. Atext for biologists, material scientists, and geologists, Plenum pres, New York, 1981 673 p.

[24] P. E. Stutzman, J.R. Clifton. Sample preparation for scanning electron microscopy. In: Proc. 21st Int. Conf. Cement Microscopy. Las Vegas, 1999, pp. 10–22.

[25] A.K. Crumbie. SEM microstructural studies of cementitious materials: Sample preparation of polished sections and microstructural observations with backscattered electron images – artefacts and practical consedirations, in: Proc. 23rd Int. Conf. Cement Microscopy, Albuquerque, 2001 320-341.

[26] H.S. Wong, M.K. Head, N.R. Buenfeld, Pore segmentation of cement-based materials from backscattered electron images. *Cem. Concr. Res*. 2006; 36: 1083-1090.

In: Computational Materials
Editor: Wilhelm U. Oster

ISBN: 978-1-60456-896-7

Chapter 9

CLEAVAGE FRACTURE CRYSTALLOGRAPHY

J. Flaquer and A. Martín-Meizoso[1]
CEIT and Tecnun (University of Navarra)
P. Manuel Lardizábal 13, 20018 San Sebastián, Spain

ABSTRACT

Cleavage fracture of cubic materials takes place usually along {001} planes. In a cubic crystal there are three sets of <001> orientations available for cleavage. The crystal will cleave along the best oriented among the three. The distribution of the best oriented cleavage plane is computed by simulation for a random polycrystalline material. It is proved that cleavage fracture is intrinsically unstable. Beside a simple and effective procedure is described for computing the distribution of observed tilt (dihedral) angles in 2D sections and/or projections distributed randomly in 3D.

Keywords: cleavage, tilt dihedral angle, stereology, distribution function, EBSD

1. INTRODUCTION

Fracture of steels and other cubic materials takes place along {001} planes at low enough temperatures by cleavage. It is well known that the scatter of the toughness is very small when fracture takes place at very low temperatures, as in the case of fracture at -196°C (77 K, usually when the test-pieces are submerged in liquid nitrogen [1]) and also that cleavage is controlled by the stress [2-4]. Usually a very thin fatigue crack is introduced in the test-piece before the fracture test. This fatigue crack samples the material intersecting a large number of crystals, as schematically shown in figure 1.

[1] e-mail: ameizoso@ceit.es

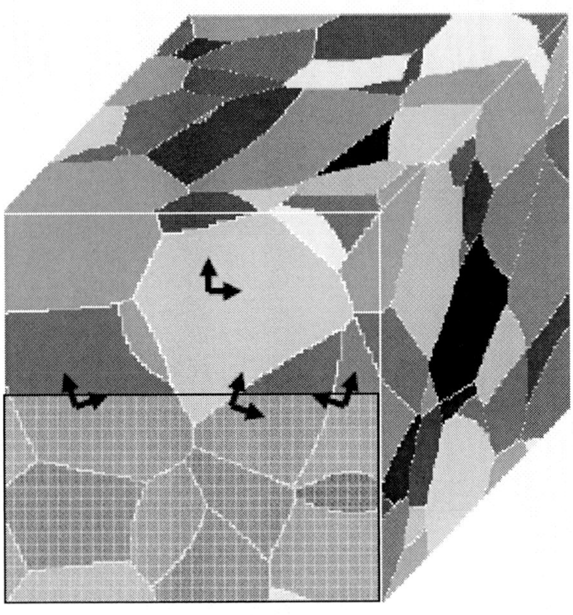

Figure 1. Schematic of a fatigue crack cutting through a polycrystalline material.

If a homogeneous isotropic polycrystalline material is assumed, it is possible to compute the distribution of their crystal orientations and particularly its cleavage planes with respect to the normal to the fatigue crack plane. A computer code has been developed to work out the distribution of the best oriented crystal orientation from the set of three possible orientations for cubic crystallites. Typically the number of grains intersected by a fatigue crack is about 500, for a standard fracture test-piece with 25 mm thickness.

After crack nucleation, the crack propagates from a first grain into a second one, where the crack branches in different steps (river marks) to accommodate a different orientation of the second grain.

There are three degrees of freedom to specify a crystal orientation. Six will be needed to specify the orientation of two different crystals. Two more will be needed if we want to specify a sectioning and observation plane (for example in a metallographic section). Even if the observation orientation is fixed, still six degrees of freedom will be required.

Note that x-ray EBSD (electron back scatter diffraction) patterns recognize the crystal orientations and may draw the boundaries among them, depending on their mismatched angles. Usually the coloured angle represents the orientation of the observed section versus the normal to the observed surface direction (other references are also possible).

To orientate identically two different crystals we need to twist one of them by the total misorientation angle [5]. This misorientation angle may be decomposed in two different twists: the first one to match the orientation of the cleavage fracture planes (after this twist {001} planes become coplanar), that is the tilt angle; and a second twist to orientate these two {001} planes identically, and that is the twist angle. From a cleavage fracture point of view, any misorientation represents a barrier for crack propagation, but most probably this barrier is much smaller if the crystals have coplanar cleavage fracture planes. So, the most important role is played by the tilt component of the misorientation and the twist component may be neglected.

2. Model Description

One particular crystal is specified by using typically the three Miller indices or the three Euler angles. The orientation of one particular (001) plane is specified by using two angles: θ, ϕ. Let us assume the first angle, θ, with respect to the normal to the fatigue crack plane (loading direction in mode I, and z axis). Subindices 1 to 3 will used to represent every one of the three {001} planes, see figure 2. For all the crystals these orientations should be uniformly distributed over a hemisphere. A uniform distribution is generated using

$$\theta_1 = \mathrm{acos}(1 - rand) \tag{1}$$

where *rand* stands for a random number uniformly distributed between 0 and 1 (most computer codes provides for such a pseudo-random number generator). The uniformity of the distribution is easy to probe if we remember that if a sphere, of a given radius R, is cut by two parallel planes, a spherical zone of two bases is obtained; the amount of surface s cut between both planes depends only on the distance from one to the other cutting planes, h.

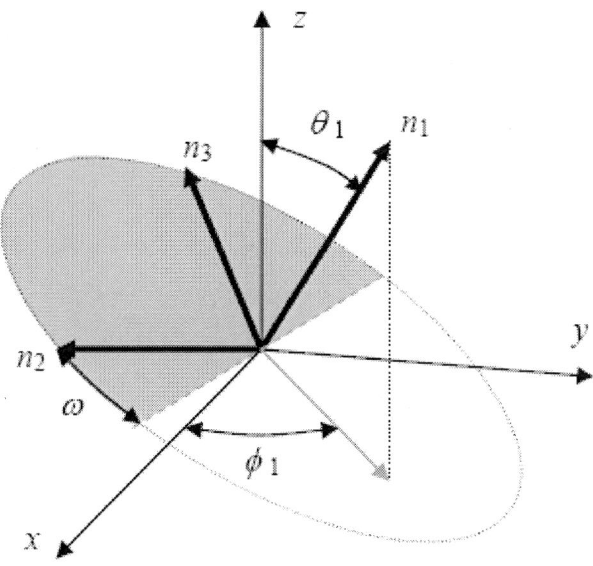

Figure 2. Symbols and axes used to specify a cubic crystal orientation. The part of the circle, with normal n_1, above the $z = 0$ plane is shown in grey colour.

$$s = 2\pi Rh \tag{2}$$

No matter if both planes are close to the equatorial plane or to one pole. Differentiating Eq. (2), *ds* scales linearly with *dh*. If the sphere is chosen of $R = 1$ and the z axis is chosen coincident with the normal to the cutting planes, then $dh = dz$, and therefore $\cos\theta_1$ distributes uniformly.

The second angle, for this first cleavage plane orientation is generated as

$$\varphi_1 = 2\pi \cdot rand \tag{3}$$

Note that this second *rand* number is, in general, a different one from that used in Eq. (1). Obviously the projections n_1 on to a normalized sphere (of radius = 1) can obtained by

$$\begin{aligned} x_1 &= \sin\theta_1 \cdot \cos\varphi_1 \\ y_1 &= \sin\theta_1 \cdot \sin\varphi_1 \\ z_1 &= \cos\theta_1 \end{aligned} \tag{4}$$

A third random number should be used to create n_2 and n_3, the second and third cleavage plane orientations, respectively; both being orthogonal to n_1. For convenience an auxiliary angle ω is introduced

$$\begin{aligned} \omega &= \frac{\pi}{2} rand \\ \theta_2 &= \operatorname{acos}(\sin\theta_1 \cdot \sin\omega) \\ \varphi_2 &= \operatorname{atan}\frac{\cos\varphi_1 \cdot \cos\omega - \cos\theta_1 \cdot \sin\varphi_1 \cdot \sin\omega}{-\sin\varphi_1 \cdot \cos\omega - \cos\theta_1 \cdot \cos\varphi_1 \cdot \sin\omega} \\ \theta_3 &= \operatorname{acos}(\sin\theta_1 \cdot \cos\omega) \\ \varphi_3 &= \operatorname{atan}\frac{-\cos\varphi_1 \cdot \sin\omega - \cos\theta_1 \cdot \sin\varphi_1 \cdot \cos\omega}{\sin\varphi_1 \cdot \sin\omega - \cos\theta_1 \cdot \cos\varphi_1 \cdot \cos\omega} \\ x_2 &= \sin\theta_2 \cdot \cos\varphi_2 \\ y_2 &= \sin\theta_2 \cdot \sin\varphi_2 \\ z_2 &= \cos\theta_2 \\ x_3 &= \sin\theta_3 \cdot \cos\varphi_3 \\ y_3 &= \sin\theta_3 \cdot \sin\varphi_3 \\ z_3 &= \cos\theta_3 \end{aligned} \tag{5}$$

We shall retain the orientations projected on to the top hemisphere from the two possible cleavage plane normals (the opposite normal being projected on to the lower hemisphere). Figure 3 shows the projection of 5,000 crystals, every one with three possible {001} plane orientations.

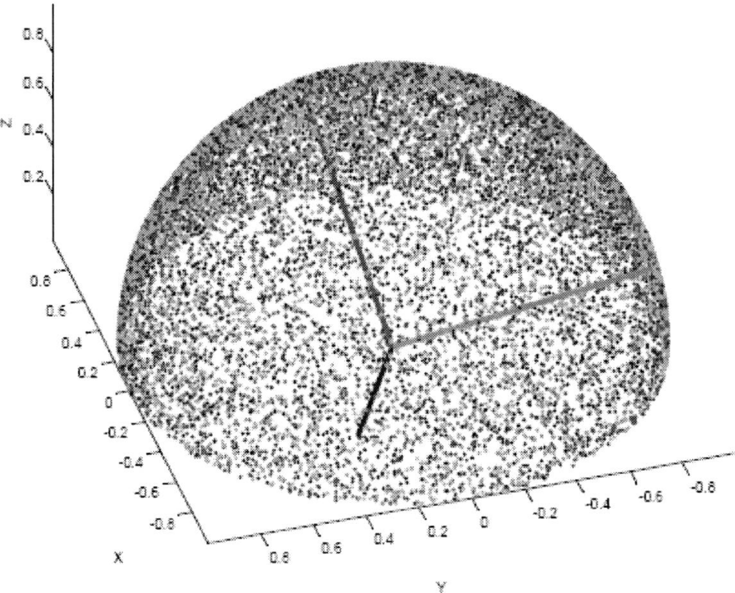

Figure 3. Five thousand cubic crystals, oriented at random, with their 15,000 {001} cleavage fracture planes. Different colours are used for the different {001} planes in each crystals. The trihedron of one particular crystal is also shown.

Figure 3 shows an evenly distribution of orientations on the whole hemisphere and also the different colours (used for subindices 1 to 3) are uniformly distributed. A particular trihedron is also shown, as solid lines from the origin; by using MatLab® it is possible to rotate the figure and to check for the trihedron perpendicularity.

For a particular crystal we shall retain the best oriented plane for cleavage fracture, which is the one with the largest projection on the loading direction (vertical in the figures). Up to now an analytical approach was possible, but to select the best from the three perpendicular orientations a simulation is probably the only simple and straight way.

Figure 4 shows the obtained result from the 5,000 crystals considered in figure 3. Note that the retained orientations are no longer uniformly distributed in figure 4 (and also that none of the three colours is preferred, as it should be).

3. Distribution of Orientations for Cleavage Planes

Figure 5 shows the distribution for the best oriented cleavage plane with respect to the loading direction and also macroscopic fracture plane, measured by the misorientation angle θ. Quite a number of crystals are close to the tensile direction. The largest miss-orientation is $\theta_{max} = 54.74° = acos\ (1/\sqrt{3}) = 0.9553$ radians, corresponding to crystals with the <111> direction lying along the vertical axis.

A polynomial regression of the cumulative density function (*CDF*) versus the misorientation angle θ is

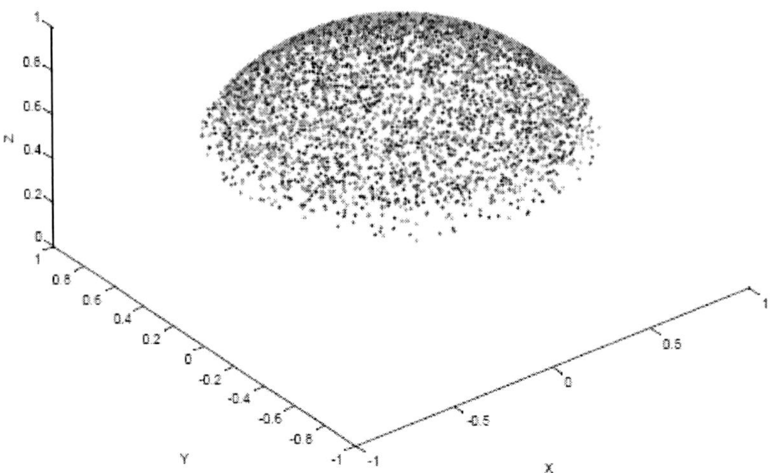

Figure 4. Distribution of the best oriented {001} for a tension in the vertical direction. Only the best oriented plane is kept for each of 5,000 randomly distributed cubic crystals.

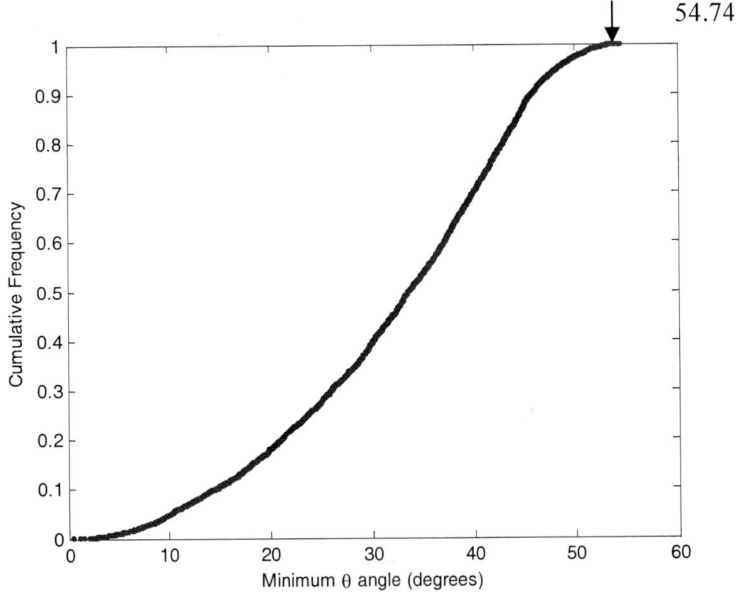

Figure 5. Cumulative Distribution Function of the misorientation angle with respect to the tensile direction (for the best oriented cleavage plane) for a homogeneous cubic polycrystalline material.

$$CDF = 1322\theta^{10} - 5530\theta^{9} + 9752\theta^{8} - 9451\theta^{7} + 5498\theta^{6}$$
$$- 1972\theta^{5} + 434.5\theta^{4} - 58.17\theta^{3} + 6.119\theta^{2} - 0.1772\theta \qquad (6a)$$

For computational purposes, Eq. (6a) is best written, in a telescopic mode, as:

$$CDF = 10^3(((((((((1.322\theta - 5.530)\theta + 9.752)\theta - 9.451)\theta + 5.498)\theta \\ -1.972)\theta + 0.4345)\theta - 5.817\times10^{-2})\theta + 6.119\times10^{-3})\theta - 1.772\times10^{-4})\theta \quad (6b)$$

where θ is given in radians ($0 \leq \theta \leq 0.9553$ radians).

We think that it is probably more useful the opposite: polynomial regression of the misorientation angles versus the cumulative density function, *CDF*, now written as r

$$\theta = -1832r^{10} + 9833r^9 - 22{,}650r^8 + 29{,}290r^7 - 23{,}350r^6 + 11{,}860r^5 \\ -3831r^4 + 765.1r^3 - 89.89r^2 + 6.762r \quad (7a)$$

$$\theta = 10^4(((((((((-0.1832r + 0.9833)r - 2.265)r + 2.929)r - 2.335)r + 1.186)r \\ -0.3831)r + 7.651\times10^{-2})r - 8.989\times10^{-3})r + 6.762\times10^{-4})r \quad (7b)$$

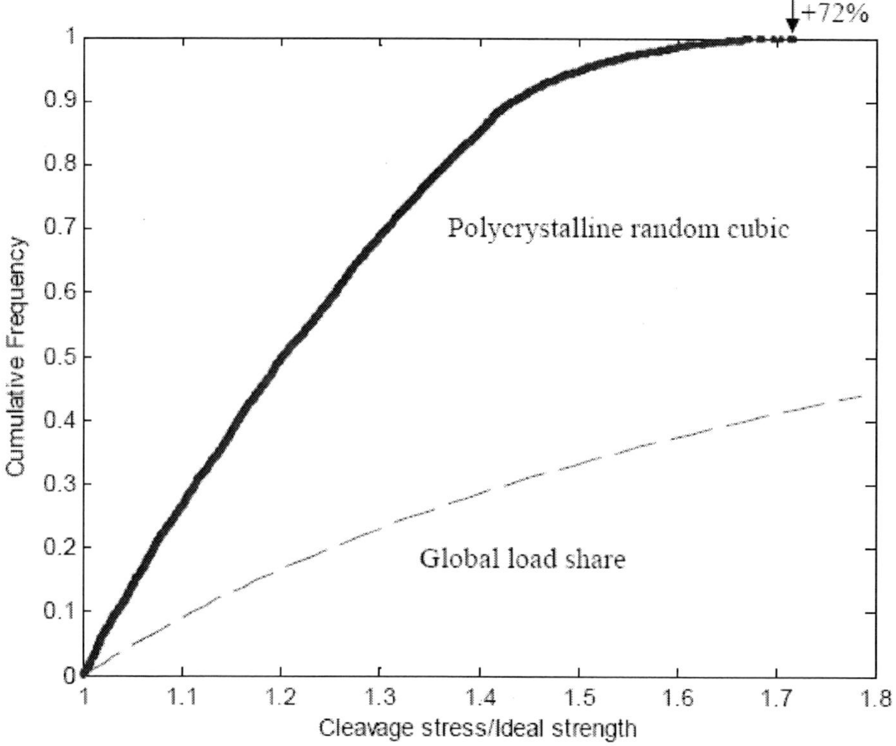

Figure 6. Cumulative Distribution Function of the required tension to cleave individual grains. The dashed line stands for a global share of the load on the remaining crystals at the crack front. Cleavage fracture is both intrinsically unstable and catastrophic, under a constant applied load.

For simulation purposes, it is possible to directly generate misorientation angles, with the right distribution, introducing a random number as r in Eq. (7) uniformly distributed between 0 and 1. The angle θ results in radians. Equations (6) and (7) are polynomials of degree ten (with no independent term) written in both: the standard and the best computing way [6].

Figure 6 shows the required tension to fracture the different crystals along a crack front. The stress in the horizontal axis is normalized by the stress required to break the best oriented crystal (with a cleavage plane coincident with the crack plane).

Figue 7 is a detail of the lower left corner of figure 6, showing the fracture of the first fifty weakest crystals. It is shown that when three crystals are broken, the other crystal resistances are located to the left of the best possible redistribution of the load. There is nearly no chance for arresting a pure cleavage fracture in cubic metals.

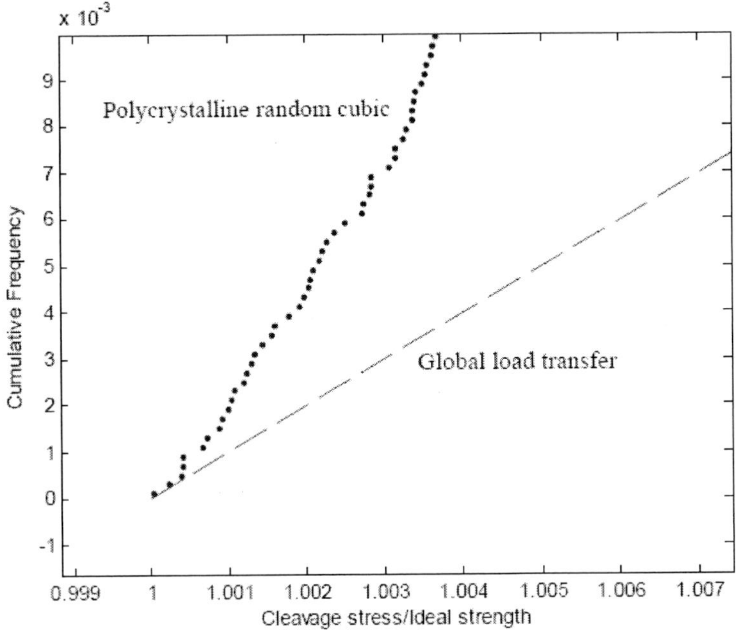

Figure 7. Detail of Fig. 6 showing the region of the weakest crystals and the global share hypothesis (dashed line).

4. Analysis of Misorientation

Up to now we were dealing with the orientation for a single crystal. From now on, we are interested on the angle between two crystals. The new problem is stated as: known the true tilt angle (dihedral angle) between two different cleavage fracture planes, θ, which is the statistical distribution for the observed angles on planar sections (or projections)?

The observed angle always ranges from 0 to 2π, for any given θ angle.

Let us assume the observation direction being coincident with the z axis, see figure 8.

The normal to the first (001) face goes from point 0 through point 1 to the surface of the sphere of unit radius. Its angle with respect to the $z = 0$ plane is α. We can write out

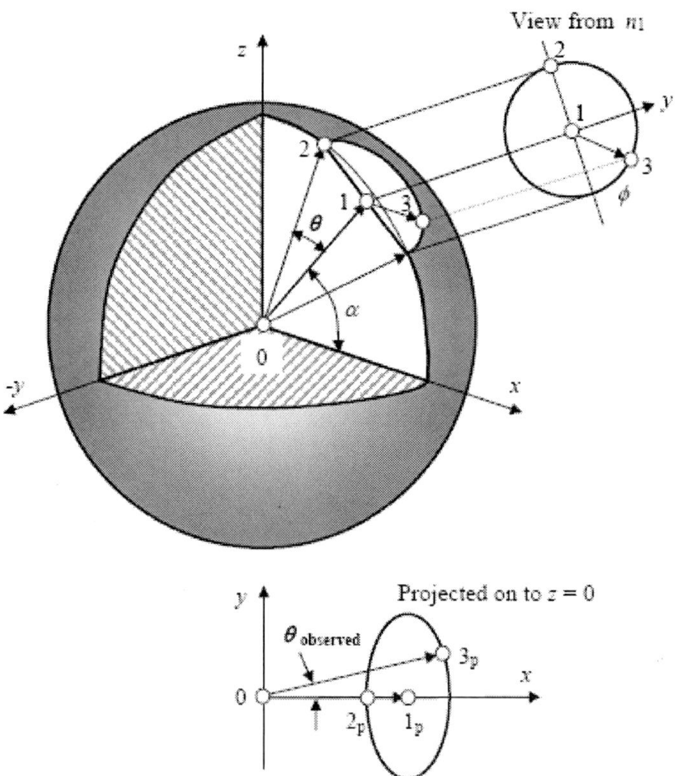

Figure 8. Definition of the axis and angles used. True angle, θ, between n_1 and n_3 and the angle observed, $\theta_{observed}$, when looking from z. (For clarity point 1 is actually drawn under the surface).

$$
\begin{aligned}
x_1 &= \cos\alpha \cdot \cos\theta \\
y_1 &= 0 \\
z_1 &= \sin\alpha \cdot \cos\theta
\end{aligned} \tag{8}
$$

The distribution of this first normal should be uniformly distributed on the whole sphere, but the angle α is not uniformly probable (it was its cosine). To account for this feature we weight each crystal in the following simulations, according to its α, as

$$\omega_\alpha = 2\pi \cos\alpha \tag{9}$$

The second (001) face forms an angle θ with the first one. Therefore its normal is located along the cone with axis trough 01 and passing through point 2. Their coordinates are

$$x_2 = \cos(\alpha + \theta)$$
$$y_2 = 0 \quad (10)$$
$$z_2 = \sin(\alpha + \theta)$$

The second normal intersects the sphere at any point at the intersection between the sphere and the cone. Using the angle ϕ, of uniform probability,

$$x_3 = \cos\alpha \cdot \cos\theta + \sin\alpha \cdot \sin\theta \cdot \cos\varphi$$
$$y_3 = \sin\theta \cdot \sin\varphi \quad (11)$$
$$z_3 = \sin\alpha \cdot \cos\theta - \cos\alpha \cdot \sin\theta \cdot \cos\varphi$$

The sign of z_3 is irrelevant for the purposes of the calculations. The observed angle θ is obtained projecting the points 1 and 3 on to the $z = 0$ plane and measuring the angle $1_p 0 3_p$

$$\theta_{observed} = \operatorname{atan}\left(\frac{y_3}{x_3}\right) = \operatorname{atan}\left(\frac{\sin\theta \cdot \sin\varphi}{\cos\alpha \cdot \cos\theta + \sin\alpha \cdot \sin\theta \cdot \cos\varphi}\right) \quad (12)$$

Note that the angle ϕ is uniformly distributed, and so is its probability, but it would not be the case once projected on the $z = 0$ plane.

A small computer code has being written using Matlab® 7.0 for doing the computations. Up to 200 α angles have being used regularly spaced between 0 and π/2 (the first one is chosen at random, close but different from 0°). Another 200 φ angles have being used (regularly distributed between 0 and π radians, the first one is also chosen at random) for computing the observed angles, using eq. (12). Figure 9.a shows the results obtained for the observed angle versus true angles, for the larger angles and figure 9.b for the smallest angles.

From figures 9.a and 9.b it is obvious that any angle may be observed forming any other angle, between 0° and 180° (already mentioned), but its most frequently observed value (its mode) is identical to its true angle; see figures 10.a and 10.b (detail) representing the probability distribution functions.

The logarithmic singularity of the density function at the dihedral angle was already noted by Reeds and Butler [7,8]. It is not so obvious in figure 10, but it was proved by DeHoff [9] and Butler and Reeds [7,8] that the average observed angle in random sections is also the true/dihedral angle.

Figure 10 was computed by numeric derivation of the results in figure 9. A second degree polynomial was fitted to vectors of the 801 closest points. The derivative is computed at the central point as that for the fitted parabola. Numerical derivation will never capture the density function singularity at the true/dihedral angle, so vertical arrows are drawn to remember this feature.

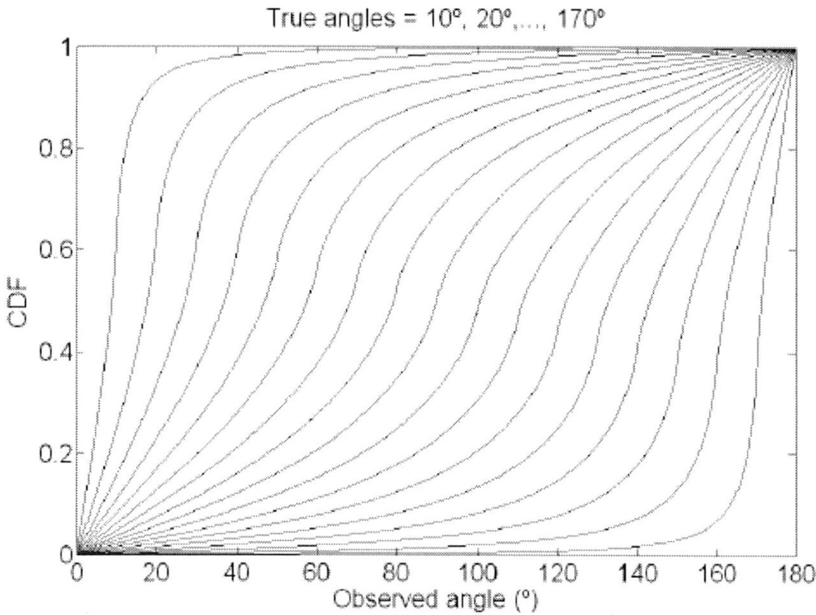

Figure 9.a. Distribution of the observed dihedral angles. Vertical tangents are observed at true angles.

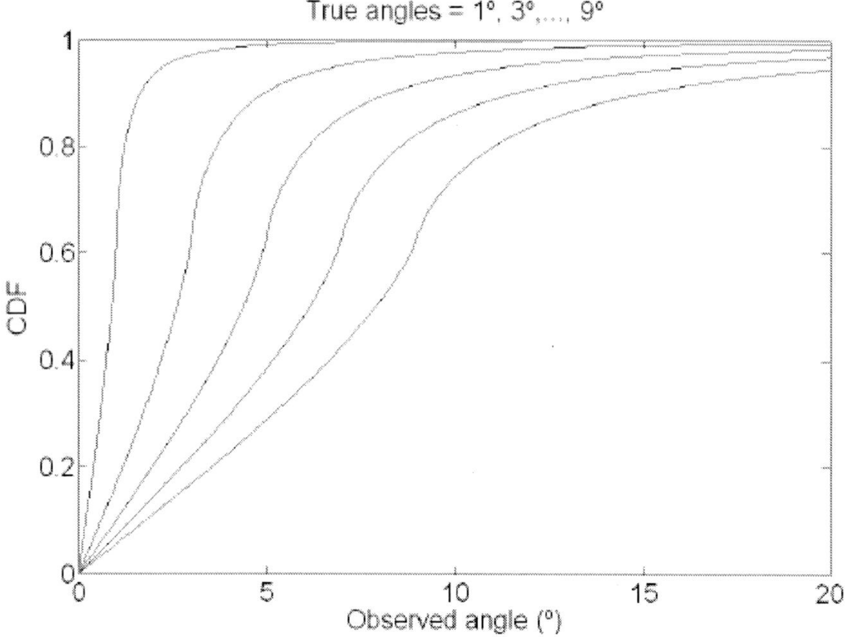

Figure 9.b. Detail of the Cumulative Distribution Function for dihedral angels smaller than 10°.

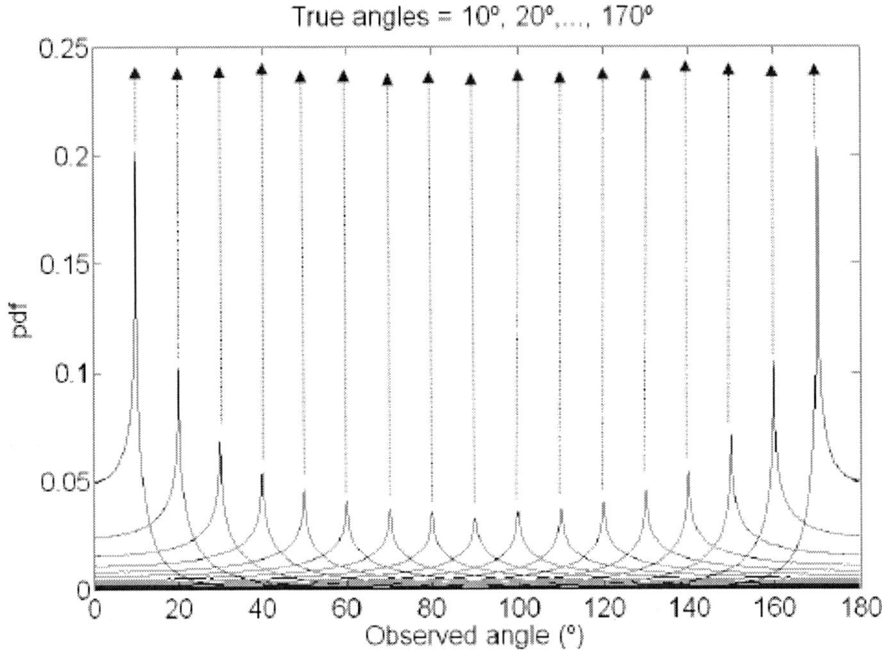

Figure 10.a. Probability distribution functions of the observed angles for different dihedral angles.

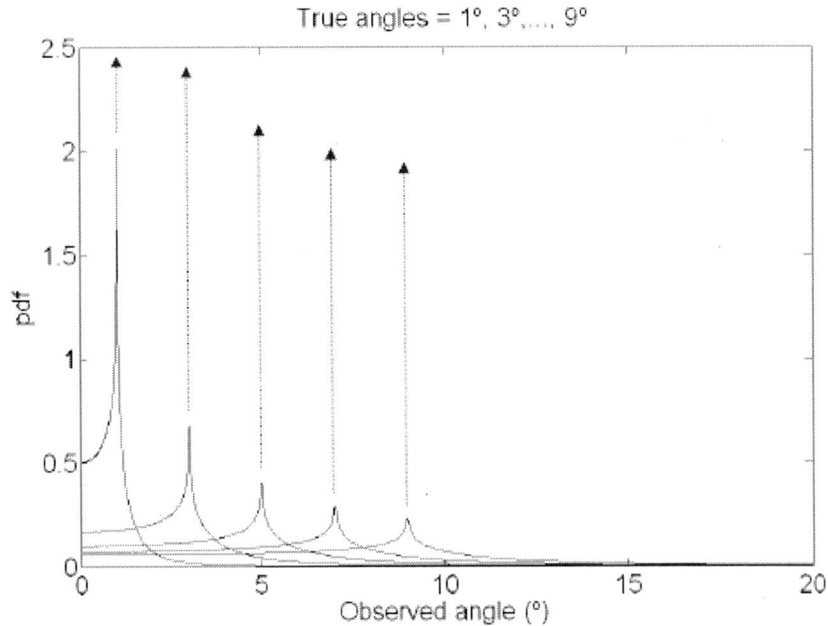

Figure 10.b. Detail of density functions for dihedral angles smaller than 10°. Arrows are used to represent the singularity at the true angles.

5. Tilt Misorientation for Random Cubic Crystals

If a cubic symmetry is imposed to the crystals, then the maximum tilt angle between two cleavage facets ({001} orientations) is 54.74° (acos(1/√3) = 0.9553 rad) and distributes as detailed elsewhere [10]. Figure 11 represents the cumulative distribution for the mismatched angle, θ, for 100,000 randomly oriented cubic crystals and the observed angle in random sections.

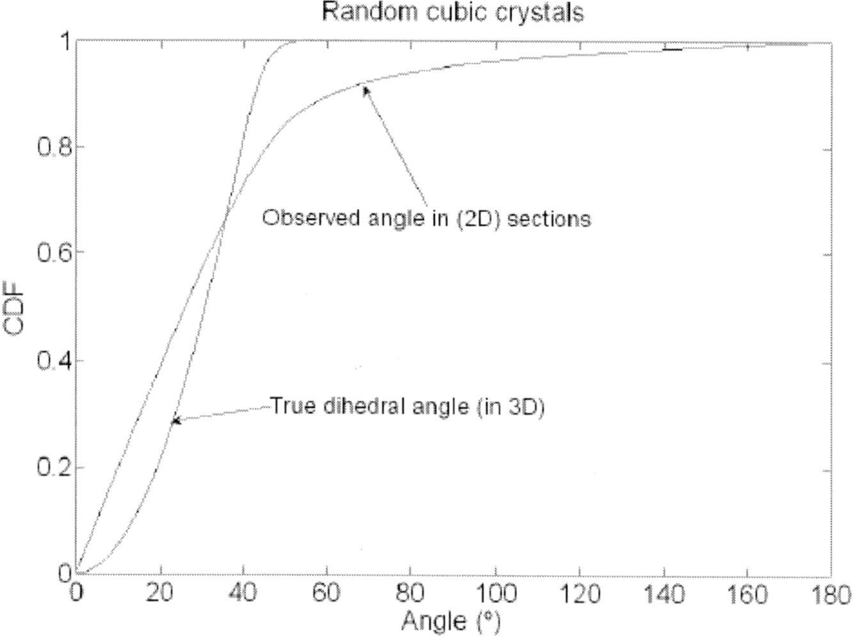

Figure 11. Cumulative distribution function for cubic crystals of dihedral angle and observed in random sections.

For every crystal and tilt θ, an α angle is randomly generated, and weighted according to eq. (9); a second φ angle is chosen at random (it is uniformly distributed, so no weight is applied), and eq. (12) is used to computed the observed angle, θ_{observed}.

6. Conclusion

From figure 6 it is concluded that for the largest possible tilt misorientation, the <111> is oriented parallel to the tension, the toughness improvement is a 72%. That will be the case for an ideal material, completely textured with the <111> orientation parallel to the tensile direction.

On figures 6 and 7 it is also represented the case of a global share of the load, after fracture of a given number of grains on the fatigue crack front. Let us imagine that a 10% of the grains on the fatigue crack front are broken, then the load will be transferred to the neighbouring grains. In the best of the possible cases this will be done in a global way [11], so the extra load is distributed on all the remaining intact crystals along the crack front. Then a 10% of extra stress is applied on them and many more grains will be broken (on the plot, a 25% of the grains along the crack front will break). There is little hope for a stable crack extension under constant applied load, as far as the dashed line (global load transfer) is located to the right of the distribution of crystal resistances to cleavage fracture. Only at the very beginning of cleavage fracture it will be possible stable crack extensions (poppings), see detail in figure 7. There is not such a thing as a resistance curve for cleavage. Once cleavage fracture is triggered there is nearly no chance for arresting (if no plastic contribution is considered).

Multiple initiation sites are frequently observed, from the previous fatigue crack front, when test-pieces are tested at very low temperatures. No inclusion or large precipitate is observed as initiator of cleavage fractures, and a small dispersion of the results is obtained (in steels, $K_{IC} \approx 27.5$ MPa$\sqrt{m}$). All these features are in good agreement with the distribution of orientations obtained for the cleavage planes.

Figure 12 represents the expected density function for randomly oriented cubic crystals. Even when the numeric derivative uses 2001 points to average the value of the density function, it is rather noisy for the smaller angles. A tougher smoothing is possible, but it is plotted as it is to remark that the results are expected to be rather noisy for the smaller angles. It is easy to explain it looking at figure 10.

The average observed angle in section is 31.9° and its standard deviation 28.3°. Note that the distribution is clearly not a Gaussian like.

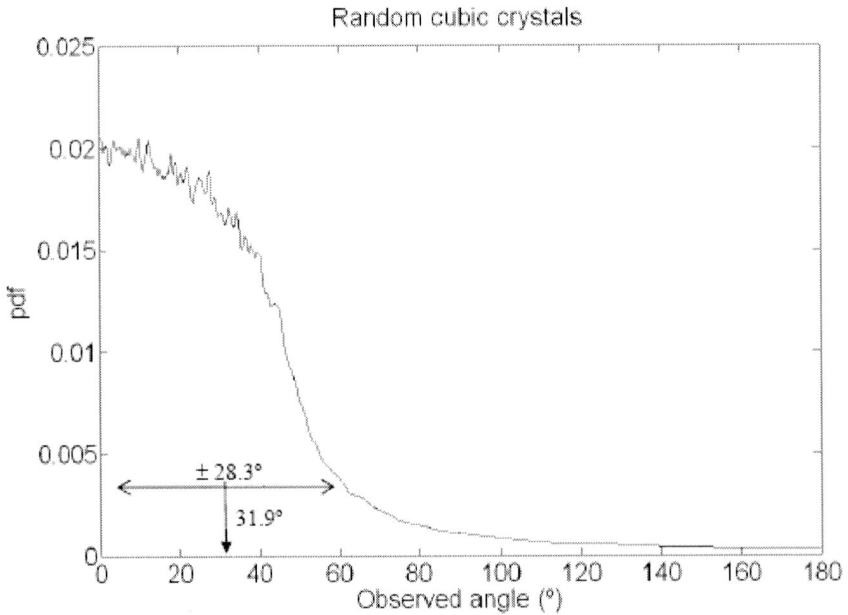

Figure 12. Density function for the observed tilt angle of randomly oriented cubic crystals.

5. ACKNOWLEDGEMENTS

This work has been performed within the Framework of the European project: "Quantitative correlation of fracture toughness and microstructure for high strength steels and their welds (355-890 MPa)" (Microtough); funded by ECSC steel research programme (Contract no. 7210-PD/312).

REFERENCES

[1] Martín-Meizoso, A., Ocaña-Arizcorreta, I., Gil-Sevillano, J., Fuentes-Pérez, M. "Modelling Cleavage Fracture of Bainitic Steels" *Acta Metall. Mater.* 1994, *42*, 2057-2068.

[2] Anderson, T. L. *Fracture Mechanics. Fundamentals and Applications*; 2nd edt., CRC Press, Boca Ratón, 1995.

[3] Broek, D. *Elementary Engineering Fracture Mechanics*; 4th edt., Noordhoff International Publishing, La Hague, The Netherlands, 1986.

[4] Ritchie, R.O., Knott, J. F. and Rice, J. R. "On the relationship between critical tensile stress and fracture toughness in mild steels" *J. Mech. Phys. Solids* 1973, *21*, 395-410.

[5] Mackenzie, J. K. and Thomson, M. J. "Some Statistics Associated with the Random Disorientation of Cubes" *Biometrika* 1957, *44*, 205-210.

[6] Press, W. H., Flannery, B. P., Teukolsky, S. A. and Vetterling, W. T. *Numerical Recipies. The Art of Scientific Computing*; Cambridge University Press, Cambridge, 1986.

[7] Butler, J. P., Reeds, J. A. "Stereology of dihedral angles I: First two moments" *SIAM J. Appl. Math.* 1987, *94*, *3*, 670-677.

[8] Reeds, J. A., Butler, J. P. "Stereology of dihedral angles II: Distribution function" *SIAM J. Appl. Math.* 1987, *47*, *3*, 678-688.

[9] DeHoff, R. T. "Estimation of dihedral angles from stereological counting measurements" *Metallography* 1986, *19*, 209-217.

[10] Horálek, V. "Stereology of dihedral angles" *Applications of Mathematics* 2000, *45*, *6*, 411-417.

[11] Martín Meizoso, A., Martínez Esnaola, J. M., Daniel, A. M., Sánchez, J. M., Puente, I. and Elizalde, M. R. "Load Transfer in Ceramic Matrix Composites" *Int. J. of Fracture* 1996, *76*, R55-60.

In: Computational Materials
Editor: Wilhelm U. Oster
ISBN: 978-1-60456-896-7

Chapter 10

THEORETICAL SIMULATION ON MOLECULAR ELECTRONIC MATERIALS AND MOLECULAR DEVICES

Jianwei Zhao[1]*, Yanwei Li[2] and Hongmei Liu[1]

[1]School of Chemistry and Chemical Engineering, Nanjing University, Nanjing 210093 P. R. China
[2]Department of Material and Chemical Engineering, Guilin University of Technology, Guilin 541004, P. R. China

ABSTRACT

Over the past decade, there has been remarkable progress in the studies of molecular electronics. In this chapter, we will survey the recent theoretical research in this field. In principle, two theoretical strategies, namely static and dynamic approaches are employed in the theoretical modeling. For static approach, an *in-situ* static theoretical calculation by considering the influence of electric field is introduced. The *in-situ* simulation results displayed that both the geometric and electronic structures of the conjugated molecular materials are sensitive to the electric field. On the other hand, the electronic transportation through molecular wires has been studied intensively in these years by using the non-equilibrium Green's functional formulism combined with density functional theory (NEGF-DFT), which may directly give the current-voltage character to compare with the experimental measurement. A series of typical molecular wires have been studied to compare with the experimental transportation behavior. Based on the chain-length dependence of conductance for each series of molecular wires, the attenuation factor, β, has been obtained and compared with experimental data where applicable. The β value has also been quantitatively correlated to the molecular HOMO-LUMO gap. Following the study of molecular wires, molecular rectification has become the focus in the field of molecular devices. A series of asymmetrically substituted conducting molecular wires have been studied with the same method. The results demonstrated that the fully-conjugated molecular wire with asymmetric substitution has minor rectification. Therefore, other rectification mechanisms are essentially required.

1. INTRODUCTION

The miniaturization of electronic components for the construction useful devices is an essential feature of modern technology. Their miniaturization allows the fabrication of ultra-densely integrated circuits and faster processors. The size of components in integrated circuits is shrinking at an exponential rate and presently close to the limit of both laws of physics and the cost production for the conventional silicon-based devices. Therefore, alternative materials for the nano-scale components of future electronic devices must be developed. Molecules and chemical systems are an obvious and promising choice [1-3].

Molecular electronics can be defined as technology utilizing single molecules, small groups of molecules, carbon nanotubes, or nanoscale metallic or semiconductor wire to perform electronic functions [3]. Conceptually different from conventional solid-state semiconductor electronics, molecular electronics allow chemical engineering of organic molecules with their physical and electronic properties tailored by synthetic methods, bringing a new dimension in design flexibility that does not exist in typical inorganic electronic materials. The conventional semiconductor devices are fabricated from the "top-down" approach that employs photolithography and etch techniques to pattern a substrate. This approach has become increasingly challenging as feature size decreases to nanometer scale due to the difficulty of controlling the electronic properties of semiconductor structures [4]. In contrast, molecular electronics are fabricated from the "bottom-up" approach that builds small structures from the atomic, molecular, or single device level (figure 1) [5].

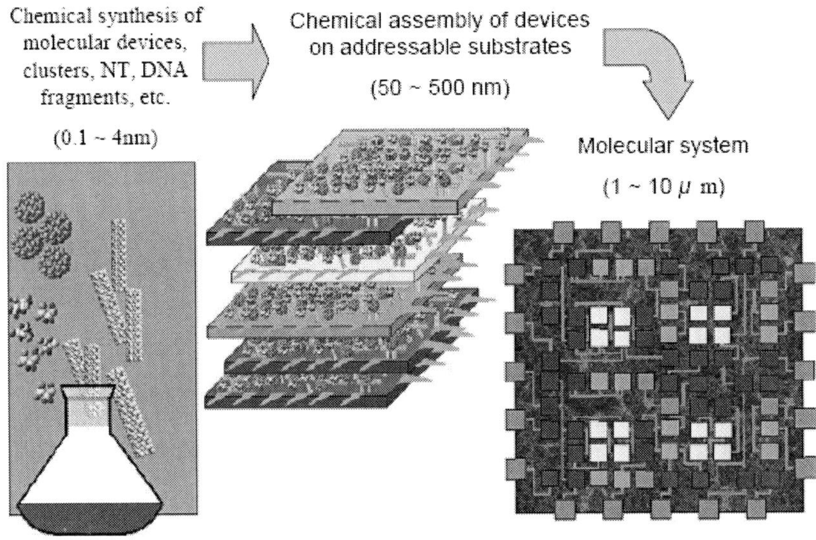

Figure 1. Schematic illustration of the "bottom up" approach for fabricating molecular electronics.

Chemical synthesis may produce large quantities of nano-scale molecules with the same uniformity but at significantly less cost than other batch-fabrication processes such as microlithography. Molecules are naturally small, and their abilities of selective recognition and binding can lead to cheap fabrication using self- assembly. In addition, they offer tenability through synthetic chemistry and control of their transport properties due to their conformational flexibility.

The idea of molecular electronics comes from a farsighted paper by Aviram and Ratner in 1974 [1], predicting that single molecules with a donor-spacer-acceptor structure could be used as diodes when place between two electrode. Today, molecular electronics is emerging as an alternative to silicon-based electronics for building integrated devices with the advent of nanotechnologies, such as scanning probe microscopy (SPM), and molecular self-assembly, which made molecular electronics in a reality. Remarkable progress in this field has been made in the last decade, as researchers have developed ways of growing, addressing, imaging, manipulating, and measuring small groups of molecules connecting metal leads [6]. Several prototype molecular electronic devices such as molecular wires, molecular diodes, molecular switches, and resonant tunneling diodes (RTD) at real molecular level have been demonstrated [7]. In parallel, there has been significant theoretical activity toward developing the description of non-equilibrium transport through molecules [8].

Experimental investigations on molecular electronics have been carried out by using mechanically controllable break junctions, nano-junctions, and scanning probe microscopy (SPM) as shown in figure 2 [9-12]. Together with the experiments, a large number of theoretical studies have also been reported [13-18]. Theoretical work can be broadly classified into two categories: one focuses on the geometric and electronic structures [13-15], and the other on the electron transport properties based on non-equilibrium Green's function (NEGF) formulism [16-18]. The theoretical work focusing on the geometric and electronic structures of molecular electronic materials or molecular devices can be defined as static theoretical investigation and the theoretical work focusing on the electron transport properties as dynamic theoretical investigations.

Several types of molecules, such as conjugated hydrocarbons, porphyrin oligomers, carbon nanotubes and DNA, have been suggested as the promising families of molecular electronic materials. They all have the same key requirements. The most obvious is that they must be electron or hole conducting in order to carry a current though the circuit. Thus, the electronic material provides a pathway for transport of the electrons from one reservoir to another that is more efficient than electron transport through space. By proper functionalization of the molecular electronic materials one can further design various molecular electronic devices, such as molecular diodes [19], molecular field effect transistors (MFET) [20], molecular switches [21], and resonant tunneling diodes (RTD) [22]. Conjugated molecules (see figure 3) comprising alternating single and double (or triple) carbon-carbon bonds can conduct electrons through their π-system, and this has been the basis of many molecular electronic materials and molecular devices. Therefore, this review is mainly focus on the theoretical simulations on these conjugated molecular electronic materials and relevant molecular devices.

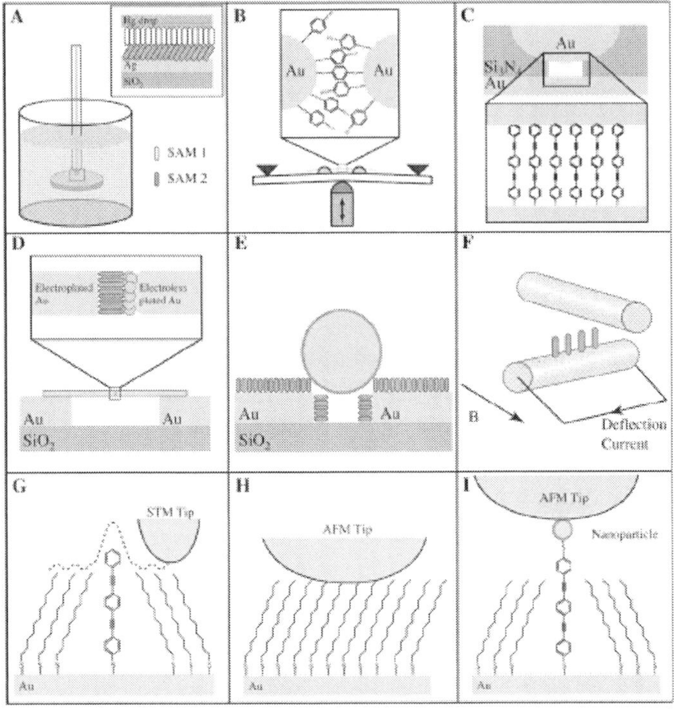

Figure 2. The various techniques used to measure electronic properties of molecules. (A) Hg drop junction; (B) Mechanically controlled break junctions; (C) Nanopore; (d) Nanowire; (E) Nanoparticle bridge; (F) Crossed wires; (G) Scanning tunneling microscope (STM); (H) Contact conductive atomic force microscope (CAFM); (I) Nanoparticle coupled CAFM.

The purpose of this chapter is to provide a broad basis for the area of theoretical simulations on molecular electronic materials and molecular devices. The heart of this chapter is section 2 and 3, which review both the theoretical method associated with electrical properties of molecular electronic electronics and some of the calculations that have been reported. In section 2, a more likely “in-situ” statical theoretical method, which considers the external electric field influence during the simulations, are described. In section 3, we introduce the dynamic theoretical investigations on molecular electronics by using the recently developed non-equilibrium Green’s function approach in combination with density-functional theory (NEGF-DTF), which is an attractive way to deal with the complexities introduced by the coupling between the discrete states of the molecule and the continuous states of the electrode. It should be noted that the conductance is not the conductance of a molecule, but the conductance of a composite system containing the molecule, two interfaces, and two electrodes. Replacing any of the molecules, interfaces and electrodes would change the transportation. Finally, we will draw some conclusions in Section 4. Due to the rapid growth in this field, this review cannot be considered as exhaustive. Instead, it should be considered as a support for researchers approaching theoretical simulations on molecular materials and devices.

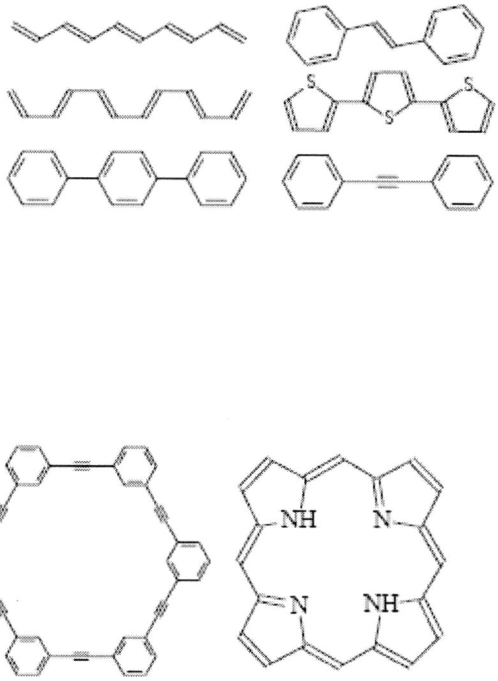

Figure 3. Typical conjugated molecules comprising alternating single and double (or triple) carbon-carbon bonds.

2. STATIC THEORETICAL SIMULATIONS

Knowing the geometric and electronic structures of the molecular electronic materials is the first step for understanding the structure-property relationship. The most commonly used approaches for the static modeling of the molecular electronic materials and devices are *ab initio* Hartree-Fock (HF) [23] and density functional theory (DFT) [24] methods. The electrical properties of a molecule can be analyzed from the geometric and electronic structures obtained from theoretical calculations. For example, if a molecule has more stationary states under specific conditions, it may be used as molecular switches or storage devices [25,26]. In a first-order approximation, the barrier for electron transfer is proportional to the energy difference between HOMO and LUMO, known as HOMO-LUMO energy gap [15]. Therefore, one may intuitively infer that electron transport through the molecules with narrow HOMO-LUMO gap should be easier than through those with large HOMO-LUMO gap. Redistribution of molecular orbitals is also excellent indicator of the molecular electron transport [27]. A conducting channel is a molecular orbital that is fully delocalized along the two metallic contacts; conversely, a non-conducting channel is localized, which can not communicate between two electrodes.

2.1. Brief Introduction to Hartree Fock and Density Functional Theories

Hartree-Fock theory The Hartree Fock theory makes use of a Slater determinant wavefunction

$$\psi_{HF}(\vec{x}_1,...\vec{x}_N) = \frac{1}{\sqrt{N!}} \begin{vmatrix} x_1(1) & ... & x_N(1) \\ . & . & . \\ . & . & . \\ x_1(N) & ... & x_N(N) \end{vmatrix} \tag{1}$$

where the spin-orbitals are orthonormal

$$\int x_i^*(1) x_j(1) dr = \delta_{ij} \tag{2}$$

The Hartree Fock energy can be represented by

$$E_{HF} = \int \psi_{HF}^* \widehat{H} \psi_{HF} d\tau = K + V_{ne} + V_{ee} \tag{3}$$

Where

$$K = \sum_i -\frac{1}{2} \int x_i^*(\vec{x}) \nabla^2 x_i(\vec{x}) d\vec{x} \quad \textit{Kinetic energy}$$

$$V_{ne} = \sum_i \int x_i^*(\vec{x}) v(\vec{r}) x_i(\vec{x}) d\vec{x} \quad \textit{Nuclear potential}$$

$$\textit{with } v(\vec{r}) = -\sum_\alpha \frac{Z_a}{|\vec{r} - \vec{R}_\alpha|}$$

$$V_{ee} = \frac{1}{2} \sum_i \sum_j (J_{ij} - K_{ij}) \quad \textit{Electronic energy}$$

J_{ij} is the *Coulombic integral*

$$J_{ij} = \iint x_i^*(\vec{x}) x_i(\vec{x}) \frac{1}{|\vec{r} - \vec{r}'|} x_j^*(\vec{x}') x_j(\vec{x}') d\vec{x} d\vec{x}'$$

K_{ij} is the *Exchange integral*

$$K_{ij} = \iint x_i^*(\vec{x}) x_j(\vec{x}) \frac{1}{\left|\vec{r} - \vec{r}'\right|} x_j^*(\vec{x}') x_i(\vec{x}') d\vec{x} d\vec{x}'$$

The Hartree-Fock energy is sometimes written as

$$E_{HF} = \sum_i H_i + \frac{1}{2} \sum_{ij} (J_{ij} - K_{ij}) \tag{4}$$

where H_i is the *core* Hamiltonian

$$H_i = K_i + V_i$$

with $V_i = \int x_i^*(\vec{x}) v(\vec{r}) x_i(\vec{x}) d\vec{x}$

Hartree-Fock theory uses a linear combination of atomic functions to construct molecular orbitals (LCAO-MOs). These atomic functions are referred to as the *basis set*.

$\psi_i(\vec{r}) = \sum_k C_{ik} \phi_k(\vec{r})$ and

$$x_i(\vec{x}) = \psi_i(\vec{r})\alpha \text{ or } x_i(\vec{x}) = \psi_i(\vec{r})\beta \tag{5}$$

where $\psi_i(\vec{r})$ and $\phi_k(\vec{r})$ are an electron wavefunction and a basis function, respectively. x_i is a spin orbital with spin α or β.

The Hartree-Fock equation is written as the following matrix equation:

$$FC = SC\varepsilon \tag{6}$$

where $F_{ij} = H_{ij} + J_{ij} - K_{ij}$ is the Fock matrix; C is a matrix of orbital coefficients C_{ik}; S is a matrix of the overlap between basis functions; ε is a matrix of energy eigenvalues.

Density functional theory Density functional theory makes use of two ideas. The first one is the energy function:

$$E[\rho] = F[\rho] + \int \rho(\vec{r}) v(\vec{r}) d^3 r \tag{7}$$

The second is the mapping between density and potential:

$$\rho(\vec{r}) \leftrightarrow v(\vec{r}) \tag{8}$$

Using the variational principle

$$\delta\left(E[\rho]-\mu\left[\int\rho(\vec{r})d\vec{r}-N\right]\right)=0 \tag{9}$$

$$\frac{\delta E}{\delta\rho}\Big|_{\rho_o}=\mu=-\chi=\left(\frac{\partial E}{\partial N}\right)_o \tag{10}$$

where μ = chemical potential, χ = electronegativity

The Mulliken definition of electronegativity is

$$\chi=\frac{IP+EA}{2} \tag{11}$$

where

IP ≡ *E* (*N-1*)- *E* (*N*) *ionization potential*
EA ≡ *E* (*N*) - *E* (*N* + 1) *electon affinity*
In terms of *N*

$$\begin{aligned}\chi&=\frac{1}{2}\left[E(N)-E(N-1)+E(N+1)-E(N)\right]\\&=\frac{E(N+1)-E(N-1)}{2}\end{aligned} \tag{12}$$

This is just a finite difference approximation to

$$\frac{\partial E}{\partial N}=lim_{\Delta N\to 0}\frac{E[N+\Delta N]-E[N]}{\Delta N} \tag{13}$$

Similarly, the hardness is

$$\eta=\left(\frac{\partial^2 E}{\partial N^2}\right)\approx\frac{1}{2}(IP-EA) \tag{14}$$

The ionization potential and electron affinity can also be expressed in terms of χ and η:

$$IP=\chi+\eta \tag{15}$$

$$EA=\chi-\eta \tag{16}$$

DFT methods are more attractive because they include the effects of electron correlation.

2.2. Static Theoretical Investigations on the Molecular Electronic Materials

Linear conjugated hydrocarbon molecular electronic materials Conjugated hydrocarbon molecules are the prototype of molecular electronic materials, and many other conjugated materials can, in principle, be derived by the structure modification. Oligo(phenylene ethynylene) (OPE) molecule which possesses not only ideal conductivity but also ideal rigidity, as well as its derivates are promising materials for a wide variety of applications in molecular electronic devices [28]. For example, in addition to wire-like properties this class of phenylene ethynylene oligomers, by proper functionalization, can behave as molecular resonant tunneling diodes (MRTDs) [22], molecular switches/storage devices [29-31], and digital computation [32]. Therefore, this kind of molecules has been intensively studied theoretically.

For molecular electronic materials, conformational change is directly related to the electron transport performance [27]. Zhou [28] observed that there exists a transition temperature (~25 K) below which the molecular current of a tolane-like molecule is rather small, but a sharp increase of two orders of magnitude in molecular current above the transition temperature (see figure 4). Seminario et al. [33] calculated the electronic structures of tolane molecule with different conformations by using DFT techniques at the B3PW92/6-311G** level and revealed the mechanism of this transition behavior.

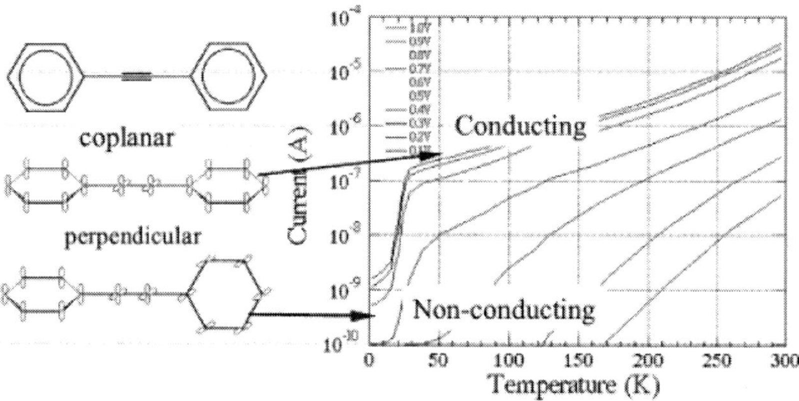

Figure 4. Current-voltage (I-V) characteristics at different biases (0.1-1.0 V) of a Au/Ti/ethyl-substituted 4,4'-di(phenylene-ethynylene)-benzothiolate/Au junction. A sharp decrease in conductivity is observed around 25 K.

The calculated results show that the relative angle between two benzene rings in each tolane molecule determines its conductivity, being maximum at 0° (planar conformation) and minimum at 90° (perpendicular conformation). At 10 K, phenyl rings in tolane molecules show very little tendency to rotate and perpendicular tolanes are more stable than parallel tolane molecules, resulting in smaller conductivity; at 30 K, the rotate rings are able to freely rotate with respect to each other, allowing the tolane molecules to be planar at some instant in time, consequently reaching higher conductivity. This theoretical simulation shows that, in addition to wire-like properties, these classes of phenylene-ethynylene oligomers can behavior as full current controllers because the electronic transport properties vary significantly based on the molecular conformations.

To correlate the relationship between the structure of molecular backbone and charge transport properties, Kushmerick et al. [34] performed experimental and theoretical investigations on two particular classes of molecular wires OPE and oligo(phenylene vinylene) (OPV) (see figure 5).

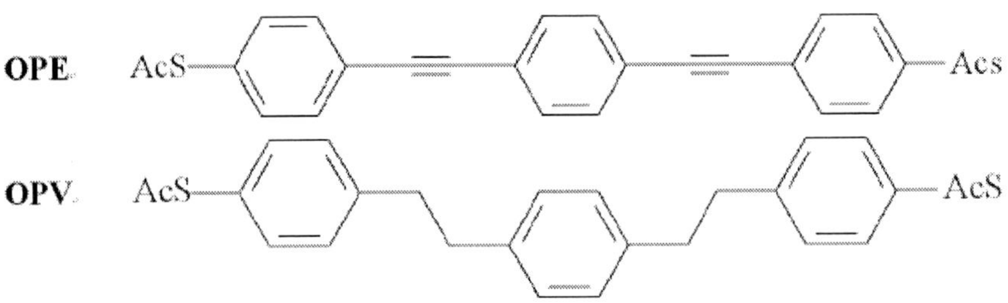

Figure 5. Molecular structures of the oligo(phenylene ethynylene) (OPE) and oligo(phenylene vinylene) (OPV) molecular wires.

The current-voltage (I-V) result by a crossed-wire tunnel junction experiment show that the OPV is more conductive than the OPE. Theoretical calculations at the B3LYP/6-31G* level for the two molecular wires highlighted a second important contribution to molecular wire conductance other than the conformational change of the OPE molecule, which disrupt the π-conjugation. The calculated HOMO-LUMO gap for the OPV (3.12 eV) is smaller than that of the OPE (3.51 eV). This difference of the energy gap can be attributed to the bond length alternation of the two molecules. The greater bond length alternation in OPE causes a larger HOMO-LUMO gap relative to the OPV system. Since the transport is dominated by charge carrier tunneling inside the HOMO-LUMO gap at low applied bias [15], the smaller HOMO-LUMO gap of the OPV leads to higher conductance. Therefore, the degree of bond length alternation needs to be considered to fully understand differences in charge transport across π-conjugated molecular wires.

Seminario et al. [15] performed DTF calculations to explain the electrical behavior of a substituted OPE resembling a resonant tunneling diode (RTD) as shown in figure 6. To understand the RTD properties, the HOMO-LUMO gap and the spatial distribution of the LUMO of OPE molecules in neutral and charged (with charges -1 and -2) states were calculated at the B3PW91/6-31G* level. The results showed that for neutral case (charge is zero) there is no connection of the LUMO between the two terminals of the molecule and the HOMO-LUMO gap is 3.48 eV. However, for the case of charge = -1, the LUMO extends over the whole molecular and the corresponding HOMO-LUMO gap decreases to 1.17 eV. For the case of charge=-2, the LUMO gets localized again and the HOMO-LUMO gap is 1.02 eV. Given the conduction is through the LUMO, this charges of the spatial distribution of the LUMO as well as the HOMO-LUMO gap can be quantitatively used to explain the RTD behavior of the substituted OPE molecule. The very small molecular current of the substituted OPE at low voltages (less than 1.74 V) can be due to the localized LUMO and the larger HOMO-LUMO gap of the OPE in neutral state. Once the molecular becomes charged by one electron upon reaching 1.74 eV, the LUMO becomes delocalized over the whole molecular backbone and the conduction will occur, increasing the current until the molecule becomes charged with two electrons. At this voltage, 2.37 eV, the LUMO corresponding to charge -2 gets localized, resulting in a concomitant current drop to practically zero.

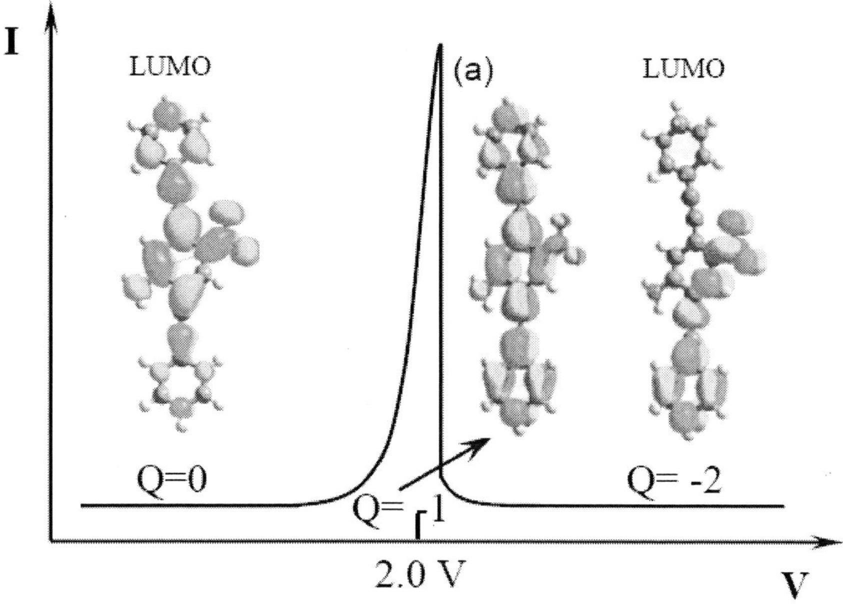

Figure 6. Experimental current-voltage (I-V) characteristics of the molecular resonant tunneling diode (MRTD). The charge Q (in electrons) determines distinct conduction channels (see the LUMO plots in different charge state) triggered by the bias voltage.

Majumder et al. [35] investigated the geometric and electronic structures of polyphenyl-based conjugated molecules (donar-spacer-accepter), potential candidates for molecular rectifying devices, by *ab initio* quantum mechanical calculations at the B3LYP/6-311G* level. The electron transport in these molecules was analyzed based on the spatial distribution of the frontier molecular orbitals. The results show that the HOMO is always localized on the donor side, while the LUMO is localized on the acceptor side. Moreover, the localization of the LUMO on the donor side depends on the number of substituents and the length of the spacers, leading to different potential drops across the molecules. In the case of monosubstituted donor-acceptor complexes, the potential drop is 1.56 and 2.05 eV for methylene and dimethylene spacers, respectively. The potential drop for the disubstituted donor-acceptor complex separated by the dimethylene group is found to be 2.76 eV. The significant increase in the potential drop by ~ 40% for the disubstituted donor-acceptor complex as compared to the monosubstituted complexes indicates that it can be used for wide range of bias voltages.

Porphyrin molecules (PH_2) are also one of the most promising materials for future application in molecular electronics due to the rigid geometric configuration, highly conjugated structure [36], and chemical stability [37]. Most important, porphyrin can be coordinated by different metals that will affect on the electron transport through the porphyrin molecule [38]. Another key advantage to the use of porphyrins as molecular materials is the large size of the monomeric unit, which is about 16.2 Å [39]. Therefore, the linear tetramers have a span of about 56±2 Å. This in itself is large enough to span the expected gap in an electrical circuit. Mizuseki et al. [40] optimized the structure of a junction in a porphyrin oligomers to mimic a p-n junction of a solid-state device by B3LYP/6-311G method. By analyzing the spatial distribution of the HOMO and LUMO, they conclude that that a full planar structure (fully conjugated) does not exhibit rectifying properties because these molecular orbitals are delocalized over the whole molecule. While the LUMO+3 for Zn–Ni in the non-conjugated oligomer forms the localized donor side (Ni porphyrin) and LUMO+1 forms the localized acceptor side (Zn porphyrin). While, HOMO is localized on free-base porphyrin between donor and acceptor porphyrins. These results show that this localization is related to the degree of conjugation between the porphyrins. Based on the above analysis, they inferred that two different metal porphyrins in a non-conjugated porphyrin would show a rectifying function.

Cyclic conjugated hydrocarbon molecular electronic materials Other than the linear molecular materials, the formation of macrocycles is an alternative way for realizing a variety of applications in molecular electronics [41]. Recently, conjugated macrocycle compounds have gathered great interests because of their potential applications in materials science [42], in host-guest systems [43], and as theoretical analysis models [44]. Phenylene-acetylene macrocycle (PAM) is the model molecule of this kind of materials. By changing the size of the macrocycles and incorporating functional coordination groups, various ions or molecular clusters with size ranging from several angstroms to several nanometers can be introduced and consequently tune the electronic properties of the conjugated macrocycles [45].

We have investigated the geometric and electronic structures of a series of PAM (from 3PAM to 10 PAM as shown in figure 7a) molecules systematically by *ab initio* HF/6-31G* calculations [46]. Two important quantum chemical tools, namely, conformational analysis [47,48] and strain-energy analysis [49], are used for understanding these features of macrocycles.

To understand the structure domination of the conjugated PAMs, the torsional barrier of the two phenyl rings as well as the bond bending energy produced by the distortion of θ_i (sp^2-sp^2-sp) or θ_j (sp^2-sp-sp) within the structural unit, tolane, have been evaluated by HF/6-31G* calculations (see figure 7b and c). In order to obtain the torsional potential energy surface (PES) drawing, the energy values computed for different conformers were fitted by the following six-term truncated Fourier expansion [48],

$$E(\phi) = \sum_{i=1}^{6} \frac{1}{2} V_i (1 - \cos i\phi) + V_0 \tag{17}$$

where ϕ is the torsional angle between the two benzene rings; $E(\phi)$ is the relative energy at the torsional angle ϕ; V_0 is relative energy at ϕ=0°.

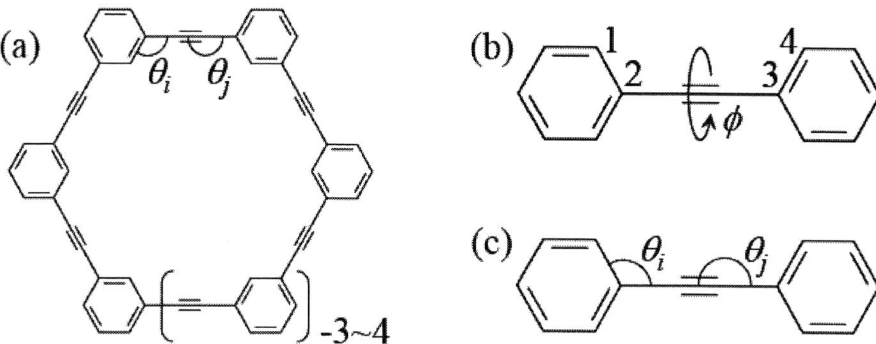

Figure 7. (a) Chemical structure of phenylene-acetylene macrocycles (PAM). (b) and (c) Chemical structure of tolane molecule. ϕ is the torsion angle formed by the planes of the phenyl rings. $\phi = 0°$ denotes the planar form. θ_i and θ_j denote the bond bending angle of (sp^2-sp^2-sp) and (sp-sp-sp^2), respectively.

Due to the molecular symmetry, the torsional potential energy surface (PES) of the tolane molecule is symmetric with respect to 90°, i.e., perpendicular conformation. Three specific points can be found on the PES profile, i.e., two energy minima at 0° and 180° and one energy maximum at 90°. The torsional barrier height is 1.77 $kJmol^{-1}$ by HF/6-31G* calculation. This low torsional barrier of tolane molecule are originated from the existence of -C≡C- linkage between the two phenyl rings. The long rigid -C≡C- linkage (~4.0 Å) can dramatically reduce the steric hindrance, favoring the planar conformation with enhanced conjugation. The PES for bond bending energy were draw according to equation (18a) or (18b) [50], separately. On the PES, the energy minimum of the conformation with θ_i^0 =120° or θ_j^0 =180° was referred to as *zero*.

$$E(\theta_i) = K_i(\theta_i - \theta_i^0)^2 \tag{18a}$$

$$E(\theta_j) = K_j(\theta_j - \theta_j^0)^2 \tag{18b}$$

where $E(\theta_i)$ and $E(\theta_j)$ are respectively the relative energies at the bond angle θ_i and θ_j. K_i and K_j are force constants, which will be assumed to be same for all the PAMs in the further ring strain energy analysis. By fitting the energy values with equation (18a) and (18b), we obtain the K_i of 0.165 $kJmol^{-1}degree^{-2}$, which is about three times larger than K_j (0.040 $kJmol^{-1}degree^{-2}$), indicating that bond angle θ_i(sp^2-sp^2-sp) is more rigid than θ_j (sp^2-sp-sp).

Geometric optimizations at the HF/6-31G* level were performed to predict geometric structure of the series of PAMs and the results were listed in table 1.

Table 1. Optimized geometries of the series of PAMs (from 4PAM to 9PAM) calculated at HF/6-31G* level

Conformation	4PAM	5PAM	6PAM	7PAM	8PAM	9PAM
Planar	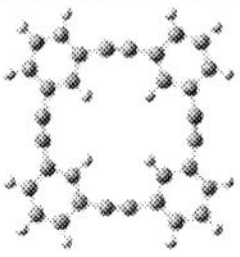	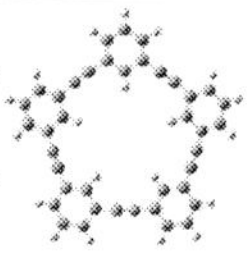	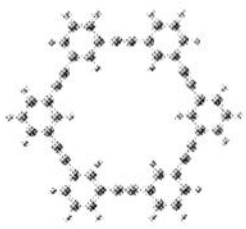	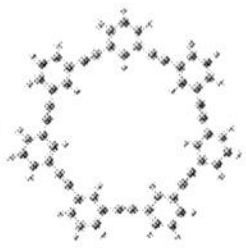	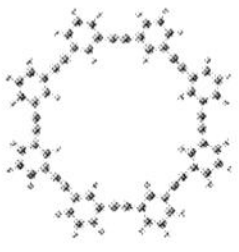	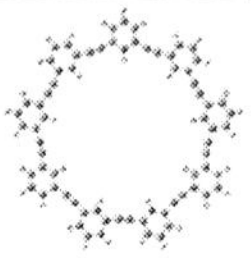
Boat	——	——	——		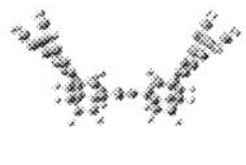	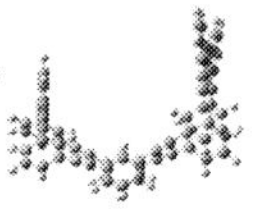
Chair	——	——	——	——		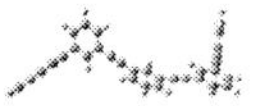

For the smallest PAM, i.e., 3PAM, a slight distorted conformation is observed, which is consistent with the results from single crystal X-ray crystallographic analysis by Oda [51]. This distortion is originated from the repulsive interactions of the three congested hydrogen groups within the macrocycles (the H···H distance is ~2.1 Å as determined by HF/6-31G* calculation). For the bigger PAMs (from 4PAM to 6PAM), The HF/6-31G* method predict only planar conformations. However, when the size of the macrocycles increases to 7PAM or larger, more stable forms are predicted. For 7PAM, HF/6-31G* calculation gives planar (4.51 $kJmol^{-1}$ higher than the planar form) and boat-like (the most stable form) forms. For 8PAM and 9PAM, the HF/6-31G* calculation gives planar, boat-like (the most stable form), and chair-like conformations. As for 10PAM, the HF/6-31G* calculations predict perfect planar conformation and one more stable nonplanar "two-hexacycle"conformation (99.83 $kJmol^{-1}$ lower than the planar form), in which two phenyl rings distort toward the centre of the macrocycle (see figure 8). This form is not a perfect plane due to the steric interaction of the two distorted phenyl rings. Obviously, both the angle strain and torsional strain can be decreased as much as possible by adopting this "two-hexacycle" conformation.

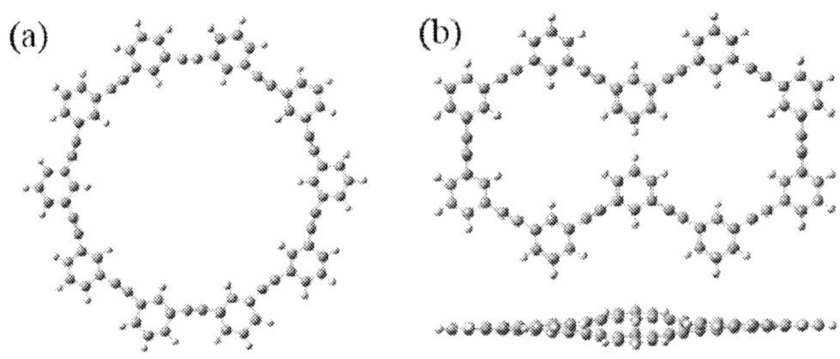

Figure 8. Optimized (a) planar and (b) nonplanar conformations of 10PAM at HF/6-31G* level. In part (b) the upper represents top view and the lower represents side view.

It should be mentioned that comprehensive understanding of the geometry of PAM is very important in exploring the material properties, for example, the self-association behavior of PAMs that depends on the molecular geometry [52-55]. The planar and rigid form of PAM exhibits the strong tendency to self-associate, whereas the flexible and nonplanar form of PAM decreases the self-association, because the π-interactions among the individual aromatic rings are not favored [56]. Moreover, detailed studies on the geometric properties of PAMs can give us inspiration in design and rationalization of novel molecular electronics. According to the cavity diameter of the macrocycles, one can introduce various metals, ions or clusters, allowing preparation of many useful molecular electronic devices, such as molecular switches [57], molecular field effect transistors (FETs) [58], and molecular storage devices [59].

Ring strain energy analysis has been proven as a quite useful tool in understanding the chemical properties, in particular, chemical activity/stability of cyclic compounds [60,61]. In general, we can evaluate the ring strain energy (E_{RS}) following two theoretical strategies, namely by *subtracting down* as proposed by Dudev and Lim [62] and by *summing up* as proposed by us [46]. According to the proposal by Dudev and Lim, the E_{RS} of a given n-membered ring relative to another structurally related r-membered reference can be calculated by the following formula,

$$E_{RS}^{sub} = E_n - E_r - (n-r)E_X \tag{19}$$

where E_X denotes the differential fragment energy evaluated from strain-free molecules; E_n and E_r are the *ab initio* energies of the n-membered ring and the r-membered reference molecule, respectively. As for PAMs, we define the meta-phenylene-acetylene unit as the differential fragment. The most appropriate candidate for evaluating E_X seems to be strain-free 6PAM. Thus, the E_X can be obtained from the *ab initio* energy of 6PAM divided by the number (n=6) of meta-phenylene-acetylene units. Here, 6PAM is used as the reference molecule as well.

Alternatively, we propose a novel approach to the evaluation of E_{RS} for this particular series of macrocycles by summing up each bond bending energies ($E(\theta_i)$ and $E(\theta_j)$) and torsional energy ($E(\phi_i)$). This approach is valid because the E_{RS} is mainly contributed by bond angle strain and dihedral angle strain in the existence of long and rigid -C≡C- linkage within the PAMs. Then, the E_{RS}^{sum} is expressed as equation (20),

$$E_{RS}^{sum} = \sum_{i=1}^{2n} E(\theta_i) + \sum_{j=1}^{2n} E(\theta_j) + \sum_{i=1}^{n} E(\phi_i) \tag{20}$$

where $E(\theta_i)$ and $E(\theta_j)$ are the contributions from the bond bending energies of θ_i and θ_j, respectively, as expressed in equation (18a) and equation (18b). $E(\phi_i)$ donates the torsional energy of dihedral angle(ϕ) as expressed in equation (17). n is the number of phenyl rings within the investigated PAM.

For the perfect planar conformation, all the dihedral angles are *zero*. According to equation (20), one can expect that the E_{RS}^{sum} is entirely contributed by the bond bending energies, i.e., $E(\theta_i)$ and $E(\theta_j)$. Table 2 lists the tension angles (defined as $\Delta\theta_i = \theta_i$-120, $\Delta\theta_j$ = 180-θ_j) and the E_{RS} calculated by both methods. It can be seen that θ_j exhibits more distortion than θ_i does. According to equation (17), we obtain the $E(\theta_i)$ is about two times larger than $E(\theta_j)$. For comparison, the E_{RS}^{sub} is also listed in table 2. The two methods give very similar results with the only exception of 3PAM. Considering the small size of 3PAM, we attribute the pronounced difference to the strong interactions of the three congested phenyl rings. The E_{RS} of 3PAM and 4PAM are up to 190.6 and 59.4 kJmol^{-1} as calculated by *subtracting down* method, which can be used to interpret the difficulty in synthesis via alkyne metathesis method [63]. With the increase of ring size from planar 7PAM to 10PAM, E_{RS} increases dramatically, resulting in the more stable nonplanar conformations. The above discussions show that the nonplanar character of 7PAM or larger PAM is mainly caused by the higher bond bending energy.

Table 2. Ring strain energies, E_{RS} (in kJmol^{-1}), and tension angles, $\Delta\theta_i$ and $\Delta\theta_j$ (in degree), of the series of planar PAMs calculated at HF/6-31G* level

Macrocycle	$\Delta\theta i$	$\sum_{i=1}^{2n} E(\theta_i)$	$\Delta\theta j$	$\sum_{i=1}^{2n} E(\theta_j)$	E_{RS}^{sum} *	E_{RS}^{sub} **
3PAM (nearly planar)	-7.2	51.3	21.2	108.4	159.7	190.6
4PAM (planar)	-3.7	18.1	10.6	36.1	54.2	59.4
5PAM (planar)	-1.4	3.2	4.2	7.1	10.3	11.5
6PAM (planar)	0.0	0.0	0.0	0.0	0.0	0.0
7PAM (planar)	1.31	4.0	3.1	5.4	9.4	9.0
8PAM (planar)	2.2	12.8	-5.4	18.8	31.5	30.9
9PAM (planar)	2.8	23.3	-7.8	44.0	67.3	61.6
10PAM (planar)	3.3	35.9	-8.6	59.5	95.4	98.5

* calculated from equation (20). ** calculated from equation (19).

As for the nonplanar PAMs, all the tension angles ($\Delta\theta_i$ and $\Delta\theta_j$) are almost *zero*. Therefore, the E_{RS} is almost entirely dominated by the dihedral angle energies ($E(\phi)$). Table 3 gives the E_{RS} and dihedral angles for each nonplanar PAM and shows good coincidence between the two methods. With the increase of ring size, the E_{RS} increases as well, but not very much. In addition, the E_{RS} of chair-like conformation is slightly higher than that of the boat-like form for the same size macrocycles, e.g., 8PAM and 9PAM, causing more stable boat-like conformation for such conjugated macrocycles. It is worthy to note that the chair-like conformations possess larger torsional angles than boat-like form does.

Here, we would like to give a simple comparison between the two methods. The first method requires performing *ab initio* calculations on each conformer. Obviously, when the ring size is big, it becomes ineffective. On the contrary, the new method partially avoids the heavy computation, and can predict the ring strain energy of the macrocycles with any size. More importantly, as proven by table 3, the contribution of each tension or torsion angle can be deconvoluted from the total E_{RS} in this method, allowing us to note those unstable parts in the molecule. This may facilitate the design of new synthesis route. It should be mentioned that several minor effects are neglected in this method. For example, this treatment underestimates the effect from the structural conjugation. Moreover, it neglects the distortion of the unit, phenyl rings in this example, also leading to the underestimation of the E_{RS}.

Table 3. Ring strain energies, E_{RS} (in kJmol^{-1}), and torsional dihedral angles (in degree) of 7PAM, 8PAM, 9PAM, and 10PAM in nonplanar conformations calculated at HF/6-31G* level

Macrocycle	Dihedral angle*										E_{RS}^{sum}	E_{RS}^{sub}
	1	2	3	4	5	6	7	8	9	10	**	***
7PAM (boat)	33	200	258	388	429	507	524	—	—	—	42	45
8PAM (boat)	288	288	288	288	288	288	760	760	—	—	58	54
8PAM (chair)	330	330	330	330	778	778	778	778	—	—	88	82
9PAM (boat)	42	128	135	210	245	762	764	890	894	—	76	73
9PAM (chair)	034	034	034	783	783	783	783	783	783	—	102	98
10PAM****	22	22	56	56	56	56	79	79	79	79	02	-14

* Torsional dihedral angles of adjacent phenyl rings within each PAM. ** Calculated from equation (20). *** Calculated from equation (19). ****.

The energy levels of molecular orbitals (MOs), especially those of the highest occupied molecular orbital (HOMO) and lowest unoccupied molecular orbital (LUMO) as well as their spatial distribution, are excellent indicators of many molecular properties, such as chemical activity [64] and electronic transportation [8,65,66]. Figure 9 shows the HOMO and LUMO levels as well as their gaps of the series of PAMs in planar conformation calculated at HF/6-31G* level (3PAM is also given in figure 9 because of its nearly planar conformation). It is noteworthy that an obvious odd-even difference can be identified in the planar conformation of such kind of conjugated macrocycles. With the increase of macrocycle size, the LUMO increases monotonously, while the HOMO increases with a zigzag manner. When we gaze at the PAMs with either odd or even number of the phenyl rings, the LUMO-HOMO gap decreases monotonously. However, it shows a zigzag manner entirely. This feature is dramatically different from the linear molecular wires such as polyacetylene and polythiophene, whose LUMO-HOMO gaps decrease monotonously with the increasing chain length [67]. Another interesting phenomenon observed in the results is that all the HOMO of the PAMs with odd number of the phenylene-acetylene units, i.e., 3PAM, 5PAM, 7PAM, and 9 PAM, are doubly degenerated. The different conformations of 7PAM, 8PAM, and 9PAM also lead to different HOMO and LUMO energy levels. Among all the conformations, the planar form shows the narrowest LUMO-HOMO gap, which is in good agreement with the fact that the planar conformation favors the highest conjugation. For 8PAM and 9 PAM, the LUMO-HOMO gaps of the three kinds of conformations increase with the order of planar, boat-like, and chair-like form, which agrees with the extent of the conjugation as evidenced by the torsional angles (see table 3).

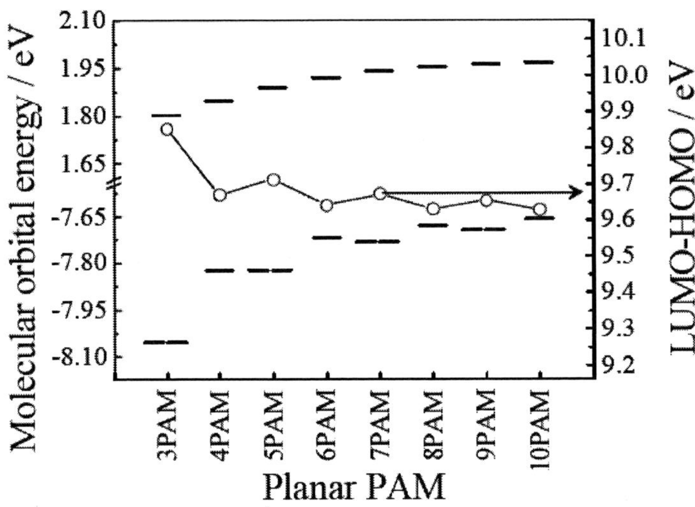

Figure 9. Comparison of the HOMO and LUMO energy levels and the LUMO-HOMO gaps for 3 PAM, 4 PAM, 5 PAM, 6 PAM, 7 PAM, 8PAM, 9PAM, and 10PAM in planer conformation calculated at HF/6-31G* level.

Table 4 illustrates the spatial distribution of the HOMO and LUMO for the series of PAMs (from 3PAM to 10PAM) in planar conformation calculated at HF/6-31G* level. The spatial distribution of HOMO also shows odd-even dependence. The HOMO of PAMs with even number of phenyl rings (4PAM, 6PAM, 8PAM, and 10PAM) is more delocalized than those PAMs with odd number of phenyl rings (3PAM, 5PAM, 7PAM, and 9PAM). Moreover, the LUMO of all the planar PAMs is delocalized over the whole molecular backbone. Table 5 presents the spatial distribution of the HOMO and LUMO for the PAMs in nonplanar conformations calculated at HF/6-31G* level. Both the HOMO and LUMO for 7PAM, 8PAM, and 9 PAMs in nonplanar conformations are not fully delocalized. In particular, the boat-like conformation has a more delocalized character than the chair-like form. Due to the "two hexacycle" configuration, the nonplanar 10PAM possesses fully delocalized HOMO and LUMO.

Table 4. Spatial distribution of the HOMO and LUMO for the series of planar PAMs (from 3PAM to 10 PAM) calculated at HF/6-31G* level. The electron density contours of the HOMO and LUMO are 0.02 e/au^3

MO	3PAM	4PAM	5PAM	6PAM	7PAM	8PAM	9PAM	10PAM
HOMO								
LUMO								

It is known that the charge state may affect the geometric and electronic structures of a molecular system [15]. We have also performed theoretical calculations on the geometric and electronic structures of another typical conjugated macrocycle, diethynylbenzene macrocycles (DBM, see figure 10), in different charge states (neutral, cationic, and anionic states) by the B3LYP method with 6-31G* (for the neutral and cationic state DBMs) and 6-31+G* (for the anionic state DBMs) basis sets.

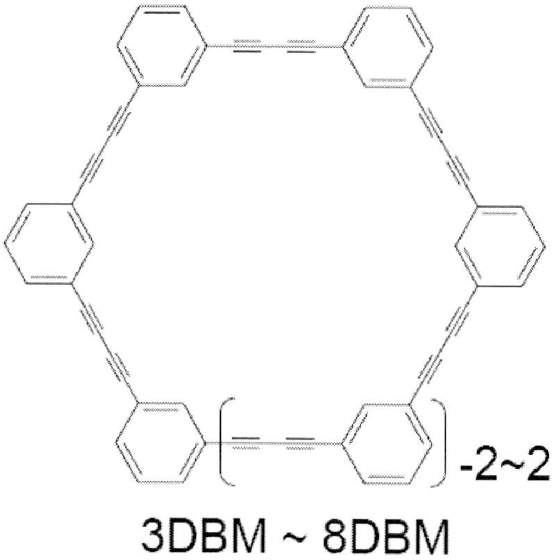

Figure 10. Chemical structures of diethynylbenzene macrocycles (DBMs).

The results show that the 3DMB, 4DBM, 5DMB, and 6DBM have only planar conformation in all charge states. However, when the macrocycle size is 7DBM or larger, more stable conformations, i.e. planar, boat-like, and chair-like conformations, are observed (table 6). The most stable conformation of the larger DBMs (7DBM and 8DBM) depends on the charge state of the macrocycles. For 7DBM, in neutral state it shows planar and boat-like conformations and the most stable conformation is the boat-like form; while in cationic and anionic states, the 7DBM gives only planar conformation. For 8DBM, it gives planar, boat-like, and chair-like conformations in all the three charge states, through the noplanar form are not same for in different charge state. In neutral and anionic states, the most stable conformation for 8DBM is the boat-like form; while for the cationic state, the most stable conformation is the planar form.

Similar to the case of PAMs, with the increase of molecular size, the HOMO increases sequentially, while the LUMO decreases with a zigzag manner for the DBMs in neutral case. The clear feature of odd-even difference for the LUMO-HOMO gap is very similar to the case of PAMs [46]. Among all the conformations, the planar form exhibits the narrowest LUMO-HOMO gap due to the highest conjugation. For the series planar DBMs in cationic state, both the HOMO and LUMO increases with the increase of molecular size. Contrary to the neutral case, for cationic state the HOMO of the DBMs with odd number of phenyl rings (i.e. 3DBM, 5DBM, and 7DBM) are doubly degenerate. With the increase of molecular size,

Table 5. Spatial distribution of the HOMO and LUMO for nonplanar 7PAM, 8PAM, 9PAM, and 10PAM calculated at HF/6-31G* level. The electron density contours of the HOMO and LUMO are 0.02 e/au^3

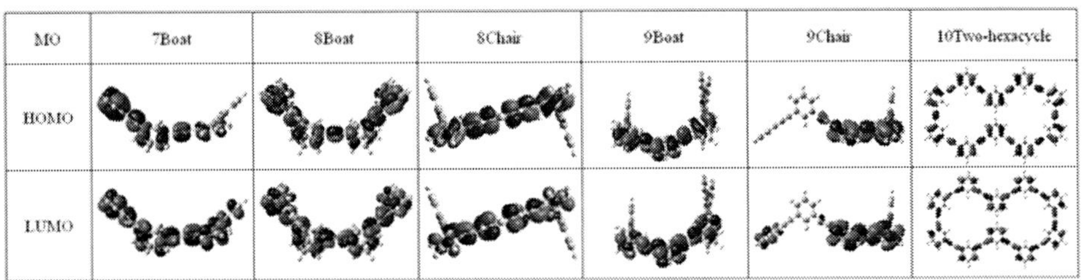

the LUMO-HOMO gap decreases with a zigzag manner. The LUMO-HOMO gaps increase in the order of: planar < boat-like < chair-like. For the series planar DBMs in anionic state, both the HOMO and LUMO decreases with the increase of molecular size, but the LUMO-HOMO gaps show a zigzag decreases with the increase of molecular size. Moreover, the LUMOs of the DBMs with even number of phenyl rings (i.e. 4DBM, 6DBM, and 8DBM) are doubly degenerate. The LUMO-HOMO gaps increase in the order of: chair-like < boat-like < planar, which is contrary to the neutral can cationic cases.

One possible reason for this odd-even difference is that the molecules with even units in neutral and cationic states have higher conjugation than the molecules with odd units. Therefore, the molecules with even units show lower LUMO-HOMO gap than the molecules with odd units. For the anionic state, the additional electron may increases the conjugation of the molecules with odd units and decreases the conjugation of the molecules with even units, resulting in a reverse evolution of the LUMO-HOMO gap with the molecular size increasing as compared to the case of neutral and cationic states.

2.3. "In-Situ" Theoretical Simulation on Molecular Electronic Materials

Although those studies focusing on the geometric and electronic structures are demonstratively useful, most of them seem to be less concerned the surroundings, where the molecular device works. It should be noted that any component in the electronic device must be subjected to a considerable external electric field (~ 10^9 Vm^{-1}) [22, 68-70]. Under this particular condition, the molecular geometry as well as the electronic structure, which plays a crucial role in determining the conductance of molecular wire, is doubted to be the same as in the zero electric field. Therefore, a detailed study of the electric field effect on the electronic and geometric structures of molecular wire is highly desired for precisely understanding the molecular electrical properties. Considering the real working conditions, we proposed an *in-situ* theoretical approach to the device modeling and investigated a wide variety of molecular materials, such as polyacetylene and its isomers [71-73], polythiophene [74,75], oligo-polyphenylene [76], diphenylacetylene [7], and asymmetrically substituted polyacetylene [77] molecules by this method.

Table 6. Optimized geometries of 7DBMs and 8DBM in neutral (calculated at B3LYP/6-31G* level), cationic(calculated at B3LYP/6-31G* level), and anionic states(calculated at B3LYP/6-31+G* level)

DBM	Neutral state	Cationic state	Anionic state
7DBM	Planar form 0.972 kJmol^{-1}	Planar form 0.972 kJmol^{-1}	Planar form 0.972 kJmol^{-1}
	Boat-like form 0.0 kJmol^{-1}	——	——
8DBM	Planar form 9.087 kJmol^{-1}	Planar form 1.130 kJmol^{-1}	Planar form 0.0 kJmol^{-1}
	Boat-like form 0.0 kJmol^{-1}	Boat-like form 0.0 kJmol^{-1}	Boat-like form 0.752 kJmol^{-1}
	Chair-like form 1.302 kJmol^{-1}	Chair-like form 5.771 kJmol^{-1}	Chair-like form 5.207 kJmol^{-1}

Linear polyacetylene molecular systems All-trans polyacetylene (PA) is a prototype of the conjugated structure for molecular electronic materials, and many other conjugated molecules can, in principle, be derived by the structure modification of it (see figure 3).

Therefore, we selected PA as the model molecules to explore the electric effect on the geometric and electronic structures of molecular electronic materials. It is well known that the performance of a molecular wire is predominated by many factors, such as the nature of the molecule itself [78], the interface between the molecule and the electrode [79], the electrode material [80], and the electrode shape [81]. Since considering all these factors looks impossible, here, we give a simplified model (figure 11a) that a PA molecule bridges two chemically inert electrodes, meaning the effect of the electrode materials has been neglected.

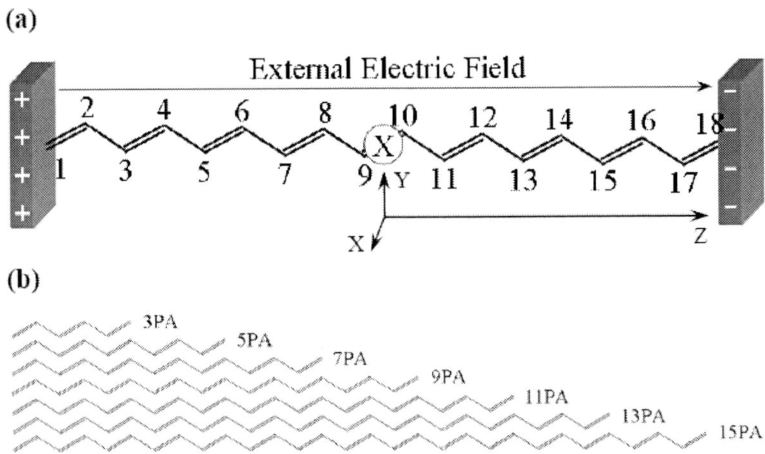

Figure 11. (a) The simplified model of the polyacetylene (PA) molecular wire. The electric field is aligned along the two terminal carbon-carbon inter-atomic vector C1→C18. The dummy atom X is located in the center of bond C9-C10 and (b) the sketches of the series of PAs studied in this work.

This assumption is valid under some particular conditions, especially when the conjugation chain is long or the coupling between the metal electrode and the conjugation backbone is impeded. Theoretical modeling has been achieved as follows. Prior to the introduction of electric field, all PAs were fully optimized at HF/6-31G* level of theory. Then, the two terminal carbon atoms were fixed in space to simulate the connection to the electrodes as shown in figure 11a. All the other geometric parameters were, then, optimized at the same level of theory in the application of uniform external electric field. A uniform electric field ranging from zero to 2.57×10^{9} Vm^{-1} and aligned along the two terminal carbon-carbon inter-atomic vector was applied to the model molecules, which may reasonably represents the working condition of the molecular electronic device [22,68-70]. To investigate the chain length effect, a series PA (H-(CH=CH)n-H, referred to as nPA, n=3, 5, 7, 9, 11, 13, and 15 as shown in figure 11b) are considered in this work.

The equilibrium geometry of all-*trans* PA without electric field (*zero* electric field) shows a coplanar conformation with C_{2h} symmetry as expected. After the application of electric field, the molecular symmetry is destroyed, although the coplanar conformation remains for all the model molecules. Figure 12 shows the evolution of the bond length with respect to that of *zero* electric field for a representative example, 9PA, under various electric fields. When the electric field increases, the carbon-carbon single bonds become shorter and the double bonds become longer, resulting in a decreased bond length alternation (BLA, the average of the difference in the length between the adjacent carbon-carbon bonds in the polyacetylene).

However, the electric field-dependence of the bond length is not identical for all the bonds. The maximum deviation occurs in the central part for both the single and double bonds due to the better conjugation as compared with those toward the end of the molecule. It is known that the molecular reconfiguration is directly related to the redistribution of the charges on each atom. Figure 13 presents the deviation of Mulliken atomic charges on each carbon atom under different electric field. With the increasing of electric field, the Mulliken atomic charges on the carbon atoms with odd number decrease and the Mulliken atomic charges on the carbon atoms with even number increase due to the easier polarization of the double bond than the single bond in the conjugated backbone. Further quantitative analyses show that the Mulliken atomic charges on all the carbon atoms evolve linearly with the electric field. Similar to the bond length case, the maximum variation of the Mulliken atomic charges also occurs in the central part due to the easy polarization of the π-conjugation in molecular center as discussed before. In order to know the significance of the electron migration along the conjugation chain, the net charge of each double bond is analyzed in this work as shown in figure 13. After the introduction of electric field, the double bonds toward positive end are negatively charged, therefore, a charge separation can be observed in the molecule. As expected, this feature also shows great electric field-dependence, i.e. the higher the external electric field, the more the electron migration. However, the electron migration along the conjugation chain is almost one order of magnitude lower than that on each double bond (see the y-axis). One can conclude, under the electric field interaction, the charge redistribution mainly occurs on each double bond.

As a kind of soft material, PA is so flexible that it can be distorted by the introduction of the environmental perturbation. The reconfiguration of the series of PA molecules has been evident from our theoretical calculations. While the electric field increases, the molecular wire bends down (negative direction for Y axis) in the π-conjugation plane from the original straight configuration (figure 14). A simple definition of the bend angle (□C1-X-C2n, the dummy atom X is located in the midpoint of the central C=C bond and the C2n is the terminal carbon atom) is used to characterize the bend amplitude of the molecular wires. It can be seen that the bend angle is decreased with increasing the electric field for all molecules. While the conjugation length increases, the sensitivity of molecular bending increases too. However, the sensitivity tends to be identical when the conjugation chain is 11PA or longer. Since the molecular wires are not typically rigid, the electric field-caused molecular bending is suggested to be taken into account in the design of the molecular electronic device. The bending of the molecular wire is, possibly, originated from the interaction between the external electric field and the induced dipole moment of each double bond, because there is an angle between them.

A quantitative correlation between the variation of the molecular property and the applied electric field is of great importance for designing novel molecular electronic device. Figure 15 shows the good linear relationship between the double bond length and the square of electric field in the concerned electric field range. This linear dependence varies depending on the location of the double bonds. The largest slope corresponds to that in the central part of the conjugation chain. As expected, the largest slope shows obvious chain length dependence. Importantly, the slopes of the central bond yield a linear dependence of the chain length as shown in figure 16. Since the molecular geometry relaxation in electric field follows an energy gradient, and, therefore, the change of the interaction energy could be an indicator of the electric field-dependence of many molecular properties. As calculated by the method

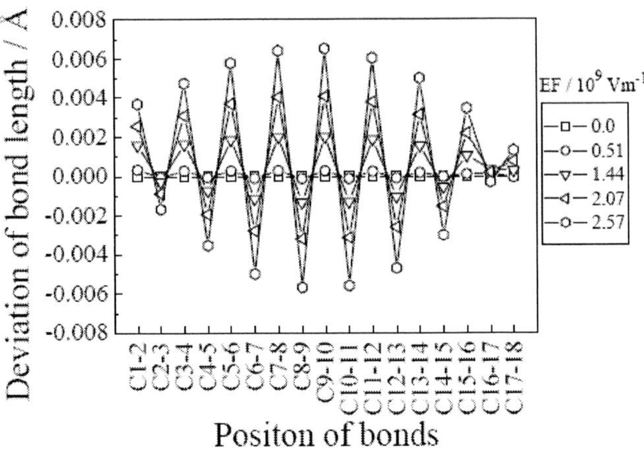

Figure 12. Geometric deviation of the bond lengths of 9PA under various electric field (see chemical structures in figure 181 for bond codes). The bond lengths of the PA under *zero* electric field are referred to as *zero*.

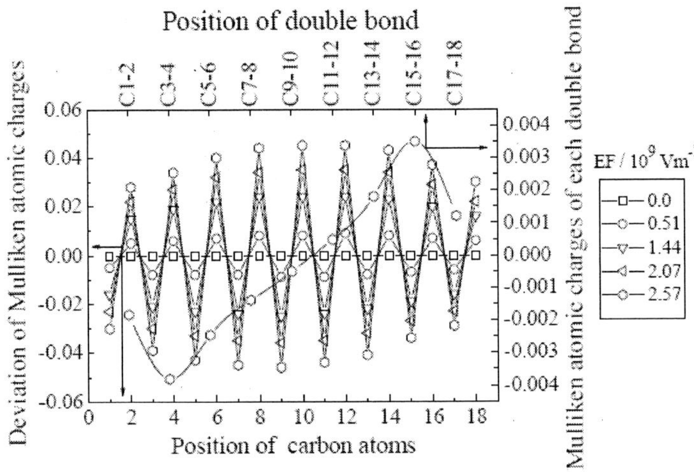

Figure 13. Deviation of the Mulliken atomic charge of 9PA under various electric field (see chemical structures in figure 18a for atom codes and bond codes). The Mulliken atomic charges of the PA under *zero* electric field are referred to as *zero*.

proposed in present work, the induced dipole moment of a double bond is about ten times of the induced dipole moment of the whole molecule (figure 13). The interaction between electric field and induced dipole moment of each double bond is a most important factor in determination of molecular energy, which may consequently predominates the stable molecular geometry. It is well known that the induced dipole moment is proportioned to the electric field. Then, we can expect that the interaction energy is proportional to the square of electric field. Although a simple quantitative relationship between single bond lengths and the external electric field has not been found, after analyses among the data, we found that single

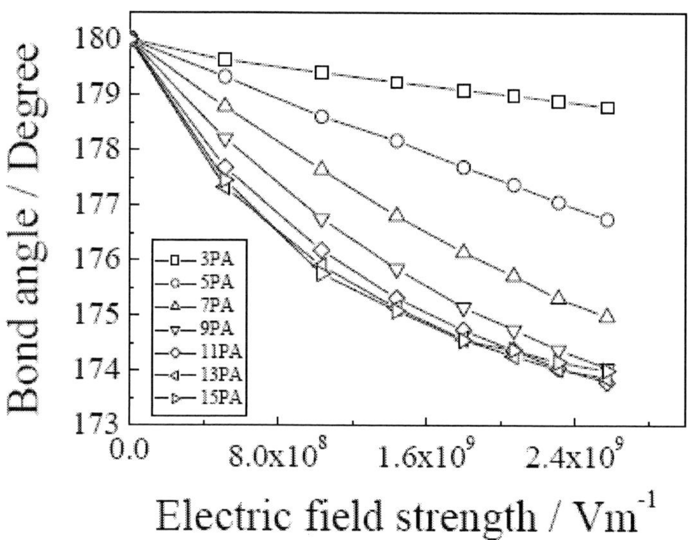

Figure 14. Electric field effect on the selected bond angles (C1-X-C2n) which are used to character the bend amplitude of PAs.

bond lengths evolve linearly with the electric field to the power of α (α is non-integer). The value for α varies from 1.60 to 3.05 corresponding to the chain length from 3PA to 15PA. The non-linearity between the variation of single bond and the square of electric field also proves that the interaction between electric field and induced dipole moment is predominant in the electric field-induced molecular evolution.

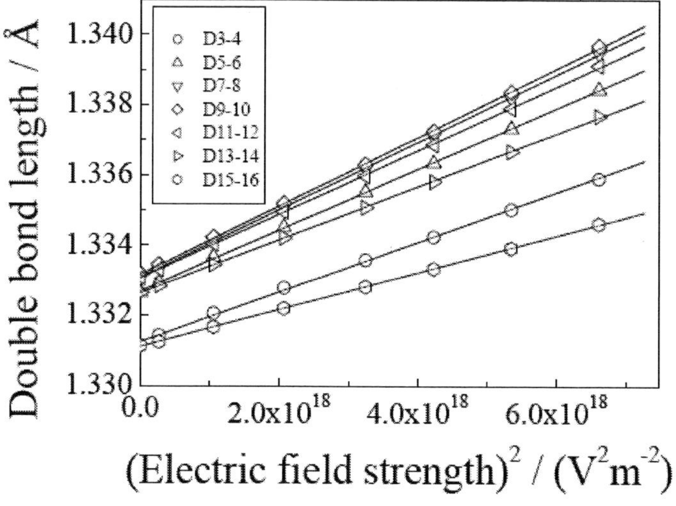

Figure 15. Correlation between the double bond lengths and the square of electric field for 9PA.

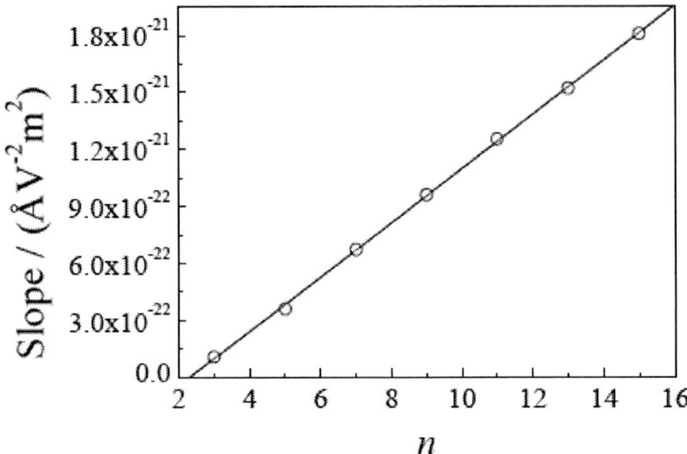

Figure 16. Linear relationship of the largest slopes and the number of double bonds (n).

Owning to the delocalized π-electrons of conjugated organic molecules, when an external electric field is applied, their charge distributions are easily modified and consequently lead to the changes of the dipole moment and the SCF energy. The results show that the induced dipole moment increases linearly with the electric field though. As expected, the electric field-dependence of the dipole moment varies with the chain length, i.e. the longer of the chain length is, the greater the induced dipole moment of the molecule.

According to the finite-field approach [82], the energy (E) of centrosymmetric molecule perturbed by a static uniform electric field (F) in the longitudinal direction can be written as:

$$\mu(F) = \alpha F + \frac{1}{2!}\beta F^2 + \frac{1}{3!}\gamma F^3 + \cdots, \tag{21a}$$

$$E(F) = E_0 - \frac{1}{2}\alpha F^2 - \frac{1}{4!}\gamma F^4 - \cdots, \tag{21b}$$

where E_0 is the energy of the molecule in the absence of the external electric field; α is the longitudinal component of the dipole polarizability; and γ is the longitudinal component of the second dipole hyperpolarizablility. For the series of PA molecules the numerical values of α were obtained from a least square fit of the SCF energy as function of the external electric field using a second order polynomial in electric field because the effect of higher order polynomials is negligible.

The energy difference between the highest occupied molecular orbital (HOMO) and the lowest unoccupied molecular orbital (LUMO), known as HOMO-LUMO gap (denoted as HLG here), is a crucial parameter in determination of the conductance of molecular wires [15,27]. It is predominated not only by the nature of the molecule, but also the surroundings. Especially, when a considerable external electric field is applied, the HLG of the molecular wire is expected to vary. In order to study, and eventually to be able to modulate the electrical properties of the molecular wire, it is important to understand the details how the HLG responses to the external electric field. Increasing the chain length, LUMO and HOMO move toward each other almost symmetrically, resulting in a narrowed HLG. For short conjugated molecular wires, 3PA and 5PA for example, the electric field-dependence of HOMO and LUMO is insensible. However, obvious electric field-dependence is observed when the conjugation chain is 7PA or longer.

It is well-known that many properties of the homologous compounds of straight molecules are the function of reciprocal of the number of the carbon atoms [83]. Figure 17 shows the chain length dependence of the HLG of the series of PA molecules. At *zero* electric field, the HLG evolves linearly with the reciprocal of the number of double bonds ($1/n$) as expected. By a simple extrapolation, the HLG of the infinite chain length ($(HLG)_{\infty}$) can be obtained. For the series of nPAs, a linear extrapolation to infinite chain length yields a HLG of approximately 6.81 eV. This result is slightly less than that predicted by semi-empirical Complete Neglect of Differential Overlap (CNDO) calculation (7.46 eV) [84] and *ab initio* "exact exchange" LCAO calculation (7.24 eV) [85]. Nevertheless, all these values are much larger than the experimental measurement of 1.35 eV [86] and 1.8 eV [87] for all-*trans* polyacetylene.

A simple comparison of the HLG evolutions among the series of PAs is very instructive for understanding the electric conductance of the molecular wire. When the electric field increases from *zero* to 2.57×10^9 V/m, the HLG is decreased by 0.052, 0.231, 0.619, 1.253, 2.086, 3.046, and 4.077 eV for 3PA, 5PA, 7PA, 9PA, 11PA, 13PA, and 15PA, respectively. Nearly 80 times difference has been found between 3PA and 15PA, indicating that the HLG for longer conjugation is easier to be modulated by the external electric field.

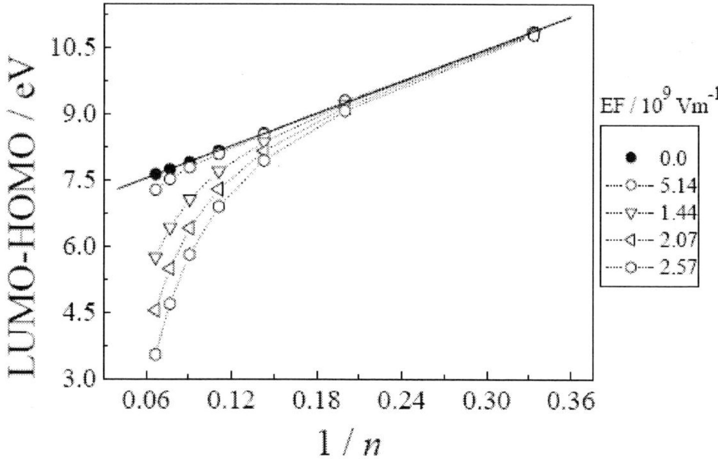

Figure 17. HOMO-LUMO gaps as functions of the reciprocal of the number of the double bond ($1/n$) at various electric field.

Figure 18 illustrates the evolution of HLG for the series of PAs as functions of the (electric field)^2. The HLG decays almost linearly with the (electric field)^2 for all the PAs, though the linearity is less pronounced for 13PA and 15PA. As mentioned before, this unideal feature is possibly caused by the chain bending under the electric field interaction. The sensitivity of the electric field-dependence of HLG can also be identified via the slopes of the lines in figure 5c. The insert shows the linear relationship between the slopes and the square of the number of double bonds (n^2). Almost a linear relationship is observed between them.

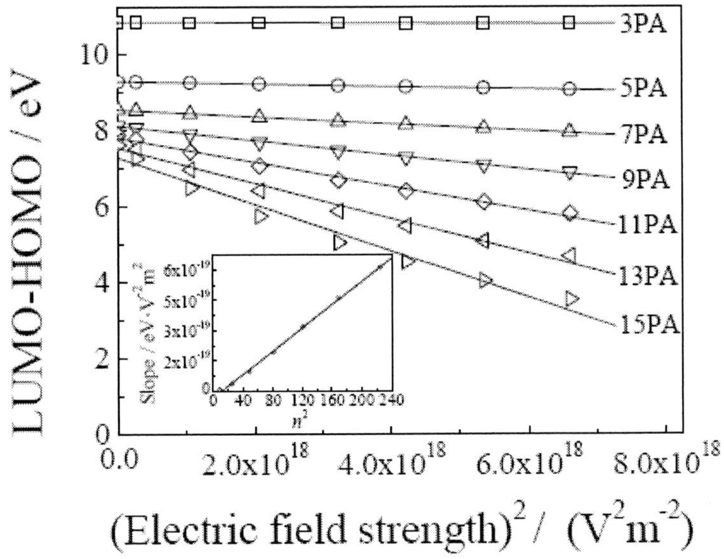

Figure 18. HOMO-LUMO gaps as functions of square of electric field for the series of PAs. The inset shows the linear relationship of the slopes in figure 28 and the square of the number of double bonds (n^2).

As discussed above, the HLG is the function of both *electric field* and chain length, *n*, therefore, providing a general expression of HLG is necessary. By detailed analysis, we can use the following formula to describe figure 5b and figure5c in the concerned electric field range:

$$HLG = k_1/n + (HLG)_\infty - k_2 \times n^2 \times (EF)^2 \tag{22}$$

where k_1 and k_2 are constants, which can be obtained from the slopes in figure 17 and the insert of figure 18, respectively. To the best of our knowledge, this is the first time that the chain length and electric field are used to determine quantitative feature of HLG of the polyacetylene molecular wire. In most cases, the electron transport barrier is directly correlated to the HLG [15]. This general HLG expression may give insight into understanding many electron transfer behaviors of the molecular wire, and further facilitate the design of the novel molecular electronic devices. In the previous studies, the absolute value of the electron tunneling barrier height of single molecules and molecular layers has been treated as a constant for a wide variety of materials. These can be true for the saturated alkyl molecules [88], because the electric field-dependence of HLG looks insensible for them. However, as proven in present study, this rule cannot be valid for the conjugated molecular wire. The longer the chain length is, the more pronounced the HLG variation.

Changes in the spatial distribution of molecular orbitals (MO), especially those of the HOMO, LUMO and a few in their neighborhood, are excellent indicators of many molecular properties [89]. Table 7 shows the spatial distribution of HOMO-2, HOMO-1, HOMO, LUMO, LUMO+1, and LUMO+2 for 15PA under various magnitudes of external electric field. At *zero* electric field, the LUMO, HOMO, and their neighboring MOs (LUMO+2, LUMO+1, HOMO-1, and HOMO-2) are delocalized throughout the whole molecule symmetrically.

Table 7. Spatial distribution of LUMO+2, LUMO+1, LUMO, HOMO, HOMO-1, and HOMO-2 orbitals for 15PA under different magnitude of electric field. The electron density contours of the molecular oribtals are 0.02 e/au^3

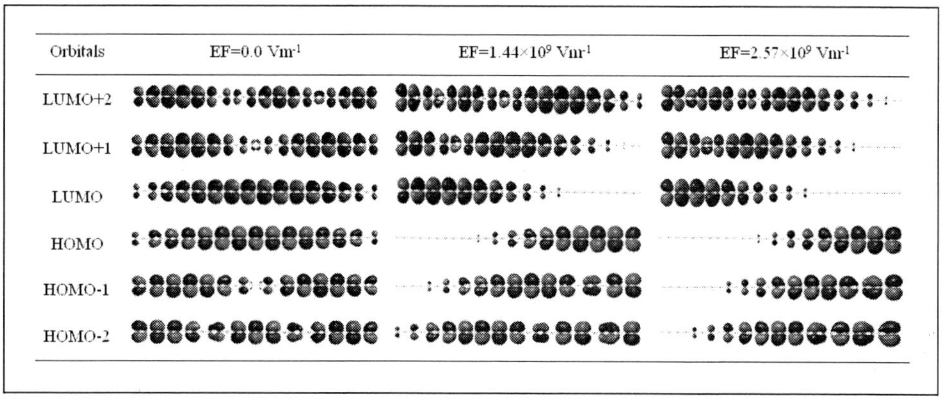

Orbitals	EF=0.0 Vm^{-1}	EF=1.44×10^9 Vm^{-1}	EF=2.57×10^9 Vm^{-1}
LUMO+2			
LUMO+1			
LUMO			
HOMO			
HOMO-1			
HOMO-2			

However, with the increase of electric field, HOMO and LUMO change oppositely from fully delocalized to a partially localized form. Meanwhile, HOMO-2, HOMO-1, and HOMO move to the negative potential side, whereas LUMO, LUMO+1, and LUMO+2 move to the positive one. We also find whose MOs closer to the HOMO and LUMO shift more significantly. This feature also shows great chain length dependence, i.e. the longer the chain length, the more obvious the shifts of the spatial distribution of the MOs.

We also investigated the geometric and electronic structures of the cis-transoid polyacetylene (Ct-PA) and the trans-cisoid polyacetylene (Tc-PA) under the interaction of electric field by using the same method for all-trans PA. Similar results to the case of all-trans PA are also observed for the Ct-PA and Tc-PA.

Linear molecular systems with torsional conformations For the molecules with torsional conformations, such as OPE molecules, one must consider electric field effect on the torsional barrier as well as the electronic structure of the molecules. To study these characters of the molecule under the interaction of electric field, we select tolane molecule, which can be viewed as the structural units of the OPE molecular systems, as model molecule and performed DFT calculations at the B3LYP/6-311+G** level. The electric field ranging from *zero* to 2.57×10^9 Vm^{-1} is defined as uniform and aligned along the main molecular axis of tolane (as shown in figure 19), which may reasonably represent the working conditions of the molecular materials [22,68-70]. Fully relaxed potential energy surface (PES) scan is carried out in redundant internal coordinates by changing the torsional angle (dihedral angle θ, C1-C2-C3-C4 as defined in figure 19) by a 10° step between 0° and 180°. On the PES, the energy of planar conformation is referred to as *zero*.

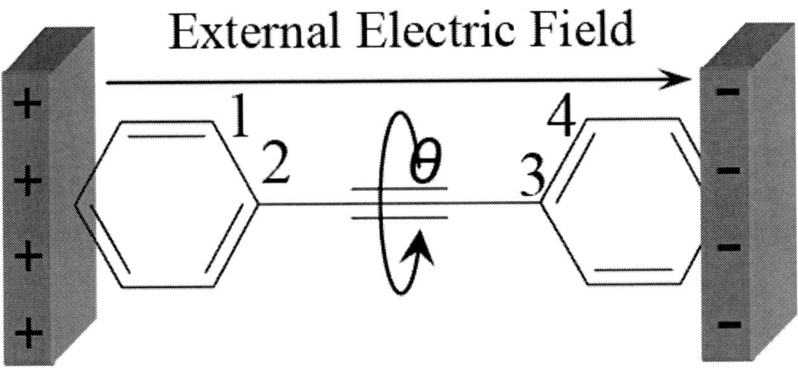

Figure 19. Schematic illustration of the simplified model of the tolane molecular wire for conformational analysis. The electric field is aligned along the two terminal carbon-carbon inter-atomic vector.

Figure 20 shows the series of torsional potential energy surface (PES) curves for tolane calculated by B3LYP/6-311+G** method under various electric field. In *zero*-electric field, the PES gives the barrier height of 3.30 $kJmol^{-1}$. When the external electric field is applied, the torsional PES curves exhibit similar feature as that of *zero*-electric field case. However, the torsional barrier ($\Delta E^{*}=E_{90°}-E_{0°}$) slightly increases with the increase of electric field, indicating that the rotation around the acetylene linkage becomes a little more difficult. With the electric field increasing from *zero* to 2.57×10^{9} Vm^{-1}, the maximum increment of ΔE^{*} is 1.07 $kJmol^{-1}$ as predicted by B3LYP/6-311+G** calculations.

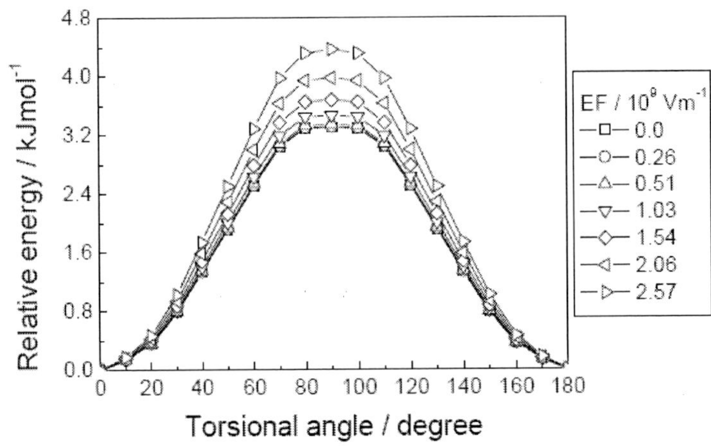

Figure 20. Torsional potentials of DPA around the central acetylene linkage under various electric field calculated at B3LYP/6-311+G** level.

By detailed analysis of the results, we note that the torsional barrier, ΔE^{*}, increases linearly with the square of the external electric field as shown in figure 21. These features can be understood by the interaction between the external electric field and the induced dipole moment. These results are very instructive for understanding the electron transport behavior of molecular wires. In particular, the low torsional barrier height favors a wide distribution of the molecular conformations under the ambient condition. In the case that each conformer corresponds to an *I-V* curve, the actual measured conductance of a molecule should correspond to an appropriate statistical average of different configurations.

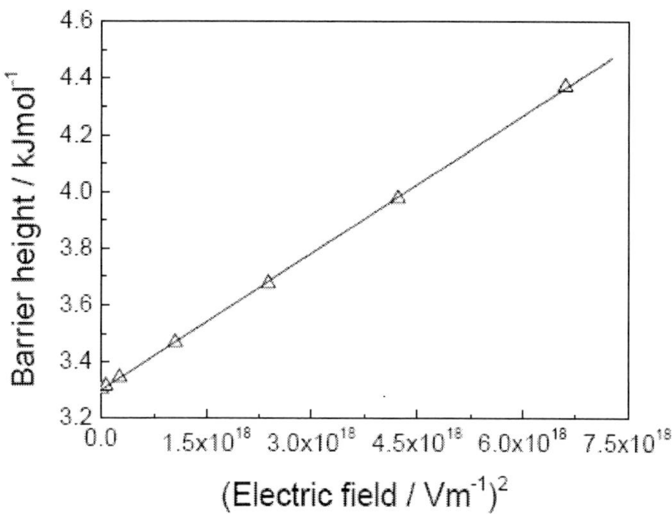

Figure 21. Correlation between the torsional barrier and the electric field calculated at B3LYP/6-311+G** level.

Figure 22 shows the molecular dipole moment as a function of the external electric field for the various conformers. It can be seen that the induced dipole moment increases with the external electric field with an excellent linear manner. Moreover, the molecular torsion results in a decrease of the induced dipole moment at a fixed external electric field. This decrease of the dipole moment can be intuitively understood from the mobility of π-electrons. The planarity of tolane with high conjugation should facilitate the π-electrons mobility under the influence of the external electric field; while the torsion of the phenyl rings destructs the molecular conjugation and consequently blocks the mobility of π-electrons along the whole molecular backbone in the external electric field, leading to a decrease of the induced dipole moment.

The numerical values of α for the series of tolane with different torsional angle can be obtained from linear fit of the data by equation (21a) with a first order polynomial, because the effect of higher order polynomials is negligible (see figure 22). Figure 23 shows the α as a function of the torsional angles (θ) calculated at B3LYP/6-311+G** level. As expected, the planar conformation possesses the maximum value of α. With the increase of the torsional angle from 0° to 90°, the α decreases and the perpendicular form shows the minimum α. By detailed analysis, we find that the value of α evolves linearly with the square of $\cos(\theta)$ (see inset in figure 23), though the reason is not clear at this stage yet. The full θ-dependent α in this case can be expressed as equation (23):

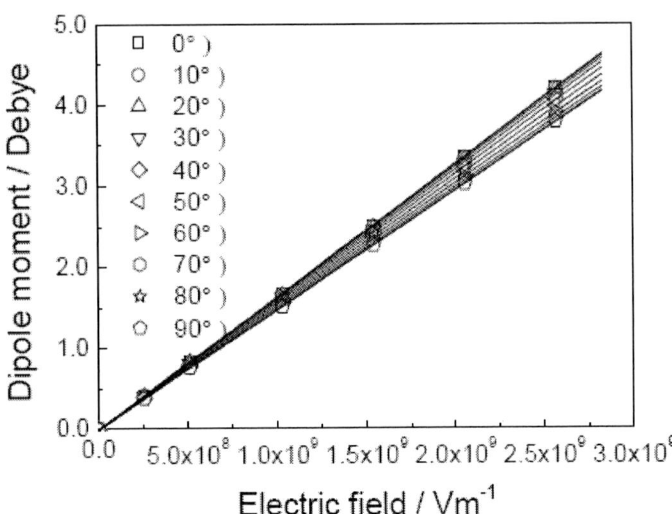

Figure 22. Correlation of the molecular dipole moment and the electric field for the various conformers calculated at B3LYP/6-311+G** level.

$$\alpha = f(\theta) = k_1 + k_2 \cos^2 \theta , \tag{23}$$

where k_1 is a constant correlated with the molecule and k_2 is the slope of the line in the inset of figure 23. According to equation (21) and equation (24), we can obtain the linear relationship between the $E(F)$ and the square of external electric field, since the effect of higher order polynomials is negligible. Thus, combined with equation (21b) and (23), the torsional barrier, ΔE^*, cab be written as equation (24),

$$\begin{aligned} \Delta E^*(F) &= E(F)_{90°} - E(F)_{0°} \\ &= (E_0(90°) - \frac{1}{2} f(90°)F^2) - (E_0(0°) - \frac{1}{2} f(0°)F^2) \\ &= k_3 + k_4 F^2 \end{aligned} \tag{24}$$

$$k_3 = (E(90°) - E(0°)),\ k_4 = \frac{1}{2}(f(0°) - f(90°)) \tag{25}$$

Obviously, equation (25) shows a linear relationship between the torsional barrier (ΔE^*) and the square of the external electric field, which conforms the result shown in figure 20.

In order to study and eventually to be able to predict the electrical properties of the molecular wire, it is important to understand the details how the HLG responds to the external electric field. Figure 24 shows the evolutions of the HOMO and LUMO energy levels as a

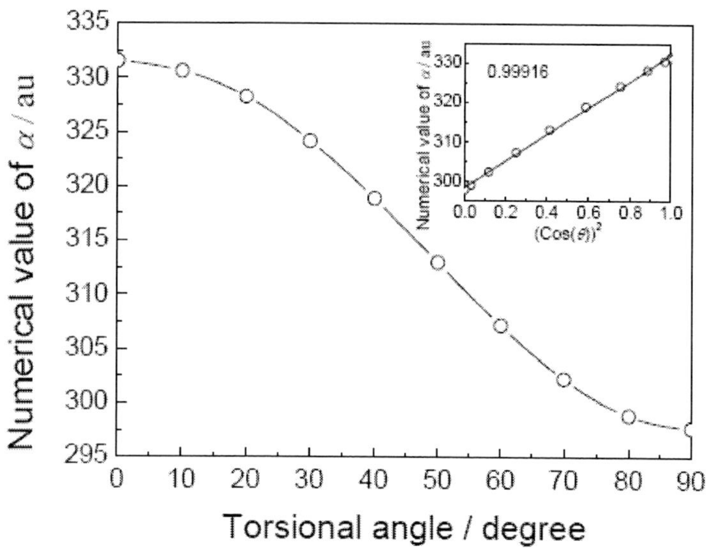

Figure 23. Evolution of the first-order derivative, α, of the induced dipole moment for DPA as a function of the torsional angle calculated at B3LYP/6-311+G** level. Inset: correlation between the value of α and the torsional angle.

function of torsional angle under various electric field. With the increase of torsional angle, HOMO decreases, whereas LUMO increases. Both of them shift away from each, leading to increase of HLG. In contrast, the introduction of electric field increases HOMO energy and decreases LUMO energy, leading the HOMO-LUMO separation decreases. In addition, the LUMO shows somewhat more electric field dependent than the HOMO does.

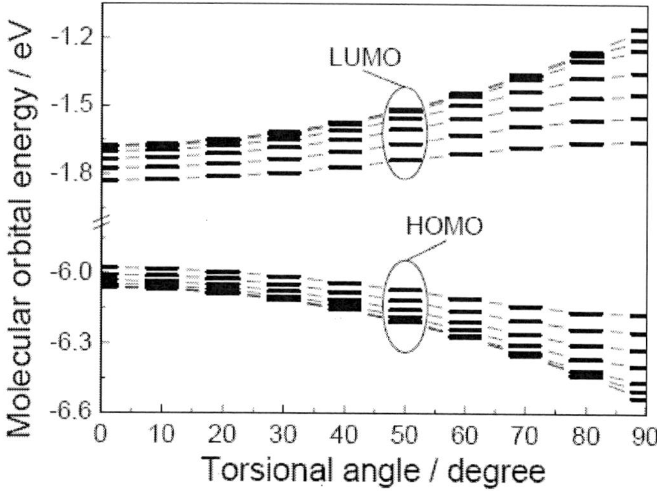

Figure 24. Evolution of HOMO and LUMO energy levels as a function of the torsional angle under various electric field calculated at B3LYP/6-311+G** level. The curves correspond to *zero*, 0.26×10^9, 0.51×10^9, 1.03×10^9, 1.54×10^9, 2.06×10^9, 2.57×10^9 Vm^{-1}.

Figure 25 illustrates the evolutions of HLG as a function of the torsional angle under various external electric field. It is interesting to find that at *zero* electric field, the HLG evolves linearly with the square of torsional angle. When the torsional angle increases from *zero* to 90°, the HLG is increased by 0.99 eV, which is consistent with that obtained at B3PW91/6-311G** (1.01 eV) by Seminario and Tour [33]. The electric field effect on HLG also shows torsional angle dependence, i.e. the larger the torsional angle, the more decrease of HLG with the external electric field. For example, with the electric field increasing from *zero* to 2.57×10^9 Vm^{-1}, the HLG is decreased by 0.24, 0.28, 0.43, and 0.85 eV for the tolane with torsional angle of 0°, 30°, 60°, and 90°, respectively, indicating the perpendicular conformation is more susceptible to change in its HLG under the external electric field. According to semi-empirical electron tunneling model, such as Simmons model [90], may give intuitive understanding of the electronic transport probability, which is determined by the barrier length and barrier height. Since the barrier for electron transfer is directly proportional to the HLG in a first-order approximation [15], one can infer that the planar tolane should yield current higher than the perpendicular one, which is in consistent with the theoretical *I-V* results calculated by NEGF-DFT method [27].

Table 8 compares the spatial distribution of HOMO and LUMO for five selected conformers under various electric field. At *zero* electric field, both the HOMO and LUMO for each conformer are fully delocalized on the whole molecular backbone.

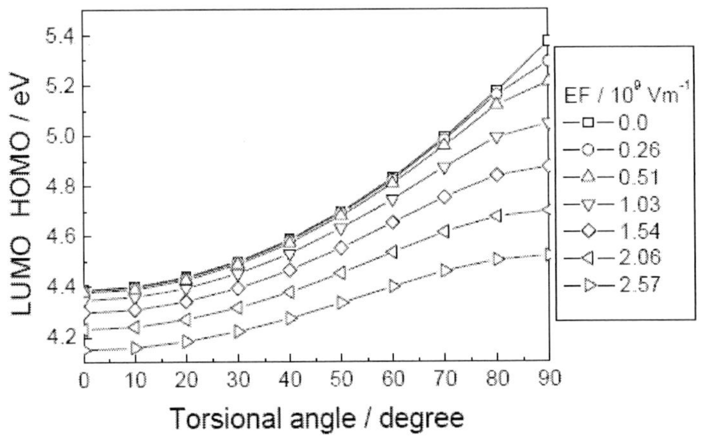

Figure 25. Evolution of LUMO-HOMO gap as a function of the torsional angle under various electric field calculated at B3LYP/6-311+G** level.

The introduction of the external electric field tends to move HOMO to the low potential side and LUMO to the high potential side. This electric field effect also shows great torsional angle dependence. When the torsional angle is less than 30°, the electric field-dependence of HOMO and LUMO seems insensible. However, when the torsional angle is 60° or larger, obvious electric field dependence is observed. In particular, while the torsional angle is 90° (perpendicular conformation), the HOMO is practically localized on central -C≡C- linkage and the phenyl ring at low potential side, while the LUMO exhibits an opposite character. Both of them almost keep constant under various *nonzero* electric field. As we know that a

Table 8. Comparison of the spatial distributions of HOMO and LUMO for selected tolane conformers under various electric field calculated at B3LYP/6-311+G level. The electron density contours of the HOMO and LUMO are 0.02 e/au^3**

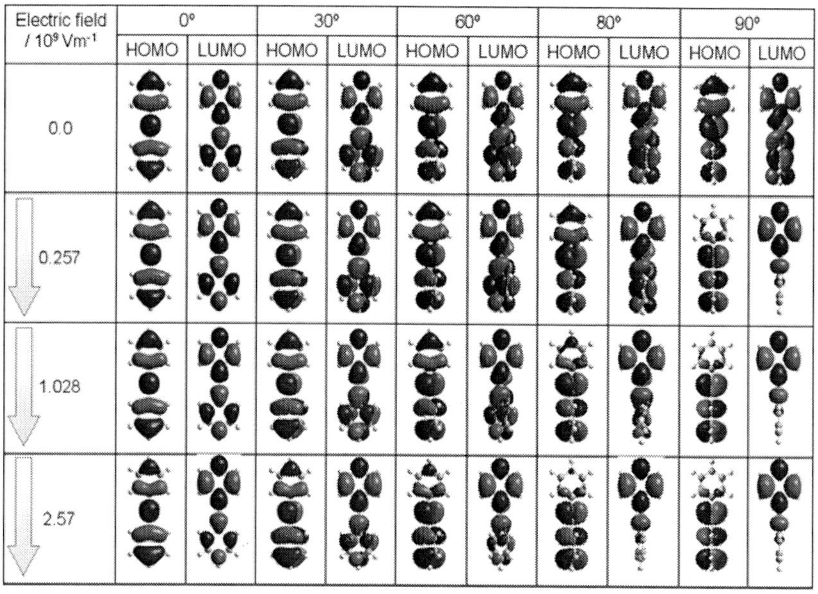

Electric field / 10^9 Vm^{-1}	0°		30°		60°		80°		90°	
	HOMO	LUMO	HOMO	LUMO	HOMO	LUMO	HOMO	LUMO	HOMO	LUMO
0.0										
0.257										
1.028										
2.57										

conducting channel is a molecular orbital that is fully delocalized along the whole molecule; conversely, a nonconducting channel is a localized molecular orbital, which cannot connect both ends of the molecule attached to the metallic contacts. Since a conducting channel is a molecular orbital that is fully delocalized while a nonconducting channel is a localized molecular orbital [27], this radical change of the spatial distribution of the HOMO and LUMO in going from a planar to a perpendicular conformation under the electric field can be another important evidence to explain the much higher current of the planar tolane than that of the perpendicular tolane [27,89].

Functionalized linear conjugated hydrocarbon molecules It is well-recognized that the electrical properties of molecular electronic devices are closely related to the electronic structures of the bridge molecule. One way to tune the molecular electronic structures is to introduce asymmetry in the wire, so that under an external electric field the induced dipole moment can be aligned to adjust the electron transport across the molecule [19]. Therefore, an understanding of the molecular geometric and electronic structures after the introduction of specific substituents is of great importance not only for elucidating the electron transport across the molecule but also for the design and realization of novel molecular electronics. In order to give theoretical evidences of the molecular electrical properties, we investigated a series of OPE wires substituted by different groups (as shown in figure 26) by the in-situ theoretical method as mentioned above. All calculations were carried out at the B3LYP/6-31G* level of theory.

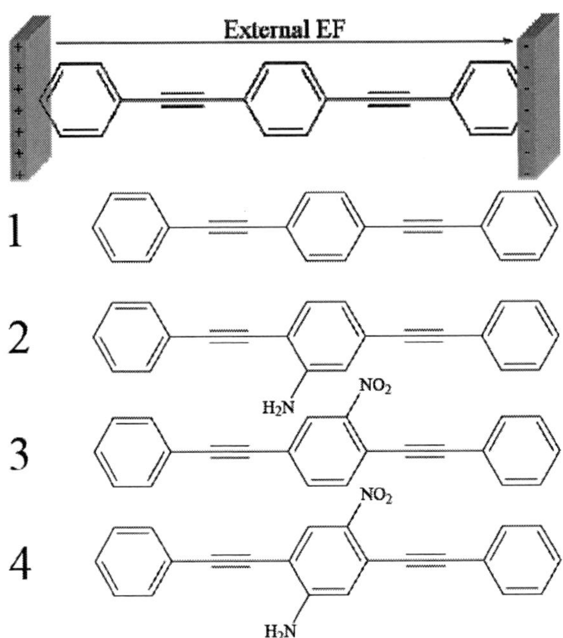

Figure 26. Top: Schematic presentation of the simplified model of the molecular wires. Bottom: Chemical structures of the 1-4 molecular wires

The electric field ranging from *zero* to 0.257 V/Å is defined as uniform and aligned along the two terminal carbon-carbon inter-atomic vector. For the convenience of discussion, we define the electric field pointing to the left side of the molecule as L-direction electric field, and R-direction electric field when the electric field points to the right side.

The evolutions of the HOMO and LUMO energy levels and HOMO-LUMO gap for the series of molecules under the external electric field are shown in figure 27, in which the negative electric field corresponds to L-direction electric field and the positive electric field corresponds to R-direction electric field. It can be seen from figure 27 that the shift of the energy level depends on both the substituents and the electric field direction. For molecule 1, the HOMO increases and LUMO decreases, leading to a decreased HLG, which is consistent with other conjugated molecules, such as polyacetylene, oligothiophenes as reported by us [71,73-75]. Both HOMO and LUMO of molecule 1 evolve symmetrically with respect to *zero* electric field due to the molecular symmetry. When the $-NH_2$ group is introduced (molecule 2), the LUMO remains the symmetrical evolution, however, the HOMO shows more increase under L-direction than under R-direction electric field, indicating the asymmetric effect from the $-NH_2$ group mainly occurs on the HOMO. Contrast to the case of molecule 2, the $-NO_2$ substituted OPE (molecule 3) shows an asymmetrical evolution of the LUMO with respect to *zero* electric field, while the HOMO almost keeps the symmetrical feature. The LUMO decreases monotonously with the increase of L-direction electric field. However, under R-direction electric field, the LUMO shows a maximum at 0.206 V/Å. Similar to the substituents effect on the geometric structure, a combined effect of both substituents on the HOMO and LUMO energy levels is observed in molecule 4. However, the asymmetrical feature of the HOMO with respect to *zero* electric field is somewhat weakened and the maximum of LUMO under the R-direction electric field shifts to lower electric field than the case for molecule 3, signifying that the $-NO_2$ group dominates electronic structure for molecule 4 under the electric field.

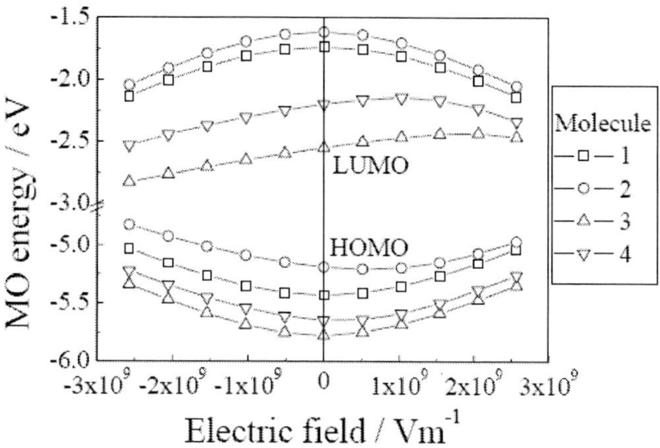

Figure 27. Electric field effect on HOMO and LUMO energy levels for the four molecules (shown in Figure 37). The negative EF and positive EF correspond to the L-direction and R-direction EF, respectively.

For comparison, we show the evolution of HLG as a function of electric field for molecules 1, 2, 3, and 4 in figure 28. All the substituted molecules give asymmetrical characters under the electric field. The $-NO_2$ substituted OPE (molecule 3) shows more pronounced asymmetry than the $-NH_2$ substituted OPE (molecule 2). While electron transport is dominated by charge carrier tunneling inside the HOMO-LUMO gap (non-resonant), we may expect a difference of current between forward and backward bias on the molecular wire. Obviously, the introduction of the $-NO_2$ substituent should facilitate this electrical asymmetry.

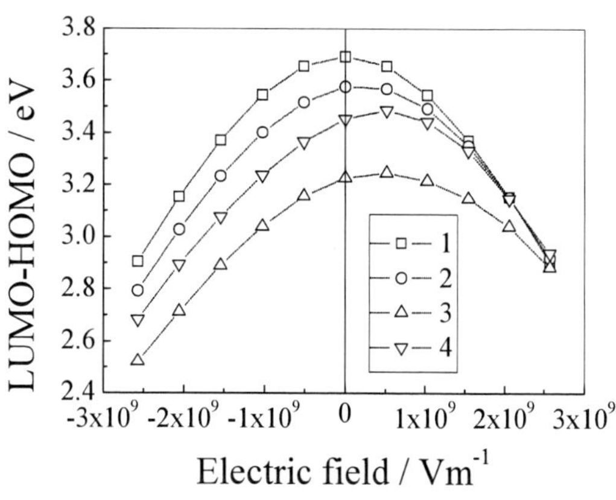

Figure 28. Electric field effect on HOMO-LUMO gap for the four molecules (shown in Figure 37). The negative EF and positive EF correspond to the L-direction and R-direction EF, respectively.

Table 9 illustrates the spatial distribution of the HOMO and LUMO for molecule 1, 2, 3, and 4. For molecule 1, both the HOMO and LUMO are fully delocalized along the whole molecular backbone. The introduction of the -NH_2 group (molecule 2) tends to slightly localize the HOMO, whereas no effect on the LUMO. In contrast, when the -NO_2 group is introduced (molecule 2), the HOMO remains, while the LUMO becomes localized. In other words, the -NH_2, a known electron-donating group, tends to localize the HOMO, while the -NO_2, a known electron withdrawing group, generates a localized LUMO.

Table 9. Spatial distribution of the HOMO and LUMO for the four model molecules (shown in figure 37) under zero electric field. The electron density contours of the HOMO and LUMO are 0.02 e/au^3

molecule	HOMO	LUMO
1	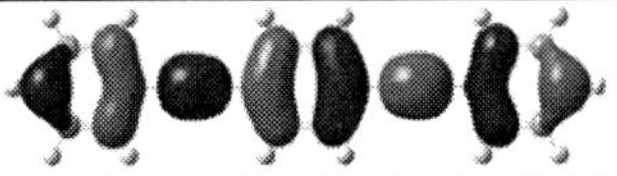	
2		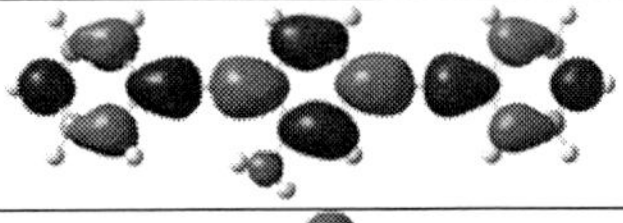
3	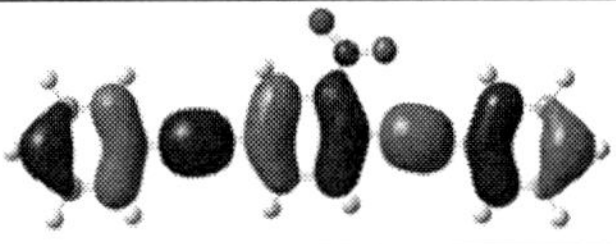	
4		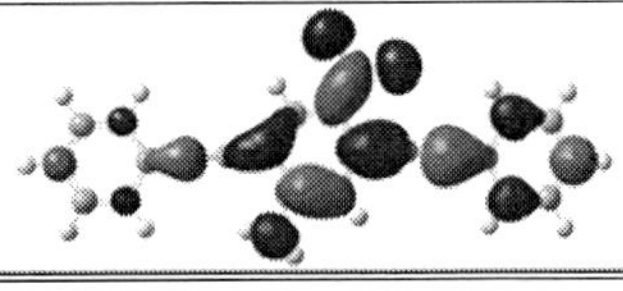

Moreover, the effect of the -NO_2 group is much larger than the corresponding effect of the -NH_2 substituent. A combined effect of both the -NH_2 and -NO_2 groups is observed in molecule 4, in which the HOMO is almost unmodified and the LUMO is greatly localized as in molecule 3. Based on these results, it can be concluded that the electron-withdrawing (-NO_2) group plays the most significant role in determining the electrical properties in such molecules. Similar to case of other conjugated molecules [71,73-75], the HOMO tends to move to the low potential side, while the LUMO tends to move to the high potential side under the interaction of electric field. Thus, both the HOMO and LUMO for molecule 1 and 2 become more localized under both the L and R-direction electric field. However, for the molecule 3 and 4 in which the LUMO is not extended to the terminal rings, different electric field will produce distinct spatial distribution of the LUMO. As a representative example, table 10 presents the spatial distribution of the LUMO of the disubstituted OPE (molecule 4) under various external electric field. Under the L-direction electric field, the LUMO becomes more localized on the electron-withdrawing group side. On the contrary, the R-direction electric field drives the LUMO from the -NO_2 group side delocalized across the molecular backbone, and then tends to make it localized on the other side. In a modern understanding of electronic transportation [18~20,36], the details of the electron transfer paths can be deconvoluted by considering each electron transmission tunnel. Generally, energy levels such as HOMO and LUMO can be potentially the transmission tunnels. A conducting channel is a molecular orbital that is fully delocalized along the molecular backbone; conversely, a nonconducting channel is a localized molecular orbital, which cannot connect both ends of the molecule to the metallic contacts.

Therefore, we can expect molecule 3 and 4 may have asymmetric current-voltage behavior, rectification and negative differential resistance as possible, due to the change of the spatial distribution of the frontier molecular orbitals, which is in agreement with the I-V results obtained by NEGF-DFT method as reported by us [91]. This further confirms the validility of our analysis by this in-situ method to molecular electronic materials. The above observations provide us an avenue to tune the molecular electronic structures by introducing specific functional groups and consequently bring the molecule some new electrical properties, such as molecular diode as reported by us [92].

Table 10. Spatial distribution of the LUMO of molecule 4 (shown in figure 37) under various external electric field. The electron density contours of the HOMO and LUMO are 0.02 e/au^3

EF/Vm^{-1}	L-direction EF	R-direction EF
0.0	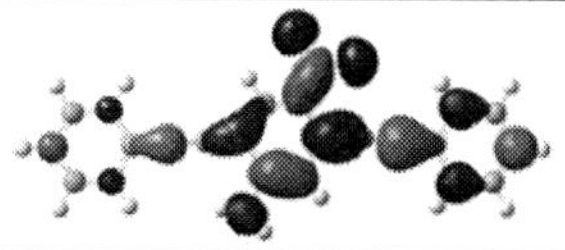	
5.14×10^8		
1.03×10^9		
2.57×10^9		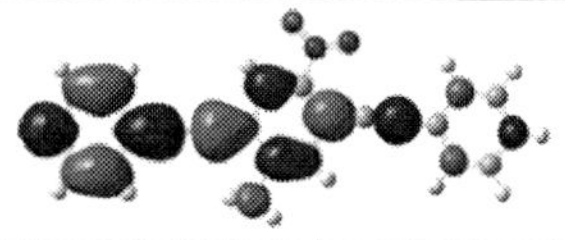

3. Dynamic Theoretical Investigation

With the aim to reveal the fundamental structure property relations for electron transfer in molecular electronic devices, electron transport through single molecule has been addressed in a number of recent theoretical papers [93]. Since recent theoretical development provides further insight into the mechanism of the electron transport. In particular, the widely used non-equilibrium Green's function formalism (NEGF) method [16-19] is a valid approach to investigate the characteristics of electron transport through organic molecules, taking electron correlation into account in an implicit and computationally expedient manner. The NEGF method can give the density matrix and charge distribution on the molecule, and the electrostatic potential can be determined from the NEGF-based charge distribution. In many studies, the potential is used in the framework of DFT in order to obtain electronic structures for a molecule connected to electrodes with finite bias [16]. Based on Density Functional Theory (DFT) and non-equilibrium Green's function (NEGF) method, the dynamic investigation on molecular electronic devices have been performed.

3.1. Theories of Electron Transfer in Molecular Junctions

Electron transport in molecular electronic devices is different from that in semiconductor mesoscopic devices in two important aspects: (1) the effect of the electronic structure and (2) the effect of the interface to the external contact. A rigorous treatment of molecular electronic devices will require the inclusion of these effects in the context of an open system exchanging particle and energy with the external environment [94]. This calls for combining the theory of quantum transport with the theory of electronic structure starting from the first-principles.

The quantum-transport theory that better adapt to microscopic theory is the NEGF formalism, developed by Keldysh (1965) and Kadanoff and Baym (1962) and applied to devices by Caroli *et al.* (Caroli et al 1971, 1972, Combescot 1971). This method also allows solving systematically for interactions within the electron propagator under non-equilibrium. Therefore, scattering mechanisms due to electron-phonon interactions or many-body electron–electron correlations can be taken into account, at least in principle, in the evaluation of the current. For the purpose of the present review, we give a brief overview of the main concepts of the NEGF theory which will serve us to introduce the relevant quantities of the formalism.

The contact self-energies Σ_1 and Σ_2 arise formally out of partitioning an infinite system and projecting out the contact Hamiltonians. When an isolated molecule with discrete energy levels is contacted to leads to make an infinite composite system, the energy-dependent one-particle retarded Green's function of the complete system is expressed in an appropriate basis set as

$$G(E) = [(E + i0^+)S - F]^{-1} \tag{26}$$

where S is the overlap matrix and F is the Fock matrix for the whole system. F incorporates the effect of external fields, the electronic kinetic energy, electron-nuclear attractions, as well as electron-electron interactions, which could in principle include Coulomb, exchange and correlation effects. The poles of this Green's function lie near the real energy axis, and represent the energy levels for the infinite system. To extract just the device part of G involving the device overlap matrix S_{dd} and the device Fock matrix F_{dd} , we utilize the fact that for a matrix

$$G = (ES - F)^{-1} \tag{27}$$

The NEGF formalism provides a suitable method for calculating the density matrix ρ for systems under non-equilibrium conditions. A tutorial description of this formalism can be found in reference [95]. Here we will simply summarize the basic relations that can be used

(27) to obtain ρ, given the molecular Fock matrix F, the self-energy matrices Σ_1, Σ_2, and the contact electrochemical potentials μ_1 and μ_2 and (28) to obtain the electron density n($\vec{r}$) and current I from the self-consistent density matrix ρ. For open systems with a continuous density of states, the density matrix can be expressed as an energy integral over the correlation function i$G^{<}(E)$, which can be viewed as an energy-resolved density matrix

$$\rho = \int dE[-iG^{<}(E)/2\pi] \tag{28}$$

The correlation function is obtained from

$$-iG^{<} = G(f_1\Gamma_1 + f_2\Gamma_2)G^{+} \tag{29}$$

Where $f_{1,2}(E)$ are the Fermi functions with electrochemical potentials $\mu_{1,2}$

$$f_{1,2}(E) = \left(1 + \exp\left[\frac{E - \mu_{1,2}}{k_B T}\right]\right)^{-1} \tag{30}$$

G is the Green's function matrix (energy-dependent) in a non-orthogonal basis

$$G = (ES - F - \Sigma_1 - \Sigma_2)^{-1} \tag{31}$$

where S is the overlap matrix. The broadening functions are the anti-Hermitian components of the self-energy

$$\Gamma_{1,2} = i[\Sigma_{1,2} - \Sigma_{1,2}^{+}] \tag{32}$$

The NEGF equation for ρ can be interpreted as the filling of broadened energy levels (eigenvalues of F) by two separated contacts, with GF_iG^{+} representing the density of states (DOS) contribution from the ith electrodes, and f_i representing the Fermi function of the ith electrode.

The converged density matrix is used to obtain the total number as well as the spatial distribution of electrons using

$$\text{N=traces}(\rho\text{S}), \tag{33}$$

$$n(\vec{r}) = \sum_{m,n} \rho_{mn}\varphi_m(\vec{r})\phi_n(\vec{r}) \tag{34}$$

where $\phi_{m,n}(\vec{r})$ represent molecular basis functions. The converged density matrix may also be used to obtain the terminal current. For coherent transport, we can simplify the calculation of the current by using the transmission formalism where the transmission function

$$T(E) = trace[\Gamma_1 G \Gamma_2 G^+] \tag{35}$$

is used to calculate the terminal current,

$$I = \frac{2e}{h}\int_{-\infty}^{\infty} dET(E)(f_1(E) - f_2(E)) \tag{36}$$

3.2. Theoretical Investigation On Molecular Wires

Geometry-correlated conductivity of molecular wires All the static theoretical investigations about molecular wires mentioned above demonstrate that the electric field affects the geometry structure and electric structure significantly. On the basis of the static investigation, a more likely in-situ dynamic theoretical simulation by DFT and NEGF formalism was carried out to study the electronic transport behavior of the typical molecular wire, 4,4'-(1,4-phenylenedi-2,1-ethynediyl)bis-benzenethiol (It drops two end hydrogen atoms when bridged two gold electrodes and is referred to as TB) [96]. Figure 29 displays the optimized geometry of the Au-TB-Au without electric field and with electric field. TB remains a planar conformation with D_{2h} symmetry and is perpendicular to one side of the gold triangles under *zero*-field. The sulfur atoms are at the center of the triangles and the distance. However, all the above features change when the electric field is applied. Since the rotational barrier of the TB around the long axis as estimated from the rotational potential energy profile is very low (3 $kJmol^{-1}$), the molecule turns to a configuration parallel to the base side and is bended towards the normal of conjugation plane even at 1.03×10^8 Vm^{-1}. Furthermore, most triple and double bonds are elongated and single bonds shrink with increasing the electric field, inferring a better conjugation of the system.

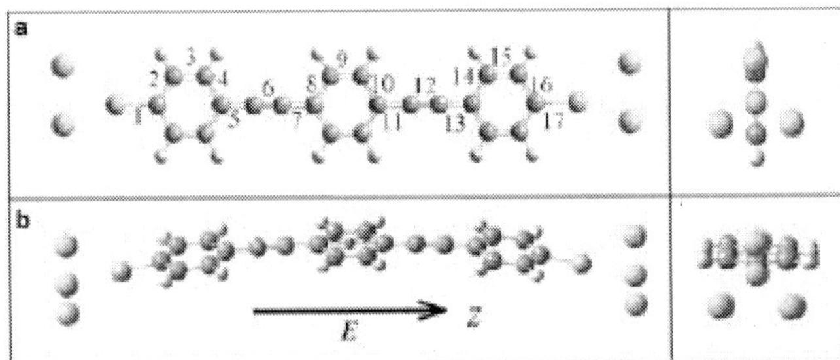

Figure 29. (a) Optimized structures of 4,4'-(1,4-phenylenedi-2,1-ethynediyl)bis- benzenethiol molecular wire under a neutral condition and (b) external electric field of 4.11×10^8 Vm^{-1}

The current calculations have been carried out for both the optimized and frozen Au-TB-Au systems. The I-V curves are compared in figure 30. In the low bias region, typically less than 1.5 V, both optimized Au-TB-Au and frozen Au-TB-Au exhibit slow current increase as well as flat differential conductance. As the bias increases over 1.5 V, the currents of both them increase rapidly. The I-V behavior of the optimized Au-TB-Au looks more sensitive to the bias compared with the frozen one. The larger current for the optimized Au-TB-Au in the high bias region arises from the lower HLG, since HLG is directly correlated with the electron transfer barrier [15]. Increasing bias may lead to the considerable decrease of the electron transfer barrier for the optimized Au-TB-Au. Therefore, the relaxation of the molecular geometry together with the Au/S interface under the interaction of electric field may lead to an obvious change of the electronic structures, such as spatial distributions of HOMO and LUMO, and the energy gaps, which result in a different transportation behavior from that of the frozen model. Carefully considering the geometric optimization in electric field is necessary for the design of novel molecular electronic devices.

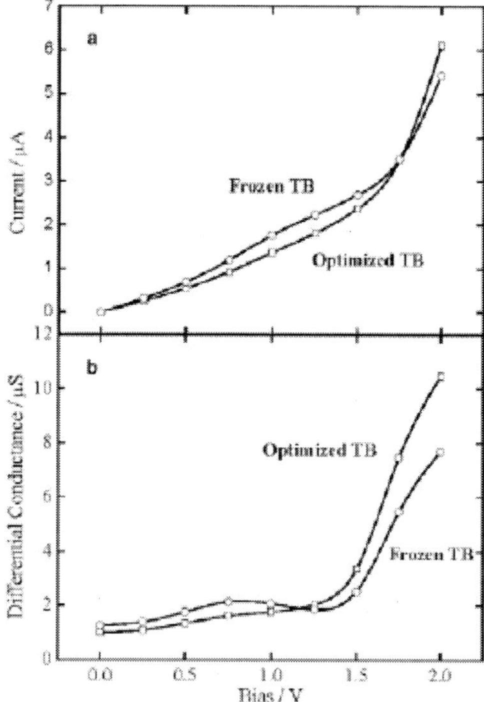

Figure 30. The calculated I-V curves (a) and The differential conductance (b) of 4,4’-(1,4-phenylenedi-2,1-ethynediyl)bis- benzenethiol molecular wire.

Torsional Effect on the conductivity of molecular wires To correlate the molecular conformation with the electron transport, we investigated the torsional effect on the conductivity of molecular wire. The I-V curves for a series of tolanes with different torsional angles were calculated by the NEGF-DFTmethod [27]. The results show that the planar conformation exhibits the highest conductance. With the increase of torsional angle, the molecular conductance decreases. In particular, when the torsional angle is 90° (perpendicular conformation), the conductance is the smallest. This change of the molecular conductance with molecular conformation can be intuitively understand from the HOMO-LUMO gap as well as the spatical distribution of the frontier molecular orbitals as discussed in section 3.2 (figure 25 and table 8). In a modern understanding, the details of the electron transfer paths can be deconvoluted by considering each electron transmission tunnel, which can be analyzed from the transmission spectrum. Figure 31a illustrates the transmission spectra for the tolane with torsional angle of 0°, 10°, 30°, 50°, 70°, and 90° at zero bias. With increasing the torsional angle, the transmission spectrum decreases radically (figure 31b). This may lead to much smaller current for the conformer with larger torsional angle. For example, at 1 V, the planar conformation yields a conductance of about 8.8 μS, while the perpendicular conformation yields a conductance of about 0.14 μS. Seminario et al.[89] also calculated the I-V characteristic of the planar and perpendicular tolane molecules using a DFT approach at the B3PW91/LANL2DZ level of theory combined with a Green function formalism, and obtained the corresponding conductance of 20.4 μS and 0.36 μS at 1 V for the planar and perpendicular conformations, respectively. The above results suggest that molecular torsion is very crucial in determining the molecular conductance. In particular, for molecular systems with low torsional barrier, the molecules favor a wide distribution of the molecular conformations under the ambient condition. Therefore, the actual measured conductance of a molecule should correspond to an appropriate statistical average of different configurations.

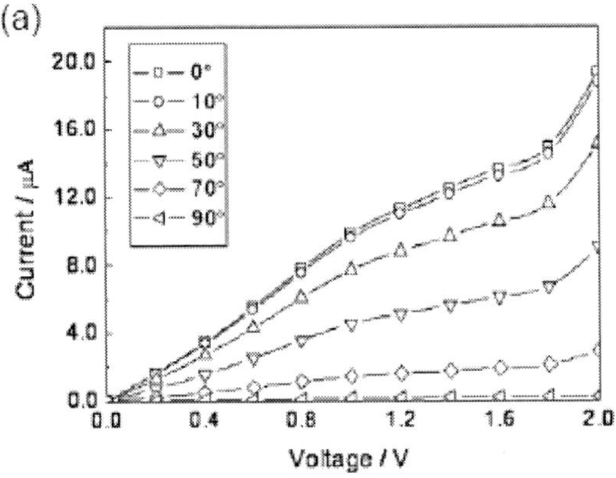

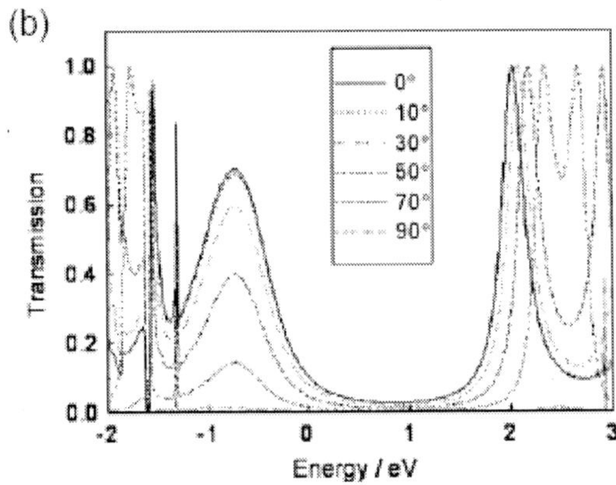

Figure 31. (a) The calculated I-V curves for tolane molecules with different torsional angle. (b) The zero-bias transmission spectra of tolane molecules with different tosional angle.

Length-dependent Conductance of Molecular Wires To control the electron transport through the molecular wires, one must know the tunneling attenuation factor for the specific tunneling bridge. The conductance of a wide variety of molecular junctions has been studied theoretically through the DFT and NEGF method, such as alkane, twisted polyphenyl (t-PP), planar polyphenyl (p-PP), poly furan (PF), oligo phenylene ethynylene (OPE), oligo phenylene vinylene (OPV), cis-trans-polyacetylene (ct-PA), and all trans-polyacetylene (t-PA). The calculation model is displayed in figure 32a. Based on the chain-length dependence of conductance for each series of molecular wires, the attenuation factor, β, has been obtained and compared with experimental data.

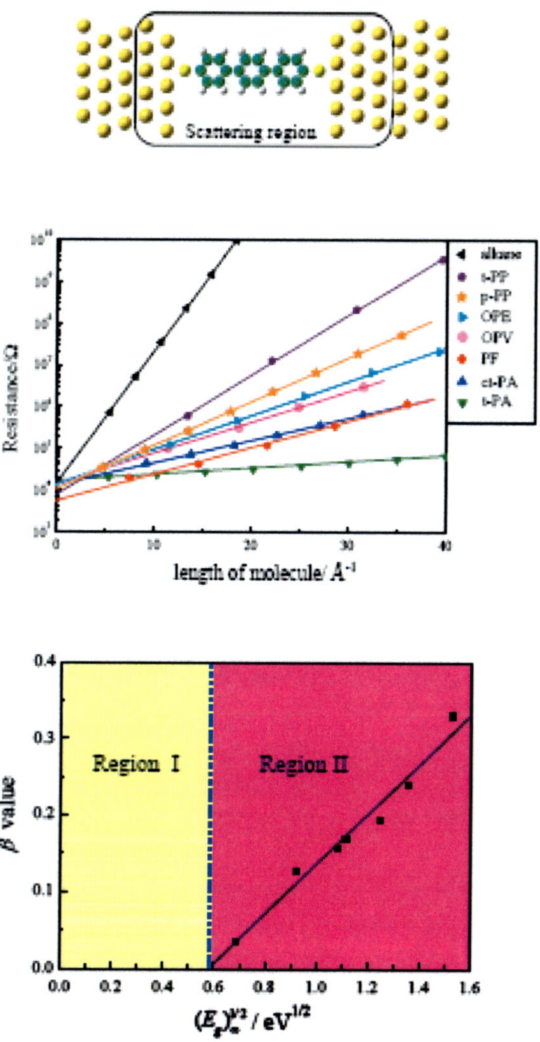

Figure 32. (a) Schematic illustration of the simplified model of the metal- molecule-metal junction (p-PP junction). (b) Resistance versus length plots for gold/molecule/gold junctions. The left triangle, pentagon, star, right triangle, circle, diamond, up triangle, and down triangle, represent alkane, t-PP, p-PP, OPE, OPV, PF, ct-PA, and t-PA, respectively. (c) β as a function of $(E_g)_\infty^{1/2}$ for a series of conjugated systems (t-PP, p-PP, OPE, OPV, PF, ct-PA, and t-PA).

Figure 32b shows the semilog plot of the junction resistance versus the molecular length of alkane, t-PP, p-PP, OPE, OPV, PF, ct-PA, and t-PA respectively. All the junction resistances increase exponentially with molecular chain length as given by the empirical equation (37) [97],

$$R = R_0 \exp(\beta d) \quad (37)$$

which is consistent with previous experimental studies, such as alkane and polyphenyl junctions. The fitting line extrapolated to zero angstrom gives the prefactor R_0 and the tunneling attenuation factor β. Due to the saturated hydrocarbon structure, the alkyl chain has the largest β value (0.73 Å^{-1}), which is very close to the value theoretically predicted by Kaun *et al.* around 0.76 Å^{-1}. Because the large resistance makes saturated alkane a poor candidate for molecular wires, conjugated molecules with alternating single and double bonds or delocalized π-electrons are most suitable. Since β depends strongly upon the intrinsic structure of the molecules, β values of saturated systems and conjugated systems significantly differ from one another, varying from 0.73 Å^{-1} to 0.036 Å^{-1}, and following the order of alkane >> t-PP > p-PP > OPE > OPV > PF > ct-PA > t-PA.

The β value has also been quantitatively correlated to the molecular HOMO-LUMO gap with the following empirical linear relationship:

$$\beta = -0.19 + 0.32(E_g)_\infty^{1/2} \quad (38)$$

The β value decreases with the order of t-PP, p-PP, OPE, OPV, PF, ct-PA, and t-PA, which is consistent with the order of the $(E_g)_\infty$. Through the comparison of $(E_g)_\infty$ instead of conductance, the length-dependent attenuation factor and the general feature of conductivity of some indeterminately-conjugated systems can be obtained approximately. This provides useful information about the relationship between molecular structure and transport behavior.

The electron transport behavior in the organic molecule is different from crystalline wire. The molecular resistance increases with the molecular length as displayed in figure 32b, corresponding to the region II of figure 32c. It is clearly shown that β is non-zero as the energy gap is larger than 0.35 eV. In other words, the resistance increases exponentially with the molecular length if the Fermi-energy of the electrodes is situated inside the energy gap of the wire. This linear relationship also infers that when the energy gap is small enough, the β value tends to be zero. In such cases, the electron transport may follow different mechanism. Most likely, when the Fermi-energy level of the electrodes resonances with the HOMO of the molecular bridge, the electron hopping could be predominant in the electron transportation. Therefore, the conductance shows less dependence on the molecular chain length, and the resistance is mainly contributed by the contact to the electrodes. In this chain length-independent region the molecule is expected to be more promising to serve as molecular wire since the conductance of the molecule is constant, which is similar to the characteristic of superconductor. The conjugated systems on two sides of this threshold behave quite differently. Above the threshold conjugated molecular wires behave normally, while below it they exhibit superconductive-like properties. This threshold will be useful for the future design of molecular wires and molecular conductors.

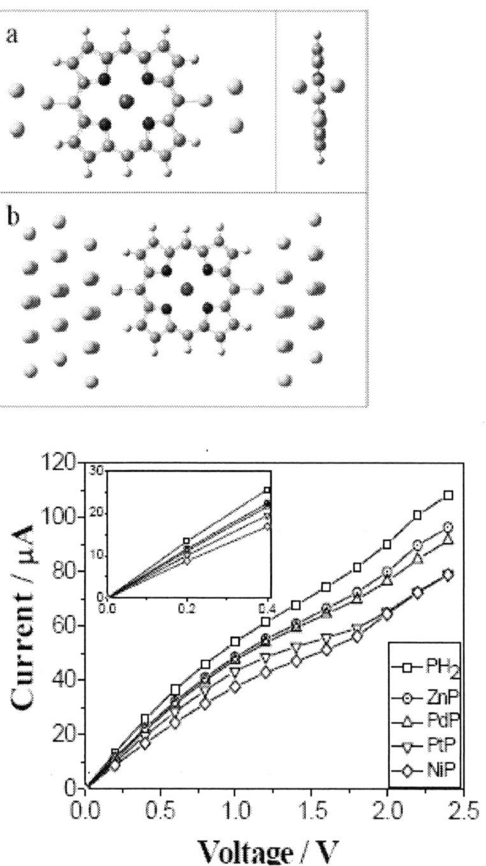

Figure 33. (a) the optimized structure of porphyrin molecule attached to two gold clusters (Top). Schematic illustration of a porphyrin molecule sandwiched between two gold (111) electrodes (Bottom). (b) The current-voltage curves of porphyrin and metalloporphyrin molecular wires.

Effect of center-metal on the conductivity of molecular wires All the above investigations about molecular wires are based on linear conjugated molecules. Porphyrin molecules (PH_2) are one of the most promising materials for molecular wires because of the rigid geometric configuration, highly conjugated structure, and chemical stability. The electron transport via a series of porphyrin molecular wires coordinated with various metal cations has been investigated with the metal-molecule-metal junctions by DFT and NEGF calculation, as displayed in figure 33a. The metal cation effect on the electron transport has been compared and discussed in terms of the electronic structures. From the I-V curves in figure 33b, we can conclude that among the series of model molecules, porphyrin is most conductive. The I-V curves of ZnP and PdP are very close to each other. For PtP, the curve falls on that of former two at low bias (less than 0.4V); however, when the bias increases to 2.0V, the curve tends to close to that of NiP. The conductance value at zero bias decreases in the order of PH2 > ZnP > PdP > PtP > NiP. The high conductivity of the porphyrin can be attributed to the higher conjugation and wider electron transfer path. Since the lone pair electrons of nitrogen atom coordinate with the metal ions, all metalloporphyrins show smaller conductance, though the difference is not so obvious. The results demonstrate that the coordination of metal cations weakens the π conjugation of the porphyrin and decreases the molecular conductance.

3.3. Theoretical Investigation on Molecular Diodes

Three decades have been passed since the first proposal of molecular rectification by Aviram and Ratner (A-R) in 1974 [1],where it was noted that a molecule, containing a donor (D) with low ionization potential and an acceptor (A) with high electron affinity, separated by a saturated bridge (D-σ-A), the electrons transfer inelastically through the σ-bridge while two electrodes are biased. Driven by the first theoretical proposal, a number of molecular rectifiers have been designed [19]. With the aim to reveal the fundamental structure property relations for electron transfer in molecular rectifier, electron transport through single molecule has been addressed in a number of recent theoretical papers. Since recent theoretical developments provide further insight into the mechanism of the electron transport. Based on Density Functional Theory (DFT) and non-equilibrium Green's function (NEGF) method, the molecular rectification have been investigated.

The proposal and mechanism of D-σ-A molecular diode Based on the classic knowledge of the semiconductor physics, the electron-donating group serves as the *n*-doping and the electron-withdrawing group serves as the *p*-doping to form a *p-n* junction. Most of the investigations focused on the Donor-bridge-Acceptor type of molecular rectifier(figure 34a), which is driven by the first theoretical proposal in 1974 [1], molecule rectifier could be achieved with a D-σ-A molecule, where D is a good one-electron donor with relatively low first ionization potential I_D, σ is some saturated covalent sigma bridge, and A is a good one electron acceptor with relatively high electron affinity A_A, when this molecule is placed between two appropriate metal contacts M_1 and M_2. The purpose of σ is to decouple the molecular orbitals of the donor moiety D from the molecular orbitals of the acceptor moiety A. Of course, this is approximate: the molecular orbitals belong to the whole molecule, but they often are more localized on one moiety than the other. If the decoupling between D and A is complete, then intramolecular electron transfer becomes impossible. The molecular ground state of D-σ-A has a relatively lower dipole moment, and can be written as D^0-σ-A^0, while the first excited state is much more polar, has a higher dipole moment, and can be written as the zwitterionic or betaine state D^+-σ-A^-. It is likely that resonant transfer would be possible (figure 34b) when a forward (left-to-right) bias was applied. The Fermi energy E_F of M_2 is resonant with the LUMO of the A moiety or part (which is close to the negative of the electron affinity A_A of the A moiety), and the HOMO of the D moiety (which is close to the negative of the ionization potential I_D of the D moiety) is in resonance with E_F of the metal M_1 (upon the application of a positive bias V onto M_1). The intramolecular electron transfer would be an inelastic tunneling from the excited electronic state D^+-σ-A^- to the ground electronic state D-σ-A. The mechanism would consist of two resonant electron transfers across metal organic interfaces, followed by (or simultaneous with) an inelastic downhill intramolecular electron transfer, which achieves, overall, the migration of one electron from M_2 to M_1 [1,19].

$$M_1 1 D^0\text{-}\sigma\text{-}A^0 1 M_2 \rightarrow M_1^{\square} 1 D^{\square}\text{-}\sigma\text{-}A^0 1 M_2 \rightarrow M_1^{\square} 1 D^{+}\text{-}\sigma\text{-}A^{-} 1 M_2^{\square} \rightarrow M_1^{\square} 1 D^0\text{-}\sigma\text{-}A^0 1 M_2^{\square}$$

In contrast, imagine now that a "reverse" voltage bias has been placed upon the system in Figure 34, with the higher voltage on the right-hand gold contact and the lower voltage on the left-hand contact driving up the Fermi energy on the left and depressing it on the right [98].

In analogy to the forward bias case described above, for the electrons in the external left-hand contact to begin to flow from left to right through the molecule, the reverse voltage bias must be sufficient to raise the Fermi energy of the gold contact on the left so that it is at least as high as the energy of the LUMO orbitals in the left-hand, donor portion of the molecule. However, in the reverse bias case, the amount of voltage that must be applied is considerably greater than in the forward bias case in order to raise the Fermi energy of the contact sufficiently to exceed the LUMO energy of the adjoining portion of the molecule. Simply applying the same amount of voltage in the reverse direction as is used to induce a current in the forward direction is insufficient to allow electrons to tunnel from the left contact into the LUMO energy levels of the molecule.

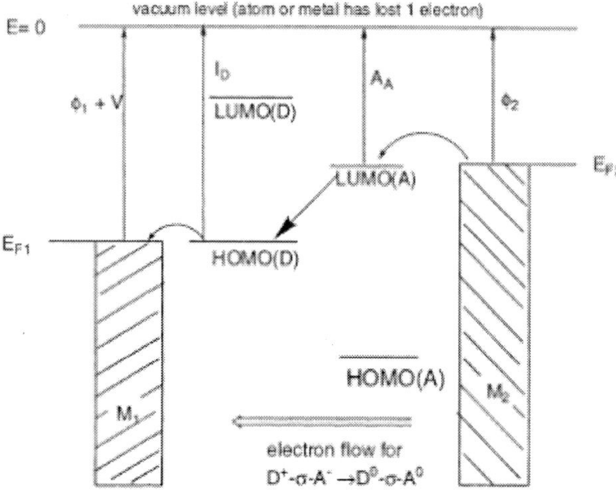

Figure 34. (a) Chemical structure of the Donor-bridge-Acceptor type of molecular rectifier. (b) Schematic illustration of the energy levels of the Donor-bridge-Acceptor type of molecular rectifier when a forward (left-to-right) bias was applied.

Thus, more voltage must be applied in the reverse (right-to-left) direction than in the forward (left-to-right) direction in order to get electrons to flow through the molecule. This is the classic behavior of a rectifying diode. Thus, this behavior and this symbol may be associated with the molecule. Furthermore, the reverse bias tends to drive up the energy of the LUMO's on the donor portion of the molecule relative to the LUMO's on the acceptor to the right. As depicted in figure 34, this increases the separation of the lowest lying unoccupied orbitals on the two sides of the molecule to a value greater than the unbiased value ΔE_{LUMO}, rather than decreasing their energy separation. This makes it difficult at moderate bias voltages to bring the energy levels of the donor-half LUMO's in coincidence with the low-energy unoccupied manifold that is localized on the acceptor half to its right. Consequently, tunneling through the central barrier from left to right, from donor to acceptor, is impeded rather than enhanced by a reverse bias.

Theoretical calculation on molecular diode From the theoretical analysis of Ratner, Metzger, Ellenbogen and Love, the A-R rectifier can be an effective rectifier in principle. Stokbro *et al.* [66] showed the advantage of evaluating possible candidates with advanced modeling tools prior to experimental verification. They have investigated the rectifying properties of a single-molecule D-σ-A diode proposed by Ellenbogen and Lowe [98] with the recently developed NEGF-DFT method. Two phenylene-ethynylene segments are connected via a dimethylene bridge and coupled to gold electrodes via thiolate bonds. An amino and a nitro side group are attached on the D and A part, respectively. The results demonstrated that although the electronic states are localized either in the D or in the A part of the molecule, the rectification is quite weak and symmetrical I-V characteristics was observed. This was due to two effects: (1) a resonant state was formed at both bias polarities, that is, no asymmetry; and (2) the transmission coefficient of the resonances was well below 1, due to electric field induced localization of the resonant levels. One model was proposed to obtain better diode characteristics, which use other electroactive substituents to obtain a much smaller gap between the HOMO of the D part and the LUMO of the A part, so that resonant conditions can be obtained for a much smaller bias.

It is known that the introduction of electron-donating and electron-withdrawing groups may lead to asymmetric electronic structure. Based on the static investigation of OPE wires [27], the asymmetric current-voltage characteristic of asymmetrically substituted conducting molecular wires has also been studied (figure 35a). To get the molecular rectification, the electron-donating group $-NH_2$ and the electron-withdrawing group $-NO_2$ are placed on the different positions of the molecular wire. The dependences of spatial distribution and lowest unoccupied molecular orbital LUMO energy level on the applied voltage have been found playing dominating but opposite roles in controlling the rectification behavior. In the tested bias range, since the shift LUMO energy level is more important, the electrons transfer more easily from donor to acceptor through the molecular junction in general (figure 35b).

Since the asymmetrically substitution on the conjugated wire could induce asymmetric electron transport. To design a molecular rectifier based on conjugated systems, we improved the design on basis of A-R rectifier by replacing the σ-connection with π-bridge (double and triple bonds) as shown in figure 36. Through the symmetrically substitution by one or two electron donating group $-NH_2$ and electron withdrawing group $-NO_2$, we get the donor-π-bridge-acceptor (D-π-A) rectifiers [99], noted as D1, T1, D2 and T2 respectively. The electron transport through the molecules was investigated by NEGF and DFT theory. The I-V curves of four D-π-A systems are displayed in figure 37a, which are almost symmetric. figure 37b shows the plot of rectification ratio (R) versus bias, where R is defined as the positive current over the negative one at the same bias. The R is variable with a maximum value close to 2.0, indicating a very slight molecular rectification for these conjugated systems. Even two electron donating and withdrawing groups were applied, the rectification ratio reaches to 2.5. With a detailed analysis of the transportation behavior and transmission spectra, we concluded that the simply substituted π-conjugated molecular wire may realize high current than the σ-connection. However, the rectification ratio is also poor. In addition, there is no obvious improvement of the rectification effect by increasing the asymmetry substituted groups, novel design should be considered.

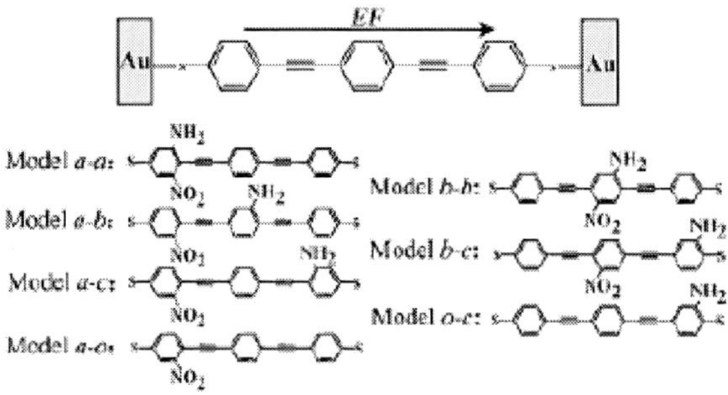

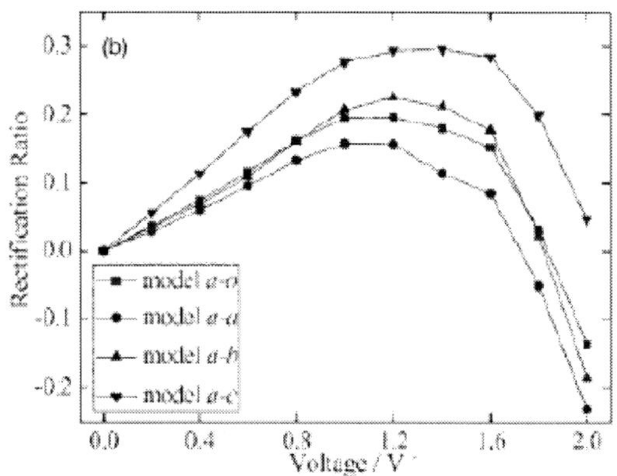

Figure 35. (a) Schematic diagram of the modelling of the Tour wire series. The phenyl rings from left to right, are numbered as *a*, *b*, and *c*, respectively. For convenience in comparison, the first letter of the model's name denotes the $-NO_2$ position and the second the $-NH_2$ position. (b) Rectification ratio as a function of the applied bias for the series of model molecules as shown in figure 34 (a).

For this investigated implementation of simply substituted conjugated molecules, we found weak rectification. This is due to the π bridge can not decouple the molecular orbitals of the donor moiety D from the molecular orbitals of the acceptor moiety A, the molecular orbitals belong to the whole molecule. For example, although the HOMO and LOMO shows slight localization, they distribute almost on the whole molecule. Therefore a resonant state was formed at both bias polarities, that is, no obvious rectification.

Staykov *et al.*[100] also studied the alternative D-π-A diode in contrast to the A-R rectifier with the same method. An asymmetric I-V diagram was observed, with a rectification ratio of 7. Above studies indicated that asymmetric substitution can induce rectification, but the rectification is not efficient and the rectification depends on the bridge between the electron donor group and the electron withdraw group.

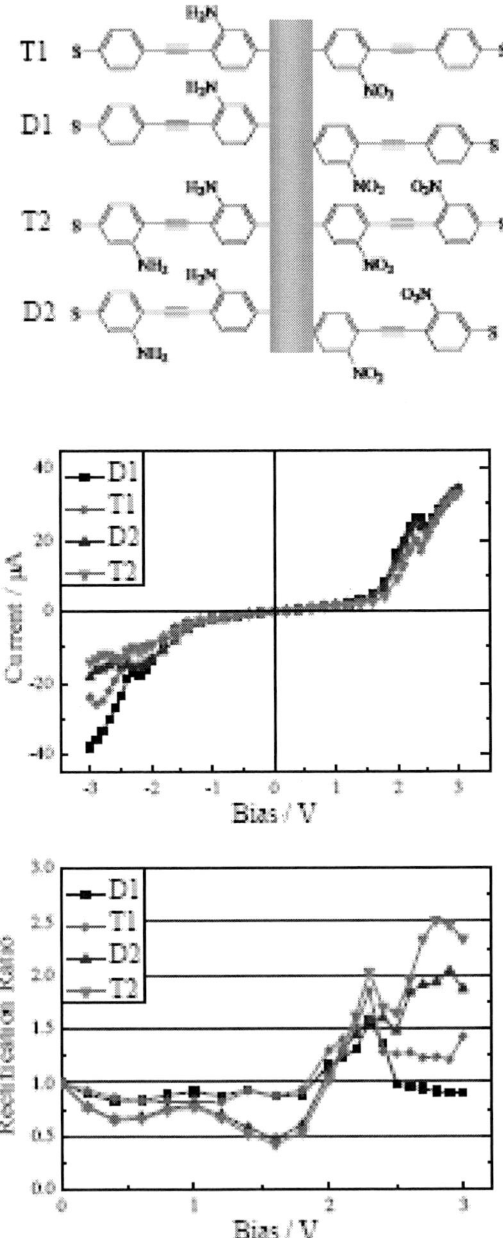

Figure 36. Chemical structures of the series of substituted D-π-A molecular systems. (b) The I-V curves of four D-π-A systems. (c) Rectification ratio (R) versus bias (R is defined as the positive current over the negative one at the same bias).

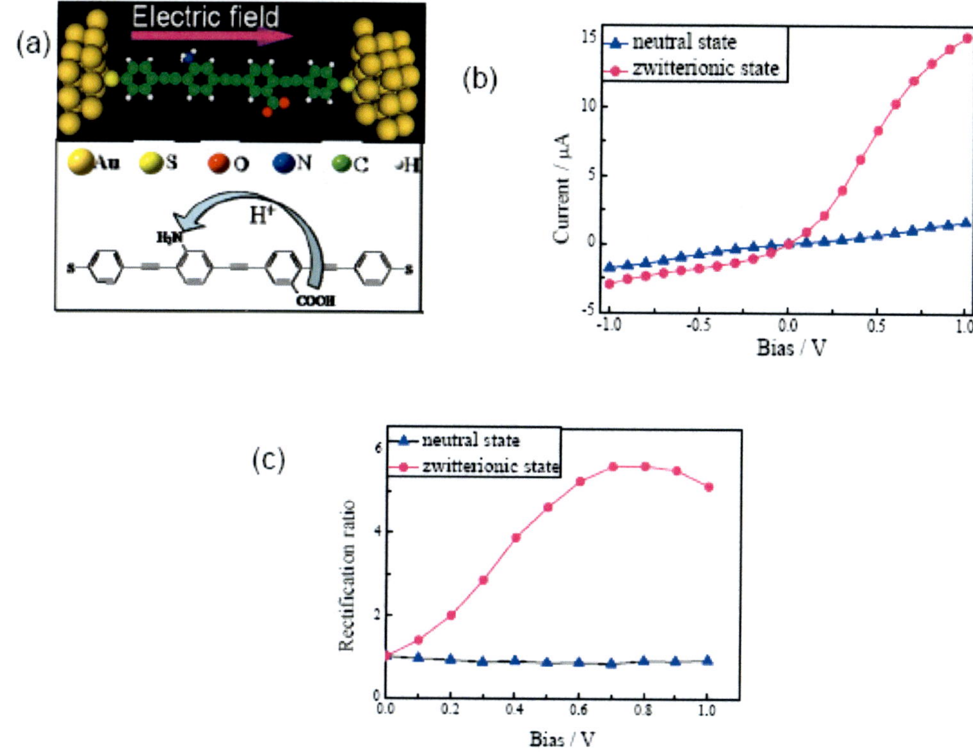

Figure 37. (a) Chemical structure of the molecular rectifier, OPE molecular wire substituted by carboxylic and the amino groups. (b) The calculated I-V curves for the systems in neutral and zwitterionic states. (c) Rectification ratio of the systems in neutral and zwittenionic states.

Our work [101] revealed a promising approach to design the molecular electronics, which is based on the perspective of chemistry. The electron transfer associated with chemical process may be an alternative for this purpose. In a designed proton transfer system, the rectification can be easily controlled by the different chemical environment. Importantly, the adjustment of the environmental condition is independent on the electron transfer. We have investigated the molecular rectifier, OPE molecular wire substituted by carboxylic and the amino groups (figure 38a), using the NEGF-DFT formalism. In the system without protonation, the rectification is hard to be distinguished and the current at negative bias is only a little larger than that at positive bias. However, when the environment, for example,the pH value of the solution, is changed, the proton transfer takes place. The current at 1.0 V could be 6 times as large as that at the corresponding negative bias, showing a pronounced rectification effect (figure 38b and 38c). It demonstrated that a high current and a great rectification ratio can be achieved simultaneously if an appropriate chemical reaction is introduced into the conjugated system and provided a promising approach to design novel molecular rectifier.

In the above devices, there is a built-in asymmetry in the structure of the molecule, which leads to asymmetric current-voltage characteristics. Taylor *et al.* studied the rectification in symmetric molecules with asymmetric electrode coupling through first-principles electron transport calculations, in this model (shown in figure 39), the molecule anchored to the electrode with one end of chemisorption and the other end of physisorption. Asymmetric electrode coupling can result in an asymmetric potential profile along the length of the molecule [103].

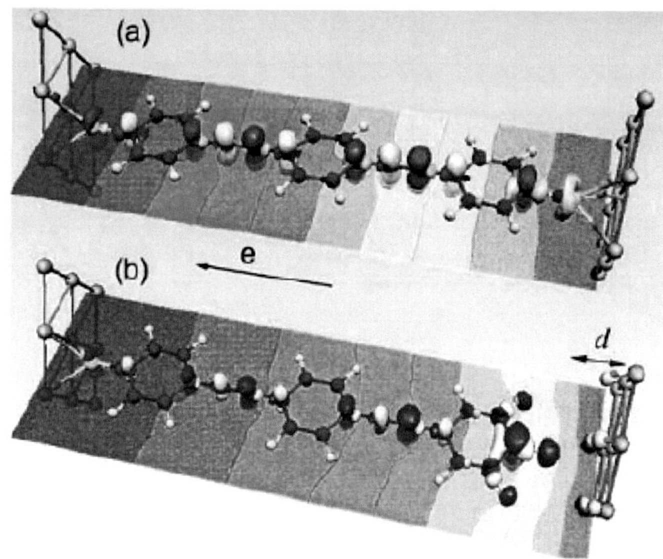

Figure 38. (a) Geometry of molecule A connected to two Au(111) surfaces. Isosurface of induced density and contour plot of induced potential at $eV_b = \mu_L - \mu_R = -1$eV. In the contour plot, red (blue) represents low (high) electrostatic potental and high (low) effective potential. For the isosurface, blue represents accumulation of positive charge (holes) and gray represents negative charge (electrons). (b) Similar to (a) but for molecule B at d=2.7 Å. From Taylor et.al.[93]

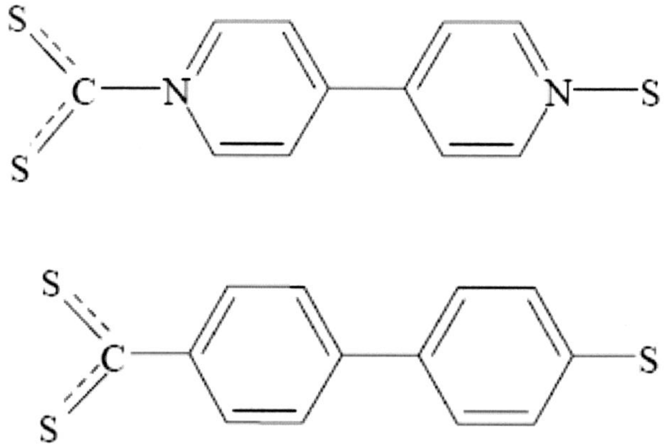

BPC

TBCT

Figure 39. (a) Chemical structure of the C4-(4′-pyridyl)-peridium-1-carbodithioate (BPC) and 4′-thiolatobiphenyl-4-dithiocarboxylate (TBCT) molecules. From Kosov et.al.[101,102]

If the potential profile is asymmetric, the molecular level can line up differently in positive and negative bias, resulting in rectification. When a bias is applied to the system, the S atoms couple differently to the bands of the left and right Au(111) surfaces, resulting in induced charges of different orbital characters on the S atoms. In model with one end of physisorption, the electric field increases near the right electrode and there is an additional drop at the right electrode due to the difference in dielectric constant of the molecule and the vacuum tunnel barrier. Therefore the asymmetric electrode coupling can induce asymmetric electron transport. However, it is important to note that one cannot obtain unlimited rectification by decreasing the molecule-electrode coupling. It is proposed that the resonances in such systems are generally quite broad and can significantly influence the rectification properties. A simple model was introduced, which takes these factors into account to explain the rectification saturation in such systems. Both the potential profile and the width of the transmission resonances must be taken into account to understand the electron transport in this kind of devices.

The asymmetric anchoring group of the molecule also could induce asymmetric electron transport. Kosov *et al.*[104,105] explored the role of dithiocarbamate anchoring group on transport properties by DFT and NEGF calculations, rectification behavior was predicted for 4-(4′-pyridyl)-peridium-1-carbodithioate (BPC) [105] and 4′-thiolatobiphenyl-4-dithiocarboxylate (TBCT) molecules (figure 40a), which have asymmetric anchoring groups. Although the rectification ratio is not large, about 2~3, the results indicated that the BPC and TBCT exhibit a new mechanism of rectification. It is different from the standard coupling-strength picture proposed by Taylor *et al.*[93], that is rectification from one resonance could be reduced by the opposite contribution from another resonance. As illustrated in the right part of figure 43c, the two transmission peaks A and B which contributed to the current respond oppositely on the applied voltage. both peaks A and B could contribute to the rectification in the same direction. As a prototype molecular rectifier, the results demonstrated that anchoring group plays a key role in the electron transport.

4. Conclusion

Molecular electronics uses single molecules as functional units in electronics, which is distinguished from the conventional silicon-based electronics and organic tin film electronics that use bulk materials and bulk-effect electron transport. It provides a bottom-up approach. to produce molecular scale functional devices. There have been many significant advances in the field of molecular electronic materials and molecular electronic devices. Since the first molecular diode proposed by Aviram and Ratner in 1974, tremendous advances have been made in the synthesis, characterization, and fabricating molecular electronic devices with molecular materials. The aim of this review is to present a broad basis for the area of theoretical simulations on molecular electronic materials and molecular devices. Two typical theoretical approaches, i.e., static and dynamic approaches, as well as the related theoretical simulation results for molecular electronic materials and molecular electronic devices have been introduced. The in-situ theoretical simulations on molecular electronic materials demonstrate that both the geometric and electronic structures of molecule are sensitive to the applied electric field. Therefore, considering the electric field effect is of great importance not only for the understanding of electron transport behavior of the molecular electronic material integrated in the circuits but also for the rational design high performance molecular electronic devices. For the dynamic theoretical investigations, the electron transport behavior and the mechanism of the molecular wires have been discussed in terms of the molecular geometric and electronic structures and the transmission spectra. Based on the investigation on molecular wires, mechanism of the typical molecular device, molecular diode, have been introduced. In particular, from the perspective of chemistry, when an appropriate chemical reaction is introduced into the conjugated system, pronounced rectification can be achieved. Through much progress in theoretical simulations on molecular electronic material and molecular electronic devices have been made, however many fundamental issues on electron transportation through molecules are still lacking. To get the realization of the molecular electronic devices, a highly collaborative, interdisciplinary arena, and a broad view will be required from all involved. Chemists, physicists, material scientists, and electrical and chemical engineers must work together to understanding the many factors of the research problems that are being uncovered as the field grows. We believe there is a very bright future for molecular electronic materials and molecular electronic devices becoming practical.

REFERENCES

[1] Aviram, A.; Ratner, M. A. *Chem. Phys. Lett.* 1974, *15,* 277-283.
[2] Joachim, C; Gimzewski, J. K.; Aviram, A. *Nature* 2000, *408,* 541-548.
[3] Carroll, R. L.; Gorman, C. B. *Angew. Chem. Int. Ed.* 2002, *41*, 4378-4400.
[4] Packan, P. L. *Science* 1999, *285*, 2079-2081.
[5] Service, R. F. *Science* 2001, *293*, 782-785.
[6] Tour, J. M. *Acc. Chem. Res.* 2000, *33*, 791-804.
[7] Fu, L.; Cao, L.; Liu, Y.; Zhu, D. *Adv. Coll. Interface Sci.* 2004, *111*, 133-157.
[8] Nitzan1, A.; Ratner, M. A. *Science* 2003, *300*, 1384-1389.
[9] Mantooth, B. A.; WEISS, P. S. *Proc. IEEE* 2003, *91*, 1785-1802.
[10] James, D. K.; Tour, J. M. *Chem. Mater.* 2004, *16*, 4423-4435.
[11] Kushmerick, J. G.; Blumb, A. S.; Longc, D. P. *Anal. Chim. Acta* 2006, *568*, 20-27.
[12] Chen, F.; Hihath, J.; Huang, Z. F.; Li, X. L.; Tao, N. J. *Annu. Rev. Phys. Chem.* 2007, *58,* 535-564.
[13] Karzazi, Y.; Cornil, J.; Bredas, J. L. *Adv. Funct. Mater.* 2002, *12*, 787-794.
[14] Majumder, C.; Briere, T.; Mizuseki, H.; Kawazoe, Y. *J. Phys. Chem. A* 2002, *106*, 7911-7914.
[15] Seminario, J. M.; Zacarias, A. G.; Tour, J. M. *J. Am. Chem. Soc.* 2000, *122*, 3015-3020.
[16] Brandbyge, M.; Mozos, J.-L.; Ordejón, P.; Taylor, J.; Stokbro, K. *Phys. Rev. B* 2002, *65*, 165401-165417.
[17] Soler, J. M.; Artacho, E.; Gale, J. D.; García, A.; Junquera, J.; Ordejón, P.; Sánchez-Portal, D. *J. Phys. Condens. Matter* 2002, *14*, 2745-2779.
[18] Taylor, J.; Guo, H.; Wang, J. *Phys. Rev. B* 2001, *63*, 245407(1-13).
[19] Metzger, R. M. *Chem. Rev.* 2003, *103*, 3803-3834.
[20] Schön, J. H.; Meng, H.; Bao, Z. *Science* 2001, *294*, 2138-2140.
[21] Donhauser, Z. J.; Mantooth, B. A.; Kelly, K. F.; Bumm, L. A.; Monnell, J. D.; Stapleton, J. J.; Price Jr., D. W.; Rawlett, A. M.; Allara, D. L.; Tour, J. M.; Weiss, P. S. *Science* 2001, *292*, 2303-2307.
[22] Chen, J.; Reed, M. A.; Rawlett, A. M.; Tour, J. M. *Science* 1999, *286*, 1550-1552.
[23] Echenique, P.; Alonso, J. L. *Mol. Phys.* 2007, *105*, 3057-3098.
[24] Geerlings, P.; De Proft, F.; Langenaeker, W. *Chem. Rev.* 2003, *103,* 1793-1873.

[25] Choi, B. Y.; Kahng, S. J.; Kim, S.; Kim, H.; Kim, H.W.; Song, Y. J.; Ihm, J.; Kuk. Y. *Phys. Rev. Lett.* 2006, *96*, 156106 (1-4).
[26] Collier, C. P.; Mattersteig, G.; Wong, E. W.; Luo, Y.; Beverly, K.; Sampaio, J.; Raymo, F. M.; Stoddart, J. F.; Heath, J. R. *Science* 2000, *289*, 1172 -1175.
[27] Li, Y. W.; Zhao, J. W.; Yin, X.; Liu, H. M.; Yin, G. P. *Phys. Chem. Chem. Phys.* 2007, *9*, 1186-1193.
[28] Chen, J.; Reed, M. A. *Chem. Phys. Lett.* 2002, *281*, 127-145.
[29] Hu, W.; Nakashima, H.; Furukawa, K.; Kashimura, Y.; Ajito, K.; Liu, Y.; Zhu, D.; Torimitsu, K. *J. Am. Chem. Soc.* 2005, *127*, 2804-2805.
[30] Reed, M. A.; Chen, J.; Rawlett, A. M.; Price, D. W.; Tour, J. M. *Appl. Phys. Lett.* 2001, *78*, 3735-3737.
[31] Li, C.; Zhang, D.; Liu, X.; Han, S.; Tang, T.; Zhou, C.; Fan, W.; Koehne, J.; Han, J.; Meyyappan, M.; Rawlett, A. M.; Price, D. W. Tour, J. M. *Appl. Phys. Lett.* 2003, *82*, 645-647.
[32] Tour, J. M.; Kosaki, M.; Seminario, J. M. *J. Am. Chem. Soc.* 1998, *120*, 8486-8493.
[33] Seminario, J. M.; Zacarias, A. G.; Tour, J. M. *J. Am. Chem. Soc.* 1998, *120*, 3970-3974.
[34] Kushmerick, J. G.; Holt, D. B.; Pollack, S. K.; Ratner, M. A.; Yang, J. C.; Schull, T. L.; Naciri, J.; Morre, M. H.; Shashidhar, R. *J. Am. Chem. Soc.* 2002, *124*, 10654-10655.
[35] Majumder, C.; Mizuseki, H.; Kawazoe, Y. *J. Phys. Chem. A*, 2001, *105*, 9454-9459.
[36] Anderson, H. L. *Chem. Commun.* 1999, *23,* 2323-2330.
[37] Tsukada, A.; Osuka, A. *Science* 2001, *293*, 79-82.
[38] Xu, Z.; Li, N.; Jin, X.; Li, Y. W.; Liu, H. M.; Zhao, J. W. *Chem. Lett.* 2007, *36*, 1278-1279.
[39] Robertson, N.; McGowan, C. A. *Chem. Soc. Rev.* 2003, *32*, 96-103.
[40] Mizuseki, H.; Belosludov, R. V.; Farajian, A. A.; Igarashi, N.; Kawazoe, Y. *Mater. Sci. Eng. C* 2005, *25*, 718-721
[41] Baszczyk, A.; Chadim, M.; Hänisch, C.; Mayor, M. *Eur. J. Org. Chem.* 2006, 3809-3825.
[42] Venkataraman, D.; Lee, S.; Zhang, J..; Moore, J. S. *Nature* 1994, *371*, 591-593.
[43] Hosokawa, Y.; Kawase, T.; Oda, M. *Chem. Commun.* 2001, 1948-1949.
[44] Sinnokrot, M. O.; Valeev, E. F.; Sherrill, C. D. *J. Am. Chem. Soc.* 2002, *124*, 10887-10893
[45] Zhao, Y.; Truhlar, D. G. *J. Am. Chem. Soc.* 2007, *129*, 8440-8442.
[46] Li, Y. W.; Zhao, J. W.; Yin, X.; Yin, G. P. *Chem. Phys. Chem.* 2006, *7*, 2593-2600.
[47] Zhao, J. W. *Chem. Phys. Lett.* 2002, *351*, 481-485.
[48] Karpfen, A.; Choi, C. H.; Kertesz, M. *J. Phys. Chem. A* 1997, *101*, 7426-7433.
[49] Bach, R. D.; Dmitrenko, O. *J. Am. Chem. Soc.* 2004, *126*, 4444-4452.
[50] Westheimer, F. H. Special Publication No 8. The Chemical Society, London, 1957.
[51] Kawase, T.; Ueda, N.; Oda, M. *Tetrahedron Lett.* 1997, *38*, 6681-6684.
[52] Zhang, J.; Pesák, D. J.; Ludwick, J. L.; Moore, J. S. *J. Am. Chem. Soc.* 1994, *116*, 4227-4239.
[53] Zhao, D.; Moore, J. S. *J. Org. Chem*. 2002, *67*, 3548-3554
[54] Kawase, T.; Tanaka, K.; Shiono, N.; Seirai, Y.; Oda, M. *Angew. Chem. Int. Ed.* 2004, *43*, 1722-1724.
[55] Desiraju, G. R.; Gavezzotti, A. *Acta. Cryst.* 1989, *B45,* 473-482.
[56] Shetty, A. S.; Zhang, J. S.; Moore, J. S. *J. Am. Chem. Soc.* 1996, *118*, 1019-1027.

[57] Bedard, T. C.; Moore, J. S. *J. Am. Chem. Soc.* 1995, *117*, 10662-10671.
[58] Park, J.; Pasupathy, A. N.; Goldsmith, J. I.; Chang, C. Yaish, Y.; Petta, J. R.; Rinkoski, M.; Sethna, J. P.; Abruna, H. D.; McEuen, P. L.; Ralph, D. C. *Nature* 2002, *417*, 722-725.
[59] Liu, Z. M.; Yasseri, A. A.; Lindsey, J. S.; Bocian, D. F. *Science* 2003, *302*, 1543-1545.
[60] Zhao, Q. H..; Zhang, S. W.; Li, Q. S. *Chem. Phys. Lett.* 2005, *407,* 105-109.
[61] Greenberg, A.; Liebman, J. F. Strained Organic Molecules, Academic Press: New York, 1978.
[62] Dudev, T.; Lim, C. *J. Am. Chem. Soc.* 1998, *120,* 4450-4458.
[63] Zhang, J.; Pesak, D. J.; Ludwick, J. L.; Moore, J. S. *J. Am. Chem. Soc.* 1994, *116,* 4227-4239.
[64] I. Fleming, Frontier Molecular Orbitals and Organic Chemical Reactions; Wiley, London, 1977.
[65] McCreery, R. L. *Chem. Mater.* 2004, *16,* 4477-4496.
[66] Stokbro, K.; Taylor, J.; Brendbyge, M. *J. Am. Chem. Soc.* 2003, *125,* 3674-3675.
[67] Ma, J.; Li, S.; Jiang, Y. *Macromolecules* 2002, *35,* 1109-1115.
[68] Reed, M. A.; Zhou, C.; Muller, C. J.; Burgin, T. P.; Tour, J. M. *Science* 1997, *278,* 252-253.
[69] Cui, X. D.; Primak, A.; Zarate, X.; Tomfohr, J.; Sankey, O. F.; Moore, A. L.; Moore, T. A.; Gust, D.; Harris, G.; Lindsay, S. M. *Science* 2001, *294,* 571-572.
[70] Waldeck, D. H.; Beratan D. N. *Science* 1993, *261,* 576-577.
[71] Li, Y. W.; Zhao, J. W.; Yin, X.; Yin, G. P. *J. Phys. Chem. A* 2006, *110*, 11130-11135.
[72] Li, Y. W.; Zhang, Y.; Yin, G. P.; Zhao, J. W. *Chem. J. Chin. Univ.* 2006, *27*, 292-296.
[73] Ye, Y. F.; Zhang, M. L.; Zhao, J. W. *J. Mol. Struct. (THEOCHEM)* 2007, *822,* 12-20.
[74] Zhao, J. W.; Li, P. Li, Y. W.; Huang, Z.Q. *J. Mol. Struct. (THEOCHEM)* 2007, *808,* 125-134.
[75] Zhang, Y.; Ye, Y. F.; Li, Y. W.; Yin, X.; Liu, H. M. Zhao, J. W. *J. Mol. Struct. (THEOCHEM)* 2007, *802,* 53-58.
[76] Kan, R. R.; Liu, H. M.; Ye, Y. F.; Li, P.; Yin, X.; Zhao, J. W. *Acta Physico-Chimica Sinica* 2007, *23,* 671-675.
[77] Li, Y. W.; Zhao, J. W.; Yin, G. P. *Comput. Mater. Sci.* 2007, *39,* 775-781.
[78] Tali1, D.; Yoav, G.; Roman,K; Ilya, K.; Diana, M.; Veronica, F.; Joseph, S.; Amir, Y.; Israel, B. J. *Science* 2005, *436,* 677-680.
[79] Hipps, K. W. *Science* 2001, *294,* 536-537.
[80] Zhitenev, N. B.; Erbe, A.; Bao, Z.; Jiang, W.; Garfunkel, E. *Nanotechnology* 2005, *16,* 1-6.
[81] Ke, S. H.; Baranger, H. U.; Yang, W. *J. Chem. Phys.* 2005, *123,* 114701(1-8).
[82] Buckingham, A. D. *Adv. Chem. Phys.* 1967, *12,* 107-142.
[83] Weimer, M.; Hieringer, W.; Della Sala, F.; Gorling, A. *Chem. Phys.* 2005, *309,* 77-87.
[84] Yamabe, T.; Tanaka, K.; Terama-e, H.; Fukui, K.; Imamura, A.; Shirakawa, H.; Ikeda, S. *J. Phys. C: Solid State Phys.* 1979, *12,* L257.
[85] Kertesz, M.; Koller, J.; Azman, A. *J. Chem. Phys.* 1977, *67,* 1180.
[86] Shirakawa, H.; Ito, T.; Ikeda, S. *Polym. J.* 1973, *4,* 460.
[87] Chung, T. C.; Feldblum, A.; Heeger, A. J.; MacDiarmid, A. G. *J. Chem. Phys.* 1981, *74,* 5504.
[88] Seminario, J. M.; Yan, L. Int. J. Quantum Chem. 2005, 102, 711-723.

[89] Seminario, J. M.; Derosa, P. A. *J. Am. Chem. Soc.* 2001, *123,* 12418-12419.
[90] Simmons, J. G. *J. Appl. Phys.* 1963, *34,* 1793-1803.
[91] Yin, X.; Liu, H.; Zhao, J. *J. Chem. Phys.* 2006, *7,* 094711.
[92] Li, Y. W.; Yin, G. P.; Yao, J. H.; Zhao, J. W. *Compu. Mater. Sci.* 2008, *in press*.
[93] Taylor, J.; Brandbyge, M.; Stokbro, K., *Phys. Rev. Lett.* 2002, *89*, 138301.
[94] Xue, Y. Q.; Datta, S.; Ratner, M. A., *Chem. Phys.* 2002, *281*, 151-170.
[95] Datta, S., Superlattices Microstruct. 2000, 28, 253-278.
[96] Yin, X.; Li, Y.; Zhang, Y.; Li, P.; Zhao, J., *Chem. Phys. Lett.* 2006, *422*, 111-116.
[97] Magoga, M.; Joachim, C. *Phys. Rev. B* 1997, *56*, 4722-4729.
[98] Ellenbogen, J. C.; Love, J. C. *Proc. IEEE* 2000, *88*, 386-426.
[99] Liu, H.; Zhao, J.; Yin, X.; Boey, F. Y. C.; Zhang, H., 2008, submitted.
[100] Staykov, A.; Nozaki, D.; Yoshizawa, K., *J. Phys. Chem. C* 2007, *111*, 11699-11705.
[101] Liu, H.; Zhao, J.; Yin, X.; Boey, F. Y. C.; Zhang, H., 2008, submitted.
[102] Troisi, A.; Ratner, M. A., *J. Am. Chem. Soc.* 2002, *124*, 14528-14529.
[103] Datta, S.; Tian, W.; Hong, S.; Reifenberger, R.; Henderson, J. I.; Kubiak, C. P., *J. Phys. Chem. B* 1999, *103*, 4447-4451.
[104] Li, Z. Y.; Kosov, D. S., *J. Phys. Chem. B* 2006, *110*, 9893-9898.
[105] Li, Z. Y.; Kosov, D. S., *J. Phys. Chem. B* 2006, *110*, 19116-19120.

In: Computational Materials
Editor: Wilhelm U. Oster
ISBN: 978-1-60456-896-7

Chapter 11

NUMERICAL SIMULATION OF THE CURING PROCESS OF RUBBER ARTICLES

Mir Hamid Reza Ghoreishy*
Iran Polymer and Petrochemical Institute, PO Box 14965/115, Tehran, Iran

ABSTRACT

This work is devoted to the numerical modeling of the curing (vulcanization) process of rubber articles. The main aim is to give an overall prospectus of the research works carried out in this field as well as practical approaches to solve an industrially scaled problem. The chapter begins with an introduction in which the aim and scope of the work is briefly described. A comprehensive review of the related published works can be found in next section. It is tried to cover all research papers from the beginning of the research in this area up to recent publications. The governing equations of the heat transfer and kinetics of the rubber curing reaction as well as finite element method used for the solution of these equations are then given and discussed in detail. The main focus is on the derivation of the working equations in conjunction with boundary conditions encountered in a typical rubber curing process. The solution algorithm and the commercial and in housed developed computer codes are drawn in the next part. In order to give a better understanding of topics presented in this chapter, two practical examples of the finite element simulation for the curing process of rubber goods including a metal reinforced rubber slab and a truck tire in the mold are presented and finally the conclusion as well references are given.

1. INTRODUCTION

The process of the vulcanization of rubber compounds in a curing process has a crucial influence on quality and service performance of the final product. This is mainly because that during the manufacturing process, the desired shape and physical properties of the rubber article are determined at this stage. This step is very energy consuming and thus its

* M.H.R.Ghoreishy@ippi.ac.ir

optimization not only helps to produce high quality products but also gives rise to more economical process. Due to the low thermal conductivity of the rubber and also using time-varying operating conditions, every point inside the rubber has a unique temperature history during the curing cycle. This leads to non-uniform distributions for temperature and state-of-cure and therefore a simple cure evaluation curve at constant temperature is generally found to be inadequate to predict the degree of cure. There are two methods to tackle this problem. The first technique is to directly measure the temperature as a function of time by the use of inserted thermocouples into critical points, especially at thick sections. Using an appropriate kinetic equation, these temperature histories are then converted to state of cure (degree of cure) and thus the necessary time for the completion of a cure cycle is determined. This method is costly and very time-consuming since for each experiment as least one rubber article must be damaged to find the exact locations of thermocouples. Consequently, rubber industries are seeking alternative approaches based on the use of computer simulations to predict the distributions of temperature and state of cure without any requirements for experimental measurement of temperature profiles. Several studies have been carried out in which both finite difference and finite element methods are used to numerically solve the governing equations.

The chapter begins with an introduction in which the aim and scope of the work is briefly described. A comprehensive review of the related published works can be found in next section. It is tried to cover all research papers from the beginning of the research in this area up to recent publications. The governing equations of the heat transfer and kinetics of the rubber curing reaction as well as finite element method used for the solution of these equations are then given and discussed in detail. The main focus is on the derivation of the working equations in conjunction with boundary conditions encountered in a typical rubber curing process. The solution algorithm and the commercial and in housed developed computer codes are drawn in the next part. In order to give a better understanding of topics presented in this chapter, two practical examples of the finite element simulation for the curing process of rubber goods including a metal reinforced rubber slab and a truck tire are presented and finally the concluding remarks as well references are given.

2. Literature Review

The idea of solving heat transfer equation either analytically or using numerical technique in rubber curing process was first emerged in 1967 by Gehman [1]. However, the first published practical work in this field belongs to Ambelang and Prentice (1972) [2] where they reported the application of computer programming for the prediction of heat transfer and curing kinetics. Their model was created for the shoulder zone of a tire and the finite difference technique was used to solve the heat equation. Thermal properties were considered to be constant and a series of parametric study on the ambient temperature and shoulder thickness were carried out using the developed model.

Later in 1980, Prentice and Williams [3] developed a more realistic numerical model for the vulcanization process of a pneumatic tire. Their model was consisted of a two-dimensional heat transfer equation in Cartesian coordinates in conjunction with an empirical integral equation for the determination of the state of cure (SOC) from temperature profile.

They have also included the bladder and mold into the model and both heating and cooling stages in mold and ambient temperature were taken into account. The energy (heat) equation was solved using the finite difference technique. The main drawbacks of this model were: 1) the finite difference was selected as the solution technique 2) only the shoulder zone of the tire was considered in the model (the lay-out of the tire was selected form Ambelang and Prentice [2]) and 3) despite the axisymmetric geometry of the tire and mold, a rectangular coordinate system was chosen for the description of the heat equation. However, the comparison between experimental measurements and the computed temperature profile showed that there were very good agreements between them. In another attempt, Schlanger (1983) [4] developed a one-dimensional numerical model for the simulation of the tire vulcanization process. Isayev et al. (1988, 1991) [5,6] combined the flow and heat equations in conjunction with a kinetics model for the simulation of the rubber injection process. They have used finite element method for the solution of the flow equations while the finite difference technique was employed to solve the energy equation. Their model was used to simulate both filling and post-filling stage where in the latter case; the state of the cure was predicted at different location in the mold.

The simulation of the tire curing process by the finite element method was first published by Toth et al. (1991) [7]. It was a relatively comprehensive work in which they used the ABAQUS code in conjunction with a user subroutine (HETVAL) to solve the coupled problem of heat transfer and cure problem in a single numerical computer code. The model was two-dimensional thus the flow in circumferential direction was ignored. The kinetics of the rubber compounds were studied by a moving die rheometer (MDR) and the variation of the thermal properties were also taken into consideration. The finite element model developed in this work was two-dimensional axisymmetric. The mold and bladder were not considered in the model so that the boundary temperature needed in the simulation were determined using the thermocouples inserted between tire-mold and tire-bladder, respectively. Marzocca (1991) [8] developed a one-dimensional finite element model for the simulation of curing process in a rubber cylinder. A first order kinetics model was also assumed for the curing reaction. Han et al. (1996) [9] developed a two-dimensional axisymmetric finite element model for the simulation of the vulcanization process of tires in molds. The working equations were solved using an in-house developed code. It was tried to take the most important aspects of the process such as complex geometry of tire, mold and bladder, time-varying boundary condition, heat and cooling steps and complicated cure kinetics model to taken into account. The numerical results were also compared with experimental measurements. They have used this model in another work (1999) [10] in conjunction with an optimization algorithm to determine the optimal cure steps for product quality. Nazockdat et al. (2000) [11] used the finite difference technique to solve the heat and cure kinetics equations. They have used this procedure to simulate the continuous curing process of conveyor belt. Ghoreishy (2001) [12] developed a complete two-dimensional axisymmetric finite element model for the simulation of a passenger tire curing process using HSTAR code (a module in COSMOS/M finite element package). The cure kinetics equations were solved using an in-house developed computer program. The computed results were also compared with experimentally measured temperature profile. The bladder and mold were also included in that model. Tong and Yan (2003) [13] and more recently Yan (2007) [14] reported on development of two-dimensional axisymmetric finite element model or the simulation of the tire curing process. Having compared with other published works on two-dimensional models, there is no novelty in their works.

The development of the finite element modeling of curing process in a three-dimensional framework, have been extensively studied by Ghoreishy and Naderi (2005) [15-17]. They created a complete and comprehensive three-dimensional finite element model for the simulation of the rubber curing process. An in-house developed computer code written in Visual Basic was used to simulate both a rubber article and a truck tire with complex tread pattern. The former was a molded rubber goods with two inserted metal pieces. Both the mold and rubber were considered. The numerical results were also compared with experimentally measured temperature profiles. It was shown that ignoring the flow in third direction (or solving the heat equation in two or one dimensions) gives rise to significant errors in predicting temperature and extent of cure reaction for complex rubber articles such as tires with tread patterns.

3. Governing Equations of Heat Transfer Process

The equation of change of energy in a rubber curing process is the conduction heat transfer equation in solid materials. In this section, we review this equation for two-dimensional axisymmetric and three-dimensional coordinate systems. There are so many good references on this topic which can be used for further reading. (See for example, [18-20]).

The governing equation of the transient heat conduction in two-dimensional cylindrical coordinate (r,z) (axisymmetric) and three-dimensional Cartesian coordinate systems are, respectively, given as:

$$\rho C\frac{\partial T}{\partial t}=\frac{1}{r}\frac{\partial}{\partial r}\left(rk\frac{\partial T}{\partial r}\right)+\frac{\partial}{\partial z}\left(k\frac{\partial T}{\partial z}\right)+\dot{Q} \qquad \textit{Two-dimensional axisymmetric} \tag{1}$$

$$\rho C\frac{\partial T}{\partial t}=\frac{\partial}{\partial x}\left(k\frac{\partial T}{\partial x}\right)+\frac{\partial}{\partial y}\left(k\frac{\partial T}{\partial y}\right)+\frac{\partial}{\partial z}\left(k\frac{\partial T}{\partial z}\right)+\dot{Q} \qquad \textit{Three-dimensional} \tag{2}$$

In this equation $\dot{Q}$ is the heat generation rate per unit volume of rubber due to the vulcanization reaction. (See section 5).

4. Transport Properties

There are three physical properties, density, heat capacity and thermal conductivity (ρ, k, C) in Eq. (1) and (2) which we refer them to as transport properties. The variation of these properties with solution variables (such as temperature and degree of cure) is a very important aspect of this topic since the success of a solution technique in solving of heat equation strongly depends on the type of the assumed equation forms that describe these parameters.

It is generally known that the density of the rubber compounds decreases with increase of temperature. On the other hand, heat capacity of rubber is believed to be increased with increasing of temperature. In addition, it is also dependent on the crosslink density and thus the effect of the degree of cure (extent of reaction) should also be taken into consideration. Consequently, two different approaches are usually adopted. In the first technique, a constant product of density times heat capacity (ρC) is assumed (see reference [3]). In the second approach (see for example, reference [9]), however, a constant density is assumed while the variation of the heat capacity with temperature and degree of cure is considered as:

$$C_m = C_u(1-\alpha) + C_c\alpha \qquad (3)$$

where C_m, C_u and C_c are the heat capacities of the partially cured $(0 < \alpha < 1)$, uncured $(\alpha = 0)$ and fully cured $(\alpha = 1)$, respectively. The dependency of the heat capacity on temperature is also expressed by:

$$C_i(T) = c_1 + c_2 T \qquad i = u, c \qquad (4)$$

In some cases the variation of density with degree of cure is assumed to be described by a relation similar to Eq. (3) as:

$$\rho_m = \rho_u(1-\alpha) + \rho_c\alpha \qquad (5)$$

Density is usually measured by a density gradient column and heat capacity is determined by differential scanning calorimeter (DSC).

Thermal conductivity of rubber compounds generally decreases with increasing of temperature. In most cases, a linear relation of the form:

$$k = a - bT \qquad (6)$$

is used. It is also assumed in many literature that the difference between thermal conductivity of uncured compounds and cured rubber is negligible [3,7,9-10] so that Eq. (5) can be used during the simulation of the curing process. It is determined either directly by a thermal conductivity measuring system (such as the instrument shown in figure 1) or indirectly by determining the thermal diffusivity. The latter parameter is a combination of the three previously mentioned properties, given by:

$$D = \frac{k}{\rho C} \qquad (7)$$

This is an essential parameter for the evaluation of heat flow in transient condition such as rubber curing process. Since it is easier to experimentally measure this parameter than thermal conductivity, it is some time preferred to determine thermal conductivity by Eq. (7) rather than its direct measurement.

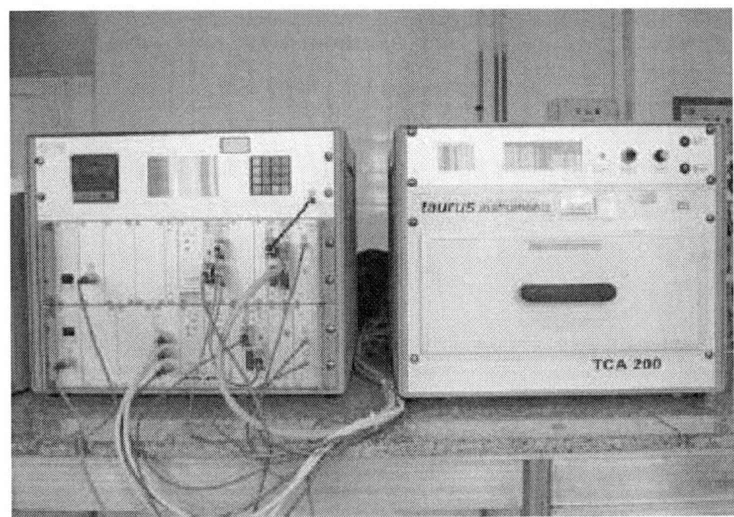

Figure 1. A fully computerized instrument for the measurement of thermal conductivity.

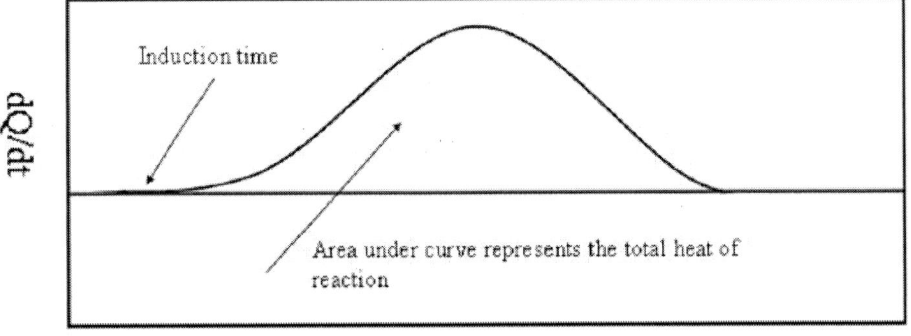

Figure 2. A typical DSC spectrum of a rubber curing reaction.

5. Heat of Reaction

The most challenging feature of the heat equation is the heat source in the right hand side of Eq. (1) or (2). The heat generated due to the vulcanization of rubber is related to the rate of cure. To derive a proper formulation for this parameter, if α denotes the state of cure (degree of cure) defined as:

$$\alpha = Q_t / Q_\infty \tag{8}$$

where Q_t and Q_∞ are the heat released up to time t and total heat of reaction, respectively, the heat generation rate per unit volume of the rubber is then calculated by:

$$\dot{Q} = Q_\infty \left(\frac{d\alpha}{dt} \right) \tag{9}$$

Total heat of reaction (Q_∞) is a material property and can be determined by either an isothermal or non-isothermal DSC experiment. Figure 2 shows a typical DSC curve from which the total heat of reaction is determined

6. Kinetics of Rubber Curing Reaction

As it is stated earlier, rubber curing process is a chemical reaction. In order to describe the extent of reaction quantitatively as function of time and temperature, a kinetic model should be proposed. It is generally known that there are three basic steps during the curing of rubber compounds. Figure 3 shows a typical torque vs. time curve obtained for a rubber compound. As it can be seen, there is usually an induction time period during which the chemical reaction does not take place and thus the measured torque is relatively remains constant. Next the curing stage starts which can be characterized by rising of torque until reaching a maximum value. At the end of this stage torque remains constant (or in some cases it goes for further increase which called marching stage). The last step is the reversion period which the cured rubber begins to degrade and thus the mechanical properties fail due to deterioration of polymer structure and destroying of cross links. The thirds step should not be occurred in an industrial process and thus we focus our attention to the first two steps i.e. induction and curing stages.

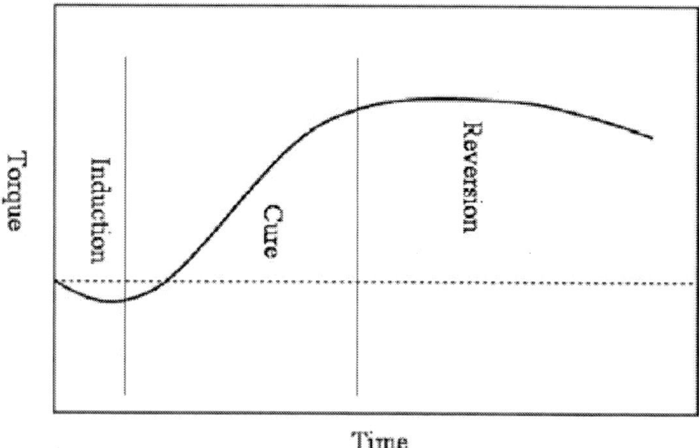

Figure 3. A typical isothermal torque rheometer curve with reversion.

6.1. Induction Time

A generally accepted form for the prediction of the induction time has been proposed in [21] with an Arrehenius-type temperature dependence as:

$$t_i = t_0 \exp(T_0 / T) \tag{10}$$

where t_0 and T_0 are material constants independent of temperature.

6.2. Curing Stage

The story of curing stage is, however, rather different form induction period and several researchers attempted to give an accurate and reliable model. The most common form of the kinetics of chemical reactions is the nth-order model given as [21]:

$$\frac{d\alpha}{dt} = k(1-\alpha)^n \tag{11}$$

The simplest form of this equation is n=1 which represents the first-order model. k in this equation is a rate constant with an Arrhenius-type temperature dependence of the form:

$$k = k_0 \exp(-E/RT) \tag{12}$$

Although the determination of the parameters of the first order model is an easy task, it cannot predict the accelerating stage of the cure reaction which normally takes place at degrees of cure less than about 0.25. A more complex model was presented by Piloyan et al. (1966) and suggested by Isayev and Deng (1988) [21] to be used for rubber curing reaction. This equation is expressed by the following form:

$$\frac{d\alpha}{dt} = k\alpha^m (1-\alpha)^n \tag{13}$$

where m and n are two parameters representing the order of reaction. This model was used by Toth et al (1991) [7], Han et al. (1996, 1999) [9-10] and recently Yan (2007) [14]. Furthermore, Kamal and Ryan (1980) [22], extended the idea of using two parameters for reaction order (m,n) to the use of two rate constants and proposed the following model:

$$\frac{d\alpha}{dt} = (k_1 + k_2\alpha^m)(1-\alpha)^n \tag{14}$$

They have successfully used this model for the prediction of the cure state of epoxy and polyester resins in injection molding process. Obviously, neither of these models can be integrated analytically to give an explicitly form for the degree of cure (α). To overcome this drawback, Kamal and Sourour (1973) [23] proposed a model in which the rate of reaction and degree of cure are, respectively given by:

$$\frac{d\alpha}{dt} = \frac{n}{k} t^{-1-n} \alpha^2 \quad (15)$$

and

$$\alpha = \frac{kt^n}{1 + kt^n} \quad (16)$$

The above mentioned model was first used for the study of cure kinetics of unsaturated polyester with styrene and further recommended by Isayev et al. (1988, 1991) [5-6] and Ghoreishy and Naderi (2005) [15-17] for rubber curing reaction as well.

The parameters in a cure kinetics model are determined in rubber industry by either a DSC or a toque rheometer (see figure 3). In a DSC experiment, the rate of reaction $(d\alpha/dt)$ is directly measured while in a torque rheometer the degree of cure (α) is obtained. In this case, the degree of cure is calculated from toque profile $(\Gamma(t))$ using the following relation:

$$\alpha(t) = \frac{\Gamma(t) - \Gamma(0)}{\Gamma(\infty) - \Gamma(0)} \quad (17)$$

where $\Gamma(0)$ and $\Gamma(\infty)$ are the initial and final recorded torque values, respectively.

6.3. Non-Isothermal Modeling

Vulcanization process of rubbery materials is a highly non-isothermal process and thus a non-isothermal cure kinetics model should be adopted in which not only the effect of time but also the influence of temperature changes on cure rate and state of cure has been taken into consideration. A suitable non-isothermal cure kinetic model should properly consider both induction time period and curing stage. For the induction time, a cumulative model [21] is usually used. It is given as:

$$\bar{t} = \int_0^t \frac{dt}{t_i(T)} \quad (18)$$

where $t_i(T)$ is the dependence of induction time to temperature described by Eq. (10). When the value of dimensionless time $\bar{t}$ becomes equal to one, the upper limit of t is considered as induction time. However, for the curing stage, several non-isothermal cure models have been reported so far [5-6,21,24-25]. The simplest form is a quasi-isothermal model in which rate constant (k) at every time step is calculated at the mean temperature between two successive time steps, i.e.

$$\alpha_i = \alpha_{i-1} + \int_{t_{i-1}}^{t_i} \left(\frac{d\alpha}{dt}\right)_{T=T_m} dt \tag{19}$$

where

$$T_m = \left(\frac{T_i + T_{i-1}}{2}\right) \tag{20}$$

In a more rigorous method, Isayev and Deng [21] proposed a total differential approach given as:

$$\alpha_i = \alpha_{i-1} + \int_{T_{i-1}}^{T_i} \left(\frac{d\alpha}{dT}\right)_t dT + \int_{t_{i-1}}^{t_i} \left(\frac{d\alpha}{dt}\right)_T dt \tag{21}$$

However, it is generally known that during the curing process the temperature at some steps (such as cooling stage) decreases with time. Our numerical studies showed that none of these models are capable of considering the effect of temperature decrease on rate of cure and state of cure thus these models fail to accurately predict the state of cure. In order to overcome to this controversy condition, two approaches are usually adopted. In the first approach, a cure kinetics equation in the form of $d\alpha/dt = k(T)f(\alpha)$ is used where $f(\alpha)$ (e.g. for Kamal and Sourour model, Eq. (15)) is expressed as:

$$\frac{d\alpha}{dt} = nk^{1/n}\alpha^{(n-1)/n}(1-\alpha)^{(n+1)/n} \tag{22}$$

Therefore, the state of cure can be computed by the solution of the ordinary differential Equation (22). However, since the right hand side of this equation is dependent on α, either a predictor-corrector technique should be used or the RHS of Eq. (22) must be evaluated at the previous time step to solve this nonlinear equation. On the other hand, since α at the start of calculation is equal to zero (initial condition), the RHS of Equation (22) cannot be expressed in terms of α and k at the previous time step (i.e., α_{i-1} and k_{i-1}) otherwise a zero profile for α is obtained. In the other method proposed by Chan, et al [24-25], a reduced time approach is used to predict the state of cure. The numerical expression for the cumulative state of cure in this method based on Kamal and Sourour model [23] is given as:

$$\left(\frac{\alpha_i}{1-\alpha_i}\right)^{1/n} = \left(\frac{\alpha_{i-1}}{1-\alpha_{i-1}}\right)^{1/n} + \left(\int_{t_{i-1}}^{t_i} k^{1/n} dt\right) \tag{23}$$

As it can be seen, in this approach, the initial condition of α (i.e. α=0 at time=t_i) does not affect on the calculation of the α at next time step. The non-isothermal rate of cure is also computed as:

$$\left(\frac{d\alpha}{dt}\right)_{non-isothermal} = \frac{\alpha_i - \alpha_{i-1}}{\Delta t} \tag{24}$$

7. Finite Element Analysis of Curing Process

As it is stated earlier, finite element method is the best choice for the solution of the most PDEs encountered in engineering problems. In this section, we first give the working equations of the finite element solution of the governing equations and then the boundary conditions existing in rubber curing process is presented. In section 8, we give a solution algorithm based on the given working equations and the practical approaches for complete solution of a rubber curing process will be discussed.

7.1. Working Equations

Solution of the governing equations (1) or (2) can be carried out by the use of the standard Galerkin technique. The equations are first multiplied by weight functions and then integrated over a typical element (see figure 4). Using the divergence theorem, the weak form is obtained. Then, by selecting an appropriate interpolation function for the solution variable (temperature), the finite element working equations is derived as:

$$[M^{(e)}]\{\dot{T}^{(e)}\} + [K^{(e)}]\{T^{(e)}\} = \{F^{(e)}\} \tag{25}$$

where $[M^{(e)}], [K^{(e)}]$ and $\{F^{(e)}\}$ are the mass matrix, stiffness matrix and load vector, respectively. These matrices and vector for axisymmetric model are given as [12]:

$$M^e_{ij} = \iint \rho c \psi_i \psi_j r dr dz \tag{26}$$

$$K^e_{ij} = \iint k\left(\frac{\partial\psi_i}{\partial r}\frac{\partial\psi_j}{\partial r} + \frac{\partial\psi_i}{\partial z}\frac{\partial\psi_j}{\partial z}\right) r dr dz \tag{27}$$

$$F^{e}_{i} = \iint \psi_i \dot{Q} r dr dz + \oint_{\Gamma^e} \psi_i k \left(\frac{\partial T}{\partial r} n_r + \frac{\partial T}{\partial z} n_z \right) r d\Gamma^{(e)} \tag{28}$$

and for three-dimensional model are expressed by [15-16]:

$$M^{(e)}_{ij} = \iiint \rho C \psi_i \psi_j dxdydz \tag{29}$$

$$K^{(e)}_{ij} = \iiint k \left(\frac{\partial \psi_i}{\partial x}\frac{\partial \psi_j}{\partial x} + \frac{\partial \psi_i}{\partial y}\frac{\partial \psi_j}{\partial y} + \frac{\partial \psi_i}{\partial z}\frac{\partial \psi_j}{\partial z} \right) dxdydz \tag{30}$$

$$F^{(e)}_{i} = \iiint \psi_i \dot{Q} dxdydz + \int_{\Gamma^e} \psi_i k \left(\frac{\partial T}{\partial x} n_x + \frac{\partial T}{\partial y} n_y + \frac{\partial T}{\partial z} n_z \right) d\Gamma^e \tag{31}$$

In these equations $\{T^{(e)}\}$ is the vector of unknown, $\{\dot{T}^{(e)}\}$ is the first derivative of the vector of unknown with respect to time, ψ is the weight (interpolation) function, n_r and n_z (for axisymmetric model) and n_x, n_y and n_z (for three-dimensional model) are the components of the unit vector normal to the boundary and $\Gamma^{(e)}$ represents the boundary of the element. In order to solve the first order ordinary matrix differential Eq. (25), the implicit-θ time stepping scheme is usually used [15-16]. The final form of the working equations in this scheme is given as:

$$\left([M]^a + \theta\Delta t[K]^a\right)\{T\}^{n+1} = \left([M]^a - \Delta t(1-\theta)[K]^a\right)\{T\}^n + \left((1-\theta)\{F\}^n + \theta\{F\}^{n+1}\right)\Delta t \tag{32}$$

Where superscripts n, n+1 refer to the time t_n and t_{n+1}, respectively. The superscript ^{a} also indicates a time level between t_n and t_{n+1} defined as:

$$\{T\}^a = (1-\theta)\{T\}^n + \theta\{T\}^{n+1} \tag{33}$$

θ is a parameter which its values varies between 0 (explicit) and 1 (full implicit) and has significant influence on both convergence and stability of the numerical results [18]. Numerical experiments showed that a θ of one is best compromise between accuracy and stability against computational cost and efforts.

7.2. Boundary Condition

To complete the mathematical model, it must be joined to proper boundary conditions. There are generally three types of boundary conditions in a rubber curing process, namely, first type or specified temperature, convection and the radiation. The latter is especially important

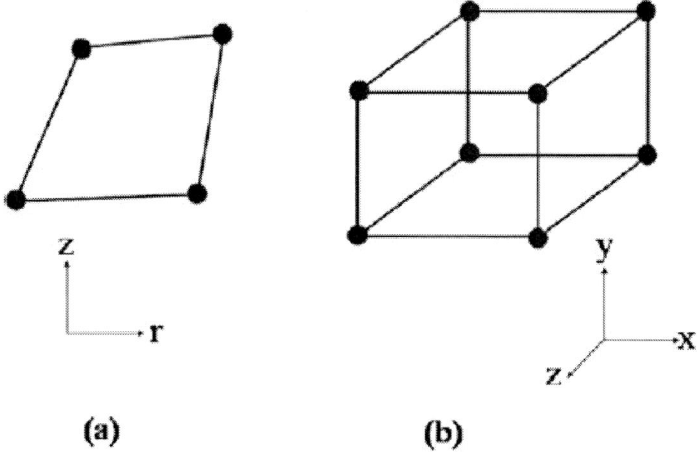

Figure 4. Elements used for the discretizaion of the domain: a) 4-noded two-dimensional axisymmetric b) three-dimensional brick.

during the post cure steps in which the rubber article is exposed to external environment. The first type boundary condition can be imposed either as a constant value or more realistically varying with time. The imposition of the convection (third type) boundary condition is somewhat different. This boundary condition represents the convection heat transfer between boundary of the domain with ambient temperature and is given by:

Axisymmetric model $$-k(\frac{\partial T}{\partial r}n_r + \frac{\partial T}{\partial z}n_z) = h(T - T_\infty) \quad (34)$$

Three-dimensional model $$-k(\frac{\partial T}{\partial x}n_x + \frac{\partial T}{\partial y}n_y + \frac{\partial T}{\partial z}n_z) = h(T - T_\infty) \quad (35)$$

Substituting of Eq. (34) into Equation (28) (for axisymmetric model) or Eq. (35) into Eq. (31) (for three-dimensional model) and using of the interpolation equation for temperature, the finite element working equations (Eq. (25)) for those parts of the model in where the convection boundary condition should be imposed are given as:

$$\left[M^{(e)}\right]\{\dot{T}^{(e)}\} + \left(\left[K^{(e)}\right] + \left[H^{(e)}\right]\right)\{T^{(e)}\} = \{F^{(e)}\} + \{P^{(e)}\} \quad (36)$$

Where

Axisymmetric model $$\begin{cases} H^{(e)}{}_{ij} = h\int_{\Gamma^e} \psi_i \psi_j r d\Gamma^{(e)} \\ P^{(e)}{}_i = h\int_{\Gamma^e} \psi_i T_\infty r d\Gamma^{(e)} \end{cases} \quad (37)$$

$$\text{Three-dimensional model} \begin{cases} H^{(e)}_{ij} = h \int_{\Gamma^e} \psi_i \psi_j d\Gamma^{(e)} \\ P^{(e)}_i = h \int_{\Gamma^e} \psi_i T_\infty d\Gamma^{(e)} \end{cases} \quad (38)$$

It should be noted that the above matrices must be computed only for those elements and boundaries that are subject to convection boundary conditions and for all other elements their values are be assumed to be zero.

The third form of the boundary condition that may arise in a rubber curing process is the radiation. The rate of heat flow by radiation (q) is governed by the relation

$$q = F_\varepsilon F_G \sigma (T^4 - T_\infty^4) \quad (39)$$

where F_ε is the emissivity function, F_G is the geometric view factor, σ is the Stefan-Boltzmann constant (5.669×10^{-8} W/m^{2}K^4), T is the absolute surface temperature of the body, and T_∞ is the absolute ambient (surrounding) temperature. If the above equation is replaced in the right hand side of either Eq. (34) or (Eq. 35), then the radiation type boundary condition is derived. As it can be seen the imposition of this boundary condition is the same as the method given for convection type condition except that the fourth order of the radiation boundary condition makes the final set of governing equations to be nonlinear. On other hand, due to the dependence of the thermal properties and also heat source to the solution variables, the working equations are already nonlinear and thus this can be treated with the same procedures adopted for them.

8. Solution Algorithm and Finite Element Code

Development of a computational strategy for the solution of the governing equations in rubber curing process based on the above described finite element working equations is not trivial and requires considering of some key factors. The most important one is the nonlinearity. Owing to the temperature-dependent of the thermal conductivity and the physical properties (such as heat capacity) and also dependency of heat generation per unit volume on the rate of cure (see Eq. 9), the set of the governing equations are nonlinear and thus an iterative approach is required. Two mostly accepted techniques for the solution of the nonlinear equations are Picard and Newton-Raphson methods. The ability of either technique in obtaining converged and correct solution is highly dependent on the initial estimate and the degree of nonlinearity. In some cases, it may be necessary to start from Picard's technique and after a few iterations switches to Newton-Raphson method. On the other hand, since the final set of working equations are transient, the proper choice of θ is very important. In most cases, a value of θ equal to 1 (backward difference) is reported to give the best converged and stable results.

There are generally two approaches adopted by researchers to solve the governing equations of rubber curing process (heat transfer and curing kinetics) by the finite element method. In the first approach, a computer code is developed from scratch to perform the required calculations. For example, Ghoreishy and Naderi [15-17] developed a computer program written in Visual Basic and used it for a series of cure simulations. The main advantage of this method is to implement the curing kinetics calculations in an integrated environment with finite element solution of the heat transfer equations. A single computer code not only analyzes the heat distribution in the rubber article but also it predicts both the state of cure and rate of cure and thus the heat source in heat equation (Eq. 1 or 2) is directly computed. However, development of such code is very time consuming and the cost is also high. In the second approach, researchers try to use commercial codes for their calculation. A number of powerful finite element codes such as ABAQUS, ANSYS, ADINA and COSMOS/M are available in the market and can be used for this purpose. Although this technique reduces the computational cost and effort associated with the first method, the problem of computing of the rate of cure and heat source as well as state of cure remains unsolved. Therefore, a supplementary calculating system is necessary for this part of calculation. The ability to include extra codes (as a user defined subroutine) into a finite element code is a great advantage among the different computer programs. For instance, ABAQUS has two different gateways to perform this task. HETVAL and UMATHT are two user subroutines which can be tailored based on the selected kinetics and implemented into the finite element model (see next section for a practical example). Pre- and post-processing stages are also crucial. Rubber articles have usually complex shapes and thus a CAD based pre-processing environment speeds up the process of providing initial modeling data to a great extent. Post-processing helps in better understanding and interpreting of results. Drawing contour and filled plots as well as x-y representation of the profiles of the temperature and state of cure are the most important features that a versatile post processor should support them. If the first approach has been adopted for the simulation of the rubber curing process i.e. using an in-house developed code, special translators for export and import of the inputs and outputs are required to be also written. This is because that the computer programs which perform pre- and post-processing tasks are generally very complex and thus it is not economical and practical to be developed individually.

9. Practical Examples

In this section, two practical examples on the simulation of the curing process are given. The first example is the simulation of the heat transfer and cure reaction for a metal reinforced rubber slab and the second one is the cure simulation of a truck tire in the mold. In each case, we begin with the problem description and then the computational strategy used to solve the problem is described. Finally, the results are presented and discussed.

9.1. A Metal Reinforced Rubber Slab

Problem Description: The article that is selected to be simulated has a cubic geometry and comprised of two similar rubber blocks with a metal plate inserted between them. Figure 5 shows a 3D representation of this article with the corresponding dimensions. The physical and thermal properties of the rubber and metal are given in table 1. It is assumed that the kinetics of the rubber curing reaction is described by the Kamal and Sourour equations given by Eqs. (15) and (16). The parameters of the cure and induction time models are recorded in table 2. The upper, lower and sides of the article are maintained at 423K (150°C) and thus a first type boundary condition can be used to represent the mold temperature. The initial temperature is 298K (25°C). The main aim of this simulation is to find the temperature and state of cure (α) inside the article during molding time.

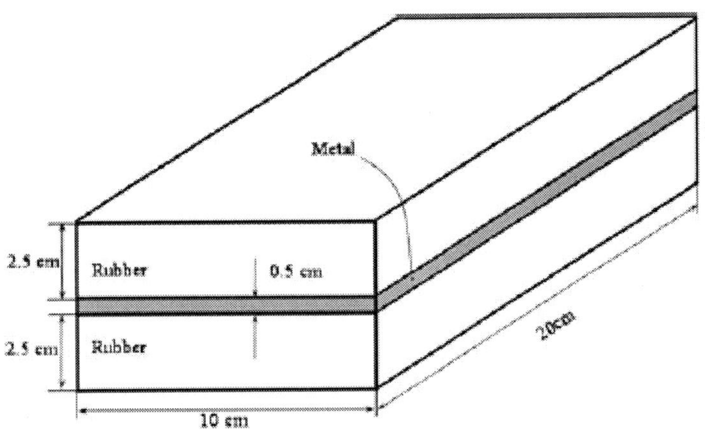

Figure 5. Schematic drawing of the rubber article in example.

Table 1. Physical and thermal properties of the materials used in the first example

Component	Density (kg/m^3)	Heat capacity (J/kgK)	Thermal conductivity (W/mK)
Rubber	1020	1980	0.47456-0.00072T
Metal	7700	460	50

Table 2. Cure model parameters of the material used in the first example

t_0 (s)	T_0 (K)	k_0	n	E (J/gmole)	Q_∞ (J/m^3)
7.55×10^{-13}	13080	1.1×10^{13}	2.7	1.5×10^5	2.2×10^7

Computational Strategy: Due to the symmetric configuration, only 1/8 of the article is considered in the model. Figure 6 shows part of the product selected for the simulation. The temperatures of those parts of the selected geometry which are in contact with the mold are set to 423K and the rest are assumed to have a zero flux boundary conditions. ABAQUS/CAE, ABAQUS/Standard and ABAQUS/Viewer are selected at the pre-processor, processor and post-processor, respectively [26]. In order to carry out the calculations required for the kinetics of the rubber curing reaction, a user-defined subroutine (HETVAL) was written and linked to the main code. This subroutine is called during the solution stage at all material calculation points. Figure 7 shows the flow chart of the HETVAL code and its interaction with the main processor (ABAQUS/Standard). Four solution dependent state variables were defined to interchange data between the computer code and the subroutine.

These are dimensionless induction time (Eq. 18), rate of cure, state of cure and rate of heat generation per unit volume at the previous step, respectively. The latter variable was required for the calculation of the rate of change of $\dot{Q}$ per temperature which is needed to define a correct Jacobian matrix. The geometry of the model was descritized into 1200 (1000 and 200 for rubber and metal parts, respectively) 8-noded elements (brick element) with total number of nodes equal to 1848 (figure 8). Although the use of higher order elements (like 20/27-noded quadratic elements) may give more accurate results on temperature field, these elements bring difficulty in computing of state of cure as they are computed at integration points. Consequently, linear elements are preferred when heat transfer and kinetics equation are intended to be solved simultaneously. Total solution time was selected to be 1500s with step time equal to 1s and thus 1500 time step are required to complete the simulation.

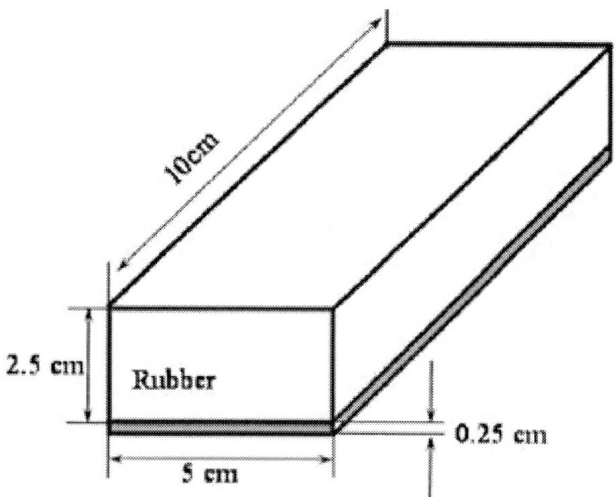

Figure 6. Solution domain selected for the rubber article.

Results and Discussion: The above mentioned model was analyzed using ABAQUS version 6.7 on a PC computer. Figures 9a to 9j, respectively, show the temperature distribution in the model at 10 consecutive time steps with an interval of 150s. As it is expected due to the low thermal diffusivity of the rubber (see table 1), there are distinguishing differences between temperatures at surfaces and core of the article. The corresponding distributions of the state of cure are shown in figures 10a to 10j, respectively. Again, at the middle zone of the article, the degree of the cure is much lower than the skin zones. This means that a non-uniform distribution of the degree of cure existed after the end of cure cycle which obviously affects the physical and mechanical properties of the product. It should be noted that the lower layer of the elements in these plots belong to the metal part and thus a zero degree of cure is reported. In order to show the profiles of the variable inside the article, the variations of the temperature and degree of cure with time for five locations (see figure 11) are shown in figures 12 and 13, respectively. The lowest profile belongs to the Node 982 which is located at the center of the article and has maximum distance from mold surfaces. Contrary, the highest profile is for the Node 406 located at the minimum distance (among the five selected points) from the mold temperature. The other profiles are between these extremes.

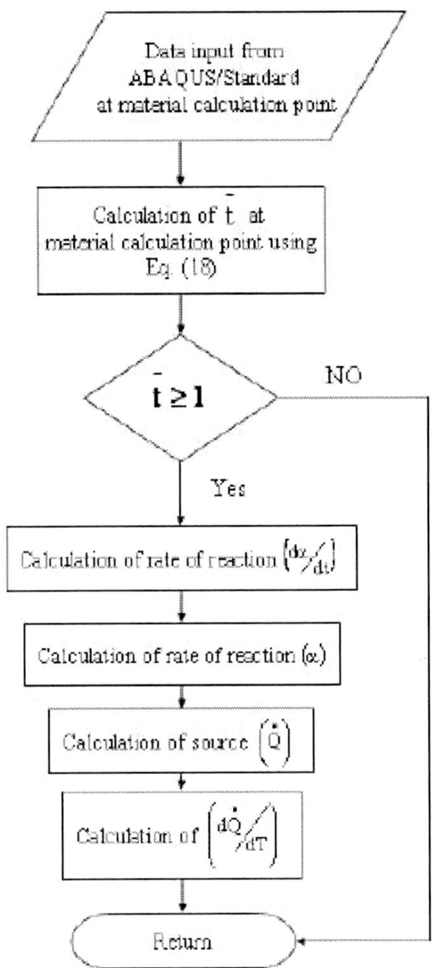

Figure 7. Flow chart of the HETVAL subroutine used in ABAQUS.

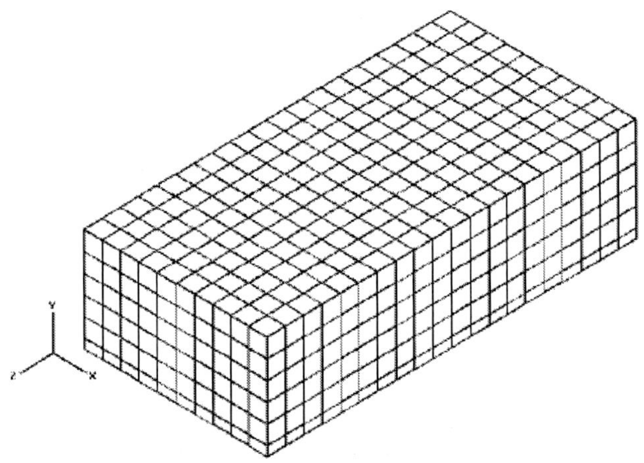

Figure 8. Finite element mesh of the rubber slab.

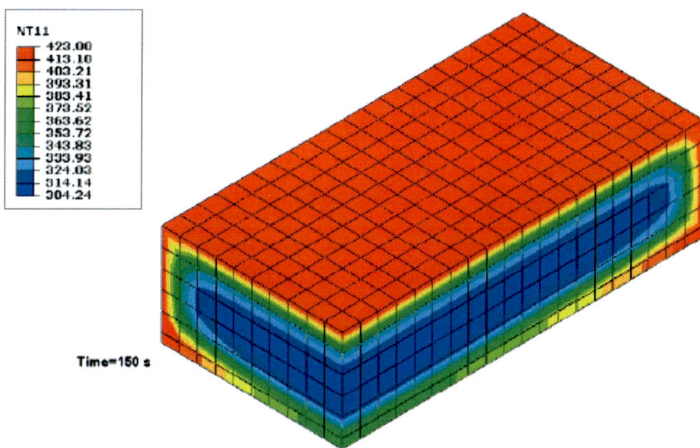

Figure 9a. Temperature field in the rubber article at time=150s.

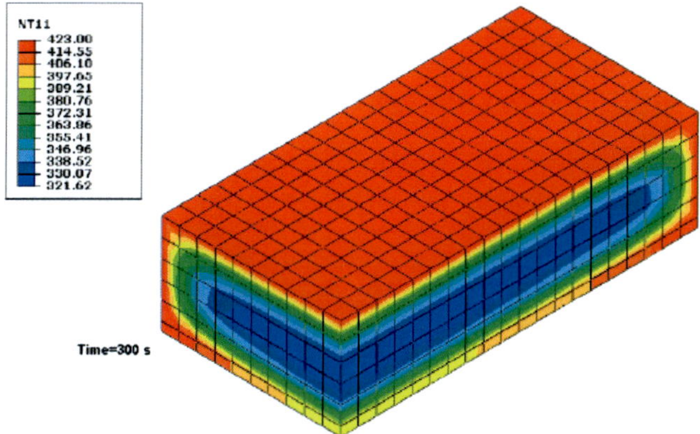

Figure 9b. Temperature field in the rubber article at time=300s.

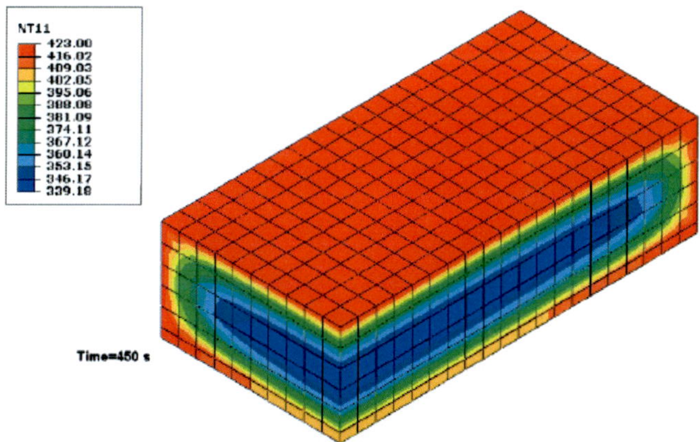

Figure 9c. Temperature field in the rubber article at time=450s.

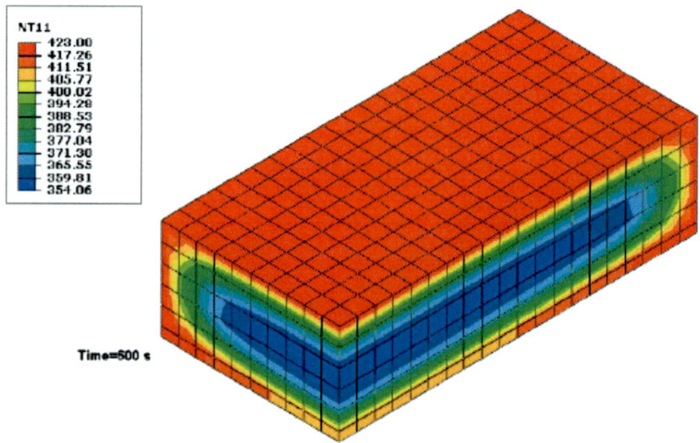

Figure 9d. Temperature field in the rubber article at time=600s.

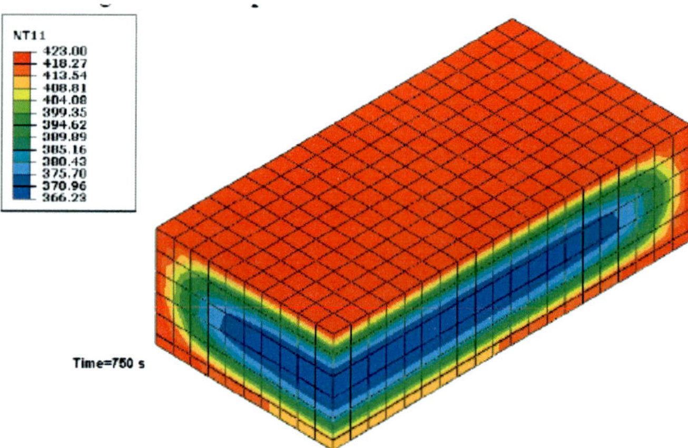

Figure 9e. Temperature field in the rubber article at time=750s.

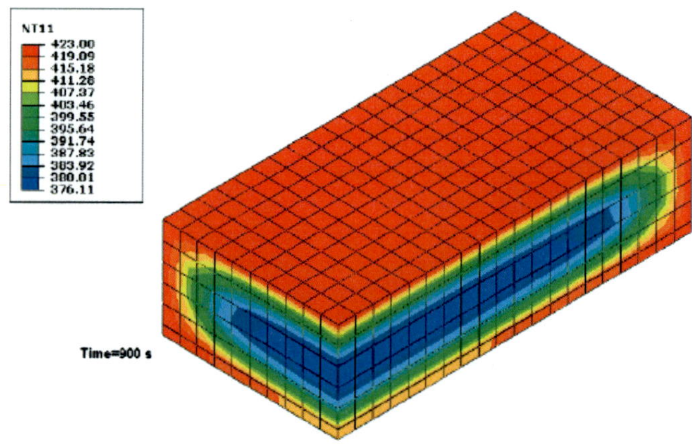

Figure 9f. Temperature field in the rubber article at time=900s.

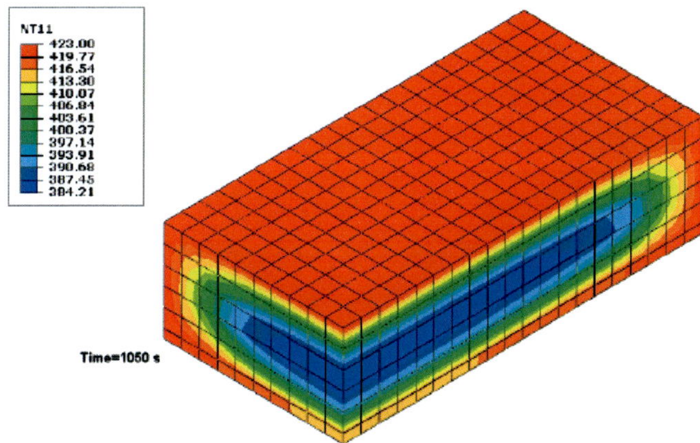

Figure 9g. Temperature field in the rubber article at time=1050s.

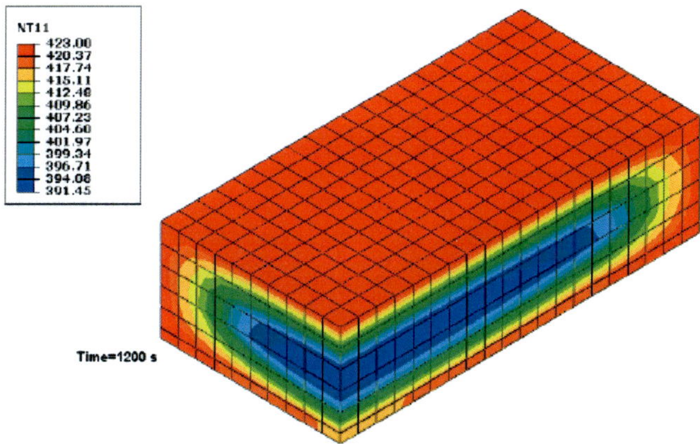

Figure 9h. Temperature field in the rubber article at time=1200s.

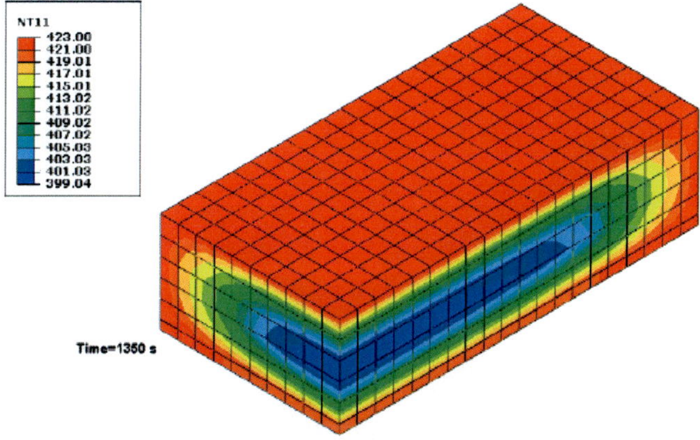

Figure 9i. Temperature field in the rubber article at time=1350s.

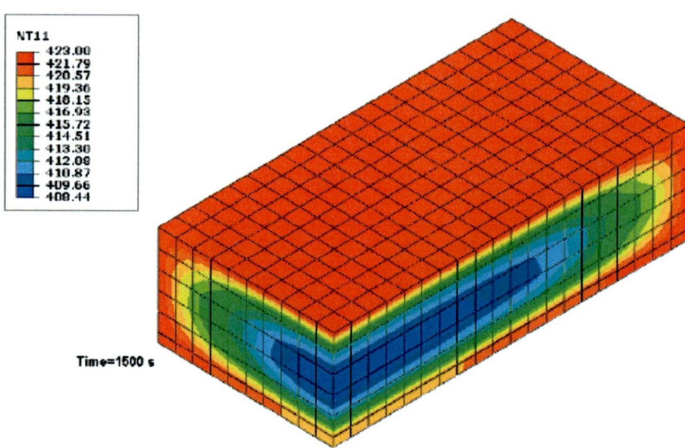

Figure 9j. Temperature field in the rubber article at time=1500s.

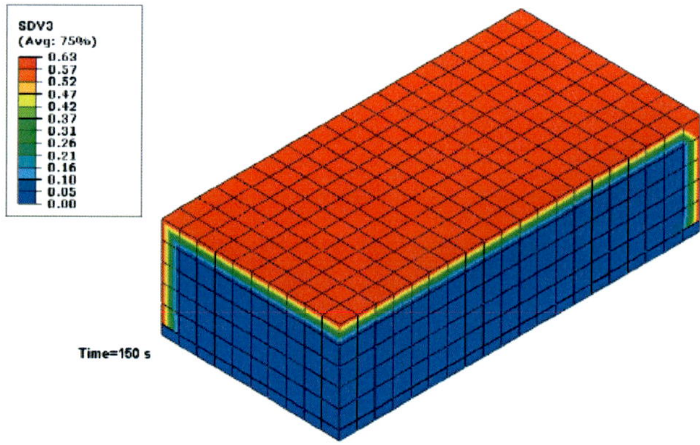

Figure 10a. Distribution of the degree (state) of cure at time=150s.

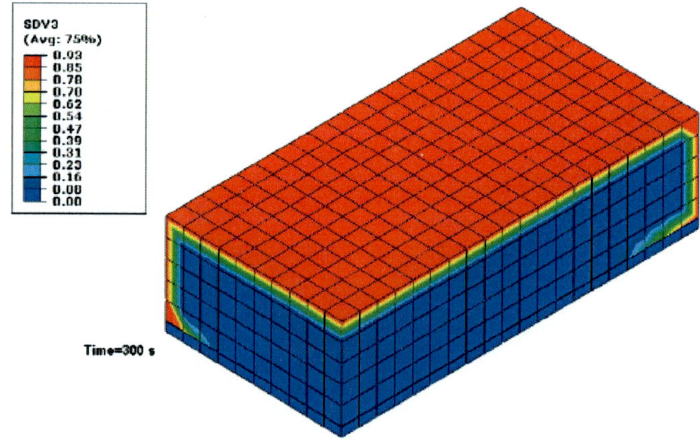

Figure 10b. Distribution of the degree (state) of cure at time=300s.

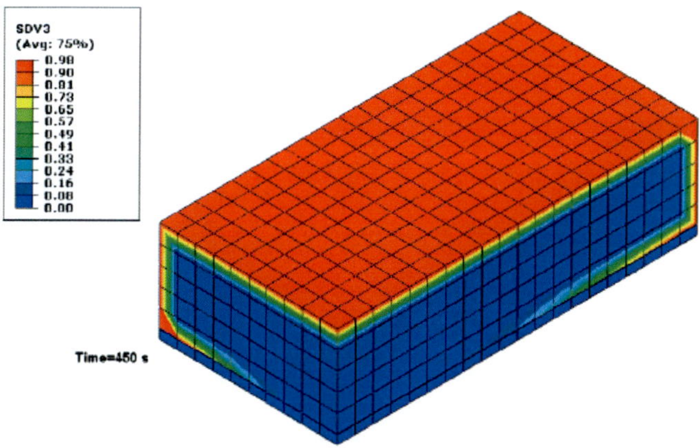

Figure 10c. Distribution of the degree (state) of cure at time=450s.

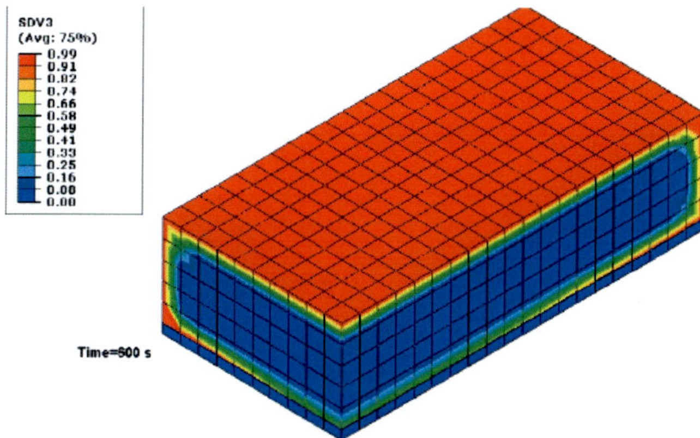

Figure 10d. Distribution of the degree (state) of cure at time=600s.

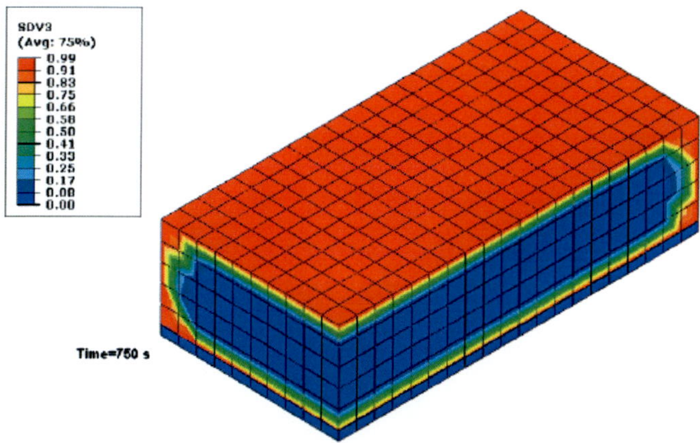

Figure 10e. Distribution of the degree (state) of cure at time=750s.

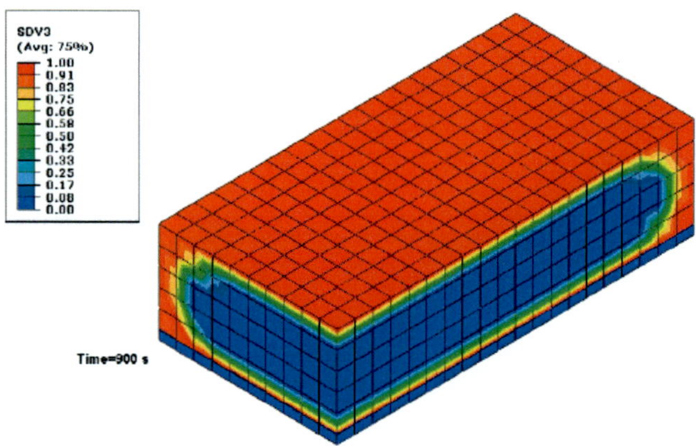

Figure 10f. Distribution of the degree (state) of cure at time=900s.

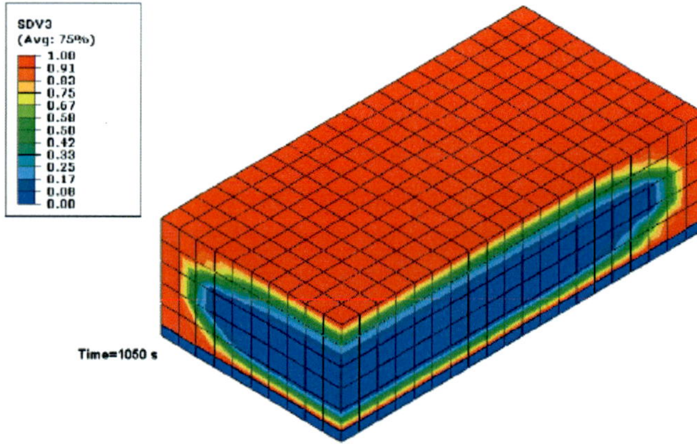

Figure 10g. Distribution of the degree (state) of cure at time=1050s.

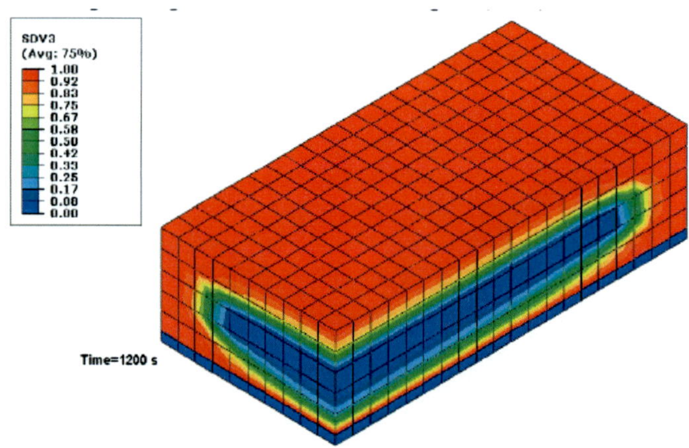

Figure 10h. Distribution of the degree (state) of cure at time=1200s.

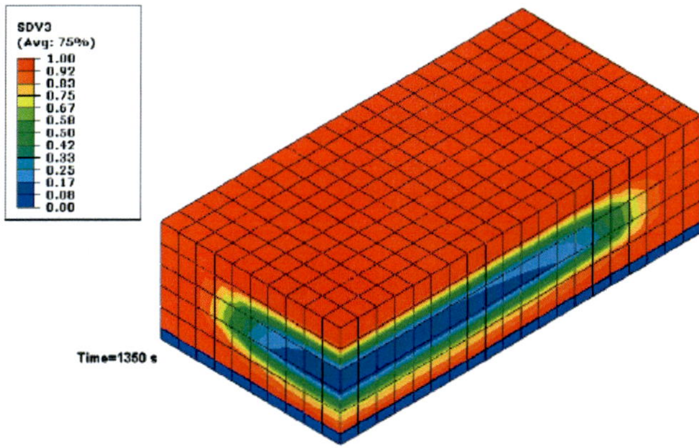

Figure 10i. Distribution of the degree (state) of cure at time=1350s.

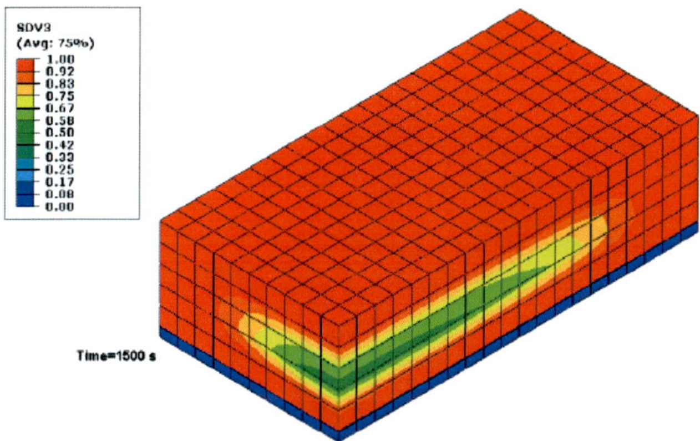

Figure 10j. Distribution of the degree (state) of cure at time=1500s.

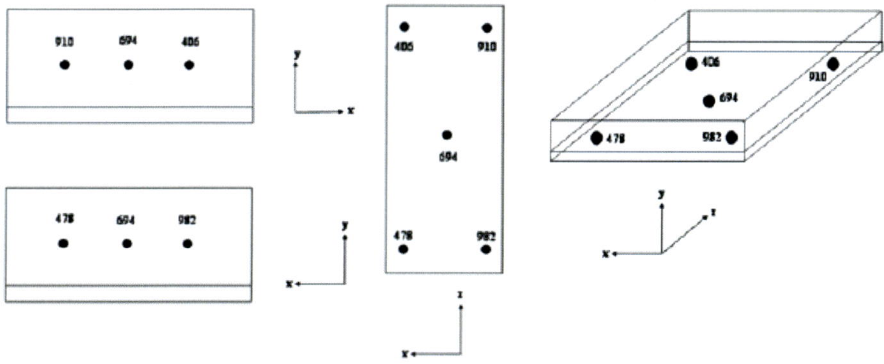

Figure 11. Spatial locations of the sampling nodes for the representation of the profiles temperature and degree of cure (views have drawn with different scale).

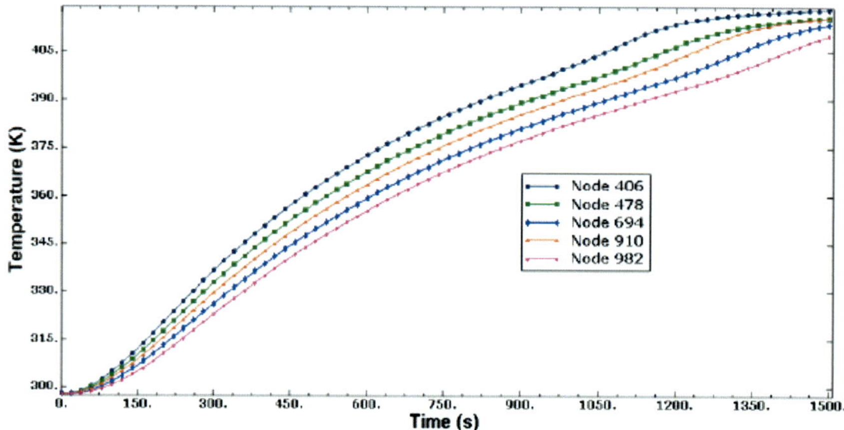

Figure 12. Variation of the temperature with time at sampling points (See figure 11).

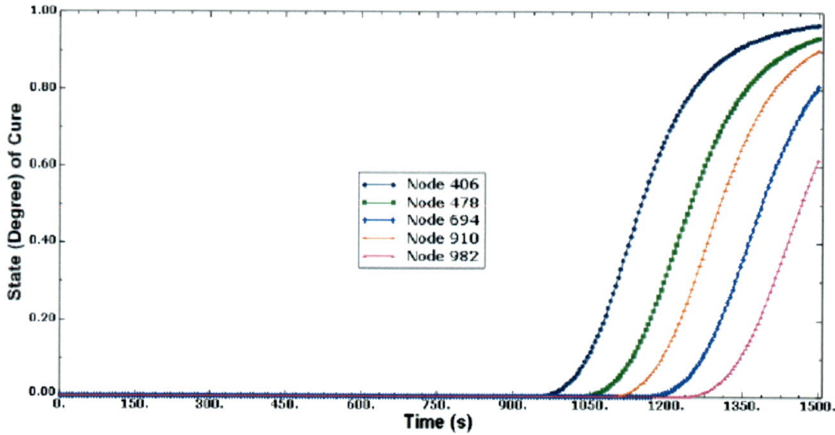

Figure 13. Variation of the degree of cure with time at sampling points (See figure 11).

9.2. Cure Simulation of a Truck Tire in Mold

Problem Description: A 12-24 truck tire with semi-rib tread pattern is used in this simulation as shown in figure 14. A non-isothermal cure kinetic model similar to one used for the first example (Kamal and Sourour with Arrehenius-type induction time model) was also used in this work. The physical and thermal properties of the different part of the tire and the parameters of the cure kinetics model are given in reference [16]. First type boundary condition with time varying temperature values was applied to the outer (tread surface) and inner surfaces of the tire, respectively.

Computational Strategy: Due to symmetry and also replication of sequences of tread blocks (periodic symmetry) only a 5^0 section of the tire as shown in figure 15 is considered in this simulation. The domain of the analysis was divided into 1216 8-noded brick elements with total number of nodes equal to 1762. This problem was analyzed using an in-house developed code written in Visual Basic [15-17]. However, the Pre-processing and post-processing stages were performed using Geostar software [27]. The simulation time for the vulcanization process was 4200s (70 min) with time step equal to 42s. Therefore, 100 time steps were required to complete the simulation.

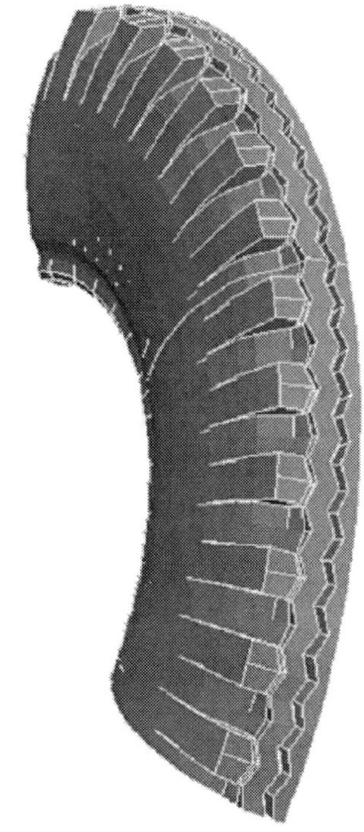

Figure 14. A 12-24 truck tire used for the simulation.

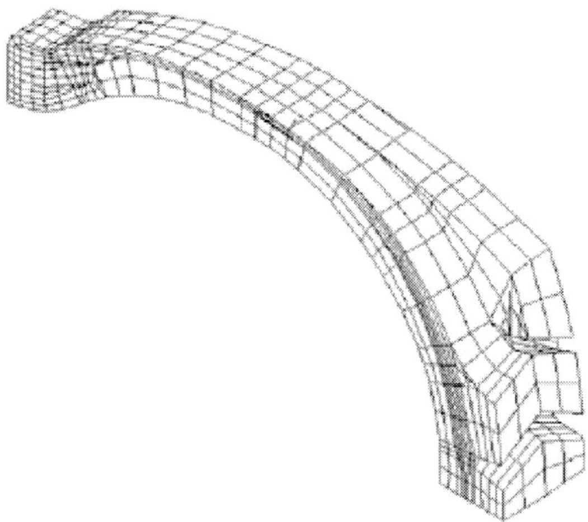

Figure 15. Finite element mesh of the section of the tire used in the study.

Results and Discussion: Six points, as shown in figure 16, have been selected for the numerical investigation. The first three points (Node 613, Node 101 and Node 118) are on the left side while the second three points (Node 679, Node 287 and Node 290) are on the right side of the model, respectively. These points are arranged in three levels as given in table 3. As it can be seen, each level indicates a certain location in tire. Level 1 (Nodes 613 and 679) is located in ply, level 2 (Nodes 101 and 287) is located in lower tread part and level 3 (Nodes 118, and 290) is located in upper tread part.

Table 3. Specification of the selected nodes for the results presentation (See figure 16)

Level	Location	Left hand side node number	Right hand side node number
1	Ply	613	679
2	Lower tread	101	287
3	Upper tread	115	290

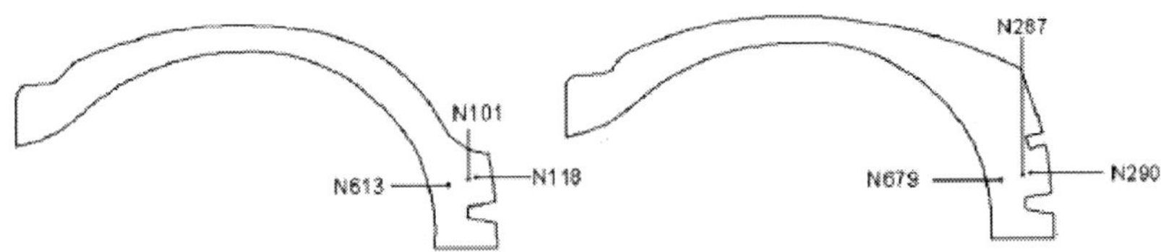

Figure 16. Locations of the sampling nodes in the tire.

Figure 17 to 19 show the variations of temperature with time for nodes located in level 1, 2 and 3, respectively. Also the variations of state of cure for these levels are shown in figures 20 to 22, respectively. In level 1, where the nodes (Nodes 613 and 679) are located in ply, minor differences in both temperature and state of cure profiles are observed (figures 17 and 20). This is due to the very negligible variations in tire geometry in circumferential direction. However, when we move from ply to tread in radial direction (from level 1 to level 2 and 3), the variation in tire geometry in circumferential direction becomes more prominent. Therefore the difference in profiles of temperature and state of cure for nodes Node 101 and Node 287 (level 2) and Node 118 and Node 287 (level 3) become greater as shown in figures 18 and 21 and figures 19 and 22, respectively. On the other hand, since the thickness of model of the tire on the left side is lower than right side, the rate of cure for those nodes located on the left side is greater than nodes located on the right side (see figure 16). Also nodes on left side have greater temperature values than nodes on right side.

In order to check the accuracy of the predicted results, the variation of temperature for two points inside a tire, located at ply and upper tread sections, were compared with experimentally measured data. This comparison is shown in figure 23, which confirms the validity and applicability of the developed model and code.

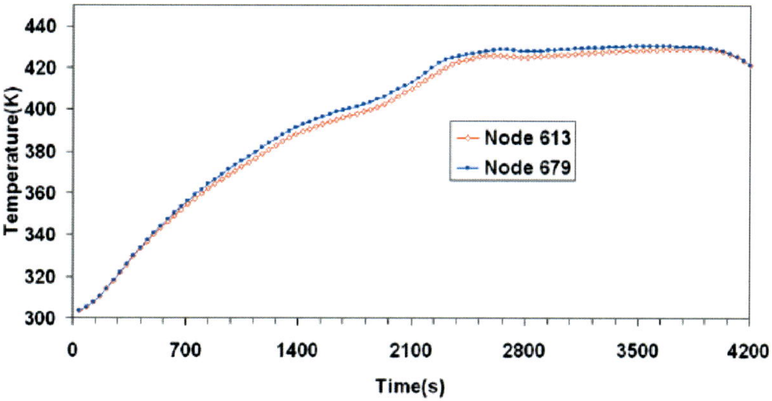

Figure 17. Variation of the temperature with time for the nodes at level 1.

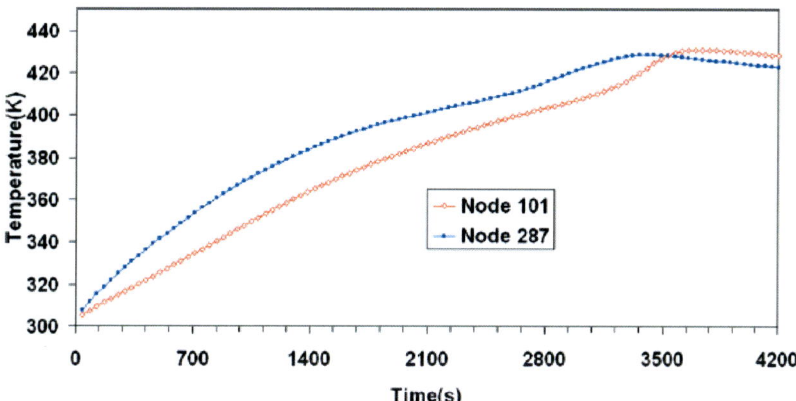

Figure 18. Variation of the temperature with time for the nodes at level 2.

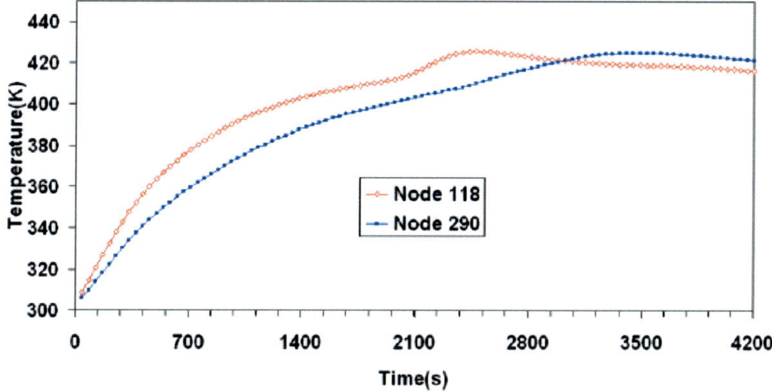

Figure 19. Variation of the temperature with time for the nodes at level 3.

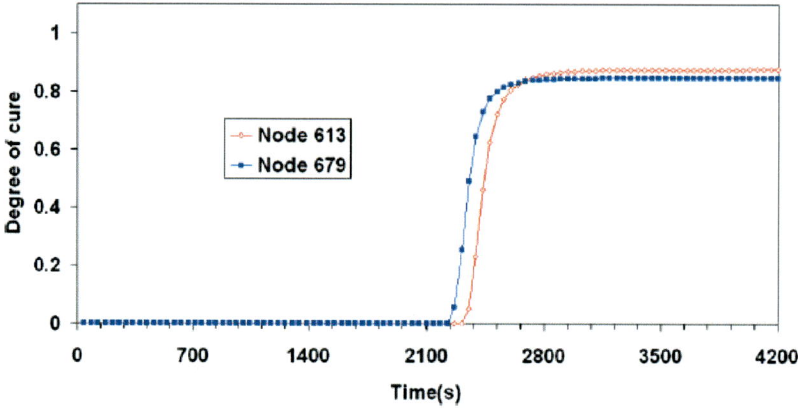

Figure 20. Variation of the degree of cure with time for the nodes at level 1.

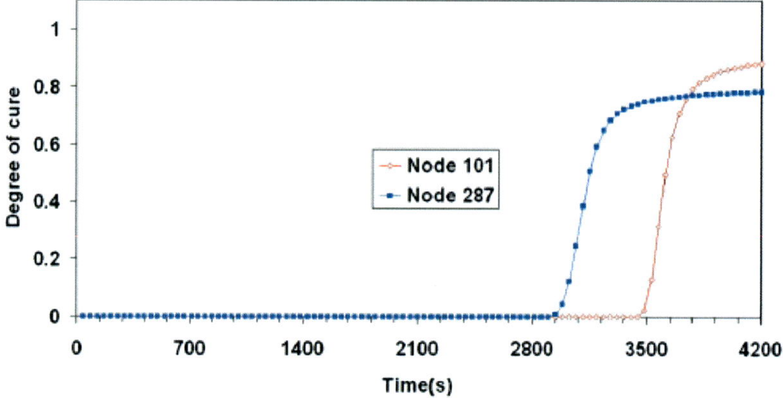

Figure 21. Variation of the degree of cure with time for the nodes at level 2.

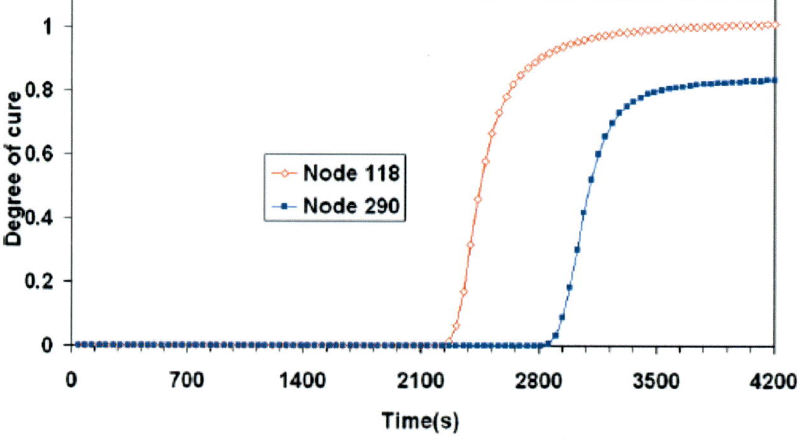

Figure 22. Variation of the degree of cure with time for the nodes at level 3.

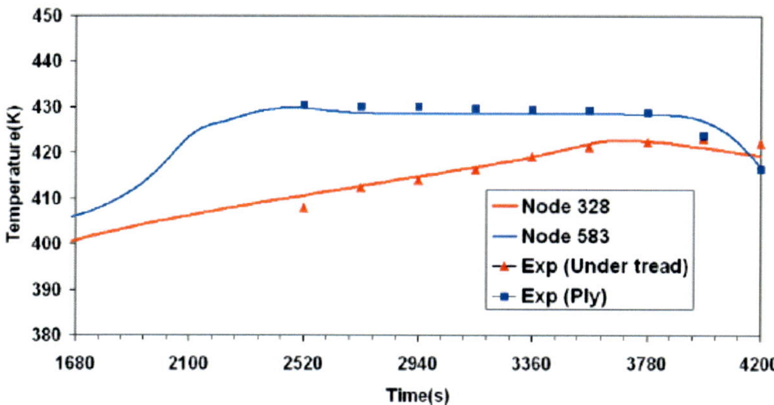

Figure 23. Comparison between predicted temperature profiles with experimentally measured data.

10. Conclusion

Optimization of the curing process is a very important stage during the technology development of the rubber articles manufacturing. This is mainly because of complex thermal behavior of rubber compounds as well as kinetics of the curing reaction that takes place inside the rubber. Owing to these features, finding of the optimum time and temperature are still carried out using traditional procedures which are based on the trial and error. Numerical simulation of the rubber curing process can significantly replace these expensive and time consuming experiments by computer runs to speed up the design step and also reduces the high expenditure which are normally required in a technology development cycle. It is tried in this monograph to cover both theoretically and practically all aspects of the numerical simulation of the rubber curing process. The mathematical formulations in conjunction with different computational strategies are studied. In order to show the applicability of the developed methods, the cure simulation of two rubber articles (a metal reinforced rubber slab and a truck tire) are presented. It is shown that the results are not only well justified but also have very good agreement with experimentally measured data that confirm the accuracy and validity of the computer simulation.

11. List Of Symbols

a,b	parameters for dependency of thermal conductivity to temperature, Eq. (6)
C	heat capacity
C_m	heat capacity of partially cured rubber compound
C_u	heat capacity of fully uncured rubber compound
C_c	heat capacity of fully cured rubber compound
c_1, c_2	parameters defining dependency of heat capacity to temperature, Eq. (4)
D	thermal diffusivity
E	activation energy
F_ε	emissivity function, Eq. (39)

F_G	geometric view factor, Eq. (39)
$\{F^{(e)}\}$	load vector
h	convection heat transfer coefficient
$[H^{(e)}]$	stiffness matrix for convection boundary conditions Eq. (36)
k	thermal conductivity
k	rate constant
k_1,k_2	parameters in kinetics model defined by Eq. (14)
k_0	parameter in the rate constant Eq. (12)
$[K^{(e)}]$	stiffness matrix
m	order of reaction
$[M^{(e)}]$	mass matrix
n	order of reaction
n_r,n_z	components of the unit vector normal to the boundary (axisymmetric)
n_x,n_y,n_z	components of the unit vector normal to the boundary (global Cartesian)
$\{P^{(e)}\}$	load vector for convection boundary condition Eq. (36)
q	heat flux
$\dot{Q}$	rate of heat generation per unit volume (heat source)
Q_t	heat released up to time t
Q_∞	total heat of reaction
r,z	axisymmetric coordinates
R	gas constant
t	time
t_i	induction time
$\bar{t}$	cumulative non-isothermal induction time (dimensionless)
T	temperature
T_m	average temperature, Eq. (20)
T_∞	ambient temperature
t_0,T_0	parameters defining dependency of induction time to temperature, Eq. (10)
x,y,z	global Cartesian coordinates
$\{T^{(e)}\}$	temperature (solution) vector (vector of unknowns)
$\{\dot{T}^{(e)}\}$	time derivative of the temperature vector

GREEK LETTERS

α	state of cure (degree of cure)
Δt	time step
$\Gamma(0)$	initial torque value, Eq. (17)
$\Gamma(\infty)$	final torque value, Eq. (17)

$\Gamma(t)$	torque profile, Eq. (17)
$\Gamma^{(e)}$	boundary of the element
θ	parameter used in implicit time stepping method
ρ	density
ρ_m	density of partially cured rubber compound
ρ_u	density of fully uncured rubber compound
ρ_c	density of fully cured rubber compound
σ	Stefan-Boltzmann constant
ψ_i	weight function
ψ_j	interpolation (shape) function

12. REFERENCES

[1] Gehman, S. D. *Rubber Chem. Technol. 1967*, 40, 36-99.
[2] Ambelang, J. C.; Prentice, G. A. *Rubber Chem. Technol. 1972*, 45, 1195-1201.
[3] Prentice G. A.; Williams, M. C. *Rubber Chem. Technol. 1980,* 53, 1023-1031.
[4] Schlanger, H. P. *Rubber Chem. Technol. 1983,* 56, 304-321.
[5] Isayev, A. I.; Sobhanie, M.; Deng, J. S. *Rubber Chem. Technol. 1988*, 61, 906-937.
[6] Deng, J. S.; Isayev, A. I. *Rubber Chem. Technol. 1991*, 64, 296-324.
[7] Toth, W. J.; Chang, J. P.; Zanichelli, C. *Tire Sci. Technol. 1991*, 19, 178-212.
[8] Marzocca, A. J. *Polymer 1991*, 32, 1456-1460.
[9] Han, I. S.; Chung, C. B.; Kim, J. H.; Kim, S. J.; Chung, H. C.; Cho, C. T.; Oh, S. C. *Tire Sci. Technol. 1996*, 24, 50-76.
[10] Han, I. S.; Chung, C. B.; Jeong, H. G.; Kang, S. J.; Kim, S. J.; Jung, H. C. *J. Appl. Polym. Sci. 1999*, 74, 2063-2071.
[11] Nazockdast, H.; Goharpey, F.; Dabir, B. *J. Appl. Polym. Sci. 2000*, 77, 2448-2454.
[12] Ghoreishy, M. H. R. Tire Technology International Annual Review 2001, 74-77.
[13] Tong, J.; Yan, X. *J. Reinf. Plast. Comp. 2003*, 22, 983-1002.
[14] Yan , X. *Polym. J. 2007*, 39, 1001-1010.
[15] Ghoreishy, M. H. R.; Naderi, G. *J. Elastom. Plast. 2005*, 37, 37-53.
[16] Ghoreishy, M. H. R.; Naderi, G. *Iran. Polym. J. 2005*, 14, 735-743.
[17] Ghoreishy, M. H. R. Tire Technology International Annual Review 2006, 84-87.
[18] Reddy, J. N.; Gartling, D. K. The Finite Element Method in Heat transfer and Fluid Dynamics; CRC Press: London, 2001.
[19] Huebner, K. H.; Dewhirst, D. L.; Smith, D. E.; Byrom, T.G. *The Finite Element Method for Engineers;* John Wiley and Sons: NY, 2001.
[20] Liu, G. R.; Quek, S. S. *The Finite Element Method A Practical Course;* Butterworth-Heinemann: Oxford, 2003.
[21] Isayev, A. I.; Deng, J. S. *Rubber Chem. Technol. 1988,* 61, 340-361.
[22] Kamal, M. R.; Ryan, M. E. *Polym. Eng. Sci. 1980*, 20, 859-867.
[23] Kamal, M. R.; Sourour, S. *Polym. Eng. Sci. 1973*, 13, 59-64.

[24] Chan, T. W.; Shyu, G. D.; Isayev, A. I. *Rubber Chem. Technol. 1993*, 66, 849-864.
[25] Chan, T. W.; Shyu, G. D.; Isayev, A. I. *Rubber Chem. Technol. 1994*, 67, 314-328.
[26] ABAQUS, Version 6.7, 2007.
[27] COSMOS/M, Version 2.85, 2003.

In: Computational Materials
Editors: Wilhelm U. Oster

ISBN 978-1-60456-896-7

Chapter 12

RELAXATION ELEMENT METHOD IN MECHANICS OF DEFORMED SOLID

***Ye.Ye. Deryugin*[1]*, G. Lasko*[1,2] *and S. Schmauder*[2]**

[1]Institute of Strength Physics and Materials Science, Siberian Branch of Russian Academy of Sciences,Tomsk, Russia

[2]Institute of Materials Testing, Materials Science and Strength of Materials (IMWF), University of Stuttgart, Stuttgart, Germany

ABSTRACT

In this chapter an original method of calculation and modelling of plastic strain localization in a loaded solid - the relaxation element method (REM) - is represented. The fundamental property of solids: "plastic deformation is accompanied by stress relaxation in local volumes" is the basis of the method. The theoretical foundation of the method is derived from the basic equations of elasticity and continuum theory of defects. For the plane-stress state, the technique of the construction of local sites of plastic deformation at the mesoscopic scale level is introduced. Examples and results of modelling the process of plastic strain localization, accompanied by the effect of Lüders band propagation and Portevin Le Chatelier effect are presented. The difficulty of the traditional description of strain localization phenomena lies in the fact that it is not possible to formulate a universal physical law of the connection between plastic deformation and stresses in the solid due to the relaxation nature of plastic deformation. Application of the REM allows to overcome this difficulty. By methods of mechanics of deformed solid it was shown that stress relaxation inside the structural element on a definite value is unambiguously connected with the change of its external shape, which according to the physical sense is not-elastic, but plastic. As a result, the stress field changes accordingly outside the considered structural element and is defined unambiguously as well. So, the structural element, having undergone plastic deformation becomes a relaxation element with its own fields of internal stresses. This stress field can be connected with the specific value of plastic deformation, ensuring corresponding changes of the external shape of the structural element. Applications of REs as mesoscopic defects, defining the relation of plastic deformation inside the RE with the stresses outside this element allow to simulate the processes of strain localization and to obtain the dependencies of the flow stress from the consequent involvement of separate structural elements of the solid into plastic deformation. The developed REM model for the plastic deformation localization operates on the principles of cellular automata. The calculation area is divided into a number of cells, playing the role of structural element (for example, grains in polycrystals). In the model each element of the simulated medium possesses the ability to switch its state by discrete steps of plastic deformation. Thus, the elements of the medium are able to increase the degree of plastic deformation one after the other and as a stress concentrator to effect the change of the stress field in the whole volume of the solid. The involvement of a structural elements into plastic deformation occurs when under the influence of external applied stress the shear stress at the center of this element achieves a critical value (e.g., according to the Mises-Tresca criterion). The interaction of the fields of internal stresses from different relaxation elements, which have undergone plastic deformation takes place automatically.

1. INTRODUCTION

The development of plastic deformation in the solid is known to proceed inhomogeneously in space and irregular in time. This property defines the process of plastic deformation localization, developing in the course of loading of structurally inhomogeneous materials. The evolution of non-homogeneous distribution of plastic deformation is realized under the influence of various stress concentrators, caused by inhomogeneity of the initial structure of the material and changing of stress state of the loaded system under prescribed boundary conditions. In the course of plastic deformation of local volumes, the continuous changing of the fields of internal stresses in the whole volume of the solid takes place. Technological and strength properties of the material depend on the character of the processes of plastic strain localization and material degradation. In this connection there is an actual problem of the description and simulation of the process of plastic strain localization and changing of the stress-strain state of a structurally-inhomogeneous medium with the site of plastic deformation under the different bondary conditions of loading. The description and simulation of these effects demand account for the multilevel character of the development of the processes in the deformed system, where the surface layer and internal interfaces effect the process of plastic deformation development and fracture [1-3]. The evolution of an inhomogeneous distribution of plastic deformation is realized under the influence of different stress concentrators (SC), and first of all under the influence of SC at the free surface of the solid. The pattern of macrolocalization of plastic deformation inherits the character of the distribution in the thin near-surface layer.

On the macrolevel one can select three types of macrolocalization of plastic deformation: Lüders band propagation, Portevin Le Chatelier Effect (PLC) and neck formation at the stage of prefracture [4-9]. For the description and simulation of the effect of intermittent flow relaxation element method (REM) developed by authors is most suitable and perspective. [10-13].

The change in stress field in the solid under loading as a result of decreasing in elastic energy in the local volume, undergone plastic deformation, lies on the basis of the method. For this purpose the notion of relaxation tensor, characterizing changing of the field of elastic stresses in the given local volume of solid as a result of its plastic deformation is used. The above approach allows to solve effectively two problems of a deformed solid (DS):

1. The description of the stress-strain state of the solid with the sites of different geometrical shape and with different distributions of the plastic deformation. Results are obtained using the known technique of continuum mechanics of DS and are represented in the form of analytical expressions for the components of the tensor of plastic deformation and stresses.
2. Simulation of the sequence of the involvement of separate structural elements into plastic deformation. Stress relaxation in local volumes changes the stress state in the whole volume of the solid. Thus, structural elements, having undergone plastic deformation in the surrounding of the elastically-deformed matrix, plays the role of the defect on a mesoscopic scale with its own field of internal stresses.

The models, developed on the basis of REM operate on the principles of cellular automata. The calculational field is devided into a number of cells, playing the role of the elements of structure (for example, grains in polycrystals). Each cell posesses the ability to switch its state by a discrete jump in stress (or plastic deformation), prescribed by a definite relaxation element, which we put at the center of a cell. A decrease in elastic energy in the cell is accompanied by an increase in stresses in the vicinity of the site of localized plastic deformation. Thus, each element of the medium is able to periodically increase the degree of plastic deformation and as the stress concentrators effect the stress state of the nearest neighbours. The state of all elements of structure changes simulataneously in a definite time interval.

The involvement of structural elements into plastic deformation is realized by prescribing of a criterion, for example, a critical shear stress τ_{cr} at the center of the cell under the influence of an external stress σ (Mises-Tresca criterion). The interaction of the fields of internal stresse of the structural elements, undergone plastic deformation occurs automatically (on the superpositional principle).

2. Stress-Strain State of the Continuum with the Site of Plastic Deformation

2.1. Plastic Deformation and Stress Relaxation

Classical experiments on relaxation in solids are realized in the mode of uniaxial loading, ensuring the constant length $l = l_0 + \Delta l$ of the loaded specimen [14]. According to the given boundary condition, the deformation averaged over the whole volume of the specimen keeps constant with time and is equal to $\varepsilon_0 = \Delta l / l_0$, where l_0 is the initial length of the specimen. The experiment shows, that the mode of constant length of the specimen after the loading above the yield stress is realized under decreasing external applied stress (Fig. 1). The physical reason of stress relaxation is the plastic deformation of the specimen. The contribution $\varepsilon^p(t)$ to the total strain of the specimen with time follows the condition:

$$\varepsilon = \varepsilon^{\mathrm{p}}(t) + \varepsilon^{\mathrm{e}}(t) = \sigma_0 / \mathrm{E}, \qquad (2.1)$$

where σ_0 – is the stress, corresponding to the elastic deformation up to the value ε_0, E – is Young's modulus. The contribution of the elastic deformation at an arbitrary instant of time t is equal to $\varepsilon^e = \sigma(t)/E$. By substituting this value into equation (2.1), we obtain the following dependence between the value of plastic $\varepsilon^p(t)$ and elastic $\varepsilon^e(t)$ deformation:

$$\varepsilon^{p}(t) = [\sigma_0 - \sigma(t)]/E = \Delta\sigma(t)/E, \tag{2.2}$$

where $\Delta\sigma(t) = \sigma_0 - \sigma(t)$ – is the value of stress drop.

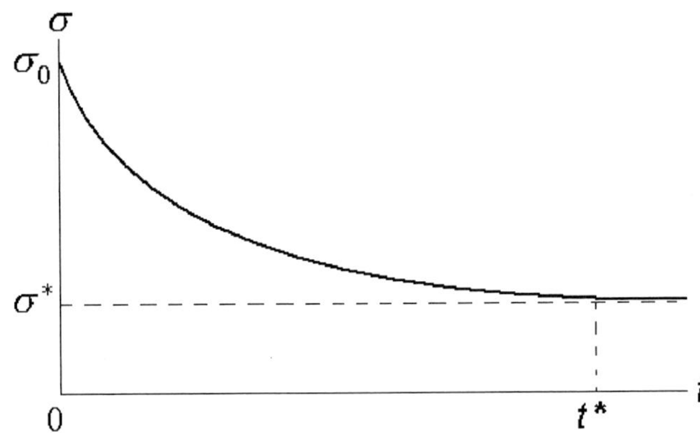

Figure 1. Changing of the external stress with time in the experiments on stress relaxation

The interrelation between the elastic and plastic deformation in experiments on stress relaxation is depicted in Fig. 2. The point A at the straight line defines the values of elastic and plastic deformation at time t, corresponding to a stress relaxation of the value $\Delta\sigma(t)$. At that time a dissipation of elastic energy takes place in a value , where

$U = 0.5\Delta\sigma\varepsilon^{p} l S = 0.5(\varepsilon^{p})^2 VE$,. l, S and V – the working length, cross-section and the volume of the specimen, respectively.

It should be noted that the value of plastic deformation is defined not by the stress $\sigma_0 - \Delta\sigma(t)$, which exists in a specimen in the considered instant of time, but by the stress $\Delta\sigma(t)$, which existed and disappeared as a result of relaxation.

In the mode of constant external load $\sigma =$ const, i.e. in experiments on creep [15], the continuous increase in the length of the speciment occurs due to the plastic deformation. In this case the stress drop takes place. The constancy of load is renewed constantly due to the fact that the dissipation of elastic energy is compensated by the work of external stress $\Delta A(t) = \sigma\varepsilon^{p}(t) l S = \sigma\varepsilon^{p}(t)V$.

In the mode of continuous loading [16] due to the stress relaxation in the local volumes of material under its plastic deformation the deviation of $\sigma - \varepsilon$ from a straight line of elastic loading takes place.

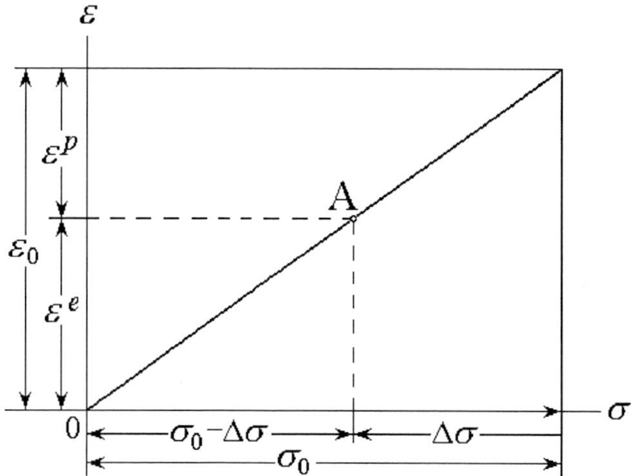

Figure 2. Interrelation between elastic and plastic deformation in experiments on stress relaxation

The examples represented prove the fact of connection between plastic deformation and stress relaxation. However, a one-dimensional approximation doesn't give an idea about the stress-strain state and on the distribution of plastic deformation in a deformed solid, since plastic deformation develops inhomogeneously in space and irregular in time.

In this connection, there is an actual problem of the description of the stress-strain state of the solid with the sites of plastic deformation. Under the site of plastic deformation we understand a stationary field of plastic deformation, localized in a separate volume of the solid. The correct statement of the problem requires to maintain a connection between the value of plastic deformation, localized in a limited region of the solid with the value of stress relaxation in it. Apparently, this connection should be represented in tensor form.

In traditional theories of plasticity and fracture, as a rule, the question about the shape of the plastic zone is considered under either, one or another limitation, imposed on the stress fields. At that time, the flow condition is prescribed and within the zone of plastic deformation the dependence of the degree of plastic deformation on the stresses is formulated, taking the macroscopic diagrams of tested specimens into account. To check the validity of such suppositions in experiments seems to be impossible. Due to this reason, the mechanics of a deformed solid up to now is phenomenological science. For the clarification of the physical mechanisms, causing changing in the mechanical characteristics, it is necessary to elaborate the technique of the constraction of the models of plasticity and fracture on the basis of the solution of an inverse problem. The Relaxation Element Method (REM) serves for such a purpose.

The difficulty in the description of the localization phenomenon lies in the fact that there is no universal physical law of the connection between plastic deformation and stresses in a solid due to the relaxation nature of plastic deformation. Application of the REM allows one to overcome such a difficulty. The solid under loading is in a thermodynamical unstable state and tries to get rid of the elastic deformation energy, stored in it. This is realized by plastic deformation, by changing in the shape of the solid, additional to the elastic formchanging. The involvement of any structural element in the local volume of solid into plastic deformation is accompanied by a stress decrease in this volume. Apparently, quantitatively this decrease should be characterized by a definite tensor- tensor of relaxation. The relaxation tensor defines the changing in the components of the stress tensor inside the structural element after its plastic deformation. If instead of the tensor of plastic deformation in the volume of the structural element one prescribes the tensor of relaxation, then the stress state inside the structural element becomes definite. In such a case the stress field beyond this structural element is defined unambiguously, since the relaxation tensor unambiguously defines the point displacements at the boundary of the structural element. At that time the formchanging of the solid as a whole changes, cause an elastic deformation of the solid as well as by contribution of plastic deformation of the considered element of microstructure. The task is reduced to the difinition of the stress field beyond the structural element, undergone plastic deformation. Thus, the structural element, undergone plastic deformation is the defect which should be considered as relaxation element of the corresponding scale level, which is characterized by a definite field of internal stresses.

The advantage of the described approach lies in the fact that the relaxation tensor unambiguously defines the stress-strain state of the solid with its structural element, undergone plastic deformation, in spite of the fact that a decrease in stresses, characterized by this relaxation tensor can be caused by different ditributions of plastic deformation inside the structural element. For the definiteness the additional displacements of the boundary of the structural element could be connected with the displacement fields inside the structural element, being adequate to such as under elastic deformation, but not conected with stresses. Corresponding derivatives define a specific field of plastic deformation. Apparently, that the field of plastic deformation, defined in such a manner will be unambiguously connected with the stress fields in the whole volume of the solid. By choosing the solution for the field of plastic deformation from the number of solutions by mentioned method, we remove umambiguity of the stress strain state of a solid.

2.2. Constitutive Equations of the Theory of Elasticity and Continuum Theory of Defects

The technque of the description of the stress-strain state of the solid is well elaborated in the continuum theory of defects. Within the framework of these theories, the mechanical state of solid with the local site of plastic deformation obeys the known constitutive equations [17-22].

In an isotropic medium the stresses σ_{ij} are interrelated with elastic deformation via Hooke's Law ε_{ij}^{e} [10]:

$$\sigma_{ij} = \frac{\mathrm{E}}{1+\nu}\left(\varepsilon_{ij}^{\mathrm{e}} + \frac{\nu}{1-2\nu}\delta_{ij}\varepsilon_{kk}^{\mathrm{e}}\right), \tag{2.3}$$

where E - is Young's modulus, ν - is Poisson's ratio, and δ_{ij} - is Kroneker's delta, which is equal 1, when $i = j$ and equal to 0, if $i \neq j$.

Continuum models are based on the assumption of the strict fulfillment of the compatibility condition for the displacement field in an arbitrary point of the deformed solid [17, 19, 21]. This condition is written down in the form of differential equations for the components of the tensor of deformation, connected with the displacement field by the known relations:

$$\varepsilon_x = \partial u_x / \partial x, \;\; \varepsilon_y = \partial u_y / \partial y, \;\; \gamma_{xy} = \partial u_x / \partial y + \partial u_y / \partial x.$$

For 2-Dimensional problems the compatibility condition is represented in the form of equation

$$\frac{\partial^2 \varepsilon_y}{\partial x^2} + \frac{\partial^2 \varepsilon_x}{\partial y^2} = \frac{\partial^2 \gamma_{xy}}{\partial x \partial y}. \tag{2.4}$$

In continuum representation the tensor of total deformation ε_{ij} is represented as a sum of the tensor of elastic and plastic deformation ε^{p}_{ij} as a function of t [20-22]:

$$\varepsilon_{ij} = \varepsilon^{p}_{ij}(t) + \varepsilon^{e}_{ij}(t). \tag{2.5}$$

The presence of plastic deformation ε^{p}_{ij} brakes the compatibility condition for elastic deformation ε^{e}_{ij}. Joint fulfillment of the conditions (2.4) and (2.5) means, that the incompatibility of the tensor of plastic deformations ε^{p}_{ij} is fully compensated by the incompatibility of the tensor of elastic deformations ε^{e}_{ij} and vice versa.

The equation of force euilibrium in the volume of a solid in its general form is represented in the form

$$\sigma_{ij,j} + f_i = 0. \tag{2.6}$$

where the f_i- function represents the volume forces. In the case of the presence of the site of plastic deformation, the function f_i represents the volume forces, caused by the presence of the field of internal stresses. According to the postulates of continuum theory of defects the components of the given volume forces are represented in the following manner [22]:

$$f_i = -C_{ijkl}\varepsilon^{p}_{kl,j}. \tag{2.7}$$

Here C_{ijkl} – are elastic constants of the material. The «minus» sign is connected with the dissipation of elastic energy. In such a manner the relaxation nature of plastic deformation reveals itself. It is connected with disappeared stresses. That means if the elastic characteristics of a material are known, than in principle the field of plastic deformation is defined. There is a special technique [23, 24] of the solutionof of equations of the type (2.6).

There are enough equations for an unambiguous definition of the connection between the local field of plastic deformations with the stresses in the whole volume of a solid. That means, if the field of plastic deformation is prescribed in the local volume of the solid, than in principle by continuum theory of defects one can define the stress/strain state in the whole volume of the solid. REM simplifies [10-13] the solution for such a problem. Given below are the specific examples.

2.3. Continuum Model of the Site of Plastic Deformation

Under the influence of the external applied stress σ in the continuum at elastic deformation in general the inhomogeneous stress field arises σ_{ij}, caused by the inhomogeneity of structural constituents of solid. Plastic deformations of a separate structural element of the medium (for example of the grain of polycrystal) change the prescribed stress field. The site of plastic deformation is the source of perturbations of the stress field as a result of irreversible plastic deformation, being realized in the local (limited) volume of the solid. Structural elements of the medium are devided by the interfaces of grains and phases. Therefore, the model of the site of plastic deformation in continuum supposes the presence of the boundary beyond which the material in an instant of time is deformed elastically. Therefore, a model of the site of plastic deformation in the continuum presumes the presence of the boundary beyond which the material in the given instant of time is deformed elastically.

Let us consider an example of the isotropic medium with the site of plastic deformation which in Fig. 3 for definiteness is depicted in the form of an ellipse. Let the change of the stress field in the site of plastic deformation due to its plastic deformation is characterized by the tensor of plastic deformation $\Delta\sigma_{ij}$. By prescribing the value $\Delta\sigma_{ij}$ we umambiguously define the stress state not only in the site of plastic deformation (it is equal there to $\sigma_{ij} - \Delta\sigma_{ij}$ by definition) but change the field of stresses $\Delta\sigma^*_{ij}$ beyond its boundary.

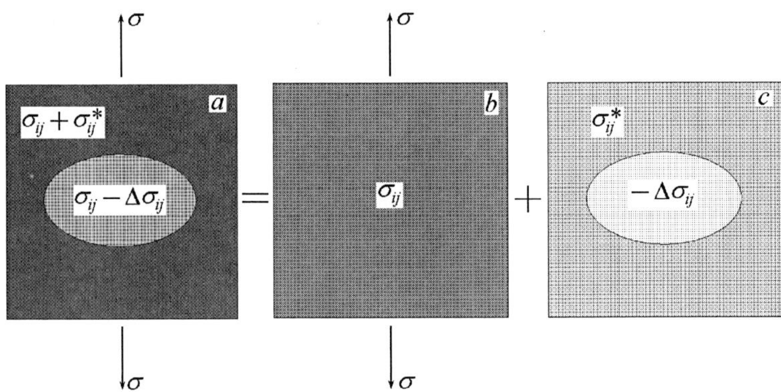

Figure 3. The scheme of the site of plastic deformation

Irreversibility of plastic deformation means, that relief of external load is accompanied only by elastic deformation of the material. Therefore, an elastic solution can be represented in the form of the superposition of two simpler solutions, one of which is simply the stress field σ_{ij} in the solid before its plastic deformation (Fig.4*b*). Without the stress field σ_{ij}, which existed under the loading in the solid before its plastic deformation of the selected region, the volume outside the site can be considered as a biconnected domain, at the external boundary of which there are no stresses, but at the boundary of the site of plastic deformation the stresses operate, prescribed by the tensor of relaxation $\Delta\sigma_{ij}$ (Fig .4*c*). In such a case a stress field $\Delta\sigma^*_{ij}$ under prescribed boundary conditions is defined by the solution of the standard problem of the linear theory of elasticity [17-19]. In such a way, prescribing the $\Delta\sigma_{ij}$ value unambiguously defines the change of the stress state inside and outside the site of plastic deformation (PD) independently on the field σ_{ij}, which existed in the solid without the site of plastic deformation. Defined in such a manner, $\Delta\sigma^*_{ij}$ inhomogeneous stress field in its essence is the field of internal stresses of defects in the solid, which is the element of structure, undergone plastic deformation.

The analysis performed shows that stress relaxation in a local volume of the solid is accompanied by an arising of the field of internal stresses $\Delta\sigma^*_{ij}$ outside the site of plastic deformation. The components of the stress tensor and their derivatives possess breaks.

In the presence of the site of plastic deformation the compatibility condition (2.3) obeys everywhere beyound the site, i.e. the material there is deformed elastically.

When approaching the boundary of the site from the external site, one can define its full displacements u_i, according to Hooke's law. Hence, the relaxation tensor $\Delta\sigma_{ij}$ unambiguously defines not only the deformed state of the system, but the changing of the geometrical shape of the site of plastic deformation. For the fullfilment of the compatibility condition at the boundary of the site of plastic deformation, it is necessary that the same displacements of the boundary are ensured by approaching it from the internal site. The contribution of elastic deformation to the given displacements, caused by the stress field in the site, at that time is definitely determined by Hooke's Law. Hence, the additional displacements of the boundary $u^p_i = u_i - u^e_i$, unambiguously characterize the contribution of plastic deformation in displacements, u_i.

However, there exist a lot of solutions for the tensor of plastic deformation, ensuring given plastic displacements u^p_i of the boundary of the site of plastic deformation. At that time, the choice doesn't influence the stress state and the change in the shape of the deformed system as a whole. In continuum theory of defects [20-22], unambiguity is ensured due to the additional condition of force equilibrium in the volume of solid (2.5). In the site of plastic deformation formally the equation of equilibrium of stress fields is written in the form of equations

$$\sigma_{ij,j} + f_i = \sigma_{ij,j} - C_{ijkl}\varepsilon^{\mathrm{p}}_{kl,j} = \sigma_{ij,j} - \Delta\sigma_{ij,j} = 0. \tag{2.8}$$

The comparison of equation (2.7) with equation (2.5) shows, that the tensor of relaxation defines the action of volume forces, i.e. $\Delta\sigma_{ij,j} = C_{ijkl}\varepsilon^{\mathrm{p}}_{kl,j}$. Plastic displacements of the boundary of the site $u^{\mathrm{p}}_i = u_i - u^{\mathrm{e}}_i$ are the boundary conditions, under which the solution of the equations unambiguously defines the field of plastic deformation in the site considered. The view of equations (2.6) testifies to the fact that from a number of solutions one postulates, which is identical to the displacement field under its elastic deformation u^p_i. The difference lies only in the fact that this additional displacements inside the site of plastic deformation should not be connected with stresses. Under such a condition with accounting of the law of elastic deformation for the isotropic medium, the tensor of relaxation is connected with the field of plastic deformation in the following manner:

$$\Delta\sigma_{ij} = \frac{\mathrm{E}}{1+\nu}\left(\varepsilon^{\mathrm{p}}_{ij} + \frac{\nu}{1-2\nu}\delta_{ij}\varepsilon^{\mathrm{p}}_{kk}\right). \tag{2.9}$$

Transformation of equation (2.9) gives the equation of the tensor of plastic deformations:

$$\varepsilon^{\mathrm{p}}_{ij} = \frac{1}{\mathrm{E}}\left[(1+\nu)\Delta\sigma_{ij} + \nu\delta_{ij}\Delta\sigma_{kk}\right] \tag{2.10}$$

Here ν - is the Poisson's ratio.

In such a way, by the method of linear theory of elasticity and continuum theory of defects a specific connection between the field of plastic deformation (or tensor of relaxation) with the field of internal stresses is maintained.

2.4. Examples of the Sites of Plastic Deformation

Apparently, that the stress state of the continuum with the site of plastic deformation depends on the fact which components of the stress tensor change as a result of relaxation, i.e. from the components of the tensor of relaxation. However, in any case, a decrease in elastic energy in the site of plastic deformation will be accompanied by an increase in a stresses beyound it.

As a simple example let us consider the stress state of the plane with the site of plastic deformations of round shape under the operation of external applied tensile stresses along the coordinate axis 0y. On a mesoscopic level in such an approximation can one imitate the influence of a separate grain of the polycrystal, undergone plastic deformation in surrounding of elastically deformed matrix?

Let us consider the case of a RE, the tensor of relaxation of which is characterized by non-zero component $\Delta\sigma_y = \Delta\sigma$ along tensile axis $0y$. That means that as a result of plastic deformation only the normal component of the tensor of stresses σ_y is relaxed, directed along the tensile axis. Lets call it tensor of relaxation of the first type. Then let us consider a RE of the second type, the tensor of relaxation of which is equal to $\Delta\sigma_{xy} = \Delta\tau$, characterizing relaxation of tangential stresses along the scheme of pure shear.

2.4.1. The Site of Plastic Deformation of the First Type

The scheme of loading is represented in Fig. 4. The general solution (*a*) can be represented in the form of superposition of two separate solutions: homogeneous stress field $\sigma - \Delta\sigma(b)$ and stress field for the plane under external tensile stress $\Delta\sigma(c)$, in the round region of which there exist no stresses. The last solution is known as Kirsch's problem for the plate with a round hole [17, 25].

In the system of coordinates at the center of the circle and the $0y$-axis along the tensile axis, beyond the site of plastic deformation, Kirsch's solution is characterized by the components

$$\Delta\sigma_y = \frac{\Delta\sigma a^2}{2r^2}\left(3\left[1-\frac{2y^2}{r^2}\right]+\left(1-\frac{8y^2x^2}{r^4}\right)\left(\frac{3a^2}{r^2}-2\right)\right)+\Delta\sigma;$$

$$\Delta\sigma_x = \frac{\Delta\sigma a^2}{2r^2}\left(\left[1-\frac{2y^2}{r^2}\right]-\left(1-\frac{8y^2x^2}{r^4}\right)\left(\frac{3a^2}{r^2}-2\right)\right); \quad (2.11)$$

$$\Delta\sigma_{xy} = \frac{\Delta\sigma a^2 yx}{r^4}\left(3-\frac{2(3a^2+4y^2)}{r^2}+\frac{12a^2y^2}{r^4}\right).$$

Here $r^2 = x^2 + y^2$.

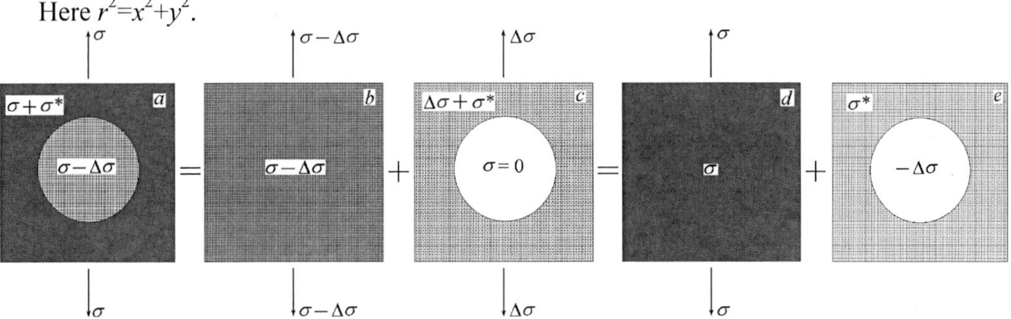

Figure 4. Representation of the solution in the form of superposition of simplier solutions:
1 – homogeneous field $\Delta\sigma-\sigma$ (*b*) and Kirsch's solution (*c*);
2 – homogeneous field $\sigma(d)$ and the field of internal stresses σ^* (*e*).

The considered example testify to the fact that Kirsch's solution has a more deeper physical sense than the simple one as a stress in the plane with round hole. This solution defines also the fields of stresses beyond the sites of plastic deformation in elastic continuum.

As it is seen, the absense of stresses in a local region doesn't mean the absense of the material. In our statement such a state is caused by plastic deformation of the material in the local region of round shape and the region without stresses can not be considered as a region where there is no material. Since by definition plastic deformation is irreversible, the material of the site under unloading will be deformed elastically as in the rest of the volume. Therefore, without the homogeneous stress field $\Delta\sigma$ (Fig. 4*d*) Kirsch's solution defines an inhomogeneous fields of internal stresses in the unloaded material σ^* (*e*), caused by the presence of the site of plastic deformation. Inside the site then the stress is equal to $\Delta\sigma_y = -\Delta\sigma$, i.e. the material is in the state of compression $-\Delta\sigma$ along the tensile axis.

Equation (2.12) without stresses $\Delta\sigma$ for the component $\Delta\sigma_y$ characterizes the field of internal stresses of the defect in the solid. Such a defect is the local volume of a round shape, as a result of relaxation the stresses drop in it by the value $\Delta\sigma$. Shown in Fig. 5 are the patterns of the spatial distributions of all components of the field of stresses σ^*. Maximum and minimum values of the component $\Delta\sigma_y$ are equal to $\Delta\sigma_{y\max} = 2\Delta\sigma$ and $\Delta\sigma_{y\min} = -\Delta\sigma$, respectively. As a result of stress relaxation inside the circle stress concentrations are observed at the boundary of the circle.

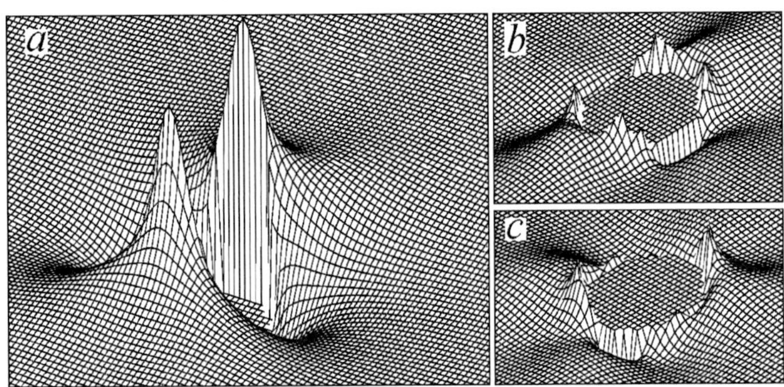

Figure 5. Distribution of the components of the field of internal stresses σ^*: (*a*) $\Delta\sigma_y$ (*b*) $\Delta\sigma_x$, (*c*). $\Delta\sigma_{xy}$

Let us characterize the field of plastic deformation in the site, corresponding to stress relaxation in the value $\Delta\sigma$ in the round region for the case of tension. This field ensures the full displacements of the points of the circle in Fig .4*a-c* under the absence of the stresses in the site of PD. According to the solution of Kirsch's problem, the components of displacements of an arbitrary point (x_0, y_0) at the circle are equal:

$$u_y = 3y_0\Delta\sigma/\mathrm{E}, \quad u_x = -x_0\Delta\sigma/\mathrm{E}. \tag{2.12}$$

These displacements define the boundary conditions for the field of plastic deformation.

It was mentioned above that according to the postulates of continuum theory of defects (2.6), plastic deformation (2.8), is chosen on the rule of elastic deformation (see section 2.2), satisfying the displacements boundary conditions (2.13) of the site. Apparantly, these boundary conditions are satisfied by the field of displacements with the following components:

$$u_y = 3y\Delta\sigma/\mathrm{E}, \quad u_x = -x\Delta\sigma/\mathrm{E}. \tag{2.13}$$

The derivatives of the components of the fields of displacements (2.13) on the corresponding coordinates define the following components of the tensor of plastic deformation:

$$\varepsilon_y^p = 3\Delta\sigma/\mathrm{E}, \quad \varepsilon_x^p = -\Delta\sigma/\mathrm{E}, \quad \varepsilon_{xy}^p = 0. \tag{2.14}$$

An increase in the external stress in $\Delta\sigma$ defines the elastic deformation of the circle which is characterized by the components:

$$\varepsilon_y^e = \Delta\sigma/\mathrm{E}, \quad \varepsilon_x^e = -\nu\Delta\sigma/\mathrm{E}, \quad \varepsilon_{xy}^e = 0, \tag{2.15a}$$

where ν - is Poisson's coefficient. Average values ν for the solids in the majority of cases lie within the limits $\nu \approx 0.33$.

After stress relaxation in the value $\Delta\sigma$ a decrease in elastic deformation of the circle takes place. This decrease is defined by the components, represented by equations (2.15a).

A comparison shows, that plastic deformations not only compensate the disappeared contribution from plastic deformations, but introduce additional contributions, which two times exceed the deformations (2.15a). At that time along 0*y*-axis the length increases, and along 0*x-axis* the width decreases. Thus, the contribution of plastic deformation of an element into the change of circle practically three times exceeds the contribution of the elastic deformation, which disappeared due to stress relaxation.

At the boundary of the site the components of the tensor of plastic deformation change in a jump from the values (2.14) to zero, but the components of the tensor of elastic deformation vice versa from 0 to the definite final values, prescribed by equations (2.10) at $r \to a$. Nevertheless, when transforming through the boundary of the site the displacements doesn't have jumps. A cooperative action of elastic and plastic deformation satisfies the condition of continuity conservation (compatibility) of material on the boundary of the site of plastic deformation.

2.4.2. The Site of Plastic Deformation of the Second Type

Let us consider another case, when stress relaxation occurs on the scheme of pure shear in the conjugate directions at an angle of 45° with respect to the tensile axis (Fig. 6).

Stress relaxation of pure shear can be realized by superposition of two separate solutions, correspondingly for the case of tensile loading along the 0*y*-axis and for the case of compression along the 0*x*-axis (Fig. 6).

The case of tensile loading is considered above and is described by the equations (2.11). Analogeously, an inhomogeneous stress field for the case of compression along 0*x*-axis is observed.

Let us consider equation (2.12), where the sign of stress $\Delta\sigma$ is assumed to be negative and perform the replacement of the coordinates from *x* into *y* and *y* into –*x*. Summing the stress components with corresponding components in equations (2.11), for the field of internal stresses of the site of plastic deformation of pure shear (i.e. without external stresses $\Delta\sigma$ along $0y$- and $-\Delta\sigma$ along $0x$- axis we obtain the final result

$$\Delta\sigma_y = \frac{\Delta\sigma a^2}{r^2}\left(2\left[1-\frac{2y^2}{r^2}\right]+\left(1-\frac{8y^2x^2}{r^4}\right)\left(\frac{3a^2}{r^2}-2\right)\right),$$

$$\Delta\sigma_x = \frac{\Delta\sigma a^2}{r^2}\left(2\left[1-\frac{2y^2}{r^2}\right]-\left(1-\frac{8y^2x^2}{r^4}\right)\left(\frac{3a^2}{r^2}-2\right)\right), \tag{2.16}$$

$$\Delta\sigma_{xy} = \frac{2\Delta\sigma a^2 xy}{r^4}\left(\frac{3a^2}{r^2}-1\right).$$

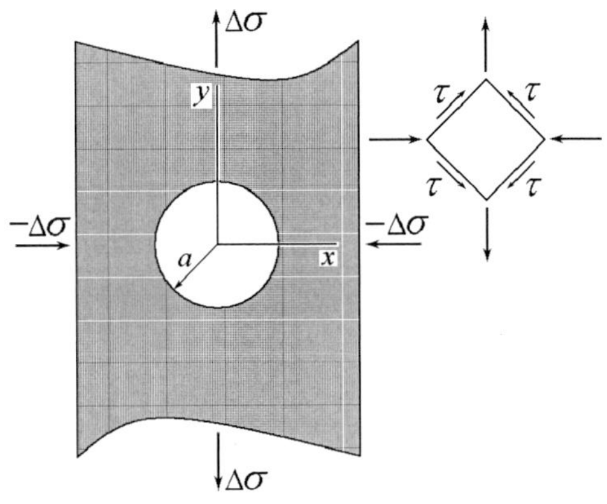

Figure 6. Boundary conditions for the calculations of stresses in the plane with a circular relaxation element of pure shear.

Shown in Fig. 7 are the patterns of spatial distributions of all components of the field of internal stresses in the system of coordinates in Fig. 6. A comparison with Fig. 5 reveals both qualitative and quantitative discrepancies. Here, the maximum and minimum value of $\Delta\sigma_y$ are

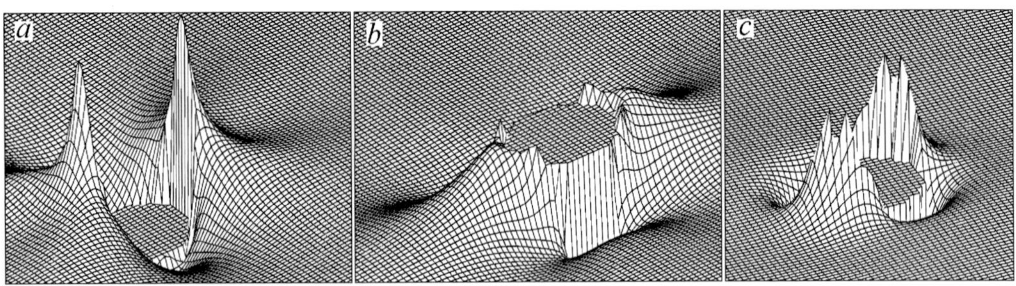

Figure 7. Distributions of the components of the field of internal stresses of a circular RE for pure shear with respect to the system of coordinates in Fig. 6: (*a*) $\Delta\sigma_y$, (*b*) $\Delta\sigma_x$, (*c*) $\Delta\sigma_{xy}$.

$\Delta\sigma_{y\max} = 3\Delta\sigma$ and $\Delta\sigma_{y\min} = -1.4\Delta\sigma$ respectively. The distribution of $\Delta\sigma_x$ component is inversed distribution $\Delta\sigma_y$, with rotation in 90°. Extreme values of the $\Delta\sigma_{xy}$ component are equal $\pm 2\Delta\sigma$.

In the present system of coordinates, rotated at an angle of 45° in an anticlockwise direction, the component $\Delta\sigma_{xy}$ is represented by the component of pure shear (Fig. 8).

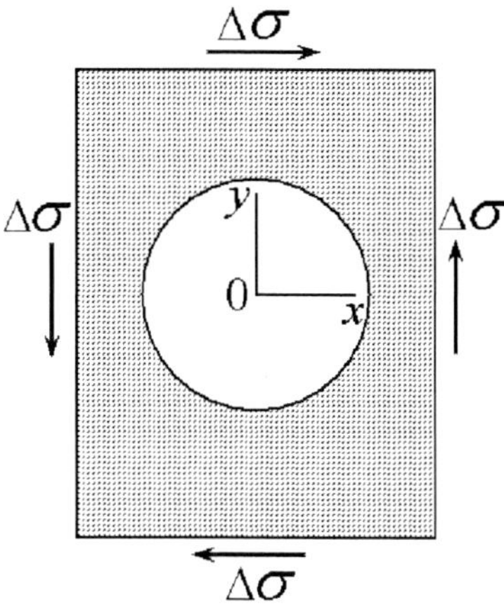

Figure 8. Boundary conditions for the relaxation of stress for the case of pure shear.

The corresponding transformations of the coordinates when rotating in 90° results in the following equations for the components of the stress tensor in the system of coordinates in Fig. 8:

$$\Delta\sigma_y = \frac{4\Delta\sigma a^2 xy}{r^4}\left(\left(\frac{3a^2}{r^2} - 2\right)\left(\frac{-2y^2}{r^2} + 1\right) - 1\right),$$

$$\Delta\sigma_x = \frac{4\Delta\sigma a^2 xy}{r^4}\left(\left(\frac{3a^2}{r^2} - 2\right)\left(\frac{2y^2}{r^2} - 1\right) - 1\right), \tag{2.17}$$

$$\Delta\sigma_{xy} = \Delta\tau = \frac{\Delta\sigma a^2}{r^2}\left(-1 + \frac{8y^2x^2}{r^4}\right)\left(\frac{3a^2}{r^2} - 2\right).$$

Corresponding distributions of $\Delta\sigma$ are shown in Fig. 9.

It is seen that beyond the site there exists an inhomogeneous stress field. The stresses undergone jumps on the round contour. Stress relaxations of pure shear in a value create a number of micro- and mesoconcentrations of stresses around itself. The components $\Delta\sigma_y$ and $\Delta\sigma_x$ in Fig. 9 testify to the fact that stress relaxations on the scheme of pure shear are accompanied by high values of stresses of overall compression and tension at the boundary of

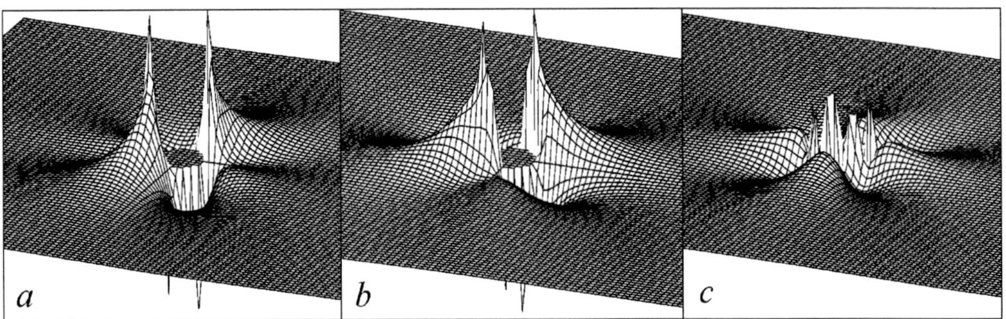

Figure 9. The field of internal stresses of a circle RE of pure shear in the system of coordinates in Fig. 8: (*a*) $\Delta\sigma_y$, (*b*) $\Delta\sigma_x$, (*c*) $\Delta\sigma_{xy}$

the RE. Elevated shear stresses, embracing the zones, commensurate with the dimension of the site (see Fig. 9*c*). Between them, at the boundary of the site, there are microconcentrators with more higher values of the stress $\Delta\tau$. However, already at the distance of 1/20 of the radius of the site, the stresses doesn't exceed the stresses of the mesoconcentrators.

The components of plastic deformation inside the RE of pure shear are caused by compression forces, and are equal to:

$$\varepsilon_y^{\mathrm{p}} = \Delta\sigma / \mathrm{E}, \quad \varepsilon_x^{\mathrm{p}} = -3\Delta\sigma / \mathrm{E}, \quad \varepsilon_{xy}^{\mathrm{p}} = 0 .$$

Together with the components (2.15) they represent the homogeneous field of plastic deformation

$$\varepsilon_y^{\mathrm{p}} = 4\Delta\sigma / \mathrm{E}, \quad \varepsilon_x^{\mathrm{p}} = -4\Delta\sigma / \mathrm{E}, \quad \varepsilon_{xy}^{\mathrm{p}} = 0 .$$

It is not easy to define, that plastic deformation of pure shear in the conjugate directions at an angle of 45° with respect to the axes x and y is calculated according to the formulae

$$\gamma^{\mathrm{p}} = \left(\varepsilon_y^{\mathrm{p}} - \varepsilon_x^{\mathrm{p}}\right)/2 = 4\Delta\sigma / \mathrm{E}.$$

Let us notice that under the condition of uniaxial tension the relaxation of shear stresses is optimally realized in direction of 45° with respect to the tensile axis only due to the relaxation of normal stresses $\Delta\sigma$ along the tensile axis. Therefore, in the given case inside the RE a field of plastic deformation will exists twice as less as it is represented in equation (2.18), i.e. :

$$\gamma^{\mathrm{p}} = \left(\varepsilon_y^{\mathrm{p}} - \varepsilon_x^{\mathrm{p}}\right)/2 = 2\Delta\sigma / \mathrm{E}. \tag{2.18}$$

The obtained equations (2.9) – (2.14) fully define a stress-strain state of the simplest site of plastic deformation of pure shear of round shape.

2.5. Construction of Relaxation Elements with Gradients of Plastic Deformation

The disadvantage of the considered distribution of the stress field of the site of plastic deformation is the fact that at the boundary of the RE the components of stresses and deformations possess jumps. And though formally the condition of compatibility for the total deformation (elastic + plastic) is obeyed, the jumps of stresses mean loose of continuity of material along the boundary. The relaxation element method allows to construct and to find relaxation elements with smooth distributions of plastic deformation. Let us demonstrate if with an example of a RE of round shape.

By definition, the site of plastic deformation is the relaxation element - a defect in the continuum medium with its own fields of internal stresses. When $\Delta\sigma$ tends to a small value $d\sigma$ we obtain an elementary defect in the continuum. The presence of such a defect doesn't change the elastic properties of the medium, i.e. it doesn't effect the solution of the boundary-value problem of linear theory of elasticity. Therefore, for the internal fields of stresses of similar defects the superpositional principle is valid. This defect can be used as the element for the construction of the various fields of localized plastic deformation. Prescribing a definite distribution of REs, one can construct the sites with the gradients of plastic deformation.

Let us consider it on the example of the family of the relaxation elements of round shape, superimposed on each other with the common center (Fig. 10). Each RE in the family is characterized by a definite dimension and the value of the elementary tensor of relaxation $d\sigma$. We will assume, that the boundaries of the neighbouring elements lie at the equal distance da from each other. A qualitative view of the supposed profile of plastic deformation (smooth, differentiated) is depicted in Fig. 10 to the right in the form of the steps, the height of which gives the degree of plastic deformation $d\varepsilon^p$ defined by the RE. Let us select a near-boundary region with the width h in the selected family of REs, in which parameters of the REs are defined in the following manner:

$$a(t'') = a - ht'', \qquad d\sigma(t'') = (\beta + 1)h\,\Delta\sigma t''^{\beta} dt/a, \quad 0 \le t'' \le 1. \tag{2.19}$$

Here t''– is a variable of integration, $\Delta\sigma$ – is the prescribed value, and $\beta + 1$ – a normalization coefficient. It is seen that with an increase in t'' the dimension of the RE evenly decreases from a to $a - t$. The value of the elementary tensor of relaxation at time $d\sigma(t'')$ is smoothly increasing. The rate of growth depends on the value β. The higher β, the quicker the value $d\sigma(t'')$ increases.

The value ε^p in the central zone should smoothly decrease to zero when moving to the center of the site. For that the change of the parameters of the RE can be defined in the following manner:

$$a(t') = (a - h)(1 - t'), \; d\sigma(t') = (\beta + 1)(1 - h/a)\,\Delta\sigma(1 - t')^{\beta} dt, \; 0 \le t' \le 1. \tag{2.20}$$

Here t' – is the variable of integration, changing within the limits $0 \le t' \le 1$, β – is the parameter, defining the changing of the value of elementary stress drop $d\sigma(t')$. According to (2.20), an increase in t' defines a smooth decrease in the radious of the RE $a(t')$ and in the value $d\sigma(t')$ as the center of the site is approached. The rate of decreasing is controlled by the β-parameter. The more β increases, the quicker decreases the value $d\sigma(t')$.

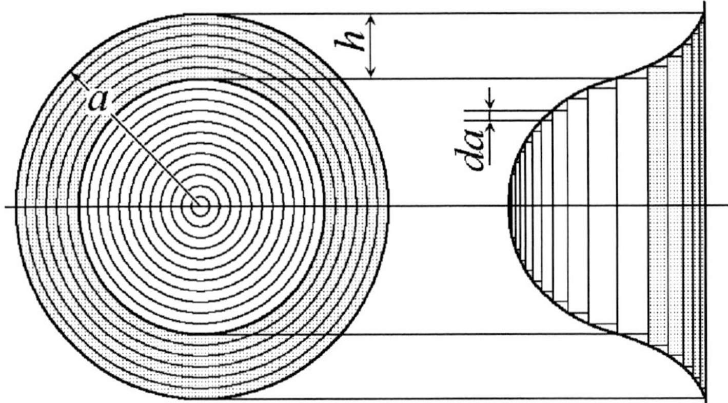

Figure 10. The scheme of the family of REs and the profile of plastic deformation in the family.

By using the parameters from eqns. (2.19) and (2.20) of the corresponding REs in equations (2.17) and (2.18), we will obtain the dependence for the elementary stress fields $d\sigma_i$ and plastic deformation $d\varepsilon_i^p$ on the variables t' and t''. Integration of such elementary fields of plastic deformations from all REs at the prescribed conditions results in the following smooth stress fields:

$$\tau(x,y)=\Delta\sigma\begin{cases}\dfrac{(\beta+1)a^2}{(\beta+3)r^2}\left[\dfrac{3(\beta+3)a^2}{(\beta+5)r^2}-2\right]\left[1-8\left(1-y^2/r_2\right)y^2/r_2\right] & if\ \ r^2\geq a^2,\\ -1+\left(\dfrac{r}{a}\right)^{\beta+1}\left\{\dfrac{\beta^2-1}{2(\beta+3)(\beta+5)}\left[1-8\left(1-y^2/r_2\right)y^2/r_2\right]+1\right\}, & if\ \ r^2\leq a^2\end{cases}. \quad (2.21)$$

and plastic deformation

$$\gamma^{\mathrm{p}}=\frac{2\Delta\sigma}{\mathrm{E}}\begin{cases}\left(1-\dfrac{h}{a}\right)\left[1-\left(\dfrac{r}{a-h}\right)^{\beta+1}\right]+\dfrac{h}{a}, & r^2\leq(a-h)^2\\ \dfrac{h}{a}\left(\dfrac{a-r}{h}\right)^{\beta+1}, & (a-h)^2\leq r^2\leq a^2.\end{cases} \quad (2.22)$$

Here r – is the distance from the center of the family of REs to the point with coordinates (x,y).

The equation (2.21) in short form is written at the value h=0.

In equation (2.22) the upper equation after the bracket defines the value of the γ^{p}-component at any point of the central zone (Fig. 9), and the lower one any point in the near-boundary zone.

Apparently, by the technique pointed out one can construct the smooth fields of plastic deformation and stresses for the RE of any type.

Shown in Fig. 11*a* is the smooth distribution of plastic deformation of pure shear (2.22) for the value β=5. The profiles of plastic deformation at different values of the β parameter are represented in Fig. 11*b*. At the distance *h* from the boundary of the site there exists a maximum gradient of plastic deformation:

$$\mathrm{grad}\varepsilon_y^p = 3\Delta\sigma(\beta+1)/\mathrm{E}a.$$

The value of the gradient of plastic deformation is seen to be defined by the β–parameter. The higher the value, the higher is the gradient of plastic deformation. For β→ ∞, we obtain the site of plastic deformation, considered above with the jump at the boundary of the site.

Shown in Fig. 12 are examples of the spatial distribution of plastic deformation γ^{p} and stress $\tau(x,y)$, for different values of β–parameter at h=0.5a. As the parameter β increases, the maximum gradients of plastic deformation and stresses increase. The case of the RE with the jump of plastic deformation (and stress), considered above is obtained as a limit case at β= ∞.

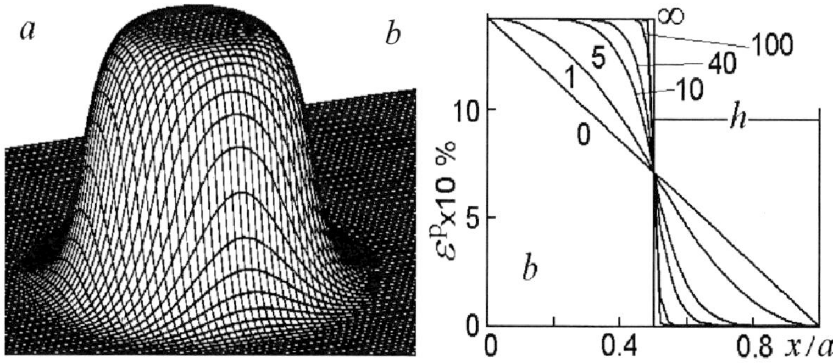

Figure 11. The distribution (*a*, β = 5) and profiles (*b*) of plastic deformation in the site at *h* = 0.5*a*.

The numbers at the curves define the values of the β –parameter.

Fig. 13 illustrates the stress field (*a*) and contour plots of pure shear (*b*) of the site of plastic deformation of pure shear for β=8 and *h*=0.2*a*. This site is seen to create elevated stresses on a mesoscopic scale at an angle of 45°with respect to the tensile axis.

Equation (2.17) was further used when simulating the jump-like propagation of the macroband of localized shear.

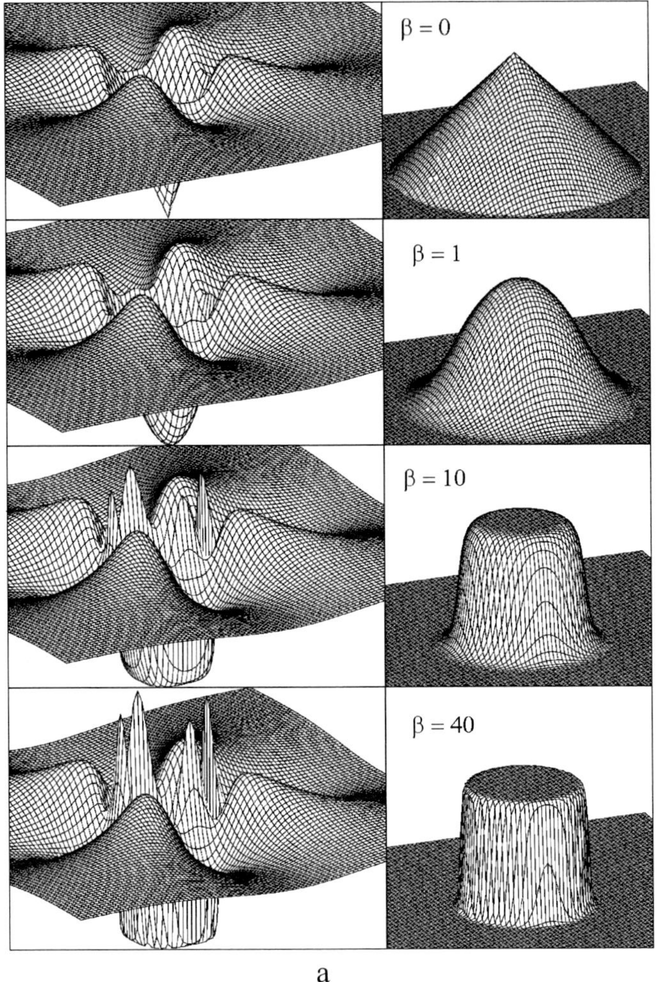

Figure 12. The spatial distribution of stress $\tau(x,y)$ (*a*) and plastic deformation of pure shear γ^{p} (*b*), for the different values of the β- parameter.

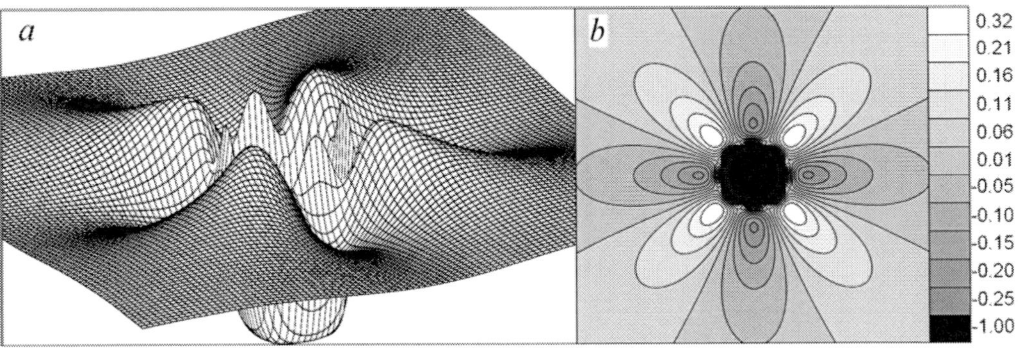

Figure 13. Stress field (*a*) and isolines of the RE of pure shear (*b*): β=8, h=0.2a.

2.6. The Stress State of the Plane with a Round Inclusion

The method, illustrated above in many cases is effective for the calculation of stresses in the continuum with the elements of structure, characterized by other elastic characteristics [27, 28]. The presence in local regions of the solid the material creates under loading with other elastic characteristics inhomogeneous stress field, being a stress concentrator of the corresponding scale. In the chapter on the superposition method the derivation of equations for all components of the stress field in the plane with a round inclusion under loading is represented. The plane-stress state is considered in the following.

The solution of the present problem is connected with the definition of boundary conditions at the contour of the inclusion. Let E_1 and ν_1 – are correspondingly Young's modulus and Poisson's ratio of the matrix and E_2 and ν_2 – correspondingly Young's modulus and Poisson's ratio of the inclusion. The scheme of loading is represented in Fig. 15*a*. The tensile stress is directed along the 0*y*- axis.

In the works of Eshelby [20, 29, 30] it was shown, that in the case of an elliptic inclusion, being oriented symmetrically with respect to the tensile axis, the stress field inside the inclusion is homogeneous with a zero σ_{xy} component. Hence, the field is homogeneous also inside the round inclusion. Let us define this stress field by the components:

$$\sigma_y^0 = k_y\sigma, \qquad \sigma_x^0 = k_x\sigma, \tag{2.23}$$

where the coefficients k_y and k_x are required to be defined.

General solutions of this boundary-value problem can be represented in the form of superposition of a homogeneous stress field (2.19) and the stress field of the plane under operation of a biaxial external load (Fig. 17*c*) with the condition of the absence of stresses in the local region of round shape. Along 0*y*-axis the tensile stress $(1-k_y)\sigma$ operates, and along 0*x* – a compression stress $-k_y\sigma$.

Deformation of the homogeneous stress field (2. 19) (Fig. 14*b*) according to Hooke's Law is characterized by the components:

$$\varepsilon_y = (k_y - k_x\nu_1)\sigma/E_1,\ \varepsilon_x = (k_x - k_y\nu_1)\sigma/E_1,\ \varepsilon_{xy} = 0. \tag{2.24}$$

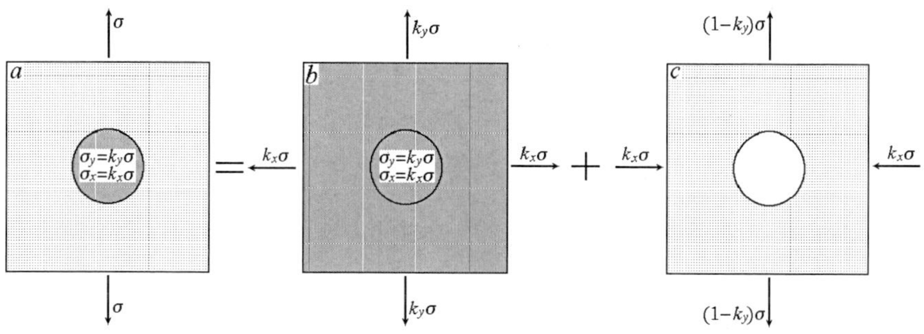

Figure 14. The schematic representation of the boundary conditions for the field of stresses of the plane with a round inclusion (*a*) in the form of superposition of the homogeneous stress field (*b*) and the stress field of the plane under the operation of two-axial external loads (*c*) and under the condition of the absence of stresses in the local region of a round shape.

Shown in Fig. 15 is the scheme of common solution for the two-axial loaded plane in the form of superposition of separate uni-axial loading, ensuring boundary conditions, represented in Fig. 14.

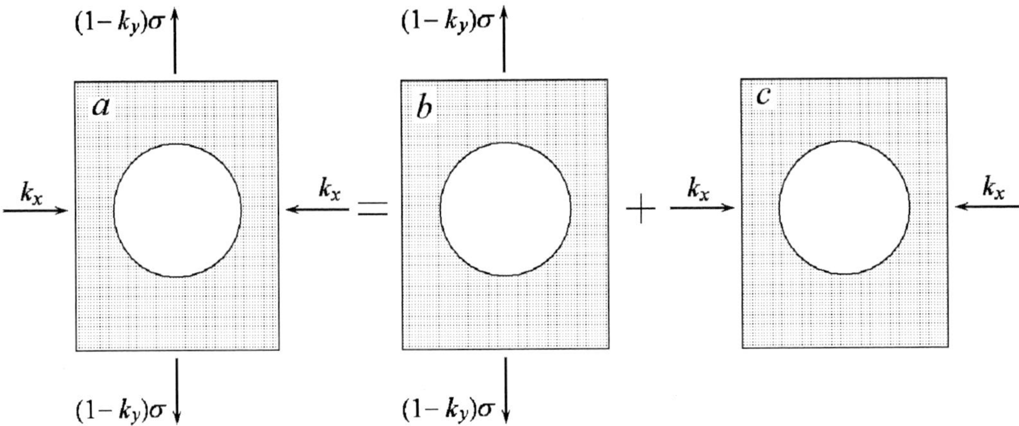

Figure 15. Schematical representation of the boundary conditions for the field of stresses of the plane under the operation of two-axial loading (a) in the form of superposition of uniaxial load (b, c), under the condition of absence of stresses in a round region.

The task about a biaxially loaded plane was already considered in section 2.4, using Kirsch's solution [17-19, 25]. In the present case these solutions results in the following results. Loading of the plane by tensile stress $(1-k_y)\sigma$ causes beyond the round contour an inhomogeneous stress field with the components

$$\sigma_y^* = \frac{(1-k_y)\sigma R^2}{2r^2}\left(3\left[1-\frac{2y^2}{r^2}\right]+\left(1-\frac{8y^2x^2}{r^4}\right)\left(\frac{3R^2}{r^2}-2\right)\right)+\sigma;$$

$$\sigma_x^* = \frac{(1-k_y)\sigma R^2}{2r^2}\left(\left[1-\frac{2y^2}{r^2}\right]-\left(1-\frac{8y^2x^2}{r^4}\right)\left(\frac{3R^2}{r^2}-2\right)\right); \qquad (2.25)$$

$$\sigma_{xy}^* = \frac{(1-k_y)\sigma R^2 yx}{r^4}\left(3-\frac{2(3R^2+4y^2)}{r^2}+\frac{12R^2y^2}{r^4}\right),$$

where R – is the radius of the inclusion, $r^2 = x^2 + y^2$ – is the distance from the center of the inclusion to the point with the coordinates (x, y).

For the case of the external compression stress, Kirsch's solution defines additional values of stress components:

$$\sigma_y^{**} = \frac{-k_x\sigma R^2}{2r^2}\left(\left[1-\frac{2y^2}{r^2}\right]+\left(1-\frac{8y^2x^2}{r^4}\right)\left(\frac{3R^2}{r^2}-2\right)\right);$$

$$\sigma_x^{**} = \frac{-k_x \sigma R^2}{2r^2}\left(\left(1-\frac{8y^2x^2}{r^4}\right)\left(\frac{3R^2}{r^2}-2\right)-3\left[1-\frac{2y^2}{r^2}\right]\right); \tag{2.26}$$

$$\sigma_{xy}^{**} = \frac{-k_x \sigma R^2 yx}{r^4}\left[5-\frac{2(3R^2+4y^2)}{r^2}+\frac{12R^2y^2}{r^4}\right],$$

The superposition of the solutions (2.25) and (2.26) together with the homogeneous stress field (2.23) (σ_y^0, σ_x^0) defines the actual stress field beyond the inclusion:

$$\sigma_y = \sigma_y^* + \sigma_y^{**} + \sigma_y^0,\ \sigma_x = \sigma_x^* + \sigma_x^{**} + \sigma_x^0 \text{ and } \sigma_x = \sigma_x^* + \sigma_x^{**} + \sigma_x^0. \tag{2.27}$$

Displacements at an arbitrary point (x_0, y_0) at the boundary of the inclusion, corresponding to the boundary conditions in Figs .15*b* and 15*c*, are defined by the equations:

$$u_y^*(x_0, y_0) = 3(1-k_y)y_0\sigma/E_1, \quad u_x^*(x_0, y_0) = -(1-k_y)x_0\sigma/E_1 \tag{2.28}$$

$$u_y^{**}(x_0, y_0) = k_x y_0 \sigma/E_1, \quad u_x^{**}(x_0, y_0) = -3k_x x_0 \sigma/E_1.$$

The components of additional displacements u_y^0, u_x^0 of the arbitrary point (x_0, y_0), caused by the homogeneous stress field (2.22), are defined by a homogeneous field of deformation (2.19):

$$u_y^0(x_0, y_0) = y_0(k_y - k_x\nu_1)\sigma/E_1,\ u_x^0(x_0, y_0) = x_0(k_x - k_y\nu_1)\sigma/E_1. \tag{2.29}$$

Actual displacements of this arbitrary point (x_0, y_0) at the boundary of inclusion is defined by summation of the corresponding components in equations (2.28) and (2.29):

$$u_x(x_0, y_0) = u_x^0(x_0, y_0) + u_x^*(x_0, y_0) + (x_0, y_0);$$

$$u_y(x_0, y_0) = u_y^0(x_0, y_0) + u_y^*(x_0, y_0) + u_y^{**}(x_0, y_0). \tag{2.30}$$

It is easy to check that the homogeneous field of deformation with the components

$$\varepsilon_x = -[1+2k_x - k_y(1-\nu_1)]\sigma/E_1, \quad \varepsilon_y = [3-2k_y + k_x(1-\nu_1)]\sigma/E_1, \quad \varepsilon_{xy} = 0, \tag{2.27}$$

satisfies the condition (2.29), where there are two unknown coefficients k_y and k_x. In equations (2.30) elastic deformations of the inclusion are expressed via elastic characteristics of the plane. Hence, the solution (2.27) is unique.

On the other hand, the homogeneous stress field (2.23) in the inclusion (Fig. 14*a*) corresponds to a homogeneous deformation, which can be expressed via the elastic characteristic of the inclusion itself:

$$\varepsilon_x = (k_x - k_y\nu_2)\sigma/E_2, \quad \varepsilon_y = (k_y + k_x\nu_2)\sigma/E_2, \quad \varepsilon_{xy} = 0. \tag{2.31}$$

By equating corresponding components in equations (2.27) and (2.28), we obtain a system of two equations with two unknown coefficients k_y and k_x, which can be written in the form:

$$k_y(E_1+2E_2) - k_x[\nu_2E_1+(1-\nu_1)E_2] = 3E_2,$$

$$k_x(E_1+2E_2) - k_y[\nu_2E_1+(1-\nu_1)E_2] = -E_2.$$

The solution of the given system defines the values of the coefficients k_y and k_x:

$$k_y = \frac{E_2[(3-\nu_2)E_1 + (5+\nu_1)E_2]}{(E_1+2E_2)^2 - [\nu_2E_1 + (1-\nu_1)E_2]^2}$$

and

$$k_x = \frac{E_2[(3\nu_2-1)E_1 + (1-3\nu_1)E_2]}{(E_1+2E_2)^2 - [\nu_2E_1 + (1-\nu_1)E_2]^2}. \tag{2.32}$$

Substituting these values in equations (2.21) – (2.23) we obtain all necessary components of stresses beyond the inclusion:

$$\frac{\sigma_y}{\sigma} = 1 + \frac{(1-k_y+k_x)R^2}{2r^2}\left[3\left[1-\frac{2y^2}{r^2}\right] + \left(1-\frac{8y^2x^2}{r^4}\right)\left(\frac{3R^2}{r^2}-2\right)\right] - \frac{k_xR^2}{r^2}\left(1-\frac{2y^2}{r^2}\right);$$

$$\frac{\sigma_x}{\sigma} = \frac{(1-k_y+k_x)R^2}{2r^2}\left[\left[1-\frac{2y^2}{r^2}\right] - \left(1-\frac{8y^2x^2}{r^4}\right)\left(\frac{3R^2}{r^2}-2\right)\right] + \frac{k_xR^2}{r^2}\left(1-\frac{2y^2}{r^2}\right);$$

$$\frac{\sigma_{xy}}{\sigma} = \frac{(1-k_y-k_x)R^2yx}{r^4}\left[3 - \frac{2(3R^2+4y^2)}{r^2} + \frac{12R^2y^2}{r^4}\right] - \frac{2k_yR^2yx}{r^4}; \tag{2.33}$$

Inside the inclusion, it is apparently

$$\sigma_y/\sigma = k_y, \quad \sigma_x/\sigma = k_x \text{ and } \sigma_{xy}/\sigma = 0. \tag{2.34}$$

Shown in Fig. 16 are the distributions of the components of the stress fields for the case of Al_2O_3 (E_2= 382GPa, ν_2=0.3) inclusion in aluminium (E_1=70GPa, ν_1=0.3) under tension [31]. It is seen, that in the inclusion the tensile stress (Fig. 16*a*) exceeds 1.4 times the external stress. Along with it near the inclusion in aluminium there are zones of lower stresses. Near the boundary of the inclusion in the local zones the components σ_x and σ_{xy} are characterized by essential positive and negative values.

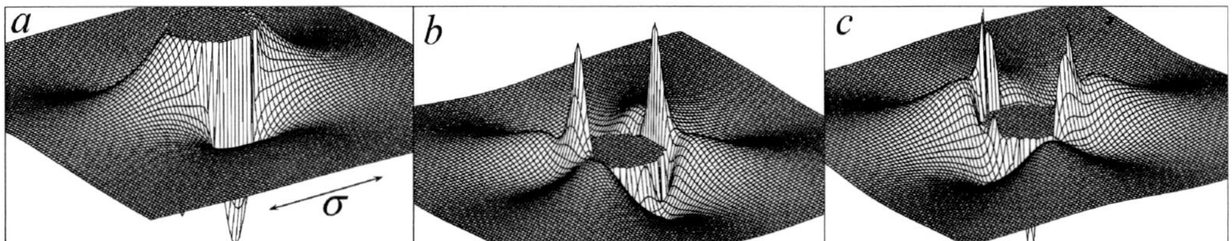

Figure 16. Distribution of the components (*a*) σ_y, (*b*) σ_x and (*c*) σ_{xy} in aluminium with a round inclusion Al_2O_3.

This practically corresponds to the case of an absolutely rigid inclusion for which the condition $E_2 \to \infty$, ν_2=0 holds. Then the equation (2.32) for the coefficients takes the simpler form:

$$k_y = \frac{5+\nu_1}{3+2\nu_1-\nu_1^2} \text{ and } k_x = \frac{1-3\nu_1}{3+2\nu_1-\nu_1^2}.$$

It is seen that in plane-stress variant the stresses can be calculated knowing the Poisson's ratio coefficient of the matrix.

The pattern inversely changes for the case of a «soft» inclusion, when $E_2 << E_1$. At that time the distribution of the stress components qualitatively match to those in the plane with a round hole. Shown in Fig. 20 is an example of the σ_y distribution, being opposite to the previous one. (E_1= 382GPa and E_2= 70GPa). It is seen that the zones of high and low stresses are exchanged.

The obtained equations (2.30) exactly coincides with that obtained in the work of A.V. Mali by selecting of a definite stress functions [32]. This fact testifies to the validity of the obtained equations.

In section 3.2 the distribution (2.30) was used when simulating strain localization in polycrystals with a solid round inclusion under tensile loading.

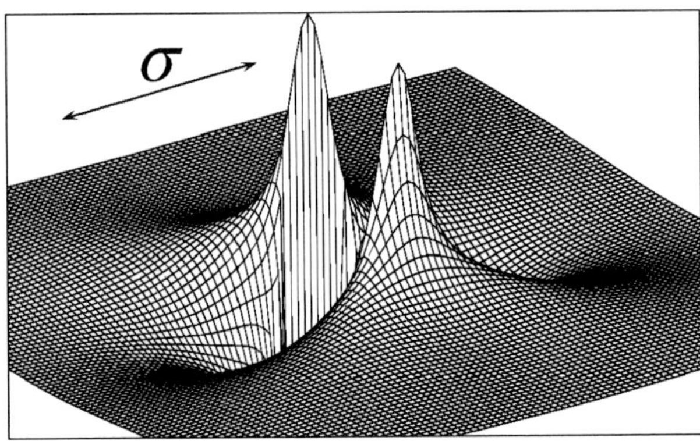

Figure 17. Distribution of the stress component σ_y in the plane with a «soft» inclusion.

2.6. The Band of Localized Plastic Deformation

One of the most interesting problems, practically and theoretically in demand in the mechanics of deformed solid is the problem of stress-strain state of the plane with the band of localized plastic deformation (LPD). By the relaxation element method a number of original solutions have been obtained for similar problems [10-13, 26].

2.6.1. The Band of Localized Plastic Deformation LPD, Constructed from Res of the 1st Type

Let us present one of the examples. The band of localized plastic deformations is composed from relaxation elements of a simple type with the field of internal stresses (2.12). Let us define for each RE an elementary value of stress drop $d\sigma = \Delta\sigma dx/2a$, where $2a$ is the diameter of the circle, and place them evenly at the distance dx from each other along $0x$-axis (Fig. 18). At such a construction in an arbitraty point (x, y) the stress drop will be equal to $nd\sigma$, where n – is a number of REs, inside which the given point hits. The stress drop in $d\sigma$, according to (2.15), is ensured by the elementary field of plastic deformation with the components

$$d\varepsilon_y^p = 3d\sigma/\mathrm{E}, d\varepsilon_x^p = -d\sigma/\mathrm{E},\ d\varepsilon_{xy}^p = 0. \tag{2.35}$$

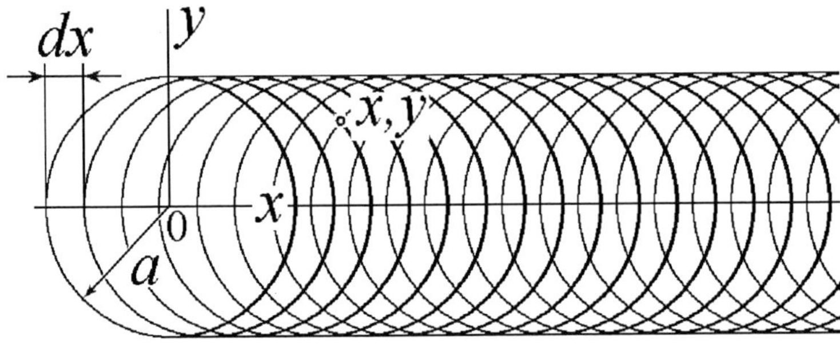

Figure 18. The scheme of construction of the band of localized deformation from a RE of the round shape.

By integration of this elementary field (2.21), we obtain the following expressions for the ε_y^p component of plastic deformation in semi-infinite band:

$$\varepsilon_y^p = \varepsilon_{y\max}^p \begin{cases} (x+\sqrt{a^2-y^2})/2a, & x^2+y^2 \le a^2; \\ \sqrt{1-y^2/a^2}, & x^2+y^2 > a^2. \end{cases} \tag{2.36}$$

Here $\varepsilon_{y\max}^p = 3\Delta\sigma/\text{E}$.

The spatial distribution of the component ε_y^p of the tensor of plastic deformation is represented in Fig. 19. The maximum value of plastic deformation $\varepsilon_y^p = \varepsilon_{y\max}^p$ is observed along the band axis. When approaching the boundary of the plastic deformation the degree of plastic deformation decreases to zero.

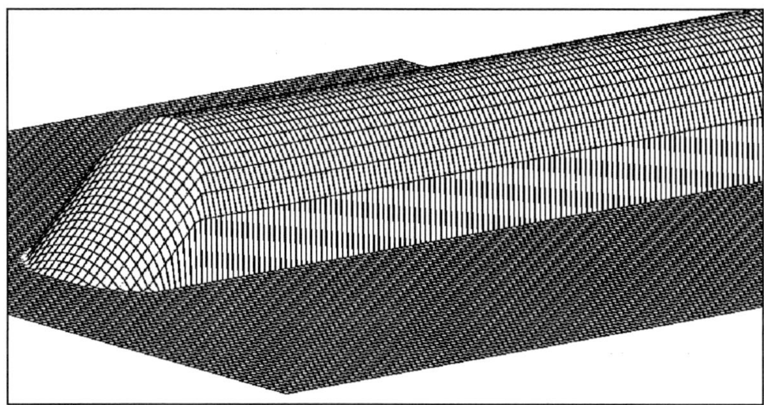

Figure 19. Semi-infinite band of localized plastic deformation.

Similarly, by integrating elementary stress fields according to equations (2.11), where $\Delta\sigma$ is replaced by equation $d\sigma = \Delta\sigma dx/2$a, we obtain the field of stresses for the half-infinite band of localized deformations. In the system of coordinates with the origin at point 0 (see Fig. 17) the components of the tensor of internal stresses of the half-infinite band of LPD is described by simple equations:

$$\Delta\sigma_y = \Delta\sigma \begin{cases} \dfrac{ax}{r^2}\left(2+\left(1-\dfrac{a^2}{r^2}\right)\left(\dfrac{4y^2}{r^2}-1\right)\right), & r^2 \ge a^2, \\ x/a, \quad r^2 \le a^2 \end{cases}$$

$$\Delta\sigma_x = \Delta\sigma \begin{cases} \dfrac{ax}{r^2}\left(2-\left(1-\dfrac{a^2}{r^2}\right)\left(\dfrac{4y^2}{r^2}-1\right)\right), & r^2 \geq a^2, \\ \sqrt{1-y^2/a^2}, & r^2 \leq a^2 \end{cases} \tag{2.37}$$

$$\Delta\sigma_{xy} = \Delta\sigma \begin{cases} \dfrac{ay}{r^2}\left(1-\dfrac{a^2}{r^2}\right)\left(3-\dfrac{4y^2}{r^2}\right), & r^2 \geq a^2. \\ 0, & r^2 \leq a^2 \end{cases}$$

The spatial distribution of the $\Delta\sigma_y$ –component is represented in Fig. 20. The field of stresses is seen to be essentially perturbed only at the end of the band. There exists stress concentrators in the front of the band. At the same time in the band at the end the stresses are lower than the average level of the external stress. This example shows that the source of the driving force for the formation of the band is the stress concentration and high stress gradients at the end of the band.

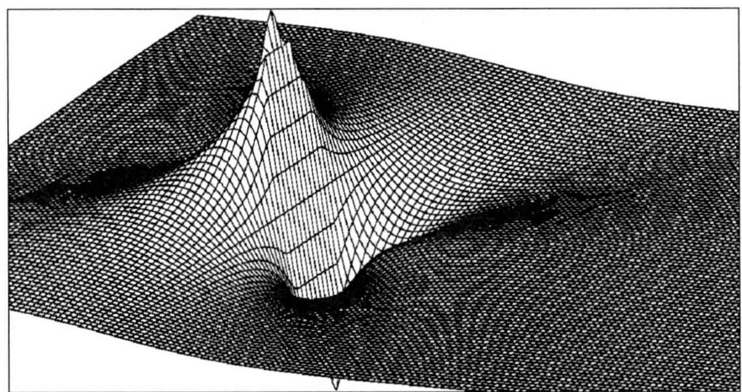

Figure 20. The field of internal stress σ_y- component of semi-infinite band of localized plastic deformation BLD of the 1st type, being oriented perpendicular to tensile axis.

2.6.2 Band of LPD, Constructed from a RE of the Second Type

Similarly, the band of LPD is constructed from the RE of pure shear, the components of which are represented by equations (2.17). Then in the system of coordinates, represented in Fig. 16 (*a*) for the LPD of finite length the components of the stress tensor are represented in the following manner:

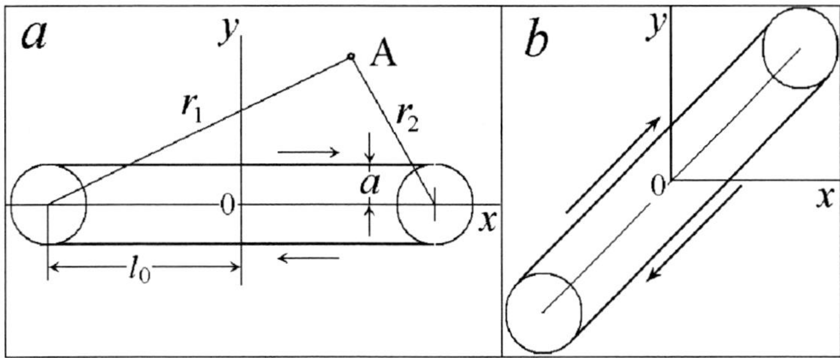

Figure 21. The scheme of the location of LPD of pure shear, directed perpendicular (a) and at an angle of 45°(b) with respect to the tensile $0y$ axis.

$$\Delta\sigma_y = \begin{cases} \dfrac{ay}{2r_2^2}\left(2-\dfrac{3a^2}{r_2^2}+4\dfrac{y^2}{r_2^2}\left(\dfrac{a^2}{r_2^2}-1\right)\right)-\dfrac{ay}{2r_1^2}\left(2-\dfrac{3a^2}{r_1^2}+4\dfrac{y^2}{r_1^2}\left(\dfrac{a^2}{r_1^2}-1\right)\right), r_2^2 \ge a^2, r_1^2 \ge a^2, \\ \dfrac{ay}{2r_2^2}\left(2-\dfrac{3a^2}{r_2^2}+4\dfrac{y^2}{r_2^2}\left(\dfrac{a^2}{r_2^2}-1\right)\right)+\dfrac{y}{2a}, r_2^2 \ge a^2, r_1^2 \le a^2, \\ \dfrac{-ay}{2r_2^2}\left(2-\dfrac{3a^2}{r_1^2}+4\dfrac{y^2}{r_1^2}\left(\dfrac{a^2}{r_1^2}-1\right)\right)-\dfrac{y}{2a}, r_1^2 \ge a^2, r_2^2 \le a^2. \end{cases}$$

$$\Delta\sigma_x = \begin{cases} \dfrac{ay}{2r_2^2}\left(\dfrac{3a^2}{r_2^2}-6-4\dfrac{y^2}{r_2^2}\left(\dfrac{a^2}{r_2^2}-1\right)\right)+\dfrac{ay}{2r_1^2}\left(6-\dfrac{3a^2}{r_1^2}+4\dfrac{y^2}{r_1^2}\left(\dfrac{a^2}{r_1^2}-1\right)\right), r_2^2 \ge a^2, r_1^2 \ge a^2, \\ \dfrac{ay}{2r_2^2}\left(\dfrac{3a^2}{r_2^2}-6-4\dfrac{y^2}{r_2^2}\left(\dfrac{a^2}{r_2^2}-1\right)\right)+\dfrac{3y}{2a}, r_2^2 \ge a^2, r_1^2 \le a^2, \\ \dfrac{ay}{2r_2^2}\left(6-\dfrac{3a^2}{r_1^2}+4\dfrac{y^2}{r_1^2}\left(\dfrac{a^2}{r_1^2}-1\right)\right)-\dfrac{3y}{2a}, r_1^2 \ge a^2, r_2^2 \le a^2. \end{cases} \tag{2.38}$$

$$\Delta\sigma_{xy} = \begin{cases} \dfrac{a(x-l_0)}{2r_2^2}\left(2-\dfrac{a^2}{r_2^2}+4\dfrac{y^2}{r_2^2}\left(\dfrac{a^2}{r_2^2}-1\right)\right)-\dfrac{a(x+l_0)}{2r_1^2}\left(2-\dfrac{3a^2}{r_1^2}+4\dfrac{y^2}{r_1^2}\left(\dfrac{a^2}{r_1^2}-1\right)\right), r_2^2 \ge a^2, r_1^2 \ge a^2, \\ \dfrac{a(x-l_0)}{2r_2^2}\left(2-\dfrac{3a^2}{r_2^2}+4\dfrac{y^2}{r_2^2}\left(\dfrac{a^2}{r_2^2}-1\right)\right)-\dfrac{x+l_0}{2a}, r_2^2 \ge a^2, r_1^2 \le a^2, \\ \dfrac{-a(x+l_0)}{2r_2^2}\left(2-\dfrac{3a^2}{r_1^2}+4\dfrac{y^2}{r_1^2}\left(\dfrac{a^2}{r_1^2}-1\right)\right)+\dfrac{x-l_0}{2a}, r_1^2 \ge a^2, r_2^2 \le a^2. \end{cases}$$

Here a and l_0 – half-width and half-length of LPD, r_1 and r_2 - are correspondingly the distance from the arbitrary point A, at which the stresses are considered, to the centers of the relaxation element at the end of the band to the right and to the left of the band (see. Fig. 21). Shown in Fig. 22 are the distributions of all components of the stress tensor according to equations.

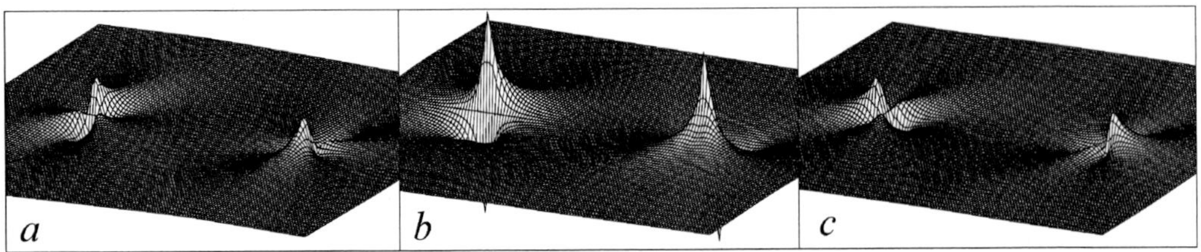

Figure 22. The distributions of the components of the field of internal stresses for the band of pure shear: (*a*) $\Delta\sigma_y$, (*b*) $\Delta\sigma_x$, (*c*) $\Delta\sigma_{xy}$.

The maximum value of shear stress under tensile loading is observed in direction at an angle of 45° with respect to the tensile axis. The relaxation of the tensile stress in the RE for the value $\Delta\sigma$ corresponds to a relaxation of the stress of pure shear by the value $\Delta\tau=\Delta\sigma/2$. Hence, the distribution $\Delta\sigma_{xy}$, represented in equations (2.38), is the field of internal stresses of localized plastic deformation of pure shear in the direction at an angle of 45° with respect to the tensile axis. In other words, the distribution in Fig. 17c corresponds to the distribution in the system of coordinates, being oriented at an angle of 45° with respect to the tensile axis. The corresponding distribution is represented in Fig. 18. The following peculiarities of the bands of localized plastic defiormation should be mentioned:

1. The stress field is highly perturbed only at the end of the band. The stress concentrator there is always observed. It favours the formation of the band of localized plastic deformation.
2. The zone of stress concentration at the end of the band is always combined with a zone of stress anticoncentration inside the band.
3. The stress concentration at the end of the band increases as the length of the band increases. It is well illustrated in Fig. 19, in which there is a surface, composed from the profiles of shear stress in the cross-section along the band of LPD as its length increases. The values of stresse are normalized by minimum value $\sigma_{xy}/\Delta\sigma = -1$. Initially V-shaped profile is transformed into the profile with two minimums and maximums at the ends of the band. When the length increases, when $l_0/a \to \infty$, $\sigma_{xy}^{max} / \Delta\sigma \to 1$. The relaxing component of the field of internal stresses is always negative. As the length of the band increases, the absolute value of this stress at the center tends to zero.

2.7. The Site of Localized Plastic Deformation of Rectangular Shape

The band of localized plastic deformation can be used as relaxation element for the construction of the shape of the sites of plastic deformation of rectangular shape. A change in the length of the sides of rectangular shape allows one to change the scale and the shape of the region of localized plastic deformation and to define corresponding stress fields in a continuum medium. Let us construct the site of plastic deformation of rectangular shape as an example of the band of LPD of the second type.

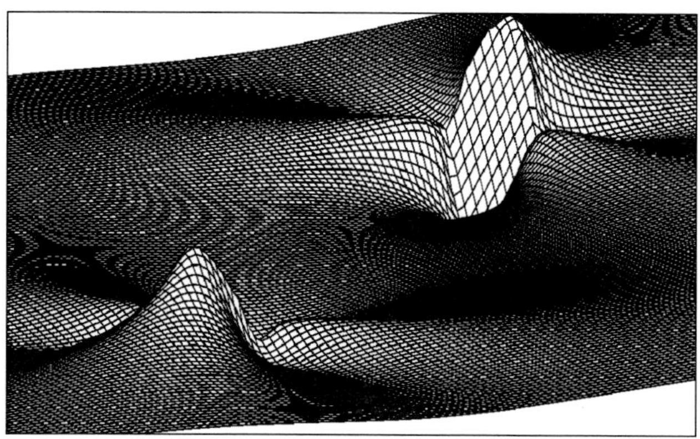

Figure 22. The stress field of the band of plastic deformation of pure shear, formed at an angle of 45° with respect to tensile axis.

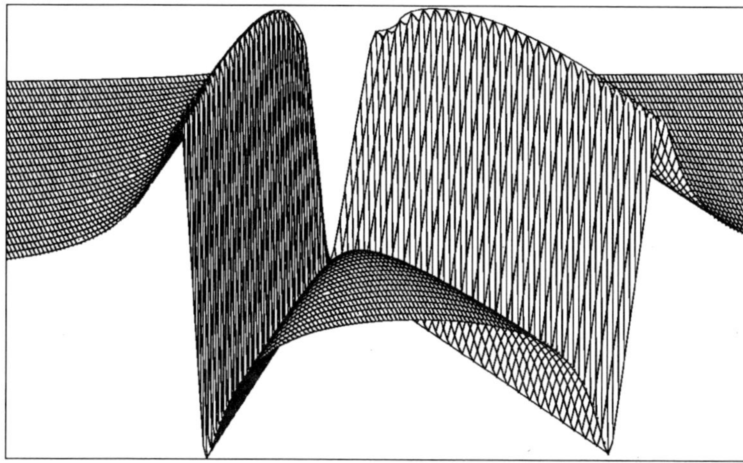

Figure 23. Changing of the profile of plastic deformation with increase in the length of the band.

Shown in Fig. 24 is the scheme of the distribution of the family of the bands of the mentioned type. The system of coordinates is located at the center of the rectangle. The dimensions Lx, Ly define the shape of rectangle with rounded angles: Lx - is the half-length of the RE (band), Ly – is the distance from the $0x$-axis to the last RE in the distribution, a - is the half-width of the RE-band or the radius of rounding of the angles of the rectangle. The variable l defines the location of any RE (in the shape of RE) in this distribution. Superposition of elementary stress fields and deformations we obtain analogously to that what we have done in item 2.6 for RE of round shape. In the figure at the angles of rectangle the shaded zones are selected in which equations for stresses and strain will be written in another form than that beyond selected zones, since in equations (2.33) for the bands of localized deformation the stresses in the zones at the end of the band differ from those in the rest of the volume.

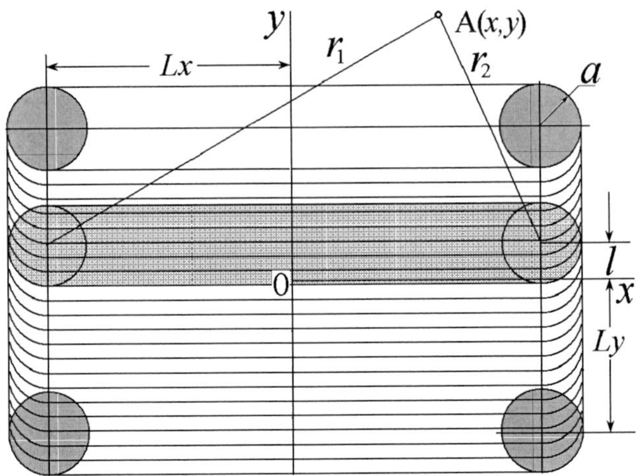

Figure 24. The scheme of distribution of the family of RE along $0y$ axis in the form of identical bands with half-length Lx, Ly – is the distance from the $0x$-axis till the last RE in the distribution.

We obtain the following results:

Distribution of the component σ_y:

Beyond the shaded zones at the angles of rectangle

$$\frac{\sigma 0_y}{\Delta\sigma}=\frac{1}{4}\ln\frac{\left[(y-Ly)^2+(x-Lx)^2\right]\left[(y+Ly)^2+(x+Lx)^2\right]}{\left[(y+Ly)^2+(x-Lx)^2\right]\left[(y-Ly)^2+(x+Lx)^2\right]}+\frac{1}{8}\left[\frac{a^2+4(x-Lx)^2}{(y-Ly)^2+(x-Lx)^2}-\frac{a^2+4(x-Lx)^2}{(y+Ly)^2+(x-Lx)^2}\right]-$$
$$-\frac{a^2}{4}\left[\frac{(x-Lx)^2}{(y-Ly)^2+(x-Lx)^2}-\frac{(x-Lx)^2}{(y+Ly)^2+(x-Lx)^2}\right]-\frac{1}{8}\left[\frac{a^2+4(x+Lx)^2}{(y-Ly)^2+(x+Lx)^2}-\frac{a^2+4(x+Lx)^2}{(y+Ly)^2+(x+Lx)^2}\right]+$$
$$\frac{a^2}{4}\left[\frac{(x+Lx)^2}{\left[(y-Ly)^2+(x+Lx)^2\right]^2}-\frac{(x+Lx)^2}{\left[(y+Ly)^2+(x+Lx)^2\right]^2}\right]=F0+F1+F2+F3+F4,\ if\left[(|y|-Ly)^2+(|x|-Lx)^2\right]\geq a^2, \tag{2.39}$$

where the functions $F0$, $F1$, $F2$, $F3$ and $F4$ in this equation a separate equations consequently are denoted as

In the round zones at the angles of the rectangle at $(y-Ly)^2+(x-Lx)^2 \leq a^2$

$$\frac{\sigma 1_y}{\Delta\sigma}=\frac{1}{4}\ln\frac{a^2\left[(y+Ly)^2+(x+Lx)^2\right]}{\left[(y+Ly)^2+(x-Lx)^2\right]\left[(y-Ly)^2+(x+Lx)^2\right]}+\frac{1}{8}\left[\frac{(y-Ly)^2+3(x-Lx)^2}{a^2}-\frac{a^2+4(x-Lx)^2}{(y+Ly)^2+(x-Lx)^2}\right]+$$
$$+\frac{0.25a^2(x-Lx)^2}{\left[(y+Ly)^2+(x-Lx)^2\right]^2}+F3+F4,\qquad if\left[(y-Ly)^2+(x-Lx)^2\right]\leq a^2, \tag{2.40}$$

at $(y+Ly)^2+(x-Lx)^2 \leq a^2$

$$\frac{\sigma 2_y}{\Delta\sigma}=\frac{1}{4}\ln\frac{\left[(y-Ly)^2+(x-Lx)^2\right]\left[(y+Ly)^2+(x+Lx)^2\right]}{a^2\left[(y-Ly)^2+(x+Lx)^2\right]}-\frac{1}{8}\left[\frac{(y+Ly)^2+3(x-Lx)^2}{a^2}-\frac{a^2+4(x-Lx)^2}{(y-Ly)^2+(x-Lx)^2}\right]+$$
$$+\frac{-0.25a^2(x-Lx)^2}{\left[(y-Ly)^2+(x-Lx)^2\right]^2}+F3+F4,\qquad if\left[(y+Ly)^2+(x-Lx)^2\right]\leq a^2, \tag{2.41}$$

at $(y-Ly)^2+(x+Lx)^2 \leq a^2$

$$\frac{\sigma 3_y}{\Delta\sigma}=\frac{1}{4}\ln\frac{\left[(y-Ly)^2+(x-Lx)^2\right]\left[(y+Ly)^2+(x+Lx)^2\right]}{a^2\left[(y+Ly)^2+(x-Lx)^2\right]}-\frac{1}{8}\left[\frac{(y-Ly)^2+3(x+Lx)^2}{a^2}-\frac{a^2+4(x+Lx)^2}{(y+Ly)^2+(x+Lx)^2}\right]+$$
$$+\frac{-0.25a^2(x+Lx)^2}{\left[(y+Ly)^2+(x+Lx)^2\right]^2}+F1+F2,\qquad if\left[(y-Ly)^2+(x+Lx)^2\right]\le a^2, \tag{2.42}$$

and at $(y + Ly)^2 + (x + Lx)^2 \le a^2$

$$\frac{\sigma 4_y}{\Delta\sigma}=\frac{1}{4}\ln\frac{\left[(y-Ly)^2+(x-Lx)^2\right]a^2}{\left[(y-Ly)^2+(x+Lx)^2\right]\left[(y+Ly)^2+(x-Lx)^2\right]}+\frac{1}{8}\left[\frac{(y+Ly)^2+3(x+Lx)^2}{a^2}-\frac{a^2+4(x+Lx)^2}{(y-Ly)^2+(x+Lx)^2}\right]+$$
$$+\frac{0.25a^2(x+Lx)^2}{\left[(y-Ly)^2+(x+Lx)^2\right]^2}+F1+F2,\qquad if\left[(y+Ly)^2+(x+Lx)^2\right]\le a^2. \tag{2.43}$$

The complete solution is represented in the form of the sum of the represented equations:

$$\sigma_y = \sigma 0_y+\sigma 1_y+\sigma 2_y+\sigma 3_y+\sigma 4_y. \tag{2.44}$$

Analogeously the distribution of the σ_x-component is defined. The difference from the distribution of $\sigma 0_y$ component is defined only in the signs before the terms, $F1$, $F2$, $F3$ and $F4$. The logarithmic term doesn't differ in anything. It can be described in the form of equation:

$$\sigma 0_x/\Delta\sigma = F1-(F2+F3+F4+F5). \tag{2.45}$$

$$\frac{\sigma 1_x}{\Delta\sigma}=\frac{1}{4}\ln\frac{a^2\left[(y+Ly)^2+(x+Lx)^2\right]}{\left[(y+Ly)^2+(x-Lx)^2\right]\left[(y-Ly)^2+(x+Lx)^2\right]}-\frac{1}{8}\left[4-\frac{3(y-Ly)^2+(x-Lx)^2}{a^2}-\frac{a^2+4(x-Lx)^2}{(y+Ly)^2+(x-Lx)^2}\right]+$$
$$+\frac{-0.25a^2(x-Lx)^2}{\left[(y+Ly)^2+(x-Lx)^2\right]^2}-F3-F4,\qquad if\left[(y-Ly)^2+(x-Lx)^2\right]\le a^2,$$

$$\frac{\sigma 2_x}{\Delta\sigma}=\frac{1}{4}\ln\frac{\left[(y-Ly)^2+(x-Lx)^2\right]\left[(y+Ly)^2+(x+Lx)^2\right]}{a^2\left[(y-Ly)^2+(x+Lx)^2\right]}+\frac{1}{8}\left[4-\frac{(y+Ly)^2+3(x-Lx)^2}{a^2}-\frac{a^2+4(x-Lx)^2}{(y-Ly)^2+(x-Lx)^2}\right]+$$
$$+\frac{0.25a^2(x-Lx)^2}{\left[(y-Ly)^2+(x-Lx)^2\right]^2}-F3-F4,\qquad if\left[(y+Ly)^2+(x-Lx)^2\right]\le a^2,$$

$$\frac{\sigma 3_x}{\Delta\sigma}=\frac{1}{4}\ln\frac{\left[(y-Ly)^2+(x-Lx)^2\right]\left[(y+Ly)^2+(x+Lx)^2\right]}{a^2\left[(y+Ly)^2+(x-Lx)^2\right]}+\frac{1}{8}\left[4-\frac{(y-Ly)^2+3(x+Lx)^2}{a^2}-\frac{a^2+4(x+Lx)^2}{(y+Ly)^2+(x+Lx)^2}\right]+$$
$$+\frac{0.25a^2(x+Lx)^2}{\left[(y+Ly)^2+(x+Lx)^2\right]^2}-F1-F2,\qquad if\left[(y-Ly)^2+(x+Lx)^2\right]\le a^2,$$

$$\frac{\sigma 4_x}{\Delta\sigma}=\frac{1}{4}\ln\frac{\left[(y-Ly)^2+(x-Lx)^2\right]a^2}{\left[(y-Ly)^2+(x+Lx)^2\right]\left[(y+Ly)^2+(x-Lx)^2\right]}-\frac{1}{8}\left[4-\frac{(y+Ly)^2+3(x+Lx)^2}{a^2}-\frac{a^2+4(x+Lx)^2}{(y-Ly)^2+(x+Lx)^2}\right]+$$
$$+\frac{-0.25a^2(x+Lx)^2}{\left[(y-Ly)^2+(x+Lx)^2\right]^2}-F1-F2,\qquad if\left[(y+Ly)^2+(x+Lx)^2\right]\le a^2.$$

The complete solution is again represented in the form of the sum all equations for the stress component σ_x:

$$\sigma_x = \sigma 0_x + \sigma 1_x + \sigma 2_x + \sigma 3_x + \sigma 4_x. \tag{2.46}$$

Shown in Fig. 21 are the distributions σ_y (*a*) and σ_x (*b*), at the prescribed relation of geometrical parameters *a/Ly/Lx*=1/2/10. The pattern qualitatively repeat the distribution for the base of the band of pure shear, which we have considered above (see Fig. 17*a, b*). The relaxation of shear stress in the band causes essential concentrations of normal stresses σ_y andσ_x at the end of the band. At that time the concentration of the component σ_x essentially exceeds that for the σ_y-component. Extreme values are observed at all angles of the rectangle.

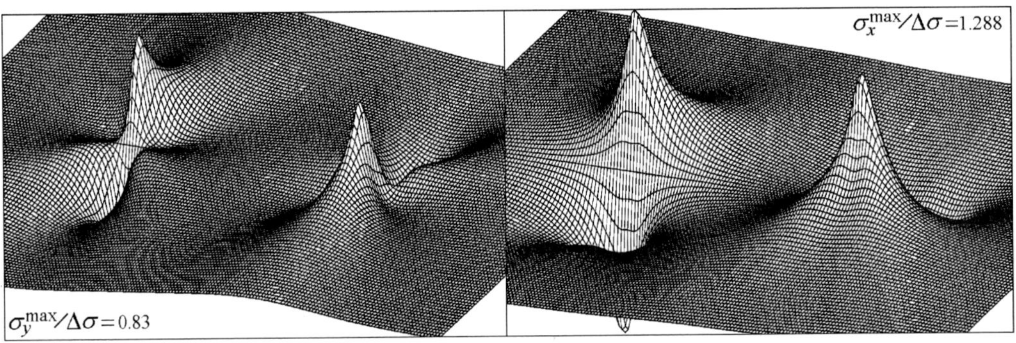

Figure 25. Distribution of stresses σ_y (*a*) and σ_x (*b*) on shear stress relaxation in the rectangular band: at a *Ly* /*Lx* =1/2/20.

The solution for the $\sigma 0_{xy}$ component is written in the following manner:

$$\frac{\sigma 0_{xy}}{\Delta\sigma} = \left[\frac{0.25(x-Lx)(y+Ly)}{(y+Ly)^2+(x-Lx)^2}\right]\left[2-\frac{a^2}{(y+Ly)^2+(x-Lx)^2}\right] - \left[\frac{0.25(x-Lx)(y-Ly)}{(y-Ly)^2+(x-Lx)^2}\right]\left[2-\frac{a^2}{(y-Ly)^2+(x-Lx)^2}\right]+$$
$$+\left[\frac{0.25(x+Lx)(y-Ly)}{(y-Ly)^2+(x+Lx)^2}\right]\left[2-\frac{a^2}{(y-Ly)^2+(x+Lx)^2}\right] - \left[\frac{0.25(x+Lx)(y+Ly)}{(y+Ly)^2+(x+Lx)^2}\right]\left[2-\frac{a^2}{(y+Ly)^2+(x+Lx)^2}\right]=$$
$$= G1+G2+G3+G4, \quad if\left[(|y|-Ly)^2+(|x|-Lx)^2\right] \geq a^2. \tag{2.47}$$

Here under the functions *G*1, *G*2, *G*3 and *G*4 the separate expressions in equation (2.33) are denoted consequently.

$$\frac{\sigma 1_{xy}}{\Delta\sigma} = F1 - \frac{(x-Lx)\sqrt{a^2-(x-Lx)^2}}{a^2} + \frac{(x-Lx)\left(Ly-y+\sqrt{a^2-(x-Lx)^2}\right)}{4a^2} + G3+G4, if\left[(y-Ly)^2+(x-Lx)^2\right] \leq a^2, \tag{2.48}$$

$$\frac{\sigma 2_{xy}}{\Delta\sigma} = \frac{-(x-Lx)\sqrt{a^2-(x-Lx)^2}}{a^2} + \frac{(x-Lx)\left(Ly+y+\sqrt{a^2-(x-Lx)^2}\right)}{4a^2} + G2+G3+G4, if\left[(y+Ly)^2+(x-Lx)^2\right] \leq a^2, \tag{2.49}$$

$$\frac{\sigma 3_{xy}}{\Delta\sigma} = G1 + G2 + \frac{(x+Lx)\sqrt{a^2-(x+Lx)^2}}{a^2} + \frac{(x+Lx)\left(Ly-y+\sqrt{a^2-(x+Lx)^2}\right)}{-4a^2} + G4, if\left[(y-Ly)^2+(x+Lx)^2\right] \le a^2, \tag{2.50}$$

$$\frac{\sigma 4_{xy}}{\Delta\sigma} = G1 + G2 + G3 + \frac{(x+Lx)\sqrt{a^2-(x+Lx)^2}}{a^2} + \frac{(x+Lx)\left(Ly+y+\sqrt{a^2-(x+Lx)^2}\right)}{-4a^2}, if\left[(y+Ly)^2+(x+Lx)^2\right] \le a^2. \tag{2.51}$$

The complete solutions represented in the form of the sum of the equations for the stress component σ_x.:

$$\sigma_{xy} = \sigma 0_{xy} + \sigma 1_{xy} + \sigma 2_{xy} + \sigma 3_{xy} + \sigma 4_{xy}. \tag{2.52}$$

The distribution of the σ_{xy} component is represented in Fig. 22. The maximum shear stress σx_y two times less than in the σ_y distribution and three times lower than in the σ_x distribution. However, the position of the concentrator at the end of the band contributes to the increase in the length of the band in the process of plastic deformation development.

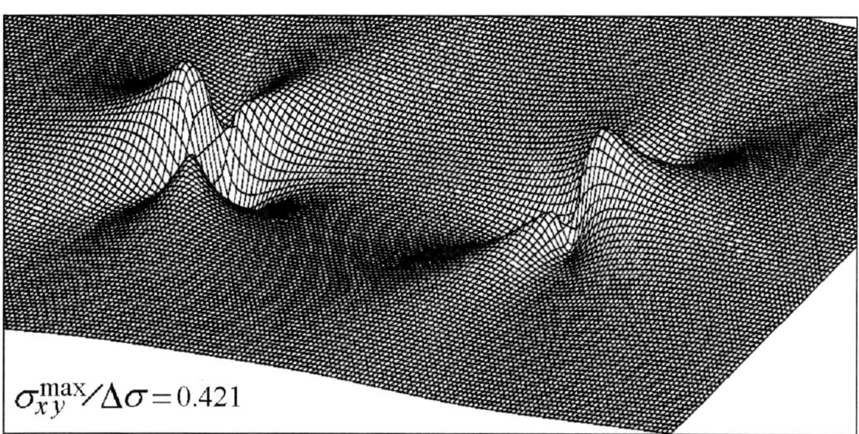

Figure 26. The distribution of σ_{xy} (*a*) stress from the relaxation of shear stress in a rectangular band: $a/Ly/Lx = 1/2/20$.

The region of shear stress relaxation will be exactly of quadratic shape, if a tends to zero. Under such a condition the equations represented above are essentially simplified and take the following form:

$$\frac{\sigma_y}{\Delta\sigma} = \frac{1}{2}\left[\frac{(x-Lx)^2}{(y-Ly)^2+(x-Lx)^2} - \frac{(x-Lx)^2}{(y+Ly)^2+(x-Lx)^2} - \frac{(x+Lx)^2}{(y-Ly)^2+(x+Lx)^2} + \frac{(x+Lx)^2}{(y+Ly)^2+(x+Lx)^2}\right] + \\ + \frac{1}{4}\ln\frac{\left[(y-Ly)^2+(x-Lx)^2\right]\left[(y+Ly)^2+(x+Lx)^2\right]}{\left[(y+Ly)^2+(x-Lx)^2\right]\left[(y-Ly)^2+(x+Lx)^2\right]}, \tag{2.53}$$

$$\frac{\sigma_x}{\Delta\sigma}=\frac{-1}{2}\left[\frac{(x-Lx)^2}{(y-Ly)^2+(x-Lx)^2}-\frac{(x-Lx)^2}{(y+Ly)^2+(x-Lx)^2}-\frac{(x+Lx)^2}{(y-Ly)^2+(x+Lx)^2}+\frac{(x+Lx)^2}{(y+Ly)^2+(x+Lx)^2}\right]+$$
$$+\frac{1}{4}\ln\frac{\left[(y-Ly)^2+(x-Lx)^2\right]\left[(y+Ly)^2+(x+Lx)^2\right]}{\left[(y+Ly)^2+(x-Lx)^2\right]\left[(y-Ly)^2+(x+Lx)^2\right]},\tag{2.54}$$

$$\frac{\sigma_{xy}}{\Delta\sigma}=\frac{1}{2}\left[\frac{(x-Lx)(y+Ly)}{(y+Ly)^2+(x-Lx)^2}-\frac{(x-Lx)(y-Ly)}{(y-Ly)^2+(x-Lx)^2}-\frac{(x+Lx)(y+Ly)}{(y+Ly)^2+(x+Lx)^2}+\frac{(x+Lx)(y-Ly)}{(y-Ly)^2+(x+Lx)^2}\right].\tag{2.55}$$

The σ_y and σ_x components in angular points have singularities. However, already at the distance of the order a from the angular points the distribution of these components are practically not different from the distribution for the RE of rectangular shape with the radius of roundess being equal a. The qualitative distribution of the σ_{xy} component remains the same. The difference is only in insignificant increase in the stress concentration. For the relationship $a/Ly\ /Lx\ =1/2/20$ the value $\sigma_{xy}^{\max}/\Delta\sigma=0.421$. For the same relationship betwenn the length of the triangle $Ly/Lx = 2/20$ at $a = 0$ the equation (2.42) defines the value $\sigma_{xy}^{\max}/\Delta\sigma=0.452$. As the length l increases the shear stress concentration at the end of the band increases as well and an asymptotic approaching to the value $\tau_{\max}/\Delta\sigma = 0.5$ takes place.

The condition of uniaxial tension optimally contributes to the shear stress relaxation τ in direction under the angle of 45° with respect to the tensile axis (Fig. 23), i.e. in the system coordinate, rotated with respect to the tensile 0y-axis by an angle $\alpha = \pi/4$. A corresponding transformation of coordinates will result to the following view of equation (2.42):

$$\frac{\tau}{\Delta\tau}=\frac{1}{2}\left[\begin{array}{l}\frac{\left(y-x-\sqrt{2}Lx\right)\left(y+x+Ly\sqrt{2}\right)}{\left(y+x+Ly\sqrt{2}\right)^2+\left(y-x-Lx\sqrt{2}\right)^2}-\frac{\left(y-x-\sqrt{2}Lx\right)\left(y+x-Ly\sqrt{2}\right)}{\left(y+x-Ly\sqrt{2}\right)^2+\left(y-x-Lx\sqrt{2}\right)^2}-\\-\frac{\left(y-x+\sqrt{2}Lx\right)\left(y+x+Ly\sqrt{2}\right)}{\left(y+x+Ly\sqrt{2}\right)^2+\left(y-x+Lx\sqrt{2}\right)^2}+\frac{\left(y-x+\sqrt{2}Lx\right)\left(y+x-Ly\sqrt{2}\right)}{\left(y+x-Ly\sqrt{2}\right)^2+\left(y-x+Lx\sqrt{2}\right)^2}\end{array}\right]\tag{2.56a}$$

The value of relaxation of the shear stress $\Delta\tau$ in the band, apparently is defined by the relaxation of the normal stress $\Delta\sigma$ along the tensile axis by the amount of $\Delta\tau = \Delta\sigma/2$.

Integrating of the elementary fields of plastic deformation of shear $d\gamma^P$=$2\Delta\sigma dl/2EL_y$ (see (2.18a)) defines the field in the rectangular coordinate system

$$\gamma^P = 2\Delta\sigma/\mathrm{E}.\tag{2.57}$$

3.

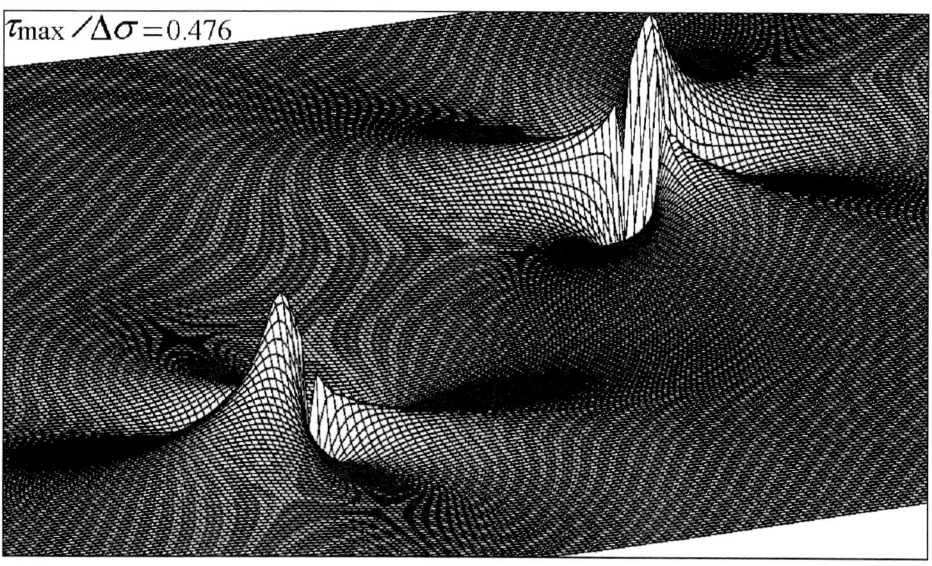

Figure 27. Distribution of shear stress field at an angle of 45° with respect to the tensile axis for the rectangular band: $Ly/Lx = 1/20$.

If the length is equal to the width of the band, it will result in a RE of quadrat shape. The distribution of shear stress τ for this case is represented in Fig. 24. It is seen that in all 4 directions at an angle of 45°with respect to the tensile axis the elevated shear stress concentration is observed which embraces the regions, commensurate with the dimension of the square itself. As distinct from the RE of a round shape, here in the distribution there is no a number of microconcentrators between given concentrators (compare with Fig. 9c).

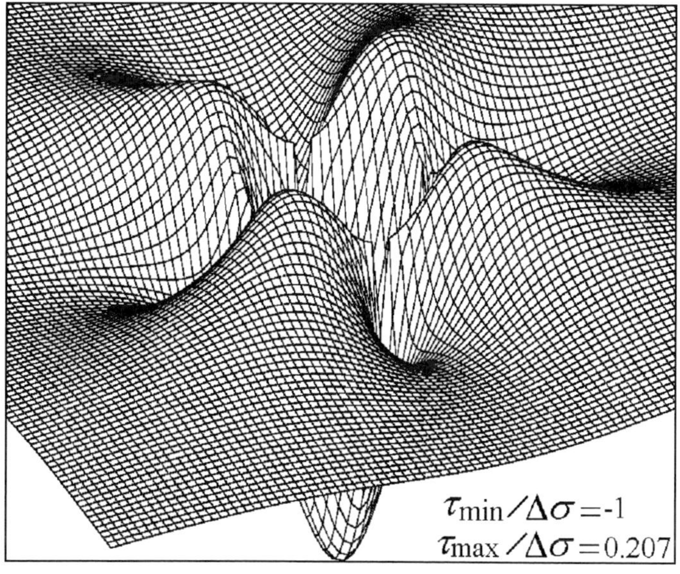

Figure 28. Distribution of shear stress at an angle of 45° with respect to the tensile axis for a RE in the form of a square.

A relaxation element of pure shear of rectangular shape can be obtained also by an other method, using Crouch's formular for the discontinuity displacement [28]. The localized plastic deformation in the continuum medium can be represented as a distribution of elementary discontinuity displacements in a limited region. Let us explain it on the scheme, represented in Fig. 25. Let in the rectangular region, being parallel to 0*x*-axis at the arbitrary distance *l* from this axis an elementary discontinuous displacment *dDx* being observed. That means, that a fixed elementary displacement *dDx* of two parallel lines with the length of 2*Lx*, located at a distance *dl* takes place. In essence the defect, arising in such an operation plays the role of a specific relaxation element. Crouch has found the solution for all components of the stress field for the discontinuity displasment of the value *Dx* at the segment of length 2*a*, assuming that this segment by itself can represent two sides of the crack. We represented the displacement discontinuity not as a displacements of the crack sides, but as an elementary shear of the points of the material, being located at an elementary distance from each other. In its physical sense this is an elementary plastic displacement in the volume of the material. A dislocation loop in crystalline lattice can be an example of it. In the cross-section, being perpendicular to the plane of the loop we observe a real displacement of the nearest closed-packed planes of a crystallite lattice by a Burgers vector. Our task is to find the solution for the integral sum of elementary discontinuity displacements with the condition that the resulting value of the displacements of the opposite sides of the rectangle is equal to *Dx* (Fig. 29).

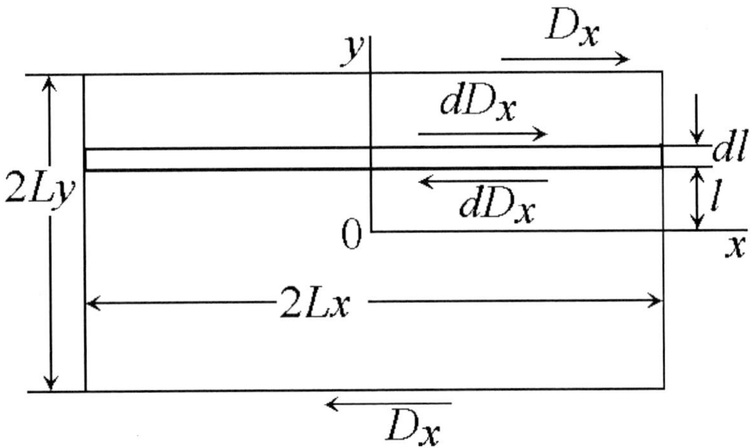

Figure 29. The scheme of an elementary discontinuous displacement in the local volume of rectangular shape.

For the elementary discontinuity displacement let us write down Crouch's solution for the displacement field in the following manner:

$$d\sigma_x = -2GD_x\left(2f_{,xy} + (y-l)f_{,xyy}\right)dl,$$
$$d\sigma_y = 2GD_x(y-l)f_{,xyy}dl,$$
$$d\sigma_{xy} = -2GD_x\left(2f_{,yy} + (y-l)f_{,yyy}\right)dl,$$

(2.58)

here comma means differentiation on corresponding coordinates, G – is the shear modulus, l – is the variable of integration, $dD_x = D_x dl$ – is the elementary discontinuously displacements. The function $f(x,y)$ is equal to

$$f(x,y)=\frac{1}{4\pi(1-\nu)}$$
$$\left[y\left(\operatorname{arctg}\frac{y-l}{x-a}-\operatorname{arctg}\frac{y-l}{y+a}\right)-\frac{x-a}{2}\ln\left[(x-a)^2+(y-l)^2\right]+\frac{x+a}{2}\ln\left[(x+a)^2+(y-l)^2\right]\right], \tag{2.59}$$

where ν - is Poisson's Ratio.

The derivatives of the function as follows:

$$\begin{aligned}
f_{,xy}&=f_{,yx}=\frac{1}{4\pi(1-\nu)}\left[\frac{y-l}{(x-a)^2+y^2}-\frac{y-l}{(x+a)^2+y^2}\right],\\
f_{,xx}&=-f_{,yy}=\frac{1}{4\pi(1-\nu)}\left[\frac{x-a}{(x-a)^2+(y-l)^2}-\frac{x+a}{(x+a)^2+(y-l)^2}\right],\\
f_{,xyy}&=-f_{,xxx}=\frac{1}{4\pi(1-\nu)}\left[\frac{(x-a)^2-(y-l)^2}{\left[(x-a)^2+(y-l)^2\right]^2}-\frac{(x+a)^2-(y-l)^2}{\left[(x+a)^2+(y-l)^2\right]^2}\right],\\
f_{,yyy}&=-f_{,xxy}=\frac{2y}{4\pi(1-\nu)}\left[\frac{x-a}{\left[(x-a)^2+(y-l)^2\right]^2}-\frac{x+a}{\left[(x+a)^2+(y-l)^2\right]^2}\right].
\end{aligned} \tag{2.60}$$

Integration of the equations (2.42) with respect to the variable l from $-Ly$ to $+Ly$ results in equations for the components of a stress tensor, matching the corresponding equations, derived by the relaxation element method with accuracy to the multuiplier. The value $\Delta\sigma$ matches the value $D_xG/\pi(1-\nu)$, defined by the formulas (2.42)-(2.44). For comparison reasons let us represent the formulae for σ_{xy}:

$$\frac{\sigma_{xy}}{\Delta\sigma}=\frac{GD_x}{2\pi(1-\nu)}\left[\frac{(x-Lx)(y+Ly)}{(y+Ly)^2+(x-Lx)^2}-\frac{(x-Lx)(y-Ly)}{(y-Ly)^2+(x-Lx)^2}-\frac{(x+Lx)(y+Ly)}{(y+Ly)^2+(x+Lx)^2}+\frac{(x+Lx)(y-Ly)}{(y-Ly)^2+(x+Lx)^2}\right]. \tag{2.61}$$

The comparison with equation (2.42) shown, that value $D_xG/\pi(1-\nu)$, defined by the equations (2.42)-(2.44) correspond to the value $\Delta\sigma$ in the relaxation element method. In the case of the relaxation element the influence of the shear stress relaxation on the field of internal stresses has been considered. In the second case- the influence of plastic deformation of shear is defined by the value of displacements D_x. In both cases the same result are obtained.

When using the derived equations one should take into account the following. In the last case we considered the scheme of a simple shear, when plastic shear occurs only along one direction (Fig. 25, along the 0x-axis). The σ_{yx} component at that time will be equal to zero, i.e. the stress tensor is antisymmetric. The component σ_{yx} will not be equal to zero, if the plastic shear Dy at the opposite sides of the rectangule is applied along the 0y-direction. One can make sure, that the equation for σ_{yx} will be the same as for σ_{xy}. Only instead of the coefficient D_x it will be the coefficient D_y in it. Shear components of the tensor of plastic deformation $\varepsilon_{xy}= D_x/L_y$ and $\varepsilon_{yx}= D_y/L_x$ will not be equal to each other in this case. Pure shear will take place under the condition

$$D_x/D_y = L_y/L_x. \tag{2.62}$$

At that time in the rectangle a homogeneous field of plastic deformation of pure shear (2.43) will be observed.

2.5. Relaxation Element of Square Shape with the Gradients of Plastic Deformation

As in the case of REs of round shape, let us construct a RE with gradients of plastic deformation for the RE to avoid the peculiarities of stresses and strains at the boundary of the RE. The technique of construction is analogoous to that described in section 2.5. As in the case of a RE of round shape let us construct Re with the gradients of plastic deformation for the RE of square shape with the aim to escape the peculiarities of stresses and deformations at the boundary of the RE. The technique of construction is analogous to the one described in section 2.5.

Let us devide the square area with the site being equal to 2a into a family of quadrats with the common center a with the sites $a(t) = at$, where t is the integration variable. At t 0 $a(t) = 0$, when $t=1$ $a(t) = a$ (Fig. 26). For each quadrat in the family let us prescribe the elementary value of stress relaxation

$$d\sigma(t) = (\beta + 1)\Delta\sigma t^{\beta} dt, \quad 0 \le t \le 1. \tag{2.63}$$

The equation (2.47) prescribes the bigger value $d\sigma$, with increasing of the site of the quadrat. That means that as a result we obtain plastic strain distribution with the maximum gradient at the RE boundary, β–parameter defines the velocity of increase in $d\sigma(t)$ with increase in the t variable. The value (β +1) – is the normalization coefficient. By substitution of the values $a(t)$ in the equations (2.40)-(2.42), replace the value $\Delta\sigma$ by $d\sigma(t)$ and integrate with respect to the t variable.

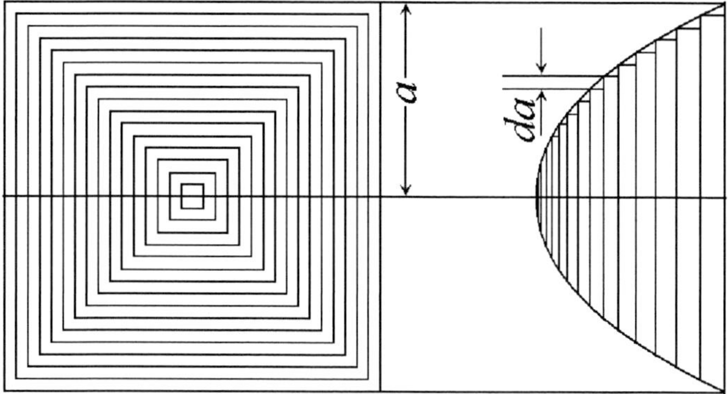

Figure 30. The scheme of a RE of quadratic shape and the profile of plastic deformation in the central cross-section of the given family

Let us represent the result for the stress σ_{xy}, which we further will use when simulating the process of plastic strain localization in polycrystals. After some transformations the equation for σ_{xy} will be written in the following form:

$$\frac{\sigma_{xy}}{\Delta\sigma}=\frac{1}{4}\int_0^1\left[\frac{(\beta+1)(x+y)^2\left[2\left(x^2+y^2\right)+4a^2t^2\right]}{\left(y^2+x^2+2a^2t^2\right)^2-4(x-y)^2a^2t^2}+\frac{(\beta+1)(x-y)^2\left[2\left(x^2+y^2\right)+4a^2t^2\right]}{\left(y^2+x^2+2a^2t^2\right)^2-4(x+y)^2a^2t^2}\right]t^{\beta}dt-1\cdot \tag{2.64}$$

The integration of the elementary fields of plastic deformation of shear ε_{xy} will result in the following distribution of ε_{xy} in quadrat:

$$\varepsilon_{xy}^{p}=\frac{2\Delta\sigma}{\mathrm{E}}\begin{cases}1-\left(\dfrac{|x|}{a}\right)^{\beta+1}, & |x|\geq|y|,\\ 1-\left(\dfrac{|y|}{a}\right)^{\beta+1}, & |y|\geq|x|.\end{cases}$$

The components ε_x and ε_y at that time, are equal to zero. In the case of a simple shear the component ε_{yx} is also equal to zero $\varepsilon_{yx}=\varepsilon_{xy}$.

Analogeously, the components σ_x and σ_y are calculated. Let us notice that when integrating with respect to the stress fields σ_x and σ_y the singularities disappear at the vertexes of square.

Shown in Fig. 27 are the stress distributions σ_{xy} (*a*) and the plastic shear deformation γ^{p} (*b*). The comparison with the distributions of REs of round shape will reveal a number of qualitative discrepancies. The difference in the distribution of plastic deformation, is caused by the difference in shape between circle and square.

The stress fields in the case of relaxation of the shear stresses in the square region are more smoother. When integrating according to the stress fields σ_x and σ_y at the edges of the square the singularities disappear. A homogeneous field of plastic deformation in the circle defines the saame stress drop in all points of the circle (see Fig. 11, β =40). The homogeneous stress field in the square region causes an inhomogeneous stress distribution in that region (see Fig. 27, β =100). Minimum stress is observed in the center of the square. As the side of the square is approached, the stress gradually increases.

The relaxation elements, represented above with the shape of circle and square have been used for the simulation of the involvement of separate structural elements of loaded material into plastic deformation.

3. Simulation of the Effects of Intermittent Flow in Polycrystals

3.1. Peculiarities of the Simulation by the Relaxation Element Method

Using of relaxation elements as the defects, characterising the interrelation betwen plastic deformation with stresses allows to simulate the process of strain localization and to obtain the dependencies of flow stress on the sequence of the involvements of separate structural elements into plastic defromation. Models, developed on the basis of REM operate on principles of cellular automata [33, 34]. The calculation field is divided into a number of cells, playing the role of elements of structure (for example, grains in polycrystals). Each element of the modelled medium possesses the ability to switch its state by discrete jumps in plastic deformation, prescribed by a definite relaxation element. In such a manner, an element of structure is able to increase discretely the degree of plastic deformation and as stress concentration to effect the stress state of the whole volume of solid. The involvement of the structural elements into plastic deformation is realized by definite transition rules (for example at the moment of achieving of a critical value of shear stress). The interaction of the stress fields from different structural elements, undergone plastic deformation, occurs automatically.

When interpreting the results of a simulation one should take into account, that the stress state of a deformed system is controlled only by incompatible plastic deformation, connected with the stresses in the volume of a solid. The relaxation elements in the present case play the role of defects, responsible for the field of plastic deformation. However, it was experimentally found, that the absolute majority of deformation defects at the stage of the developed plastic deformation disappear as a result of annihilation, disappearing at internal interfaces of the structure and exposing at the free surface of the solid. The stresses, caused by these defects will also disappear. What remains is the corresponding field of plastic deformation, not connected with stresses, which satisfy the compatibility condition. Thus, in a general case, one should not neglect the compatible plastic deformation. The problem lies in the fact, that the compatible plastic deformation can not be represented by a definite analytical function of coordinates. The same formchanging of a solid can be realized by a number of variants of mass transfer of the material, not influencing the stress field. That means that the Relaxation Element method can correctly calculate the change of stress fields and plastic deformation in the volume of material with time. For the definition of the total plastic deformation, including the contributions from the defects, exposed at the surface of the material, the additional conditions are necessary. They can be formulated when calculating

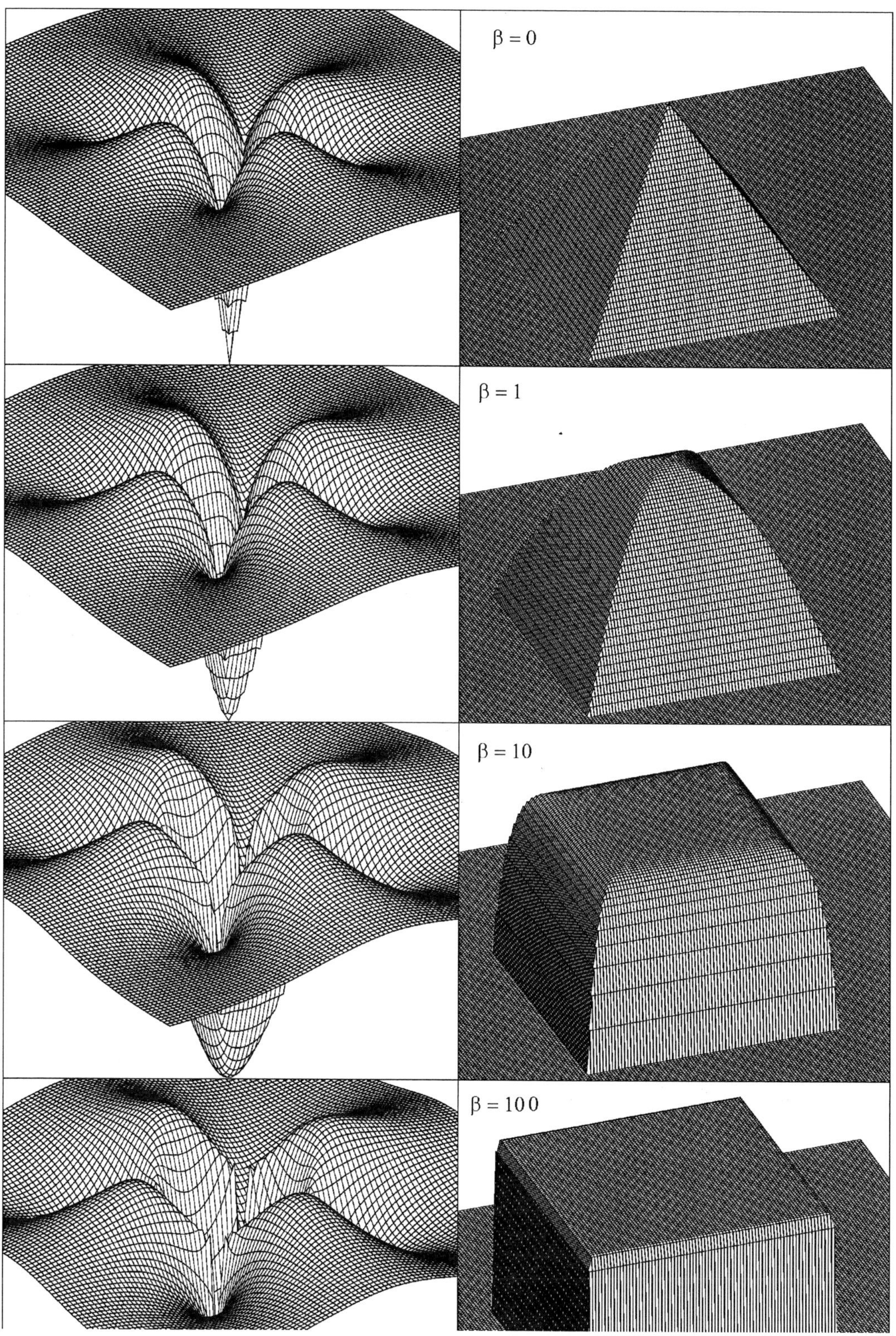

Figure 31. Spatial distributions of stress $\sigma_{x,y}$ (*a*) and plastic deformation of shear (*b*), for the different values of the β –parameter.

the specific structures, taking the known quantitative data from the experiment. In other words, it is necessary to know the prehistory of the development of plastic deformation.

3.2. Modelling of Localization of Deformation in Polycrystals by the Relaxation Element Method

When simulating with the relaxation element method a modelled medium is represented as an conglomerate of a discrete elements, playing the role of structural elements. Each element of the modelled medium has the ability to switch its state by discrete jumps of plastic deformation. The structural element with the field of plastic deformation is the relaxation element itself. With a definite degree of accuracy, this field can be characterized, using the type of RE considered above. By placing into the structural element a definite relaxation element we define along with that corresponding changing in the field of internal stresses in the solid. Thus, the interaction of the fields of internal stresses from the elements of structure, undergone plastic deformation is laid down on the basis of the RE method.

Let us consider an example of the simulation of plastic strain localization in polycrystals under loading, using the relaxation element in the form of a circle. In this case we use an approximation, assuming that on the mesolevel, each crystallite, involved into plastic deformation in the polycrystalline conglomerate can be considered as a relaxation element of round shape. As a criterion of the crystallite involvement into plastic deformation, serves a critical shear stress τ_{cr} in one of two possible directions which was prescribed in each crystallite by the generator of random numbers. The directions imitate the orientation of slip planes in the crystallite lattice. The second slip system was directed at an angle of $\pi/3$ with respect to the first one. Let us define boundary conditions by the assumption that the transition of the crystallite from elastic into the plastic deformed state occurs at the minimum external applied tensile stress σ. That means that the probability of the involvement into plastic deormation is excluded for all crystallites except only one. According to this condition, as new elements with the coordinates (x_n, y_n) transfer into the plastic state under external stress

$$\sigma_n(x_n, y_n) = \frac{2(\tau_{\text{cr}} - \sum_{i=1}^{n-1} \Delta\tau(x_i, y_i))}{1 + 2\Delta\tau(x_n, y_n)/\Delta\sigma}, \tag{3.1}$$

where the sum represents the contribution of all previous REs to the shear stresses.

The contribution of the i–th RE to the effective shear stress from the n- RE is calculated according to the equation

$$\Delta\tau(x_i, y_i) = (\Delta\sigma_y^i - \Delta\sigma_x^i)\sin\alpha_n \cos\alpha_n + \Delta\sigma_{xy}^i(\cos^2\alpha_n - \sin^2\alpha_n), \tag{3.2}$$

where α_n – is the angle between the allowed direction and the axis of loading, the components $\Delta\sigma_x^i$, $\Delta\sigma_y^i$ and $\Delta\sigma_{xy}^i$ are calculated, using the equations (2.12) for REs of the first type, at the values $\Delta\sigma = 50$MPa and $\tau_{cr} = 50$MPa. The calculation field consists of 25x50 points – centers of virtual crystallites, which in Fig. 27 are depicted in the form of hexagons.

Shown in Fig. 27 is the sequence of the patterns of the crystallites involvement into plastic deformation. It is seen that under the present conditions, the model predictes the selforganization of the bands of localized shear. From the very beginning of plastic flow the mesobands divide the material into fragments. The external applied stress (3.1) oscillates near some average value (Fig. 28). Each stress drop is associated with the formation of a separate mesoband embracing several grains.

A different pattern is observed in the polycrystal with rigid inclusions (Fig. 29). The plastic deformation starts at the boundary of the inclusions and is located near it. The macrobands in the conjugate direction of maximum tangential stresses originates from the inclusion. The contribution of additional stress fields from the inclusion result into a lower flow stress of the polycrystal. The development of plastic deformation is realized in this case by the formation of mesobands, causing oscillations of the external stress.

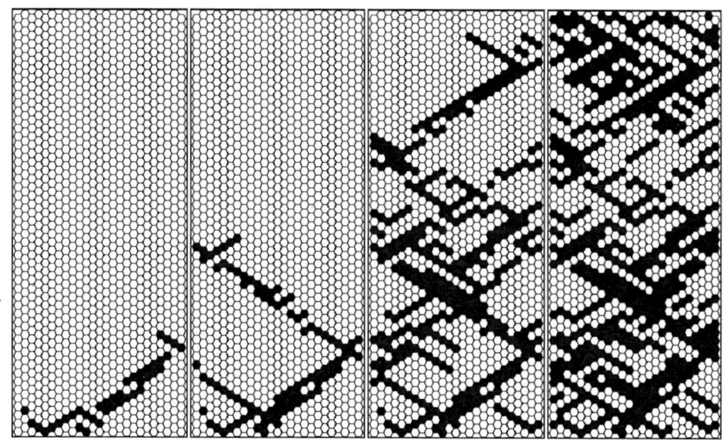

Figure 32. Self-organization of the bands of localized plastic deformation in the modelled 2D-polycrystal under vertical tensile loading

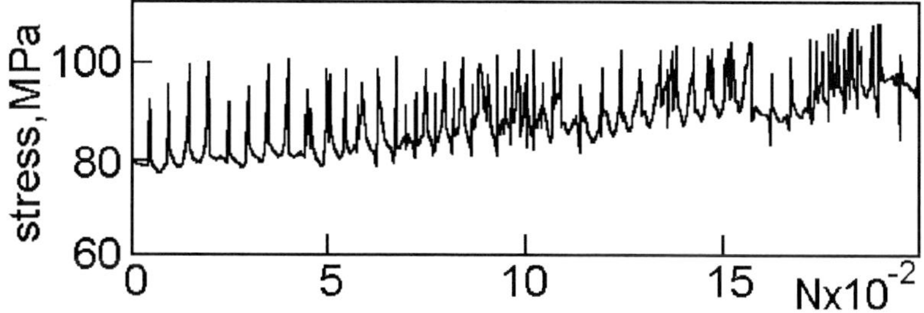

Figure 33. Dependence of the external stress on the number of crystallite, involved into plastic deformation.

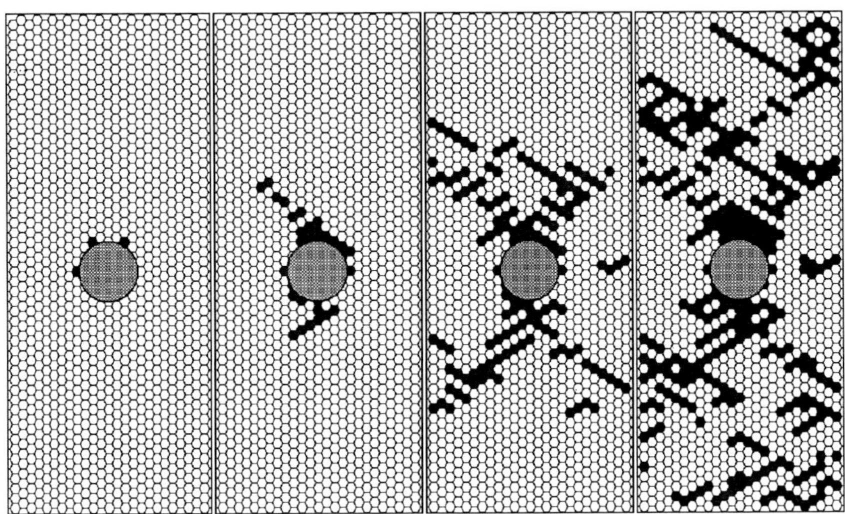

Figure 34: Self-organization of the bands of localized plastic deformation in polycrystal aluminium with a rigid circular inclusion under tensile stress.

Thus, modelling by the relaxation element method reveals qualitative and quantitative distinctions of the developments of plastic deformation in polycrystal structures without the inclusion and with a rigid one. The model presented predicts the jump-like dependence of the external stress on the number of acts of structural element involvements into plastic deformation. Its value (external stress) defines the onset of plastic flow in the grain, where a critical value of shear stress (according to Tresca criterion) is achieved. Since plastic deformation of each new grain changes the field of internal stresses, then the value of external stress oscillates within corresponding limits. Therefore, the stress diagram changes in jumps with increase in the number of grains involvement into plastic deformation.

The presented results of the simulation by the relaxation element method points out to the necessity of further improvement of the model. A simplified approach doesn't allow to describe a temporal evolution of strain localization. Besides that, to measure the value of $\sigma(x,y)=\sigma_{\min}$, according to the equation (3.1), is not possible. That is why any further improvement of the model of plastic strain localization is realized with accounting of the real boundary conditions of loading.

3.4. The Influence of the Rigidity of the Testing Device

Theoretical models of plasticity are developed in assumption of a definite and as a rule simple boundary condition of loading. The changing of the applied load with time is controlled by the change of plastic deformation. It is not possible to realize in practice precise theoretical boundary conditions. Therefore, the loading diagram of the same material depends essentially not only on the mode of loading (tension, compression, bending, torsion), geometrical shape and the dimension of the specimen but on the technical characteristics of the testing device. One of the most important technical characteristics of the machine is the rigidity modulus M. It is defined as a force which is necessary to apply to the punch of the machine in order to shift in it in 1 mm at the rigid coupling of clamps. Such a displacement it possible due to elastic deformations of the parts of the machine from the punch to the clamp. Depending on the M value the machines conditionally are divided in two classes: into rigid with a big value and in a soft with a small value of rigidity modulus. Rigid machines are very sensitive to the quick change in the rate of plastic deformation and react on it by a drop in the load. (Fig. 29*a*). The stress-strain diagrams, obtained on the soft mashine have stair-case type. (Fig. 30*b*).

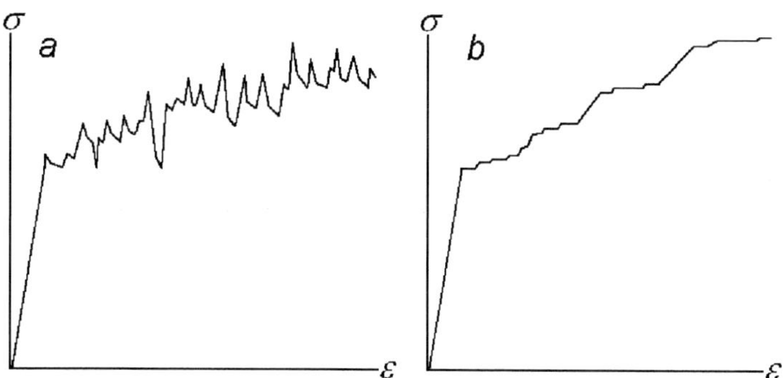

Figure 35. The view of the loading diagram under tensile loading with application of «rigid» (*a*) and «soft» (*b*) testing.

A constant rate of the movement of the punch of the testing machine v_0 under tensile loading defines a velocity of the clamps of the mashine in the unloaded state. In the loaded state in a course of deformation of the specimen, the elastic deformation of the intermediate parts of the machine takes place. Shown in Fig. 31, the parts of the machine which contribute to elastic displacement are depictued in the form of springs.

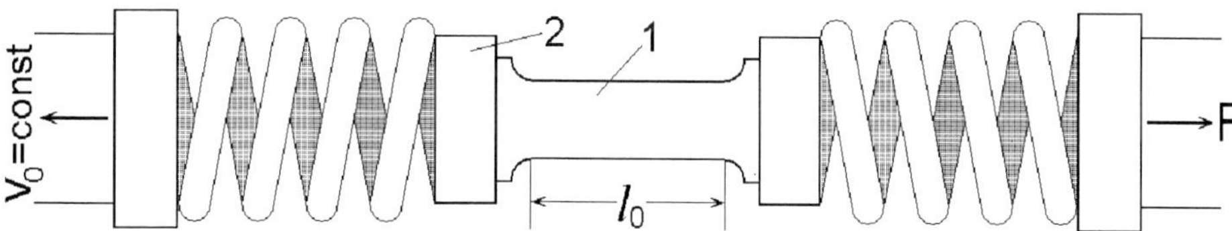

Figure 36. The scheme: 1 – sample, 2 – clamp.

According to definition, the modulus of rigidity of the machine is equal to

$$M=F/\Delta l_m, \quad (3.3)$$

where Δl_m – is the displacement of the clamps with respect of each other, caused by the elastic deformation of the machine itself. Experimentally, the modulus of rigidity can be defined in the following manner. The clamps of the machine are coupled, excluding the possibility to move with respect to each other. Then the loading is switching on till the nominal magnitude of the force F_{nom}, after which the loading is stopped. Next, it is necessary gradually to uncouple the clamps, to measure the distance Δl_m between them and calculate the modulus of rigidity of the machine M, according to equation (3.5).

During the time Δt the punch of the machine will move in a distance

$\Delta l^* = v_0 \Delta t.$

At the rigid coupled clamps it will match to the force

$\Delta f^* = \mathrm{M}\Delta l^* = v_0 \Delta t.$

If the specimen is deformed in the clamps, then due to the deformation the distance between clamps will increase in

$$\Delta l = [\Delta\varepsilon^{e}(\Delta t) + \Delta\varepsilon^{p}(\Delta t)]l, \quad (3.4)$$

where l - is the effective length of the working part, but $\Delta\varepsilon^{e}(\Delta t)$ and $\Delta\varepsilon^{p}(\Delta t)$ – respectively contribution of elastic and plastic deformation of the specimen. In the same length Δl will decrease «the effective length»of the imaginary spring, representing elastic deformation of the intermediate parts of the machine. As a result, the decrease in the force applied to the clamps will take place in a value:

$\Delta f = \mathrm{M}\,\Delta l.$

The resulting changing of the force will be equal to

$$\Delta \mathrm{F} = \Delta f^* - \Delta f = \mathrm{M}(\Delta l^* - \Delta l) = \mathrm{M}(v_0 \Delta t - \Delta l). \quad (3.5)$$

Then the elastic deformation of the specimen is equal

$$\Delta\varepsilon^{e}(\Delta t) = \Delta \mathrm{F}/(\mathrm{SE}) = \mathrm{M}(v_0 \Delta t - \Delta l)/(\mathrm{SE}), \quad (3.6)$$

where S – is the area of the cross-section of the working part of the specimen, E – is Young's modulus.

From the equations (3.6) and (3.8) the change in the length of the specimen in time Δt takes place :

$$\Delta l = l[\Delta\varepsilon^{e}(\Delta t) + \mathrm{M}v_0\Delta t/(\mathrm{SE})]/[1+\mathrm{M}l/(\mathrm{SE})]. \quad (3.7)$$

Plastic deformation of the specimen is equalv to

$\Delta\varepsilon^{p}(\Delta t) = \dot{\varepsilon}^{p}\Delta t,$

where $\dot{\varepsilon}^{p}$ – is the average rate of plastic deformation in a prescribed time interval Δt.

hence,

$$\Delta l = l\Delta t[\Delta\varepsilon^{e}(\Delta t) + \mathrm{M}v_0/(\mathrm{SE})]/[1+\mathrm{M}l/(\mathrm{SE})]. \quad (3.8)$$

By substituting of the value Δl (3.9) in equation (3.7), we obtain the change in the force, applied to the clamps in a time Δt:

$$\Delta F(\Delta t) = M\Delta t(v_0 - v^p)/[1 + Ml/(SE)]. \quad (3.9)$$

here $v^p = \dot{\varepsilon}^p l$ – is the rate of specimen length change due to its plastic deformation.

The sign depends on the difference of the rates of the Punch of the machine and the rate of the change of the length of the specimendue to its plastic deformation v^p. At $v^p = v_0$ the flow plateau will be observed, and at a $v^p > v_0$ – a decrease in external load. The higher the rigidity of the machine M, the higher is the value of the external drop of the load.

The Portevin-le Chatelier Effect is connected with periodical spontaneous arising of the band of localized shear. The accumulation of plastic deformation in the band proceeds so effective, that due to this time the velocity of the increase in the length of the specimen due to its plastic deformation passes ahead the velocity of the movement of the punch of the testing machine. For example, in Al/Al_2O_3 alloys the decrease in external load occurs during 1-2s. The dependence of the change in external load ΔF (3.10) in that time interval on the average rate of plastic in the given time interval for the given alloy is presented in Fig. 32. The amplitude of oscillations of external stress at the pointed parameters of the experiments is seen to be 4MPa if the velocity of plastic deformation of specimen two times passes ahead of the rate of movement of the punch of the machine.

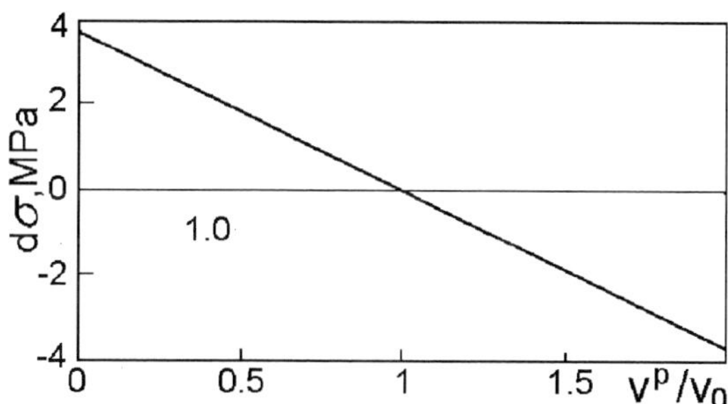

Figure 37. Dependence of the changing of external load on the relative rate of plastic deformation v^p/v_0: E=70GPa, l=18mm, M=1.3x103r'kN/mm,Δt=2s, v_0=5x10-2mm/s.

Equation (3.9) can be transformed for the calculation of the increment of external stress in time interval dt during which the stress relaxation in the value takes place in the crystallite undergone plastic deformation

$$d\sigma = \frac{M(v_0 dt - 2\pi a^2 \Delta\sigma / bE)}{(b + M/E)}. \quad (3.10)$$

Here v_0 – is the velocity of the movement of the punch of the machine, a - is the radius of the crystallite, S, l_0 and b – is the cross-section, length and the width of the working part correspondently.

3.3. Accounting for Edge Effects

More exact accounting of the boundary conditions of loading in the model is planned to perform because of the following circumstances. The equations represented above for the stress fields are valid for an infinite plane. At that time at the lateral boundaries of the modelled specimen there exist normal and tangential stresses, which should be absent according to the boundary conditions of loading.

Removal of shearing stresses have been performed, using the method of reflection [28] following the scheme, depicted in Fig. 33. A system is considered the, consising of three REs, one of which is actual and two others are fictitious. The last creates the fields of inverse stresses. The proposed scheme of RE-position is practically fully compensates the shear components at the edge of the specimen (Fig. 30). It is necessary to take into account the following: An exact compensation of shear stresses is possible only in the case, if two similar defects (real and fictitious) are located symmerically with respect to the half-plane boundary. The edge of the specimen (for example, the right one in Fig. 34) could be considered as a boundary of the half-plane without the presence of other fictitious RE from the opposite (left edge of the specimen=which breaks the strict obeying of the condition of zero tangential stresses at the edge of the half-plane. The stress field is essentially perturbated only in the vicinity being commensurable with the dimension of the RE itself (see Fig. 5, 9, 12, 24). Therefore, if the width of the specimen is much bigger than the dimension of the RE, then the influence of the second fictitious RE becomes negligibly small.

If the RE is exposed at the edge of the polycrystal then there is no need in the presence of a fictitious RE, since the edge of the specimen coinsides with the axis of symmetry of the RE itself, along which the tangential component of stress tensor is equal to zero.

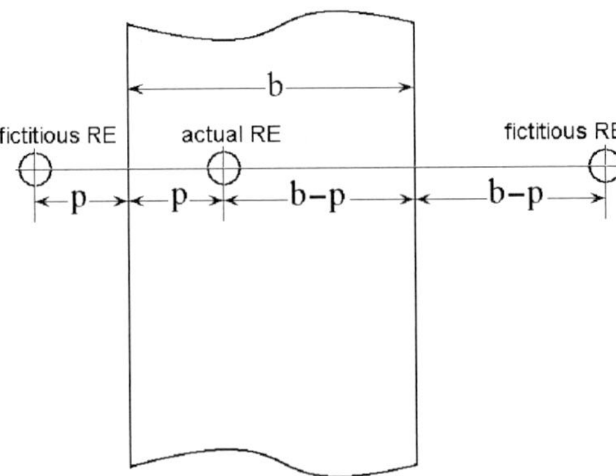

Figure 38. The scheme of compensation of shear stresses at the edge of the specimen

Shown in Fig. 35 is the result of compensation of tangential stresses at the edges of the specimen according to the proposed scheme for the RE of the second type, the components of the fields of internal stresses of which is defined by equations (2.16). The case corresponds to the relation of the dimensions of the diameter of the RE to the width of the specimen as 1/10. RE is located at the distance of half-diameter from the edge. The relaxation element is seen to cause at the edge of the specimen the components of tangential stresses of significant value (Fig. 34a). Application of technique proposed allows easily and simply to compensate for these stresses up to practically negligible values (Fig. 31b). At that time, as seen, the stress distribution near the given edge undergo essential change.

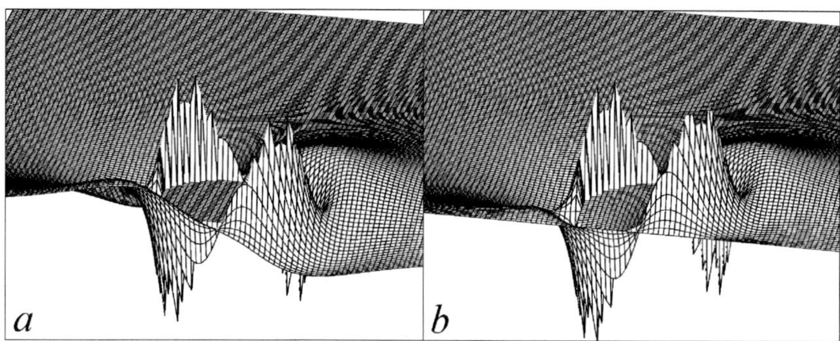

Figure 39. The distribution of shear stresses in the specimen σ_{xy} before(a) and after (b) compensation of them at the edge of the specimen..

For the releasing of normal stresses at the edges of the specimen, the following techniques have been applied. In the present case the edge of the specimen was considered to be a half-plane. A RE creates at the edge of this half-plane normal stresses. A fictitious RE, located symmerically with respect to the boundary of the half-plane, increases these stresses two times. This distribution is symmetrical with respect to the straight line, connecting the centers of real and imaginary relaxation element.

In the mechanics of deformed solids, the Flahmans problem on concentrated force, applied perpendicularly to the surface of an elastic isotropic half-plane, is known [28]. In two-dimensional problems, the dimension of stress is defined as the force, divided by length (N/m).

According to the known formula of Flahmans [28], in the system of coordinates, depicted in Fig. 32, under the influence of an elementary force $dF_x = \Delta\sigma_x dl$, operating at the boundary of the half-plane, in the volume of material, operating at the boundary of the half-plane, in the volume of the material the field of stresses with the components arises

$$d\sigma_x = \frac{2}{\pi}\sigma_x \frac{x^3}{\left(x^2+t^2\right)^2}dt,\quad d\sigma_y = \frac{2}{\pi}\sigma_x \frac{xt^2}{\left(x^2+t^2\right)^2}dt,\quad d\sigma_{xy} = \frac{2}{\pi}\sigma_x \frac{x^2 t}{\left(x^2+t^2\right)^2}dt. \tag{3.11}$$

Here dl – is the elementary segment of the half-plane, in which the stress $\Delta\sigma_x$. In order to compensate the normal stresses, created by the RE at the boundary of the half-plane, it is necessary to prescribe the same distribution of the load at the plane but with the sign reversed. In the system of coordinates in Fig. 32, according to the second equation in the system (2.16), the RE of the second type creates the following distributions of normal stresses $\Delta\sigma_x$ at x=0, which we can write down with the opposite sign:

$$\Delta\sigma_x(l) = \frac{-\Delta\sigma a^2}{p^2+t^2}\left(2\left[1-\frac{2t^2}{p^2+t^2}\right]-\left(1-\frac{8p^2t^2}{\left[p^2+t^2\right]^2}\right)\left(\frac{3a^2}{p^2+t^2}-2\right)\right), \tag{3.12}$$

where l –is the variable, defining at the boundary of the half-plane a location of a specific point in the distribution $\Delta\sigma_x(l)$ (Fig. 35), $\Delta\sigma$ – is the value, defining the degree of shear stress relaxation in the RE (see equations (2.16)).

Substitution of the function $\Delta\sigma_x(l)$ in equations (3.3) and their integrations within the limits–L till L, where L – is the length of the specimen, defines with enough accuracy the changing of the corresponding components of the stress field in the plane specimen and the compensation of the normal stress at the edge of the specimen. This is very good illustrated in Fig. 36, where the distribution of the normal σ_x stresses are represented in the specimen before (*a*) and after (*b*) their compensation at the edge of the specimen. It is seen that the approach applied really results in releasing of normal stresses at the edge of the specimen. In such a manner an essential changing of the character of distribution takes place.

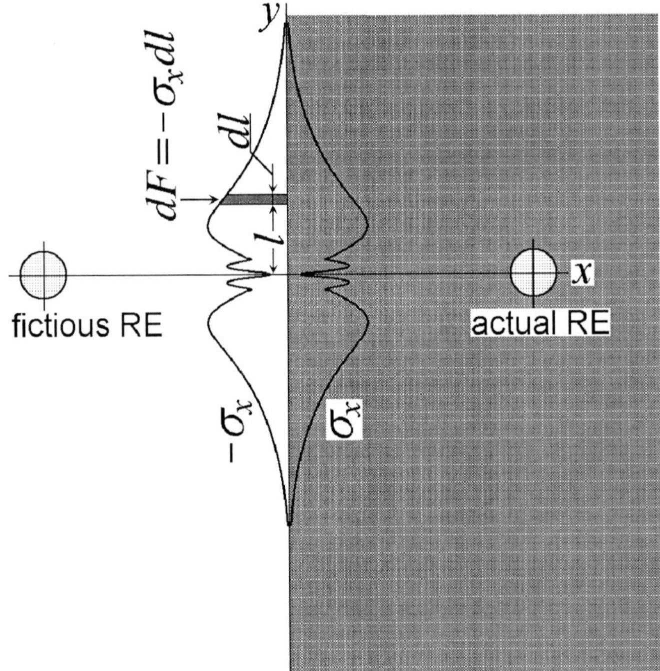

Figure 40. The scheme of the compensation of normal stresse at the edge of the specimen

In a similar way, the analytical expressions have been derived, defining the change in the stress fields in the specimen with accounting of the gradients of plastic deformation as a result of releasing of normal and tangential stresses at the edge of the specimen, which were used further in simulations.

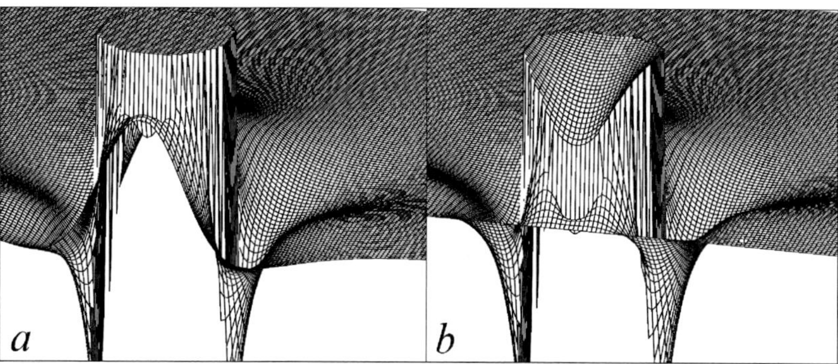

Figure 41. Distribution of normal stresses □x in the specimen before (a) and after (b) their compensation at the edge of the specimen

3.4. Influence of Periodical Boundary Conditions

The examples represented show, that the models, elaborated on the basis of the relaxation element satisfy the properties of cellular automata:

- Locality of rules. Only surrounding elements effect the new state of the structural element
- The number of possible states of the element is finite. For obtaining the new state a finite number of operations is required.
- The state of all elements changes simultaneously, at the end of the cycle but not during the calculation. Therefore the order of the element involvement into plastic deformation during one cycle is defined unambiguously..
- Classical cellular automata posess another property-homogeneity of the system. No one element can differ from another in any peculiarity. However, in practice, the calculation field is limited by the finite number of elements. As a result, an edge effects appears. In order to avoid it one introduces the so-called periodic boundary conditions.

With the aim of verification of the reply of the model to different limitations, imposed on the edge of the specimen, the simulation of the plastic strain localization have been performed for three cases:

1. Without releasing of normal and tangential stresses at the edge of the specimen, caused by the presence of RE,
2. With application of periodic boundary conditions.
3. With releasing of the normal snd tangential stresses at the edges of the specimen.

The initial parameters of the model for all cases were the same. A calculational field consisted of 10x40 calculational points. A RE of the second type was selected as a structural element with the field of internal stresses (2.12), with the radius *a*, being equal to the half of the distance between the points. When simulating in the present case the expressions (3.10) have been used.

Shown in Fig. 35 is the result without accounting of the edge effect. Limitations of the calculation field result in the reflection of the band of localized shear from the edge of the specimen (*a, b*). Further the rapid transfer of plastic deformation along the formed bands of localized shear takes place (*b, c*). Due to this mechanism, the increase in the width of the bands takes place. At the later stage of simulations, the volumes before the formed bands are involved into plastic deformation (*г*). All the calculation field is filled by relaxation elements. The partial fragmentation of the material (*д*) takes placealong some boundaries. After filling out of the whole space with relaxation elements the stage of high workhardening comes.

Another structure of the bands of localized shear is obtained in the case of the application of the periodic boundary conditions (Fig. 37). Artificial shift of the relaxation element from one edge to another at the beginning results to the formation of the parallell bands of localized shear (*a, b*). Then these bands extended and the mesobands of conjugate direction are formed between them (*c, d*). Finally, unordered structure is formed. (*e*).

In the case of accounting for the edge effects following the technique, described in the previous paragraph, we obtain a repid transfer of plastic deformation from one element into another on the mechanism of Lüders band (Fig. 38). A macroscopic deformtion for all three cases starts at the same stress and is accompanied by decreasing in stress (Fig. 39). After the whole volume is spanned by plastic deformation, the stage of sharp work-hardening is observed. In the intermediate interval the loading diagramms differ. In the case of not accounting of edge effects (curve 1) after the flow «tooth » the stage of weak work hardening comes. On average as the space is filling out with plastic deformation the flow stress increases. Account for the edge effect by releasing of the normal and tangential stresse at the edge of the specimen is not accompanied by essential work-hardening (Curve 3). Using of periodical bondary conditions results at the beginning to the weak work hardening. Transfer to the stage with high strong hardening occurs already before the whole volume will be embraced with plastic deformation (curve 2).

Comparisons with the real loading diagrams show, that the full coincidence is observed in the case of accounting of edge effects by releasing of normal and tangential stresses at the edges of the specimen. A sharp flow tooth and the flow plateau is observed at the initial stage of macroplastic deformation of many polycrystalline materials. The experience shows that the flow plateau is connected with the propagation of Lüders bands. The case of non-accounting of the edge effect qualitatively correctly reflects the behaviour of polycrystalline materials, the surface of which from both sides is blocked by high-strength coatings. Than, really, plastic deformation is developing on the mechanism of reflecting waves [35]. Accounting to the periodic boundary conditions they do not reflect any ordered mechanisms of the onset of macroplastic flow.

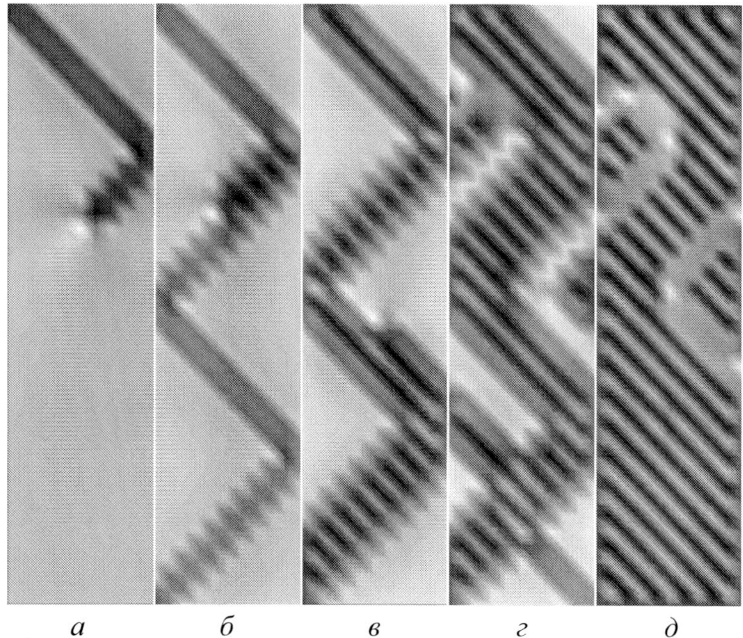

Figure 42. Formation of the band structures without accounting of edge effects

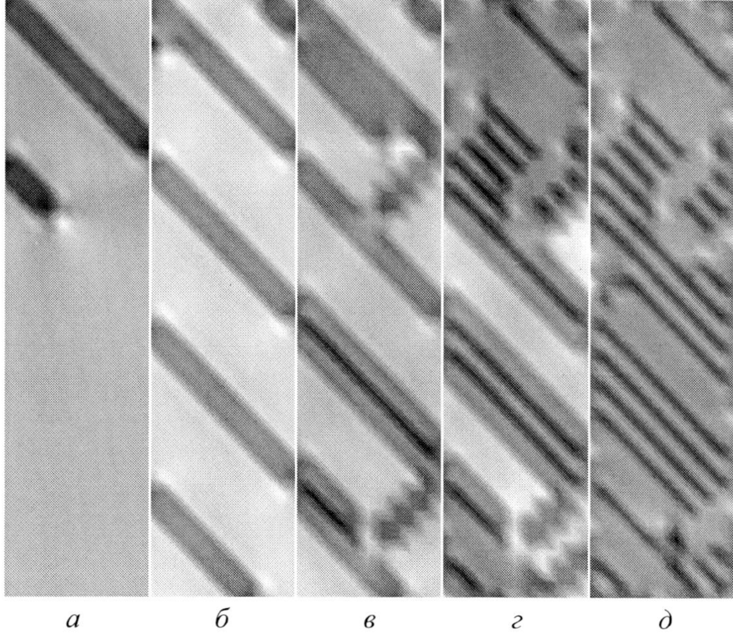

Figure 43. Formation of the band structures with accounting of the periodic boundary conditions

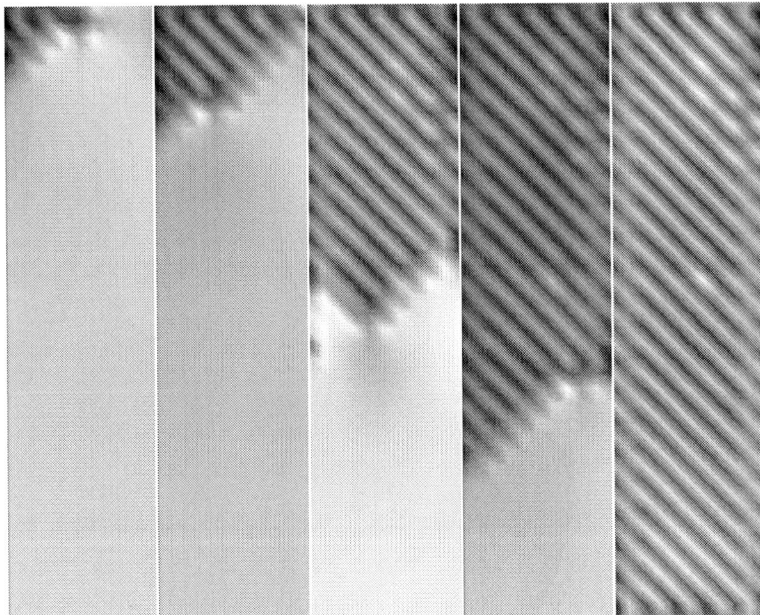

Figure 44. Formation of the band structures with accounting of the releasing of normal and tangential stresses at the edges of the specimen.

The analyse performed show that when simulating the effect of plastic strain localization the account for periodic boundary conditions doesn't result in the practically important results. In practice, when solving definite tasks the necessity arises ro refuse from some of the properties of the classical models of cellular automata.

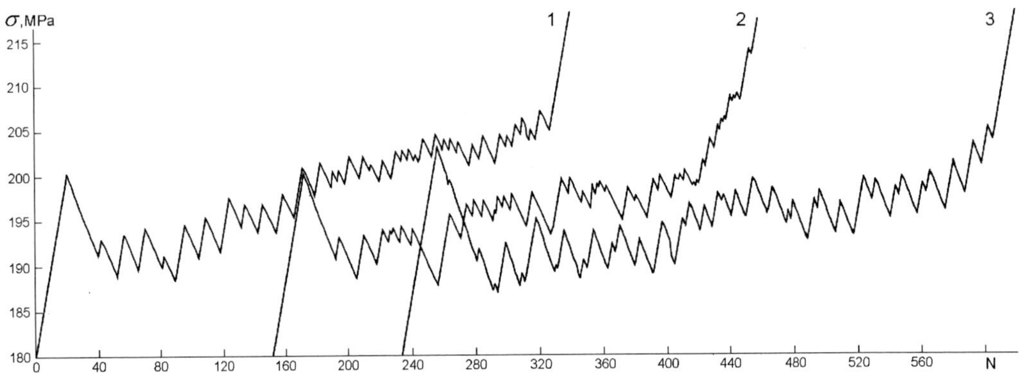

Figure 45. Modelled diagrams of loading of the specimen without accounting of the edge effects (1), with accounting of periodical boundary conditions (2) with accounting of the releasing of the normal and tangential stresses at the edges of the specimen (3).

3.4. Simulation of the Jump-Like Propagation of the Macroband of Localized Shear

When simulating the process of crystallites involvement into plastic deformation the relaxation elements of other types have been used. Equation (2.19) was put as the basis of the algorithm, described in item 3.2, for the stress distribution of pure shear in conjugate directions at an angle of 45° with respect to the tensile axis. Such a simplification is justified by the fact that in any case the front of localized shear is usually oriented at an angle of maximum tangential stresses.

The calculation field is represented in the form of a matrix from 10x50 points – the centers of crystallites. Onset of plastic deformation was initiated at the edge of the calculation field by relaxation elements RE represented in Fig. 10 and equations (2.19) and (2.20) at the values h=0 and β=6. This grain created a non-homogeneous field of shear stresses (2.19) in the volume of the polycrystal. The field of plastic deformation in the crystallite at h=0 is characterized by the tensor with the components

$$\varepsilon_y(x,y) = 2\frac{\Delta\sigma}{E}\left[1-\left(\frac{r}{a}\right)^{\beta+1}\right],\ \varepsilon_x(x,y) = -\varepsilon_y(x,y),\ \varepsilon_{xy}(x,y) = 0, \qquad (3.14)$$

By the formula (3.1), the minimum external stress was calculated, at which at the center of any crystallite the shear stress achieves its critical value τ_{cr} = 50MPa, according to the Tresca's criterion. The sum in the numerator in Eqv. 3.4 defines the contributions from the previous relaxation elements in shear stress. Minimum values of this function $\sigma(x,y)_{min}$ correspond to the minimum external stress, at which there is a possibility of involvement into plastic deformation of a single crystallite. The coordinates of the point, in which $\sigma(x,y)=\sigma(x,y)_{min}$ have been found and a new relaxation element of considered type was put there. In such a manner, this crystallite obtains a discrete value of plastic deformation (3.4) and creates around it a corresponding field of internal stresses (2.19). Thus, the crystallite effects the stress field in the whole volume of the solid. Interaction of the fields of internal stresses from the crystallites, undergone plastic deformation, together with the external applied stress result in the formation of meso- and macrobands of localized plastic deformation.

Shown in Fig. 34 is the result of the simulation of the process of strain localization. It is seen that under the operation of the changing the inhomogeneous stress field in the volume of the polycrystal a consequent development of the band structure of type B occurs, being typical for a clearly pronounced PLC-effects. The jump-like propagation of the process of strain localization along the working part of the specimen in the form of macrobands of localized shear occurs.

The formation of a separate macro-band starts from the initiation at the edge of the specimen of the mesoband , with the width of one crystallite diameter (see frames 1, 3, 6, 9, 10), which goes through the whole cross-section of the specimen (frames 2, 4, 5, 7), oriented at an angle of 45° with respect to the tensile axis. The development of mesobands occurs spontaneously without any increase in the external stress. Further, the expansion of this band occurs on the mechanism of Lüders band propagation. (frames 1-5) by consequent involvement of the grains along the front of initially formed mesoband. A set of three mesobands composes a fully formed macroband (frame 5). At a definite distance from the macroband, near the edge of the specimen a zone of increasing tangential stresses is created (dark background at the edge of the specimen in frame 8). The achieving of the critical value τ_{cr} defines the initiation at the edge of the specimen and development of a new macroband of localized shear (frames 6, 9, 10). The process of initiation and development of a band of localized plastic deformation repeats periodically.

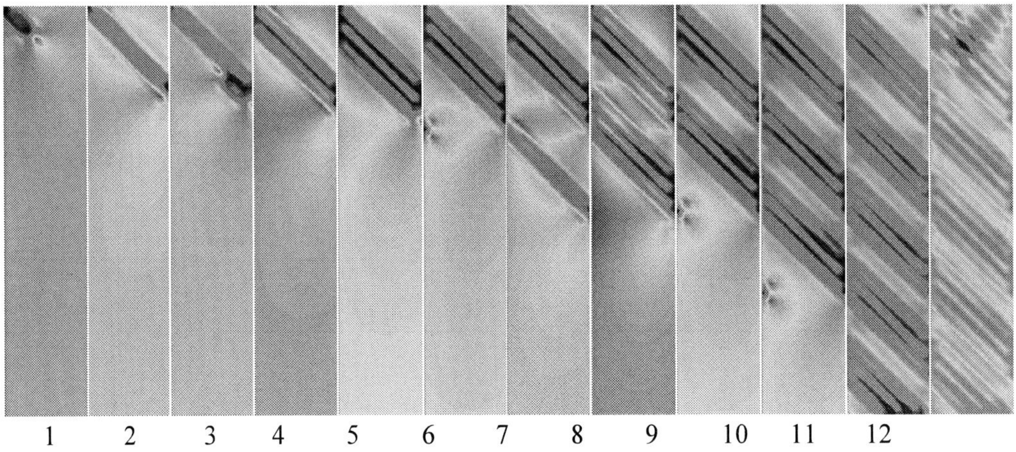

Figure 46. Consequent formation of meso- and macrobands of localized deformation.

After the process of strain localization achieves the opposite end of the specimen (frame 11), the repeated formation of the macrobands takes place, but already in conjugate direction of the maximum shear stresses (frame 12).

Shown in Fig. 35 is the dependence of external stress on the number of grain involvements into plastic deformation. In fact, this dependence represents by itself a loading diagram of the modelled polycrystal, i.e. each *N*- act defines a definite quantum (3.4) of plastic deformation.

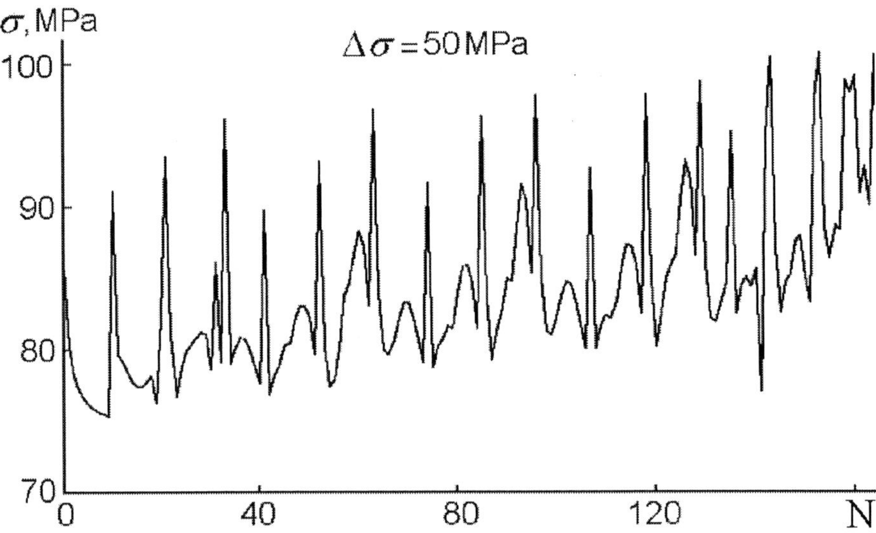

Figure 47. Effect of interrupted flow at the curve of the dependency of external stress on the number of grains, involved into plastic deformation..

Each peak is connected with the initiation of a mesoband of localized deformation at the edge of the specimen. The formation of mesobands occurs spontaneously at the decreasing external stress. The first mesoband from three, composing the macroband requires for its initiation the higher external stresses than others. The lowest external stress corresponds to the initiation of the second mesoband.

In the course of the development of deformation the onset of the formation of a new macroband occurs at higher external stresses in comparison with the previous band of localized deformation (BLD). Thus, the change in the field of internal stresses in the chosen mode of intermittent flow results in the effect of work-hardening.

A similar result is obtained for a polycrystal with a noticeably larger number of grains. (50x200). (Fig. 28). Besides the high-frequent oscillations of external stress, here it is observed a long-periodical modulation of stress, connected with arising and formation of new macrobands of localized shear.

The evolution of the development of strain localization in the modelled polycrystal qualitatively repeats the regularities of the band formation in alloys with clearly pronounced Portevin-le Chatelier (PLC) effect at the stage of jump-like propagation of the band along the working part of the specimen.

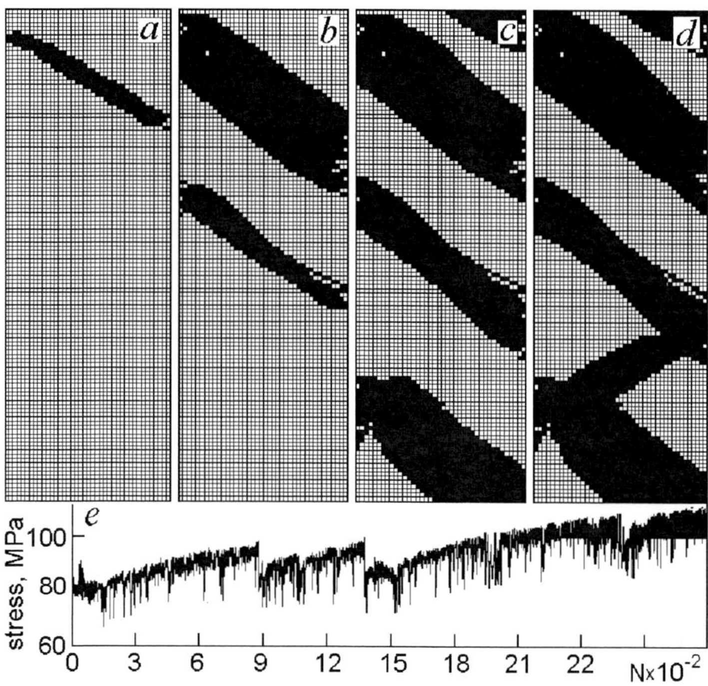

Figure 48. Jump-like propagation of the process of localized deformation (*a-d*), and jump-like loading diagramm (*e*).

Modelling of the Effects of Intermittent Flow in Polycrystalline Materials

In the REM each act of element involvement into plastic deformation is associated with a definite time interval, which we define from the physical reasonings. So, for example in the macroband of the polycrystals with the dimension of the crystallites of 40 μм there are not less than $4x10^3$ grains. The band is formed in 1-2s. Hence, a separate grain is involved into plastic deformation for the time of order $5x10^{-4}$s.

When simulating the localization of deformation in polycrystals under tensile loading we used the equation (3.10). The involement of a grain into plastic deformation was considered to proceed consequently. The time of relaxation *dt* of the grain involvement into plastic deformation was defined by the value which corresponds to the one in the macroband of localized shear of the composite $Al/10\%Al_2O_3$. The involement of a separate grain into plastic deformation occurs when in its center $\tau = \tau \geq \tau_{cr}$, i.e. τ exceeds the critical shear stress, which we associate with the stress of unpinning dislocations [4 - 6, 8, 36-38]. The parameter $\Delta\sigma$ includes in itself the mechanisms of plastic defor-mation, i.e. the ability of the material for plastic form changing:

$$\Delta\sigma = 2(\tau_{max} - \tau_0),$$

where τ_{max} is the maximum shear stress in the direction at an angle 45° (in accordance with the Mises-Tresca criterion), τ_0 is the stress of free movement of dislocations. The σ-ε-diagram was constructed by summing the increment $d\sigma$, using the obtained expressions.

Thus, the expression (3.11) includes in itself the following characteristics of the material

- E is Young's modulus
- *dt* is the relaxation time of a separate structural element when involving into plastic deformation,
- a is the size (radius) of structural element,
- τ_{cr} is the critical stress of dislocations unpinning
- τ_0 is the resistance of unpinning (free).

Along with them there are characteristics, defining the boundary conditions of loading:

- v_0 velocity of movement of the punch of the machine
- b is the width of the working part of the specimen
- M – is the rigidity modulus of the testing machine.

The influence of the rigidity of the tensile machine on the loading diagrams of low-carbon steel with E = 210000 MPa [31] is depicted in Fig. 39.

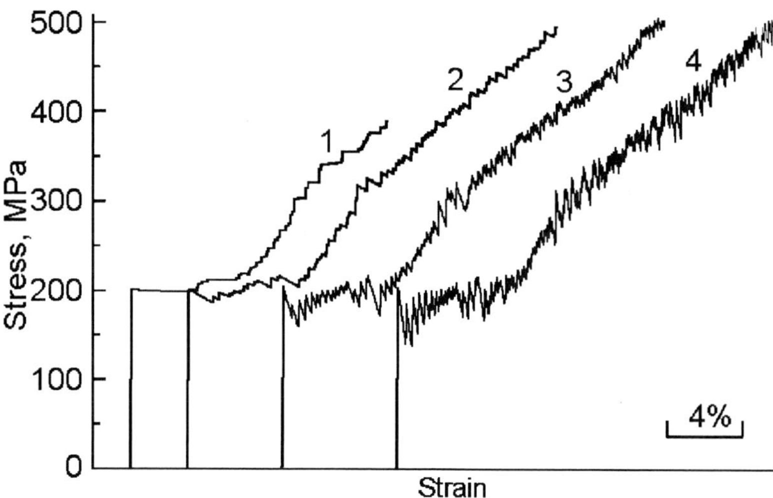

Figure 49. The influence of the rigidity modulus of the machine M on the type of loading diagrams: M, kN/mm = 1.3 x10^2 (1), 1.3 x10^3 (2), 1.3 x10^5(3), 1.3 x10^8(4).

The curve 1 for the «soft» mode of loading (M = 1.3x$10^2$3kN/mm) has stair-case type. As the rigidity M of the machine increases from 1.3x10^2 to 1.3x10^8 kN/mm the curve takes a more saw-tooth shape. The yield drop appears and grows. The amplitude of external stress oscillations increases. At all curves the flow plateau is obsered, after which the stage of work-hardening follows, caused by the increase in internal stress field from relaxation elements. The flow plateau is formed on the mechanism of the Lüders band propagation, when the crystallites are involved into plastic deformation, consequently filling up the working part of the specimen. Repeated involvement of the grains into plastic deformation occurs at higher external applied stresses.

The rate of loading exerts much influence upon the σ–ε-curves. The less the rate of loading, the more pronounced is the effect of intermittent flow (Fig. 33). The increase in the loading curve results in a decrease in the amplitude of the oscillations of the external stress. Starting from the definite rate of loading, there exist no oscillations of external load at the curves (curve 5). A further increase in the rate of loading results in disappearing of the yield drop and the flow plateau (Fig. 33). The flow stress increases and the effect of a sharp yield stress disappears (curve 7 and 8 in Fig. 34).

Along with the influence of the rigidity of the machine and the rate of loading, defining the boundary conditions of loading, the effect of the characteristics of the material itself have been considered at the same other parameters of the model. Shown in Fig. 41 are the σ–ε–curves for the different values of τ_{cr}. If the dislocations are not blocked by atoms of admixture (when $\tau_{cr} = \tau_0$), than the phenomenon of intermittent flow is not observed (low curve). As the τ_{cr} increases, the Portevin-le Chatelier Effect arises and is enhanced. At that time the flow stress, flow plateau and the amplitude of the peaks at the loading diagrams increase.

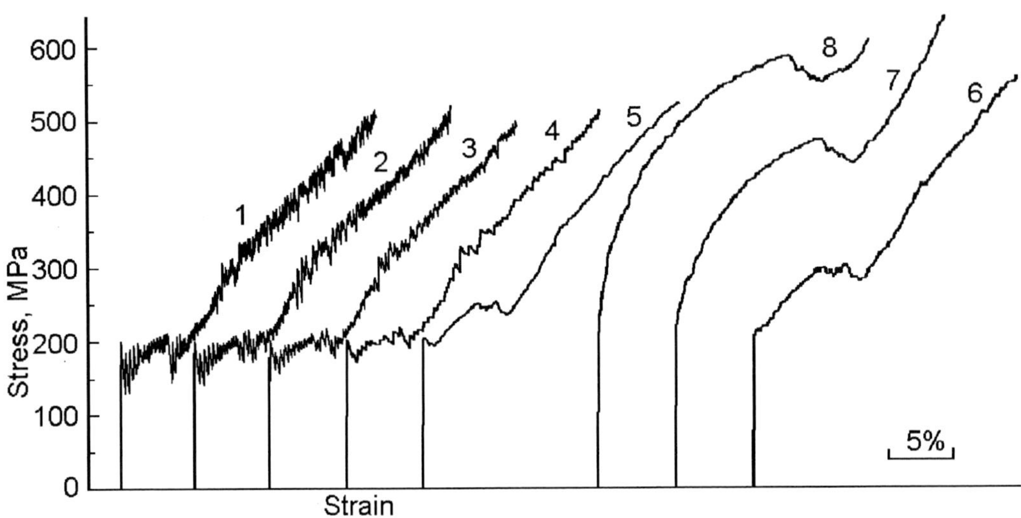

Figure 50. The influence of the velocity of the free movement of punch of the testing machine v_0 on the type of loading diagrams: v_0, mm/s = 1(1), 10 (2), 20 (3), 30 (4), 40 (5), 50 (6), 80 (7), 110 (8).

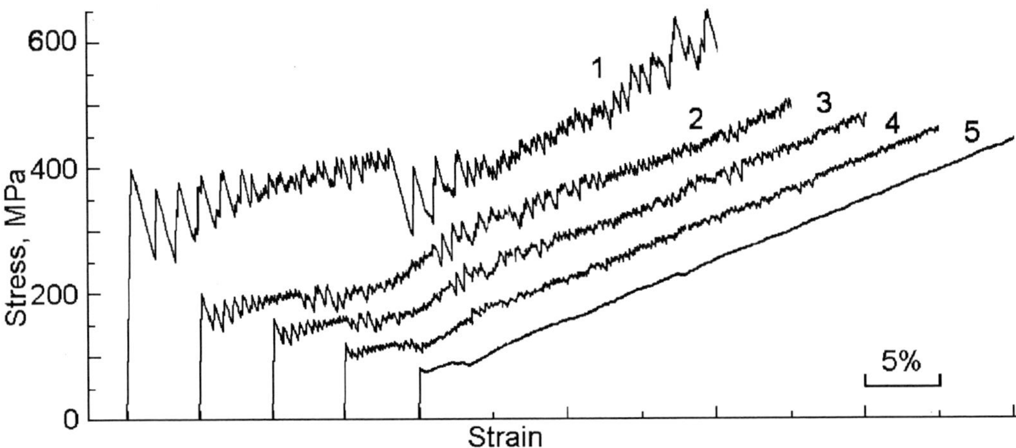

Figure 51. Influence of critical shear stress τ_{cr} on the form of loading diagrams:

τ_{cr} = 200(1), 100(2), 80(3), 60(4), 40(5).

The performed simulations allowed us to reveal qualitative and quantitative changing curves of loading, depending on the characteristics of the material itself and on the boundary conditions of loading. The obtained characteristics of the changes of the qualitative and quantitative loading diagrams when varying the parameters of the model, are in agreement with known experimental findings. [4, 6 - 8, 39 - 42].

4. CONCLUSION

In the present work the possibilities of a new method - the relaxation element method (REM) are demonstrated. With the help of the this method the examples of calculation of stress-strain state of elastic plane with the sites of plastic deformation of the various geometrical shapes and dimensions have been shown. The examples of the construction of the sites with the gradients of plastic deformation are given. The stress-strain state of the plane with round inclusions is described. An analytical description of the plane with the band of localized shear of different orientations with respect to the tensile axis have been performed. The original analytical solution for the field of internal stresses of the band of localized shear of rectangular shape have been obtained, the specific case of which is the site of homogeneous plastic deformation of square shape.

The possibilities of simulations with the presented method of the effects of plastic strain localization have been discussed. The influence of the rigidity of the testing device on the effects of interrupted flow have been analyzed. The different techniques of accounting of the edge effect in the model of cellular automata have been considered.

The important results, testifying to the high predictive possibilities and advantage of the proposed method have been obtained. They are in a good agreement with the known experimental data:

- The effect of intermittent flow is the consequence of the formation of meso- and macrobands of localised deformation;
- In the changing field of stresses the jump/like displacement of the process of strain localiyation along the working part of the specimen takes place;
- An increase in the rate of loading supresses the effect of interrupted flow;
- A saw/like type of the σ-ε-curves is typical for the rigid mode of loading (the device with the high value of rigiditz mosulus устройство), in the soft mode of σ-ε loading curves have stair case shape;
- The folloing peculiarities of the development of strain localisation have been elucidated:
- The macroband formation occurs on the mechanism of Lüders band propagation and is accompanied by the decrease in external stress.
- The structure of a separate macroband consists of a number of mesonbands, being oriented along the direction of maximum tangential stresses.
- The accumulation of the field of internal stresses in the volume of a polycrystal results in the effect of strain hardening.

The necessary condition of the PLC-effect manifestation is the fact that the stress of onset of plastic deformation of structural element is essentially higher than that at which the following plastic deformation can proceed. Sufficient is the condition at which the velocity of length increase of the specimen due to its plastic deformation periodically exceed the velocity of the movement of the clamps of the mechine in unloaded state.

4. Conclusion

In the present work the possibilities of a new method - the relaxation element method (REM) are demonstrated. With the help of the this method the examples of calculation of stress-strain state of elastic plane with the sites of plastic deformation of the various geometrical shapes and dimensions have been shown. The examples of the construction of the sites with the gradients of plastic deformation are given. The stress-strain state of the plane with round inclusions is described. An analytical description of the plane with the band of localized shear of different orientations with respect to the tensile axis has been performed. The original analytical solution for the field of internal stresses of the band of localized shear of rectangular shape have been obtained, the specific case of which is the site of homogeneous plastic deformation of square shape.

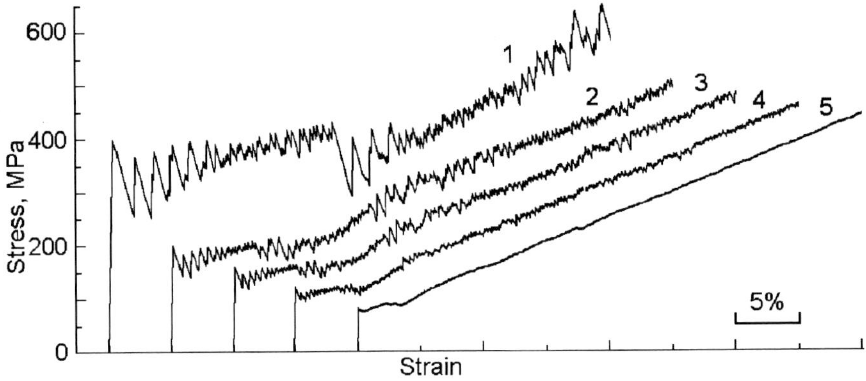

Fig. 52. Influence of critical shear stress τ_{cr} on the form of loading diagrams:

τ_{cr} = 200(1), 100(2), 80(3), 60(4), 40(5).

The possibilities of simulations with the presented method of the effects of plastic strain localization have been discussed. The influence of the rigidity of the testing device on the effects of interrupted flow has been analyzed. The different techniques of accounting of the edge effect in the model of cellular automata have been considered.

The important results, testifying to the high predictive possibilities and advantage of the proposed method have been obtained. They are in a good agreement with the known experimental data:

- The effect of intermittent flow is the consequence of the formation of meso- and macrobands of localised deformation;
- In the changing field of stresses the jump/like displacement of the process of strain localiyation along the working part of the specimen takes place;
- An increase in the rate of loading supresses the effect of interrupted flow;
- A saw/like type of the σ-ε-curves is typical for the rigid mode of loading (the device with the high value of rigiditz mosulus устройство), in the soft mode of σ-ε loading curves have stair case shape;

The following peculiarities of the development of strain localisation have been elucidated:

- The macroband formation occurs on the mechanism of Lüders band propagation and is accompanied by the decrease in external stress.
- The structure of a separate macroband consists of a number of mesonbands, being oriented along the direction of maximum tangential stresses.
- The accumulation of the field of internal stresses in the volume of a polycrystal results in the effect of strain hardening.

The necessary condition of the PLC-effect manifestation is the fact that the stress of onset of plastic deformation of structural element is essentially higher than that at which the following plastic deformation can proceed. Sufficient is the condition at which the velocity of length increase of the specimen due to its plastic deformation periodically exceeds the velocity of the movement of the clamps of the mechine in unloaded state.

ACKNOWLEDGEMENTS

The work is funded by the German Research Foundation (DFG), project DFG Schm 746/52-2, 746/52-3 and the Russian Foundation for Basic Researches (RFBR), project № 07-08-00144.

REFERENCES

[1] Sutton, A.P.; Balluffi, R.W. *Interfaces in Crystalline Materials*; Clarendon Press: Oxford, UK, 1995.

[2] Panin V.E., Proc. Int. Conf. MESOMECHANICS 2000: Role of Mechanics for Development of Science and Technology; Tsinghua University Press: Beijing, 2000, 1, 127–142.

[3] Panin, V.E; Egorushkin V.E.; Panin A.V. *Phys mesomech* 2006, *9 (3)*, 9-22.

[4] Estrin, Y; Kubin L.P. In *Spatial Cupling and Propagative Plastic Instabilities: Continuum Models of Materials with Microstructure*; Mühlhaus, H.-B.; Ed.; John Wiley&Sons Ltd., 1995; pp 395-450.

[5] Klose, F.B.; Ziegenbein, A.; Weidenmüller, J.; Neuhäuser, H.; Hähner, P. *Comput Mat Sci* 2003, 26, 80-86.

[6] Zhang, Q.; Jiang, Z.; Jiang, H.; Chen, Z.; Wu, X. *Int J Plasticity, 2005, 21*, 2150-2173.

[7] Deryugin, E.E; Panin B.E.; Schmauder S.; Storozhenko I.V. *Phys Mesomech,* 2001, 4 (3) 35-47.

[8] Krishtal, M.M., *Phys mesomech* 2001, *7 (5),* 5-45.

[9] Panin, V.E.; Grinyaev, Yu.V.; Danilov, V.I. *Structural level of plastic deformation and fracture*; Nauka: Novosibirsk, RF, 1990 (in russian).

[10] Deryugin, Ye. *Relaxation element method*; Nauka: Novosibirsk, RF, 1998 (in russian).

[11] Deryugin, Ye.Ye.; Lasko, G.V. In *Physical mesomechanics and computer-aided designing of materials*; Panin, V.E.; Ed.; Nauka: Novosibirsk, RU, 1995; Vol. 1, pp 131-161.

[12] Deryugin Ye.Ye., Lasko G.V., Schmauder S. Relaxation element method, Computational Materials Science 11 (3), (1998) 189-203.

[13] Deryugin, Ye. Ye. *Computational Material Science* 1999, *19*, 53-68.

[14] Stress relaxation testing. Symposium on Mechan. Testing, American Soc. for Testing and Materials, Kansas City Mo., 24, 25 May 1978, ed. Fox A. Philadelphia, Pa.: American Soc. for Testing and Materials, (1979) 216 p.

[15] Creep and stress relaxation in metals / Ed. by I. A. Oding. Academy of Sciences of the U.S.S.R. All-Union Institute of Scientific and Technical Information. Edinburgh: Oliver & Boyd, X (1965) 377S.

[16] Atlas of stress-strain curves / ed. by Howard E. Boyer. Metals Park, Ohio: ASM International, XX (1987), 630 p.

[17] Timoshenko, S.P.; Goodier, J.N. *Theory of elasticity*; 3-rd edition; McGraw Hill: New York, NY, 1970.

[18] Hahn, H.G. *Elastizitätstheorie*: *Grundlagen der linearen Theorie und Anwendungen auf eindimensionale, ebene und räumliche Probleme*; B.G. Teubner: Stuttgart, 1985.

[19] Musheliŝvili, N.I. *Some Basic Problems of the Mathematical Theory of Elasticity*; P. Noordhoff Ltd.: Groningen-Holland, 1953.

[20] Eshelby, J. D. *Continuum theory of defects*, *Solid State Physics*, N.Y., 1956; 3.

[21] de Wit, R. In *Fundamental aspects of dislocation*; Simons, J.A.; R. de Wit; Bullough, R.;Eds; Nat. Bur. Stand.; (US) Spec. Publ. 317, 1970, vol. I, pp 651-673.

[22] Likhachev, V.A.; Volkov, A.E.; Shudegov V. E. *Continuum heory of defects*; Publishing house of Leningrad university: leningrad,, RU, 1986, 232 p.

[23] Morse, P.M; Feshbach, H. *Methods of mathematical physics*; McGraw-Hill: New York, Vol. 1, 1953.

[24] Tewary, V.K.; Wagoner, R.H.; Hirth J.P. *J Mater Res 1989,* 113 (4).

[25] Kirsch, G. *Zantralblatt Verlin Deutscher Ingenieure* 1898, 42, 797-807.

[26] Deryugin, Ye.Ye.; Lasko, G.V.; Smolin, Yu. E *Russian Physician J* 1995, 38(5), 449-454.

[27] Levin, V.A. *Many-folded superposition of great deformations in elastic and viscous-elastic bodies*; Fizmatgiz: Moscow, RU, 1999.

[28] Crouch, S.L.; Starfield, A.M. *Boundary element methods in solid mechanics*; George Allen & Unwin: London, UK, 1983.

[29] Eshelby, D.E. *Definition of the stress field, which was creating by elliptical inclusion*, Proc. Roy. Soc., 1957 (1226).

[30] Eshelby, D.E., *Elastic field outside the elliptical inclusion*, Proc. Roy. Soc., 1959, A 252, (1271).

[31] Physical values: Handbook, Energoatomizdat, Moscow, RF, 1991.

[32] A.V. Mali, S.J. Singh, *Deformation of Elastic Solids*, Prentice Hall: New York, NY, 1992.

[33] Gould, H.; Tobochnik, Ya. *An introduction to computer simulation methods. Part. 2 Application to physical systems;*. Mir: Moscow, 1990.

[34] Lasko, G.V.; Deryugin, Ye.; Schmauder S. *LNCS* 2004, 3305, 375-384.

[35] Panin, V.E.; Deryugin, Ye.; Schmauder, S. *The Physics of Metals and Metallography* 2003, 96, Suppl. 1, S1-S15.
[36] Rizzi, E; Hähner, P. *Int J Plasticity* 2004, 20, 121-165.
[37] Zang, S.; McCormick, P.G.; Estrin, Y. *Acta Materialia* 49, 2001, 1087-1094.
[38] Hirth, J.P.; Lothe, J. *Theory of dislocations*, 2nd ed.; Wiley: New York, NY, 1982, 857p.
[39] Pink, T.; Bruckbauer, P.; Weinhandl, H. *Scripta Marerialia*, 1998, 38(6), 945-951.
[40] Toyooka, S.; Widiastuti, R.; Zhang, Q.; Kato, H. *Jpn J Appl Phys* 2001, 40 (2A), 873-876.
[41] Dillon, O.W. *J Mech Phys Solids* 1963, Vol. 11, 289-304.
[42] Panin, V.E.; Grinyaev, Yu.V. *Phys mesomech* 2003, V.6. № 4, 9-36.

INDEX

A

Abelian, 310, 311
absolute zero, 322
absorption, 30, 203, 206, 208, 209, 210, 317
absorption spectra, 30, 317
acceptor, 387, 395, 396, 432, 434, 435
accessibility, 353
accidental, 80, 353
accounting, 488, 524, 528, 530, 532, 533, 534, 541
accuracy, 46, 134, 271, 338, 360, 456, 472, 475, 517, 522, 530
acetylene, 396, 397, 399, 401, 414
ACF, vii, 1, 13, 14, 15, 17, 18, 19, 20, 22
achievement, 329
acoustic, 67, 79, 192, 198, 207, 209, 214, 215, 250, 324
acoustic emission, 79
acoustical, 232
activation, 42, 44, 266, 268, 269, 279, 282, 283, 284, 286, 475
activation energy, 42, 44, 266, 268, 282, 283, 284, 475
adaptation, 77
adiabatic, 75, 195
adjustment, 332, 437
adsorption, 32
aerospace, 59
AFM, 191, 210
Ag, 18, 185, 217, 238, 268, 269, 286
age, 212, 228, 352, 360, 366
ageing, 57
aggregates, 352
aging, 33, 77
aging process, 77
air, 57, 58, 59, 64, 66, 121, 270, 320, 353
aircraft, 87, 88, 93, 94, 95, 108
alcohol, 356
algorithm, xi, 86, 355, 445, 446, 447, 455, 535
alkali, 178
alkaline, 185, 193, 211, 235
alkane, 429, 430
Allah, 114
alloys, ix, 52, 57, 61, 63, 65, 66, 67, 79, 80, 84, 120, 126, 129, 173, 174, 269, 285, 287, 289, 290, 309, 312, 322, 325, 326, 327, 527, 537
alpha, 116
alternative, 52, 192, 321, 336, 386, 387, 396, 435, 437, 446
alternatives, 55
aluminium alloys, 57, 66, 79
aluminum, 53, 57, 58, 62, 66, 68, 70, 79, 84, 94, 101, 103, 104, 110, 115, 116, 117, 322, 324, 503, 524
ambient pressure, 210, 235, 239, 243
amino, 434, 437
amino groups, 437
amorphous, vii, 1, 2, 33, 39, 43, 44, 45, 47, 126, 127, 128, 133, 174
amplitude, 50, 61, 78, 79, 196, 205, 208, 213, 321, 407, 409, 527, 539
AMS, 88, 89, 116
angular momentum, 305, 306, 308
anharmonicity, 136, 143, 150, 156, 163, 171, 234, 242
anisotropy, 210, 267, 268, 269, 285
annealing, 33, 43, 64, 67, 78, 85, 98, 109, 286
annihilation, 72, 80, 292, 520
anomalous, 139, 152, 165, 211, 212, 228, 246, 325
antiferromagnetic, 179, 190, 191, 210, 211, 213, 227

application, 51, 52, 59, 65, 119, 174, 180, 182, 183, 188, 191, 209, 235, 236, 237, 238, 239, 240, 245, 255, 256, 266, 335, 349, 354, 395, 406, 432, 446, 525, 531, 532
argument, 127
arithmetic, 273
aromatic rings, 399
arrest, 74, 77, 324
Arrhenius dependence, 269
Arrhenius law, 266, 269, 284
Asia, 175
aspect ratio, ix, 319
assessment, 321, 340, 348, 349
assignment, 197
assumptions, 129, 267, 269, 293, 304, 329
ASTM, 111, 114, 322, 340, 348, 354, 367
asymmetry, 209, 419, 421, 434, 435, 437
asymptotic, 3, 267, 514
asymptotically, 28
atmosphere, 320
atmospheric pressure, 320
atomic distances, 195
atomic force, 388
atomic force microscope, 388
atomic orbitals, 304
atomic positions, 214, 251
atoms, ix, 2, 3, 4, 8, 10, 120, 121, 126, 127, 128, 133, 174, 178, 179, 187, 194, 195, 196, 197, 198, 199, 200, 202, 204, 205, 208, 214, 217, 222, 223, 226, 227, 230, 236, 237, 239, 241, 242, 246, 247, 248, 251, 252, 254, 255, 256, 265, 266, 267, 268, 270, 271, 272, 273, 276, 277, 279, 280, 282, 283, 284, 285, 304, 306, 308, 309, 312, 406, 407, 411, 426, 438, 539
ATP, 200
attacks, 352, 363
austenitic stainless steels, 322, 323, 326
autocorrelation, vii, 1, 2, 13, 18
automata, xii, 480, 481, 520, 531, 534, 541
autoradiography, 267
availability, 254, 321
averaging, 13, 128, 266, 306, 339, 340

B

backscattered, 354, 366, 367
band gap, ix, 289, 308, 309, 310, 311, 312, 314, 315
bandwidth, 178, 181, 188
barrier, 370, 389, 396, 397, 412, 413, 414, 415, 416, 418, 426, 427, 428, 434, 439
barriers, 58, 85, 184, 328
basis set, 295, 304, 308, 309, 391, 403, 424
beams, 204
behavior, ix, x, xi, 30, 53, 67, 78, 84, 133, 181, 189, 230, 237, 238, 239, 243, 244, 251, 270, 286, 319, 320, 321, 327, 328, 339, 344, 352, 360, 366, 385, 393, 394, 399, 414, 423, 426, 430, 433, 434, 435, 439, 475
Beijing, 543
bending, 59, 64, 65, 75, 76, 79, 94, 208, 209, 212, 215, 217, 220, 222, 228, 230, 231, 237, 238, 243, 256, 323, 396, 397, 399, 407, 411, 524
beneficial effect, 69
benefits, 65, 183, 354
benzene, 393, 396
Bessel, 3
bias, 60, 342, 394, 395, 421, 424, 426, 428, 431, 432, 433, 434, 435, 436, 437, 438
biaxial, 61, 499
binding, 121, 182, 195, 205, 290, 386
binding energy, 205
biochemistry, 208
bladder, 447
blocks, 270, 271, 272, 273, 415, 459, 470
boiling, 58, 321, 323
Boltzmann constant, 205, 271, 458, 477
bonding, 79, 121, 195, 197, 208, 300, 308
bonds, 190, 300, 308, 387, 389, 406, 426, 434
Bose, 178, 191, 198, 241
Bose-Einstein, 191, 198, 241
bosons, 178
bottom-up, 386, 439
boundary conditions, xi, 2, 69, 86, 273, 290, 445, 446, 455, 456, 458, 460, 476, 480, 487, 488, 490, 499, 500, 501, 522, 524, 528, 531, 532, 533, 534, 538, 539, 540
boundary surface, 340
brass, 62
Brazilian, 257
breads, 92
breathing, 190, 222, 230, 231
Brillouin light scattering, 203, 209
buildings, 321
bulk materials, 439

C

CAD, 459
CAE, 460
calcium, 352, 360
calibration, 59, 321
candidates, 185, 395, 434
capillary, 353
carbide, 65, 325, 326
carbides, 323, 327, 328, 331, 348

carbon, ix, 57, 61, 64, 74, 76, 79, 92, 98, 99, 111, 113, 116, 289, 290, 320, 322, 325, 327, 386, 387, 389, 406, 407, 411, 414, 420, 538
carbon atoms, 406, 407, 411
carbon nanotubes, 386, 387
carboxylic, 437
carrier, 181, 182, 190, 394, 421
Cartesian coordinates, 195, 446, 476
case study, x, 351
cast, 57, 65, 66, 84, 101, 356
casting, 352
catalysis, 185
cation, 188, 203, 211, 431
cavitation, 58, 328
cavities, 272, 328, 331
CEA, 349
cell, 52, 81, 86, 133, 185, 186, 195, 200, 201, 217, 220, 226, 227, 228, 230, 236, 239, 255, 271, 272, 273, 332, 337, 338, 481
cement, x, 351, 352, 353, 354, 355, 356, 360, 361, 362, 363, 364, 365, 366, 367
ceramic, 57, 119
ceramics, 57, 66, 67
channels, 395
chemical content, 237
chemical engineering, 386
chemical properties, 266, 304, 399
chemical reactions, 452
chemical stability, 395, 431
chemical structures, 408
chemical vapor deposition, 290
chemisorption, 290, 437
China, 96, 175, 385
chromium, 64, 192, 326
Chromium, 98, 101
cis, 413, 429
classes, 269, 393, 394, 525
classical, 40, 205, 209, 281, 296, 328, 340, 534
classification, 197, 205, 239, 241
cleaning, 65
cleavage, ix, x, 319, 321, 324, 326, 327, 328, 329, 333, 340, 341, 342, 344, 348, 369, 370, 372, 373, 374, 376, 381, 382
closure, 66, 244
clusters, ix, 2, 289, 290, 306, 308, 309, 312, 313, 314, 316, 317, 327, 396, 399, 431
Co, 7, 90, 94, 185, 246, 248, 251, 252, 254, 329
coatings, 532
cobalt, 61, 194, 246, 247, 254
codes, xi, 54, 308, 321, 371, 408, 445, 446, 459
cohesion, 125
coil, 67, 69, 72, 77
collisions, vii, 1, 2, 5, 8, 9, 10, 13, 14, 28
combined effect, 64, 173, 420, 422
combustion, 320, 321
community, 178, 180, 328
compatibility, 484, 485, 487, 491, 495, 520
compensation, 528, 529, 530, 531
complement, 255
complex systems, 86, 255
complexity, 69, 321
compliance, 323
components, viii, 49, 51, 59, 60, 64, 67, 72, 73, 74, 75, 102, 120, 121, 125, 197, 251, 320, 333, 334, 337, 340, 349, 386, 425, 456, 476, 481, 484, 485, 487, 488, 489, 490, 491, 492, 493, 494, 495, 499, 500, 501, 502, 503, 504, 505, 506, 507, 508, 514, 516, 517, 518, 519, 523, 528, 529, 530, 535
composites, 67, 340
composition, 178, 193, 211, 213, 352
compound semiconductors, 290
compounds, viii, 80, 177, 179, 180, 183, 185, 191, 192, 202, 211, 212, 217, 227, 238, 245, 246, 247, 290, 309, 396, 399, 411, 449
compressibility, 122, 123, 145
compressive strength, 353, 354, 361, 363, 364
Compton scattering, 205
computation, 123, 124, 125, 127, 128, 174, 193, 240, 251, 300, 304, 393, 400
computer simulations, 268, 446
computing, x, 85, 120, 121, 133, 369, 375, 378, 459, 461
concentration, 37, 38, 125, 178, 181, 185, 187, 191, 193, 211, 213, 220, 227, 239, 243, 244, 246, 247, 267, 506, 508, 512, 514, 515, 520
concrete, 352, 353, 363, 366
condensation, vii, 1, 7, 8, 9, 10, 12, 23, 24, 25, 27, 28, 29, 30, 31, 267, 320
condensed matter, 178, 180, 183, 203, 291
conditioning, 57
conductance, xi, 385, 388, 394, 404, 410, 411, 415, 426, 427, 429, 430, 431
conduction, 129, 188, 270, 300, 303, 339, 394, 395, 448
conductive, 388, 394, 431
conductivity, 45, 179, 182, 188, 190, 194, 393, 426, 427, 430, 431, 446, 448, 449, 450, 458, 460, 475, 476
confidence, 232, 243, 343, 344
confidence interval, 343, 344
confidence intervals, 343
configuration, 195, 356, 395, 402, 407, 426, 460
conformational analysis, 396, 414
conformity, 364
Congress, 87

conjecture, 26
conjugation, 272, 394, 396, 397, 401, 403, 404, 406, 407, 411, 415, 426, 431
consensus, 212
conservation, 203, 296, 308, 339, 491
constant rate, 18, 41, 42, 525
construction, vii, xi, 185, 197, 271, 272, 273, 325, 352, 386, 479, 495, 504, 508, 518, 541
construction materials, 352
consulting, 367
consumption, 320
contamination, 57
continuity, 124, 338, 345, 491, 495
control, 57, 60, 72, 74, 183, 356, 357, 360, 361, 363, 365, 387, 429
convection, 456, 457, 458, 476
convergence, 20, 329, 456
conversion, 237
cooling, 271, 272, 273, 320, 321, 447, 454
copper, 84, 266, 267, 285, 287
correction factors, 343
correlation, viii, 18, 26, 59, 60, 70, 81, 119, 122, 126, 127, 128, 133, 134, 135, 137, 139, 141, 142, 143, 145, 148, 149, 152, 155, 156, 161, 162, 165, 167, 168, 169, 174, 177, 178, 180, 181, 201, 205, 206, 211, 214, 243, 255, 286, 297, 298, 299, 300, 303, 363, 364, 365, 383, 392, 407, 417, 423, 424, 425
correlation function, viii, 26, 119, 126, 127, 128, 133, 134, 135, 137, 139, 142, 143, 145, 148, 149, 155, 156, 161, 162, 165, 168, 169, 174, 205, 206, 303, 425
correlations, 123, 205, 217, 296, 300, 302, 366, 424
corrosion, viii, 49, 58, 64, 65, 66, 67, 74, 84, 323, 326
corrosive, 64, 65, 66
cosine, 377
costs, 57, 194, 300, 326, 348
Coulomb, 135, 141, 148, 155, 161, 167, 178, 179, 181, 202, 291, 306, 424
Coulomb energy, 179, 181
Coulomb interaction, 202
coupling, 179, 181, 182, 188, 189, 190, 192, 204, 210, 211, 213, 234, 247, 255, 256, 388, 406, 437, 439, 524
covalent, 300, 432
covering, 73, 266, 271, 340
crack, viii, ix, 49, 51, 52, 53, 61, 64, 65, 67, 74, 75, 76, 77, 78, 79, 81, 89, 319, 322, 323, 326, 327, 331, 335, 344, 346, 347, 348, 369, 370, 371, 375, 376, 381, 382, 516
cracking, 326
CRC, 348, 383, 477
creep, 63, 67, 80, 265, 482
critical points, 446
critical temperature, 188, 192, 234
critical value, xii, 327, 332, 345, 480, 520, 524, 535
cross links, 451
cryogenic, 320, 321, 322, 323
crystal lattice, 126, 192, 270, 324
crystal structure, 63, 180, 186, 192, 198, 237, 255, 267, 268, 306
crystal structures, 180, 267, 306
crystalline, vii, 2, 31, 34, 36, 37, 39, 41, 43, 44, 45, 56, 76, 80, 84, 120, 126, 127, 128, 135, 142, 149, 155, 162, 168, 174, 184, 430, 516
crystalline solids, 126
crystallites, vii, 2, 33, 34, 36, 37, 38, 39, 41, 45, 370, 522, 523, 534, 535, 538, 539
crystallization, vii, 1, 2, 33, 34, 35, 36, 38, 39, 40, 41, 42, 44, 45, 46, 47, 72, 285, 328
crystallization kinetics, vii, 1, 2, 33, 38, 40, 41, 44, 45, 47
crystals, 56, 72, 79, 85, 126, 191, 193, 195, 199, 201, 205, 210, 227, 271, 282, 290, 316, 327, 340, 369, 370, 371, 372, 373, 374, 375, 376, 381, 382
cuprate, 246
cuprates, 178, 180, 227, 246
curing, xi, 363, 445, 446, 447, 448, 449, 450, 451, 452, 453, 454, 455, 456, 458, 459, 460, 475
curing process, xi, 363, 445, 446, 447, 448, 449, 450, 451, 454, 455, 456, 458, 459, 475
cycles, 50, 67, 70, 76, 78, 79, 80, 321

D

damping, 71, 72, 230
decomposition, 121, 333
decompression, 324
decoupling, 127, 128, 432
defects, xi, 51, 67, 80, 182, 269, 342, 479, 484, 485, 486, 487, 488, 490, 495, 520, 528, 544
definition, 183, 187, 191, 195, 392, 407, 486, 490, 495, 499, 520, 525
deformability, 67
deformation, viii, xi, 49, 53, 57, 61, 62, 66, 70, 74, 78, 84, 85, 186, 266, 272, 286, 287, 327, 328, 331, 339, 349, 479, 480, 481, 482, 483, 484, 485, 486, 487, 488, 489, 490, 491, 492, 494, 495, 496, 497, 498, 501, 502, 504, 505, 506, 508, 509, 513, 514, 516, 517, 518, 519, 520, 521, 522, 523, 524, 525, 526, 527, 530,

531, 532, 534, 535, 536, 537, 538, 539, 541, 543
degenerate, 123, 187, 188, 190, 197, 214, 217, 220, 248, 255, 403, 404
degradation, 335, 480
degrees of freedom, 178, 179, 239, 255, 370
delivery, 52, 57
Denmark, 91
density fluctuations, 131
density functional theory, x, 300, 312, 385, 389
dependent variable, 54
deposition, vii, 1, 2, 3, 8, 12, 14, 22, 31, 290
deposition rate, 31
derivatives, 201, 227, 338, 484, 487, 491, 517
desensitization, 66
designers, 321
desorption, 3, 29
deviation, 25, 32, 139, 165, 174, 266, 269, 284, 407, 408, 482
DFT, x, 308, 309, 312, 385, 389, 392, 393, 413, 418, 423, 424, 426, 428, 429, 431, 432, 434, 437, 439
diamond, 209, 300, 306, 354, 356, 429
dielectric constant, 201, 439
dielectric function, 122, 152, 174
differential approach, 454
differential equations, 484
differential scanning, 449
differential scanning calorimeter, 449
differentiation, 7, 517
diffraction, 26, 43, 67, 84, 126, 205, 211, 243, 370
diffusion, ix, 2, 3, 4, 265, 266, 267, 268, 269, 270, 273, 274, 276, 277, 278, 279, 280, 281, 282, 283, 284, 285, 286, 287
diffusion mechanisms, ix, 265, 266, 268, 269, 270, 276, 277, 283, 284
diffusion permeability, 268, 269
diffusion process, 266, 267, 268, 269, 279
diffusion rates, 285
diffusivity, 267, 269, 449, 461, 475
dihedral angles, 379, 380, 383, 399, 400, 401
dimensionality, 41, 182, 271
diodes, 182, 290, 387, 393
dipole, 207, 208, 209, 308, 407, 408, 409, 410, 414, 415, 416, 417, 419, 432
dipole moment, 207, 208, 209, 407, 408, 409, 410, 414, 415, 416, 417, 419, 432
Dirac delta function, 337
direct measure, 450
disaster, 321
discontinuity, 516
discs, 80
dislocation, ix, 56, 63, 64, 66, 72, 76, 77, 78, 79, 83, 84, 85, 86, 265, 268, 269, 274, 275, 276, 277, 279, 280, 284, 287, 327, 328, 516, 544
dislocations, viii, ix, 49, 53, 57, 64, 84, 85, 265, 267, 268, 269, 273, 274, 275, 276, 277, 278, 279, 280, 281, 282, 284, 327, 328, 538, 539, 545
disorder, 126, 135, 142, 149, 155, 162, 168, 188, 241, 256
dispersion, viii, 119, 120, 126, 128, 131, 134, 136, 137, 138, 144, 145, 150, 151, 157, 158, 163, 164, 169, 170, 174, 193, 194, 197, 200, 206, 207, 210, 217, 220, 221, 222, 228, 230, 231, 238, 239, 250, 251, 255, 256, 382
displacement, 128, 174, 181, 182, 195, 197, 199, 201, 215, 226, 227, 234, 241, 256, 277, 280, 323, 337, 338, 345, 346, 347, 484, 488, 516, 524, 525, 541
distortions, viii, 177, 188, 189, 192, 210, 213, 217, 235, 236, 237
distribution function, 200, 206, 330, 340, 369, 380, 381
divergence, 336, 349, 455
DNA, 387
donor, 387, 395, 396, 432, 433, 434, 435, 436
dopant, 187, 211, 212, 217, 238
doped, ix, 177, 181, 183, 189, 190, 192, 193, 211, 212, 220, 223, 226, 227, 237, 243, 244, 245, 246, 255, 256
doping, ix, 177, 181, 185, 187, 189, 190, 191, 193, 194, 210, 211, 212, 220, 223, 225, 227, 236, 237, 239, 243, 255, 256, 432
dosage, 354, 362
double bonds, 406, 407, 410, 411, 412, 426, 430
dry ice, 58
DSC, 449, 450, 451, 453
ductility, 81, 323, 324, 325, 327, 328
duplex stainless steels, 326
durability, 57, 321, 352, 353
duration, 56, 57, 60, 72, 77, 269, 273, 277
dust, 58, 320
dynamical properties, 193, 194, 212, 225

E

earth, 180, 185, 193, 211, 227, 235, 239
EBSD, 369, 370
ECSC, 383
elastic constants, 201, 325, 485
elastic deformation, 482, 483, 484, 485, 486, 487, 488, 490, 491, 501, 525, 526
elasticity, xi, 209, 334, 337, 338, 349, 479, 487, 488, 495, 544

electric field, x, 178, 183, 202, 385, 388, 404, 406, 407, 408, 409, 410, 411, 412, 413, 414, 415, 416, 417, 418, 419, 420, 421, 422, 423, 426, 427, 434, 439
electrical conductivity, 179
electrical properties, 388, 389, 404, 410, 416, 419, 422, 423
electrical resistance, 183
electrodes, 185, 388, 389, 406, 424, 425, 426, 430, 431, 432, 434
electromagnetic, 30, 182, 204
electron beam, 204
electron charge, 203
electron density, 121, 181, 402, 404, 413, 419, 422, 423, 425
electron diffraction, 43
electron gas, 129, 130, 174, 294, 299
electron microscopy, 43, 275, 367
electronegativity, 392
electronic materials, 386, 387, 388, 389, 393, 396, 405, 423, 439
electronic structure, x, 304, 385, 387, 389, 393, 395, 396, 403, 404, 406, 413, 419, 421, 423, 424, 427, 431, 434, 439
electronic systems, 211
electron-phonon, viii, 177, 181, 190, 191, 192, 210, 211, 213, 237, 238, 241, 246, 247, 254, 255, 256, 424
electron-phonon coupling, 190, 192, 210, 211, 213, 247, 255, 256
electrons, 121, 127, 129, 130, 173, 178, 179, 180, 181, 187, 188, 189, 191, 192, 194, 195, 210, 236, 255, 290, 291, 293, 294, 298, 300, 301, 302, 304, 305, 306, 308, 309, 387, 395, 410, 415, 425, 430, 431, 432, 433, 434, 438
electroplating, 58
elliptical inclusion, 544
email, 177
emission, 79, 206
encouragement, 257
endurance, 69
energetic characteristics, 270
energy density, 76, 296, 320, 339
energy efficiency, 58
energy-momentum, 308
engines, 79
England, 175
entropy, 68, 199
environment, viii, 49, 64, 66, 67, 77, 188, 208, 304, 437, 459
environmental conditions, 352
epitaxy, 2, 290
epoxy, 354, 452
equating, 502
equilibrium, x, 194, 195, 200, 205, 255, 269, 296, 336, 385, 387, 388, 406, 423, 424, 432, 487
erosion, 74
estimating, 328
estimator, 342, 343
estimators, 342, 349
Eulerian, 54
Europe, 89
europium, 183, 192
evolution, vii, 1, 2, 19, 20, 32, 36, 37, 47, 52, 79, 349, 352, 404, 406, 409, 411, 420, 421, 480, 524, 537
excitation, 30, 191, 194, 207, 209, 222, 230, 241, 298, 303, 304
execution, 355
exertion, 355
exfoliation, 66
expansions, 267
experimental condition, 267
exposure, 52, 59, 66, 74, 79, 104, 115, 191
external environment, 424, 457
external influences, 274
extrapolation, 411

F

fabrication, 386
failure, ix, 51, 59, 64, 79, 85, 182, 253, 265, 319, 321, 323, 328, 329, 330, 332, 335, 340, 341, 342, 344, 345, 348, 349, 353
family, 185, 213, 495, 496, 509, 510, 518, 519
fatigue, viii, 49, 50, 51, 52, 53, 57, 58, 60, 61, 62, 63, 64, 65, 66, 67, 68, 69, 70, 72, 73, 74, 75, 76, 77, 78, 79, 80, 81, 84, 86, 116, 323, 369, 370, 371, 381, 382
faults, 286
fax, 1, 319
feeding, vii, 1, 3, 8, 9, 10, 24
FEM, 52, 54, 69, 70, 71, 75, 84, 86, 329, 337
Fermi, 122, 129, 130, 296, 316, 425, 430, 432, 433
Fermi energy, 432, 433
fermions, 179
ferrite, 323, 326
ferromagnetic, viii, 177, 179, 185, 188, 189, 190, 191, 192, 210, 211, 228, 235, 237, 325
ferromagnetism, 183, 190, 227, 234, 241, 255
ferromagnets, 183
fiber, 340
fibers, 352
field theory, 349

film, vii, 1, 2, 3, 4, 5, 10, 12, 13, 14, 15, 20, 22, 23, 25, 26, 28, 29, 30, 31, 32, 33, 36, 37, 38, 39, 40, 41, 42, 43, 44, 45, 47, 356, 439
film formation, 29
film thickness, vii, 1, 2, 4, 20, 23, 25, 28, 29, 30, 31, 32, 33, 36, 38, 39, 40, 41, 42, 45
films, 2, 12, 13, 14, 18, 22, 29, 30, 31, 33, 38, 43, 44, 45, 46, 184, 285, 290
finite element method, x, xi, 54, 55, 86, 89, 319, 329, 445, 446, 447, 455, 458
finite volume, 54, 337
finite volume method, 54, 337
first principles, 247, 248, 250, 251, 252
fixation, 273
flexibility, 58, 386, 387
flexural strength, 354, 355, 361, 363, 365
flow, xii, 50, 58, 64, 70, 71, 268, 326, 329, 334, 335, 348, 361, 365, 433, 447, 448, 450, 458, 460, 479, 480, 483, 520, 523, 524, 527, 532, 536, 537, 539, 541
flow rate, 50, 58
flow value, 361, 365
fluctuations, 131, 204, 205, 208, 268, 279
fluid, 121, 122
focusing, 52, 387, 404
force constants, 202, 397
forecasting, 352
Fourier, 18, 205, 206, 396
fracture, ix, x, 52, 54, 67, 76, 77, 78, 84, 85, 86, 281, 319, 321, 322, 323, 326, 327, 328, 329, 330, 331, 333, 336, 340, 342, 344, 345, 348, 349, 369, 370, 373, 375, 376, 381, 383, 480, 483, 543
fractures, 382
fragmentation, 532
France, 88, 91, 92, 93, 94, 95, 96, 99, 100, 101, 103, 104, 105, 106, 107, 108, 110, 112, 114, 116, 117, 177
free energy, 195, 199, 271
free volume, 243
freedom, 178, 179, 183, 210, 239, 255, 370
frequency distribution, 206
friction, 65, 70, 72
frustration, 212
fuel cell, 185
fulfillment, 484, 485
Full Width at Half Maximum (FWHM), 50, 222
fullerenes, ix, 289, 290
functional approach, 296
functionalization, 387, 393
furan, 429

G

G4, 512
GaAs, ix, 289, 290, 312
GaP, 290
gas, x, 84, 320, 322, 324, 348, 476
gas turbine, 84
gases, 322
Gaussian, vii, 1, 18, 26, 32, 316, 382
gel, 353, 360
gene, 126
generalization, 126, 195
generalizations, 126
generation, 53, 68, 85, 300, 321, 448, 451, 458, 461, 476
geometrical parameters, 512
Germany, 87, 88, 91, 92, 93, 94, 95, 96, 97, 98, 99, 100, 101, 102, 103, 104, 105, 106, 108, 109, 110, 111, 112, 113, 114, 115, 116, 479
GFP, 10, 11, 12, 13
glass, 57, 58, 63, 64, 66, 120, 129, 131, 134, 135, 136, 137, 138, 139, 140, 141, 142, 143, 144, 145, 146, 147, 148, 149, 150, 151, 152, 153, 154, 155, 156, 157, 158, 159, 160, 161, 162, 163, 164, 165, 166, 167, 168, 169, 170, 171, 172, 173, 174
glasses, viii, 119, 120, 121, 124, 125, 126, 131, 132, 133, 134, 174
God, 289
gold, 426, 429, 431, 433, 434
government, iv
grades, 273, 323
grain, ix, 57, 62, 67, 85, 86, 265, 266, 267, 268, 269, 270, 271, 272, 273, 275, 276, 277, 278, 279, 280, 281, 282, 283, 284, 285, 286, 287, 322, 327, 328, 370, 486, 488, 524, 535, 536, 538
grain boundaries, ix, 85, 265, 266, 267, 268, 269, 270, 271, 273, 276, 279, 281, 282, 283, 284, 285, 286, 287, 327, 328
grain boundary structure, 266, 269, 286
grains, ix, xii, 85, 266, 267, 268, 269, 272, 273, 275, 282, 319, 327, 328, 370, 375, 381, 480, 481, 486, 520, 523, 524, 535, 536, 537, 538, 539
graph, 360
gravity, 63, 121
ground state energy, 301, 304
grouping, 327
groups, 183, 250, 252, 306, 353, 386, 387, 396, 398, 419, 422, 423, 434, 435, 437, 439
growth, vii, viii, ix, 1, 2, 4, 5, 7, 8, 9, 10, 12, 14, 15, 16, 18, 19, 20, 23, 24, 28, 29, 30, 31, 32,

34, 41, 42, 45, 46, 47, 49, 52, 53, 65, 67, 74, 76, 77, 81, 89, 282, 319, 321, 330, 331, 335, 340, 346, 347, 348, 388, 495
growth modes, 4
growth rate, viii, 2, 5, 19, 29, 42, 46, 52, 65, 77
guidance, 203
Gujarat, 119, 120

H

Hamiltonian, 178, 181, 182, 291, 292, 293, 294, 295, 300, 302, 307, 308, 391
handling, 51, 86
hanging, 495
hardness, 57, 59, 60, 61, 62, 63, 64, 65, 66, 67, 68, 70, 71, 72, 74, 75, 77, 79, 83, 84, 202, 325, 392
harmonics, 306
healing, 76
heart, 235, 388
heat, xi, 58, 64, 68, 133, 139, 165, 226, 232, 233, 254, 266, 320, 321, 323, 339, 352, 445, 446, 447, 448, 449, 450, 451, 457, 458, 459, 461, 475, 476
heat capacity, 133, 254, 448, 449, 458, 475
heat release, 451, 476
heat transfer, xi, 58, 321, 445, 446, 447, 448, 457, 458, 459, 461, 476
heating, 42, 43, 68, 193, 272, 447
height, vii, 1, 2, 5, 13, 14, 22, 23, 25, 26, 27, 28, 29, 32, 34, 36, 45, 60, 397, 412, 414, 418, 495
helium, 320, 323
hemisphere, vii, 1, 4, 5, 371, 372, 373
heterogeneous, 286, 353
heterogeneous systems, 286
heterostructures, 183
hexagonal lattice, 180
high pressure, 58, 194, 210, 237
high temperature, ix, 64, 67, 68, 78, 79, 80, 81, 177, 185, 189, 199, 210, 227, 255, 265, 269, 275, 277, 279, 281, 282, 283, 286, 328
high-frequency, 217
Hilbert space, 301
histogram, 222, 360
Holland, 544
HOMO, xi, 385, 389, 394, 395, 396, 401, 402, 403, 404, 410, 411, 412, 413, 416, 417, 418, 419, 420, 421, 422, 423, 427, 430, 432, 434, 435
homogeneity, 362, 531
homogenous, 296, 299, 365
host, 309, 396
HSP, 63
Hubbard model, 181
humidity, 352
hybrid, 300
hybridization, 188
hybrids, 300, 301
hydrate, 360
hydrates, 352
hydration, 352, 353, 354, 356, 360, 363, 365, 366, 367
hydro, 320, 387
hydrocarbon, 393, 396, 419, 430
hydrocarbons, 320, 387
hydrodynamic, 65
hydrogen, 58, 66, 208, 308, 398, 426
hydrogen atoms, 426
hydrostatic pressure, 61, 236
hydrostatic stress, 332, 333
hydroxide, 352
hypothesis, 26, 376

I

ice, 58, 491
ICE, 110
id, 312
identification, 208, 336, 341, 345
identity, 174, 340
image analysis, x, 57, 351, 354, 355, 360, 363, 364, 366
images, 43, 354, 355, 363, 366, 367
imaging, ix, 210, 289, 354, 355, 387
impact analysis, 69
impact energy, 322, 326
implementation, x, 52, 182, 351, 435
impurities, 58, 320
in situ, 43
in transition, 178, 179, 181
inactive, 217
incentive, 290
inclusion, 134, 135, 141, 143, 145, 149, 155, 161, 162, 167, 169, 281, 303, 312, 332, 382, 424, 499, 500, 501, 502, 503, 504, 523, 524, 544
incompatibility, 485
India, 119, 177, 257
indicators, 401, 412
indices, 270, 327, 371
indium, 290
induction, 64, 451, 452, 453, 454, 460, 461, 470, 476
induction period, 452
induction time, 451, 452, 453, 454, 460, 461, 470, 476
industrial, 86, 290, 325, 451

industrial application, 86, 290, 325
industry, 52, 58, 59, 72, 453
inelastic, 120, 126, 197, 203, 204, 206, 209, 211, 217, 220, 221, 222, 228, 230, 239, 240, 247, 253, 334, 335, 337, 338, 432
inert, 304, 406
inertia, 70
infinite, 33, 39, 40, 70, 195, 267, 298, 300, 411, 505, 506, 528
information processing, 182
information technology, 183
infrared, 30, 192, 197, 201, 203, 207, 208, 209, 211, 212, 238, 247, 248
infrared spectroscopy, 212
inhomogeneity, 480, 486
initiation, ix, 51, 64, 76, 77, 79, 111, 279, 319, 323, 326, 327, 328, 344, 348, 382, 535, 537
injection, 447, 452
innovation, 116
inorganic, 178, 386
InP, 290
INS, 217, 230, 231, 239, 253
insight, 86, 203, 412, 423, 432
inspiration, 399
instabilities, 329, 349
instability, 136, 143, 150, 156, 163, 171, 230, 256, 269
insulation, 320, 325
insulators, 179, 180, 181, 210, 211, 227
integrated circuits, 386
integration, x, 16, 133, 183, 271, 319, 338, 340, 349, 461, 495, 505, 517, 518, 519
integrity, 60, 97
interaction, viii, xii, 120, 121, 129, 177, 178, 181, 182, 188, 190, 191, 192, 194, 197, 203, 205, 209, 210, 228, 229, 237, 241, 244, 247, 254, 255, 256, 293, 303, 304, 332, 398, 407, 408, 409, 411, 413, 414, 422, 427, 460, 480, 481, 520, 522
interaction effect, 182, 190
interaction effects, 182, 190
interactions, 86, 120, 121, 129, 135, 141, 148, 155, 161, 167, 174, 178, 181, 182, 189, 190, 193, 194, 197, 202, 203, 210, 212, 213, 229, 235, 238, 246, 248, 270, 398, 399, 400, 424
interdisciplinary, 440
interface, 33, 36, 41, 45, 321, 338, 353, 406, 424, 427
interference, 205, 206
intermetallic compounds, 286
internet, 88
interstitial, 185, 188, 189, 268, 269
interval, 50, 266, 267, 344, 461, 481, 526, 527, 532, 538
intrinsic, 180, 184, 194, 204, 235, 247, 248, 321, 327, 338, 353, 430
inversion, 74, 211, 214, 306, 307
investigations, 65, 93, 97, 114, 366, 393
ionic, 180, 187, 195, 202, 203, 211, 235, 255, 290
ionicity, 202
ionization, 392, 432
ions, 128, 129, 130, 180, 181, 185, 186, 187, 188, 191, 195, 202, 203, 211, 213, 222, 230, 235, 267, 290, 396, 399, 431
Iran, 289, 445, 477
iron, 65, 66, 324, 327
irradiation, 33, 65
island, vii, 1, 2, 3, 4, 5, 8, 9, 10, 12, 13, 14, 15, 16, 18, 22, 23, 28, 30, 31, 33, 47
island density, 12, 31
isomers, 404
isothermal, 42, 44, 45, 120, 131, 132, 451, 453, 454, 455, 470, 476
isothermal crystallization, 42, 45
isotope, 206, 212
isotopes, 266, 267
isotropic, 15, 81, 132, 201, 267, 334, 370, 484, 486, 488, 529
Israel, 442
I-V curves, 426, 427, 428, 431, 434, 436, 437

J

Jacobian matrix, 461
Japan, 91, 92, 93, 94, 95, 96, 97, 98, 99, 102, 103, 104, 105, 107, 108, 109, 110, 111, 112, 114, 116
J_c, 328
joints, 102, 325
judge, 174
Jun, 259
Jung, 477

K

kernel, 340
kinetic curves, 28, 40, 41, 46
kinetic energy, 58, 72, 85, 178, 189, 271, 291, 296, 297, 339, 424
kinetic model, 451, 453, 470
kinetics, vii, xi, 1, 2, 9, 11, 12, 13, 22, 24, 27, 30, 31, 33, 38, 39, 40, 41, 42, 44, 45, 46, 47, 266, 285, 445, 446, 447, 452, 453, 454, 458, 460, 470, 475, 476

Kirchhoff, 331
knots, 268
Kolmogorov, vii, 1, 2, 33, 47

L

Lagrangian, 54, 291
land, 322
language, 316
lanthanide, 180
lanthanum, 187, 193, 227, 255
large-scale, 287
laser, 58, 81, 84, 290
lasers, ix, 289
lattice, viii, 18, 21, 22, 120, 126, 177, 178, 179, 180, 181, 182, 186, 187, 189, 190, 191, 192, 193, 194, 195, 197, 199, 202, 203, 205, 206, 209, 210, 211, 212, 213, 215, 225, 227, 230, 234, 235, 236, 237, 238, 240, 241, 242, 243, 247, 248, 254, 255, 256, 268, 271, 309, 313, 325, 340, 516, 522
lattice parameters, 236
lattices, 126, 186, 268
law, vii, xi, 1, 2, 5, 7, 8, 9, 10, 12, 13, 14, 15, 16, 20, 22, 24, 28, 30, 38, 41, 66, 81, 120, 133, 139, 165, 266, 269, 284, 308, 309, 334, 339, 479, 483, 487, 488
laws, 71, 291, 296, 308, 339, 386
leaks, 320
LED, 182, 290
LEED, 45, 46
lifetime, 3
ligand, 180
light emitting diode, 182
light scattering, 209
likelihood, 329
limitation, 483
limitations, x, 60, 203, 217, 319, 329, 531
linear, vii, viii, 2, 18, 28, 34, 41, 44, 57, 69, 79, 80, 81, 86, 121, 177, 180, 194, 207, 216, 227, 231, 238, 255, 256, 266, 271, 325, 337, 338, 342, 360, 391, 396, 401, 407, 411, 412, 415, 416, 419, 430, 431, 449, 461, 487, 488, 495
linear dependence, 266, 407
linear law, 28
linkage, 397, 399, 414, 418
liquefaction, 320, 322
liquefied natural gas, ix, 319, 320, 326, 348
liquid helium, 323
liquid nitrogen, 323, 369
liquid phase, 321
liquids, 123, 126, 127
LNG, 320, 321, 322, 323, 325, 326, 348
localised, 541
localization, xi, 190, 191, 348, 395, 396, 434, 435, 479, 480, 483, 503, 519, 520, 522, 524, 531, 534, 535, 536, 537, 538, 541
location, 51, 69, 75, 77, 407, 447, 472, 507, 509, 530
London, 175, 176, 257, 258, 259, 260, 261, 262, 263, 286, 441, 442, 477, 544
long distance, 191
losses, 31, 68
low temperatures, 178, 183, 189, 191, 193, 210, 227, 266, 282, 283, 324, 325, 327, 369, 382
low-temperature, 213, 323, 328, 344
LPG, 322
lubrication, 58, 65
LUMO, xi, 385, 389, 394, 395, 396, 401, 402, 403, 404, 410, 411, 412, 413, 416, 417, 418, 419, 420, 421, 422, 423, 427, 430, 432, 433, 434
lying, 135, 143, 149, 155, 163, 169, 180, 188, 223, 232, 304, 373, 434

M

M1, 432
machines, 59, 525
magnesium, 67, 77, 78, 84, 103
magnesium alloys, 67, 78, 103
magnetic, 65, 178, 179, 180, 181, 182, 183, 184, 185, 189, 190, 191, 192, 194, 206, 209, 210, 211, 213, 228, 235, 243, 246, 299, 325
magnetic field, 178, 182, 183, 184, 185, 191, 210, 235, 243, 299, 325
magnetic materials, 183
magnetic moment, 179
magnetic properties, 179, 189, 211
magnetic sensor, 184, 185
magnetic structure, 191
magnetization, 185, 325
magnetoelectronics, 182
magnetoresistance, viii, 177, 179, 180, 182, 183, 184, 185, 189, 192, 193, 210, 211, 227, 235
magnetostriction, 325
management, 367
manganese, 179, 183, 185, 187, 191, 192, 211, 331, 332
manganites, viii, 177, 180, 181, 183, 184, 185, 187, 188, 189, 190, 191, 192, 193, 202, 210, 211, 212, 217, 220, 226, 227, 234, 235, 236, 237, 238, 241, 243, 244, 255
manifold, 72, 434
manifolds, 72
manufacturer, 354

manufacturing, 52, 349, 445, 475
mapping, 391
market, 459
MAS, 101, 113
mass transfer, 520
material degradation, 480
materials science, 119, 180, 208, 210, 396
matrix, vii, 1, 67, 120, 128, 182, 189, 196, 197, 198, 199, 241, 292, 294, 295, 299, 301, 302, 303, 329, 331, 335, 391, 423, 424, 425, 455, 456, 461, 476, 481, 488, 499, 503, 535
MBE, 290
MDR, 447
measurement, xi, 52, 54, 72, 206, 244, 253, 323, 367, 385, 411, 446, 450
measures, 59, 60, 327, 353
mechanical behavior, 339, 360, 366
mechanical properties, x, 265, 270, 351, 352, 353, 354, 360, 361, 362, 363, 365, 366, 451, 461
mechanical stress, 180
mechanical treatments, 51
media, 58, 185, 331, 348
medicine, 210
melting, 81, 126, 265, 266, 277, 327
melting temperature, 126, 266, 277
melts, 120
memory, 182, 184
mercury, 353, 367
mesoscopic, xi, 424, 479, 481, 488, 497
metal ions, 179, 181, 431
metal oxide, viii, 177, 178, 179, 181, 182, 193, 227, 246
metal oxides, viii, 177, 178, 179, 181, 182, 193, 227
metallography, 331
metalloporphyrins, 431
metals, ix, 53, 58, 59, 61, 63, 67, 74, 75, 81, 85, 119, 120, 121, 129, 178, 180, 183, 188, 193, 210, 211, 265, 266, 269, 270, 271, 274, 277, 284, 285, 286, 287, 327, 339, 349, 376, 395, 399, 544
methane, 320
methylene, 395
microscope, x, 30, 273, 349, 351, 353, 356, 367, 388
microscopy, 275, 353, 387
microstructure, viii, x, 26, 37, 39, 49, 52, 54, 62, 64, 65, 72, 73, 77, 78, 84, 320, 323, 326, 340, 348, 352, 353, 354, 356, 360, 362, 365, 366, 383, 484
microstructure features, 353
microvoid, ix, 319, 345
microwave, 290
microwave radiation, 290
migration, ix, 265, 266, 269, 271, 276, 277, 279, 281, 282, 284, 286, 407, 432
miniaturization, 386
MIP, 353
MIT, 259
mobility, 266, 283, 290, 415
MOCVD, 290
modeling, x, xi, 54, 69, 71, 89, 203, 285, 321, 338, 341, 344, 345, 348, 385, 389, 404, 406, 434, 445, 448, 459
models, vii, x, 2, 3, 18, 36, 47, 53, 55, 56, 72, 74, 77, 81, 84, 86, 181, 195, 212, 246, 268, 269, 271, 321, 328, 351, 354, 363, 364, 366, 396, 448, 452, 454, 460, 481, 483, 484, 524, 531, 534
modulation, 537
modules, 270
modulus, 18, 120, 129, 130, 131, 132, 200, 271, 334, 336, 342, 343, 345, 482, 484, 499, 517, 524, 525, 526, 538, 539
moisture, 352
molar volume, 200
mold, xi, 445, 447, 448, 459, 460, 461
molecular beam, 290
molecular beam epitaxy, 290
molecular dynamics, ix, 265, 269, 270, 272, 287
molecular orbitals, 304, 389, 391, 395, 396, 401, 412, 423, 427, 432, 435
molecular structure, 208, 430
molecular-beam, 2
molecules, 206, 207, 209, 304, 386, 387, 388, 389, 393, 394, 395, 396, 399, 404, 405, 406, 407, 410, 411, 412, 413, 419, 420, 421, 422, 423, 428, 430, 431, 434, 435, 437, 438, 439
molybdenum, 266
momentum, 120, 121, 126, 136, 143, 150, 156, 163, 171, 192, 203, 204, 205, 210, 303, 305, 306, 308
monolayers, 9, 30
monomeric, 396
Monte-Carlo, 73, 123
morphology, vii, 1, 2, 12, 14, 18, 28, 32, 47
Moscow, 1, 47, 285, 316, 544
motion, 57, 127, 196, 199, 230, 241, 269, 271, 287, 291
motivation, x, 319, 328
movement, 66, 72, 179, 268, 327, 328, 525, 527, 538, 540, 541
multiplier, 335

N

nanocrystals, 72, 313
nanometer scale, 271, 386
nanometers, 396
nanostructures, 177
nanotechnologies, 387
nanotubes, ix, 289, 290, 386, 387
naphthalene, 354, 365
NASA, 89
natural, ix, x, 7, 36, 319, 320, 326, 348
natural gas, ix, x, 319, 320, 326, 348
Nb, 185
Nd, 185, 235
neck, 480
neglect, 520
Netherlands, 257, 383
network, 185, 255
neutrons, 192, 194, 203, 204, 205, 206, 210, 217
New York, 88, 175, 257, 258, 259, 316, 348, 349, 367, 442, 544, 545
Newton, 458
next generation, ix, 289
Ni, ix, 43, 79, 101, 103, 109, 115, 116, 198, 262, 265, 267, 270, 277, 278, 279, 282, 283, 284, 285, 287, 322, 323, 325, 326, 327, 396
nickel, 61, 286, 320, 321, 322, 323, 325, 326, 348
nitrogen, 323, 327, 369, 431
nodes, 340, 461, 469, 470, 472, 473, 474
noise, 58
non-crystalline, 120
non-destructive, 54
nonequilibrium, 286
non-linearity, 409
non-uniform, 446, 461
normal, 57, 62, 73, 135, 139, 142, 149, 155, 162, 165, 168, 197, 202, 207, 214, 272, 336, 343, 370, 371, 372, 376, 377, 378, 426, 456, 476, 489, 494, 512, 514, 528, 529, 530, 531, 532, 534
normal distribution, 73, 343
normalization, 198, 495, 518
normalization constant, 198
novelty, 448
nuclear, 65, 195, 305, 309, 310, 348
nuclear charge, 305
nuclear reactor, 65
nucleation, 2, 5, 7, 9, 11, 12, 13, 18, 19, 28, 29, 31, 34, 37, 39, 41, 67, 77, 81, 327, 330, 331, 348, 370
nuclei, 6, 8, 18, 19, 37, 41, 45, 46, 206, 291, 306, 309
nucleus, 6, 15, 34, 291, 306, 309, 327
numerical tool, 52

O

observations, 30, 45, 84, 327, 367, 423
Ohio, 544
oil, 58, 65
oligomer, 396
oligomers, 387, 393, 396
one dimension, 179, 448
on-line, 74
operator, 59, 60, 291, 294, 295, 304, 305, 306, 339
optical, viii, x, 26, 31, 67, 177, 179, 180, 182, 190, 207, 209, 214, 222, 228, 230, 231, 232, 249, 250, 290, 351, 367
optical lattice, 179
optimization, 53, 74, 84, 85, 86, 427, 446, 447
organic, 290, 386, 410, 423, 430, 432, 439
orientation, 267, 268, 271, 344, 370, 371, 372, 373, 376, 381, 522
orthogonality, 125
orthorhombic, 180, 181, 185, 186, 187, 190, 210, 211, 213, 215, 217, 227, 229, 237, 238, 255
oscillations, 135, 136, 141, 143, 148, 150, 155, 156, 161, 163, 167, 171, 523, 527, 537, 539
oscillator, 199, 201, 241
oxidation, 64
oxide, 183, 185, 192, 194, 246, 247, 254, 327
oxides, viii, 177, 178, 179, 182, 185, 192, 193, 202, 203, 211, 227, 246
oxygen, 54, 178, 185, 187, 188, 190, 203, 212, 215, 216, 217, 222, 223, 225, 228, 230, 232, 238, 244, 248, 251, 255, 256, 322, 327
oxygen sensors, 185

P

PAA, viii, 119, 120, 174
Pacific, 175
pairing, 187
paramagnetic, viii, 177, 189, 190, 192, 210, 211, 235, 325
parameter, 7, 11, 12, 26, 29, 36, 70, 74, 75, 76, 84, 124, 129, 189, 212, 213, 216, 234, 242, 256, 271, 296, 326, 329, 330, 331, 332, 333, 340, 345, 410, 449, 450, 456, 476, 477, 495, 497, 498, 518, 521, 538
Paris, 88, 91, 92, 93, 94, 95, 96, 97, 98, 99, 100, 101, 102, 103, 104, 105, 106, 107, 108, 109, 110, 111, 112, 114, 116, 117
particle density, 301, 303

particles, 65, 67, 178, 192, 204, 205, 328, 329, 331, 344
partition, 55
passenger, 447
Pb, 185, 193, 194, 211, 227, 235, 237, 243, 244, 245, 246, 327
PCF, viii, 119, 126, 135, 142, 148, 155, 162, 168
PDC, 119, 120, 126, 129, 133, 134, 135, 139, 141, 143, 145, 149, 152, 155, 157, 161, 163, 165, 167, 169, 172, 174, 217, 228, 230, 239
PDEs, 455
percolation, 2
periodic, 19, 120, 179, 197, 304, 306, 309, 470, 531, 532, 533, 534
periodic table, 304, 306, 309
periodicity, 127, 196
permeability, 268, 269
permit, 322
perovskite, 185, 186, 187, 192, 193, 194, 202, 210, 211, 213, 217, 227, 234, 235, 236, 241, 243
perovskite oxide, 193, 194, 202
perovskites, 183, 185, 186, 190, 192, 213, 217, 229, 234
perturbation, 407
perturbations, 486
pH, 437
phase diagram, 190, 227, 235, 243, 255
phase shifts, 223
phase transformation, 33, 190
phase transitions, 192, 194, 212
Philadelphia, 544
phone, 319
phosphorus, 290
photoelectron spectroscopy, 267
photolithography, 386
photoluminescence, ix, 289
photon, 204, 205, 206, 207
photons, 191, 192, 194, 206, 207, 209
physical and mechanical properties, 265, 461
physical mechanisms, 483
physical properties, 30, 187, 189, 193, 194, 197, 205, 227, 235, 247, 445, 448, 458
physicists, 2, 126, 178, 205, 440
physics, viii, 49, 178, 189, 194, 203, 255, 266, 290, 386, 432, 544
planar, 13, 376, 393, 396, 397, 398, 399, 400, 401, 402, 403, 404, 413, 415, 418, 419, 426, 427, 429
Planck constant, 199
plasmons, 205
plastic, viii, x, xi, 49, 53, 54, 55, 57, 61, 62, 63, 68, 69, 70, 72, 74, 78, 80, 83, 84, 85, 108, 320, 326, 327, 329, 331, 333, 335, 338, 341, 349, 356, 382, 479, 480, 481, 482, 483, 484, 485, 486, 487, 488, 489, 490, 491, 492, 494, 495, 496, 497, 498, 504, 505, 506, 508, 509, 513, 514, 516, 517, 518, 519, 520, 521, 522, 523, 524, 526, 527, 530, 531, 532, 534, 535, 536, 538, 539, 541, 543
plastic deformation, viii, xi, 49, 53, 57, 61, 62, 74, 78, 85, 327, 479, 480, 481, 482, 483, 484, 485, 486, 487, 488, 489, 490, 491, 492, 494, 495, 496, 497, 498, 504, 505, 506, 508, 509, 513, 514, 516, 517, 518, 519, 520, 521, 522, 523, 524, 526, 527, 530, 531, 532, 534, 535, 536, 538, 539, 541, 543
plastic strain, xi, 54, 63, 69, 70, 72, 80, 83, 84, 331, 341, 479, 480, 518, 519, 522, 524, 531, 534, 541
plasticity, 54, 56, 66, 68, 70, 79, 80, 86, 265, 323, 326, 334, 335, 339, 349, 483, 524
plasticization, 75, 79
play, viii, 49, 53, 55, 72, 76, 81, 178, 185, 192, 210, 211, 228, 234, 241, 283, 323, 520
PLC, 480, 535, 537, 541
pleasure, 257
point defects, 269, 285, 290
Poisson, 6, 15, 23, 50, 120, 131, 132, 334, 345, 484, 488, 491, 499, 503, 517
Poisson distribution, 6, 15, 23
Poisson ratio, 50
Poland, 69, 88, 91, 92, 93, 94, 97, 98, 99, 100, 101, 102, 103, 104, 105, 106, 109, 110, 112, 113, 116, 349
polar coordinates, 128
polarizability, 202, 203, 207, 410
polarization, 201, 203, 204, 211, 300, 407
pollution, 58
polycrystalline, x, 184, 227, 228, 230, 234, 265, 266, 286, 320, 326, 339, 348, 369, 370, 374, 522, 532
poly-crystalline, 126
polyester, 354, 356, 452, 453
polymer, 65, 451
polymer structure, 451
polynomial, 55, 373, 375, 378, 410, 415
polynomials, 375, 410, 415, 416
poor, 63, 78, 430, 435
population, vii, 2, 37, 38, 343
population density, 37
pore, x, 351, 352, 353, 354, 356, 360, 363, 364, 365, 366, 367
pores, 66, 353, 360
porosity, 26, 33, 344, 352, 353, 360, 363, 366
porous, 331, 348, 349, 353, 363, 365

porous media, 331
porphyrins, 396
portability, 134
powder, 67
power, 13, 16, 20, 22, 38, 66, 182, 339, 409
prediction, 60, 71, 72, 81, 157, 330, 335, 345, 348, 446, 452
pressure, ix, 57, 58, 61, 65, 177, 178, 187, 188, 191, 193, 194, 209, 210, 213, 228, 235, 236, 237, 238, 239, 241, 242, 243, 244, 245, 246, 255, 256, 270, 272, 312, 320, 321, 323, 325, 340, 348
probability, vii, ix, 1, 2, 5, 6, 13, 14, 15, 16, 19, 22, 28, 32, 34, 58, 188, 200, 206, 265, 274, 277, 284, 329, 330, 342, 343, 345, 348, 349, 378, 418, 522
probability distribution, 22, 28, 32, 200, 378
probe, 43, 203, 204, 205, 208, 209, 371, 387
process control, 73
processing stages, 459, 471
production, 58, 60, 320, 352, 386
program, 447, 459
programming, 446
propagation, xi, 52, 55, 64, 65, 75, 76, 77, 85, 127, 131, 201, 207, 209, 323, 326, 327, 328, 335, 370, 479, 480, 497, 532, 535, 537, 539, 541
property, xi, 65, 73, 183, 208, 270, 304, 349, 354, 389, 407, 423, 432, 451, 479, 480, 531
protection, 321
prototype, 185, 210, 387, 393, 405, 439
pseudo, viii, 119, 120, 305, 306, 334, 336, 337, 339, 371
purification, 290
PVP, 88
pyramidal, 22

Q

quanta, 209
quantum, 191, 199, 209, 241, 290, 291, 306, 316, 395, 396, 424, 536
quantum chemistry, 290, 306
Quantum Field Theory, 316
quantum mechanics, 291
quartz, 18
quasiparticle, 300, 303
quasiparticles, 290

R

radiation, 182, 204, 208, 209, 290, 456, 458
radioactive isotopes, 266, 267
radius, 3, 4, 5, 6, 8, 15, 20, 32, 33, 34, 50, 73, 124, 127, 130, 135, 142, 149, 155, 162, 168, 180, 187, 193, 235, 236, 237, 243, 246, 255, 271, 272, 371, 372, 376, 494, 500, 509, 514, 527, 531, 538
Raman, 192, 193, 194, 197, 201, 203, 206, 207, 208, 209, 211, 212, 214, 215, 216, 217, 223, 225, 228, 229, 230, 232, 237, 238, 239, 243, 244, 247, 248, 249, 256, 289
Raman and Brillouin scattering, 209
Raman scattering, 192, 201, 206, 207, 208, 209, 237, 248
Raman spectra, 211, 215, 223, 289
Raman spectroscopy, 203, 206, 208, 211, 212
random, vii, x, 1, 5, 9, 13, 14, 15, 17, 18, 19, 21, 22, 23, 26, 28, 34, 73, 126, 182, 184, 206, 369, 371, 372, 373, 375, 378, 381, 522
random access, 182, 184
random numbers, 73, 522
range, ix, 56, 57, 60, 69, 73, 123, 133, 135, 142, 149, 155, 162, 168, 179, 184, 188, 189, 191, 192, 193, 198, 202, 203, 204, 205, 208, 210, 212, 213, 222, 227, 228, 229, 230, 234, 235, 237, 238, 240, 242, 246, 248, 253, 254, 265, 277, 282, 283, 284, 290, 291, 320, 321, 323, 325, 327, 328, 340, 395, 407, 412, 434
rare earth, 185, 193, 211, 227, 235, 239
RAS, 265, 286
rat, 70, 72, 81, 84, 86
Rayleigh, 207
reaction order, 452
reading, 448
reality, 202, 387
recall, 214
reciprocity, 336
recognition, 386
recovery, 63, 328
recrystallization, 63, 67, 78, 265
rectification, xi, 385, 423, 432, 434, 435, 437, 439, 440
red shift, 193, 222, 223
redistribution, 376, 407
reference system, 238
reflection, 65, 209, 528, 532
reflectivity, 230
refrigeration, 185
regression, 63, 73, 343, 373, 375
regular, 18, 19, 285, 336
regulations, 277
relationship, x, 7, 44, 52, 64, 80, 214, 255, 333, 336, 351, 353, 354, 360, 361, 363, 364, 365,

366, 383, 389, 394, 407, 408, 410, 412, 416, 430, 514
relationships, 63, 66, 131
relaxation, xi, 52, 53, 64, 73, 74, 76, 77, 78, 79, 80, 81, 84, 85, 86, 202, 266, 271, 272, 273, 407, 427, 479, 480, 481, 482, 483, 484, 485, 487, 488, 489, 490, 491, 492, 493, 494, 495, 504, 507, 508, 512, 513, 514, 516, 517, 518, 519, 520, 522, 524, 527, 528, 529, 530, 531, 532, 534, 535, 538, 539, 541, 544
relaxation effect, 80, 81
relaxation model, 86
relaxation rate, 79, 81
relaxation time, 538
reliability, 60, 74, 94, 109
REM, xi, 479, 480, 481, 483, 486, 520, 538, 541
replication, 470
reservoir, 387
resins, 452
resistance, 43, 51, 62, 64, 65, 79, 84, 85, 116, 183, 184, 211, 323, 326, 382, 423, 430, 538
resistive, 43, 44, 182
resistivity, 43, 44, 180, 183, 184, 185, 188, 189, 228, 235, 237
resolution, 203, 267, 275
resources, 85, 320
response time, 182
retardation, 354, 365
retention, 64
rhombohedral, ix, 177, 186, 187, 190, 193, 194, 203, 210, 211, 212, 213, 214, 215, 216, 217, 218, 219, 220, 221, 222, 223, 224, 225, 226, 227, 229, 230, 231, 232, 233, 234, 235, 236, 237, 238, 239, 243, 244, 255, 256
rigidity, 120, 131, 132, 393, 524, 525, 527, 538, 539, 541
rings, 393, 396, 397, 398, 399, 400, 401, 402, 403, 404, 415, 422, 435
risk, 321, 322
robust design, 73
robustness, 73
rolling, 84
Romania, 89
room temperature, viii, 119, 126, 184, 194, 205, 210, 213, 227, 229, 231, 239, 242, 243, 254, 325, 326
root-mean-square, 271
Rössler, 263
rotations, 208, 256
roughness, vii, 1, 2, 14, 17, 18, 19, 21, 22, 23, 24, 25, 26, 28, 29, 30, 31, 32, 33, 50, 51, 52, 58, 61, 62, 63, 64, 65, 67, 69, 73, 74, 75, 76, 77, 80, 81, 84, 86, 354
rubber, xi, 445, 446, 447, 448, 449, 450, 451, 452, 453, 455, 456, 457, 458, 459, 460, 461, 462, 463, 464, 465, 466, 475, 477, 478
rubber compounds, 445, 447, 449, 451, 475
Russia, 1, 265, 479
Russian, 47, 316, 544
Russian Academy of Sciences, 1, 479

S

SAE, 52, 88
safety, 52, 320, 321, 322, 333
salt, 312, 316
sample, 63, 205, 209, 312, 342, 343, 525
sampling, 222, 360, 469, 470, 472
sapphire, 31
satellite, 207
saturation, 25, 26, 59, 80, 439
savings, 60, 304
scalar, 304, 331
scandium, 180
scanning electron microscopy, 367
scatter, ix, 72, 206, 319, 329, 369, 370
scattered light, 206, 207
scattering, 120, 126, 178, 182, 190, 192, 197, 201, 203, 204, 205, 206, 207, 208, 209, 211, 212, 216, 217, 220, 221, 222, 225, 228, 229, 230, 234, 237, 239, 240, 247, 248, 252, 253, 424
Schrödinger equation, 298
search, 246, 269
secular, 128
segmentation, 355, 356, 360, 363, 367
selecting, 174, 455, 503
self, 3, 90, 107, 291, 293, 295, 523, 524
self-assembly, 387
SEM, 353, 366, 367
SEM micrographs, 353
semiconductor, ix, 182, 289, 290, 303, 386, 424, 432
semiconductors, ix, 178, 183, 289, 290, 300, 302
sensitivity, 67, 209, 228, 340, 407, 411
sensitivity analysis, 106
sensors, ix, 182, 184, 185, 289
separation, 58, 195, 266, 407, 417, 434
series, xi, 63, 122, 180, 185, 186, 316, 385, 396, 397, 398, 399, 400, 401, 402, 403, 404, 406, 407, 410, 411, 412, 414, 415, 419, 420, 427, 429, 431, 435, 436, 446, 459
service loads, 52, 53
shape, vii, xi, 1, 15, 17, 18, 21, 22, 28, 55, 57, 59, 69, 73, 74, 75, 86, 130, 135, 141, 142, 148, 149, 155, 161, 162, 167, 168, 194, 329, 338,

349, 406, 445, 477, 479, 481, 483, 487, 488, 490, 494, 495, 499, 504, 508, 509, 513, 514, 515, 516, 518, 519, 520, 522, 524, 539, 541
sharing, 247
shear, xii, 61, 75, 76, 209, 324, 327, 328, 332, 334, 335, 480, 481, 489, 491, 492, 493, 494, 497, 498, 506, 507, 508, 509, 512, 513, 514, 515, 516, 517, 518, 519, 520, 521, 522, 523, 524, 527, 528, 529, 530, 532, 535, 536, 537, 538, 540, 541
shear deformation, 519
shell, 213, 235, 248
shock, 116, 117
shortage, 241
short-range, ix, 202, 203, 228, 238, 265, 284, 290
shoulder, 246, 446, 447
shoulders, 222, 232, 253
Si3N4, 45
SIC, vii, 2, 33, 34, 35, 36, 37, 38, 40, 41, 42, 45, 46
side effects, 352
sign, 179, 195, 378, 485, 491, 527, 529
signs, 511
silica, 43, 57, 366
silicate, 360
silicon, 386, 387, 439
silver, 30, 267, 286
similarity, 126, 215, 227, 304
simulation, ix, x, xi, 52, 53, 54, 69, 70, 81, 85, 86, 193, 203, 238, 270, 280, 286, 308, 309, 310, 311, 312, 319, 328, 333, 345, 348, 366, 369, 373, 375, 385, 393, 426, 439, 445, 446, 447, 448, 449, 459, 460, 470, 471, 475, 480, 520, 522, 524, 531, 535, 544
simulations, ix, x, 52, 56, 57, 62, 68, 73, 75, 85, 86, 268, 284, 287, 309, 319, 321, 344, 345, 352, 377, 387, 388, 439, 446, 459, 530, 532, 540, 541
sine, 67
sine wave, 67
Singapore, 47, 176
single crystals, 56, 290
singular, 336
singularities, 514, 519
sites, xi, 18, 120, 125, 181, 182, 185, 188, 189, 191, 193, 211, 214, 215, 247, 248, 251, 269, 327, 344, 348, 382, 479, 481, 483, 489, 495, 508, 518, 541
skewness, 26
skin, 461
Sm, 185
smoothing, 382
SOC, 446
sodium, 247, 251, 256
software, 75, 203, 471
solid oxide fuel cells, 185
solid state, 126, 178, 182, 205, 386, 396
spacers, 395
space-time, 307, 308
Spain, 117, 369
spatial, 13, 15, 17, 18, 21, 54, 56, 205, 206, 285, 293, 294, 344, 394, 395, 396, 401, 402, 412, 413, 418, 419, 422, 423, 425, 427, 434, 490, 492, 497, 498, 505, 506
spatial frequency, 18
species, 254, 304
specific heat, 120, 133, 134, 139, 140, 145, 146, 147, 152, 153, 154, 157, 159, 160, 165, 166, 167, 171, 172, 193, 194, 199, 200, 225, 227, 228, 232, 234, 242, 243, 251, 254, 256
spectral analysis, 207
spectroscopy, ix, 209, 267, 289
spectrum, 18, 74, 79, 83, 128, 131, 198, 201, 206, 207, 208, 209, 232, 243, 246, 427, 450
speed, 70, 324, 475
SPF, 69
spheres, 50, 68, 271
spin, viii, 177, 178, 179, 181, 182, 183, 186, 187, 188, 189, 190, 191, 192, 194, 206, 210, 227, 246, 294, 295, 296, 299, 303, 308, 390, 391
springs, 60, 67, 69, 72, 75, 77, 525
stability, 52, 68, 72, 79, 134, 186, 197, 210, 212, 286, 395, 399, 431, 456
stabilization, 69, 79, 236, 272
stabilize, 180, 191
stable crack, 382
stages, vii, 1, 2, 8, 10, 12, 13, 14, 29, 79, 266, 331, 354, 363, 447, 451, 459, 471
stainless steel, 57, 65, 72, 79, 322, 323, 325, 326
stainless steels, 65, 325, 326
standard deviation, 343, 382
standardization, 59
standards, 59, 72, 367
statistical analysis, 52, 75
statistics, x, 191, 271, 320, 329
steel, ix, 57, 60, 62, 63, 64, 65, 66, 67, 68, 70, 72, 73, 74, 77, 78, 79, 83, 84, 319, 320, 321, 322, 323, 325, 326, 328, 329, 340, 341, 344, 345, 348, 349, 383, 538
steric, 397, 398
stiffness, 226, 333, 335, 455, 476
STM, 388
STO, 308, 312
stochastic, viii, 49, 54, 72, 73
storage, 182, 185, 320, 321, 348, 389, 393, 399

strain, xi, 50, 58, 59, 60, 61, 63, 68, 69, 70, 72, 75, 76, 77, 80, 81, 83, 84, 85, 86, 234, 240, 286, 323, 327, 331, 333, 334, 335, 336, 338, 339, 340, 341, 349, 396, 397, 398, 399, 400, 401, 479, 480, 481, 483, 484, 486, 494, 503, 504, 509, 518, 519, 520, 522, 524, 525, 531, 534, 535, 536, 537, 541
strains, 69, 336, 337, 518
strategies, x, 385, 399, 475
strategy use, 459
strength, 50, 51, 52, 57, 58, 61, 62, 63, 64, 65, 66, 67, 68, 69, 70, 72, 73, 75, 77, 79, 80, 81, 84, 103, 108, 110, 114, 116, 181, 201, 202, 210, 212, 229, 321, 322, 325, 327, 329, 332, 333, 344, 352, 353, 354, 355, 361, 362, 363, 364, 365, 366, 383, 439, 480, 532
stress fields, 483, 484, 487, 496, 503, 505, 508, 509, 519, 520, 523, 528, 530
stress intensity factor, 77, 79
stress-strain curves, 64, 544
stretching, 59, 63, 66, 72, 208, 209, 212, 215, 217, 220, 228, 230, 237, 238, 239, 243, 255, 256
strong interaction, 178, 400
structural changes, 63, 222, 223, 266
structural formation, x, 351
structural transformations, 211
structural transitions, 211
structure formation, 354
styrene, 453
subdomains, 337, 338
substances, 185, 267
substitution, xi, 42, 212, 227, 236, 385, 434, 435, 518
substrates, 12, 29, 31, 43, 97, 290
sulfur, 426
superalloys, 53, 67, 270
superconducting, 180, 246
superconductivity, viii, 177, 178, 180, 192, 227, 246
superconductor, 194, 246, 247, 254, 430
superconductors, 179, 202, 246
superposition, 61, 66, 68, 75, 487, 489, 491, 499, 500, 501, 544
supply, 339
suppression, 152
surface diffusion, 3
surface layer, 2, 33, 45, 46, 57, 63, 67, 75, 81, 480
surface modification, 86
surface properties, 52, 61
surface roughness, vii, 1, 2, 14, 17, 22, 26, 30, 31, 33, 51, 58, 62, 63, 64, 66, 67, 69, 73, 74, 76, 77, 80, 81, 84, 86
surface structure, 65
surplus, 272
survival, vii, 1, 2, 58, 329, 330
susceptibility, 64
Sweden, 262, 348
switching, 525
Switzerland, 175, 367
symbols, 220, 221, 231, 239
symmetry, ix, 120, 177, 188, 189, 190, 197, 207, 211, 213, 214, 215, 217, 220, 222, 227, 228, 229, 237, 238, 239, 248, 255, 289, 295, 296, 307, 308, 309, 310, 311, 381, 397, 406, 420, 426, 470, 528
synchrotron, 203
synthesis, 180, 386, 400, 439
systems, viii, 52, 57, 58, 86, 119, 120, 127, 136, 143, 150, 156, 163, 171, 178, 179, 180, 182, 183, 189, 192, 193, 208, 211, 212, 223, 225, 227, 234, 235, 236, 237, 243, 246, 255, 256, 294, 298, 323, 353, 386, 396, 405, 413, 424, 426, 428, 429, 430, 434, 436, 437, 439, 448, 544

T

tankers, 323
tanks, 320, 321, 322, 348
Taylor expansion, 195
technology, 119, 182, 185, 290, 352, 386, 475
TEM, 43, 45
temperature dependence, 133, 139, 165, 193, 227, 230, 234, 241, 256, 270, 325, 452
temporal, 37, 56, 524
tensile, 52, 57, 59, 61, 62, 64, 65, 66, 68, 71, 73, 75, 76, 79, 80, 81, 84, 269, 326, 327, 329, 332, 333, 341, 349, 373, 374, 381, 383, 488, 489, 490, 491, 494, 497, 499, 500, 503, 506, 507, 508, 509, 514, 515, 522, 523, 524, 525, 535, 538, 541
tensile strength, 329
tensile stress, 52, 57, 59, 61, 62, 64, 65, 66, 73, 76, 79, 80, 81, 84, 269, 327, 329, 383, 488, 489, 499, 500, 503, 508, 522, 524
tension, 102, 272, 323, 327, 330, 332, 341, 342, 345, 349, 374, 375, 376, 381, 400, 490, 493, 503, 524
terminals, 320, 394
test statistic, 343
thermal energy, 205
thermal equilibrium, 205

thermal expansion, 194, 200, 325
thermal load, 52, 77, 80, 84
thermal properties, 238, 447, 458, 459, 460, 470
thermal relaxation, 79
thermal stability, 72
thermal treatment, 51
thermal vibrations, 191
thermodynamic, viii, 58, 119, 120, 121, 131, 133, 134, 139, 145, 152, 157, 165, 195, 199, 226, 255, 256
thermodynamic parameters, 256
thermodynamic properties, viii, 119, 120, 121, 131, 199, 226
thermodynamical parameters, 226
thermodynamics, 339
thin film, vii, 1, 2, 7, 26, 33, 39, 40, 41, 45, 47, 182, 183, 184
thin films, 7, 33, 47, 182, 183, 184
Thomson, 383
three-dimensional, 69, 71, 185, 272, 448, 456, 457
three-dimensional model, 456, 457
threshold, 2, 74, 76, 353, 360, 430
threshold level, 360
thresholds, 360
tin, 125, 439
titanates, 185
Titanium, 53, 57, 66, 79, 80, 83, 84, 91, 94, 96, 102, 116, 117
Tokyo, 91, 92, 93, 94, 95, 96, 97, 98, 99, 102, 103, 104, 105, 107, 108, 109, 110, 111, 112, 114, 116
tolerance, 109, 186, 187, 227
top-down, 386
torque, 451, 453, 476, 477
total energy, 294, 296, 298, 338, 339
toughness, ix, 67, 319, 323, 326, 340, 344, 348, 369, 381, 383
traction, 337
trajectory, 279, 280
trans, 316, 405, 406, 411, 413, 429
transfer, xi, 58, 136, 143, 150, 156, 163, 171, 188, 189, 204, 210, 321, 382, 389, 412, 418, 422, 423, 427, 431, 432, 434, 437, 445, 446, 447, 448, 457, 458, 459, 461, 476, 477, 522, 532
transformation, 33, 62, 190, 287, 307, 308, 316, 338, 514
transformations, 212, 493, 519
transistors, 58, 182, 387, 399
transition, viii, ix, x, 43, 121, 125, 141, 145, 152, 161, 167, 177, 178, 179, 181, 182, 184, 188, 189, 190, 192, 193, 207, 210, 211, 213, 216, 227, 235, 243, 246, 255, 257, 277, 312, 319, 320, 321, 324, 325, 326, 327, 328, 333, 340, 345, 348, 393, 520, 522
transition metal, viii, 121, 125, 141, 145, 152, 161, 167, 177, 178, 179, 181, 182, 193, 227, 246, 257
transition metal ions, 179, 181
transition temperature, 189, 246, 324, 325, 327, 328, 340, 393
transitions, 181, 209
translation, 192, 306
translational, 120, 198
transmission, 43, 324, 422, 425, 427, 428, 434, 435, 439
transmission electron microscopy, 43, 107
transport, 120, 181, 182, 185, 189, 192, 194, 211, 228, 254, 320, 387, 389, 393, 394, 395, 412, 414, 418, 419, 421, 423, 424, 425, 426, 427, 429, 430, 431, 432, 434, 437, 439, 448
transportation, x, 385, 388, 401, 422, 427, 430, 435, 440
traveling waves, 194
trial, 475
trial and error, 475
tungsten, 266
tunneling, 182, 387, 388, 393, 394, 395, 412, 418, 421, 429, 430, 432, 434
Turkey, 351
twinning, 328
two-dimensional, 22, 69, 71, 75, 287, 446, 447, 448, 457, 529

U

UHF, 308, 309, 312
UK, 91, 92, 93, 94, 95, 97, 99, 101, 103, 104, 105, 106, 108, 111, 113, 115, 116, 543, 544
ultrasound, 58
uniaxial tension, 494, 514
unification, 268, 275
uniform, 62, 70, 192, 267, 268, 299, 329, 371, 378, 406, 410, 413, 420, 446, 461
user-defined, 460
USSR, 316
UV, 60

V

vacancies, 189, 268, 286
vacuum, 43, 58, 67, 294, 320, 439
valence, 122, 178, 183, 184, 185, 246, 304, 305, 306

validity, x, 206, 319, 323, 472, 475, 483, 503
van der Waals, 195
vanadates, 179
vapor, 3, 290, 320
variables, viii, 7, 23, 28, 31, 36, 46, 49, 59, 72, 73, 75, 86, 235, 243, 256, 448, 458, 460, 496
variance, 188
variation, ix, 43, 44, 57, 60, 65, 72, 74, 78, 81, 83, 84, 193, 197, 206, 222, 226, 232, 236, 237, 238, 239, 240, 241, 242, 256, 289, 293, 342, 407, 409, 412, 447, 448, 449, 472
vector, 18, 122, 128, 136, 173, 181, 192, 196, 198, 204, 230, 239, 272, 273, 274, 275, 336, 339, 406, 414, 420, 455, 456, 476, 516
velocity, 4, 14, 19, 23, 57, 58, 59, 62, 63, 64, 68, 69, 70, 74, 75, 120, 131, 209, 266, 267, 271, 324, 518, 525, 527, 538, 540, 541
vessels, 321, 322, 348
vibration, 191, 197, 206, 208, 209, 231, 244, 251
vibrational modes, 207, 214, 215, 239
vibrational spectrum, 208
Victoria, 287
Vietnam, 175
visible, 205, 246, 290
voids, 29, 326, 328, 330, 331, 332, 353
Volmer-Weber, 2, 47
vulcanization, xi, 445, 446, 447, 448, 450, 471

W

Wall Street Journal, 257
Warsaw, 88, 91, 92, 93, 94, 97, 98, 99, 100, 101, 102, 103, 104, 105, 106, 109, 110, 112, 113, 116, 348
water, 58, 64, 65, 180, 246, 320, 352
wave number, 122, 207, 222, 223, 232, 246
wave propagation, 55
wave vector, 122, 128, 131, 136, 173, 192, 196, 197, 198, 204, 230, 239
wavelengths, 207
wavelet, 54
wealth, 179, 183, 203
wear, 52, 86
Weibull, x, 320, 329, 330, 336, 342, 343, 344, 345, 347, 348, 349
Weibull distribution, 329
welding, 67, 81, 322
wires, x, 385, 387, 388, 394, 401, 407, 410, 414, 419, 420, 426, 427, 429, 430, 431, 434, 439
workers, 235, 266
working conditions, 404, 413
workload, 59

X

X-ray diffraction, 44, 67, 80, 84, 126, 316
x-rays, 194, 205, 209

Y

yield, 23, 44, 57, 62, 63, 67, 69, 72, 75, 77, 79, 81, 179, 197, 238, 270, 297, 326, 327, 330, 331, 334, 407, 418, 481, 539

Z

Zener, 80, 188, 210, 235, 258, 262
zinc, 65, 100, 266, 267, 306, 307, 308, 309, 312, 313, 314, 315
Zn, 67, 103, 396